Statistics for Research in Psychology

To Danielle

With love∞

Sara Miller McCune founded SAGE Publishing in 1965 to support the dissemination of usable knowledge and educate a global community. SAGE publishes more than 1000 journals and over 800 new books each year, spanning a wide range of subject areas. Our growing selection of library products includes archives, data, case studies and video. SAGE remains majority owned by our founder and after her lifetime will become owned by a charitable trust that secures the company's continued independence.

Los Angeles | London | New Delhi | Singapore | Washington DC | Melbourne

Statistics for Research in Psychology

A Modern Approach Using Estimation

Rick Gurnsey
Concordia University

Los Angeles | London | New Delhi
Singapore | Washington DC | Melbourne

FOR INFORMATION:

SAGE Publications, Inc.
2455 Teller Road
Thousand Oaks, California 91320
E-mail: order@sagepub.com

SAGE Publications Ltd.
1 Oliver's Yard
55 City Road
London EC1Y 1SP
United Kingdom

SAGE Publications India Pvt. Ltd.
B 1/I 1 Mohan Cooperative Industrial Area
Mathura Road, New Delhi 110 044
India

SAGE Publications Asia-Pacific Pte. Ltd.
3 Church Street
#10-04 Samsung Hub
Singapore 049483

Acquisitions Editor: Abbie Rickard
Content Development Editor: Morgan Shannon
Editorial Assistant: Jennifer Cline
Production Editor: Jane Haenel
Copy Editor: Christina West
Typesetter: C&M Digitals (P) Ltd.
Proofreader: Scott Oney
Indexer: Nancy Fulton
Cover Designer: Karine Hovsepian
Marketing Manager: Jenna Retana

Printed in the United States of America

Library of Congress Cataloging-in-Publication Data

Names: Gurnsey, Frederick Norman, author.

Title: Statistics for research in psychology : a modern approach using estimation / Rick Gurnsey, Concordia University.

Description: First Edition. | Thousand Oaks : SAGE Publications, [2017] | Includes bibliographical references and index.

Identifiers: LCCN 2017020842 | ISBN 9781506305189 (hardcover : alk. paper)

Subjects: LCSH: Psychology—Statistical methods. | Psychology—Research—Methodology.

Classification: LCC BF39 .G867 2017 | DDC 150.72/7—dc23
LC record available at https://lccn.loc.gov/2017020842

This book is printed on acid-free paper.

17 18 19 20 21 10 9 8 7 6 5 4 3 2 1

BRIEF CONTENTS

DETAILED CONTENTS

*Sections with asterisks contain optional or advanced material.

PREFACE

Along with other domains of human behavior and activity, research is also affected by routines of thought and, particularly, of premises. [M]any routines are very useful by providing important economy of mental effort. But there are exceptions. It does happen that some premises of our thought, acquired through traditional learning in schools and universities, have no other backing than tradition. And the longer the tradition of a commonly accepted premise, the more difficult it is, psychologically, to notice its routineness and to question its validity.

Jerzy Neyman (1974)

WHY ANOTHER STATISTICS TEXTBOOK?

Television shows, websites, and scientific journals frequently tell us that the result of some research study or other is "statistically significant." Statistical significance seems to be the scientific community's way of saying that the result is significant in the everyday sense of being important or meaningful. This is not so. Results may be statistically significant without being meaningful. Furthermore, it is increasingly evident that statistical significance is widely misunderstood not only by students and laypeople but also by researchers themselves. In fact, confusions about p-values are so widespread that in 2016, the American Statistical Association took the unusual step of issuing a statement on their meaning (Wasserstein & Lazar, 2016). Statistics instructors across the world are the ones most aware of these confusions, their consequences, and the rapidly changing attitudes toward the routine use of significance tests in research. Unfortunately, many statistics texts retain a decades-old syllabus built on traditional significance tests.

This textbook offers a syllabus based on estimation that emphasizes confidence intervals, effect sizes, and practical significance. This approach encourages us to attach meaning (practical significance) to the size of some measured effect in the context of a specific research question, and discourages reliance on the context-independent notion of statistical significance. When we have to explain the meaning of research results without using the term "statistically significant," we recognize how heavily we've leaned on it.

Instructors interested in making a shift to this syllabus must be able to do so without rebuilding their courses from scratch. A glance at the table of contents shows a reassuringly traditional sequence of topic areas. This is because estimation and significance tests rest on the same foundations, even though they differ in emphasis. The goal of this book is to explain both approaches but shift the emphasis to estimation. Rather than devoting the main body of each chapter to testing hypotheses about population parameters, with estimation mentioned as a niche alternative, the focus is squarely on estimation, followed by a discussion of how confidence intervals (and test statistics) can be used to test null hypotheses.

The material in this book is structured in a way that allows the advantages of estimation to emerge naturally. All of the standard material on significance testing is presented, but estimation is shown to be the more general method. In my experience, students find estimation to be a more intuitive and less tangled approach to statistics. Therefore, I firmly believe that students who are introduced to statistics through estimation will be better able to evaluate published research and will be better prepared to undertake meaningful research of their own.

ENGAGING STUDENTS IN STATISTICS

This book is written in an engaging conversational style that addresses the student directly. Sections include **learning checks** with questions for students to answer in order to assess their understanding before moving on to new material. There are also many **end-of-chapter exercises,** including definitions and concepts, true or false questions, calculations, and scenarios. Each chapter includes brief model reports of statistical analyses that follow American Psychological Association (APA) guidelines.

The people who developed the statistical methods routinely used in psychological research are introduced in Part 2 of the book. Among the most important of these are William Sealy Gosset, Jerzy Neyman, Sir Ronald Fisher, and Jacob Cohen. As most statistics instructors know, the so-called Fisher-Neyman hybrid model of statistical decision making would have pleased neither Fisher nor Neyman. Therefore, a historical perspective on statistics shows how some current confusions arose from blending two quite different philosophies. Unlike others, this text devotes a chapter to the problems associated with significance tests, including the file-drawer problem, p-hacking, and basic misunderstandings about p-values. The message to students is that statistical reform in behavioral research will be the responsibility, and accomplishment, of their generation.

Each chapter has a very clear through line and typically uses a single research question to aid its development. To avoid diversions from the main story, additional material is presented in end-of-chapter **appendices**. These include discussions of precision planning for estimation, power, and other technical material such as collinearity, bootstrapping, and probability density.

Many useful tools for performing statistical analyses are also discussed in appendices. The two most important tools are Excel and IBM® SPSS® Statistics.* Instructors will have different opinions about which of these is most appropriate for their classes. Students using this text as part of an advanced course may need to have experience with SPSS in the likely event that they go on to conduct an independent research project. Students in an introductory course will benefit from a good grounding in Excel, because it provides a simple way to check hand calculations and provides useful tools for working with normal and t-distributions. Furthermore, Excel is in wide use in settings beyond universities. (Personally, I use Excel far more than SPSS in daily work-related activities.) Additional appendices show students how to use **R** to put confidence intervals around statistics that have complex distributions, such as Cohen's d and R^2. Finally, there are appendices showing how to use G*Power for prospective power analysis and sensitivity analysis.

CHAPTER SUMMARIES

Part I: Introduction to Statistics and Statistical Distributions

Chapter 1: Basic Concepts

The first chapter makes the case that we all have an intuitive understanding of statistics. It then proceeds to establish the vocabulary to be used through the rest of the book. The

---

*SPSS is a registered trademark of International Business Machines Corporation.

concepts covered are scales of measurement, psychological constructs, operational measures, measurement error, populations and samples, parameters and statistics, random sampling, and the distinction between sampling error and sampling bias. Three appendices provide introductions to Excel, SPSS, and **R**.

Chapter 2: Distributions of Scores

The central concept in statistics is that of a sampling distribution, and explaining distributions of scores is a first step toward explaining sampling distributions. This chapter covers tabular and graphical characterizations of distributions of qualitative and quantitative variables, with emphasis on grouped frequency distributions. This chapter also introduces probability so that we can establish the concept of a probability distribution.

Chapter 3: Properties of Distributions

Although graphical and tabular characterizations of distributions have value, it is essential to have more succinct summaries of their properties. This chapter introduces measures of central tendency and dispersion for populations and samples, and it establishes the notation for parameters and statistics. The primary emphasis is on the mean and variance of populations and samples, but variations in shape are also discussed.

Chapter 4: Normal Distributions

The role of normal distributions in statistics cannot be overstated. This chapter provides a brief historical introduction to normal distributions and then turns to area-under-the-curve problems to deepen the notion of probability distributions. The 34-14-2 approximation is used first to connect the material covered in this chapter to the material covered in Chapter 2, and then the standard normal distribution is introduced to deal with less constrained situations. To build toward the foundations of estimation and significance testing, students learn to find the two values of z that enclose the central $(1-\alpha)100\%$ of the standard normal distribution.

Chapter 5: Distributions of Statistics

Part 1 culminates with an introduction to sampling distributions, with an emphasis on the distribution of means and the central limit theorem. Once the distribution of means has been introduced, the structure of this chapter follows that of Chapter 4. The central message is that distributions of scores and means differ mainly in their dispersion. The chapter ends with a brief discussion of the distribution of variances to make clear the notions of biased and unbiased statistics.

Part II: Estimation and Significance Tests (One Sample)

Chapter 6: Estimating the Population Mean When the Population Standard Deviation Is Known

It is a small step from knowing the distribution of sample means to using a sample mean to estimate a population mean. This chapter brings together the material developed in Part 1 to explain the basic elements of estimation using confidence intervals. For the purposes of exposition, we consider the case in which the population standard deviation is known. This assumption is made in Chapters 7 through 9 as well, to facilitate the introduction of statistical inference.

Chapter 7: Significance Tests

Significance tests are the most widely used method of statistical inference in psychology. In this chapter, significance tests are explained from the Fisherian perspective. The null hypothesis is introduced, and p-values are shown to be conditional probabilities under the null hypothesis. An important caution is made about the inverse probability error, which is the mistaken belief that a p-value is the probability that the null hypothesis is true. We avoid mention of alternative hypotheses through most of the chapter, but Fisher's curious views on these are discussed at the end.

Chapter 8: Decisions, Power, Effect Size, and the Hybrid Model

Running a significance test makes sense only if there is a high probability of rejecting a false null hypothesis. This chapter develops the concept of the alternative hypothesis, Type I and Type II errors, and power. Some historical context is given to briefly introduce the Neyman-Pearson model and to make clear how the views of Fisher and Neyman differed. The standard Fisher-Neyman hybrid is then presented. The historical perspective allows students to recognize the constituents of the hybrid and avoid some of the confusions that arise from merging these two approaches (specifically, the mistaken belief that retaining the null hypothesis means that it is true, and confusing a p-value with the probability that a Type I error has been made).

Chapter 9: Significance Tests: Problems and Alternatives

Having introduced the hybrid model of significance testing, this chapter turns to a discussion of its many problems. Following a brief description of the publication process in psychological research, there is a discussion of publication bias, the file-drawer problem, proliferation of Type I errors, p-hacking, binary thinking, and misinterpretations of statistical significance. We then review how these problems can be avoided by abandoning the language of statistical decision making and instead using estimation to support researchers' judgement about data interpretation. However, external constraints may require researchers to use significance tests, so there is a discussion of how confidence intervals can be used to test null hypotheses. Because the null hypothesis has almost no chance of being exactly true, we can use estimation to judge just how untrue it is.

Chapter 10: Estimating the Population Mean When the Population Standard Deviation Is Unknown

This chapter provides a model for the rest of the book. It opens with a discussion of t-distributions and how they connect to confidence intervals for the population mean when the population standard deviation is unknown. Confidence intervals for estimated effect sizes are then discussed. Finally, we discuss how significance tests can be conducted with confidence intervals and t-statistics.

Part III: Estimation and Significance Tests (Two Samples)

Chapter 11: Estimating the Difference Between the Means of Independent Populations

This chapter considers estimation problems relating to two independent populations. We cover confidence intervals for the difference between two independent means and the standardized difference between two population means. There is also a brief discussion of significance tests. The notion of partitioning variance is introduced here, and r^2 is presented as a second measure of effect size. Finally, we show that many statistics (t, F, and r) can be transformed to Cohen's d for the purpose of meta-analysis.

Chapter 12: Estimating the Difference Between the Means of Dependent Populations

This chapter introduces the notion of repeated measures and matched samples and shows how to estimate the difference between the means of two dependent populations. It also continues the discussion of partitioning variance and cautions about combining results from dependent and independent populations in a meta-analysis.

Chapter 13: Introduction to Correlation and Regression

Many research questions in psychology involve the association between two quantitative variables. This chapter deals with the basics of correlation and regression, showing how to compute the correlation coefficient and the regression equation. We also return to the notion of partitioning variance. Although no inferential statistics are introduced, we discuss how regression with fixed levels of a predictor variable (set by the experimenter) allows for easier causal inferences than when the predictor and outcome variables are random variables.

Chapter 14: Inferential Statistics for Simple Linear Regression

This chapter begins with the assumed bivariate distribution of scores under the regression model and then derives the sampling distribution of the regression slope (b) for the case of fixed levels of x. From this follows a discussion of a confidence interval around b. There is then a discussion of bivariate normal distributions and the computation of confidence intervals for b when both x and y are random variables. This is followed by a confidence interval for $\hat{y}$. One of the most distinctive parts of this chapter is the discussion of prediction intervals. This discussion shows that regression equations rarely have much to say about outcomes at the level of individual cases.

Chapter 15: Inferential Statistics for Correlation

A confidence interval for the sample correlation coefficient is complicated by the fact that its sampling distribution is complex. This chapter shows how to compute an approximate confidence interval using the Fisher transform and then how to compute an exact confidence interval using **R**. The chapter ends with a discussion of the generality of correlation and how regression and correlation underlie the two-independent-groups design discussed in Chapter 11.

Part IV: The General Linear Model

Chapter 16: Introduction to Multiple Regression

Multiple regression is shown to be a modest extension of simple linear regression. This chapter covers the omnibus regression model, and topics include R^2, confidence intervals for R^2, the central F-distribution, ΔR^2, and hierarchical regression. This is also the first point at which a complete explanation of degrees of freedom can be given. Finally, confidence intervals for $\hat{y}$ and prediction intervals are discussed.

Chapter 17: Applying Multiple Regression

The analysis of individual regression predictors is typically more interesting than omnibus effects. The opening section explains the meaning of partial regression coefficients and then moves on to explain how to compute confidence intervals for partial regression coefficients and standardized regression coefficients. Three applications of regression are covered, including statistical control, mediation analysis, and moderation analysis.

Chapter 18: Analysis of Variance: One-Factor Between-Subjects

This chapter presents the basic terminology of analysis of variance (ANOVA). ANOVA is shown to be a special case of multiple regression, with fixed predictors coding for groups. Although the omnibus level is discussed, the main focus is on the analysis of specific contrasts. The chapter covers confidence intervals for planned and unplanned contrasts, orthogonal contrasts and trend analysis, and standardized contrasts. An argument is made against the practice of conducing post hoc analyses following an examination of the omnibus effect. Because contrasts are the only informative aspects of an ANOVA, they should be planned and conducted without regard for the omnibus effect.

Chapter 19: Analysis of Variance: One-Factor Within-Subjects

This chapter also discusses omnibus and contrast analyses, with the greatest emphasis placed on confidence intervals for planned contrasts.

Chapter 20: Two-Factor ANOVA: Omnibus Effects

This chapter covers omnibus main effects and interactions for the two-factor between-subjects, within-subjects, and mixed designs. The approach is conceptual, showing how orthogonal vectors (and their products) can be used to decompose the variability in two-factor designs into components that can be tested for statistical significance. Importantly, variability accounted for by the vectors coding for the omnibus effects can itself be tested for statistical significance, making the case that contrasts have greater value than omnibus effects.

Chapter 21: Contrasts in Two-Factor Designs

The final chapter deals with contrasts in two-factor designs. These are conceptually far simpler than omnibus analyses and are more directly related to the questions of interest to the researcher. Confidence intervals for contrasts in the three two-factor designs are very similar to those used in one-factor designs, although there are a few special cases in the mixed design.

ONLINE SUPPLEMENTS

study.sagepub.com/gurnsey

The password-protected **Instructor Resources** site features author-created tools designed to help instructors plan and teach their course. These include an extensive test bank, chapter-specific PowerPoint presentations, and lecture notes. For even more coverage and support, visit our Instructor Resource site for an additional chapter on "Meta-Analysis" and for numerous appendices on using Excel and SPSS.

The open-access **Student Resources** site provides eFlashcards, web quizzes, access to full-text SAGE journal articles with accompanying assessments, and multimedia resources.

ACKNOWLEDGMENTS

I thank Professor Rex Kline for reading the entire draft manuscript and providing invaluable suggestions for improvements. I am profoundly grateful for his advice and encouragement throughout the writing of this book. Imagine Roger Federer taking the time to help you with your tennis game.

The SAGE team has been tremendous. Reid Hester provided steady and enthusiastic support through the initial development of the book. Abbie Rickard continued with the same through the latter stages of development and was always ready to answer questions, big and small, with incredible thoroughness and patience. The production team has been fantastic. Eric Garner always provided quick feedback and expert advice on technical matters while preparing for production. Christina West did an outstanding job of copyediting an extremely complex manuscript. And Jane Haenel expertly shepherded this project through production. Thank you all.

I am indebted to the many reviewers and accuracy checkers for the time they devoted to reviewing and critiquing initial drafts of the chapters.

Emmanuel Angel, University of Pennsylvania

Michael T. Brannick, University of South Florida

Christopher J. Ferguson, Stetson University

Lynda K. Hall, Ohio Wesleyan University

David K. Jones, Westminster College

Yasuo Miyazaki, Virginia Tech

Jason Popan, University of Texas–Pan American

Kristopher J. Preacher, Vanderbilt University

Laura Sabourin, University of Ottawa

John M. Spores, Purdue University Northwest

Devdass Sunnassee, University of North Carolina Greensboro

Zhigang Wang, Carleton University

Cengiz Zopluoglu, University of Miami

Words cannot express my gratitude to Dr. Danielle Sauvé, love of my life, for her unflagging support. Had she known what she was getting into, she might have had second thoughts about encouraging me to write this book. And to our best friends ever, David, John, and Nyssa, where would we be without you?

With such exceptional support, it is difficult to imagine that errors of any sort have found their way into this book. However, those that may be lurking within these pages are entirely my responsibility.

ABOUT THE AUTHOR

Rick Gurnsey is a professor in the Department of Psychology at Concordia University, Montreal. He received both his MSc in Computer Science and PhD in Psychology from Queen's University, Kingston. His research covers many areas of low-level vision, including texture perception, subjective contours, motion, orientation, symmetry, and changes in visual function across the visual field. Dr. Gurnsey's research has been funded by federal and provincial agencies continuously since 1989. He has taught introductory and advanced statistics for 15 years.

INTRODUCTION TO STATISTICS AND STATISTICAL DISTRIBUTIONS

1 BASIC CONCEPTS

STATISTICS IN PSYCHOLOGY

If you're reading this text, there is a high probability that you've taken an introductory course in psychology or a related discipline. Therefore, you know that psychology is a very broad field with many subdomains. Some of these are perception, cognition, social psychology, clinical psychology, personality, learning, development, motivation, and neuroscience. All professors in psychology departments were trained to do research in one of these subdomains. Typically these professors think of themselves as researchers first and foremost. If you check the website of your psychology department, you will find an entry for each faculty member (professor) in the department that describes his or her research interests. There will typically be a list of recent publications as well, so that you can get an idea of what kind of research each of your professors does.

Although faculty members have all trained in different areas of psychology, they address similar types of questions in their research. For example, asking about the effectiveness of two kinds of psychotherapy is very similar to asking about differences in brain activity following presentations of two different visual stimuli. Asking whether two drugs have different effects on the amount of food rats eat is very similar to asking whether one method of studying produces better grades than another. In these examples, the researchers obtain measurements from two groups of individuals, assess the differences between groups on these measurements, and reason about the results. Therefore, despite large differences in research issues, researchers use very similar methods to assist their reasoning. The purpose of this book is to explain these methods.

Statistics Will Improve Your Life

Many of us enter psychology precisely because we think we have no interest in topics such as statistics. It's quite common to hear psychology students say, "I'm in psychology to learn how to help other people, but I'm not interested in research." Well, if one thinks that psychology has something to teach about the best ways to help people, how did researchers determine what these are? They learn by testing different methods of helping to see which ones work best. To compare methods, you must quantify the improvements. By this I mean you will have to measure how much better people are after treatment than before. If you are treating depression, for example, you might count the depressive symptoms exhibited before and after treatment. Better still, you might measure the days of work missed before and after treatment. Now you are in a position to compare methods and see how much better (or worse) one method is than the alternatives. To make informed use of these comparisons, one has to make use of statistical methods. Therefore, if your objective is to learn the best way to help people, then it is essential for you to know how to evaluate the evidence relating treatments to outcomes. Evaluating the evidence requires understanding statistics.

There is an even more fundamental reason for studying statistics. People today have access to more information than at any time in human history. (Google it.) Some of the information available might influence important decisions we make. Check any news website and you'll find a section on health. Such sites often have links to articles with titles such as

- Fish Oil's Impact on Cognition and Brain Structure Identified in New Study
- Carrots Cut Heart Attack Risk
- Socializing Key to "Successful Aging"
- Live Longer by Avoiding Hunger Games
- Multivitamin Use Linked to Lowered Cancer Risk
- Exercise After Stroke Could Help Improve Memory: Study
- Study Finds That Free Birth Control Means Fewer Abortions and Fewer Teen Births
- Fasting Hormone Helps Mice Live Longer

Many such headlines might be relevant to us or those close to us. Some of the stories under these headlines may contain credible information about health, and others may be simply clickbait. Being able to sort signal (credible research results) from noise (clickbait) is a crucial survival skill in today's world, and this requires a basic understanding of proper research methodologies and appropriate statistical analyses. Being a better consumer of statistical information will help you to better navigate a complex world and make better decisions for yourself and those you love as well as in any career you might undertake. In my view, all of this will contribute to improving your life.

We All Have an Intuitive Grasp of Statistics

Even though many of us approach statistics with trepidation, the underlying concepts are very simple. I am certain that you already have an intuitive understanding of many of these concepts. For example, think about flipping a coin. Coins have two sides that are typically referred to as heads and tails. A fair coin is one that will land with the heads side up 50% of the time when it is flipped. We can say that the probability of coming up heads is $1/2 = .5$. Therefore, because the coin can only come up heads or tails, the probability of coming up tails is also $1/2 = .5$. Now, if I give you a fair coin, do you think it will come up heads exactly 5 times in every 10 flips? My guess is that you'd say no. You would probably say that it would often come up heads 5 times, but it might also come up heads 4 or 6 times. It might even come up heads 3 or 7 times.

From these thought experiments, I suspect that you wouldn't be too surprised if a fair coin came up heads in 6 of 10 flips, you might think it just a tad unusual for it to come up heads in 7 of 10 flips, but you would be very surprised if it came up heads in 10 of 10 flips. If these are your intuitions, then you already have a basic understanding of randomness and probability, which are concepts at the heart of statistics.

Now let's think of a slightly different situation. I have a coin that might or might not be fair. I will flip it 10 times and ask you to decide whether you think it's fair based on how many times it comes up heads. If the coin comes up heads 10 times in 10 flips, would your guess be that the coin is fair? I think you'd conclude that it is not a fair coin because your intuition is that it would be really unusual for a fair coin to come up heads 10 times in 10 flips. Of course, unusual does not mean impossible. It is certainly possible for a fair coin to come up 10 times in 10 flips, but if we are forced to make a decision, most of us would decide that the coin is not fair.

We can further illustrate important intuitions underlying statistics by thinking about jelly beans. Let's say that there is a large jar of jelly beans and you are asked to estimate the percentage of red jelly beans in the jar. You can't see the jar but you're allowed to take out a handful of jelly beans to help you make your estimate. We will call your handful a sample.

Now, if you're allowed to take a sample of jelly beans, which do you think would be most helpful: a large sample or a small sample? I'm sure you'd say that a large sample would allow you to make a better estimate.

So let's say you've reached into the jar and pulled out as many jelly beans as you can fit in one hand. If you find that 28% of the jelly beans in your sample are red, then what would be your estimate of the percentage of red jelly beans in the jar? Well, with nothing else to go on, your best estimate of the percentage in the jar is the percentage in your sample. That is, your best estimate of the percentage of red jelly beans in the jar is 28%. However, I don't think you'd bet your life savings on this estimate, because you probably recognize that if you took another sample you would almost certainly find a slightly different percentage of red jelly beans. This second sample could have more than 28% or less than 28%. As with the coin-flipping example, however, you would be very surprised to find that a second sample from the same jar produced all red jelly beans. If so, this shows that you have some intuitions about what is normal sample-to-sample variation and what is unusual sample-to-sample variation.

Finally, as gamblers like to say, let's make things interesting. Imagine that the jar of jelly beans belongs to your roommate, and she knows the percentage of red jelly beans in the jar. She wants to make the following deal with you. If your guess about the percentage of red jelly beans in the jar is within 1% of the actual percentage, she'll do the dishes for a month. If your guess is off by more than 1%, you'll do the dishes for a month. In other words, if the actual percentage of red jelly beans in the jar is 28% ± 1% (i.e., 27% to 29%), you won't have to do the dishes for a month. Would you take this deal? I don't think I would. Given my intuitions about sample-to-sample variation, I wouldn't have much confidence that the true percentage in the jar is in the interval 28% ± 1%. However, if the criterion were that the estimate be within 20% of the actual percentage, then you might be tempted. Personally, given my intuitions about sample-to-sample variation, I'd be quite confident that the actual percentage of red jelly beans in the jar is in the interval 28% ± 20%, or 8% to 48%. Maybe you wouldn't have the same confidence I have, but I think you'd agree that the wider the interval, the more confident you'd be that it includes the true percentage of red jelly beans.

You might find most of what I've said to be basic common sense. That's because most of us have an intuitive grasp of statistics. Throughout this book, we will build some relatively simple formalisms on top of these intuitions. There is surprisingly little mathematics involved in statistics. You will have to know how to add, subtract, multiply, and divide. You'll also have to know how to square numbers and take their square root. And you'll have to know the order of operations. There is not much beyond this. Statistics is more logic than mathematics.

In the rest of this chapter, we will introduce some essential concepts that will be used throughout the rest of this book. So, without further ado, we will get on with the introduction.

VARIABLES, VALUES, AND SCORES

All sciences involve measuring things, and psychology is no exception. For example, we might measure height, weight, eye color, or hair color for any individual. Height, weight, eye color, and hair color are examples of **variables**. We can say that variables are physical or abstract attributes or quantities that we wish to measure (Figure 1.1). Any variable has a range of values that it can take on. For example, possible values for height are positive numbers representing height in inches or centimeters, and possible values of weight are positive numbers representing weight in pounds or kilograms. Possible values for eye color are blue and brown, and possible values of hair color are black, brown, red, and blond. A **score** is the value that an individual has on a particular variable. For the variables height, weight, eye color, and hair color, an individual might have the scores 70 inches, 158 pounds, blue, and brown, respectively.

You may have noticed in the examples just given that variables come in different kinds. The variables height and weight, for example, can take on values that are positive numbers, whereas eye color and hair color take on values that are color names. In general, we refer to variables whose values are numbers (positive, negative, integer, or real) as **quantitative variables** because they often reflect how much of some quantity an individual possesses. Variables whose values are qualities (such as blue or brown) are referred to as **qualitative variables**.

Quantitative Variables

Ratio Scales

When we think of measurement we immediately think of measuring devices. For example, length is measured with a ruler, weight is measured with a scale, and time is measured with a clock. Such devices produce numbers that represent units on the scale in question. These numbers can be used to *order* the things being measured (e.g., from shortest to longest, or lightest to heaviest). Although length, weight, and time are measures of different physical quantities, their scales share two important properties. The first is that there is an *absolute zero* value on the scale. If we are measuring height in inches, for example, there is the theoretical possibility of the absence of the thing being measured. The same goes for weight, time, temperature, area, and so on. The second feature common to these scales is that they have a *unit of measurement*. An inch may be the unit of measurement for length, a gram may be the unit of measurement for weight, and a second may be the unit of measurement for time (Figure 1.2). Therefore, the scales we are most familiar with are divided into *intervals* that correspond to a constant physical quantity, such as an inch, a gram, or a second. Quantitative variables for which there is a unit of measurement and an absolute zero point are said to be measured on **ratio scales**.

FIGURE 1.1 ■ Individuals and Variables

iStock.com/FatCamera

An individual can be characterized by his or her scores on many variables.

Variables are physical or abstract attributes or quantities that we wish to measure. A variable can take on specific values.

A **score** is the value that an individual has on a particular variable.

Quantitative (scale) variables have values that are numbers. They often reflect how much of some quantity an individual possesses.

Qualitative variables have values that are qualities or categories. They are also referred to as nominal or categorical variables.

Ratio scales have units of measurement and an absolute zero. The ratio of two values on a ratio scale expresses their relative distances from 0.

FIGURE 1.2 ■ Three Measuring Devices

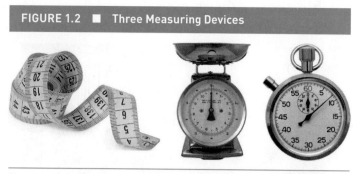

Photo credits (*left to right*): iStock.com/SvetaVo, iStock.com/Matmart, and iStock.com/Gannet77

Discrete variables have fixed values. Discrete quantitative variables have values that are integers or whole numbers.

Continuous variables have values that are real numbers.

Interval scales have equal units of measurement but no absolute zero point.

The term *ratio scale* highlights the fact that we can compare two values of the variable in terms of their *relative distance* from 0. For example, if one person is 3 feet tall and another is 6 feet tall, then the taller person is twice as tall as the shorter person, because his or her height is twice as far from zero as that of the shorter person. That is, the ratio of 6 to 3 is 2.

Quantitative variables can be **discrete** or **continuous**. For example, the variable *number of people in a classroom* can take on only positive integers as values (e.g., 27.32 is not a possible value of this variable). In other words, *number of people in a classroom* is a discrete quantitative variable. On the other hand, the variable *height* is a continuous quantitative variable and can take on any positive real number as a value. Therefore, 27.32 inches is a possible value of *height*.

Interval Scales

Not all scales having a unit of measurement have an absolute zero point. The most common examples of such scales are the Celsius and Fahrenheit scales for temperature. Both have a unit of measurement and a zero point, but the zero points do not represent the absence of heat. On the Celsius scale, 0° is the point at which water freezes; on the Fahrenheit scale, 0° is 32° below the freezing point of water. These are arbitrary points on the scale. Therefore, it is not true that an object radiating 100°C emits twice as much heat as an object radiating 50°C, because 100°C is not twice as far as 50°C from the absence of heat. The only scale of heat that has an absolute zero point is the Kelvin scale (Figure 1.3). Zero on the Kelvin scale corresponds to the absence of atomic activity. A body at 0°K radiates 0 heat.

FIGURE 1.3 ■ Ratio and Interval Scales

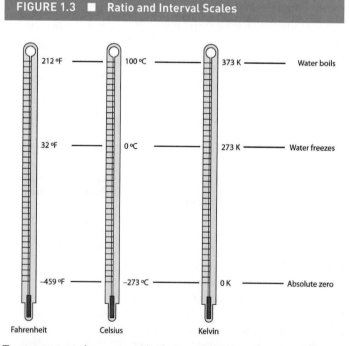

Temperature can be measured in degrees Fahrenheit, Celsius, or Kelvin. All three are equal interval scales but only the Kelvin scale is a ratio scale because it has an absolute 0 value.

Adapted from Key & Ball (2014). *Introductory Chemistry*. Retrieved from https://opentextbc.ca/introductorychemistry/.

When a scale has a unit of measurement (e.g., degrees Fahrenheit) but no absolute zero point, we say that it is an *equal interval scale*, or simply an **interval scale**. All ratio scales are equal interval scales, but not all equal interval scales are ratio scales. Few scales in the physical domain are interval scales without being ratio scales, but such scales are common in psychology, as we will see later.

Qualitative Variables

Qualitative variables are sometimes called *categorical* variables because their values can be discrete categories. Examples of such variables are sex, religious or political affiliation, or health status. Qualitative variables may also be called nominal variables because their values are simply names and so they are said to be measured on a **nominal scale**. For example, if we are interested in political affiliation, the values of our nominal (or categorical) variable may be *conservative* and *liberal*. Unlike quantitative variables, whose values are numbers, there is no natural ordering to the values of qualitative variables. For example, we can't say that blue is greater than orange or that conservative is greater than liberal. The terms *categorical* and *nominal* both capture the fact that qualitative variables are just names for particular qualities and there is no natural order to the values of such variables.

> The values of variables measured on a **nominal scale** are qualitative and discrete and have no natural ordering.

Ordinal Variables

Ordinal variables represent a final category that rests somewhere between quantitative and qualitative because they can have properties of both. Consider the variable *rank of my university*. Magazines sell many copies by creating yearly rankings of universities. The ranks are qualitative and discrete but, like quantitative variables, they are numeric (first, second, third, etc.). Furthermore, there is, by definition, a natural ordering to the ranks and to the values of ordinal variables in general. Obviously, first place ranks higher than fifth place. However, ranks do not have the property of being on an equal interval scale. The difference in quality between the first- and second-place universities is not necessarily the same as the difference in quality between the seventh- and eighth-place universities. Similarly, the difference in marathon finishing time between the first- and second-place runners is unlikely to be the same as the difference between the 45th- and 46th-place runners. Because there is a natural ordering to scores on an ordinal variable, but no unit of measurement, we say that ordinal variables are measured on an **ordinal scale**.

Sometimes ordinal variables can blur the qualitative/quantitative dichotomy. One such variable is *letter grade*. Grades at universities are often given as A, B, C, D, and F. These are categories that reflect something about academic performance. In addition, there is a natural ordering to the values of this variable; for example, A reflects better performance than B, and B better than C, yet there is no unit of measurement on this scale. However, ordinal variables, such as *letter grade*, are sometimes converted to numbers and treated as interval variables. In fact, universities do this all the time when computing grade point averages (GPAs). Often A is converted to 4 grade points, B to 3 grade points, C to 2 grade points, D to 1 grade point, and F to 0 grade points. When this conversion is done for each course, the average of these grade points produces a GPA. One has to be cautious when applying statistical analyses to ordinal variables that have been converted to quantitative variables. The equal interval property may not hold for such measures; if this is the case, then any analysis that requires the equal interval property of the variable will be invalid.

> The values of variables measured on an **ordinal scale** are qualitative and discrete and have a natural ordering, but they do not have a unit of measurement.

Stevens's Hierarchy of Scales

We've seen that the values of variables can be qualitative or quantitative, as well as discrete or continuous. Psychologist S. S. Stevens (1946) proposed a hierarchy of measurement scales to help us think clearly about the variables we encounter in psychological research. In his hierarchy, the four measurement scales were ordered as follows:

1. Ratio scales
   - Natural ordering to values
   - Units of measurement
   - An absolute zero

2. Equal interval scales
    - Natural ordering to values
    - Units of measurement

3. Ordinal scales
    - Natural ordering to values
    - No units of measurement

4. Nominal scales
    - Unordered categories

Lowest on the hierarchy are nominal scales, which represent unordered categories. Next are ordinal scales, which represent ordered categories. Then there are interval scales that have units of measurement but no absolute zero point. Finally, there are ratio scales that have units of measurement and an absolute zero point.

The largest part of this book is devoted to the analysis of quantitative variables. In statistics, the equal-interval nature of quantitative variables is their most important feature, because most statistical analyses of quantitative variables rest on the assumption that the variables are measured on an interval scale. Whether or not the scale is ratio will make no difference. There are clear cases of each of these scales, as illustrated above, but it can be difficult sometimes to figure out exactly what kind of scale we're dealing with.

Think of a questionnaire that lists 25 daily activities most people engage in at one time or another (e.g., brushing your teeth, having dinner, walking between classes, talking on the phone, looking at Facebook). For each item on this questionnaire, a respondent is asked to answer yes or no to the question "Does this activity make you happy?" A respondent's happiness might then be defined as the number of questions to which he or she answered yes.

What kind of scale are we dealing with? The variable is clearly ordinal because there is a natural ordering to the values, which can range from 0 to 25. But can we think of this variable as measured on an interval scale or even a ratio scale? Because 0 is a possible value of this variable, does this mean we are dealing with a ratio scale? If so, does this imply that happiness is completely absent from the life of a person scoring 0 on this measure? Does that make sense? Furthermore, is a person who scores 16 out of 25 twice as happy as a person who scores 8 out of 25? To me, this seems very unlikely to be true. So the idea that this is a ratio scale seems very iffy. Might the scale be an interval scale? This would imply that the difference in happiness between people who score 16 and 8, respectively, is the same as the difference between people who score 24 and 16, respectively. I'm not sure that the values of this scale can be legitimately called happiness units.

The core difficulty we face when trying to decide what kind of scale we're dealing with is the question of what is being measured. There is a fundamental difference between measuring happiness and temperature, and it is this difference that makes it hard to decide whether a variable should be characterized as ordinal, interval, or ratio. This question is examined in more detail in the next section.

MEASUREMENT

Many of the quantitative variables we've considered are physical (e.g., height, weight, temperature, and finishing times). In the social sciences, many of the measurements we make involve physical scales like these. For example, the time it takes to identify a face in a degraded image is a physical measure associated with a ratio scale. The electrical conductance of the skin is a physical measure. The number of action potentials per second relates to physical events. Measurements require a measuring device. Height can be measured with a ruler,

LEARNING CHECK 1

1. State whether the following statements are true or false.

   (a) Test scores on a multiple choice test correspond to the number of correct answers. Therefore:

      (i) test score is a quantitative variable,

      (ii) test score is a continuous variable,

      (iii) a person who scores 10 on the test knows twice as much about the topic as a person who scores 5.

   (b) Time in milliseconds is an ordinal variable.

   (c) A questionnaire about satisfaction with a psychology course has the values *very satisfied*, *satisfied*, *neutral*, *dissatisfied*, and *very dissatisfied* as possible responses. Therefore, satisfaction is measured on an equal interval scale.

2. Which of the following variables is measured on a ratio scale: (a) height in inches, (b) weight in kilograms, (c) temperature in degrees Celsius, (d) gender, (e) religious affiliation, (f) university ranking, or (g) marathon finishing time in hours?

Answers

1. (a) (i) True. (ii) False. It is discrete. (iii) False. There is no necessary connection between test scores and units of knowledge.

   (b) False. It is a scale/quantitative variable.

   (c) False. It is an ordinal variable.

2. a, b, and g.

---

weight can be measured with a scale, temperature can be measured with a thermometer, finishing time and response time can be measured with a stopwatch, skin conductance can be measured with a galvanometer, and action potentials can be counted.

Operational Measures

Many variables of interest to psychologists are not physical. Intelligence may be the clearest example of this. Although we can measure someone's temperature with a thermometer, there is no "intelligometer" that we can insert into the head to measure his or her intelligence. Intelligence is different from temperature because it is a **psychological construct**. That is, it is a distinct but abstract quantity that we believe explains some aspect of behavior. A very incomplete list of psychological constructs would include intelligence, depression, anxiety, confidence, introversion, sociability, authoritarianism, narcissism, happiness, and compassion.

One can't escape the fact that intelligence (and other psychological constructs) must be measured indirectly. A frequently used, indirect measuring method is the paper-and-pencil test. For example, we might attempt to measure intelligence by having an individual fill out a questionnaire with questions testing general knowledge, spatial reasoning ability, verbal reasoning, and so on. The speed and accuracy with which such a questionnaire is completed can be combined into a single score that is supposed to reflect the degree of intelligence possessed by an individual. We say that such a test provides an **operational measure** of intelligence. High scores are thought to capture the abstract notion of high intelligence, and low scores are thought to capture the abstract notion of low intelligence. Operational measures of intelligence may also include memory and problem-solving tasks, among other things.

A **psychological construct** is a hypothetical attribute (such as intelligence, introversion, or happiness) that is thought to explain some aspect of behavior, but which cannot be measured directly with a physical measuring device.

An **operational measure** is a tool used to measure a psychological construct. Very often, operational measures are derived from questionnaires. Operational measures may also derive from the speed and accuracy with which psychological tasks are completed.

Table 1.1 provides an example of how the concept of narcissism might be operationalized. The questions in Table 1.1 make up the 16-item Narcissistic Personality Inventory, or NPI-16 (Ames, Rose, & Anderson, 2006), which was adapted from the larger NPI-40 (Raskin & Terry, 1988). According to Ames et al. (2006), narcissism involves "a grandiose yet fragile sense of self and entitlement as well as a preoccupation with success and demands for admiration" (p. 441). For each of the 16 items on the NPI-16, a respondent is asked to check the alternative that best describes himself or herself. Browsing through the items, you can see that for each item, one of the choices is more closely associated with the description of narcissism given above. Therefore, one can count up the times a respondent chose the more narcissistic option. The number of such choices represents one's narcissism score. The higher the score, the more narcissistic the person is assumed to be. (Note that this is very similar in structure to the happiness questionnaire that we imagined at the end of the previous section.)

Validity and Reliability

There are deep and difficult issues associated with the construction of operational measures of constructs like happiness, intelligence, and narcissism. The first question is whether an operational measure is a valid measure of the psychological construct.

TABLE 1.1 ■ The NPI-16 as a Short Measure of Narcissism			
1	○ When people compliment me I sometimes get embarrassed	○	I know that I am good because everybody keeps telling me so
2	○ I prefer to blend in with the crowd	○	I like to be the center of attention
3	○ I am no better or worse than most people	○	I think I am a special person
4	○ I like to have authority over other people	○	I don't mind following orders
5	○ I find it easy to manipulate people	○	I don't like it when I find myself manipulating people
6	○ I insist upon getting the respect that is due me	○	I usually get the respect that I deserve
7	○ I try not to be a show off	○	I am apt to show off if I get the chance
8	○ I always know what I am doing	○	Sometimes I am not sure of what I am doing
9	○ Sometimes I tell good stories	○	Everybody likes to hear my stories
10	○ I expect a great deal from other people	○	I like to do things for other people
11	○ I really like to be the center of attention	○	It makes me uncomfortable to be the center of attention
12	○ Being an authority doesn't mean that much to me	○	People always seem to recognize my authority
13	○ I am going to be a great person	○	I hope I am going to be successful
14	○ People sometimes believe what I tell them	○	I can make anybody believe anything I want them to
15	○ I am more capable than other people	○	There is a lot that I can learn from other people
16	○ I am much like everybody else	○	I am an extraordinary person

Source: Ames, D. R., Rose, P., & Anderson, C. P. (2006). The NPI-16 as a short measure of narcissism. *Journal of Research in Personality, 40*(4), 440–450. Copyright (2006), with permission from Elsevier.

A measuring device is **valid** if it measures what it is supposed to measure. Generally, rulers are valid measures of length and clocks are valid measures of time. But can a 16-item questionnaire provide a valid measure of narcissism? Can a 200-item questionnaire provide a valid measure of intelligence? Can a 25-item questionnaire provide a valid measure of happiness? Does the NPI-16 measure narcissism or self-esteem? Even more serious, are happiness, intelligence, and narcissism even valid constructs? Questions like these are hugely important. The success of psychology as a discipline will depend on whether its constructs (e.g., intelligence, narcissism, anxiety, introversion, sensation seeking), as well as its operational measures of these things (e.g., responses on questionnaires), are valid.

> A measuring device is **valid** if it measures what it is supposed to measure.

Any measuring device, whether a thermometer or an intelligence scale, can be characterized in terms of its reliability. A highly **reliable** device gives very similar (if not identical) measurements each time it is applied to the same object. An unreliable device does not. If a thermometer gives a measure of body temperature that ranges from 92° to 104°F in a period of 10 minutes, we would judge it as highly unreliable. Therefore, an important question for operationalized variables is their reliability. If we think we have a valid operational definition of intelligence, we'd like to get very similar scores each time we use it to measure the same individual. If we do, the measure is reliable.

> A **reliable** measuring device gives very similar (if not identical) measurements each time it is applied to the same object.

One has to be careful with the notions of the reliability and validity of measuring instruments because these are always determined in a specific context. The NPI-16 shown in Table 1.1 might provide reliable and valid scores for North American university students but that does not mean it will provide reliable and valid scores for Argentinian cattle ranchers. Because reliability and validity are determined in a particular context, they are not properties of the measuring device itself (see Thompson, 2003). Therefore, a researcher using a measuring instrument (like the NPI-16) must ensure that reliability and validity have been established for the situation in which it is being used.

Reliability and validity are particularly important issues when we deal with operationalized measures. Determining validity in particular can be difficult. We can create reliable scales that are supposed to measure intelligence, but one can argue about whether the measurement actually reflects the thing we intend to measure. For example, an intelligence test may be culturally biased if the items on the "general knowledge" component of the test are relevant to one culture but not another.

Operational Measures and Scale of Measurement

Of course, even if we have a valid and reliable measure of some psychological concept, there is still the question of the kind of measurement scale we are dealing with. In essence, our question comes down to connecting the properties of our operational measure to the underlying psychological construct. If we think of the NPI-16, we can view it from two perspectives. The first perspective ignores the question of what psychological construct it is intended to measure and focuses only on the values of the variable. In this case, because the values range from 0 to 16, we can say that our variable is measured on a ratio scale; there is an absolute zero, and the ratio of two values expresses their relative distances from zero. It makes sense to say that one person's score is twice that of another person (e.g., 10 versus 5).

The second perspective treats the numbers of the variable as reflecting the degree of the underlying construct (i.e., narcissism) present in the individual we're measuring. In this case, the units of measurement are supposed to be units of narcissism. It is from this perspective that we encounter the challenges discussed at the end of the previous section. It is difficult, if not impossible, to say that the equal intervals on an operational measure correspond to equal intervals of the underlying psychological construct. That is, a person who scores 10 on the narcissism scale may not be twice as narcissistic as a person who scores 5. In spite of these complications, operational measures are used routinely in psychological research.

Many statistical analyses involve computing averages of variables that are assumed to be measured on an interval scale. For example, if 100 people completed the 16-item narcissism questionnaire, then the average narcissism score could be computed by summing the 100 scores and dividing by 100. Any statistical procedure that depends on this average "knows" only about the numbers and not about the underlying construct whose values the scale is supposed to represent. It is really up to the researcher to interpret such results carefully. For example, if men score 5.92 on average on the narcissism scale and women score 5.44 on average, it is up to the researcher to make sense of this difference. This means keeping in mind that the difference between 5.92 and 5.44 might represent a very small (or very large) difference in actual narcissism, assuming that narcissism is even a valid construct.

Measurement Error

It is important to note that all measurements are subject to error. Consider measuring a person's height. If you were to measure the same person several times, it is almost certain that you would come up with a slightly different measurement each time. Similarly, if you asked 10 people to measure the height of the same individual, it's almost certain that each person would come up with a slightly different answer. We call this variability in measurements **measurement error**. One of the major insights of modern science is that nothing can be measured without error. Rather, any single measurement must be seen as imperfect.

Measurement error refers to the fact that each time something is measured a slightly different score will be obtained.

Measurement error makes the question of your height really interesting. A statistician would say that your height is the average of all possible measures of your height. To make this a little more concrete, just think about measuring your height 25 times and taking the average. A statistician thinks about this by replacing 25 with ∞. So, to a statistician, your height is a theoretical quantity. Any time your height is measured, the result is really an *estimate* of this theoretical average. Because of measurement error, any single measurement is really an estimate of a theoretical average. So, when the doctor tells you that you're 5 feet, 7 inches, you think to yourself, "5 feet 7 is an estimate of my height."

LEARNING CHECK 2

1. Which of the following can be measured directly: (a) intelligence, (b) compassion, (c) length of a cod, (d) number of cows in a field, (e) race completion time, (f) anxiety, (g) blood pressure, or (h) life span of a house cat?

2. State whether the following statements are true or false.

   (a) Measurement error is inevitable.

   (b) Measurement error implies sloppily collected scores.

   (c) An operational measure is required to overcome measurement error.

   (d) Height is a valid measure of IQ.

   (e) A stopwatch is the most reliable way to measure time.

   (f) Intelligence is a psychological construct.

3. Which of the following requires an operational measure: (a) dominance, (b) introversion, (c) pupil diameter, (d) heart rate, (e) narcissism, (f) depression, or (g) time to complete a sudoku puzzle?

Answers

1. c, d, e, g, and h.

2. (a) True.

   (b) False.

   (c) False. This statement makes no sense.

   (d) False.

   (e) False. If 20 people measured the same event, there would probably be 20 very different times.

   (f) True.

3. a, b, e, and f.

POPULATIONS AND SAMPLES

At the heart of statistics is the distinction between populations and samples. A **population** is a collection of individuals that share some characteristic of interest. For example, the collection of all eligible voters represents a population. A **sample** is a subset of a population. So, the subset of voters who participated in a poll is a sample from the population of eligible voters.

We often think about populations as large collections of people. However, the term as used in statistics is not restricted to humans. A population could be

- all Canadians,
- all university students in Finland,
- all maple trees in Vermont,
- all houses built in France in 1953, or
- all worms living in my backyard.

Similarly, a sample from each of these populations is simply a subset of the population. That is, we could have a sample of trees, worms, or houses selected from the corresponding population. Any subset of a population is a sample. The population of Quebec is a sample of Canadians, as are the populations of Medicine Hat, Alberta, and Conway, Ontario. Similarly, all Canadians born in February is a sample of Canadians, as is the subset of Canadians over 6 feet tall.

The notion of a population can be refined somewhat. Typically researchers are interested in measurements (scores) taken from individuals in a population. Therefore, we might be interested in

- the heart rates of Canadians,
- the ages of university students in Finland,
- the circumferences of maple trees in Vermont,
- the values of houses built in France in 1953, or
- the lengths of the worms living in my backyard.

For our purposes, we will typically consider populations as the *scores* of individuals (that share some characteristic of interest) on a variable of interest. A sample is a subset of these scores.

In a research context, the researcher defines the population of interest. One researcher might be interested in a population defined by the ages of all university students in the United States. Therefore, the ages of all university students in Florida represent a sample from this population. Another researcher might be interested in a population defined by the ages of all university students in Florida. So, now, what was a sample for the first researcher is a population for the second researcher. Therefore, the ages of university students in Miami would be a sample from this population.

Parameters and Statistics

In most sciences we would like to know things about populations. For example, what is the average heart rate of Canadians? What is the average age of university students in Finland? What is the average circumference of a maple tree in Vermont? What is the average

A **population** comprises the scores on a variable of interest, obtained from individuals that share some characteristic of interest.

A **sample** is a subset of a population.

A **parameter** is a numerical characteristic of a population.

A **statistic** is a numerical characteristic of a sample.

Inferential statistics is the act of inferring population parameters from sample statistics.

Descriptive statistics are used to describe the characteristics of a collection of scores, without the goal of inferring something about a population parameter.

value of houses built in France in 1953? Finally, what is the average length of the worms living in my backyard? In each case, the averages in question are called **parameters**. The corresponding averages in samples are called **statistics**. A parameter is a numerical characteristic of a population. A statistic is a numerical characteristic of a sample.

So, the average heart rate of Canadians is a parameter, and the average heart rate of people in Montreal is a statistic. In almost all cases, populations of interest are too large to allow us to determine parameters directly. It would be virtually impossible to determine the heart rate of every single living Canadian. Therefore, we infer population parameters from sample statistics. **Inferential statistics** is the act of inferring population parameters from sample statistics.

The examples given at the beginning of this chapter used sample statistics (e.g., proportion of red jelly beans in a sample) to estimate a parameter (e.g., percentage of red jelly beans in a population). This kind of inference is at the heart of this entire book.

Inferential statistics are sometimes distinguished from **descriptive statistics**, which do not involve inferences about population parameters. For example, an instructor might compute the average grade on a midterm exam to provide some indication of how difficult the exam was. In doing so, he or she is simply describing the results and is not attempting to infer something about a population parameter.

LEARNING CHECK 3

1. State whether the following statements are true or false:

   (a) A statistic is to a parameter as a sample is to a population.

   (b) A statistic is a numerical characteristic of a population.

   (c) The IQs of all people in Roslyn, Washington, form a sample of the IQs of Americans.

   (d) The IQs of all people in Roslyn, Washington, can be a population.

   (e) Estimating the mean IQ of all people in Roslyn, Washington, from the mean IQ of all Americans is a typical example of inferential statistics.

Answers

1. (a) True.

   (b) False. A statistic is a numerical characteristic of a sample.

   (c) True.

   (d) True.

   (e) False. We infer parameters from statistics, not statistics from parameters.

SAMPLING, SAMPLING BIAS, AND SAMPLING ERROR

Several subsets of populations were given above as illustrations of samples. For the purposes of inferential statistics, however, most of these samples are not particularly useful. If the objective of sampling is to infer a population parameter from a sample statistic, then **simple random sampling** is essential. Simple random sampling means that all members of the population had an equal chance of being selected in the sample.

Simple random sampling means that all members of the population had an equal chance of being selected in the sample.

Simple Random Sampling

Let's say that for some unexplained reason, you might wish to estimate the average heart rate of Canadians. To do so, you would need a random sample of Canadians. A truly random sample of Canadians could be obtained (at least conceptually) in the following way. Every single Canadian is assigned a unique number between 1 and 36,560,587 (the number of Canadian citizens). These numbers are written on ping-pong balls and then placed in a (very big) bingo hopper (Figure 1.4). If we crank the hopper and then let 36 balls pop out, we have a simple random sample from our population. That is, because each number had an equal chance of being pulled from the hopper, these numbers are a random sample from the numbers 1 and 36,560,587, and the corresponding individuals represent a simple random sample of Canadians. Again, if our objective is to infer the average heart rate of Canadians, then random sampling is essential. The average heart rate of all Canadians can be estimated from the average heart rate of the 36 people in our sample.

FIGURE 1.4 ■ Random Sampling

iStock.com/popovaphoto

Sampling Bias

Random sampling is central to most of the statistical analyses that we will consider. The importance of random sampling can be appreciated when we consider the consequences of non-random sampling. Let's imagine that we wish to estimate the average fitness of American university students, and that fitness can be measured on a scale having values ranging from 0 to 100. The population of interest then is American university students. The numbers 0 to 100 are values of the variable *fitness*. To acquire a sample to study we might post advertisements around the Psychology Department of Tufts University soliciting volunteers for our study. And we might choose to study the first 64 volunteers that sign up for the study. This sample is clearly not random because not every American university student had an equal chance of being selected. Therefore, this sample is *biased*. **Sampling bias** means, simply, that not all members of the population had an equal chance of being selected in the sample.

> **Sampling bias** means that not all members of the population had an equal chance of being selected in the sample. Sampling bias can and should be avoided.

A second obvious source of bias here is that we chose to study the first 64 people who signed up. It is quite possible that the people most likely to sign up for a study of fitness are those who are—or view themselves to be—fit. Therefore, the average fitness of this sample may be higher than the average fitness of the population. Remember, we are using a statistic (the average fitness of the sample) to estimate a parameter (the average fitness of the population). Because of sampling bias our statistic is a poor *estimator* of our parameter.

Another source of bias is that sampling was done in one department of one university. There is no reason to suppose that the average fitness of students in this department of this university is a good representation of the average fitness of American university students in general.

When we are reading about studies conducted in university settings, it is worth keeping in mind that true random sampling has probably not occurred. Such studies often involve samples that are conveniently available. Such samples are often referred to as **convenience samples**. Choosing volunteers from a psychology department participant pool is an example of a convenience sample. Similarly, the first 64 individuals to sign up to participate in a study is an example of a convenience sample. If the study's objective is to estimate a parameter (e.g., average fitness) of a population (e.g., American university students, or even adults in general), then use of a convenience sample makes this estimation suspect. Sampling bias limits the conclusions that can be drawn from a study.

> A **convenience sample** is a sample that is conveniently available. It is the most common type of biased sample.

Sampling Error

In practice, a truly random sample is difficult to obtain. However, let's assume that we would like to estimate the average fitness of American university students, and we've managed to obtain a truly random sample of 25 university students. Do we expect the average fitness score of this sample to be exactly equal to the average fitness score of the population? With a little reflection you would probably say no. In fact, a sample statistic is almost never exactly equal to the population parameter, even with random sampling. The difference between a sample statistic and the population parameter it estimates is called **sampling error**. As you might imagine, sampling error tends to be large when sample size is small and decreases as sample size increases.

Sampling error must be distinguished from sampling bias (sampling error ≠ sampling bias). Sampling error is unavoidable and happens every time we estimate a parameter from a statistic. Sampling bias is avoidable in principle but may be difficult to avoid in practice.

Sampling error is the difference between a statistic and the parameter it estimates. Sampling error cannot be avoided; it is an inevitable feature of random sampling.

An Example

Whenever elections approach, news reports are filled with poll results. The purpose of a poll is to give us an idea about how the election would turn out if it were held that day. For example, in the 2016 U.S. presidential election, Hillary Clinton and Donald Trump were the Democratic and Republican nominees, respectively. On any given day leading up to the election, pollsters wanted to know what percentage of all eligible voters (that is, the population of eligible voters) would vote for Clinton. Because there are millions of people in this population, we can't ask them all whether they'd vote for her. Therefore, pollsters would choose a subset, or sample, of eligible voters and ask each member of the sample if he or she would vote for Clinton. On a typical day, it might be reported that a national poll found that 46% of eligible voters would vote for Clinton, with a *margin of error* of 3%. We hear this kind of thing all the time, but what does it mean?

The 46% mentioned above is the percentage of voters in the sample who would vote for Clinton. This percentage is used to *estimate* the percentage of voters in the population that would vote for Clinton. That is, the percentage in the sample is a statistic that estimates the corresponding parameter in the population. The margin of error associated with an estimate provides a sense of how precise the estimate is. Your intuitions probably suggest that if you took many polls (samples) of eligible voters, the percentage who'd vote for Clinton would vary somewhat from poll to poll (sample to sample), just as the percentage of red jelly beans would vary from sample to sample. When pollsters mention a margin of error (3% in our example), they are expressing something about the degree of variation they would expect to see from sample to sample. In other words, the margin of error is related to sampling error. In our example, the pollsters are saying they are confident that the true percentage of all eligible voters who would vote for Clinton is within the interval 46% ± 3%. In other words, they are confident that the true percentage is between 43% and 49%. (As it turned out, Clinton received 48.2% of the 136 million votes cast, and Trump received 46.2% of the votes cast.)

A PREVIEW OF WHAT'S AHEAD

We will end this chapter with a look ahead to where we're going, and we will also have a look at the tools we use in statistical analysis.

Estimation Versus Significance Testing

At the beginning of this chapter we noted that there are a lot of commonalities in the kinds of questions researchers ask in psychology. The most common questions are (i) how do samples drawn from two populations differ on some variable of interest? and (ii) what does this difference imply about the populations from which the samples were drawn?

LEARNING CHECK 4

1. State whether the following statements are true or false.

   (a) A random sample from the psychology department participant pool at the University of California, Los Angeles (UCLA) is a random sample of American university students.

   (b) Sampling error means that there was some kind of mistake in the way a sample was chosen.

   (c) Sampling bias can be avoided with truly random sampling.

   (d) Sampling error can be eliminated by using large samples.

Answers

1. (a) False (for so many reasons).

   (b) False. Sampling error is unavoidable, even with random sampling.

   (c) True.

   (d) False. Sampling error decreases as sample size increases but it cannot be eliminated.

Two approaches to these questions go by the names *significance testing* and *estimation*. To illustrate the difference between significance testing and estimation, let's think about red jelly beans a final time. Imagine two large barrels of jelly beans that we will call populations. For some reason we want to know how these two populations differ in the percentage of red jelly beans they contain.

The purpose of a significance test, in this example, is to answer the simple question of whether the two populations have the same or different percentages of red jelly beans. A significance test starts with the assumption that both barrels have the same percentage of red jelly beans. (In Chapter 7 we will see that this assumption is called the *null hypothesis*.) Then a sample (handful) is taken from each barrel and the percentage of red jelly beans in each sample is determined. If the difference in the percentage of red jelly beans in the two samples is unusually large given the assumption that the two populations have the same percentage of red jelly beans, then the difference is declared *statistically significant*. In this case, the "researcher" decides that the two populations contain different percentages of red jelly beans.

The focus of estimation is not on *whether* two populations differ but on *how much* they differ. In our example, the objective would be to estimate the difference in the percentage of red jelly beans in the two populations. The difference in percentages in the samples is the best estimate of the difference in percentages in the two populations. Because no estimate is perfect (there is always sampling error), the estimate is accompanied by an interval and an expression of the confidence we have that the interval contains the true difference. For example, we might say that our best estimate is that one population has 10% more red jelly beans than the other population, with a margin of error of 2%. That is, we're pretty sure that the true difference is 10% ± 2%. In other words, we're pretty sure that the true difference is between 8% and 12%.

For more than 70 years, psychologists have relied on significance tests to analyze their data. However, over those years many psychologists and statisticians have questioned the value of significance testing, and they've argued that significance testing has actually damaged psychology as a discipline. These arguments have gathered force in the last 15 to 20 years, and we will review them in Part II of this book. Those who criticize significance tests point to estimation as a very healthy alternative. Therefore, throughout this book we will emphasize estimation procedures. We will also see that estimation is the more general procedure, because one can do significance tests with estimation procedures.

Tools

There are many computer tools available to assist with statistical calculations. Each of these serves a different purpose, and we will draw on these at different points throughout the book. In the following sections, four of these tools are described: Microsoft Excel, IBM SPSS, **R**, and G*Power. Brief introductions to Excel, SPSS, and **R** are given in appendices at the end of this chapter. It is not necessary to read all of these at once. For example, your instructor may not want to use Excel, so you may not find the Excel appendix useful at this time. Or there may be no need to look at the **R** appendix right away because we won't use **R** until later in the book.

Excel

Microsoft Excel is a widely available computer application that has fantastic support for statistical analysis. I use many tools for statistical analyses, but Excel is the one I use most often. In many ways, Excel is like a super powerful hand calculator that lets you store calculations and revise them easily. There are many example calculations given in the following chapters. If you have Excel open as you read, you can enter the formulas into it to make sure that you understand the steps in a calculation. Furthermore, there are lots of practice questions within the chapters. You should do these because there is a good chance that similar questions will show up on a midterm or final exam. If you know how to use Excel, you can easily double-check your answers as you study. Most of the data sets used in chapters, learning checks, and end-of-chapter exercises are available as Excel files at study .sagepub.com/gurnsey

No matter what kind of computer you have, there is a very good chance that you have Excel installed on it. However, there are two alternatives with (almost) exactly the same functionality. One is OpenOffice, which can be downloaded and installed for free. More information about installation is available at the OpenOffice website (openoffice.org).

A second alternative is Numbers for Macintosh, which is essentially the Apple version of Excel. If you have a Macintosh computer, it is almost certain that Numbers came preinstalled. With very few exceptions, everything done in Excel can be done in exactly the same way in Numbers.

SPSS

SPSS stands for Statistical Package for the Social Sciences and is widely used by researchers in psychology. SPSS is owned by IBM, and student versions can be leased at a very modest cost. It differs from Excel in that statistical analyses can be specified through point-and-click menus, then executed with the click of a button. Whereas Excel requires you to specify the steps in a calculation, SPSS hides most of this from you and just provides the results of the analysis, often in great detail. Of course, this makes it easier to get results, but it allows you to get an answer without knowing what computations were performed or why. Chapter appendices will explain how to use SPSS to conduct many of the analyses we cover. We will use SPSS much more in Part IV, where we deal with complex computations.

R

There is a freely available statistical application called **R** that is used by many scientists (r-project.org). **R** is incredibly powerful and can be used to do all the calculations described in this book. For the purposes of this book, the main advantage of **R** is that it allows us to do calculations that are difficult or impossible to do in Excel or SPSS. Unfortunately, the **R** environment can be a challenge to learn, so we will devote less space to describing it than we devote to Excel and SPSS. However, some important calculations can be done easily in **R**, so it will play an important role in Parts II, III, and IV of this book.

G*Power

A final, freely available tool we'll make use of is G*Power (gpower.hhu.de/en.html). I won't say anything about G*Power at this point because understanding why we use it requires a deeper understanding of significance testing. I will say that G*Power is an amazingly useful tool, and researchers can be grateful to the G*Power team (Faul, Erdfelder, Buchner, & Lang, 2009; Faul, Erdfelder, Lang, & Buchner, 2007) for developing it and making it available for all to use. We will first see G*Power in Chapter 10 and then in many subsequent chapters.

A Word of Advice

It is easy to think that a statistics course is a math course and that to succeed, one mainly needs to memorize formulas and know how to do calculations. Of course, calculations are an important part of statistics, but to excel in statistics you need to understand the logic behind the calculations. When I meet with my students at the beginning of each term, I tell them that the first question on the first midterm will be: What is a sampling distribution? I also tell them that the following is a good answer to memorize: A **sampling distribution** is a probability distribution of all possible values of a sample statistic based on samples of the same size.

> A **sampling distribution** is a probability distribution of all possible values of a sample statistic based on samples of the same size.

Of course, this definition may not mean much to you at this point. However, the first part of this book (Chapters 1 to 5) is all about developing the concept of a sampling distribution. If by the end of this book you understand the notion of a sampling distribution, then you will have understood the most important concept in statistics. That is why I also put this question on the final exam.

There are three other concepts that derive from sampling distributions that you should pay close attention to as you work through the book. By the end of Part II, you should be able to answer the following questions:

- What is a p-value?

- What does it mean for a result to be statistically significant?

- What does it mean to have 95% confidence in an interval?

Return to these questions often. If you can answer them correctly by the end of the course, statistics will make complete sense to you.

SUMMARY

Science involves *measurement* of physical or abstract quantities. These physical or abstract quantities are called *variables*, and variables can take on specific *values*. When we make a measurement the result is the *score* for the individual or entity that was measured. We distinguish between *quantitative* and *qualitative* variables. The values of quantitative variables are numbers that are associated with *equal interval* scales. When the scale has an absolute zero value, it is a *ratio* scale. Quantitative variables can be *discrete* or *continuous*. The values of qualitative variables are abstract qualities that typically have no natural ordering. A special class of qualitative variables (called *ordinal* variables) are discrete, and they have a natural ordering but are not measured on an equal interval scale. However, ordinal variables are sometimes converted to numbers and treated as though measured on an equal interval scale.

The measurements made in science require measuring devices. Many physical quantities can be measured directly, but *psychological constructs* must be *operationalized* and measured indirectly. Whether measures are direct or indirect, all are subject to *measurement error*, meaning that every time a measurement is made it will differ to some degree from all previous measures. This is inevitable. Measuring devices that produce small measurement errors are *reliable* and those that don't are unreliable. A measuring device that measures what it is supposed to measure is *valid*, and those that don't are not valid.

A *population* is a set of individuals that share some characteristic of interest. A population can also be the scores on some variable for a set of individuals that share some characteristic of interest. A *sample* is any subset of a population. A *parameter* is a number that describes the scores in a population, and a *statistic* is a number that describes the scores in a sample drawn from a population. When we use a sample statistic to *estimate* a population parameter we are doing *inferential statistics*. For inferences to be valid our sample must be a *random sample* from the population, meaning that all members of the population must have had an equal chance of being selected as part of the sample. If not all members of the population had an equal chance of being selected as part of the sample, then our sample is a *biased sample*. *Convenience samples* are among the most frequent biased samples in psychological research. Sample statistics rarely equal the parameter they estimate, and the difference between a sample statistic and the corresponding population parameter is called *sampling error*. Sampling error is an inevitable part of inferential statistics.

KEY TERMS

continuous variable   6
convenience sample   15
descriptive statistics   14
discrete variable   6
inferential statistics   14
interval scale   6
measurement error   12
nominal scale   7
operational measure   9

ordinal scale   7
parameter   14
population   13
psychological construct   9
qualitative variable   5
quantitative variable   5
ratio scale   5
reliable   11
sample   13

sampling bias   15
sampling distribution   19
sampling error   16
score   5
simple random sampling   14
statistic   14
valid   11
variable   5

EXERCISES

Definitions and Concepts

1. Define variables, values, and scores.

2. What is a quantitative variable?

3. What is a qualitative variable?

4. What is the difference between discrete and continuous variables?

5. Explain the relationship between equal interval scales and units of measurement.

6. Define a ratio scale.

7. Explain how an ordinal variable shares characteristics with qualitative and quantitative variables.

8. What is a psychological construct?

9. What is an operational measure?

10. What does measurement error refer to?

11. What does it mean to say that a measuring device is reliable?

12. What does it mean to say that a measuring device is valid?

13. Define the concepts of populations and samples.

14. Define the concepts of parameters and statistics.

15. Give an example of inferential statistics.

16. What does it mean to say that a sample is a simple random sample from a population?

17. What is sampling bias?

18. What is a convenience sample?

19. What is sampling error?

True or False

State whether the following statements are true or false.

20. All variables whose values are numbers are quantitative variables.

21. Eye color is an ordinal variable.

22. A questionnaire allows the responses *strongly disagree*, *disagree*, *neither agree nor disagree*, *agree*, and *strongly agree*. Therefore, this variable is associated with an equal interval scale.

23. If a scale has units of measurement, then it is an equal interval scale.

24. Determination is a psychological construct.

25. Measurement error is inevitable.

26. The Celsius scale is a ratio scale.

27. Time to solve a sudoku puzzle is a valid measure of intelligence.

28. Estimating the average monthly income of Australians from a random sample of monthly incomes of Australians is an instance of inferential statistics.

29. Simple random sampling is essential for valid inferences from samples to populations.

30. A statistic is computed from all scores in a population.

31. A convenience sample is an instance of sampling error.

32. The worms living in a backyard in Cleveland, Ohio, represent a sample of all worms living in North America.

33. A parameter is a numerical characteristic of a sample.

34. The average annual income of working Britons is £26,000, but the average annual income in a random sample of working Britons is only £23,000. This is an example of sampling bias.

35. Measurement error is more like sampling error than sampling bias.

36. A jar contains 100 black beans and 100 white beans. A handful of beans drawn from this jar has 15 black beans and 19 white beans. This illustrates sampling error.

37. The worms living in a backyard in Charleston, South Carolina, represent a random sample of all worms living in North America.

Scenarios

38. Explain how you would obtain a simple random sample of 10 students in your class.

39. A researcher at a well-known university chose a random sample of 30 students from his department's participant pool and measured how long it took each student to complete a crossword puzzle. The 30 numbers in the table below are the scores he obtained. He found that it took 253.4 seconds (s) on average to finish the puzzle.

307	212	274	270	213
245	267	283	241	274
226	207	259	290	212
248	308	223	268	261
307	212	274	270	213
258	264	198	226	292

(a) Is this a discrete or continuous variable?

(b) What kind of scale characterizes this variable?

(c) Let's say the researcher concludes that university students in general can solve this crossword puzzle in 253.4 s on average. What can you say about his conclusion?

APPENDIX 1.1: INTRODUCTION TO EXCEL

This appendix introduces some basic functions in Microsoft Excel and then shows how to use them to compute your GPA. If you have used Excel before, you will probably not find much that's surprising here. However, if you don't know how a GPA is calculated, you may find the discussion below useful.

Excel Basics

An Excel file is called a *workbook*, and a workbook may contain several *worksheets*. By default, when a workbook is opened for the first time (created), a blank worksheet appears. The essential feature of an Excel worksheet is an array of cells, with rows indexed by numbers (1, 2, 3 . . .) and columns indexed by letters (A, B, C . . .), as shown in Figure 1.A1.1. Each cell is indexed by a unique combination of column and row indices (e.g., cell **C6**). Worksheet cells can contain many things, including text, dates, numbers, and formulas of many sorts.

At the bottom of Figure 1.A1.1, you will see *tabs* named **GPA**, **Final Grade**, **Histogram**, and **Sheet1**. The tab labeled **Sheet1** is highlighted, which means that this is the current worksheet. The other worksheets (**GPA**, **Final Grade**, and **Histogram**) contain Excel calculations that we will discuss later. Clicking on these other tabs will reveal the content of the corresponding worksheets.

Although the array of cells is constant in all versions of Excel, there are differences in the *user interface* between Macintosh and PC versions of Excel and between different versions of Excel for both Macintosh and PC. For example, the top part of Figure 1.A1.1 shows the user interface for a recent version of Excel for Macintosh; the interface might be somewhat different in another version of Excel. In the top part of Figure 1.A1.1, there are a number of icons that permit quick access to Excel functionality. Some of these may be self-explanatory and some may not. (Clicking on the disk icon, second from the left, saves changes made to the workbook.)

Below these icons are a number of tabs (**Home**, **Insert**, **Page Layout**, **Formulas** . . .). Clicking on a tab (e.g., **Home**) reveals a *ribbon* of icons and options. In Figure 1.A1.1, the **Home** ribbon has been chosen and a number of drop-down lists (e.g., for choosing fonts and font sizes) and icons are revealed. These are segregated by function to control (i) how clipboard items are pasted into cells, (ii) font styles, (iii) the alignment of cell contents, and (iv) number formatting. Clicking a different ribbon tab (e.g., **Insert**) will reveal a new set of icons relating to the ribbon's function.

The ribbon philosophy is that functions within Excel should be *discoverable*. So, if you would like to

---

FIGURE 1.A1.1 ■ An Excel Worksheet

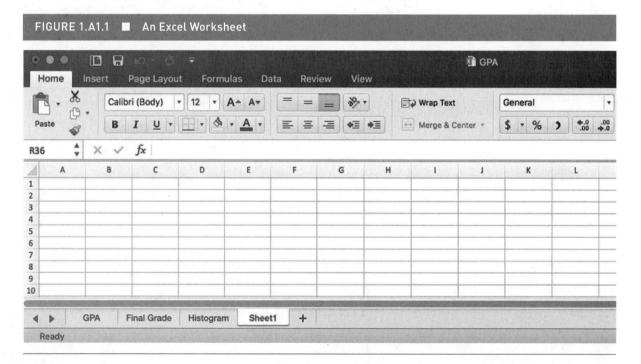

make a graph, you can click the **Insert** ribbon and find lots of options relating to graphs, among other things. If you are interested in a particular formula, you can click the **Formulas** ribbon, and lots of options relating to formulas will appear. Most of these icons will be self-explanatory, and you can click on them and discover, through experimentation, how they work.

We will typically have little to say about ribbons. Our focus will be on the array of cells that comprise the unchanging and important part of a worksheet. Even though the Excel interface may look different across platforms and versions, these differences will (typically) have no effect on how we use Excel.

Computing Your GPA

We will introduce computations in Excel by considering the question of how to compute your GPA. To make the discussion of GPA concrete, we will consider a university student who has completed six courses and received the letter grades A, B, C+, A+, A−, and D. At many universities, these letter grades (scores on an ordinal variable) would be converted to the following numbers: 4.0, 3.0, 2.3, 4.3, 3.7, and 1.0, respectively. We will call these numbers *grade points*. To compute the average of these grade points, we could simply sum them and divide by 6. Computing a GPA is just a bit more complicated than this because each course may be worth a different number of credits. In some universities, one-semester courses are worth 3 credits and two-semester courses are worth 6 credits. Much greater variety is possible, however. Some courses may be worth 2 credits, and others worth .5 or 4 credits. These differences do not affect how a GPA is computed. However, it is important to recognize that courses worth more credits contribute more heavily to the GPA.

To compute a GPA, we take into account both the grade points and the number of credits associated with each course. We say that GPA is a *weighted sum* of grade points. For our example, let's say that the first five courses in our example are worth 3 credits and the last course is worth 6 credits. Therefore, the student has earned 5 * 3 + 6 = 21 credits. For this example, we would compute the GPA as follows:

$$\text{GPA} = \frac{3}{21}(4) + \frac{3}{21}(3) + \frac{3}{21}(2.3)$$
$$+ \frac{3}{21}(4.3) + \frac{3}{21}(3.7) + \frac{6}{21}(1).$$

The numbers in parentheses are the grade points, and each fraction represents the proportion of total credits associated with a course. For example, the first course is worth 3/21 or 1/7 of the total credits, and the last course is worth 6/21 or 2/7 of the total credits. These fractions sum to 1, so we say that GPA is computed as a weighted sum.

Because each fraction has 21 in its denominator, an equivalent (and simpler) way to compute GPA is to (i) multiply each grade point by its corresponding credit, (ii) sum these products, and then (iii) divide this sum by the sum of the number of credits earned. This produces the following:

$$\text{GPA} = \frac{3(4) + 3(3) + 3(2.3) + 3(4.3) + 3(3.7) + 6(1)}{21}$$
$$= \frac{57.9}{21} = 2.76.$$

That's all there is to computing a GPA.

It would be a useful exercise to figure out what this student's GPA would have been if the course in which she got a grade point of 4.3 (i.e., A+) had been worth 6 credits and the course in which she got a grade point of 1 (i.e., D) had been worth 1 credit. (The answer is 3.23.) Courses with greater weight (more credits) contribute more heavily to the GPA than courses with less weight (fewer credits).

Computing Your GPA in Excel

The GPA calculation described above is shown in Figure 1.A1.2. Notice that the **Home** ribbon is not shown in this figure. Also note that the **GPA** tab at the bottom has been clicked to reveal the contents of this worksheet. Cells **B2** to **F2** are labels that I've typed in for the content below. The contents of these cells (**B2:F2**) are text. Some formatting has been done, which is not relevant to our calculations. However, you should be able to discover how to format text by using the contents of the **Home** ribbon in your version of Excel.

The column labeled **Grade** shows the six letter grades mentioned above. These are just for reference and are not part of a calculation. The column labeled **Grade Point** shows the grade points associated with the letter grades. The column **Credits** shows the number of credits associated with each course. These are numbers that you would enter yourself, and they involve no calculations.

The first calculations appear in the column labeled **GP*Credits**, and the formulas used to do these calculations are shown in the adjacent column labeled **Formulas**. For the first course, the product of **Grade Point** and **Credit** is 4 * 3 = 12, as shown in cell **E3**. Cell **F3** (in column **Formulas**) shows how this product was computed. To enter a formula into a cell, we first type the equal sign (e.g., in cell **E3**). To form the product of **Grade**

FIGURE 1.A1.2 ■ GPA Calculation

	A	B	C	D	E	F
1						
2		Grade	Grade Point	Credits	GP*Credits	Formulas
3		A	4.00	3	12.00	=C3*D3
4		B	3.00	3	9.00	=C4*D4
5		C+	2.30	3	6.90	=C5*D5
6		A+	4.30	3	12.90	=C6*D6
7		A-	3.70	3	11.10	=C7*D7
8		D	1.00	6	6.00	=C8*D8
9						
10						
11				Sum of Credits	21.00	=SUM(D3:D8)
12				Sum of GP*Credits	57.90	=SUM(E3:E8)
13						
14				GPA	2.76	=E12/E11

Point and **Credit**, we next click on cell **C3**, which makes cell **C3** appear immediately following the '=' sign in E3. (*Note:* The single quotes here and elsewhere are just used to delimit the characters to be typed. They are never typed as part of a formula.) We then type in '*' through the keyboard and then click cell **D3**. When these three things are done, the content of cell **E3** will look exactly like what is shown in **F3**. When we press return or enter, the formula will disappear and its value (12) will show up in the cell. (The number in cell **E3** has been formatted, and you may be able to discover how to do this through the **Home** ribbon in your version of Excel.)

We could repeat this procedure for the remaining five courses, but there is a much easier way to achieve the same result. If you click once on cell **E3**, it will be surrounded by a colored rectangle with a little square in the bottom right corner. Click and hold the mouse on this little square and then drag the mouse down to cell **E8**. Doing this will copy the formula in cell **E3** into cells **E4** to **E8**, exactly as shown to the right in the **Formulas** column. When you release the mouse, you will see that the formula has been applied to **Grade Point** and **Credit** for each course, and the result is what you see in Figure 1.A1.2. With a few simple steps, we have computed the product of **Grade Point** and **Credit** for each of the six courses.

To compute GPA, we must first sum **GP*Credits** and **Credits**. These sums, shown in cells **E11** and **E12**, were computed with the Excel function **SUM**. To compute the sum of **Credits**, we first type '=' (without the quotes) into **E11**. We then type '**SUM(**' after the equal sign. At this point, Excel is waiting to find out what to sum. Therefore, click on cell **D3** and hold the mouse down. While you hold the mouse down, drag it to cell **D8**. When this is done, **D3:D8** will appear after the opening parenthesis, as shown in **F11**. Now enter '**)**' and press return or enter. The

sum of the credits is shown in **E11**. We follow the same procedure to calculate the sum of **GP*Credits** in cell **E12**.

Finally, to compute the GPA, we simply divide the sum of **GP*Credits** in cell **E12** by the sum of **Credits** in cell **E11**. This calculation is shown in cell **E14**. The formula to achieve this is shown to the right in cell **F14**.

GPA is an important part of your academic record, so it is important to know how it is calculated and how grade points and credits affect it. You can play with the GPA calculator we just created to get a sense of these things. You should also consult the undergraduate calendar for your university to see how letter grades are converted to grade points. And you should check the credits associated with each of your courses. If you know the credit value of each course, then you will know which courses to devote the most time to if you are squeezed for time to study.

Finally, Table 1.A1.1 shows a typical printout from a student record. Not all of the information here is relevant to computing a GPA. As practice, use the information in Table 1.A1.1 to compute this student's GPA in Excel. The correct answer is 3.77.

Once you have computed the GPA for the data in Table 1.A1.1, try to discover how to use the Excel function **SUMPRODUCT** and the function **SUM** in a single cell to compute GPA. Here's a big hint: For the data in Figure 1.A1.2, the solution would be '= SUMPRODUCT(C3:C8, D3:D8)/SUM(D3:D8)'.

TABLE 1.A1.1 ■ Practice Data for Computing GPA

Course	Number	Term	Section	Credits	Letter Grade	Grade Point
PSYC	485	/4	3	2	B	3.0
APSS	316	/2	2	3	B	3.0
BIOL	100	/2	1	5	B+	3.3
PSYC	322	/2	1	3	A+	4.3
BIOL	388	/4	2	4	A+	4.3
APSS	412	/4	5	3	A+	4.3
APSS	256	/4	3	2	B	3.0
BIOL	335	/4	2	6	A+	4.3
BIOL	233	/2	1	3	A-	3.7
ITAL	322	/2	1	5	A	4.0
PSYC	258	/2	51	4	A+	4.3
PSYC	305	/4	5	4	B	3.0
PSYC	332	/4	5	3	B+	3.3
APSS	214	/2	51	5	A+	4.3
PSYC	311	/4	2	3	A+	4.3
ITAL	621	/2	1	7	B+	3.3

APPENDIX 1.2: INTRODUCTION TO SPSS

SPSS is among the most widely used statistical packages in psychology and the social sciences. It is almost certain that professors in your department use it as a routine part of their research. If you do not have SPSS installed on your computer, the SPSS website (www-03.ibm.com/software/products) provides links to sites that sell yearly SPSS licenses for approximately $110.00 (USD). When you purchase a license you will receive instructions about how to install it on your computer. The material in this appendix assumes that you have a student version of SPSS installed on your computer, or that you have access to a computer with SPSS installed. SPSS is available for Macintosh and PC, and these versions are almost indistinguishable. The images in this section were captured from a Macintosh version of SPSS, but you will have no trouble connecting these to the dialogs in the PC version.

SPSS allows one to perform many statistical analyses, but we will use just a subset of these. Our introduction to SPSS will be highly selective, with the objective of providing enough information to be able to carry out standard statistical analyses. Once you understand the logic of SPSS in one context, you will find it quite easy to learn its other features. For those who would like a more extensive introduction to SPSS, I recommend *Discovering Statistics Using IBM SPSS Statistics* by Andy Field (2013).

Data Editor

When SPSS is launched (whether on Mac or PC), you are presented with the Data Editor window where you can enter and save the data that you wish to analyze. The Data Editor provides two views of the data. At the bottom of Figure 1.A2.1, you will see that a tab labeled Data View has been highlighted, so we are looking at the data in Data View. Figure 1.A2.1 shows that I've pasted the data that was used in Appendix 1.1 on Excel. Like Excel, the main part of Data View is an array of cells indexed by columns and rows. The rows are numbered but the columns initially have no names. Each column in the Data Editor is a *variable*. When data are pasted into this window (as I've done in Figure 1.A2.1), SPSS will name these columns VAR00001, VAR00002, VAR00003, and so on; these names have been changed as described below.

The second view of the data is Variable View. When the Variable View option is chosen (clicked on), the window will switch and each column shown in Data View now corresponds to a row in Variable View (see Figure 1.A2.2). Variable View allows us to control properties and formats of our variables. These

FIGURE 1.A2.1 ■ Data Editor in Data View Mode

properties are shown in the columns in Variable View (i.e., Name, Type, Width, Decimals, Label, Values, Missing, Columns, Align, Measure, and Role). A few of these options will be described briefly.

The column in Figure 1.A2.2 labeled Name allows you to provide a meaningful name for the numbers in each column. As noted, SPSS originally names these variables VAR00001, VAR00002, and VAR00003. These names are useful for distinguishing variables in analyses, but I thought the names Letter_Grade, Grade_Points, and Credits would be better reminders of the meaning of the variables in each column. (We will discuss the column GPxCredits later.) Therefore, in Variable View, I changed VAR00001 to Letter_Grade, VAR00002 to Grade_Points, and VAR00003 to Credits. To do this, I simply selected the existing name (e.g., VAR00001), entered a better name (e.g., Letter_Grade), and then hit return to finish the job. When choosing names for your variables, you will find that they cannot contain spaces or special characters, such as ':', '*', or '$'.

Many of the remaining options in Variable View control the formatting of data in the Data View window. Let's look at the entries in the Type column. The first variable, Letter_Grade, is set to String, which means that the values of Letter_Grade are strings of letters and not numbers. The second and third variables (Grade_Points and Credits) are set to Numeric, which means that the values of these variables are numbers, not Dates or letter Strings.

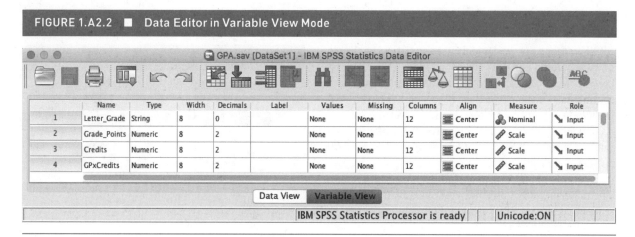

FIGURE 1.A2.2 ■ Data Editor in Variable View Mode

The full range of data types available can be seen by clicking the data type (e.g., Numeric, row 2, column Type), and then clicking again on the ellipsis (. . .) that appears. When this is done, the Variable Type dialog in Figure 1.A2.3 pops up. This dialog shows additional formatting options and data types. The desired data type is selected by clicking the appropriate radio button (◉). In this case, the data type Numeric has been chosen, and to the right there are two formatting controls that can be adjusted. The first is a Width, which is a number from 1 to 40 that specifies the maximum number of digits used to display numbers. The second formatting choice is Decimal Places, which specifies the number of digits *shown* to the right of the decimal point. (As in Excel, SPSS stores many digits following the decimal point, but Decimal Places only affects how many decimal places are displayed.) The settings in Variable Type are shown in the columns Width and Decimals in the Variable View window.

We will skip over most of the remaining columns in Variable View and just mention the columns labeled Align and Measure. The Align column allows you to specify the alignment of numbers in the columns in the Data View window. I've chosen to center the numbers, but you can right or left justify them if you prefer. The Measure column allows you to specify the nature of your data. The choices are Scale, Ordinal, and Nominal, as shown in Figure 1.A2.4. Letter_Grade is specified to be a nominal variable, and Grade_Points and Credits are set to scale.

The Transform and Analyze Menus

To compute a GPA, we (i) multiply each grade point by its corresponding credit, (ii) sum these products, and then (iii) divide this sum by the sum of the number of credits earned. (If you haven't read Appendix 1.1, this material is covered in the section "Computing Your GPA.") We will do this in two steps in SPSS. First, through the Transform menu item we will choose the Compute Variable . . . dialog to compute a new variable that is the product of Grade_Points and Credits, which

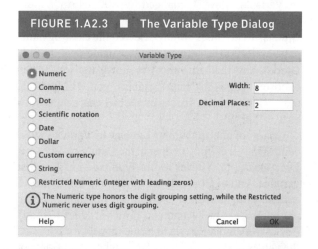

FIGURE 1.A2.3 ■ The Variable Type Dialog

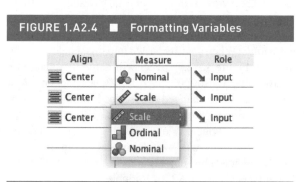

FIGURE 1.A2.4 ■ Formatting Variables

will be called GPxCredits. Through the Analyze menu item, we will then sum both Credits and GPxCredits.

THE TRANSFORM MENU

If you look back to Figure 1.A2.1, you will see at the top the menu items File, Edit, View, Data, Transform, and Analyze. (There are a few more to the right of Analyze that are not shown because they won't be used in this book.) Each of these menus provides access to specific functionality. If you click the Transform menu, a number of options will appear, the first of which is Compute Variable . . . Throughout this book we will use the convention Transform→Compute Variable . . . to denote how dialogs are accessed through the SPSS menus. In this example, Transform→Compute Variable . . . means click on the Transform menu and choose the Compute Variable . . . option. We will next see how the variable GPxCredits, shown in Figure 1.A2.1, was created.

When the Compute Variable . . . option is chosen, the dialog box shown in Figure 1.A2.5 appears. This dialog allows us to compute new variables. In this case, we want to compute a variable called GPxCredits. In the top left of Figure 1.A2.5 (Target Variable:), you will see that this new variable name has been typed in. Below this (on the left) is a panel showing the existing variables. Our goal is to compute the product of Grade Points and Credits, so we click on these in turn and then use the arrow (➡) to move them into the box titled Numeric Expression: at the top right. Between the two variable names is the multiplication symbol '*', which can be entered with the keypad as shown in the center of Figure 1.A2.5. When the OK button is pressed, SPSS will create a new variable called GPxCredits and store in it the product of Grade Points and Credits. When you check the Data Editor, you will find this new variable and its scores.

THE ANALYZE MENU

To compute GPA, we must sum Credits and Grade Points. We can do this with functions in the Analyze menu: Analyze→Descriptive Statistics→Descriptives. . . . This means that under the Analyze menu there is an item called Descriptive Statistics, which itself provides several options, one of which is Descriptives. . . . When the Descriptives . . . dialog appears (Figure 1.A2.6a), it shows the existing variables on the left (Grade Points, Credits, GPxCredits). Credits and GPxCredits have been transferred into the Variable(s): region with the ➡ button.

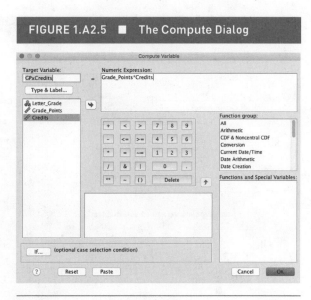

FIGURE 1.A2.5 ■ The Compute Dialog

We must now tell SPSS which descriptive statistics we wish to compute. This is done by clicking on the Options button at the top right of the Descriptives . . . dialog. Figure 1.A2.6b shows that many options are available, many of which won't mean much at this point. The option we want is Sum, so we check the box (✓) next to it and uncheck all of the other check boxes. We click Continue to return to the Descriptives . . . dialog, and there we click OK to proceed with the analysis.

The Statistics Viewer (Output Window)

The result of our analysis is shown in Figure 1.A2.7. Results are (almost) always shown in the Statistics Viewer. (If you click on Window in the SPSS menu bar, you will now see that there is a Data Editor window open as well as a Statistics Viewer window.) In the box labeled Descriptive Statistics, we see that the sum of Credits is 21 and the sum of GPxCredits is 57.9. Of course, these are exactly the same sums computed in Excel. This is as far as we can go in the GPA calculation in SPSS. To compute the GPA, you will have to divide GPxCredits by Credits using a calculator or Excel.

SPSS Versus Excel

It might seem that we've taken a lot of steps in SPSS only to arrive at an incomplete calculation of GPA. This is only partially true. As you become more familiar with SPSS, you will be able to do these calculations

FIGURE 1.A2.6 ■ Descriptive Statistics: Sum

(a)   (b)

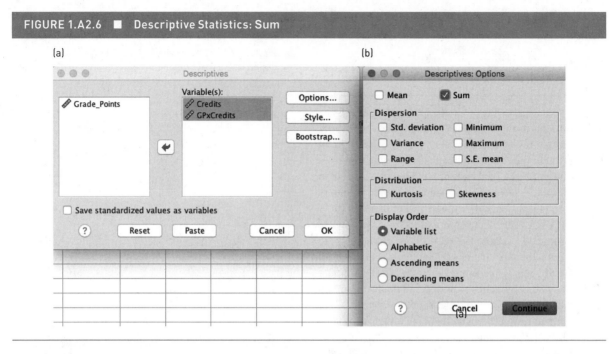

FIGURE 1.A2.7 ■ Output Window

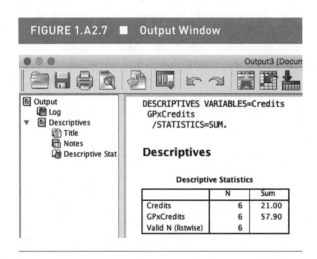

more quickly. Furthermore, in later chapters, we will encounter statistical calculations that are done far more quickly in SPSS than in Excel. This appendix shows only the first steps in the calculations, and you will have to know how to enter and format data before you can do the more advanced calculations.

However, it is also worth noting that each statistical platform has its strengths and weaknesses. Some things are done more efficiently in Excel and others more efficiently in SPSS. For the simple GPA calculation, Excel is probably the more appropriate platform. For multiple regression (Chapter 16), SPSS is more efficient. In Appendix 1.3, we will introduce **R**, which is far more efficient for calculations that we will see in Parts II, III, and IV.

APPENDIX 1.3: AN INTRODUCTION TO R

R is a widely used, free statistics package that can be downloaded from r-project.org. If you follow this link, it will take you to the **R** main page that is shown in Figure 1.A3.1. From there, you can download and install **R** on your computer. **R** is a fantastic resource for researchers because it allows us to do all of the analyses available in SPSS and more. Furthermore, researchers can write their own statistical analysis packages in the **R** programming language and share these with other researchers.

When **R** is launched, you will be presented with a window that looks like the one shown in Figure 1.A3.2. Most of the text is just legal-ish stuff that tells you that you're using freeware. The text also tells you how to get a demonstration of **R** functionality and how to get help.

The most important part of the **R** window is at the bottom where the **R** prompt (>) is shown. When you type things into the prompt, **R** evaluates them and returns a result. In its simplest use, **R** is like a calculator.

FIGURE 1.A3.1   ■   The R Project Page

The R Project for Statistical Computing

[Home]

Download

CRAN

R Project

About R
Logo
Contributors
What's New?
Mailing Lists
Reporting Bugs
Development Site
Conferences
Search

R Foundation

Foundation
Board
Members
Donors
Donate

Documentation

Manuals
FAQs
The R Journal
Books
Certification
Other

Links

Bioconductor
Related Projects

Getting Started

R is a free software environment for statistical computing and graphics. It compiles and runs on a wide variety of UNIX platforms, Windows and MacOS. To **download** R, please choose your preferred CRAN mirror.

If you have questions about R like how to download and install the software, or what the license terms are, please read our answers to frequently asked questions before you send an email.

News

- R version 3.3.1 (Bug in Your Hair) has been released on Tuesday 2016-06-21.
- R version 3.2.5 (Very, Very Secure Dishes) has been released on 2016-04-14. This is a rebadging of the quick-fix release 3.2.4-revised.
- **Notice XQuartz users (Mac OS X)** A security issue has been detected with the Sparkle update mechanism used by XQuartz. Avoid updating over insecure channels.
- The R Logo is available for download in high-resolution PNG or SVG formats.
- useR! 2016, will take place at Stanford University, CA, USA, June 27 - June 30, 2016.
- The R Journal Volume 7/2 is available.
- R version 3.2.3 (Wooden Christmas-Tree) has been released on 2015-12-10.
- R version 3.1.3 (Smooth Sidewalk) has been released on 2015-03-09.

We type in mathematical expressions, and **R** evaluates them and then returns the answer. For example, I typed in the mathematical expression `4+3+2.3+4.3+3.7+1`, and **R** evaluated it and returned the answer. The numbers I typed into the prompt are the six grade points that we used as examples in the previous two appendices. (If you haven't read Appendix 1.1, this material is covered in the section "Computing Your GPA.") The sum of these six numbers is then printed out to the window. **R** shows that the sum of these six numbers is 18.3.

To compute GPA, we first sum the product of `Grade Points` and `Credits`. To compute the sum of these products, I typed in `3*4+3*3+3*2.3+3*4.3+3*3.7+3*1`, and **R** returned 54.9. I realized that I had made a mistake; the last product should have been 6*1. Having to retype all of this would be tedious, so I pushed the up arrow on my keyboard and **R** showed me the last line I had entered: `3*4+3*3+3*2.3+3*4.3+3*3.7+3*1`. I changed the 3 to a `6`, pressed return, and **R** displayed the sum

I wanted, 57.9. Next, I entered `3+3+3+3+3+6` to get the sum of credits; **R** returned 21. Finally, knowing the sum of Grade Points and Credits (57.9) and the sum of Credits (21), I can enter `57.9/21` at the prompt to find GPA and **R** returns 2.757143. So, in just three lines, I have computed a GPA. Computing GPA this way could be quite tedious and error prone when there are many more grades to combine.

R can be used more efficiently by defining variables. For example, at the **R** prompt, one can type

```
> sumCredits <- 3+3+3+3+3+6
```

The symbol <- is called the assignment operator. Using it in this way creates a variable called `sumCredits` that stores the sum of the six credits. When we make this assignment, **R** does not display any output; however, if we type `sumCredits` at the prompt, **R** displays the value it has stored in `sumCredits`:

```
> sumCredits
[1] 21
```

Single values can be stored in variables, and collections of values can also be stored in variables. For

FIGURE 1.A3.2 ■ The R Project Window

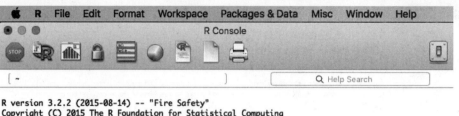

example, to store the six credits in a variable, we would enter the following command at the **R** prompt:

```
> credits <- c(3, 3, 3, 3, 3, 6)
```

The c in this case stands for combine. Again, nothing is returned when we make this assignment, but if we type credits at the prompt, **R** displays

```
> credits
[1] 3 3 3 3 3 6
```

In mathematics, we call a single row or column of numbers a vector. Therefore, we will call credits a vector. We can define gradePoints as a vector in a similar way:

```
> gradePoints <- c(4, 3, 2.3, 4.3, 3.7, 1)
```

To compute GPA, we have to multiply gradePoints by credits and then sum these products. The following command computes the product of our two vectors and then stores the result in a new variable called GPxCredits:

```
> GPxCredits <- gradePoints*credits
```

When we type GPxCredits at the prompt, **R** displays the vector stored in GPxCredits:

```
> GPxCredits
```

```
[1] 12.0 9.0 6.9 12.9 11.1 6.0
```

We saw these products in Appendices 1.1 and 1.2.

The final step of computing GPA is to divide the sum of GPxCredits by the sum of credits. This can be done in a single command as follows:

```
> sum(GPxCredits)/sum(credits)
[1] 2.757143
```

If we had wanted to store this result for later reference, we could have used the assignment statement as follows:

```
> GPA <- sum(GPxCredits)/sum(credits)
```

This concludes our brief introduction to **R**. Because **R** is freely available, you can download and use it at any time. **R** is far more powerful than you might expect from our simple introduction, and there are many online guides to **R** that you may wish to explore. In this book, we will use **R** only for advanced computations that can't be done in Excel or SPSS. But keep in mind that all computations that can be done in Excel and SPSS can also be done in **R**.

2 DISTRIBUTIONS OF SCORES

INTRODUCTION

Imagine that you received a grade of 80% on your first statistics midterm. Would you be happy with this? Is 80% a good grade? Well, it all depends. It would be a great mark if it was the highest in the class and a terrible mark if it was the lowest in the class. You can't interpret a grade without knowing where it falls in the distribution of grades.

Distribution is one of the most frequently used terms in statistics. A distribution conveys the relative frequency with which values of a variable occur in a sample or population. A distribution also conveys the relative standing of scores within a sample or population. In this chapter, we will show how distributions can be conveyed in tabular and graphical form. There are two reasons to examine distributions. First, from a strictly descriptive point of view, distributions can provide some sense of the order in a large set of scores, as in the example above. Second, from an inferential point of view, distributions provide the crucial link between statistics and parameters. We will encounter distributions in all subsequent chapters, so now is a good time to establish the basics.

DISTRIBUTIONS OF QUALITATIVE VARIABLES

Frequency Tables for Qualitative Variables

We've noted that psychology is a large and varied area of study, so the mention of psychology brings many different things to mind. Students entering psychology may have quite different interests and quite different expectations about what they will be studying. Let's imagine a psychology department in a large North American university. The researchers in this department are concentrated in the following five areas: clinical psychology, neuroscience, developmental psychology, cognition, and health psychology. This year the department admitted 512 new students and each was asked to choose which of the five areas listed above best represents his or her interest in psychology. There are two standard ways to represent the results of this survey. We could create either a **frequency table** or a *bar graph*. Table 2.1 shows the survey results in a frequency table.

The variable in this case is *area preference*. It is a qualitative variable (also called a categorical or nominal variable) and its values are Clinical, Neuroscience, Developmental, Cognition, and Health Psychology. In Table 2.1, the number of the 512 students who chose the corresponding area as their preferred area is listed under the heading f (frequency). We refer to these numbers as **raw frequency counts** (or tallies). The total number of students

A **frequency table for a qualitative variable** conveys the number or proportion of scores in a sample or population having each value of a variable.

Raw frequency counts (or tallies) represent the number of scores in a sample or population having a particular value or falling in a given interval.

who provided responses (i.e., 512) is shown at the bottom of the column. The proportion of students preferring each of the areas is shown under the heading *p* (proportion). Proportions are obtained by dividing the number of students preferring a given area by the total number of students surveyed: e.g., 151/512 = .29, when rounded to two decimal places. Proportions convey the **relative frequency** of occurrence of each value of the variable. It is clear from Table 2.1 that students are quite evenly split between Clinical, Neuroscience, and Developmental Psychology but they have less interest in Health and Cognitive Psychology.*

Bar Graphs

A frequency table is a perfectly cromulent way to convey the relative frequency of occurrence of each value of the variable. The same information can be conveyed graphically using a **bar graph,** as illustrated in Figure 2.1. The *x*-axis in Figure 2.1 shows area preference and the *y*-axis shows the proportion of students choosing each of the five areas. In general, graphical representations are preferred over tables because they provide an immediate visual characterization of the relationship between values and frequency. That is, we can judge which bar is tallest more quickly than we can find the largest number in a column of numbers. A detail to keep in mind when making a bar graph is that the bars should not touch each other. We will return to this point later when we look at histograms.

In psychology, we can use frequency tables and bar graphs to show the distribution of scores on qualitative variables such as sexual orientation, religious affiliation socioeconomic status, and career type, to name just a few. Because there is no natural ordering to the values of a qualitative variable, the order in which they're listed in the table or bar graph is arbitrary.

Relative frequencies represent the proportion of scores in a sample or population having a particular value or falling in a given interval.

A **bar graph** is a graphical depiction of the information in a frequency table. Each value of the variable is represented by a bar, and the height of each bar represents the number or proportion of scores having that value.

TABLE 2.1 ■ Area Preferences

Area Preference	f	p
Clinical	151	.29
Neuroscience	142	.28
Developmental	124	.24
Health	51	.10
Cognition	44	.09
Total	512	1.00

Note: Area preferences for 512 newly admitted psychology students; frequency (*f*), proportion (*p*).

FIGURE 2.1 ■ Bar Graph of Area Preference

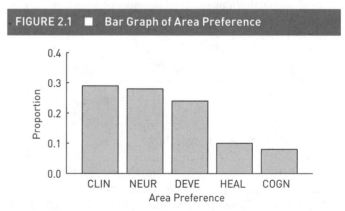

Proportion of 512 students preferring each of five areas of psychology. CLIN = clinical psychology; COGN = cognitive psychology; DEVE = developmental; HEAL = health psychology; NEUR = neuroscience.

---

*The great thing about an education is that it opens your eyes to new things. These (hypothetical) students may have little interest in cognitive psychology when they enter college because they don't really know what it is. Once they are exposed to the fascinating phenomena covered in cognitive psychology, their minds may well change. After all, that's what minds are for: changing.

LEARNING CHECK 1

1. A pollster was interested in social network preferences of Americans. A random sample of ninety-six 18- to 23-year-old Americans was chosen and asked about their social network preferences. The possible responses were Facebook (F), Twitter (T), or other (O). (The data in this table, and most others in the book, are available at study.sagepub .com/gurnsey.)

O	F	F	F	T	F	F	F	O	F	F	F	O	F	F	F	T	F	F	F	O	F	F	F
T	F	F	F	T	F	F	T	F	F	T	F	T	F	F	F	T	F	F	T	F	F	T	F
F	F	F	F	F	T	F	F	F	F	O	F	F	F	F	F	F	F	T	F	F	F	O	F
F	F	F	T	F	F	F	F	O	T	F	F	F	F	F	T	F	F	F	F	O	T	F	F

(a) How many individuals chose Facebook?

(b) How many individuals chose Twitter?

(c) How many individuals identified something besides Facebook or Twitter?

(d) From the data above, create a frequency table with columns labeled Social network preferences, f, and p.

(e) Plot the data above in a bar graph with axes properly labeled.

Answers

1. (a) 72. (b) 16. (c) 8.

(d) Frequency table of social network preferences.

Social network preferences	f	p
Facebook	72	.75
Twitter	16	.17
Other	8	.08
Total	96	1.00

(e) Histogram of social network preferences.

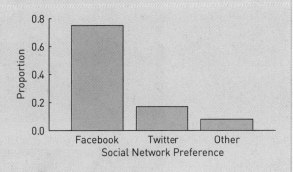

DISTRIBUTIONS OF DISCRETE QUANTITATIVE VARIABLES

We now turn to quantitative variables. Recall from Chapter 1 that quantitative variables can be discrete or continuous. Discrete quantitative variables are typically integers, and continuous quantitative variables are typically real numbers. The simplest situation to consider is the distribution of a discrete quantitative variable with a small number of values, so this is where we'll start.

Let's consider the grades on a pop quiz in an introductory statistics course. The quiz consisted of 10 multiple-choice questions, so the values of the variable are the numbers 0 to 10 correct. There were 80 students in the class, and their quiz scores are shown in

Table 2.2. It is rather hard to get an overall impression of these grades. Was the quiz hard? Was it easy? Was it moderately difficult? To answer these questions we would have to see how these grades are "spread out," or in the terminology we've been developing, we would need to see how these grades are *distributed*. What could we do to see the order in these data? Well, sorting them might help. This has been done in Table 2.3. Each column is sorted from lowest to highest, and the last score in a given column is lower than or equal to the first score in the column to its right. Sorting the grades in this way helps because we can see clearly that there are fewer 4s and 10s than there are 7s. We can do better than this, however, by making a *frequency table*.

Frequency Tables for Discrete Quantitative Variables

A **frequency table** for a discrete quantitative variable conveys the number or proportion of scores in a sample or population having specific values of the variable. The frequency table may also convey the number or proportion of scores at or below a given value of the variable.

Table 2.4 is another example of a **frequency table**. To create this frequency table, we first determine the maximum and minimum scores in the collection. We then list all values of the variable between the maximum and the minimum in the left-hand column. Values are

TABLE 2.2 ■ Scores From a Statistics Quiz									
6	7	9	8	8	7	7	7	7	6
6	6	7	8	6	7	7	8	4	7
7	6	8	7	6	7	7	6	8	8
5	7	5	7	10	8	8	7	10	6
7	7	7	6	6	5	8	7	7	6
6	6	7	6	9	9	6	8	7	7
8	8	6	7	8	8	7	5	6	7
6	7	7	7	6	6	8	7	7	8

TABLE 2.3 ■ Scores From a Statistics Quiz (Sorted)									
4	6	6	6	7	7	7	7	8	8
5	6	6	6	7	7	7	7	8	8
5	6	6	7	7	7	7	8	8	8
5	6	6	7	7	7	7	8	8	9
5	6	6	7	7	7	7	8	8	9
6	6	6	7	7	7	7	8	8	9
6	6	6	7	7	7	7	8	8	10
6	6	6	7	7	7	7	8	8	10

Note: Sorted from lowest to highest.

always listed from largest to smallest, for reasons that will soon become clear. We don't need to include values greater than the maximum or lower than the minimum, which in this case are 10 and 4, respectively. However, we can't skip values that happen not to occur. For example, if there had been no occurrences of 9 in our data set, we must still have an entry for 9 in our table.

Once the values have been listed, we count the number of instances of each. These counts are shown in the column labeled f (for frequency). You should be able to confirm that these entries are correct. There is 1 occurrence of 1 and there are 4 occurrences of 5, 21 occurrences of 6, and so on. The number of scores counted (i.e., $n = 80$) is shown at the bottom of this column. Now we are starting to get a much clearer picture of how grades were distributed on this quiz. The most frequently occurring grade was 7, and grades of 6 and 8 also occurred quite frequently. On the other hand, grades of 4, 5, 9, and 10 were quite infrequent.

Now, if you were a student in this class, you would want to know how your score compares with those of the rest of the class. One way to do this is to count how many people scored the same or lower than you. Such numbers are called **cumulative frequencies**, and they're listed in the column "Cumulative f" (cumulative frequency). Consider the value 4. This is the lowest value in the table, and there is only one instance of a 4 in this data set. Because only one person scored 4 or lower, we enter a 1 in the row for 4 in the cumulative frequency column. There were four instances of 5 in our table. This means that there were $1 + 4 = 5$ scores at or below 5 in our data set, so we enter a 5 in the row for 5 in the cumulative frequency column. There were 21 scores of 6, so there were $1 + 4 + 21 = 26$ scores at or below 6. Therefore, we enter 26 in the row for 6 in the cumulative frequency column, and so on. Because 10 was the highest score in our set of 80 scores, 80 scores are at or below 10.

Raw frequency counts (f) and cumulative frequency counts (Cumulative f) are useful, but the meaning of either of these numbers depends on how many scores we're dealing with. A cumulative frequency of 80 means one thing if $n = 80$, but quite another if $n = 800$. Therefore, it is conventional to express frequencies and cumulative frequencies as proportions, by dividing them by the number of scores in the data set (n). Consider the value 10 in Table 2.4. Only 2 of 80 scores in the set were 10, which corresponds to a proportion of $p = 2/80 = .025$. (Please note that proportions will always be rounded to two decimal

> A **cumulative frequency** is the number of scores at or below a given value of a variable.

Value	f	Cumulative f	p	P
10	2	80	.03	1.00
9	3	78	.04	.98
8	17	75	.21	.94
7	32	58	.40	.73
6	21	26	.26	.33
5	4	5	.05	.06
4	1	1	.01	.01
	$n = 80$			

TABLE 2.4 ■ Frequency Table of 80 Quiz Scores

Note: Frequency (f), cumulative frequency (cumulative f), proportion (p), and cumulative proportion (P).

A **cumulative proportion (P)** is the proportion of scores at or below a given value of a variable.

The **percentile rank** of a score is its cumulative proportion multiplied by 100; i.e., (*P**100%).

places in tables.) The fourth column in Table 2.4 shows the proportion (*p*) of scores occurring for each value of the variable. Similarly, to obtain the **cumulative proportion (*P*)** for cumulative frequencies, we divide the number of scores at or below a given value by the number of scores in the set. For example, 26 scores are at or below 6, which means the proportion of scores at or below 6 is 26/80 = .325. Of course, proportions can be expressed as percentages by multiplying them by 100. Thus, 32.5% of the 80 scores were at or below 6. The cumulative percentage associated with a score is often called the **percentile rank** of the score.

In statistics, notation is critical. In this case, it is absolutely essential to recognize that *p* and *P* denote very different quantities. Lower-case *p* is the proportion of scores having a particular value. Upper-case *P* is the proportion of scores at or below a given value. It may seem evil that a difference in capitalization is so important. However, this notation is almost universal in the statistics literature, so it is good to get used to it at the outset. With time, *p* and *P* will become as distinctive as X and O.

Histograms

A **histogram** of a discrete quantitative variable is a graphical depiction of the number or proportion of scores in a set having a specific value of a variable.

Figure 2.2 plots the values shown in the left column of Table 2.2 against the proportion of scores having each value. The graph in Figure 2.2 is called a **histogram**. Histograms are very similar to bar graphs but there are two differences. First, qualitative variables have no natural ordering, so how the categories are arranged on the *x*-axis of a bar graph is arbitrary. It would change nothing if the area preferences in Figure 2.1 were reversed or shown in random or alphabetical order. This is not true for quantitative variables. There is a natural ordering to the values of quiz score, so they must be placed in this natural order on the *x*-axis of a histogram. Think about how hard it would be to read Figure 2.2 if the *x*-axis were labeled 4, 10, 9, 5, 8, 6, 7. This order would put the proportions in ascending order, but the sense of how scores are distributed in the set would be lost.

The second difference between histograms and bar graphs is that there is no space between the bars of a histogram as there is between the bars of a bar graph (see Figure 2.1). We adopt this convention to convey the fact that the levels of the variable have a natural order. When the bars touch, this conveys the continuity of adjacent values of the variable.

The example we just worked through involved a variable with a conveniently small number of values. Our approach would be quite different if we had 50 scores that ranged from 0 to 100. The approach we would take in this situation is very similar to the approach we would take when dealing with continuous variables, which will be described in the next section. Therefore, as frequency tables for continuous variables are explained, keep in mind how this would apply to discrete quantitative variables having a large number of values.

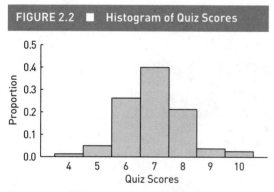

FIGURE 2.2 ■ Histogram of Quiz Scores

Histogram of 80 scores on a statistics quiz.

DISTRIBUTIONS OF CONTINUOUS VARIABLES

Much of what we've learned about frequency tables for discrete quantitative variables applies to continuous quantitative variables, but there are some interesting differences. Let's continue thinking about statistics grades, but this time they will be final grades that are computed from a number of different sources. Your final grade may be derived

LEARNING CHECK 2

1. A literature professor was interested in whether his students were familiar with the characters in J. R. R. Tolkien's *Lord of the Rings* (*LOTR*). He made a list of 24 names. Twelve were from *LOTR* and 12 were not. Each of the 48 students in his class answered yes or no to each of the 24 names. The professor counted the number of the 12 *LOTR* characters to which each of his students said yes. The table below shows the data he collected.

6	2	5	2	2	6	8	4	3	3	12	5
4	2	3	3	4	3	5	7	5	4	5	12
6	7	0	4	4	0	3	5	6	3	5	5
6	12	8	3	2	5	4	2	8	4	4	8

(a) From the data above, create a frequency table that is formatted like Table 2.4. Use the following column heads and fill in the appropriate values: Value, f, Cumulative f, p, and P.

(b) Plot the data in a histogram with axes properly labeled.

(c) Can you think of any reason that the professor's test was not a good one? (*Hint:* How might you get a perfect score?) Can you think of a way to fix the problem?

Answers

1. (a) Frequency table of LOTR quiz-scores

Value	f	Cumulative f	p	P
12	3	48	.06	1.00
11	0	45	.00	.94
10	0	45	.00	.94
9	0	45	.00	.94
8	4	45	.08	.94
7	2	41	.04	.85
6	5	39	.10	.81
5	9	34	.19	.71
4	9	25	.19	.52
3	8	16	.17	.33
2	6	8	.13	.18
1	0	2	.00	.04
0	2	2	.04	.04
	$n = 48$			

(b) Histogram of LOTR quiz-scores

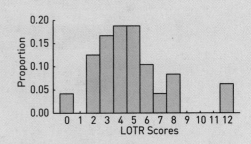

(c) Saying yes to all 24 names on the list means you would earn a perfect score, even if you knew nothing about *LOTR*. A better test would be to count up the number of times a participant correctly said yes to an *LOTR* character and correctly said no to a non-*LOTR* character. Now the maximum score would be 24, and we'd expect a score of approximately 12 if a student was guessing, or exactly 12 if a student said yes to all 24 names.

from your midterm grades, your final exam grade, your lab assignments, and possibly participation credits. Each of these contributions (numbers) is scored on an interval scale. When your professor combines all of these numbers to obtain a final grade, it will almost certainly be a real number. A collection of such grades is shown Table 2.5. Again, it is hard to get an overall impression of these grades. To get a sense of whether the course was hard, easy, or a joke, we again need to see how the scores are distributed.

As before, sorting the scores can reveal some order. This has been done in Table 2.6. Each column is sorted from lowest to highest, and the last score in a given column is lower than or equal to the first score in the column to its right. Sorting the grades in this way helps, because it becomes clear that there are fewer scores in the 40s and 90s than there are in the 70s. It might be useful to stop and think for a moment about what difficulties we might face if we tried to make a frequency table for these scores in the same way that we made a frequency table for the quiz-score data.

The scores in Tables 2.5 and 2.6 are real numbers rounded to one decimal place. If we were to make a frequency table from the numbers in Table 2.6 in the same way we did for discrete variables, we would have to list all numbers between the minimum and maximum scores in the data set. From Table 2.6 we can see that the maximum score is 96.9, and the minimum score is 38.9. How many real numbers, rounded to one decimal place, are there between 38.9 and 96.9? Well, this is the same as asking how many numbers there are between 389 and 969. (That is, the decimal was shifted one place to the right.) The answer is $969 - 389 + 1 = 581$. That's a lot of numbers. If we proceed to make a frequency table with 581 entries, we'd find that (i) the table would stretch across many pages, (ii) most of the entries would be empty (remember there are only 60 scores and 581 rows), (iii) those entries that are not empty have only one or two scores (look at Table 2.6 and you'll see that very few numbers occur twice), and therefore (iv) the table would give us no sense of how the scores are distributed. How will we deal with this?

TABLE 2.5 ■ Final Grades From an Introductory Statistics Course

95.3	88.0	67.6	72.2	76.0	70.5	83.6	69.9	87.7	70.7
87.2	68.0	69.6	73.6	52.2	73.7	60.3	85.3	70.9	96.9
56.6	76.2	54.0	85.8	84.8	45.7	71.7	72.8	63.1	67.5
57.1	87.3	74.4	80.7	82.4	89.2	60.9	89.9	69.6	42.0
59.3	76.4	49.6	64.3	71.9	70.0	72.1	75.3	66.6	67.3
59.9	74.1	66.7	78.4	72.1	88.7	38.9	58.6	54.5	66.7

TABLE 2.6 ■ Final Grades From an Introductory Statistics Course (Sorted)

38.9	54.5	60.3	66.7	69.6	71.7	73.6	76.2	84.8	88.0
42.0	56.6	60.9	67.3	69.9	71.9	73.7	76.4	85.3	88.7
45.7	57.1	63.1	67.5	70.0	72.1	74.1	78.4	85.8	89.2
49.6	58.6	64.3	67.6	70.5	72.1	74.4	80.7	87.2	89.9
52.2	59.3	66.6	68.0	70.7	72.2	75.3	82.4	87.3	95.3
54.0	59.9	66.7	69.6	70.9	72.8	76.0	83.6	87.7	96.9

Note: Sorted from lowest to highest.

Grouped Frequency Tables

We use a *grouped frequency table* when we have a large (potentially infinite) number of possible scores. In such tables we define **intervals** and count the number of scores that fall in each interval. This sounds straightforward, but there are some rules that should be followed. Table 2.7 is a grouped frequency table. It contains a great deal of information, which we will break down bit by bit so that we can see the rules and logic that generated it. Then we will use these rules to build a somewhat different version of the table. Before we begin this process you should note that the main difference between Tables 2.4 and 2.7 is that the rows correspond to values of the variable in Table 2.4, whereas the rows correspond to intervals of the variable in Table 2.7. An interval is a range of values of a quantitative variable. Intervals in Table 2.7 serve the same role as categories in Table 2.1 and values in Table 2.4.

*An **interval** is a range of values of a quantitative variable.*

Number of Intervals and Interval Width

The first decisions to make when constructing a grouped frequency table concern the number and widths of the intervals. It is important that our intervals all have the same width, otherwise the table would be very difficult to interpret. The width of the intervals should be an integer number of the units of measurement. In the present example, we are treating grades as the units of measurement. In other examples, the units of measurement might be inches, milliseconds, or IQ points.

The number of intervals we have is determined by how many scores we have. Typically we aim to have 5 to 20 intervals. A small number of intervals is appropriate when we have very few scores, and a large number of intervals is appropriate when we have many scores.

Interval width depends on (i) the number of intervals we wish to have and (ii) the range of scores in the data set. The range is

$$range = maximum - minimum.$$

In our example, the range is $96.9 - 38.9 = 58$. We then divide the range by the approximate number of intervals we'd like to have. So, let's say we'd like to have about 10 intervals. This means we'd be aiming for an interval width of approximately $58/10 = 5.8$. Because our interval width must be an integer, we will have to choose one close to 5.8. We have some choice about which integer we choose, but we prefer widths that are natural, or intuitive. In my view, 5 and 10 would be good choices, whereas 6 and 9 are less intuitive. In Table 2.7, the interval width is 10, but it

TABLE 2.7 ■ Grouped Frequency Distribution of 60 Final Grades						
Score Limits	Real Limits	Midpoint	*f*	Cumulative *f*	*p*	*P*
90–99	89.5–99.5	94.5	3	60	.05	1.00
80–89	79.5–89.5	84.5	12	57	.20	.95
70–79	69.5–79.5	74.5	22	45	.37	.75
60–69	59.5–69.5	64.5	12	23	.20	.38
50–59	49.5–59.5	54.5	8	11	.13	.18
40–49	39.5–49.5	44.5	2	3	.03	.05
30–39	29.5–39.5	34.5	1	1	.02	.02
			n = 60			

Note: Frequency (*f*), cumulative frequency (cumulative *f*), proportion (*p*), and cumulative proportion (*P*)

could have just as well been 5. By choosing an interval width of 10 we will have fewer than 10 intervals. If we'd chosen an interval width of 5, we'd have had more than 10 intervals.

Score Limits

The **score limits** of an interval are the minimum and maximum whole values of the units of measurement that define the interval.

Intervals in a grouped frequency table are typically determined by so-called **score limits**. Score limits are expressed as whole numbers in the *units of measurement*. We've noted that the units of measurement in this example are grades. We have already determined that our intervals will be 10 units wide so there are 10 distinct grades in each interval. The lower score limit of each interval must be a multiple of the interval width. The highest interval must include the highest score in the distribution and the lowest interval must include the lowest score in the distribution. For our example, the score limits of the lowest interval are [30–39], and they include the lowest score in the distribution, which is 38.9 (see Table 2.6). The score limits of the highest interval in our example are [90–99], and they include the highest score in the distribution, which is 96.9.

Real Limits

When constructing a grouped frequency table, we have to assign scores to intervals. If the scores were whole numbers, we'd have no problem. A score of 90 would go in the highest interval [90–99] and a score of 89 would go in the second highest interval [80–89]. Things are not so simple when the scores are real numbers. Consider the score 89.9. Should it go in the interval [80–89] or the interval [90–99]? If these were grades from your class, you would argue, with justification, that it is most reasonable for 89.9 to go in the interval [90–99], because it is closer to the lower score limit of this interval than to the upper score limit of the interval [80–89]. Another way to put this is that if 89.9 were rounded to the next whole number, it would become 90, and not 89.

The **real limits** of an interval are the minimum and maximum real values of scores that define the interval.

This brings us to the column in Table 2.7 labeled "real limits." The **real limits** show the limits of all real numbers (values of the variable) included in each interval. Notice that for each interval, the real limits go from half a unit (grade) below the lower score limit to half a unit above the upper score limit. For example, the real limits of [90–99] go from 89.5 to 99.5. Because the score 89.2 is below the lower real limit of [90–99], it goes in the interval below (i.e., [80–89]). Because 89.9 is greater than the lower real limit of [90–99], it goes in this interval. Table 2.7 shows that there are three scores in the interval [90–99]. From Table 2.6, you can see that these three scores were 89.9, 95.3, and 96.9.

There is an inevitable question here about what to do when a score falls exactly on a real limit. For example, what to do with 39.5, which is simultaneously the upper real limit of the lowest interval and the lower real limit of the next interval. First, this happens very, very rarely in the real world of data collection. When real numbers are not rounded (as they are in Tables 2.5 and 2.6), it is very rare for a number to have 5 in the first decimal place and only zeros in the remaining decimal places. However, it helps to have a rule to rely on should this happen in an exercise or practice question. So, here it goes: we will always put a score that falls on a real limit boundary in the higher interval. This is completely arbitrary, but it won't have much practical consequence, and, if we're dealing with things like grades, we're giving the benefit of the doubt to the student.

Interval Midpoints

An **interval midpoint** is the middle value of an interval.

The middle points of intervals are called **interval midpoints** (duh!). Midpoints are defined as

$$\frac{maximum - minimum}{2},$$

where maximum is the upper real limit and minimum is the lower real limit.

Therefore, we can describe the real limits of an interval as

$$midpoint \pm \frac{width}{2}.$$

In Table 2.7, this would mean that the real limits of intervals are defined as

$$midpoint \pm \frac{10}{2}.$$

The midpoints are evenly spaced so that

$$midpoint_2 = midpoint_1 + width.$$

You can see that the lowest midpoint in Table 2.7 (third column) is 34.5. Therefore, the second lowest midpoint is 34.5 + 10 = 44.5. Interval midpoints can be useful when constructing histograms because they require less space to print than do score limits or real limits. Therefore, one consideration in the choice of interval width is that odd widths are better than even widths. The midpoint of an interval with odd width (e.g., 5) will be an integer, whereas the midpoint of an interval with even width will end with .5 (e.g., 34.5).

Relative Frequencies

I've devoted several paragraphs to explaining intervals, interval widths, score limits, real limits, and midpoints. However, we don't want to lose sight of our objective, which is to express the scores in Tables 2.5 and 2.6 in a way that conveys how they are *distributed*. Column f in Table 2.7 shows the number of scores falling in each of the seven intervals in the table. It is now clear how our scores are distributed. There are very few extreme scores (i.e., less than 50 or greater than 90), so most of the scores in the set (54 of them) lie between 49.5 and 89.5. Furthermore, we see that the most frequently occurring scores fall in the interval whose score limits are 70–79 (real limits, 69.5–79.5; midpoint, 74.5).

We can express the frequency counts (tallies) as proportions (or percentages). There are 60 scores in the set, so dividing each number in f by $n = 60$ gives us the proportion of scores in each interval. This proportion expresses the *relative frequency* with which scores fall in each interval. Please note that these proportions have been rounded to two decimal places.

As we saw before, it is often important to know the relative standing of scores in a set. The column labeled "Cumulative f" shows the number of scores at or below each interval. There is one score in the interval [30–39], and there are two scores in the interval [40–49]. Therefore, there are three scores (1 + 2 = 3) at or below the interval [40–49]. There are eight scores in the interval [50–59]; therefore, there are 11 scores (1 + 2 + 8 = 11) at or below the interval [50–59], and so on. These cumulative frequencies can be converted to cumulative proportions (column "P") by dividing them by the total number of scores in the set. Therefore, for each interval we can determine the proportion of the distribution falling at or below each interval. Of course, 100% ($P = 1$) of the distribution falls at or below the highest interval. We can now see why we list intervals (or values) from highest to lowest in frequency tables. Doing so makes intuitive sense because the first entry in the table is the one with 100% of the distribution at or below it.

Desirable Features of Grouped Frequency Tables

Table 2.7 was constructed to illustrate the components of a grouped frequency distribution. We will now summarize some general rules that emerged from this discussion.

1. The interval width should be an integer number of the units of the variables. The units might be centimeters, inches, pounds, seconds, milliseconds, or nanoseconds, to name just a few. Although it would be possible to have an interval width of 3.1415 inches, for example, such a width would make it impossible to specify intuitive score limits.

2. The interval width should be familiar. In Table 2.4, the interval width is 10, which is familiar, or intuitive. In my view, 2, 5, 10, 20, 25, 50, or 100 are also good; of course, which of these you choose depends on the range of the scores in your set. So, 100 is a sensible interval width if your scores range from 0 to 2000 but not if they range from 0 to 200.

3. All lower score limits should be multiples of the interval width. For example, if we had chosen an interval width of 10 for a set whose lowest score is 12, the lowest interval should not be 12 to 21. Rather, the lowest interval should be 10 to 19.

4. The real limits of the highest interval should include the highest score, and the real limits of the lowest interval should include the lowest score.

5. All intervals should be continuous with each other. For example, even if the interval [40–49] (from Table 2.7) contained no scores, that interval must be included in the table.

Procedure for Making a Grouped Frequency Table

From the review above we can outline the steps to follow when making a grouped frequency table.

Step 1. Determine the range of scores in your set by subtracting the smallest from the largest. For the data shown in Tables 2.5 and 2.6, the range is $96.9 - 38.9 = 58$.

Step 2. Determine a reasonable number of intervals, which will provide a rough estimate of interval width. To do so, divide the range by possible numbers of intervals (e.g., 7, 10, 15) to get a sense of what interval widths are appropriate. In the case of the data in Table 2.6, these widths would be 8.4, 5.9, and 3.9. None of these widths is "intuitive," but 8.4 is close to 10, and 3.9 is close to 5. Both 10 and 5 seem like reasonable choices. We chose 10 for Table 2.7, but for this example we'll choose 5.

Step 3. Define the score limits, making sure that

(a) the lowest real limits contain the lowest score in the data set,

(b) the highest real limits contain the highest score in the data set, and

(c) all lower score limits are multiples of the chosen interval width.

Step 4. Arrange your intervals in the table from highest to lowest, with the highest at the top.

Step 5. Fill in columns as needed. You may choose all columns shown in Table 2.7 or a subset of these, as suits your purposes.

Table 2.8 shows a grouped frequency table for the data shown in Table 2.6. In contrast with Table 2.7, there are fewer columns because many of those in Table 2.7 are redundant and were shown only for illustration. Unlike Table 2.7, the interval width in Table 2.8 is 5. This makes 13 intervals rather than 7. This table provides a nice summary of the distribution of scores in Tables 2.5 and 2.6. However, you might choose to show the real limits or the interval midpoints rather than the score limits. You might also choose to show raw frequencies rather than proportions; in my view, proportions more effectively convey information because the units are standardized and thus can be compared across data sets.

Graphical Depiction of Frequencies

We noted earlier that graphs convey the distribution of scores as well as or better than tables. Figure 2.3 plots the data from Table 2.5 as six different histograms. Each histogram corresponds to a grouped frequency distribution having a different interval width (i.e., 40, 20, 10, 5, 2, and 1). The x-axis in each panel shows the interval midpoints. In some cases, only a subset of midpoints is shown for readability. Interval midpoints have been used on the x-axis because they require less space than the real limits, or the score limits. The y-axes show the proportion of scores falling in each interval.

Figures 2.3c and 2.3d correspond to the grouped frequency distributions in Tables 2.7 and 2.8. Both show a reasonable characterization of the data in Table 2.6. In my view, a width of 10 best conveys the data from Table 2.5, but this is largely an aesthetic judgment. Others might think a width of 5 better conveys the distribution of scores. Therefore, a large part of graph making requires judgments that can't be captured by a set of rules that will work in all possible cases. However, there should be little disagreement that an interval width of 40 (Figure 2.3a) is far too wide, and an interval width of 1 (Figure 2.3f) is far too narrow. Both interval widths yield a poor characterization of the data.

TABLE 2.8 ■ Grouped Frequencies		
Score Limits	p	P
95–99	.03	1.00
90–94	.02	.97
85–89	.15	.95
80–84	.05	.80
75–79	.08	.75
70–74	.28	.67
65–69	.12	.38
60–64	.08	.27
55–59	.08	.18
50–54	.05	.10
45–49	.02	.05
40–44	.02	.03
35–39	.02	.02

Note: Proportion (p), cumulative proportion (P).

LEARNING CHECK 3

1. I want to make a grouped frequency table from the following numbers: [4.659, 5.135, 5.811, 6.346, 6.792, 6.801, 7.273, 7.479, 9.179, 9.269, 9.377, 9.61, 10.476, 10.705, 12.005, 12.971]. If the interval width is 3, then

(a) Create a frequency table that is formatted like Table 2.8 from the data above. Use the following column headings and fill in the appropriate values: Score Limits, p, and P.

(b) Plot the data as a histogram.

Answers

1. (a) Grouped frequency distribution of 16 scores.

Score Limits	p	P
12–14	.13	1.00
9–11	.38	.88
6–8	.38	.50
3–5	.13	.13

(b) Histogram of 16 scores.

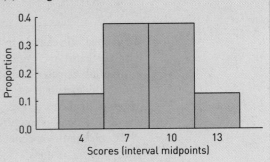

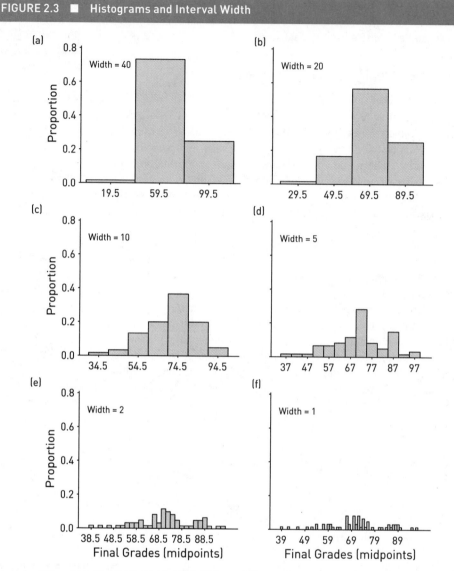

FIGURE 2.3 ■ Histograms and Interval Width

Interval widths of (a) 40, (b) 20, (c) 10, (d) 5, (e) 2, and (f) 1. Values on the *x*-axis represent interval midpoints.

PROBABILITY

We started this chapter by noting that "distribution" is one of the most frequently used terms in statistics. *Probability* is another frequently used term. In fact, probability may be used just as often as distribution because a huge part of statistics deals with so-called probability distributions. In this section, we will develop a specific notion of probability that is at the heart of everything to follow. We first distinguish between subjective and objective notions of probability.

If you heard someone say "The Blue Jays probably won't make the post-season this year," you would understand this to mean that in the speaker's judgment, it is unlikely that the Blue Jays will make it into the post-season. [The Blue Jays are a major league baseball team from Toronto, Ontario. They won the World Series 2 years in a row in the early 1990s but have made it into the post-season (playoffs) only twice since then.] The speaker is expressing his subjective judgment about the likelihood of a specific outcome. This is an example of **subjective probability**. You would be expressing a subjective probability if you were to say, "I'm probably going to ace this course."

Objective probabilities are numerical expressions of the likelihood of some event occurring. The definition of probability used in this book is related to the answer to the following familiar question: If I flip a fair coin, what is the probability that it will come up heads? Most people would say the probability is $1/2 = 0.5$, recognizing that half the time a flipped coin will come up heads, and half the time it will come up tails. This answer is what statisticians call the *frequentist*, or *long-run* notion of probability. Central to the frequentist view of probability are the ideas of *proportions*, outcomes, events, and sampling experiments. To understand these concepts, we will consider two familiar situations in which they apply.

Coin Flipping

Typically the two sides of a coin are referred to as heads and tails. We can think of heads and tails as two values of a qualitative variable. (When qualitative variables have two values, we call them *dichotomous* variables.) Each time you flip a coin there are two possible **outcomes**, heads and tails. If the coin is fair, then both outcomes are equally likely. Therefore, flipping is one way to select one of the two possible outcomes randomly. We call the random selection of one of the possible values of the variable a **sampling experiment** (or **trial**). In a sampling experiment, each outcome has an equal chance of being selected.

iStock.com/asafta

If one of the outcomes is defined as a *success*, we can count the number of successes ($N_{success}$) in a given number of sampling experiments (N_{SE}). The **proportion of successes** in N_{SE} sampling experiments is $p = N_{success}/N_{SE}$. For example, if a success is defined as "coming up heads," then 14 successes in 25 sampling experiments means that the proportion of successes is $p = N_{success}/N_{SE} = 14/25 = .56$.

Rolling a Die

A die is a cube with one to six dots painted its sides. (The plural of die is dice.) We can think of a die as a qualitative variable with six values. Therefore, each time you roll a die, the six possible outcomes are the six values of the variable: 1, 2, 3, 4, 5, or 6. The act of rolling a die is essentially to select one of the six possible outcomes randomly. If a success is defined as "rolling a 1," then 6 successes in 60 sampling experiments means that the proportion of successes is $p = N_{success}/N_{SE} = 6/60 = .1$.

iStock.com/malerapaso

Events

A success was defined above as a specific outcome in a sampling experiment. An **event** is one or more outcomes in a sampling experiment. For example, we can define a success as "rolling a 3 or a 6." In this case, the event is 3 or 6. We can compute the proportion of times an event occurs in the same way that we computed the proportion of times an outcome occurs. Let's imagine that in 75 rolls of a die, there are fourteen 3s and sixteen 6s. In this case, we would have $14 + 16 = 30$ successes, because the event is "3 or 6." Therefore, the proportion of successes is $p = N_{success}/N_{SE} = 30/75 = .4$.

Frequentist Definition of Probability

Proportions, events, and sampling experiments are critical to the frequentist, or long-run, notion of probability. To this point we have considered the proportion of times that events occur in a fixed number of sampling experiments. For example, if we obtained 14 heads (successes) in 25 coin flips (sampling experiments), then the proportion of successes is $p = .56$. However, our intuitive understanding of coin flipping tells us that heads should occur on 50% of flips, not 56% of flips. Therefore, our intuition is based on the idea of an *infinite* number of flips, or sampling experiments. If a fair coin is flipped (theoretically) infinitely many times, then it will come up heads on exactly 50% of the flips. That is, in the long run, a flipped coin will come up heads exactly 50% of the time. So the **probability** of coming up heads is $p = .5$.

From the frequentist perspective, the **probability** of an event is the proportion of times the event would occur if the same sampling experiment were repeated infinitely many times.

With these observations in mind, we can now say that the probability of an event is the proportion of times the event would occur in an infinite number of identical sampling experiments. Again, if a sampling experiment is the flip of a coin and a success is "coming up heads," then the probability of a success is the proportion of times heads would occur in an infinite sequence of flips. In this case, the probability of heads is .5. If a sampling experiment is the roll of a die and success is defined as rolling a 1, then the probability of rolling a 1 in an infinite number of sampling experiments is $1/6 = .1667$ when rounded to four decimal places.

Mutually Exclusive and Independent Events

In this section we have discussed the probability of outcomes and events. There are several interesting rules of probability that provide answers to the following questions: (i) What is the probability that two rolled dice will both come up 6? (ii) What is the probability of drawing two aces in a row from a deck of cards? (iii) What is the probability that a randomly chosen university student is a female studying psychology? Appendix 2.3 (available at study.sagepub.com/gurnsey) provides a brief introduction to these rules. So that we can move on to discussing probability distributions, we will skip over these rules. However, even if it is not required in your course, you may find this material interesting.

Two important concepts in probability are *mutual exclusivity* and *independence*. These are discussed in more detail in Appendix 2.1, but they will be defined here briefly.

Mutually exclusive events are events that cannot co-occur.

Two events are **mutually exclusive** if they cannot co-occur. For example, a flipped coin can come up heads or tails, but not both. Therefore, the possible outcomes of a coin flip are mutually exclusive. Similarly, a randomly chosen individual may have an IQ of 100 or 115, but not both. A randomly chosen car may be a Ford or a Toyota, but it can't be both, so these outcomes are mutually exclusive.

Events are **independent** if the occurrence of one does not affect the probability of the other occurring.

Two events (or outcomes) are **independent** if the occurrence of one does not affect the probability that the other will occur. For example, if two coins are flipped, the outcomes are independent because if one coin comes up heads it has no effect on whether the other coin will come up heads. Or, if the same coin is flipped twice, coming up heads on the first flip has no effect on the probability of it coming up heads on the second flip.

If I draw an ace from a deck of cards and then put it back in the deck (drawing *with replacement*), this has no effect on the probability that the next card drawn will be an ace; the probability is 4/52 that the next card drawn will be an ace because all four aces are in the deck. In other words, draws done with replacement are independent events.

Not all events are independent. If I draw an ace from a deck of cards and don't put it back in the deck (drawing *without replacement*), this does affect the probability that the next card drawn will be an ace; the probability is 3/51 that the next card drawn will be an ace, because one of the aces has been removed. Draws done without replacement are *dependent* events.

LEARNING CHECK 4

1. The statement "It will probably rain tomorrow" involves a subjective probability. [True, False]

2. I have rolled a die 500 times and the following table shows the number of times each of the six faces came up.

Die face	1	2	3	4	5	6
Frequency	87	78	88	78	92	77

(a) What proportion of times did the die come up 2?

(b) What proportion of times did the die come up 5?

(c) What proportion of times did the die come up 2 or 5?

(d) What proportion of times did the die not come up 6?

3. What is the probability that a flipped coin will come up heads or tails?

Answers

1. True.

2. (a) $p = 78/500 = .156$.

   (b) $p = 92/500 = .184$.

   (c) $p = (78 + 92)/500 = .34$.

   (d) $p = (87 + 78 + 88 + 78 + 92)/500 = .846$. Or, $1 - 77/500 = .846$.

3. $p = .5 + .5 = 1$.

PROBABILITY DISTRIBUTIONS

Although our discussion of probability may have seemed like a digression, probability is intimately related to our use of distributions in statistics. In this section, we will see that frequency tables can be thought of as **probability distributions**.

> A **probability distribution** conveys the probability that a randomly selected score will have a given value or fall in a given interval.

Grouped Frequency Tables, Random Sampling, and Probability

Table 2.8 showed the distribution of 60 scores. We can use it to answer a simple question: What's the probability that a randomly chosen score from this distribution will fall in the interval 80 to 84? Of course, by this we mean, what is the probability that the score will be within the real limits of 79.5 to 84.5? By posing the question this way, we have defined the *event* as "*in the interval [80–84]*." Looking back to Table 2.8 we see that 5% of scores fall in this interval. Drawing a single score from this distribution is a sampling experiment. If we repeat this sampling experiment infinitely many times with replacement, then 5% of the time the score will fall in the interval 80 to 84. We could equally say that the *probability* is .05 that a randomly chosen score will fall in the interval [80–84].

Remember the connection that we're making here. We've said that a distribution conveys the relative frequency with which each value of a variable (or range of values of a variable) occurs in a sample or population. This is equivalent to saying that a distribution shows the proportion of scores falling in any interval. When we specify the probability that a randomly chosen score will fall in a given interval, we mean the proportion of times a randomly chosen score would fall in that interval in an infinite number of identical sampling experiments.

We could describe another event as "*not in the interval [80–84]*." So, what is the probability that a randomly chosen score will not fall in the interval 80 to 84? Again, by this we mean, What is the probability that the score will be outside the interval with real limits of 79.5 to 84.5? Well, if 5% of the distribution falls in the interval [80–84], then 95% of the distribution falls outside this interval. Put another way, the probability is .95 that a randomly chosen score will fall outside the interval [80–84].

Given any distribution of scores, we can define any number of events. Events might be "score is greater than x," "score is less than x," "score is between x and y," or "score is outside the interval x to y." In the immortal words of Donald Trump, "this is huge." With this background established, we can now say that a probability distribution conveys the probability that a randomly selected score will have a given value or fall in a given interval.

This definition tells us that any relative frequency distribution is a probability distribution. For example, Table 2.1 and Figure 2.1 represent the same probability distribution. If we repeatedly choose individuals at random from this distribution (with replacement), then the probability of drawing a student whose area preference is Clinical would be .29. Table 2.4 and Figure 2.2 are also probability distributions. If we repeatedly choose scores at random from this distribution, then the probability of drawing a score of 7 would be .40. Tables 2.7 and 2.8 are probability distributions, as are all of the histograms in Figure 2.3.

Properties of Probability Distributions

There are some basic features of probability distributions that we will make use of in subsequent chapters. The first is that the probability of any event A is between 0 and 1. The symbol A can denote any event, such as a coin coming up heads, a rolled die coming up 4, a drawn card being red, or a randomly chosen university student being in psychology. The constraint that the probability of event A must be between 0 and 1 can be expressed as follows:

$$0 \leq p(A) \leq 1, \tag{2.1}$$

where $p(A)$ should be read as "the probability of A." If we think of coin flips, then the two possible outcomes are heads and tails, so $p(heads) = .5$ and $p(tails) = .5$, both of which are between 0 and 1.

The second property of a probability distribution is that the sum of all probabilities is 1. So, n mutually exclusive events can be denoted $A_1, A_2 \ldots A_n$, where n can be any integer. For example, if we are thinking about drawing cards from a deck, then n would be 52. The second property of a probability distribution can be stated as follows:

$$p(A_1) + p(A_2) + \ldots + p(A_n) = 1. \tag{2.2}$$

In any distribution there are many possible mutually exclusive events. Again, in the case of coin flips, the two possible events are heads and tails, and according to equation 2.2, $p(heads) + p(tails) = 1$. For qualitative variables the events are categories (such as clinical, developmental, neuroscience, etc.). There is a probability associated with each of these categories, and the sum of all these probabilities is 1. For discrete quantitative variables, the events might be whole numbers. The probability associated with each value of the variable is a number between 0 and 1, and the sum of all probabilities is 1. The same logic holds for grouped frequency distributions. The probability of a score falling in a given interval is between 0 and 1, and the sum of all such probabilities is 1.

Probability Density Functions

In the next few chapters, we will spend a lot of time discussing populations comprising scores on continuous variables. We will not represent these distributions using histograms (as in Figure 2.3), because it is impossible to record all scores in the population. Instead we will make use of **probability density functions** (*pdf*s). Figure 2.4 shows a hypothetical distribution of heights measured in inches. You can think of this as the distribution of heights of all American women. The interpretation of this figure should be somewhat intuitive. There are more scores around the center of the distribution, and the number of scores decreases as we move away from the center. If we think about the number of scores in each 1-inch interval, then we could talk about the *density* of scores per inch. When we use the term *density*, we mean the number of things per unit measure. Examples could be the number of people per square mile, the number of neurons per cubic millimeter, or the number of scores per 1-inch interval.

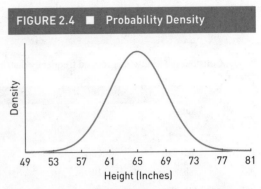

FIGURE 2.4 ■ Probability Density

The line depicts a probability density function that relates height (in inches) to the abstract quantity, density.

> A **probability density function** plots the density of scores at each value of a continuous variable. In statistics, density is defined at a point.

The statistician's use of the term *density* differs somewhat from our intuitive use of the term. For statisticians, density is the proportion of scores in an interval divided by interval width. That is, density (*d*) is defined as $d = p/w$, where *p* is the proportion of scores in an interval and *w* is the width of the interval. If we think about the distribution of a continuous variable, then the proportion of scores in an interval decreases as the interval width narrows. However, dividing by interval width compensates for the reduced proportion of scores. As interval width becomes extremely narrow, *d* converges to a very specific value. In the limit, density is defined at a point.

Therefore, for every value of *x* (where *x* might be height), there is a single density. You may be perplexed by the idea of density being defined at a point, because this means an interval having zero width. If you simply accept this new sense of the term, you will have no trouble following what is presented in subsequent chapters. If you would like to deepen your understanding, you can consult Appendix 2.4 (available at study.sagepub.com/gurnsey).

Density functions often have relatively simple mathematical definitions. From these we can determine the proportion of a distribution in any interval defined by two values of *x*, no matter how small or large this difference is. For example, the proportion of the distribution, shown in Figure 2.4, between 57 and 61 inches is .1359, or 13.59%. (We'll see in Chapter 4 how we know this.) Therefore, the proportion of the distribution outside this interval is $1 − .1359 = .8641$, or 86.41%. If we were to draw an individual at random from this distribution, the probability is .1359 that the person's height will be in the interval 57 to 61. As before, this means that in an infinite series of identical sampling experiments, 13.59% of the time the score we choose will be in the interval 57–61. Because density functions can be used to determine the probability of a score falling in a given interval with random sampling, they are referred to as probability density functions. One implication of this is that the area under the entire curve is 1. (If you've taken calculus, this means that the integral of this function is 1.) This fact is the analogue of what is shown in equation 2.2, which states that the sum of all probabilities in a frequency table is 1.

LEARNING CHECK 5

1. Consider the following grouped frequency distribution.

Score Limits	p	P
95–99	.01	1.00
90–94	.03	.99
85–89	.06	.96
80–84	.10	.90
75–79	.15	.80
70–74	.17	.65
65–69	.17	.48
60–64	.14	.31
55–59	.09	.17
50–54	.05	.08
45–49	.02	.03
40–44	.01	.01
35–39	.00	.00

(a) What is the probability that a randomly drawn score will fall in the interval 60 to 64?

(b) What is the probability that a randomly drawn score will fall in the interval 60 to 74?

(c) What is the probability that a randomly drawn score will fall below 89?

(d) What is the probability that a randomly drawn score will fall outside the interval 45 to 84?

2. State whether the following statements are true or false.

(a) The table to the left is a probability density function.

(b) Probability density function defines density at a point.

(c) We can use a probability density function to determine the proportion of scores in any interval.

Answers

1. (a) .14.

   (b) .48.

   (c) .96.

   (d) .11.

2. (a) False. Probability density is defined at a point.

   (b) True.

   (c) True.

SUMMARY

In this chapter, we've seen that a *distribution* conveys the *relative frequency* with which each value of a variable occurs in a sample or population, or the relative frequency with which a score falls in a given *interval*. The distribution of a qualitative variable can be conveyed in a *frequency table* that shows the *proportion* of individuals falling in each category. This distribution can also be conveyed as a *bar graph*, where the height of each bar shows the proportion of scores in each category. The distribution of a discrete quantitative variable can also be shown in a *frequency table* or graphically conveyed as a *histogram*, where the height of each bar shows the proportion of scores associated with each level of the variable. Because there is a natural ordering to quantitative variables, a

frequency table can also show *cumulative proportions* (i.e., the proportion of scores at or below a given value of the variable). If the quantitative variable has a large number of possible values, then the distribution of scores can be conveyed in a *grouped frequency table*, in which values of the variable are replaced with *intervals*. Intervals have *score limits* and *real limits*, and all intervals in a grouped frequency table have the same *interval width*. The grouped frequency table shows the proportion of scores falling in each interval, and this information can also be conveyed using a histogram.

On the *frequentist* view, *probability* is the proportion of times an *event* occurs in an infinite series of identical *sampling experiments*. Frequency distributions, bar graphs, and histograms can be viewed as *probability distributions* because they convey the proportion of times that a randomly chosen score would have a particular value, or fall in a particular interval. The probability distribution of a continuous variable is conveyed by a *probability density function*. For each value of the variable there is an associated *density*. Density is an abstract quantity defined at a point. The area under a probability density function is 1. The proportion of a probability density function between any two values of x can be determined.

KEY TERMS

bar graph   32
cumulative frequency   35
cumulative proportion (P)   36
event   45
frequency table for a discrete
  quantitative variable   34
frequency table for a qualitative
  variable   31
histogram   36

independent events   46
interval   39
interval midpoint   40
mutually exclusive events   46
outcome   45
percentile rank   36
probability   46
probability density function   49
probability distributions   47

proportion of successes   45
raw frequency count   31
real limits   40
relative frequency   32
sampling experiment   45
score limits   40
subjective probability   45
trial   45

EXERCISES

Questions with asterisks (*) are based on material from Appendix 2.3.

Definitions and Concepts

1. What is a frequency table?

2. What are raw frequency counts (or tallies)?

3. How are relative frequencies different from raw frequencies?

4. What is a bar graph and how does it differ from a histogram?

5. What is the difference between relative frequency and cumulative relative frequency?

6. Explain why the concept of an interval does not pertain to qualitative variables.

7. How do the real limits of an interval relate to the interval width and interval midpoint?

8. What are the score limits of an interval?

9. What is a sampling experiment?

10. What is the outcome of a sampling experiment?

11. Is an event different from an outcome? Explain.

12. How does a proportion relate to probability?

13. What does it mean to say that two events are mutually exclusive?

14. What does it mean to say that two events are independent?

15. *Give an example of a conditional probability.

16. Is there a difference between a relative frequency distribution and a probability distribution? Please explain.

True or False

State whether the following statements are true or false.

17. A frequency table can be thought of as a probability distribution.

18. A frequency table can be thought of as a probability density function.

19. A value of 19.4999 falls in an interval with score limits of 20–29.

20. The probability of drawing a face card from a deck of 52 cards is .2308.

21. The probability of drawing a red ace from a deck of 52 cards is .0385.

22. The probability of drawing the ace of spades or the ace of hearts from a deck of 52 cards is .0385.

23. For large samples with many different values of the variable, we should have a small number of intervals when constructing a frequency table.

24. In 50 flips of a fair coin it came up heads 28 times. Therefore, the probability of coming up heads is .56.

25. If the interval width is 10, then 35–44 is a possible interval for the lowest score limits.

26. In a frequency table for a qualitative variable, values of the variable should be listed from largest to smallest.

27. The only difference between a bar graph and a histogram is the labeling of the y-axis.

28. The range of 100 scores is 50; therefore, 20 would be a reasonable interval width.

29. A collection of reaction times ranges from .936 seconds to 2.301 seconds. Therefore, a reasonable interval width would be 20 milliseconds.

30. A collection of IQs range from 75 to 141. If interval width is 15, then our grouped frequency distribution would have 5 intervals.

Scenarios

31. The following table represents responses of a random sample of eligible voters in the United States. Each voter was asked if he or she was a Democrat (D), Independent (I), or Republican (R). Use the data in the table to create a frequency table, with column headers Value, f, and p. Create a bar graph that summarizes the data.

D	I	R	D	I	D	D	D	R	R	I
R	R	D	D	R	D	R	R	R	R	D
D	R	R	D	R	R	D	D	I	D	D
R	R	D	R	R	R	I	D	R	D	R
D	R	D	R	D	D	D	R	I	I	I
D	R	R	D	R	D	D	R	I	R	I
R	D	I	R	R	R	R	D	R	I	D
D	D	D	D	R	D	I	R	R	I	I
R	D	D	D	I	I	I	R	D	D	D
R	I	D	R	R	D	D	I	R	D	D
R	D	D	R	D	D	R	R	I	D	R
R	D	D	R	D	I	I	D	R	D	D
I	R	D	R	D	D	I	I	D	R	I
I	D	I	I	R	D	D	R	R	R	R
I	R	D	R	D	I	D	D	D	D	R

32. Create a grouped frequency table with an interval width of 10 for the following collection of final grades. Use the following column headers: Score limits, f, p, and P. Plot your frequency table as a histogram, making sure to properly label your axes.

62.4	48.5	61.4	51.7	49.3	52.4	60.8	73.3
50.7	57.1	52.3	40.7	34.5	45.2	41.4	56.6
38.1	43.9	51.6	40.8	46.5	53.9	57.7	61.0
44.1	38.8	62.7	50.8	35.6	55.9	51.0	41.5
41.5	38.6	43.6	55.8	56.7	63.7	61.5	46.3
68.8	53.9	55.6	70.7	42.9	48.4	34.3	47.8
45.1	48.6	56.7	44.7	50.8	32.9	48.4	58.8
56.0	48.4	48.1	51.5	45.6	43.3	40.3	47.8

33. A set of 150 scores has a maximum and minimum of 65.8 and 14.1, respectively. What is the range in this set of scores? If you were creating a frequency table, how many intervals would you choose? What would be the interval widths?

34. For the preceding question, what are the interval midpoints?

35. What is the difference between a proportion and a probability?

36. What is the probability of each of the following events:

(a) drawing a 2 from a deck of cards;

(b) drawing a 2 or a King from a deck of cards;

(c) rolling a 2 or a 4 in a single roll of a die;

(d) rolling a 1 or a 5 in a single roll of a die;

(e) flipping a coin and having it come up heads;

(f) flipping a coin and having it come up heads or tails;

(g) randomly choosing a voter from the table that you created in Question 31 and finding that the voter is a Republican;

(h) randomly choosing a final grade from the table in Question 32 and finding that it is in the interval 50–54;

(i) a randomly drawn card will be a 1 or a 3;

(j) *in two independent draws from a 52-card deck, the first card will be a 5 and the second card will be a 7;

(k) *a card drawn from a 52-card deck will be a Queen or red; and

(l) *in two successive draws from a 52-card deck, the first card will be a Queen and the second will be a King when sampling is without replacement.

APPENDIX 2.1: GROUPED FREQUENCY TABLES AND HISTOGRAMS IN EXCEL

In the body of the chapter, we discussed the rules to follow when computing a grouped frequency table. In this appendix, we will discuss some very useful functions in Excel for creating grouped frequency tables and histograms.

Grouped Frequency Tables

Figure 2.A1.1 shows an array of cells in Excel. Cells **A1:J6** contain the 60 scores from Table 2.5. Below the data are two columns labeled **Values** and **Formulas**. The cells below **Values** contain numbers computed with either the **COUNTIF** or **COUNTIFS** functions, which will be described below. The cells below **Formulas** show the formulas that were typed into the cells to their left.

COUNTIF *and* COUNTIFS

In cell **C9**, we see the value 11. This is the number of the 60 scores falling below 59.5. This was computed using the **COUNTIF** function, which is used in the following way:

$$\text{COUNTIF(range,criterion).}$$

Range is a range of cells in the worksheet and **criterion** specifies a criterion that cells in the specified range must satisfy to be counted. (You can get more information about Excel functions through the Excel Help menu.) In cell **D9**, you can see how **COUNTIF** was used to count the number of scores below 59.5.

FIGURE 2.A1.1 ■ Computing a Grouped Frequency Table in Excel

	A	B	C	D	E	F	G	H	I	J
1	95.3	88.0	67.6	72.2	76.0	70.5	83.6	69.9	87.7	70.7
2	87.2	68.0	69.6	73.6	52.2	73.7	60.3	85.3	70.9	96.9
3	56.6	76.2	54.0	85.8	84.8	45.7	71.7	72.8	63.1	67.5
4	57.1	87.3	74.4	80.7	82.4	89.2	60.9	89.9	69.6	42.0
5	59.3	76.4	49.6	64.3	71.9	70.0	72.1	75.3	66.6	67.3
6	59.9	74.1	66.7	78.4	72.1	88.7	38.9	58.6	54.5	66.7
7										
8			Values	Formulas						
9			11	=COUNTIF(A1:J6,"<59.5")						
10			3	=COUNTIF(A1:J6,"<=49.5")						
11			8	=C9-C10						
12			8	=COUNTIFS(A1:J6, "<59.5",A1:J6, ">=49.5")						
13	49.5	59.5	8	=COUNTIFS(A1:J6, "<"&B13,A1:J6, ">="&A13)						
14										
15			10							
16			99.5							
17			89.5	=C16-$C$15						
18			79.5	=C17-$C$15						
19			69.5	=C18-$C$15						
20										
21	Lower	Upper	Count	Formulas						
22	89.5	99.5	3	=COUNTIFS($A$1:$J$6,"<"&B22,$A$1:$J$6,">="&A22)						
23	79.5	89.5	12	=COUNTIFS($A$1:$J$6,"<"&B23,$A$1:$J$6,">="&A23)						
24	69.5	79.5	22	=COUNTIFS($A$1:$J$6,"<"&B24,$A$1:$J$6,">="&A24)						
25	59.5	69.5	12	=COUNTIFS($A$1:$J$6,"<"&B25,$A$1:$J$6,">="&A25)						
26	49.5	59.5	8	=COUNTIFS($A$1:$J$6,"<"&B26,$A$1:$J$6,">="&A26)						
27	39.5	49.5	2	=COUNTIFS($A$1:$J$6,"<"&B27,$A$1:$J$6,">="&A27)						
28	29.5	39.5	1	=COUNTIFS($A$1:$J$6,"<"&B28,$A$1:$J$6,">="&A28)						

Because **COUNTIF** is a function, we first enter '=', just as we did in Appendix 1.1. After '=', we type in '**COUNTIF**' followed by an opening parenthesis. To specify the range of scores, we click on cell **A1**, hold the mouse down, and then drag to cell **J6** and release the mouse. Now type in a comma to separate **range** from **criterion**. The criterion is an expression in quotes. As shown in cell **D9**, the criterion is "**< 59.5**". Once this expression is typed in, close the parentheses and hit return. Now **COUNTIF** does its job and counts the number of scores having values less than 59.5.

In row **10**, the **COUNTIF** function has been used to count the number of scores having values *less than or equal* to 49.5. As shown in cell **C10**, there are three scores satisfying this criterion.

In row **11**, the number of scores in the interval 49.5 to 59.5 is given by the difference between the number of scores that are less than 59.5 and the number less than or equal to 49.5. This is the difference between **C10** and **C9**, as shown in cell **C11**. Note that in this example, we are following the rule given in the chapter: if a score falls exactly on a real limit (e.g., 49.5 and 59.5), it goes in the upper interval. For example, if a score equals 49.5, it goes in the interval 49.5 to 59.5. However, if a score equals 59.5, it would not go in the interval 49.5 to 59.5; it would go in the next interval.

Cell **C12** shows another way of counting scores between 49.5 and 59.5 using the **COUNTIFS** function. As the name suggests, the **COUNTIFS** function counts the number of scores satisfying multiple criteria. It has the following form:

COUNTIFS(range₁, criterion₁, range₂, criterion₂...).

That is, there may be many range-criterion pairs. Cell **D12** shows that the first range-criterion pair has range **A1:J6** and criterion "**<59.5**". The second range-criterion pair has range **A1:J6** and criterion "**>=49.5**". This means that scores are counted only if they are greater than or equal to 49.5 and less than 59.5. That is, both criteria have to be met for scores to be counted.

When we compute a grouped frequency table, we count the scores in many intervals, so it would be nice to be able to compute the interval limits in cells and then refer to these from within **COUNTIFS**. Next we will show how to compute the real limits of intervals, and we will then show how to connect these to **COUNTIFS**.

Absolute References

Cell **C15** contains the number 10, which will be the interval width for the grouped frequency table that we will build. Below, in cell **C16**, is the number 99.5, which is the upper real limit of the highest interval. In cell **C17**, I initially typed '= **C16-C15**' to compute the upper real limit of the second highest interval. We've seen before that we can avoid repeatedly typing formulas into cells by selecting a cell and then clicking and dragging its contents to other cells (see Appendix 1.1). Our goal here is to drag the contents of cell **C17** down so that we subtract 10 from the contents of the cell above. This would produce all the upper real limits of our intervals. However, if you drag the contents of **C17** to **C18**, the result will be '= **C17-C16**'. This is not what we want. Rather, we want the cell to contain '= **C17-C15**'. No matter where we drag **C17**, the quantity being subtracted should always be the number in cell **C15** (the interval width).

We use something called *absolute referencing* to ensure that wherever the formula in cell **C17** is copied, the second cell referred to in the formula will always be **C15**. To do this, we select cell **C17**, then click on the '**C15**' that appears as part of the formula. On a Macintosh, we then type ⌘t (that is, press the command key and the 't' key at the same time). This will cause the formula in cell C17 to appear as '**C16-$C$15**'. (To achieve this result on a PC, we use the F4 key rather than ⌘t.) The '**$**' symbol preceding '**C**' and '**15**' ensures that no matter where cell **C17** is dragged, the second cell referred to in the formula will always be column **C**, row **15**. This is shown in cells **C18** and **C19**. (There are other ways to use absolute referencing. For more information, search for "switch between relative and absolute references" in Excel Help.) Absolute referencing is one of the most useful features of Excel, and one that many people are unaware of.

Using Cell Content to Specify Criteria

The numbers below the headings **Lower** and **Upper** in Figure 2.A1.1 were created as described above using absolute referencing. The goal now is to determine the number of scores falling in each of these intervals. We can do this with the **COUNTIFS** function. In cell **C22**, the following formula has been entered (as shown in cell **D22**):

= COUNTIFS($A$1:$J$6,"<"&B22,$A$1:$J$6,"> ="&A22)

There are two range-criterion pairs. In both cases, the range is **A1:J6** and these ranges have been made absolute references so that no matter where this formula is dragged, it will always point to the same range of scores in the worksheet.

The interesting change to **COUNTIFS** concerns how the criteria are specified. The first criterion is '**"<"&B22**'. This specifies that scores must have a value that is less than the content of cell **B22**. The '**<**' is in double quotes, as before. The change is that the quantity in question is the content of a cell. The ampersand ('**&**') in the above formula tells Excel which cell contains the criterion number. Therefore, the first criterion says that to be counted, a score must be less than the contents of cell **B22**. The second criterion works in the same way. It says that to be counted, a score must be greater than or equal to the contents of cell **A22**.

When cell **C22** is dragged down to **C28**, all formulas will change appropriately. The ranges will always specify cells **A1:J6**. And the criteria will adjust their references to the real limits of the intervals according to the rows they are in.

Histograms

Excel offers a very large number of ways to create graphs, which Excel refers to as *charts*. In this section, we will create a histogram based on the data computed above.

These frequencies are shown in Figure 2.A1.2 and have been highlighted by clicking and dragging the mouse.

The charting options are reached through the **Insert** ribbon. Therefore, the first thing to do is click on the **Insert** tab to reveal the **Insert** ribbon. Figure 2.A1.2 shows a subset of the icons in the **Insert** ribbon, starting with **Recommended Charts**. Immediately to the right of this are icons for several kinds of charts, including **Column graphs**, **Line graphs**, **Pie charts**, **Bar graphs**, **Scatter plots**, and more. I've clicked on the icon for a **Column** graph and seven options were presented; these include **Clustered Column**, **Stacked Column**, and **100% Stacked Column**, among others. (I recommend avoiding 3-D graphs. They may look cool but they will irritate your audience because they are difficult to read.)

To create a histogram we select **Clustered Column**, which is highlighted in Figure 2.A1.2. Selecting this option immediately produces the bar graph from the data that have been highlighted (under **Frequencies**).

FIGURE 2.A1.2 ■ Creating a Histogram in Excel

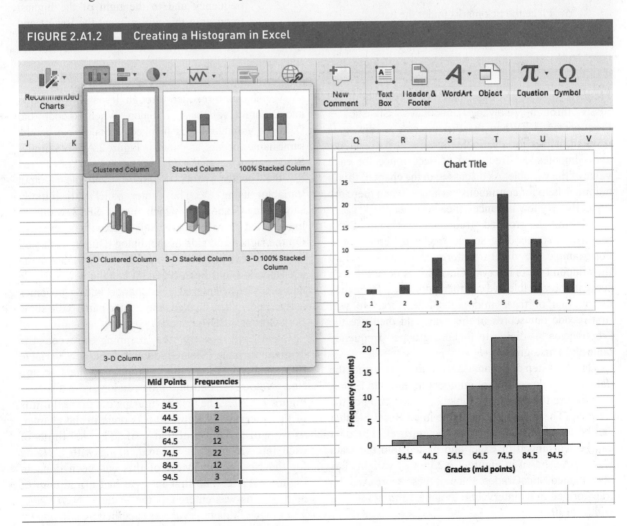

This is the graph with narrow bars and grid lines. So, as easily as that we've created a histogram.

The default histogram just created doesn't look great. In fact, it is a bar graph and not a histogram. Below the default graph is one that has been formatted to look much better. It would take too much space to describe how all of this formatting was done. To learn more, you can discover some of the formatting options yourself. I will provide a few hints in point form.

- Double-click on one of the bars in the graph. A menu will appear that allows you to
  ○ Change the separation between the bars.
  ○ Change the color of the bars to solid colors.
  ○ Make the lines around the bars thicker and darker.
  ○ Remove the shadows from the bars.

- Double-click on the *x*-axis to
  ○ Place the tick marks under the bars.
  ○ Darken the axis line.

- Double-click anywhere on the figure to open a dialog that allows you to
  ○ Change the font size.
  ○ Remove the frame around the graph.

- Click on one of the grid lines (horizontal lines) and press delete to remove them.

- Click once on the chart and notice that a **Chart Design** tab appears. Click on the **Chart Design** tab to reveal the **Chart Design** ribbon. The leftmost icon is **Add Chart Element**, and choosing this allows you to add axis titles, among other things.

- Control-click (right click on a PC) on the *x*-axis to find out how to
  ○ Change the categories on the *x*-axis to the mid-points shown to the left of the **Frequencies**.
  ○ Add blank entries to the left of the lowest frequency and to the right of the highest frequency; this will prevent the lowest and highest bars from being cut in half.

APPENDIX 2.2: GROUPED FREQUENCY TABLES AND HISTOGRAMS IN SPSS

Frequency distributions can be computed in SPSS through Analyze→Descriptive Statistics→ Frequencies. . . . When the Frequencies . . . dialog is invoked, it asks for a variable (or variables) to analyze and computes the frequency of occurrence for each value of the variable. As discussed in the chapter, this is not much help for continuous variables, when there are typically very few instances of each value.

Making a grouped frequency table in SPSS requires first recoding scores into a different form. For example, we might transform each raw score into the midpoint of its interval. Once this is done for each score in the variable, we can then compute a frequency distribution of these midpoints. Therefore, we will first recode our scores in this way, and then use the Frequencies . . . dialog to make a grouped frequency table and a histogram.

The first step of the analysis requires arranging the data as a single column of numbers in the Data Editor; we will use the data from Table 2.5 in the body of the chapter. (These data are available in an Excel file that can be found at study.sagepub.com/gurnsey. The data can be rearranged in Excel and pasted into the Data Editor.) As shown in Figure 2.A2.1a, this variable has been named StatsGrades. Each of these scores will be transformed to its interval midpoint assuming interval widths of 10.

Variables are recoded through Transform→Recode into Different Variables . . . dialog. When Recode into Different Variables . . . is invoked, a dialog box of the same name appears, as shown in Figure 2.A2.1b. Initially StatsGrades was in the left panel, but it was transferred to the center panel (Numeric Variable→Output Variable) using ⇨. The center panel will initially show StatsGrades→?, which means SPSS wants to know what to call the variable that will be created. On the right-hand side of the dialog (Output Variable), there is a text box into which the new variable name (SGRecoded) has been typed. When the new variable name has been entered, the Change button becomes active and, when clicked, the center panel will show StatsGrades→SGRecoded.

Next, we specify the mapping between the original variable (StatsGrades) and the new variable (SGRecoded). To do this we click on the Old and New Values . . . button. This dialog is shown in Figure 2.A2.1c. The task of figuring out the real limits of the intervals of width 10 was done in the chapter, so we will make use of that work. The real limits of each interval must be entered, along with a Name for the New Value. Most of this has been done in Figure 2.A2.1c. The first interval had a midpoint of 34.5, so this was entered as the Value in New Value. The range of this interval was specified by clicking on

FIGURE 2.A2.1 ■ Recoding a Variable in SPSS

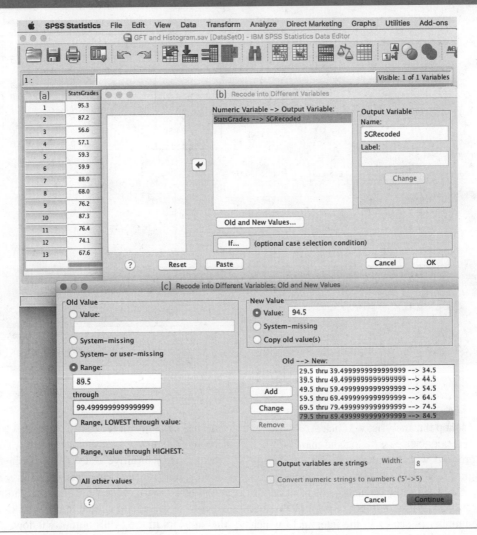

the Range radio button (◉) and entering the real limits of the interval. For this first interval, the lower limit was 29.5 and the upper limit was entered as 39.49̇. With these three things entered, I clicked the Add button and this mapping showed up as the first line in the sub-panel labeled Old→New.

The first mapping required the most typing; all subsequent mappings can be created from an existing mapping. Notice that in Figure 2.A2.1c in the last line in the Old→New sub-panel, the mapping 79.5 thru 89.49 → 84.5 has been highlighted. This causes these numbers to appear in the ◉ Value and ◉ Range text boxes on the left. In Figure 2.A2.1c, these numbers have been changed to 94.5, 89.5, and 99.49, which define the top interval; only one digit was changed in each number. Now when the Add button is pressed,

this new mapping appears in the Old→New sub-panel. All mappings except the first were created this way. Once the mappings have been entered, we click **Continue** to return to the Recode into Different Variables . . . dialog, and there we click **OK** to proceed with the analysis. This produces a new variable called SGRecoded.

As mentioned before, the grouped frequency distribution is computed via the Analyze→Descriptive Statistics→Frequencies . . . dialog. This dialog is shown in Figure 2.A2.2a. The variable SGRecoded has been moved (via ⇨) into the Variable(s): region. One of the options in this dialog is to create a chart. Clicking on the Charts . . . button produces the dialog Frequencies: Charts. As shown in Figure 2.A2.2b, the Histograms: option has been chosen by clicking the radio button

FIGURE 2.A2.2 ■ The Frequencies Dialog

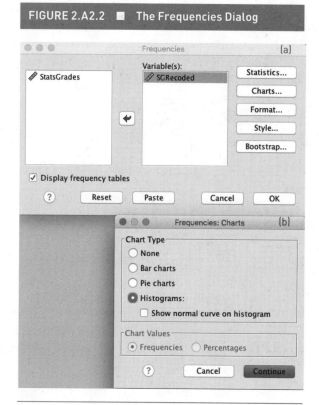

FIGURE 2.A2.3 ■ Frequency Table and Histogram

(a)

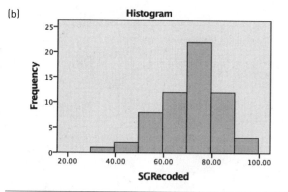

SGRecoded

		Frequency	Percent	Valid Percent	Cumulative Percent
Valid	34.50	1	1.7	1.7	1.7
	44.50	2	3.3	3.3	5.0
	54.50	8	13.3	13.3	18.3
	64.50	12	20.0	20.0	38.3
	74.50	22	36.7	36.7	75.0
	84.50	12	20.0	20.0	95.0
	94.50	3	5.0	5.0	100.0
	Total	60	100.0	100.0	

(b)

(◉). Having set up the analysis, we click Continue to return to the Frequencies . . . dialog and there we click OK to proceed with the analysis. The results are shown in Figure 2.A2.3a.

Figure 2.A2.3a shows a frequency distribution for the recoded values, and the number of each of these is equal to the number of scores in the interval to which the midpoint corresponds. The numbers in the column labeled Frequency are exactly the same as those shown in Table 2.7 and in Figure 2.A1.1. Unfortunately, the midpoints are arranged from smallest to largest, but we can live with that. The column labeled Percent shows the percentage of the 60 scores in each interval, and the column labeled Cumulative Percent shows the percentage of scores at or below the upper real limit of each interval.

Below the grouped frequency table in Figure 2.A2.3b is a histogram of the counts. If you double-click on the histogram, you will be taken to a new window that provides a degree of control over the figure. These controls won't be covered in this book but, as with Excel, most of these are discoverable.

This material above shows how to create a frequency table in SPSS and provides information that parallels the material in Appendix 2.1. However, histograms

can be created more easily in SPSS through the Graphs→Legacy Dialogs→Histogram . . . dialog. This dialog is very intuitive and you should have no trouble creating a histogram. However, SPSS may not produce a histogram with the desired interval widths or limits. To control these, you should double-click on the figure that appears in the SPSS output window, and this will bring you to the Chart Editor shown in Figure 2.A2.4a. The circled icon in the Chart Editor allows you to control many properties of the figure. When it is double-clicked, the Properties dialog pops up (Figure 2.A2.4b) and Binning is one of the properties that can be controlled.

Switching from Automatic binning to Custom binning allows you to specify the desired Interval width and the lower real limit of the lowest interval (Custom value for anchor). When these values are specified, click the Apply button at the bottom of the Properties dialog and a histogram with the desired interval widths and limits will be produced. Notice that the bars of the histogram show (by default) the number of scores falling in each interval. These are exactly the same frequency counts that we recovered in the chapter (Table 2.7), in Appendix 2.1 using Excel, and earlier in this appendix using SPSS.

FIGURE 2.A2.4 ■ Creating a Histogram

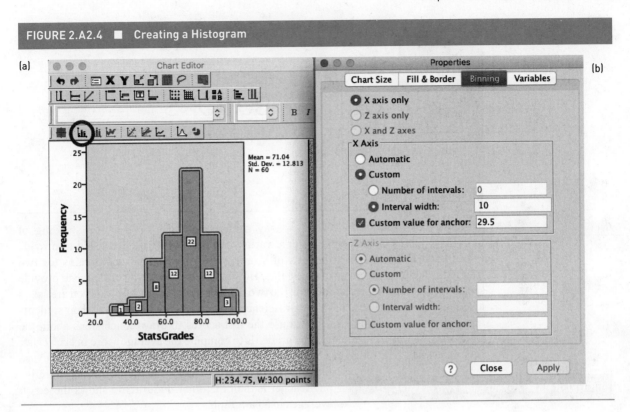

3

PROPERTIES OF DISTRIBUTIONS

INTRODUCTION

Frequency distributions allow us to say something about the relative standing of individual scores or intervals on some variable of interest. If we are asked to provide information about two distributions of scores, then showing two histograms (or two frequency tables) is one way of doing this. However, it is often more effective to provide succinct summary information about the two distributions. For example, if you are asked to assess the effectiveness of two neuroscience instructors, you might have their students take the same standardized test and use the test scores to draw conclusions about the teachers. One way to summarize the grades is to compute the average score in each class. If one class had an average grade of 70 and the other 73, then you might be inclined to use this difference as a measure of how much more effective one teacher is than the other. Figure 3.1a shows these two averages in a bar graph.

We have to be careful about drawing conclusions from averages alone. If it turned out that the grade range was 42 to 95 in the first class and 45 to 89 in the second, then we would recognize that there is considerable *variability*, and hence overlap, in the two distributions of grades. Figure 3.1b shows the distribution of individual grades in the two classes; each circle represents one student's test grade. Therefore, a difference of 3 percentage points between the class averages might not seem particularly impressive given the variability within the two distributions of grades. We will see that averages convey information about distributions succinctly, but measures of variability are required to help us evaluate the difference between two averages.

We saw in Chapter 1 that inferential statistics involves inferring parameters of distributions from the statistics of samples. In this chapter we will discuss averages and variability in greater detail because these are two of the most important parameters of distributions in inferential statistics.

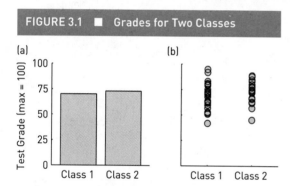

FIGURE 3.1 ■ Grades for Two Classes

Data from two classes on a standardized test of neuroscience knowledge. (a) The means from the two classes, which were 70 and 73, respectively. (b) The 30 test scores for each of the two classes.

CENTRAL TENDENCY

Central tendency is a number that is typical of the scores in a distribution.

The most common way to summarize a distribution of scores is to provide a single number that is typical of the scores in the distribution. We call this number a measure of **central tendency**. In statistics we use the term *average* to mean central tendency. We might compare two communities by referring to differences in average income, average age, average level of education, and so on. We might characterize the psychological stress associated with different occupations by noting the average number of depressive symptoms in each. We can use the average activity in two brain regions to infer something about how they differ in function.

Although statisticians use the word *average* synonymously with central tendency, in everyday use, the term *average* means the *arithmetic mean*. However, the arithmetic mean is just one of three commonly used measures of central tendency. The other two are the *median* and the *mode*. In the following sections, we will discuss these three measures of central tendency and comment on the situations in which each provides the most appropriate characterization of a sample or a population of scores.

The Mean

The most common measure of central tendency is the arithmetic mean. You learned about the mean in elementary school. If five people had grades of 41, 64, 55, 79, and 81 on a test, then the mean is 64. We obtain the mean of a set of numbers by adding them up and then dividing this sum by the number of scores in the set.

We will denote a collection (or set) of numbers with a symbol. For example, we could write

$$y = \{41, 64, 55, 79, 81\}.$$

In this case, the symbol y stands for the five numbers in the set. (There is nothing special about the symbol y. We could just as easily have denoted this set of scores with the symbols x, g, q, or *Bilbo*.) To calculate the mean of the scores in y, we first sum them. We use the following expression to indicate the sum of a set of numbers:

$$sum = \Sigma y. \tag{3.1}$$

The Greek letter Σ (capital S) means sum. Evaluating equation 3.1 for $y = \{41, 64, 55, 79, 81\}$ shows that

$$sum = \Sigma y = \Sigma\{41, 64, 55, 79, 81\} = \{41 + 64 + 55 + 79 + 81\} = 320.$$

Because the mean of a set of scores is the sum divided by the number of scores in the set, we can define the mean as follows:

$$m = \frac{\Sigma y}{n}. \tag{3.2}$$

In this case, n refers to the number of scores in y (i.e., $n = 5$), and m is the letter we use to denote the mean. Therefore, equation 3.2 says that the mean is

$$m = \frac{\Sigma y}{n} = \frac{\Sigma\{41, 64, 55, 79, 81\}}{5} = \frac{\{41 + 64 + 55 + 79 + 81\}}{5} = \frac{320}{5} = 64.$$

So, equation 3.2 simply provides a succinct way of referring to a mathematical operation that we're already familiar with.

The mean is sometimes thought of as a balance point in a set of numbers. If we were to subtract the mean (i.e., 64) from each score in y, we would have the following:

$$y - 64 = \{-23, 0, -9, 15, 17\}.$$

The sum of the negative numbers is –32, and the sum of the positive numbers is 32. Therefore, –32 + 32 = 0. Whenever we subtract the mean from all numbers in the set, some of these differences will be positive and some will be negative. The sum of the *absolute values* of the negative numbers will always be the same as the sum of the positive numbers. Therefore, the positive and negative differences "balance."

Notation for Parameters and Statistics

We saw in Chapter 1 that the mean of a population is a parameter, and the mean of a sample is a statistic. The mean is computed in exactly the same way for populations and samples, but we use different symbols to distinguish between parameters and statistics. Throughout this book, parameters will be denoted using Greek symbols and statistics will be notated using Roman characters. Therefore, the mean of a population will be denoted as

$$\mu = \frac{\sum y}{N}. \tag{3.3}$$

The Greek letter μ (pronounced "mew") corresponds to the Roman letter m. Equation 3.3 says that the mean (μ) of the population of scores is the sum of all the scores in the population divided by the number of scores (N) in the population. In some situations where there are several variables under consideration, we may add a subscript to μ (e.g., μ_y, μ_x, μ_{men}, or μ_{women}) to make explicit which population mean we're talking about.

If we draw a sample from a population, we would define the mean of the sample as

$$m = \frac{\sum y}{n}, \tag{3.4}$$

where n is the number of scores in the sample. Inferential statistics involves estimating population parameters with sample statistics. Therefore, m estimates μ.

Throughout this book, I will use the symbol m to denote the sample mean, and I'll add subscripts when necessary to distinguish between the means of different samples (e.g., m_y, m_x, m_{men}, or m_{women}) as was done for the means of different populations. This notation maintains consistency between statistics and parameters.

However, you may find that other sources (e.g., books or websites) denote the means of samples using $\bar{x}$ or $\bar{y}$, or possibly $\overline{Bilbo}$. That is, putting a bar over the name given to the sample of scores indicates the mean of the sample. This is a perfectly legitimate and traditional way of denoting the mean. However, this "bar" notation is not applied to other sample statistics, so I find it somewhat inconsistent.

Another way of denoting a statistic is to put a hat ($^\wedge$) over the parameter it estimates. For example, $\hat{\mu}$ is sometimes used for the sample mean. We read this as "mew-hat" and interpret it as a sample mean that estimates a population mean (μ). This is a very elegant way to connect statistics and parameters. However, it is used less often in sources that you're likely to read, so I will stick with the convention of denoting statistics with lowercase Roman letters and parameters with lowercase Greek letters. Very occasionally, when I am desperate for symbols, I will use the hat notation.

The Median

The **median** is the number that divides an ordered set of numbers into two groups of equal size, such that half of the numbers are greater than the median and half are less than the median. To determine the median, we must first put our scores in order from smallest to largest. For example, if

$$y = \{9, 1, 7, 4, 3, 2, 3, 4, 5, 6, 8, 6, 7\},$$

The **median** is the number that divides an ordered set of numbers into two groups of equal size.

the scores are as follows when they are put in order:

$$y = \{1, 2, 3, 3, 4, 4, 5, 6, 6, 7, 7, 8, 9\}.$$

The position of the median is given by

$$\text{Median Position} = \frac{n+1}{2}.$$

This definition will have different implications for samples with *odd* or *even* numbers of scores. In this case, there are 13 scores in y so the median is in position $(13 + 1)/2 = 7$. Therefore, 5 is the median because it is the seventh (middle) score of the sorted list. Six of the 13 scores are above 5 and six are below 5.

If there is an *even* number of scores in a set, then the median is the mean of the middle two numbers. For example, if

$$y = \{1, 4, 3, 2, 3, 4, 5, 6, 7, 6, 7, 8\}$$

when put in order we have

$$y = \{1, 2, 3, 3, 4, 4, 5, 6, 6, 7, 7, 8\}.$$

In this example, there are 12 numbers in y so the median is in position $(12 + 1)/2 = 6.5$. This means that the median position is half way between positions 6 and 7. The sixth number is 4 and the seventh is 5, so we average these two numbers to produce the median. That is, the median is $(4 + 5)/2 = 4.5$.

We won't provide Greek and Roman notation for the median for two reasons. First, we will use the median infrequently in this book. Second, if we were to use characters to notate the median, they would probably be μ and m, which we've already used for the mean. Therefore we will just skip notating the median so that we don't use up symbols unnecessarily. The same is true for the mode, which is the last measure of central tendency we'll discuss.

The Mode

The **mode** is the most frequently occurring score in a set. For example, if

$$y = \{1, 2, 3, 3, 4, 4, 5, 6, 6, 6, 6, 6, 7, 7, 8\},$$

> The **mode** is the most frequently occurring score in a set.

then the mode is 6 because it is the most frequently occurring score. Unlike the mean and median, the mode can also be computed for qualitative variables. For example, if

$$y = \{\text{dog, cat, dog, parrot, cat, cat, parrot, dog, cat, cat, cat}\},$$

then the mode is cat. There are six instances of cat, three instances of dog, and two instances of parrot.

Which Measure of Central Tendency to Use?

When we are dealing with quantitative variables (such as height, income, and time), the mean and the median are the most frequently used measures of central tendency. However, the mean and median differ in important ways. The mean always changes when one of the numbers in the set changes, because all numbers in the set contribute to the mean.

However, the median typically doesn't change (or change much) when one of the numbers in the set changes, because only one or two numbers in the set determine the median. For example, if $y = \{1, 2, 3\}$, then the mean and median are both 2. However, if y changes to $y = \{1, 2, 9\}$, the mean changes to 4, whereas the median remains 2.

In some cases, the median is a better measure of central tendency than the mean. Consider the annual incomes of 10 employed people, chosen at random, in Washington, DC. Let's say their annual incomes (in thousands of dollars) are

$$i = \{29, 32, 40, 45, 1200, 33, 42, 49, 28, 29\}.$$

In this case, the mean income is $152,700 per year, which gives the impression that the people in Washington, DC, are generally living quite comfortably. However, we see that this large mean income is a consequence of one person who makes $1,200,000 per year. If we were to compute the median income, we'd find it to be $36,500 per year. There are 10 incomes in this set, so the median lies between the fifth and sixth incomes:

$$i = \{28, 29, 29, 32, \mathbf{33}, \mathbf{40}, 42, 45, 49, 1200\}.$$

The middle two incomes are $33,000 and $40,000, so their mean is $36,500. In this case, the median gives a much better sense of central tendency than the mean.

The mode is rarely used to characterize the central tendency of quantitative variables. Rather, it is used to characterize qualitative variables. For example, if we categorized people based on their political preferences (e.g., Conservative, Liberal, Moderate), the mode would be the category that has the largest number of occurrences in the sample or population.

Most of the analyses we discuss in this book will involve quantitative variables. Therefore, the mean and the median are both reasonable measures of central tendency. However, in most of the cases we'll consider, the mean and median will be very similar. For this reason (and others that will become apparent as we move along), the mean will be our preferred measure of central tendency.

LEARNING CHECK 1

1. If $y = \{1, 3, 6, 6, 4, 5, 7, 2, 4, 6\}$, compute (a) the mean, (b) the median, and (c) the mode.

2. Which measure of central tendency best characterizes the following set of numbers $x = \{49, 42, 40, 45, 1, 43, 42, 49, 48, 49\}$?

Answers

1. (a) 4.4. (b) 4.5. (c) 6.

2. The median. All but one of the numbers is in the "40s," but the remaining number is what we will call an outlier. The mean is 40.8 and the median is 44.

Dispersion refers to how spread out scores are in a distribution of a quantitative variable.

DISPERSION (SPREAD)

Dispersion refers to how spread out scores are in distributions of quantitative variables. Consider two sets of scores, y_1 and y_2, where $y_1 = \{1, 3, 5, 7, 9\}$ and $y_2 = \{3, 4, 5, 6, 7\}$. Both

y_1 and y_2 have a mean (and median) of 5, but the scores in y_1 are more widely dispersed (i.e., more spread out) than those in y_2.

As we saw in the opening section of this chapter, dispersion is critically important because it helps guide judgements about central tendency. For example, imagine that a researcher wanted to assess differences in math skills between men and women, so she gave the same test to a random sample of 5 men and 5 women. The mean score on the math test was 75 for the 5 men and 80 for the 5 women. From these means, you might be inclined to think that men generally score more poorly than women. However, if the numbers from which these means were derived were

$$men = \{35, 75, 80, 85, 100\} \text{ and}$$

$$women = \{70, 75, 80, 85, 90\},$$

then I think you would question what can be inferred from the difference of 5 points in the sample means. The grades are quite widely dispersed for both men and women, and particularly so for men; one man got the highest score overall (100) and another got the lowest score overall (35). Therefore, characterizing the dispersion in a set of numbers will have important consequences for how we view differences in central tendency.

Before describing different measures of dispersion, we will take a moment to comment on outliers. The math scores for men in the example above contain an unusually low score of 35. This score is an example of an **outlier**. As the term suggests, an outlier is a score that is unusually far from the central tendency of the distribution. Outliers can be perplexing because it is often difficult to say why they occur. In the case of the math score of 35, it could be that the person in question was truly bad at math, and so the score is a valid reflection of his math ability. However, it might be that the outlier is not a valid reflection of his math ability. It could be that he (i) was very ill during the test, (ii) was not wearing his glasses, or (iii) did not bring his calculator to the test. It could also be that the outlying score was the result of a data entry error; the person entering the grades may have mistaken an 8 for a 3 (i.e., the correct grade was 85) or was simply distracted for a moment and entered the wrong number. Any of these explanations for the outlier would undermine our ability to judge differences in math abilities between men and women because the outlier is not a legitimate reflection of math skill. So, should this outlier be included in the analysis or not?

Outliers are scores that deviate by an unusual amount from the central tendency of a distribution.

Dealing with outliers often involves subjective judgements of whether they are valid or invalid measures of the quantity of interest (e.g., math skill). If they are judged invalid, they may be excluded from the analysis. In this simple example, excluding 35 from the analysis would change the mean of men from 75 to 85. A researcher who decides to exclude an outlier from data analysis in a published report must provide justification for doing so and must also show what the excluded values were. This allows readers to judge for themselves if the decision to exclude is justified and to draw their own conclusions about the data.

In the next section, we will introduce the three most commonly used measures of dispersion. These are the *range*, the *variance*, and the *standard deviation*.

The Range

The **range** is the difference between the largest and smallest scores in a set. In $y_1 = \{1, 3, 5, 7, 9\}$, the minimum (min) score is 1 and the maximum (max) is 9. Therefore, we could characterize the range as a pair of numbers [min, max], where the first number is the minimum and the second is the maximum. In our example, we can say the range of y_1 is [1, 9]. We can also characterize the range as the difference between the largest and smallest numbers in the set. So, if the minimum score is 1 and the maximum is 9, then the range could be expressed as $max - min = 9 - 1 = 8$. This means that the maximum value is 8 units above the minimum value.

The **range** is the difference between the largest and smallest scores in a set.

Like the median, the range does not depend on the values of all the scores in a set. The median is based on the middle number, or middle two numbers, which makes it insensitive to outliers. Conversely, because the range is based on only the minimum and maximum, it is highly sensitive to outliers. For example, the following two sets

$$a = \{1, 2, 2, 2, 2, 2, 2, 2, 2, 10\}$$

$$b = \{1, 2, 3, 4, 5, 6, 7, 8, 9, 10\}$$

have the same range [1, 10], despite being quite different collections of numbers. The score of 10 seems to be an outlier in a. Because the range is highly affected by outliers, it would seem better to have a measure of dispersion computed from all scores in a population or sample, rather than just two.

The Variance

In a population, the **variance** is the mean squared deviation from the mean.

The **variance** is a measure of dispersion that takes account of all scores in a population or sample. The variance is a kind of mean and so we use the same notation that we used previously to define the mean.

The Population Variance

The variance in a population of scores is defined as follows:

$$\sigma^2 = \frac{\Sigma(y-\mu)^2}{N}. \tag{3.5}$$

σ^2 is pronounced "sigma squared." You might wonder why we use the symbol σ (the Greek letter "s") as part of the notation for something called the variance; the Greek symbol for v might seem a better choice. There is actually a good reason for this notation, but we'll have to wait a bit for the explanation.

There are two differences between equation 3.3, which defines the mean of a population, and equation 3.5, which defines the variance of a population. In both cases we sum numbers, but in the case of equation 3.5 (the variance) we sum the *squared deviations from the mean* rather than scores themselves. To see this, we will unpack equation 3.5. Let's say that $y = \{1, 3, 5, 7, 9\}$ is a (very small) population of scores. The mean of these scores is $\mu = 5$; therefore, the $(y - \mu)$ part of equation 3.5 tells us to subtract 5 from each score in y:

$$(y - \mu) = \{1, 3, 5, 7, 9\} - 5 = \{-4, -2, 0, 2, 4\}.$$

A **deviation score** is the difference between a score and the mean of the sample or population from which it came.

We refer to the numbers in $(y-\mu)$ as **deviation scores**, because they represent deviations from the mean. Knowing what $(y-\mu)$ is, we then ask what $(y-\mu)^2$ is. The exponent (2) tells us to square each number in $(y-\mu)$. If we do this, we find that

$$(y - \mu)^2 = \{-4, -2, 0, 2, 4\}^2 = \{16, 4, 0, 4, 16\}.$$

A **squared deviation score** is the square of a deviation score.

Therefore, the numbers in $(y-\mu)^2$ are **squared deviation scores**, or simply *squared deviations*. Looking back at equation 3.5, we see that to compute the variance, we must sum these squared deviations, as follows:

$$\Sigma(y - \mu)^2 = \Sigma\{-4, -2, 0, 2, 4\}^2 = \Sigma\{16, 4, 0, 4, 16\} = 40.$$

The **sum of squares** (**ss**) is the sum of all squared deviation scores in a sample or population.

It is common in statistics to refer to this sum of squared deviations as simply the **sum of squares**, or *ss* for short. So, in this example *ss* = 40. (Sums of squares are important, and we will see them in many later chapters.)

Equation 3.5 states that the variance is the mean squared deviation from the mean. This means we must divide the sum of squares (ss) by the number of squared deviation scores that have been summed. We can now use equation 3.5 to calculate the variance of $y = \{1, 3, 5, 7, 9\}$ as follows:

$$\sigma^2 = \frac{\sum(y-\mu)^2}{N} = \frac{ss}{N} = \frac{40}{5} = 8. \tag{3.6}$$

Therefore, for the population $y = \{1, 3, 5, 7, 9\}$, the variance is $\sigma^2 = 8$. That is, the mean (average) squared deviation from the mean is 8.

The Variance for a Sample

We have been treating $y = \{1, 3, 5, 7, 9\}$ as a very small population of scores. However, if $y = \{1, 3, 5, 7, 9\}$ were a sample from a larger population, then a minor complication arises. There are two ways of defining the variance of a sample. The first definition simply mirrors the definition of the population variance:

$$s_{pop}^2 = \frac{\sum(y-m)^2}{n}, \tag{3.7a}$$

where m is the sample mean and n is the sample size. We use s in our notation of the sample variance because, as noted, s is the Roman letter corresponding to the Greek letter σ. The subscript "pop" is a reminder that this version of the sample variance is computed exactly like the population variance. As we will see, this definition of the variance is almost never used in statistics.

The more frequently used definition of the sample variance is

$$s^2 = \frac{\sum(y-m)^2}{n-1}. \tag{3.7b}$$

Therefore, the sample variance for $y = \{1, 3, 5, 7, 9\}$ is

$$s^2 = \frac{\sum(y-m)^2}{n-1} = \frac{ss}{n-1} = \frac{40}{4} = 10.$$

Note that for $y = \{1, 3, 5, 7, 9\}$, ss is the same whether we're talking about a sample or a population. However, because ss is divided by $n-1$ (equation 3.7b) rather than n (equation 3.7a), s^2 will always be greater than s_{pop}^2.

If one wishes to simply *describe* the variability in a sample of scores, then equation 3.7a can be used. However, if the purpose is to *estimate* the population variance (σ^2), then equation 3.7b *must* be used. From now on, we will use equation 3.7b as the definition of the **sample variance**. Explaining why s^2 is a better estimate of σ^2 than s_{pop}^2 would lead us away from the main point of this chapter. Therefore, we'll have to wait until Chapter 5 for a full explanation of this point. However, if you would like a small preview of the connection between s^2 and s_{pop}^2, you can read the following optional section.

The **sample variance** (s^2) is the sum of squares (ss) divided by $n-1$.

*The Connection Between Two Versions of the Sample Variance

It turns out that the quantity defined in equation 3.7b could also be written as

$$s^2 = s_{pop}^2 * \frac{n}{n-1}. \tag{3.7c}$$

The following simple manipulations show the relationship between equations 3.7a, 3.7b, and 3.7c.

$$s^2 = s_{pop}^2 * \frac{n}{n-1} = \frac{\sum(y-m)^2}{n} * \frac{n}{n-1} = \frac{n\sum(y-m)^2}{(n-1)n} = \frac{\sum(y-m)^2}{n-1}.$$

The color portion of equation 3.7c is a so-called *correction factor*. When s_{pop}^2 is multiplied by this correction factor, it provides a better estimate of σ^2. It is very common for the equations defining statistics to have two parts like this. One part will look a lot like the definition of the parameter that is estimated and the second part is a correction factor that makes the statistic a better estimator.

If you think about this for a while, you might notice that the correction factor is always greater than 1. However, the size of the correction factor decreases as n increases. That is, $n/(n-1)$ gets closer to 1 as n increases. This means that s_{pop}^2 becomes a better estimate of σ^2 as the sample size increases. We'll have more to say about what it means to be a good estimator of σ^2 in Chapter 5.

The Standard Deviation

The **standard deviation** is the square root of the variance. The standard deviation is roughly the average distance (deviation) of scores from the mean.

Although the variance is important in statistics, we will see that it is not a particularly intuitive measure of variability. Instead, it is easier to describe the variability in a set of scores (in this case, a population of scores) using the square root of the variance, which we call the **standard deviation**. The standard deviation in a population is defined as follows:

$$\sigma = \sqrt{\sigma^2} = \sqrt{\frac{\sum(y-\mu)^2}{N}}, \tag{3.8}$$

where μ is the population mean and N is the population size. You may now be able to see why the variance is denoted σ^2. The Greek s (sigma, σ) was chosen to name the standard deviation, and the variance is therefore denoted as the square of the standard deviation (σ^2). The standard deviation in a sample is defined as follows:

$$s = \sqrt{s^2} = \sqrt{\frac{\sum(y-m)^2}{n-1}}. \tag{3.9}$$

We saw before that for $y = \{1, 3, 5, 7, 9\}$, the sum of squares is $ss = 40$, regardless of whether y is a population or a sample. We also saw that $\sigma^2 = 8$ and $s^2 = 10$. Therefore, the population and sample standard deviations are

$$\sigma = \sqrt{\sigma^2} = \sqrt{8} = 2.8284$$

and

$$s = \sqrt{s^2} = \sqrt{10} = 3.1623,$$

respectively.

So, how do we put into words the meaning of the standard deviation? First, remember that the variance in a population is the average squared deviation of scores from the mean, and the standard deviation is the square root of the variance. Therefore, in a population, the standard deviation is *roughly* the average deviation (distance) of scores from the mean. The same is true for samples. The standard deviation is a rough measure of the average distance of scores from the mean.

The Standard Deviation Versus the Variance

The variance and standard deviation are both useful measures of dispersion, and both are preferable to the range, in that they make use of all scores in a sample or population. However, the standard deviation is a more intuitive measure of variability than the variance because it is more directly related to differences in variability between two sets of scores. This point is most easily understood with a concrete example.

Let's consider the following two sets of scores:

$$y_1 = \{1, 3, 5, 7, 9\} \text{ and}$$
$$y_2 = \{2, 6, 10, 14, 18\}.$$

There is a very simple relationship between y_1 and y_2. Specifically, we produced y_2 by multiplying all numbers in y_1 by 2 (i.e., $y_2 = 2*y_1$). In an intuitive sense, the variability in y_2 is twice that of y_1. There are two ways to see this. First, consider the range. The range for y_1 is $9 - 1 = 8$, and the range for y_2 is $18 - 2 = 16$. This seems completely intuitive; doubling all scores in a set doubles the range.

Second, notice that the separation between pairs of numbers in y_2 is twice that of the corresponding pairs of numbers in y_1. For example, think of the first and third numbers in y_1, which are 1 and 5, and the first and third numbers in y_2, which are 2 and 10. The separation between 2 and 10 is twice the separation between 1 and 5. This is true for all corresponding pairs of numbers taken from y_1 and y_2.

So, if all separations between scores have increased by a factor of 2, one would like a measure of dispersion/variability that also changes by a factor of 2. In other words, a good measure of variability would have this property: multiplying all scores in a set by a constant c should cause the measure of variability to change by the same constant. This is true of the standard deviation (and range) but not the variance. We will illustrate this by comparing the standard deviations and variances of y_1 and y_2.

We saw in the section above that the sample standard deviation for y_1 is $s_{y_1} = 3.1623$. If you do the calculations, you will find that the sample standard deviation for y_2 is $s_{y_2} = 6.3246$. So, $s_{y_2}/s_{y_1} = 6.3246/3.1623 = 2$. Therefore, when all scores in a set are multiplied by a constant c, the standard deviation has the desirable property of changing by the same constant.

The variance is a different story. We know without doing all the calculations that the variances of y_1 and y_2 are just the squares of their standard deviations. When we square these two sample standard deviations, we find that the two sample variances are $s_{y_2}^2 = 3.1623^2 = 10$ and $s_{y_2}^2 = 6.3246^2 = 40$. So, $s_{y_2}^2/s_{y_1}^2 = 40/10 = 4$. In this example, multiplying all scores by $c = 2$ caused the standard deviation to change by $c = 2$ and the variance to change by $c^2 = 4$. In general, when all scores in a set are multiplied by a constant c, the variance has the undesirable property of changing by the square of the constant, c^2.

A USEFUL RULE

Multiplying all scores in a set by a constant c changes the standard deviation by a factor of c and the variance by a factor of c^2.

Therefore, it is far easier to reason about how two sets of scores differ in variability using the standard deviation rather than the variance. This is what it means to say that the standard deviation is a more intuitive measure of variability than the variance.

Computing the Variance by Hand

It is almost certain that on your first midterm test you will be asked to compute the variance and standard deviation of a sample or population. The best way to do this is to lay your data out in a table and compute the quantities necessary, as in Table 3.1.

The left column of Table 3.1 shows a *sample* of scores $y = \{9, 10, 6, 11, 7, 9, 12, 8\}$ in column form. Below these numbers are their sum and mean. The second column shows the deviation scores $(y-m)$. For each score in y, we have subtracted the mean $m = 9$. The final column shows the squared deviation scores $[(y-m)^2]$. Below this column is the sum of squared deviations, or $ss = 28$. To compute the *sample variance*, we divide ss by $n-1$ to get $s^2 = ss/(n-1) = 28/7 = 4$. To compute the sample standard deviation, we take the square root of s^2 to get $\sqrt{4} = 2$.

TABLE 3.1 ■ Computing *ss*			
	y	*y–m*	*(y–m)²*
	9	0	0
	10	1	1
	6	–3	9
	11	2	4
	7	–2	4
	9	0	0
	12	3	9
	8	–1	1
Sum	72	0	28
Mean	9		

If we had been asked to compute the population variance, then the mean would be $\mu = 9$ because the sample mean and the population mean are computed in exactly the same way. As a consequence, the deviation scores and squared deviation scores will be exactly the same as those computed with $m = 9$. To compute the population variance, we divide *ss* by *N* to get $\sigma^2 = ss/N = 28/8 = 3.5$. To compute the population standard deviation, we take the square root of σ^2 to get $\sqrt{3.5} = 1.8708$ when rounded to four decimal places.

Table 3.1 simply provides a convenient and orderly way to do our calculations. Using a table like this, you can keep track of your calculations and check your work easily.

Reporting Data

Let's return to the illustration that opened this chapter. The data from Figure 3.1 are presented in Table 3.2. If you compute the mean and sample standard deviation for each class, you will find that the means for Class 1 and Class 2 are 70.07 and 73, respectively, and the sample standard deviations are 12.56 and 10.33, respectively. The statistics of these two samples would be reported as follows:

APA Reporting

Two classes of 30 students each completed the same neuroscience exam. The mean for Class 1 was $M = 70.07$, $s = 12.56$, and the mean for Class 2 was $M = 73$, $s = 10.33$.

TABLE 3.2 ■ Class Grades					
Class 1			Class 2		
70	51	71	83	73	66
83	95	53	63	72	75
42	72	72	63	77	62
73	64	81	66	77	63
68	72	70	45	66	74
52	63	75	89	74	89
60	63	72	77	73	66
68	80	62	67	80	78
95	79	68	88	85	72
92	79	57	57	85	85

Note that APA style requires us to use the symbol *M* for the sample mean and *s* for the sample standard deviation. Table 4.5 in the sixth edition of the *APA Publication Manual* provides a complete list of symbols to be used for reporting statistical quantities and units of measurement. Furthermore, real numbers should generally be rounded to two decimal places.

The means and standard deviations can also be reported in a bar graph, as shown in Figure 3.2. The I-type bars about the means are called *error bars*. In this example, the error bars represent one standard deviation above and below the mean. This is 70.07 ± 12.56 for Class 1 and 73 ± 10.33 for Class 2. Therefore, Figure 3.2 summarizes both the means of the two distributions of scores and the variability within each group. This is a more concise summary of the data than showing two frequency tables or two histograms.

We should note that error bars do not always represent standard deviations. Later we will see that error bars can also represent *standard errors* or *confidence intervals*, which we will discuss in Chapters 5 and 6, respectively.

APA Reporting

Error bars are required in graphs to convey information about the variability about the means. In this case, the error bars represent ± one standard deviation, which APA requires us to denote as *s*. It is extremely important that your figure captions state what the error bars represent, because not all error bars represent standard deviations.

The Importance of the Mean and Standard Deviation

The mean and standard deviation of a population of scores are the two most important parameters discussed in this book. There is an important theorem by the Russian mathematician Pafnuty Chebyshev (1821–1894) stating that the mean and standard deviation of a distribution place important constraints on the relative standing of (almost) any score drawn from the distribution, even when we know absolutely nothing else about the distribution. This beautiful result is described in Appendix 3.3 (available at study.sagepub .com/gurnsey). I think you will find this appendix interesting, and it sets the stage for the material discussed in Chapter 4. Moreover, the elegance of Chebyshev's theorem will be more apparent after you've read Chapter 4. So, if you have the time, I think you will find it a worthwhile read.

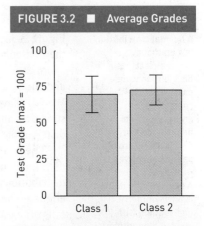

FIGURE 3.2 ■ Average Grades

Average test grades for two neuroscience classes. Error bars represent ±s.

LEARNING CHECK 2

1. If $y = \{1, 3, 6, 6, 4, 5, 7, 2, 4, 6\}$ is a sample of scores, compute (a) the range, (b) the variance, and (c) the standard deviation.

2. If all of the scores in $y = \{1, 3, 6, 6, 4, 5, 7, 2, 4, 6\}$ from question 1 were multiplied by 5, determine the values of the following without recomputing the new sum of squares: (a) the range, (b) the variance, and (c) the standard deviation.

3. Why is the standard deviation a more intuitive measure of dispersion than the variance?

Answers

1. (a) [1, 7]. (b) 3.8222. (c) 1.9551.

2. (a) [1, 7]*5 = [5, 35]. (b) 3.8222*5^2 = 3.8222*25 = 95.555. (c) 1.9551*5 = 9.7755.

3. Doubling all scores in a sample (or population) causes the standard deviation to double and the variance to quadruple. Therefore, the standard deviation captures differences in dispersion more intuitively than the variance. In general, multiplying all scores in a set by a constant c will cause the standard deviation to change by the same factor but the variance will increase by a factor of c^2.

SHAPE

The **shape** of a distribution refers to how density (or frequency) changes as a function of the values of the variable. Distributions can vary tremendously in shape, but at this point we will consider only differences in *skew*, *kurtosis*, and *modes*.

In this section we will see a large number of probability density functions. Don't lose track of the fact that probability density functions simply show us about how scores are distributed. These scores will have a mean, median, mode, variance, and standard deviation. Therefore, we will often comment on the parameters of a distribution of scores described by a probability density function.

The **shape** of a distribution refers to how density (or frequency) changes as a function of the values of the variable.

Skew

A **normal distribution** is a unimodal, symmetrical distribution whose cross-section resembles the cross-section of a bell. Hence, normal distributions are often called bell curves.

A **skewed distribution** is an asymmetrical distribution, with one tail longer than the other. When the longer tail is on the right of the peak, it is said to be **right skewed**, or **positively skewed**. Right skewed distributions are a consequence of *floor* effects. When the longer tail is on the left of the peak, it is said to be **left skewed**, or **negatively skewed**. Left skewed distributions are a consequence of *ceiling* effects.

Figure 3.3a shows a **normal distribution**. It is often referred to in popular writing as the bell curve because it looks like the cross-section of a bell. Normal distributions are symmetrical with a single peak (mode). For all normal distributions, the mean, median, and mode are all exactly the same, as shown by the inverted triangles in Figure 3.3a. We will have much more to say about normal distributions in Chapter 4, because many statistical analyses are built on the assumption that samples have been drawn from normal distributions of scores.

There are many "non-normal" distributions. One such distribution is a **skewed distribution**. An example is shown in Figure 3.3b. Unlike normal distributions, skewed distributions are not symmetrical and, consequently, the mean, median, and mode are not aligned. The point of highest density is the mode of the distribution. In Figure 3.3b, density falls off more quickly to the left of the mode than to the right. This distribution is described as **right skewed** (or **positively skewed**) because the "long tail" of the distribution is to the right of the mode (i.e., the long tail "points" to increasingly large positive numbers). Figure 3.3c shows a distribution that is **left skewed** (or **negatively skewed**) because the "long tail" of the distribution is to the left of the mode (i.e., the long tail "points" to increasingly large negative numbers).

The reasons why distributions may show skew are best illustrated with specific examples of test scores. Test scores may be skewed to the right or left, depending on the difficulty of the test. For example, if a test is extremely hard, then scores tend to be low. Because there is a minimum value of the variable (i.e., 0), scores can't go below it and so they pile up near 0. (We call this a *floor effect*.) That is, scores are more densely packed at the low end (floor) than at the high end.

Figure 3.3b represents a distribution of scores (percentages) on a test that was extremely hard. The mean of this distribution is shown by the dark blue marker on the *x*-axis. The median is shown by the light blue marker, and the mode is shown by the white marker.

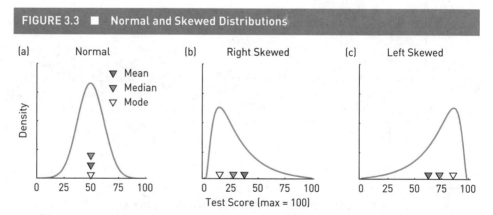

FIGURE 3.3 ■ Normal and Skewed Distributions

Three probability density functions. (a) A normal distribution. For such distributions, the mean, median, and mode are identical. In this case, all equal 50. (b) A right-skewed distribution. The mean (darker marker) is greater than the median (lighter marker) and the median is greater than the mode (white marker). (c) A left-skewed distribution. The mean is less than the median and the median is less than the mode.

When distributions are skewed to the right, the median is below the mean and the mode is below the median.

If a test is extremely easy, scores are very high but they can't exceed the maximum possible score, which is 100 in this case. (We call this a *ceiling effect*.) Therefore, scores are most densely packed close to the maximum possible score (the ceiling). The long tail now points toward negative numbers and so the distribution would be negatively skewed, or skewed to the left. In this case, the mean (dark blue marker) is below the median (light blue marker), which is below the mode (white marker).

Skewed distributions provide a good illustration of why different measures of central tendency are more appropriate than others in specific situations. Income is an example of a distribution that is skewed to the right (as in Figure 3.3b). As noted earlier, the mean of a positively skewed distribution is higher than the median. If a politician wants to tout her excellent stewardship of the economy, she might point out that the mean income in her constituency is $100,000 per year. Those less impressed with her stewardship of the economy might point out that median income in her constituency is $20,000 per year. Remember, distributions are skewed when there is a limit on how small or large the value of the variable can be. Income has a lower limit, or floor, (0) but no upper limit. Thus, income is a classic example of a right-skewed distribution.

Kurtosis

Kurtosis refers to the sharpness of the peak of a probability distribution. Figure 3.4 shows three distributions that differ in kurtosis. Figure 3.4a shows a normal distribution again. Normal distributions are said to have normal kurtosis, or to be *mesokurtic*. Figure 3.4b shows a distribution with a very sharp peak. Distributions with sharp peaks are said to be *leptokurtic*. Figure 3.4c shows a distribution that is much less peaked than the other two. Distributions with flat peaks are said to be *platykurtic*. We will rarely see distributions that are as strongly leptokurtic or platykurtic as those shown in Figure 3.4. However, one extremely important distribution that we'll see in Chapter 10

The **kurtosis** of a distribution refers to how sharp or flat its peak is.

FIGURE 3.4 ■ Variations in Kurtosis

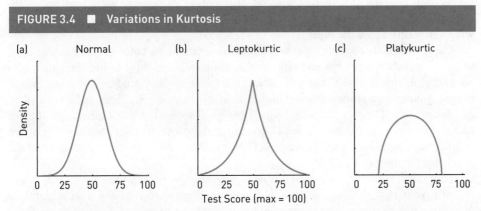

(a) A normal distribution, which is said to be mesokurtic. (b) A leptokurtic distribution. (c) A platykurtic distribution.

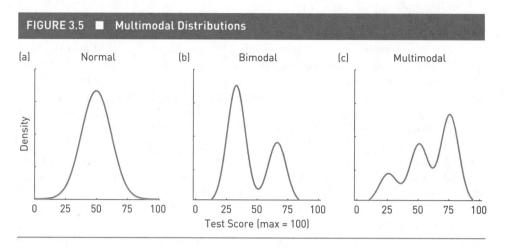

FIGURE 3.5 ■ Multimodal Distributions

(a) Normal   (b) Bimodal   (c) Multimodal

Density

Test Score (max = 100)

is leptokurtic. This distribution, called the *t*-distribution, has a sharper peak than a normal distribution.

Multiple Modes

A **multimodal distribution** has more than one peak.

All of the distributions in Figures 3.3 and 3.4 have a single peak, or mode, which is a single point of high density. Some distributions have more than one peak, and such distributions are called **multimodal distributions**. Figure 3.5b shows a *bimodal* distribution (two peaks), and Figure 3.5c shows a *trimodal* distribution (three peaks). Multimodal distributions can occur when a population comprises two or more subpopulations.

The *x*-axes in Figure 3.5 are labeled "test score." Imagine that these three distributions come from scores on a History of Psychology test given in three different classes. The first (Figure 3.5a) was a class of high school seniors who had taken no psychology courses. There is a normal distribution of scores, with a mean of 50, reflecting a relatively poor understanding of the subject, on average, but with fairly wide variability about the mean.

The second class (Figure 3.5b) was a third-year course on color vision that was open to both psychology students and engineering students. One-third of the class members were psychology students and two-thirds were engineering students. As one would expect, psychology students score higher than engineering students on a History of Psychology test. The two peaks in the distribution of scores represent the means of these two subpopulations; the higher peak (greater density) corresponds to the mean of the engineering students and the lower peak (lower density) corresponds to the mean of the psychology students.

The third class (Figure 3.5c) was a psychology proseminar involving many senior psychology students, some senior anthropology students, and a small number of accounting students. The three peaks represent the means of each of these three subpopulations. The psychology students did well on the test, the anthropology students had some trouble with the test, and the accounting students had no clue.

LEARNING CHECK 3

1. Which of the following would be most likely to produce a skewed distribution: the incomes of professional baseball players or the scores on a test that was of average difficulty?

2. Which of the following is said to be mesokurtic: a very peaked distribution, a normal distribution, or a very flat distribution?

3. Which of the following is likely to be multimodal: the scores on a very difficult test, a distribution of reaction times, or a distribution of heights for male basketball players and their spouses?

4. If a distribution is negatively skewed, which will be higher, the median or the mean?

Answers

1. The incomes of professional baseball players.

2. A normal distribution.

3. A distribution of heights for male basketball players and their spouses.

4. The median.

Computing Sample Statistics

It is important to know how to compute statistics by hand, and you will certainly have to demonstrate that you can do this on tests. However, most real-world situations involve large sample sizes and thus hand calculations are not recommended because they are slow and error prone. Appendix 3.1 shows how to compute all the statistics (and some parameters) discussed in this chapter in Excel. Appendix 3.2 shows how to compute these statistics in SPSS.

In this section, we discussed skew and kurtosis but did not show formulas for how to compute them. Normally the calculations are done by computer as described in Appendices 3.1 and 3.2. Appendix 3.3 (available at study.sagepub.com/gurnsey) provides formulas for the skew and kurtosis of populations and samples. If you are in an advanced statistics course, this material might interest you. In Appendix 3.3, you will see that most parameters of interest are ultimately related to deviation scores. Computing skew and kurtosis from deviations scores in a population is really quite straightforward. Things get a little more complex for samples, but this complexity is just an elaboration of the correction factor shown in equation 3.7c.

SUMMARY

The *mean*, *median*, and *mode* are measures of *central tendency*. Each provides a different way of summarizing a "typical score" in a sample or population. The *range*, *variance*, and *standard deviation* are measures of *dispersion*. That is, they are measures of how spread out the scores are in a sample or population. Distributions having the same mean can differ in their standard deviation and vice versa. Distributions can also differ in their *shapes*. The *normal distribution* is symmetrical and bell shaped, with the mean, median, and mode having the same value. For *skewed* distributions, one tail is longer than the other.

When the long tail points toward positive infinity, the distribution is skewed to the right, or *positively skewed*. In this case, mean > median > mode. When the long tail points toward negative infinity, the distribution is skewed to the left, or *negatively skewed*. In this case, mean < median < mode. *Kurtosis* defines how "pointy" or flat a distribution is. *Leptokurtic* distributions are pointy and *platykurtic* distributions are flat. The normal distribution is *mesokurtic*.

All populations have associated *parameters* that can be computed from the scores in the distribution. The mean of a population is computed as

$$\mu = \frac{\sum y}{N}$$

and the sample mean is defined as

$$m = \frac{\sum y}{n}.$$

The variance of a population is computed as

$$\sigma^2 = \frac{\sum (y - \mu)^2}{N},$$

and the variance of a sample is computed as

$$s^2 = \frac{\sum (y - m)^2}{n-1},$$

when used to estimate σ^2. The population standard deviation is the square root of the population variance

$$\sigma = \sqrt{\frac{\sum (y - \mu)^2}{N}},$$

and the sample standard deviation is the square root of the sample variance:

$$s = \sqrt{\frac{\sum (y - m)^2}{n-1}}.$$

The standard deviation is a more intuitive measure of dispersion than the variance because doubling the separation between all scores in a set doubles the standard deviation but quadruples the variance.

KEY TERMS

central tendency   60

deviation score   66

dispersion   64

kurtosis   73

left skewed   72

median   62

mode   63

multimodal distribution   74

negatively skewed   72

normal distribution   72

outlier   65

positively skewed   72

range   65

right skewed   72

sample variance (s^2)   67

shape   71

skewed distribution   72

squared deviation score   66

standard deviation   68

sum of squares (ss)   66

variance   66

EXERCISES

Definitions and Concepts

1. What is the central tendency of a distribution?

2. What is the mean of a distribution?

3. What is the median of a distribution?

4. What is the mode of a distribution?

5. What do we mean by dispersion?

6. What is the range of a distribution?

7. What is a deviation score?

8. What do we mean by the sum of squares?

9. Show the formula for the population variance and put into words what it represents.

10. How does the sample variance differ from the population variance? Why is the sample variance computed differently from the population variance?

11. How does the standard deviation of a population or sample relate to the variance?

12. What are the features of a normal distribution?

13. What does it mean for a distribution to be skewed?

14. Explain the term *kurtosis*.

15. What does it mean for a distribution to be multimodal?

16. The expression $\Sigma(y - m)$ will always evaluate to the same number, no matter what the scores in y are. What is the number? Please explain why this number is always obtained.

True or False

State whether the following statements are true or false.

17. In a normal distribution, the mean equals the mode.

18. In a population, the variance is the average squared deviation from the mean.

19. In a sample, the variance is the average squared deviation from the mean.

20. The sample variance computed with $n-1$ in the denominator is a poor estimator of the population variance.

21. Kurtosis is a property of distributions that cannot be computed from the scores.

22. Multiplying all scores in a sample by 10 will increase the variance by a factor of 20.

23. Adding 10 to all scores in a sample will increase the variance by a factor of 100.

24. If $y = \{1, 2, 3, 3, 3, 4, 5\}$, then the mean equals the mode.

25. If $y = \{1, 2, 3, 4, 5, 6, 7, 8, 9\}$, then the mean equals the median.

26. A very easy test will produce a positively skewed distribution of scores.

27. The mean, like the range, does not make use of all scores in a set.

28. $500,000 and $5,000,000 represent two measures of central tendency for the salaries of professional baseball players. $5,000,000 is the median and $500,000 is the mean.

29. Outliers will have more of an effect on the median than the mean.

30. Outliers will have more of an effect on the range than the variance.

Calculations

31. For each of the following five samples, compute the mean, median, sample variance, and sample standard deviation.

Sample 1	27	3	15	24	6	6	11	0	1	62	15	4
Sample 2	0.2	1.3	–1.8	0.6	0.9	–0.6	0.3	–1.2	–1.2	0.1	–0.2	1.6
Sample 3	5.2	6.3	3.2	5.6	5.9	4.4	5.3	3.8	3.8	5.1	4.8	6.6
Sample 4	9.4	8.9	9.2	11.8	10.6	9.4	9.6	9.9	10	11.3	10.3	11.4
Sample 5	21.6	22.4	22.8	22.7	22.1	21.5	20.6	20.5	21	23.4	21.3	23.5

32. If $y = [1, 2, 3, 4, 5, 6, 7]$ is a population:

    (a) What is the variance, standard deviation, and range?

    (b) If I add 5 to all scores in y (i.e., now $y = [6, 7, 8, 9, 10, 11, 12]$), what is the new mean, the new variance, and the new standard deviation?

    (c) If I multiply all scores in y by 5 (i.e., now $y = [5, 10, 15, 20, 25, 30, 35]$), what is the new mean, the new variance, and the new standard deviation? (Do this question without recomputing the sum of squares.)

APPENDIX 3.1: BASIC STATISTICS IN EXCEL

All the computations described in the body of this chapter can be performed easily in MS Excel or Numbers for Macintosh. Figure 3.A1.1 shows all the statistical functions described in the body of the chapter. In all cases, the functions are applied to the numbers in cells **A2:A6**. A description of the function is given in column **B** under the heading **Quantity**. The value of the function is given in column **C** under the heading **Values**. The instructions to compute these functions are given in column **D** under the heading **Formulas**. The only comment required here is that the function **KURT** actually calculates excess kurtosis (i.e., kurtosis – 3). To understand this point, you should review the material in Appendix 3.4 (available at study.sagepub.com/gurnsey).

FIGURE 3.A1.1 ■ Statistical Functions in Excel

	A	B	C	D
1	**Scores**	**Quantity**	**Values**	**Formulas**
2	1	sum	25	=SUM(A2:A6)
3	3	n	5	=COUNT(A2:A6)
4	5	mean	5	=C2/C3
5	7	mean	5	=AVERAGE(A2:A6)
6	9	median	5	=MEDIAN(A2:A6)
7		maximum	9	=MAX(A2:A6)
8		minimum	1	=MIN(A2:A6)
9		range	8	=C7-C8
10		sum of squares	40	=DEVSQ(A2:A6)
11		population variance	8	=VAR.P(A2:A6)
12		sample variance	10	=VAR.S(A2:A6)
13		population std. dev.	2.8284	=STDEV.P(A2:A6)
14		sample std. dev.	3.1623	=STDEV.S(A2:A6)
15		sample skew	0	=SKEW(A2:A6)
16		sample excess kurtosis	-1.2	=KURT(A2:A6)

APPENDIX 3.2: BASIC STATISTICS IN SPSS

Figure 3.A2.1a shows a Data Editor window with three variables, each with five scores. These small collections of numbers were discussed at various points in the chapter. Computing descriptive statistics for these variables is very easy in SPSS. First, we choose Analyze→Descriptive Statistics→Descriptives . . . , which produces the dialog Descriptives in Figure 3.A2.1b. The three variables (Var1, Var2, and Var3) have been moved into the Variable(s): panel. To specify which statistics are to be computed, we click on the Options . . . button in the Descriptives dialog. This produces the Descriptives: Options dialog box shown in Figure 3.A2.1c. All Descriptives options except S.E. mean have been checked (S.E. mean stands for standard error of the mean, which will be discussed in detail in Chapter 5). Once the options have been chosen, we click ⟨ Continue ⟩ to return to the Descriptives dialog, and there we click ⟨ OK ⟩ to

proceed with the analysis. The results are shown in Figure 3.A2.2.

Most of the statistics in Figure 3.A2.2 were computed at some point in the chapter. The results computed by SPSS can be compared for accuracy with those computed in the chapter. Only skew and kurtosis were not computed before. However, the SPSS output shows that all three variables have 0 Skewness, as we would expect from symmetrical distributions of scores. However, the three distributions all have (excess) Kurtosis of −1.2, which we noted in Appendix 3.4 (available at study.sagepub .com/gurnsey), is the excess kurtosis value for uniform distributions (i.e., the flattest possible distributions). Beside the measures of skew and kurtosis are quantities labeled Std. Error, meaning standard error. Each statistic has an associated standard error and, like the standard error of the mean, will be discussed in Chapter 5.

FIGURE 3.A2.1 ■ Descriptives Dialog in SPSS

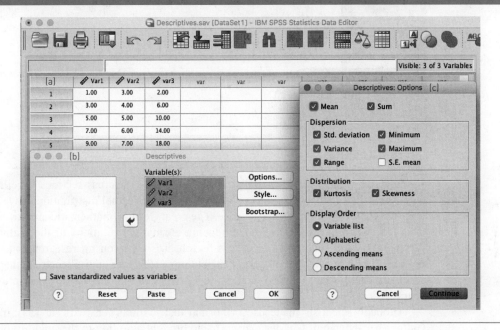

FIGURE 3.A2.2 ■ Descriptives Output in SPSS

Descriptive Statistics

	N	Range	Minimum	Maximum	Sum	Mean	Std. Deviation	Variance	Skewness		Kurtosis	
	Statistic	Statistic	Statistic	Statistic	Statistic	Statistic	Statistic	Statistic	Statistic	Std. Error	Statistic	Std. Error
Var1	5	8.00	1.00	9.00	25.00	5.0000	3.16228	10.000	.000	.913	−1.200	2.000
Var2	5	4.00	3.00	7.00	25.00	5.0000	1.58114	2.500	.000	.913	−1.200	2.000
var3	5	16.00	2.00	18.00	50.00	10.0000	6.32456	40.000	.000	.913	−1.200	2.000
Valid N (listwise)	5											

4 NORMAL DISTRIBUTIONS

INTRODUCTION

In Chapter 3, we saw that normal distributions look like the cross-sections of bells and are therefore called bell curves in the popular literature. Normal distributions have a rather long history in mathematics and science. The great German mathematician Carl Friedrich Gauss is typically credited with introducing normal distributions to the mathematical literature in 1809 (Figure 4.1). For this reason, it is common for researchers to refer to normal distributions as Gaussian distributions. However, the French mathematician Abraham de Moivre seems to have provided the same mathematical description many years earlier (Salsburg, 2001).

Although scientists made use of normal distributions for years, it was Sir Francis Galton (half-cousin of Charles Darwin) who brought attention to the ubiquity of normal distributions in biology and psychology. An example comes from his interest in heredity. In one of his studies, Galton examined the association between the heights of parents and their adult children. (This work provided the foundation for a statistical analysis known today as regression, which we will cover in Chapters 13 and 14.) The heights of the adult children in his study are shown in the histogram in Figure 4.2. The mean of the distribution was 68.09 and the standard deviation was 2.52. Overlaid on the histogram in Figure 4.2 is the probability density function of a normal distribution having the same mean and standard deviation. The normal distribution provides a reasonably good fit to the data. Galton was impressed by the fact that many variables in biology and psychology are normally distributed. (We will suggest a reason for the ubiquity of normal distributions in Appendix 5.2, available at study.sagepub.com/gurnsey.)

Normal distributions bring together the concepts covered in Chapters 2 and 3. First, normal distributions are probability distributions, so they have associated probability

FIGURE 4.1 ■ Gauss, De Moivre, and Galton

Photo credits (*Left to right*): Carl Friedrich Gauss (© Archives of the BBAW, Collections Collection, Collection of paintings, ZIMM-0001 "Carl Friedrich Gauss." Photograph by Stephan Fölske.); Abraham de Moivre (© The Royal Society); Sir Francis Galton (© National Portrait Gallery, London).

density functions. Second, each normal distribution is completely described by its mean and standard deviation, which is another reason why the mean and standard deviation were covered so extensively in Chapter 3. Third, if we know the mean and standard deviation of a variable that is (or is assumed to be) normally distributed, we immediately know the exact proportion of the distribution below any value of the variable. This will have very important consequences for inferential statistics, because normal distributions provide the foundation of many statistics used in psychology and related disciplines.

NORMAL DISTRIBUTIONS

Although we may hear people speak about *the* normal distribution, there are, in reality, infinitely many normal distributions. All normal distributions have a shape that resembles the cross-section of a bell. However, normal distributions differ in their means and standard deviations. Figures 4.3 and 4.4 illustrate these variations. Figure 4.3 shows three normal distributions that have the same standard deviation (σ) of 4 but differ in their means (μ). These means are 62 for the lighter curve, 65 for the darker curve, and 68 for the black curve. As noted in Chapter 3, the mean, median, and mode are all the same for normal distributions; therefore, the mean corresponds to the peak of the distribution. Changing only the mean of the distribution simply shifts its *position* on the x-axis.*

Figure 4.4 shows three normal distributions that have the same mean (μ) of 65 but differ in their standard deviations (σ). Therefore, the only difference between these distributions is their dispersion. These standard deviations are 3 for the lighter curve, 4 for the darker curve, and 5 for the black curve. Notice that density decreases at the peak of the distribution as σ increases and scores become more spread out. It may seem a bit odd, but the three distributions in Figure 4.4 are said to have the same shape. Your immediate response to this might be "Huh, don't they differ in kurtosis?" The answer is no, but please hold onto that thought because we'll return to it in a moment.

---

*The mathematical definition of normal distributions is described and illustrated in Appendix 4.2 (available at study.sagepub.com/gurnsey). You might find this appendix useful because it will help you to recognize and interpret this definition if you ever come across it in your readings in later years.

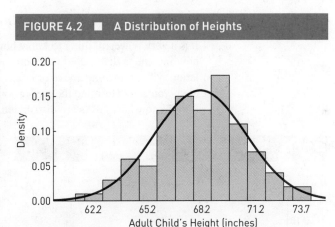

FIGURE 4.2   ■   A Distribution of Heights

A distribution of heights (*n* = 928) for the adult children of 205 parents. The bars show the density in each interval and the overlaid curve is the normal distribution having the same mean (68.09) and standard deviation (2.52) as the distribution of the 928 scores.

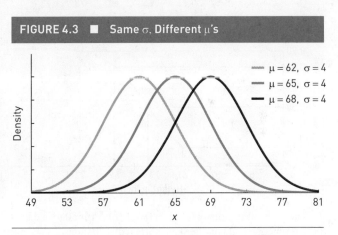

FIGURE 4.3   ■   Same σ, Different μ's

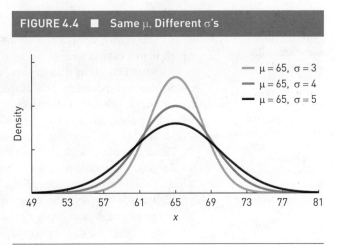

FIGURE 4.4   ■   Same μ, Different σ's

Probability and Normal Distributions

It is a very powerful thing to know how a variable is distributed. For example, if you know how income is distributed in a population, you can determine the proportion of the population with an income above or below yours. If you know how intelligence (IQ) is distributed, you can determine the proportion of the population with an IQ above or below yours. If you know how GPAs are distributed, you can determine the proportion of the population with a GPA above or below yours. In this chapter we will consider many area-under-the-curve questions related to normal distributions.

Figure 4.5a shows a probability density function (*pdf*) for a normal distribution with a mean of $\mu = 65$ and a standard deviation of $\sigma = 4$. The variable is height (measured in inches) for some population, such as the heights of adult Australian women. On the *x*-axis, we've marked off a small number of values of height in intervals of σ. For example, 69 is one σ above μ and 61 is one σ below μ. Thus, the *pdf* is broken into intervals, each of which is one σ wide. The numbers within each region indicate the *approximate* percentage of the distribution falling in each interval. Although these numbers are only approximate, they will help generate intuitions about probability and normal distributions.

FIGURE 4.5 ■ The 34-14-2 Approximation

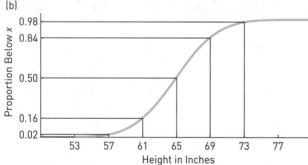

(a) The probability density function (*pdf*). (b) A cumulative distribution function (*cdf*) for the same normal distribution.

Area Under a Normal Curve: The 34-14-2 Approximation

In this section we will introduce the so-called **34-14-2 approximation** (Minium, Clarke, & Coladarci, 1999) to help generate intuitions about the proportion of a normal distribution falling within given intervals.

Figure 4.5a shows that *approximately* 34% of the distribution falls between μ and $\mu + \sigma$. Because the distribution is symmetrical, another 34% falls between μ and $\mu - \sigma$. In other words, 34% of the distribution falls between 65 (μ) and 69 ($\mu + \sigma$), and 34% of falls between 61 ($\mu - \sigma$) and 65 (μ). Therefore, approximately 68% of the distribution falls in the interval $\mu \pm \sigma$, which in this case is 65 ± 4.

Another 14% of the distribution falls in the interval $\mu + \sigma$ to $\mu + 2\sigma$, which in this case means that approximately 14% of the distribution falls between 69 and 73. Again because of symmetry, 14% falls in the interval $\mu - 2\sigma$ to $\mu - \sigma$, or 57 to 61. Therefore, approximately 96% of the distribution (14 + 34 + 34 + 14) falls in the interval $\mu \pm 2\sigma$, which in this case is 65 ± 8, or 57 to 73.

Finally, about 2% of the distribution lies above $\mu + 2\sigma$ (73), and 2% falls below $\mu - 2\sigma$ (57). You should memorize the numbers 34, 14, and 2 because soon we'll ask many questions whose answers depend on them.

Figure 4.5b conveys the same information as Figure 4.5a but in a different format. This type of graph is called a *cumulative distribution function* (*cdf*). The *y*-axis shows the proportion of the distribution falling below each value of *x*. In this example, the curve shows the proportion of the distribution falling below each value of height. For every *pdf*, there is an associated *cdf*, and all *cdf*s have this general *S*-shape.

The **34-14-2 approximation** summarizes the approximate proportion of normal distributions in intervals that are one standard deviation (σ) wide. Approximately 34% of a normal distribution falls between μ and $\mu + \sigma$, approximately 14% of a normal distribution falls between $\mu + \sigma$ and $\mu + 2\sigma$, and approximately 2% of a normal distribution lies above $\mu + 2\sigma$.

Figure 4.5b shows the following:

- approximately 2% of the distribution falls below 57 (i.e., $\mu - 2\sigma$),
- approximately 16% (i.e., $2 + 14$) falls below 61 (i.e., $\mu - \sigma$),
- exactly 50% (i.e., $2 + 14 + 34$) falls below 65 (i.e., μ),
- approximately 84% (i.e., $2 + 14 + 34 + 34$) falls below 69 (i.e., $\mu + \sigma$), and
- approximately 98% (i.e., $2 + 14 + 34 + 34 + 14$) falls below 73 (i.e., $\mu + 2\sigma$).

Although we've chosen only a few values of x to consider in detail, it should be clear that for any value of x, the *cdf* tells us the proportion of the distribution falling below x. We will deal with a larger number of x-values in later sections.

p(x) Versus P(x)

It is very common in the statistical literature to denote the proportion of the distribution below x with $P(x)$. This notation was introduced in Chapter 2, and we will continue with the same notation in this and subsequent chapters. Do not confuse $P(x)$ with $p(x)$, which can mean the proportion of scores in a distribution having the value x, or the density of scores in a distribution at value x.

LEARNING CHECK 1

1. If a normal distribution has a mean of 10 and standard deviation of 2, approximately what proportion of the distribution falls below $x = 12$?

2. If a normal distribution has a mean of 60 and standard deviation of 8, approximately what proportion of the distribution falls below $x = 52$?

3. If a normal distribution has a mean of 100 and standard deviation of 16, approximately what proportion of the distribution falls below $x = 84$?

4. If a normal distribution has a mean of 10,000 and standard deviation of 2, approximately what proportion of the distribution falls below $x = 10,004$?

Answers

1. $P(12) = .84$.

2. $P(52) = .16$.

3. $P(84) = .16$.

4. $P(10,004) = .98$.

THE STANDARD NORMAL DISTRIBUTION: z-SCORES

In Figure 4.5, the values on the x-axis are given in the units of measurement for the variable (i.e., inches). However, the 34-14-2 approximation applies to all normal distributions, not just the one with $\mu = 65$ and $\sigma = 4$. This is why we related the rule to μ and σ. A simpler way to characterize a normal distribution of scores is to convert all scores to **standard scores**, which are also referred to as **z-scores**. A z-score has the following definition:

$$z = \frac{x - \mu}{\sigma}. \tag{4.1}$$

A **z-score**, or **standard score**, expresses the distance of a score from the mean (μ) of its distribution in units of standard deviation (σ).

In words, a z-score represents the number of standard deviations (σ) that a score (x) is from the mean (μ). For the examples given in Figure 4.5, we see that 69 is one standard deviation *above* the mean, so $z = (69 − 65)/4 = 1$. Similarly, 57 is two standard deviations *below* the mean, so $z = (57 − 65)/4 = −2$.

Figure 4.6 illustrates the distribution of z-scores. We call this distribution the **standard normal distribution**. The standard normal distribution has a mean of $\mu = 0$ and a standard deviation and variance of $\sigma = \sigma^2 = 1$. Any normal distribution can be converted to the standard normal distribution by converting x-scores to z-scores using equation 4.1. Conversely, z-scores can be transformed back into the original units using the following equation:

$$x = \mu + z(\sigma). \qquad (4.2)$$

For example, if $\mu = 65$, $\sigma = 4$, and $z = 2$, then $x = \mu + z(\sigma) = 65 + 2(4) = 65 + 8 = 73$.

It is extremely important to note that converting a non-normal distribution of scores to z-scores does *not* magically turn it into the standard normal distribution. For example, if a distribution of scores is skewed or leptokurtic, transforming the scores into z-scores will produce distributions that remain skewed or leptokurtic. The transformation of scores into z-scores only changes the *scale* of the variable; it does not change the shape of the distribution. As for the thought about shape that I asked you to hold onto, we can now see why the three distributions in Figure 4.4 are said to have the same shape. If we convert the scores in all three of these distributions to z-scores, then the three distributions would be identical. That is, all three distributions can be converted to the standard normal distribution and for this reason we say that all normal distributions have the same shape.

The **standard normal distribution** is a distribution of z-scores. The standard normal distribution has a mean of $\mu = 0$, a standard deviation of $\sigma = 1$, and a variance of $\sigma^2 = 1$.

FIGURE 4.6 ■ The Standard Normal Distribution

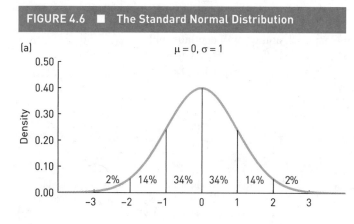

(a)

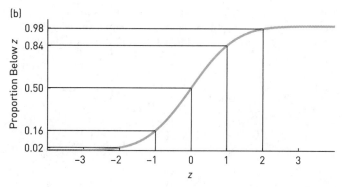

(b)

(a) The probability density function (*pdf*) for the standard normal distribution. (b) A cumulative distribution function (*cdf*) for the standard normal distribution.

AREA-UNDER-THE-CURVE PROBLEMS: APPROXIMATE SOLUTIONS

Proportion Above or Below *x*

In Chapter 2 we made grouped frequency tables and thought about questions such as "What's the probability of drawing a score at random and finding that it is in the interval [80–89]?" Using the terminology developed in Chapter 2, we defined "in the interval [80–89]" as an event and we asked about the probability of this event occurring by chance. It should be clear that Figures 4.5a and 4.6a depict in graphical form the information previously depicted in a grouped frequency table. Table 4.1 shows this connection. (Please note that some liberties have been taken here. In a real frequency table, all intervals are the same width; however, the first and last intervals are infinitely wide in Table 4.1.)

We can now ask questions about the probability of a randomly selected score, drawn from a normal distribution, falling in a particular interval. Here is one such question:

■ What is the *approximate* probability of drawing a score at random from a normal distribution with $\mu = 65$ and $\sigma = 4$ and finding that it is 57 or less?

LEARNING CHECK 2

1. If a normal distribution has a mean of 10 and standard deviation of 2, compute the z-score corresponding to $x = 12$ and state the approximate proportion of the distribution falling *below x*.

2. If a normal distribution has a mean of 60 and standard deviation of 8, compute the z-score corresponding to $x = 52$ and state the approximate proportion of the distribution falling *below x*.

3. If a normal distribution has a mean of 25 and standard deviation of 4.2, compute the z-score corresponding to $x = 33.4$ and state the approximate proportion of the distribution falling *below x*.

4. If a normal distribution has a mean of 71.62 and standard deviation of 9.73, compute the z-score corresponding to $x = 81.35$ and state the approximate proportion of the distribution falling *above x*.

5. If $z = 1$, $\mu = 1.8$, and $\sigma = .5$, what is x?

6. If $z = 1.96$, $\mu = 1.8$, and $\sigma = .5$, what is x?

7. If $z = 1.64$, $\mu = 1.8$, and $\sigma = .5$, what is x?

Answers

1. $z = 1$, $P(1) = .84$.

2. $z = -1$, $P(-1) = .16$.

3. $z = 2$, $P(2) = .98$.

4. $z = 1$, $1 - P(1) = .16$.

5. $x = 1.8 + .5 = 2.3$.

6. $x = 1.8 + 1.96(.5) = 2.78$.

7. $x = 1.8 + 1.64(.5) = 2.62$.

In this case, the event is "a score of 57 or less." One way to answer this kind of question is to convert the score we are given (i.e., 57) to a z-score and make use of the 34-14-2 approximation. We saw previously that 57 corresponds to a z-score of −2. From Figure 4.6 (or Table 4.1), we find that approximately 2% of the standard normal distribution lies below $z = -2$, and so approximately 2% of the original distribution lies below 57.

Below are a few more examples with answers. Note that these questions ask about the *approximate* probability of an event occurring. If you are asked on a test about the *approximate probability* of an event occurring, this is a signal that you can use the 34-14-2 approximation. Therefore, you know that all x-values will transform into z-scores that are integers. When you approach this kind of question, it is very useful to draw a figure, such as Figure 4.5a, so that you can make sure you're on the right track. I would suggest learning to draw normal distributions. (It's really not so hard.) In each of the questions below, I've provided a graphical depiction of the question. The figure you draw serves as a check on your answer. For example, if your figure suggests an answer of .02 and your calculated answer is .98, then you know something has gone wrong.

TABLE 4.1 ■ Z-scores		
z-Score Limits	p	P
2 to ∞	.02	1.00
1 to 2	.14	.98
0 to 1	.34	.84
−1 to 0	.34	.50
−2 to −1	.14	.16
−∞ to −2	.02	.02

Note: Grouped frequency distribution for intervals of the standard normal distribution.

■ What is the approximate probability of drawing a score at random from a normal distribution with $\mu = 65$ and $\sigma = 4$ and finding that the score is:

(a) 57 or less? When we convert 57 to a z-score, we find that it is $z = (57 - 65)/4 = -2$. Figure 4.5 shows that the approximate proportion of the z-distribution falling below $z = -2$ is $P(-2) = .02$, and this is also shown in the figure to the right. Therefore, the approximate probability is $p = .02$ that a randomly chosen score from a normal distribution will be less than 57, when $\mu = 65$ and $\sigma = 4$.

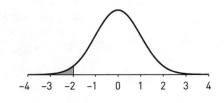

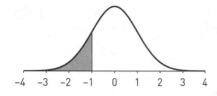

(b) 61 or less? $z = (61 - 65)/4 = -1$. Because approximately 2% of the z-distribution falls below -2 and approximately 14% falls between -2 and -1, then $2 + 14 = 16\%$ of the z-distribution falls below $z = -1$. Therefore, the approximate probability is $p = .16$ that a randomly chosen score from a normal distribution will be less than 61, when $\mu = 65$ and $\sigma = 4$.

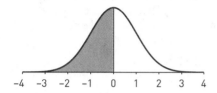

(c) 65 or less? $z = (65 - 65)/4 = 0$. In this case, 65 is the mean of the distribution, so we know right away that exactly 50% of the distribution lies below the mean and 50% lies above the mean. Therefore, the exact probability is $p = .50$ that a randomly chosen score from a normal distribution will be less than 65, when $\mu = 65$ and $\sigma = 4$.

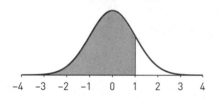

(d) 69 or less? $z = (69 - 65)/4 = 1$. Because 50% of the distribution is below $z = 0$ and another 34% (approximately) is between 0 and 1, then approximately 84% of the z-distribution falls below $z = 1$. Therefore, the approximate probability is $p = .84$ that a randomly chosen score from a normal distribution will be less than 69, when $\mu = 65$ and $\sigma = 4$.

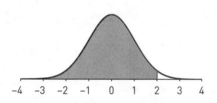

(e) 73 or less? $z = (73 - 65)/4 = 2$. Because approximately 84% of the distribution is below $z = 1$ and another 14% (approximately) is between 1 and 2, then approximately $84 + 14 = 98\%$ of the z-distribution falls below $z = 2$. Therefore, the approximate probability is $p = .98$ that a randomly chosen score from a normal distribution will be less than 73 when $\mu = 65$ and $\sigma = 4$.

We can turn the examples above around and ask what proportion of a distribution with $\mu = 65$ and $\sigma = 4$ falls *above* each of the following numbers: 57, 61, 65, 69, and 73? The answers are .98, .84, .5, .16, and .02, respectively. We find these proportions by subtracting the proportion of the distribution below each score from 1. For example, the proportion of the distribution below 57 is .02, so the proportion above is $1 - .02 = .98$.

Proportion in the Interval x_1 to x_2

Another kind of area-under-the-curve question that will be extremely important later on has the following form:

- What is the *approximate* probability that a randomly selected score from a normal distribution with $\mu = 65$ and $\sigma = 4$ will be *between* x_1 and x_2, where $x_1 = 69$ and $x_2 = 73$?

As with the previous examples, we can answer this question using z-scores (i.e., we can think of how many standard deviations x_1 and x_2 are from the population mean). In the example given, 69 corresponds to $z_1 = 1$ and 73 corresponds to $z_2 = 2$. If we consult Figure 4.6 (or Table 4.1), we see that 14% of the z-distribution lies between $z_1 = 1$ and $z_2 = 2$. Therefore, the approximate probability is $p = .14$ that a randomly chosen score from this distribution will be in the interval 69 to 73. Again, we can answer these questions by converting the scores in question to z-scores and then using the 34-14-2 approximation. Here are two further examples with answers.

- What is the approximate probability of drawing a score at random from a normal distribution with $\mu = 65$ and $\sigma = 4$ and finding a score that is:

(a) Between 57 and 61? When we convert 57 and 61 to z-scores, we find that they are $z_1 = (57 - 65)/4 = -2$ and $z_2 = (61 - 65)/4 = -1$. Figure 4.6 (and Table 4.1) shows that approximately 14% of the z-distribution falls between -2 and -1. Therefore, the approximate probability is $p = .14$ that a randomly chosen score from a normal distribution will be in the interval 57 to 61, when $\mu = 65$ and $\sigma = 4$.

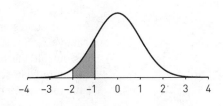

(b) Between 57 and 69? When we convert 57 and 69 to z-scores, we find that they are $z_1 = (57-65)/4 = -2$ and $z_2 = (69-65)/4 = 1$. Figure 4.6 (and Table 4.1) shows that approximately $14 + 34 + 34 = 82\%$ of the z-distribution falls between -2 and 1. Therefore, the approximate probability is $p = .82$ that a randomly chosen score from a normal distribution will be in the interval 57 to 69, when $\mu = 65$ and $\sigma = 4$.

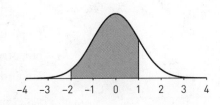

A second way to approach this kind of problem makes use of cumulative probabilities, as shown in Figures 4.5b and 4.6b or the column of Table 4.1 labeled P. Think again about the following question:

■ What is the approximate probability that a randomly selected score from a normal distribution with $\mu = 65$ and $\sigma = 4$ will be between x_1 and x_2, where $x_1 = 69$ and $x_2 = 73$?

(a) We've seen that converting 69 and 73 to z-scores yields 1 and 2. The column labeled P in Table 4.1 shows the proportion of the z-distribution at or below the z-scores. From this table, we see that approximately 98% of the z-distribution is below 2 and approximately 84% of the z-distribution is below 1.

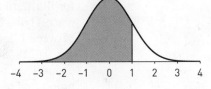

The approximate proportion of the z-distribution *between* 1 and 2 is the proportion below 2 (i.e., .98) *minus* the proportion below 1 (i.e., .84), which yields $p = .98 - .84 = .14$, or 14%.

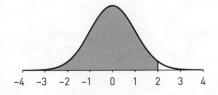

Therefore, the approximate probability is $p = .14$ that a randomly chosen score from a normal distribution, with $\mu = 65$ and $\sigma = 4$, will be in the interval 69 to 73. We will return to this method very soon because it is the method we typically use when we are asked for *exact* proportions.

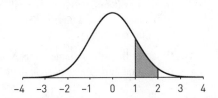

Proportion Outside the Interval x_1 to x_2

A final area-under-the-curve question has the following form:

■ What is the approximate probability of selecting a score at random from a normal distribution with $\mu = 65$ and $\sigma = 4$ and finding that it is *less than* x_1 or greater than x_2, where $x_1 = 69$ and $x_2 = 73$? In other words, what is the probability of selecting a score at random and finding that it is *outside* the interval x_1 to x_2?

A simple way to answer this question is to determine the proportion of the distribution that is in the interval x_1 to x_2 and then subtract this proportion from 1. For example, if the proportion of the distribution between x_1 and x_2 is .14, then the proportion outside this interval is $1 - .14 = .86$. Another way to answer this question would be to convert x_1 and x_2 to z-scores and then (assuming $x_1 < x_2$) add the proportion below x_1 to the proportion above x_2. So, if 84% of the distribution is below x_1 and 2% is greater than x_2, then 86% is outside the interval x_1 to x_2. Therefore, the approximate probability is $p = .86$ that a randomly chosen score from a normal distribution, with $\mu = 65$ and $\sigma = 4$, will be outside the interval 69 to 73.

In summary, we have just considered four area-under-the-curve problems:

- proportion less than x,

- proportion greater than x,

- proportion within the interval x_1 to x_2, and

- proportion outside the interval x_1 to x_2.

All of these examples involved values of $x - \mu$ that were multiples of σ. That is, when we converted our x-values to z-scores, the results were always integers (i.e., $-2, -1, 0, 1,$ or 2). We then used the 34-14-2 approximation to solve these questions.

There are two points to be made before we move on. First, on the off chance that you haven't gotten the message, 34, 14, and 2 are *approximate* values that are used for illustration. In reality, the *exact* proportion of the z-distribution (to four decimal places) lying between 0 and 1 is 0.3413. The exact proportion of the z-distribution lying between 1 and 2 is 0.1359. And the exact proportion of the z-distribution lying above 2 is 0.0228. Second, the 34-14-2 approximation does not allow us to answer questions about z-scores that are not integers. For example, from this rule we can't determine the proportion of the z-distribution that lies above $z = 1.23$. Both of these issues are dealt with in the next section.

LEARNING CHECK 3

1. What is the approximate probability of drawing a score at random from a normal distribution with $\mu = 65$ and $\sigma = 4$ and finding that it is (a) 61 or less and (b) 61 or more?

2. What is the approximate probability of drawing a score at random from a normal distribution with $\mu = 8$ and $\sigma = 1$ and finding that it is (a) 9 or more, (b) 10 or less, (c) between 8 and 10, and (d) between 6 and 7?

3. What is the approximate probability of drawing a score at random from a normal distribution with $\mu = 12$ and $\sigma = 3$ and finding that it is outside the intervals (a) 12 to 15 and (b) 6 to 18?

Answers

1. (a) $z = -1$, $P(z) = .16$, $p = .16$.
   (b) $z = -1$, $1 - P(z) = .84$, $p = .84$.

2. (a) $z = 1$, $1 - P(z) = 1 - .84 = .16$, $p = .16$.
   (b) $z = 2$, $P(z) = .98$, $p = .98$.

   (c) $z_1 = 0$, $z_2 = 2$, $p = .34 + .14 = .48$.
   (d) $z_1 = -2$, $z_2 = -1$, $p = .14$.

3. (a) $z_1 = 0$, $z_2 = 1$, $p = 1 - .34 = .66$.
   (b) $z_1 = -2$, $z_2 = 2$, $p = .02 + .02 = .04$.

THE z-TABLE

Now that we've developed some intuitions about z-scores, we will increase the range of z-scores we can use and the precision of our proportions. Table 4.2 expands on Table 4.1

TABLE 4.2 ■ The z-Table for a Selection of z-Scores Ranging From −2.59 to 2.59									
	.00	.01	.02	.03	.04	.05	.06	.07	.08	.09
−2.5	.0062	.0060	.0059	.0057	.0055	.0054	.0052	.0051	.0049	.0048
−2.4	.0082	.0080	.0078	.0075	.0073	.0071	.0069	.0068	.0066	.0064
−2.3	.0107	.0104	.0102	.0099	.0096	.0094	.0091	.0089	.0087	.0084
−2.2	.0139	.0136	.0132	.0129	.0125	.0122	.0119	.0116	.0113	.0110
−2.1	.0179	.0174	.0170	.0166	.0162	.0158	.0154	.0150	.0146	.0143
−2.0	.0228	.0222	.0217	.0212	.0207	.0202	.0197	.0192	.0188	.0183
−1.9	.0287	.0281	.0274	.0268	.0262	.0256	.0250	.0244	.0239	.0233
−1.8	.0359	.0351	.0344	.0336	.0329	.0322	.0314	.0307	.0301	.0294
−1.7	.0446	.0436	.0427	.0418	.0409	.0401	.0392	.0384	.0375	.0367
−1.6	.0548	.0537	.0526	.0516	.0505	.0495	.0485	.0475	.0465	.0455
−1.5	.0668	.0655	.0643	.0630	.0618	.0606	.0594	.0582	.0571	.0559
−1.4	.0808	.0793	.0778	.0764	.0749	.0735	.0721	.0708	.0694	.0681
−1.3	.0968	.0951	.0934	.0918	.0901	.0885	.0869	.0853	.0838	.0823
−1.2	.1151	.1131	.1112	.1093	.1075	.1056	.1038	.1020	.1003	.0985
−1.1	.1357	.1335	.1314	.1292	.1271	.1251	.1230	.1210	.1190	.1170
−1.0	.1587	.1562	.1539	.1515	.1492	.1469	.1446	.1423	.1401	.1379
−0.9	.1841	.1814	.1788	.1762	.1736	.1711	.1685	.1660	.1635	.1611
−0.8	.2119	.2090	.2061	.2033	.2005	.1977	.1949	.1922	.1894	.1867
−0.7	.2420	.2389	.2358	.2327	.2296	.2266	.2236	.2206	.2177	.2148
−0.6	.2743	.2709	.2676	.2643	.2611	.2578	.2546	.2514	.2483	.2451
−0.5	.3085	.3050	.3015	.2981	.2946	.2912	.2877	.2843	.2810	.2776
−0.4	.3446	.3409	.3372	.3336	.3300	.3264	.3228	.3192	.3156	.3121
−0.3	.3821	.3783	.3745	.3707	.3669	.3632	.3594	.3557	.3520	.3483
−0.2	.4207	.4168	.4129	.4090	.4052	.4013	.3974	.3936	.3897	.3859
−0.1	.4602	.4562	.4522	.4483	.4443	.4404	.4364	.4325	.4286	.4247
−0.0	.5000	.4960	.4920	.4880	.4840	.4801	.4761	.4721	.4681	.4641
0.0	.5000	.5040	.5080	.5120	.5160	.5199	.5239	.5279	.5319	.5359
0.1	.5398	.5438	.5478	.5517	.5557	.5596	.5636	.5675	.5714	.5753
0.2	.5793	.5832	.5871	.5910	.5948	.5987	.6026	.6064	.6103	.6141
0.3	.6179	.6217	.6255	.6293	.6331	.6368	.6406	.6443	.6480	.6517
0.4	.6554	.6591	.6628	.6664	.6700	.6736	.6772	.6808	.6844	.6879
0.5	.6915	.6950	.6985	.7019	.7054	.7088	.7123	.7157	.7190	.7224
0.6	.7257	.7291	.7324	.7357	.7389	.7422	.7454	.7486	.7517	.7549
0.7	.7580	.7611	.7642	.7673	.7704	.7734	.7764	.7794	.7823	.7852
0.8	.7881	.7910	.7939	.7967	.7995	.8023	.8051	.8078	.8106	.8133
0.9	.8159	.8186	.8212	.8238	.8264	.8289	.8315	.8340	.8365	.8389
1.0	.8413	.8438	.8461	.8485	.8508	.8531	.8554	.8577	.8599	.8621
1.1	.8643	.8665	.8686	.8708	.8729	.8749	.8770	.8790	.8810	.8830
1.2	.8849	.8869	.8888	.8907	.8925	.8944	.8962	.8980	.8997	.9015
1.3	.9032	.9049	.9066	.9082	.9099	.9115	.9131	.9147	.9162	.9177
1.4	.9192	.9207	.9222	.9236	.9251	.9265	.9279	.9292	.9306	.9319
1.5	.9332	.9345	.9357	.9370	.9382	.9394	.9406	.9418	.9429	.9441
1.6	.9452	.9463	.9474	.9484	.9495	.9505	.9515	.9525	.9535	.9545
1.7	.9554	.9564	.9573	.9582	.9591	.9599	.9608	.9616	.9625	.9633
1.8	.9641	.9649	.9656	.9664	.9671	.9678	.9686	.9693	.9699	.9706
1.9	.9713	.9719	.9726	.9732	.9738	.9744	.9750	.9756	.9761	.9767
2.0	.9772	.9778	.9783	.9788	.9793	.9798	.9803	.9808	.9812	.9817
2.1	.9821	.9826	.9830	.9834	.9838	.9842	.9846	.9850	.9854	.9857
2.2	.9861	.9864	.9868	.9871	.9875	.9878	.9881	.9884	.9887	.9890
2.3	.9893	.9896	.9898	.9901	.9904	.9906	.9909	.9911	.9913	.9916
2.4	.9918	.9920	.9922	.9925	.9927	.9929	.9931	.9932	.9934	.9936
2.5	.9938	.9940	.9941	.9943	.9945	.9946	.9948	.9949	.9951	.9952

The left column shows z-scores rounded to one decimal, and the top row shows the value in the second decimal place. Each row/column combination specifies z to two decimal places. Each table entry shows $P(z)$, the proportion of the standard normal distribution falling below z. (Values were computed in Excel by the author.)

The **z-table** shows the proportion of the standard normal distribution falling below each of a large number of z-scores.

$P(z)$ denotes the proportion of the standard normal distribution falling below a given z-score.

above. Rather than showing the proportion of the z-distribution below −2, −1, 0, 1, and 2, it shows the proportion of the z-distribution below z-scores ranging from −2.59 to 2.59. We call Table 4.2 the **z-table**, and the table entries show $P(z)$, the proportion of the z-distribution below a given value of z.

This z-table was designed to make optimal use of space on the page. The first column represents z-scores −2.5 to 2.5 rounded one decimal place. The numbers in the column header row represent the second decimal place of the z-score. From this table, we can determine that the proportion of the z-distribution below −2.50 is .0062. To see this, look at the top left corner of the table. The numbers in this row begin with −2.5. The first column corresponds to z-scores with 0 in the second decimal place. You will find that the proportion below −2.50 is .0062. If we look one column to the right, we find that the proportion below −2.51 is .0060; if we look in the third column, we see that the proportion below −2.52 is .0059. The last column shows that the proportion below −2.59 is .0048.

Let's consider a second example. The proportion below 1.6 is .9452. We find this in the first column of proportions in row 1.6. The second decimal in this case is 0. The proportion below 1.64 is found by looking in row 1.6 and moving over to the fifth column, with .04 as the column header. The cell corresponding to row 1.6 and column .04 contains the proportion .9495. This is the proportion of the z-distribution falling below 1.64.

Because the z-distribution is symmetrical, we know that the proportion of the z-distribution *above* a given z is exactly the same as the proportion *below* the z-score having the opposite sign (i.e., −1 * z). For example, the proportion *above* z = 1.64 is exactly the same as the proportion *below* −1 * 1.64 = −1.64. In both cases, the answer is .0505. Alternatively, to find the proportion above z = 1.64, you could have found the proportion below 1.64, which is .9495, and then calculated the proportion above 1.64 by subtracting .9495 from 1 to get 1 − .9495 = .0505.

Similarly, to find the proportion of the z-distribution *above* −1.96, you could simply find the proportion *below* −1 * −1.96 = 1.96, which is .975. Alternatively, to find the proportion above z = −1.96, you could have found the proportion below −1.96, which is .025, and then calculated the proportion above −1.96 by subtracting .025 from 1 to get 1 − .025 = .975.

As a final point, notice that there are two rows for 0, labeled −0.0 and 0.0. The row labeled −.0 contains z = .0 to −.09. The row labeled 0.0 contains z = .0 to .09. We will next revisit some of the questions we dealt with before to get more precise solutions.

LEARNING CHECK 4

1. What proportion of the z-distribution falls below (a) 1.5, (b) −1.92, (c) 1.64, and (d) −1.23?

2. What proportion of the z-distribution lies above (a) −2.01, (b) 2.33, (c) −.82, and (d) −.08?

Answers

1. (a) .9332. (b) .0274. (c) .9495. (d) .1093.

2. (a) .9778. (b) .0099. (c) .7939. (d) .5319.

AREA-UNDER-THE-CURVE PROBLEMS: EXACT SOLUTIONS

We will now return to the four kinds of questions we answered previously using the 34-14-2 approximation. Now, however, we will answer them using the z-table to get answers that are more precise. Furthermore, the z-table allows us to answer questions that can't be answered with the 34-14-2 approximation.

Four Rules

We can state the rules for solving area-under-the-curve-problems very succinctly using the z-table. In all cases, we will assume a normal distribution with known mean and standard deviation. We will also assume that we've transformed scores to z-scores. Here are the rules:

Rule 1. $P(z)$: The probability that a randomly chosen score will be less than z.

Rule 2. $1 - P(z)$: The probability that a randomly chosen score will be greater than z. This can also be computed as $P(-1*z)$; that is, the proportion above z is the same as that below $(-1*z)$.

Rule 3. $P(z_2) - P(z_1)$: The probability that a randomly chosen score will be between z_1 and z_2, assuming $z_2 > z_1$.

Rule 4. $1 - [P(z_2) - P(z_1)]$: The probability that a randomly chosen score will be outside of the interval z_1 to z_2, assuming $z_2 > z_1$. This can also be computed as $P(-1 * z_2) + P(z_1)$, which is the proportion above z_2 plus the proportion below z_1.

We can now put these rules into practice to solve the following problems.

Rule 1: Probability of Obtaining x or Less

- If a score is selected at random from a normal distribution with $\mu = 100$ and $\sigma = 20$, what is the exact probability that the score will be 80 or less?

    Step 1. Choose the rule. Rule 1: $P(z)$.

    Step 2. Compute z and draw an approximate figure where the region in question is shaded. $z = (80 - 100)/20 = -1$.

    Step 3. Find the proportion of the distribution below z. Locate $z = -1$ in the z-table and note the proportion of the distribution below it. In this case, the proportion is $P(-1) = .1587$.

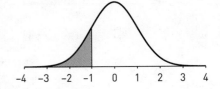

    Step 4. Report the answer. The exact probability of drawing a score of 80 or less from a normal distribution with $\mu = 100$ and $\sigma = 20$ is $p = .1587$.

■ If a score is selected at random from a normal distribution with $\mu = 100$ and $\sigma = 20$, what is the exact probability that the score will be 120 or less?

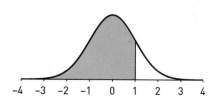

Step 1. Choose the rule. Rule 1: $P(z)$.

Step 2. Compute z and draw an approximate figure with the region in question shaded. $z = (120 - 100)/20 = 1$.

Step 3. Find the proportion of the distribution below z. Locate $z = 1$ in the z-table and note the proportion of the distribution below it. In this case, the proportion is $P(1) = .8413$.

Step 4. Report the answer. The exact probability of drawing a score of 120 or less from a normal distribution with $\mu = 100$ and $\sigma = 20$ is $p = .8413$.

■ If a score is selected at random from a normal distribution with $\mu = 100$ and $\sigma = 20$, what is the exact probability that the score will be 114.468 or less?

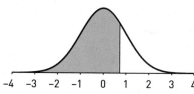

Step 1. Choose the rule. Rule 1: $P(z)$.

Step 2. Compute z and draw an approximate figure with the region in question shaded. $z = (114.468 - 100)/20 = 0.7234$.

Step 3. Find the proportion of the distribution below z. $z = 0.7234$ is not in the z-table so we choose the closest z to 0.7234. Our choices are 0.72 and 0.73. Because 0.72 is closer to 0.7234 than is 0.73, we choose 0.72. The proportion of the distribution below 0.72 is $P(0.72) = .7642$.

Step 4. Report the answer. The exact probability of drawing a score of 114.468 or less from a normal distribution, with $\mu = 100$ and $\sigma = 20$, is $p = .7642$.

Rule 2: Probability of Obtaining x or Greater

■ If a score is selected at random from a normal distribution with $\mu = 100$ and $\sigma = 20$, what is the exact probability that the score will be 114.468 or greater?

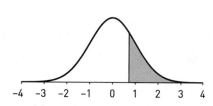

Step 1. Choose the rule. Rule 2: $1 - P(z)$.

Step 2. Compute z and draw an approximate figure with the region in question shaded. $z = (114.468 - 100)/20 = 0.7234$.

Step 3. Find the proportion of the distribution above z. $z = 0.7234$ is not in the z-table, so we choose the closest z, which is 0.72. The proportion of the distribution below 0.72 is $P(0.72) = .7642$. Therefore, the proportion above 0.72 is $1 - P(0.72) = 1 - .7642 = .2358$. (Note that this is equivalent to the proportion below $P(-1 * 0.72)$; i.e., $P(-0.72) = .2358$).

Step 4. Report the answer. The exact probability of drawing a score of 114.468 or greater from a normal distribution, with $\mu = 100$ and $\sigma = 20$, is $p = .2358$.

Rule 3: Probability of Obtaining x Within the Interval x_1 to x_2

■ If a score is selected at random from a normal distribution with $\mu = 100$ and $\sigma = 20$, what is the exact probability that the score will be between 61 and 140?

Step 1. Choose the rule. Rule 3: $P(z_2)-P(z_1)$.

Step 2. Compute z_1 and z_2 and draw an approximate figure with the region in question shaded. $z_1 = (61 - 100)/20 = -1.95$. $z_2 = (140 - 100)/20 = 2$.

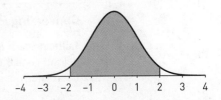

Step 3. Find the proportion of the distribution below z_1 and the proportion below z_2; then apply Rule 3. The proportion of the distribution below $z_2 = 2$ is .9772, and the proportion below $z_1 = -1.95$ is .0256. Apply Rule 3 to find $P(z_2) - P(z_1) = .9772 - .0256 = .9516$.

Step 4. Report the answer. The exact probability of drawing a score between 61 and 140, from a normal distribution with $\mu = 100$ and $\sigma = 20$, is .9516.

Rule 4: Probability of Obtaining x Outside the Interval x_1 to x_2

■ If a score is selected at random from a normal distribution with $\mu = 100$ and $\sigma = 20$, what is the exact probability that the score will be outside the interval 61 to 140?

Step 1. Choose the rule. Rule 4: $1 - [P(z_2) - P(z_1)]$.

Step 2. Compute z_1 and z_2, and draw an approximate figure with the regions in question shaded. $z_1 = (61 - 100)/20 = -1.95$. $z_2 = (140 - 100)/20 = 2$.

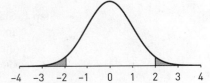

Step 3. Find the proportion of the distribution below z_1 and the proportion below z_2; then apply Rule 4. The proportion of the distribution below $z_2 = 2$ is .9772, and the proportion below $z_1 = -1.95$ is .0256. Apply Rule 4 to find $1 - [P(z_2) - P(z_1)] = 1 - [.9772 - .0256] = .0484$. [Note that this is equivalent to adding the proportion below z_1 and the proportion above z_2. To accomplish this, we add $P(z_1)$ and $P(-1 * z_2)$, which yields $P(-1.95) + P(-2) = .0256 + .0228 = .0484$.

Step 4. Report the answer. The exact probability or drawing a score outside the interval 61 and 140, from a normal distribution with $\mu = 100$ and $\sigma = 20$, is $p = .0484$.

LEARNING CHECK 5

1. What proportion of the z-distribution falls below -1.96?

2. What proportion of the z-distribution falls above 1.96?

3. What is the probability that a randomly chosen z-score will fall in the interval -1.96 to 1.96?

4. What is the probability that a randomly chosen z-score will fall in the interval -1.32 to 1.36?

5. A normal distribution has a mean of 100 and a standard deviation of 15. What is probability that a randomly chosen score from this distribution will be (a) greater than 106, (b) greater than 94, (c) between 82 and 91, and (d) outside the interval 85 to 120?

Answers

1. .025.

2. .025.

3. .95.

4. .8197.

5. (a) .3446. (b) .6554. (c) .1592. (d) .2505.

CRITICAL VALUE PROBLEMS

Converting P(x) to a Score

In the preceding sections, we asked about the proportion of a normal distribution above or below x, and the proportion inside or outside the interval x_1 to x_2. We now turn to a different but related question:

- If the mean and standard deviation of a normal distribution are known, what is the score such that a given proportion [$P(x)$] lies below it?

We approach this kind of problem using the z-table in reverse, so to speak. Rather than computing z and then finding the proportion of the distribution below it, we are given a proportion and then find the z to which this proportion corresponds.

To be concrete, if a normal distribution is defined by $\mu = 65$ and $\sigma = 8$, what is the score such that 75% [i.e., $P(x) = .75$] of the distribution lies below it? To solve this kind of problem, we need to find the z-score corresponding to $P(z) = .75$. This means we have to find the number in the z-table that is closest to .75, note the corresponding z-score, and covert z to x. Here are the steps we need to take.

Step 1. Start by looking down the first column (0 in the second decimal place) until you find two rows that bracket the proportion you're looking for. In this example, these rows correspond to 0.6 and 0.7. The proportion below 0.6 is .7257, and the proportion below 0.7 is .7580. The z-score we're looking for is between 0.6 and 0.7 and must be in the row for 0.6.

Step 2. Look along the chosen row until you find two proportions that bracket the one you're looking for. In this example, the two proportions are .7486 and .7517.

Step 3. Choose the proportion that is closest to the one you're looking for. In this example, .7486 is closer to .75 than is .7517; that is, $|.7486 - .75| = .14$ and $|.7517 - .75| = .17$.

Step 4. Find the z-score corresponding to the selected proportion. In this example, $z = 0.67$.

Step 5. Convert z to a raw score using equation 4.2: $x = \mu + z(\sigma)$. We know $z = 0.67$, $\mu = 65$, and $\sigma = 8$, so we can substitute these into equation 4.2 to find that

$$x = \mu + z(\sigma) = 65 + 0.67(8) = 65 + 5.36 = 70.36.$$

Therefore, 75% of a normal distribution with $\mu = 65$ and $\sigma = 8$ falls below 70.36.

Sketching a graph of the situation we've been asked to consider can also help for problems like this. If we're told that the z-score in question has 75% of the distribution below it, then we should recognize right away that the z-score is between 0 (which has 50% below it) and 1 (which has about 84% below it). Furthermore, the z-score should be closer to 1 than 0. If the z-score we've obtained from the z-table doesn't match this situation, then something has gone wrong.

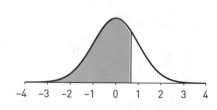

Critical Values of x and $z_{\alpha/2}$

At the end of Chapter 1 we discussed the important question of estimating population parameters from sample statistics. To illustrate the concept of estimation, we can consider a researcher who wants to estimate the mean number of text messages sent each day by 15-year-old American boys. To do so, the researcher would obtain a simple random sample of 15-year-old American boys and determine the number of texts sent by each boy on a given day. These numbers would then be averaged and this average (*m*) would be an estimate of the population parameter (μ), which is the mean number of texts sent each day in the population of 15-year-old American boys. The researcher might report that "We have 95% confidence that the mean number of text messages sent each day by 15-year-old American boys is 79.3 ± 8.6." At this point I can't explain what it means to have 95% confidence; that will have to wait until Chapter 6. However, as you read the researcher's conclusion, just mentally replace "We have 95% confidence" with "We're pretty sure."

When the researcher concluded that he's "pretty sure" that "the mean number of text messages sent each day by 15-year-old American boys is 79.3 ± 8.6," the number 79.3 is the mean number of text messages sent by boys in the sample. Therefore, 79.3 is a statistic (*m*) that estimates a parameter (μ). We saw in Chapter 1 that the number 8.6 is called the *margin of error*. It is used to define an interval around the sample mean. We have high confidence that this interval contains μ. So, the researcher could have said, "We have 95% confidence [we're pretty sure] that the mean number of text messages sent each day by 15-year-old American boys is in the interval 70.7 to 87.9."

The question now is why the researcher has high confidence that the parameter is in this interval. The answer is related to the margin of error. We will see in Chapter 6 that the margin of error is the product of two numbers. One number is called the standard error of the mean, which is discussed in detail in Chapter 5. The other part is a quantity called $z_{\alpha/2}$. This important quantity is related to the following kind of question:

■ What two values of *x* enclose the central 96% of a normal distribution?

Answering this question requires converting proportions to *z*-scores and then *z*-scores to raw scores. In Chapter 6 we'll see that $z_{\alpha/2}$ is related to the confidence we have that an interval contains a parameter. The larger $z_{\alpha/2}$ is, the greater our confidence. The following discussion will help you understand this point when we get to Chapter 6.

The question above asks about the *two* values of *x* enclosing the *central* 96% of the distribution. Therefore, these two values are the same distance from μ; one is below μ and the other is above μ, and 96% of the distribution lies between them. These two distances can be expressed as the number of standard deviations they are from μ. In other words, we can say that our two values of *x* are $\mu \pm z(\sigma)$. Therefore, to answer our question, we simply need to find the appropriate *z*-score, which, as noted above, we will call $z_{\alpha/2}$.

An Approximate Solution

Let's first consider an approximate problem that allows us to use the 34-14-2 approximation.

■ What approximate values of *x* enclose the central 96% of a normal distribution with $\mu = 65$ and $\sigma = 8$?

To answer this, we first think about the *z*-distribution and recall that approximately 96% of *z*-scores fall between –2 and 2. That is, the interval 0 ± 2 encloses (approximately) the central 96% of the *z*-distribution. These *z*-scores can be converted to raw scores as follows:

$$\mu \pm 2(\sigma) = 65 \pm 2(8) = 65 \pm 16 = [49, 81] .$$

Therefore, the interval [49, 81] encloses (approximately) the central 96% of a normal distribution with $\mu = 65$ and $\sigma = 8$.

If we say that $\mu \pm 2(\sigma)$ encloses the central 96% of a normal distribution, this means that the proportion of the distribution falling in this interval is .96. Therefore, the proportion of the distribution *not* falling in the interval $\mu \pm 2(\sigma)$ is $1 - .96 = .04$. From now on we will denote the proportion of the distribution *not* falling in a specified interval with the symbol α. In this example, $\alpha = .04$. In the general case, we can say that α is the proportion of a normal distribution falling outside the interval $\mu \pm z(\sigma)$.

From now on we will use α in our definition of intervals, centered on the mean, containing some percentage of a normal distribution. Specifically, we will refer to such intervals as the central $(1-\alpha)100\%$ of a normal distribution. That is, when α is the proportion of a normal distribution falling *outside* the interval $\mu \pm z(\sigma)$, then $1-\alpha$ is the proportion of a normal distribution falling inside the interval $\mu \pm z(\sigma)$. When $\alpha = .04$, then

$$(1-\alpha)100\% = (1 - .04)100\% = (.96)100\% = 96\%.$$

In the approximate example we worked through, we computed our interval as $\mu \pm 2(\sigma)$, where $z = 2$. We refer to this value of z as $z_{\alpha/2}$, because $(\alpha/2)100\% = 2\%$ of the z-distribution lies below -2 and $(\alpha/2)100\% = 2\%$ lies above 2. Therefore, when $\alpha = .04$, we can say that

$z_{\alpha/2}$ is a z-score for which $(\alpha)100\%$ of the standard normal distribution lies outside the interval $\pm z_{\alpha/2}$.

$$\mu \pm z_{\alpha/2}(\sigma) = 65 \pm 2(8) = 65 \pm 16 = [49, 81].$$

Remember, this solution used the 34-14-2 approximation. We will next see how to solve these problems exactly.

An Exact Solution

When we're asked what two values of x enclose the central $(1-\alpha)100\%$ of a normal distribution, our main task is to determine $z_{\alpha/2}$. We use the following three steps to find $z_{\alpha/2}$, and for this example we will assume $\alpha = .05$.

Find $z_{\alpha/2}$ when $\alpha = .05$.

Step 1. Calculate $\alpha/2$. For this example, $\alpha/2 = .05/2 = .025$.

Step 2. Find $\alpha/2$ in the z-table and note the z-score to which it corresponds. For this example, we must find .025 in the z-table. The z-score corresponding to $\alpha/2 = .025$ is -1.96.

Step 3. $z_{\alpha/2}$ is the absolute value of the z-score we found. For this example, $z_{\alpha/2} = 1.96$.

We will now work through concrete examples of finding the two values of x that enclose the central $(1-\alpha)100\%$ of a normal distribution.

■ What two values of x enclose the central $(1-\alpha)100\%$ of a normal distribution with $\mu = 65$ and $\sigma = 8$, when $\alpha = .05$?

Step 1. Calculate $z_{\alpha/2}$. $\alpha/2 = .05/2 = .025$; therefore, $z_{\alpha/2} = 1.96$.

Step 2. Use $\mu \pm z_{\alpha/2}(\sigma)$ to determine our two values of x. We do this as follows:

$$\mu \pm z_{\alpha/2}(\sigma) = 65 \pm 1.96(8) = 65 \pm 15.68 = [49.32, 80.68].$$

Step 3. State the answer. The interval [49.32, 80.68] encloses the central 95% of a normal distribution with $\mu = 65$ and $\sigma = 8$.

■ What two values of x enclose the central $(1-\alpha)100\%$ of a normal distribution with $\mu = 100$ and $\sigma = 15$, when $\alpha = .2$?

Step 1. Calculate $z_{\alpha/2}$. $\alpha/2 = .2/2 = .1$; therefore, $\alpha/2 = .1$ corresponds to a z-score of -1.28, so $z_{\alpha/2} = 1.28$.

Step 2. Use $\mu \pm z_{\alpha/2}(\sigma)$ to determine the two values of x as follows:

$$\mu \pm z_{\alpha/2}(\sigma) = 100 \pm 1.28\,(15) = 100 \pm 19.2 = [80.8, 119.2].$$

Step 3. State the answer. The interval $[80.8, 119.2]$ encloses the central $(1-\alpha)100\% = (1 - .2)100\% = 80\%$ of a normal distribution with $\mu = 100$ and $\sigma = 15$.

LEARNING CHECK 6

1. A normal distribution has a mean of 100 and a standard deviation of 15. What score corresponds to (a) $P(x) = .5000$, (b) $P(x) = .9788$, (c) $P(x) = .2296$, and (d) $P(x) = .1170$?

2. For each of the following values of α, calculate $z_{\alpha/2}$: (a) .0628, (b) .3030, (c) .0150, and (d) .4010.

3. What two values of x enclose the central $(1-\alpha)100\%$ of a normal distribution when (a) $\mu = 16$, $\sigma = 4$, and $\alpha = .101$; (b) $\mu = 25$, $\sigma = 6$, and $\alpha = .05$; (c) $\mu = 1000$, $\sigma = 40$, and $\alpha = .1336$; and (d) $\mu = 128$, $\sigma = 16$, and $\alpha = .0098$?

Answers

1. (a) $z = 0, x = 100 + 0(15) = 100.$

    (b) $z = 2.03, x = 100 + 2.03(15) = 130.45.$

    (c) $z = .74, x = 100 .74(15) = 88.9.$

    (d) $z = -1.19, x = 100 - 1.19(15) = 82.15.$

2. (a) 1.86. (b) 1.03. (c) 2.43. (d) 0.84.

3. (a) $z_{\alpha/2} = 1.64, \mu \pm z_{\alpha/2}(\sigma) = [9.44, 22.56].$

    (b) $z_{\alpha/2} = 1.96, \mu \pm z_{\alpha/2}(\sigma) = [13.24, 36.76].$

    (c) $z_{\alpha/2} = 1.50, \mu \pm z_{\alpha/2}(\sigma) = [940, 1060].$

    (d) $z_{\alpha/2} = 2.58, \mu \pm z_{\alpha/2}(\sigma) = [86.72, 169.28].$

APPLICATIONS

Let's return to the question of relative standing that we mentioned at the beginning of this chapter. Imagine that Lisa got a grade of 83% on her last stats test. Is 83% a good grade? Without knowing the mean and standard deviation of the distribution of grades, we really don't know if it is good or bad. However, if we're told that the test grades were normally distributed with a mean of 87 and standard deviation of 2, we can determine the proportion of the distribution above or below 83. When we express 83 as a standard score, we find that it is $z = (83 - 87)/2 = -4/2 = -2$. Our intuitions about z-scores should now tell us that 83% is a terrible grade on the test. By consulting the z-table, we can determine that only 2.28% of grades on the test were at or below 83%. In other words, 97.72% of grades were higher than this. It seems that Lisa's professor gives very high marks, but Lisa's 83% puts her near the bottom of her class.

Let's now consider David's grade of 67% on a statistics test given by another professor. These grades were normally distributed with a mean of 61 and standard deviation of 4. David might feel bad about his 67% compared with Lisa's 83%. Can you make David feel better? Well, yes. When 67% is expressed as a standard score, we find that it is $z = (67 - 61)/4 = 6/4 = 1.5$. When we consult the z-table, we can determine that 93.32% of grades on the test were at or below 67%. (I'd be very happy with this outcome!) David's professor gives very low marks, but David's 67% puts him near the top of his class.

Perhaps Lisa thought to herself, "At least my 83% isn't as low as poor David's 67%!" Poor Lisa! No wonder she did so poorly in statistics. If David had taken the test that Lisa took and he obtained a grade that was 1.5 standard deviations above the mean, his grade would have been $87 + 1.5(2) = 90\%$. The same point can be made the other way around. If Lisa had taken the test that David took and scored two standard deviations below the mean, her grade would have been $61 - 2(4) = 53\%$. So, z-scores allow us to compare apples and oranges by putting the values of quantitative variables on a common (standard) scale.

This ends our introduction to normal distributions. We will make heavy use of normal distributions in Chapters 5 to 9, so it is important for you to master these concepts. I highly recommend working through the end-of-chapter exercises to consolidate your understanding of z-scores and normal distributions.

Assessing Normality

Many statistical applications that we will discuss in later chapters assume that a sample of scores have been drawn from a normal distribution. The analyses are only valid if this assumption is true. Appendix 4.3 (available at study.sagepub.com/gurnsey) explains several methods that can be used within Excel and SPSS to assess this assumption. You may wish to return to this material when you work through Chapters 10 to 21.

LEARNING CHECK 7

1. In Professor Gauss's statistics class, grades are normally distributed with a mean of 72 and a standard deviation of 8. In Professor Laplace's statistics class, grades are normally distributed with a mean of 82 and a standard deviation of 6.

   (a) What grade corresponds to $z = 1$ in (i) Professor Gauss's class and (ii) Professor Laplace's class?

   (b) What grade corresponds to $z = -2$ in (i) Professor Gauss's class and (ii) Professor Laplace's class?

   (c) What proportion of the grades in Professor Gauss's class fall below 77 and what proportion of grades in Professor Laplace's class fall below 77?

Answers

1. (a) $x_{Gauss} = 72 + 1(8) = 80$; $x_{Laplace} = 82 + 1(16) = 88$.

   (b) $x_{Gauss} = 72 - 2(8) = 56$; $x_{Laplace} = 82 - 2(6) = 70$.

   (c) $z_{Gauss} = .63$, $P = .7357$; $z_{Laplace} = -.83$, $P = .2033$.

SUMMARY

A *normal distribution* is a bell-shaped distribution that is completely described by its mean (μ) and standard deviation (σ). Any normal distribution can be transformed to the *standard normal distribution* by converting all scores (x) to z-scores as follows:

$$z = \frac{x - \mu}{\sigma}.$$

The standard normal distribution is also\ called the *z-distribution*. The distribution of z-scores can be

transformed into any normal distribution by converting all z-scores (z) to x-scores as follows:

$$x = \mu + z(\sigma).$$

The *34-14-2 approximation* is used to generate intuitions about normal distributions. Approximately 34% of the z-distribution lies between $z = 0$ and $z = 1$, approximately 14% lies between $z = 1$ and $z = 2$, and approximately 2% lies above $z = 2$. Similarly, approximately 34% of the z-distribution lies between

$z = -1$ and $z = 0$, approximately 14% lies between $z = -2$ and $z = -1$, and approximately 2% lies below $z = -2$. With this rule, we can compute the approximate proportion of any distribution *below x, above x, within the interval* x_1 *to* x_2, and *outside the interval* x_1 *to* x_2.

The z-table provides a more extensive listing of z-scores between $z = -2.59$ and $z = 2.59$. Each entry in the table corresponds to the proportion of the z-distribution below z; that is, $P(z)$. Using the z-table, we can compute the exact proportion of any normal distribution *below x, above x, within the interval* x_1 *to* x_2, and *outside the interval* x_1 *to* x_2. To do so, we convert x_1 and x_2 to z_1 to z_2, and we then consult the z-table. We then use to following rules:

1. $P(z)$: The probability that a randomly chosen score will be less than z.

2. $1 - P(z)$: The probability that a randomly chosen score will be greater than z. This can also be computed as $P(-1 * z)$; that is, the proportion above z is the same as that below $-1 * z$.

3. $P(z_2) - P(z_1)$: The probability that a randomly chosen score will be between z_1 and z_2, assuming $z_2 > z_1$.

4. $1 - [P(z_2) - P(z_1)]$: The probability that a randomly chosen score will be outside of the interval z_1 and z_2, assuming $z_2 > z_1$. [This can also be computed as $P(-1 * z_2) + P(z_1)$, which is the proportion above z_2 plus the proportion below z_1.]

We can use the z-table in reverse to find a z-score when given the proportion of the distribution below it; that is, $P(z)$. Once the z-score has been found, we can plug it into the following equation to determine x:

$$x = \mu + z(\sigma).$$

We can also use the z-table in reverse to determine the two values of z that enclose the central $(1-\alpha)100\%$ of a normal distribution when given μ, σ, and α. To do this, we first determine $\alpha/2$ and then $z_{\alpha/2}$. Once we have $z_{\alpha/2}$, we simply insert the appropriate numbers into the equation

$$\mu \pm z_{\alpha/2}(\sigma)$$

to determine the two values of x that enclose the central $(1-\alpha)100\%$ of the distribution.

KEY TERMS

34-14-2 approximation 82	standard score 83	z-table 90
$P(z)$ 90	$z_{\alpha/2}$ 96	
standard normal distribution 84	z-score 83	

EXERCISES

Definitions and Concepts

1. What is a z-score?

2. What two parameters completely describe a normal distribution?

3. What is the standard normal distribution?

4. Put into words the meaning of $P(z)$.

True or False

State whether the following statements are true or false.

5. All normal distributions can be transformed to the standard normal distribution.

6. Skewed distributions can be transformed to the standard normal distribution.

7. All normal distributions have a mean of zero and a standard deviation of 1.

8. $P(z)$ conveys the proportion of the z-distribution above z.

9. $P(z_2) - P(z_1)$ conveys the proportion of the z-distribution between z_1 and z_2.

10. Approximately 14% of the standard normal distribution lies between -1 and 0.

11. If $z = 1$, then $P(z) = .8413$.

12. If $z = -1$, then $P(z)$ is approximately $.34$.

13. If $z_1 = 1$ and $z_2 = 2$, then $P(z_2) - P(z_1)$ is the proportion of a normal distribution outside the interval x_1 to x_2.

Calculations

14. What proportion of the z-distribution lies below (a) 1.6, (b) -1.6, and (c) 2.3?

15. What proportion of the z-distribution lies above
    (a) 1.33,
    (b) −0.53, and
    (c) 1.86?

16. What proportion of the z-distribution lies between
    (a) −1.22 and 1.86,
    (b) 1.22 and 1.86,
    (c) −1.96 and 1.96,
    (d) −1.64 and 1.96, and
    (e) −2.58 and 2.58?

17. What proportion of the z-distribution lies outside the interval
    (a) −1.00 to 2.58,
    (b) −.5 to .66, and
    (c) .5 to .92?

18. What is the approximate probability of drawing a score at random from a normal distribution with $\mu = 100$ and $\sigma = 10$ and finding that the individual's score is
    (a) 80 or less,
    (b) 90 or less,
    (c) 100 or less,
    (d) 110 or less, and
    (e) 120 or less?

19. What is the approximate probability of drawing a score at random from a normal distribution with $\mu = 100$ and $\sigma = 10$ and finding that the individual's score is
    (a) 80 or more,
    (b) 90 or more,
    (c) 100 or more,
    (d) 110 or more, and
    (e) 120 or more?

20. What is the approximate probability of drawing a score at random from a normal distribution with $\mu = 100$ and $\sigma = 10$ and finding that the individual's score is
    (a) between 80 and 90,
    (b) between 80 and 100,
    (c) between 90 and 120,
    (d) between 110 and 120, and
    (e) between 80 and 120?

21. What is the exact probability of drawing a score at random from a normal distribution with $\mu = 99.2$ and $\sigma = 12.8$ and finding that the individual's score is
    (a) 81.024 or less,
    (b) 91.008 or less,
    (c) 100.992 or less,
    (d) 110.976 or less, and
    (e) 120.96 or less?

22. What is the exact probability of drawing a score at random from a normal distribution with $\mu = 99.2$ and $\sigma = 12.8$ and finding that the individual's score is
    (a) 81.024 or more,
    (b) 91.008 or more,
    (c) 100.992 or more,
    (d) 110.976 or more, and
    (e) 120.96 or more?

23. What is the exact probability of drawing a score at random from a normal distribution with $\mu = 99.2$ and $\sigma = 12.8$ and finding that the individual's score is
    (a) between 81.024 and 91.008,
    (b) between 81.024 and 100.992,
    (c) between 91.008 and 120.96,
    (d) between 110.976 and 120.96, and
    (e) between 81.024 and 120.96?

24. A score was drawn from a normal distribution with $\mu = 99.2$ and $\sigma = 12.8$. What is the score if
    (a) the proportion below it is 0.1788,
    (b) 17.88% of the distribution falls below it,
    (c) the proportion below it is 0.9591,
    (d) 95.91% of the distribution falls below it,
    (e) the proportion below it is 0.4364, and
    (f) 15.15% of the distribution falls above it?

25. What two values of x enclose the central $(1-\alpha)100\%$ when $\mu = 99.2$, $\sigma = 12.8$, and $\alpha =$
    (a) .05,
    (b) .0434,
    (c) .0802,
    (d) .0990, and
    (e) .0096?

APPENDIX 4.1: NORM.DIST AND RELATED FUNCTIONS IN EXCEL

Microsoft Excel provides some really useful statistical tools that can help us solve the kinds of questions we worked through in this chapter. We will discuss four of these functions below.

NORM.DIST and NORM.S.DIST

In the Excel spreadsheet in Figure 4.A1.1, I've entered values for x, μ, and σ, as shown in cells **B2** to **B4**. The **NORM.DIST** function uses these numbers to compute either

- the density of scores in a normal distribution at the given value of x, i.e., $p(x)$, or

- the proportion of scores in a normal distribution at or below a given value of x, i.e., $P(x)$.

The **NORM.DIST** function requires us to provide x, μ, σ, and *cumulative*, as follows:

NORM.DIST(*x*,*μ*,*σ*,*cumulative*)

The numbers within the parentheses are referred to as **arguments**; they are the inputs to the function. When

> The inputs to mathematical functions are called **arguments**.

cumulative is set to 0, **NORM.DIST** returns $p(x)$; when *cumulative* is set to 1, **NORM.DIST** returns $P(x)$. $p(x)$ and $P(x)$ are calculated in cells **B7** and **B8**, respectively, in Figure 4.A1.1. When $x = 68$, $\mu = 65$, and $\sigma = 4$, the density of score at x is $p(x) =$ .0753 and the proportion of the distribution below x is $P(x) = .7734$.

The **NORM.S.DIST** function works much like **NORM.DIST** but operates on z-scores. Therefore, **NORM.S.DIST** asks us to provide z and *cumulative*, as follows:

NORM.S.DIST(*z*,*cumulative*)

As with **NORM.DIST**, **NORM.S.DIST** provides

- the density of scores at z when *cumulative* is set to 0,

- and the proportion of the standard normal (z) below z when *cumulative* is set to 1.

In cell **B5** in Figure 4.A1.1, x has been converted to a z-score in the usual way; i.e., $z = (x - \mu)/\sigma$. In this example, $z = .75$. In cell **B10**, we see the density of scores at $z = .75$; in cell **B11**, we see the proportion of the z-distribution below $z = .75$. It is important to note that the proportion of the z-distribution below $z = .75$ is exactly equal to the proportion of the normal distribution below the value of x to which z corresponds. In this example, the proportion of the z-distribution below $z = .75$ is the same as the proportion below $x - \mu + z(\sigma) = 65 + .75(4) = 65 + 3 = 68$. In both cases, $P = .7734$.

Figure 4.A1.2 shows how to use **NORM.DIST** to solve all area-under-the-curve problems that we discussed in the chapter. As before, we have $\mu = 65$ and $\sigma = 4$ in cells **B2** and **B3**. We now have two values of x, which we denote x_1 and x_2. Cell **B8** shows that the

FIGURE 4.A1.1 ■ NORM.DIST and NORM.S.DIST

	A	B	C
1	Quantities	Values	Formulas
2	x	68	
3	μ	65	
4	σ	4	
5	z	0.75	=(B2-B3)/B4
6			
7	p(x)	0.0753	=NORM.DIST(B2,B3,B4,0)
8	P(x)	0.7734	=NORM.DIST(B2,B3,B4,1)
9			
10	p(z)	0.3011	=NORM.S.DIST(B5,0)
11	P(z)	0.7734	=NORM.S.DIST(B5,1)

FIGURE 4.A1.2 ■ Four Rules and NORM.DIST

	A	B	C
1	Quantities	Values	Formulas
2	μ	65	
3	σ	4	
4			
5	x₁	61	
6	x₂	69	
7			
8	P(x₁)	0.1587	=NORM.DIST(B5,B2,B3,1)
9	P(x₂)	0.8413	=NORM.DIST(B6,B2,B3,1)
10	P(x₂)-P(x₁)	0.6827	=B9-B8
11	1-[P(x₂)-P(x₁)]	0.3173	=1-B10

proportion of this normal distribution below $x_1 = 61$ is $P(x_1) = .1587$ and cell **B9** shows that the proportion below $x_2 = 69$ is $P(x_2) = .8413$. The proportion of the distribution between these two values of x is simply $P(x_2) - P(x_1)$ as shown in cell **B10**. The proportion of the distribution outside this interval is $1 - [P(x_2) - P(x_1)]$ as shown in cell **B11**.

NORM.INV and NORM.S.INV

The **NORM.DIST** function takes x as an argument and returns a proportion. The **NORM.INV** function takes a proportion as an argument and returns x. That is, **NORM.INV** answers the question, what value of x has proportion $P(x)$ below it? We provide **NORM.INV** with the following arguments:

NORM.INV($P(x)$, μ, σ)

where $P(x)$ is the proportion of the distribution below x, and μ and σ are the mean and standard deviation of the distribution. In Figure 4.A1.3, μ and σ have the same values as in Figures 4.A1.1 and 4.A1.2. The value of $P(x)$ (cell **B5**) is .8413. In cell **B8**, we use the **NORM.INV** function to determine the value of x having 84.13% of the distribution below it when μ = 65 and σ = 4. The answer is 69.

The **NORM.S.DIST** function takes z as an argument and returns a proportion. The **NORM.S.INV** function takes a proportion as an argument and returns z. That is, **NORM.S.INV** answers the question, what value of z has proportion $P(z)$ below it? We provide **NORM.S.INV** with a single argument:

NORM.S.INV($P(z)$),

where $P(z)$ is the proportion of the distribution below z. In Figure 4.A1.3, the value of $P(z)$ (cell **B6**) is .8413. In cell **B9**, we use the **NORM.S.INV** function to determine the value of z having 84.13% of the distribution below it. The answer is 1.

In this chapter, we saw how to compute the values of x that enclose the central $(1-\alpha)100\%$ of a normal distribution of scores. In cell **B11**, α has been set to .05. Therefore, $(1-\alpha)100\% = 95$, as shown in cell **B12**. To determine the values of x that enclose the central

FIGURE 4.A1.3 ■ NORM.INV and NORM.S.INV

	A	B	C
1	**Quantities**	**Values**	**Formulas**
2	μ	65	
3	σ	4	
4			
5	P(x)	0.8413	
6	P(z)	0.8413	
7			
8	x	69.00	=NORM.INV(B5,B2,B3)
9	z	1.00	=NORM.S.INV(B6)
10			
11	α	0.05	
12	(1-α)100%	95	=(1-B11)*100
13	α/2	0.025	=B11/2
14	$z_{\alpha/2}$	1.96	=ABS(NORM.S.INV(B13))
15			
16	μ-$z_{\alpha/2}$(σ)	57.16	=B2-B14*B3
17	μ+$z_{\alpha/2}$(σ)	72.84	=B2+B14*B3

$(1-\alpha)100\%$ of a normal distribution of scores, we must compute $z_{\alpha/2}$. α/2 is calculated in cell **B13**, and $z_{\alpha/2}$ is computed in cell **B14** using **NORM.S.INV**. Note that in cell **B14**, the **ABS** function has been used. **ABS** takes a real number as an argument and returns its absolute value. To determine our two values of x, we computed $\mu - z_{\alpha/2}(\sigma)$ in cell **B16** and $\mu + z_{\alpha/2}(\sigma)$ in cell **B17**.

Practicing With These Functions

You can use the functions described in this appendix to answer questions 7 to 25 in the end-of-chapter exercises. Proficiency with these functions is extremely useful. There will be many calculations in the next few chapters that you can check with **NORM.DIST** and **NORM.INV**. Many researchers keep Excel open at all times so that we can check routine calculations in our own work and in the work of our colleagues.

APPENDIX KEY TERM

argument   101

DISTRIBUTIONS OF STATISTICS

INTRODUCTION

In Chapter 1 we saw that inferential statistics involves estimating parameters from statistics. Estimating a population mean from a sample mean is the most common instance of estimation. For example, we might estimate the mean number of *Facebook friends* North American university students have by computing the mean number of *Facebook friends* in a random sample drawn from this population. The main objective of this chapter is to show that a sample mean is a more *precise* estimate of the population mean than is a single score. And the larger the sample, the more precise the estimate. In Chapter 1 we explored our statistical intuitions about this issue with the example of drawing jellybeans from a large jar. Our intuition was that a large handful of jellybeans would provide a better estimate of the true proportion of red jellybeans in the jar than would a small handful. We can now work toward explaining why this intuition is valid.

To provide a preview of what is to come, imagine that the number of *Facebook friends* in the population of North American university students is normally distributed with a mean of $\mu = 300$ and standard deviation of $\sigma = 50$. With this information, we can use the 34-14-2 approximation from Chapter 4 to say that approximately 68% of individuals will have *scores* (number of *Facebook friends*) within the interval $\mu \pm \sigma$ (i.e., 300 ± 50).

Now for something new. If we consider all possible pairs of scores drawn from this distribution, then approximately 68% of these pairs will have *means* that fall in the interval $\mu \pm \sigma/\sqrt{2}$ (i.e., 300 ± 35.36). If we consider all possible triplets of scores drawn from this distribution, then approximately 68% of these triplets will have *means* that fall in the interval $\mu \pm \sigma/\sqrt{3}$ (i.e., 300 ± 28.87). You might be able to discern a pattern here. If so, you might be able to predict the interval that will contain approximately 68% of sample means when sample size is 4. The answer is 300 ± 25. The pattern here shows that as sample size increases, 68% of sample *means* fall in narrower and narrower intervals.

The critical new concept in this chapter is that of a *statistical distribution* (the distribution of a statistic). All statistics (e.g., mean, variance, skew, median, kurtosis, etc.) have a distribution, and statistical distributions are the foundation of inferential statistics. In fact, a course on statistics is really a course on how to use statistical distributions. Our discussion of statistical distributions begins with the distribution of sample means, because the mean is the statistic most commonly used to characterize a distribution. We also discuss distributions of variances, and we say a few things about distributions of proportions in Appendix 5.5 (available at study.sagepub.com/gurnsey).

We saw in Chapter 1 that *estimation* and *significance tests* are the most widely used statistical methods in psychology. These two methods rely on exactly the same statistical distributions but use them in different ways. The topics of the previous four chapters (populations and samples, parameters and statistics, μ and *m*, σ and *s*, and the normal distribution) laid the foundations for this chapter, and this chapter lays the foundation for the rest of the book. So, let's continue the story.

THE DISTRIBUTION OF SAMPLE MEANS

In the Facebook example that opened this chapter we considered two different questions. The first concerns the interval containing 68% of scores from the population, and the second concerns the interval containing 68% of sample means. To answer the second question, we need to know something about the *distribution of sample means*. The **distribution of means** (or **sampling distribution of the mean**) is a probability distribution of all possible sample means computed from samples of the same size. We will unpack this definition as we go along.

Before we discuss the distribution of means, we must first think about how samples are formed. Samples can be formed by drawing scores at random from a population *with* or *without* replacement. **Sampling with replacement** means that each time a score is drawn from a population, it is returned to the population before the next score is drawn. Therefore, a score from the same individual can be selected more than once in the sample. **Sampling without replacement** means that when a score is drawn from a population, it is not returned to the population before the next score is drawn. This means that a score from the same individual can be selected only once in the sample. We encountered the distinction between sampling with and without replacement when discussing independent and dependent events in Appendix 2.3 (available at study.sagepub.com/gurnsey).

When we sample with replacement, the number of possible samples of size n is N^n, where N is the number of scores in the population. This leads to an enormous number of possible samples, even for a small N. For a small population with $N = 1000$ and a smallish sample size of $n = 30$, the number of possible samples is unimaginably large. In this case, $N^n = 1000^{30}$ would be a 1 followed by 90 zeros. There aren't this many atoms in the known universe, and it would be impossible to calculate 1000^{30} means even on today's fastest computers.

If $N^n = 1000^{30}$ is an unimaginably large number of samples, then the number of possible samples one could form when sample size is $n = 64$, and N is the number of North American university students, is even larger. For practical purposes, the number is not different from infinity. Fortunately, mathematicians are able to tell us about the properties of distributions of means even if the distributions involve such large numbers of means. Therefore, the properties of statistical distributions are typically derived theoretically. However, we can illustrate these properties using small, concrete examples.

Although we've been discussing sampling with replacement, studies in psychology involve sampling without replacement. Is this difference important? Well, even though there are fewer possible samples when we sample without replacement, the actual number is still so big that there is no practical difference between sampling with or without replacement. Therefore, we will begin our discussion of the distribution of means with a concrete example involving sampling with replacement, because our points can be made more easily this way. However, everything we will say is true when sampling without replacement from populations of the size normally studied in psychology.

Let's start by considering a small population of four scores:

$$y = \{1, 3, 5, 7\}.$$

The **distribution of means** is a probability distribution of all possible values of a sample mean based on samples of the same size. The distribution of means is also referred to as the **sampling distribution of the mean**, and these terms are used interchangeably.

Sampling with replacement means that each time a score is drawn from a population, it is returned to the population before the next score is drawn.

Sampling without replacement means that when a score is drawn from a population, it is not returned to the population before the next score is drawn.

It's hard to think of a real population comprising only four scores; the population of people who have walked on the moon is greater than 4. However, for our purposes, the things that are true of this population of four scores are also true of populations comprising hundreds, thousands, millions, or billions of scores. Therefore, considering this tiny population will help us make some important points with minimal fuss.

If you take a moment to do the calculations, you will find that the mean of this population is $\mu = 4$ and the variance is $\sigma^2 = 5$. Remember, the population variance is the mean squared deviation from the mean, which is $\sigma^2 = \sum (y - \mu)^2 / N$.

Now let's think about all possible samples of size $n = 2$ that can be formed by sampling with replacement from this population. Remember, the number of possible samples (and hence the number of sample means) is $N_m = N^n$. Therefore, in this case, the number of samples is $N_m = N^n = 4^2 = 16$.

The column labeled "Sample" in Table 5.1 shows the 16 possible samples of size 2 drawn from our population. To distinguish samples, each has been given a unique name: for example, $y_1 = [1, 1]$, $y_2 = [1, 3]$, and $y_{16} = [7, 7]$. The column labeled "Sample mean (m_i)" in Table 5.1 shows the mean for each of the 16 samples. These means are all possible means for samples of size 2 drawn from our population. Therefore, these 16 sample means constitute a population (i.e., distribution) of means. And, like any population, the distribution of means itself has a mean and variance, which we'll now compute.

The mean of the distribution of means is simply

$$\mu_m = \frac{m_1 + m_2 + m_3 + m_4 + \cdots + m_{N_m}}{N_m}. \tag{5.1a}$$

That is, we add up all the sample means and divide by the number of sample means (N_m). The symbol μ_m denotes the mean of sample means. We can write equation 5.1 in the simpler (and familiar) form shown in equation 5.1b:

$$\mu_m = \frac{\sum m_i}{N_m}. \tag{5.1b}$$

The subscript in m_i is simply to make clear that we are summing a number of different means (i.e., $m_1, m_2 \ldots m_{16}$). The sum of the sample means is 64. When this sum is divided by $N_m = 16$, we find that the mean of the distribution of means is $\mu_m = 64/16 = 4$.

To summarize, for a population of N scores, there will be exactly $N_m = N^n$ samples of size n that can be formed. Each of the N_m samples has a mean, so we have a population of sample means. This population of sample means itself has a mean, which we denote μ_m.

The column labeled "Squared deviation" in Table 5.1 shows the squared deviation of each sample mean from μ_m. For the first sample, this squared deviation is $(m_1 - \mu_m)^2 = (1 - 4)^2 = 9$. From these squared deviations, we can compute the variance of a distribution of means in the usual way:

$$\sigma_m^2 = \frac{\sum (m_i - \mu_m)^2}{N_m}. \tag{5.2}$$

In equation 5.2, m_i again denotes sample means and N_m is the number of sample means. Table 5.1 shows that $\sigma_m^2 = 2.5$. (It would be worthwhile to do this calculation yourself in Excel.) As with μ_m, the subscript on σ_m^2 reminds us that it denotes the variance of a population of sample means (m).

The main point to take from this discussion is that we can conceive of all possible samples of size n drawn from a population of N scores. Each of these samples has a mean, and we often refer to this distribution of sample means as the *sampling distribution of the*

TABLE 5.1 ■ Demonstration of the Sampling Distribution of the Mean

i	Sample	Sample Mean (m_i)	Squared Deviation $(m_i - \mu_m)^2$
1	$y_1 = [1, 1]$	1	9
2	$y_2 = [1, 3]$	2	4
3	$y_3 = [1, 5]$	3	1
4	$y_4 = [1, 7]$	4	0
5	$y_5 = [3, 1]$	2	4
6	$y_6 = [3, 3]$	3	1
7	$y_7 = [3, 5]$	4	0
8	$y_8 = [3, 7]$	5	1
9	$y_9 = [5, 1]$	3	1
10	$y_{10} = [5, 3]$	4	0
11	$y_{11} = [5, 5]$	5	1
12	$y_{12} = [5, 7]$	6	4
13	$y_{13} = [7, 1]$	4	0
14	$y_{14} = [7, 3]$	5	1
15	$y_{15} = [7, 5]$	6	4
16	$y_{16} = [7, 7]$	7	9
		$\mu_m = \dfrac{\sum m_i}{N_m}$ $= \dfrac{64}{16}$ $= 4$	$\sigma_m^2 = \dfrac{\sum (m_i - \mu_m)^2}{N_m}$ $= \dfrac{40}{16}$ $= 2.5$

Note: The population is $y = \{1, 3, 5, 7\}$, and sample size is $n = 2$.

mean. A distribution of sample means itself has a mean (μ_m) and variance (σ_m^2). Our next step is to see how μ_m and σ_m^2 relate to sample size (n), as well as the mean (μ) and variance (σ^2) of the population from which the samples were drawn.

LEARNING CHECK 1

1. Let $y = \{1, 3, 5\}$ be a population of scores.

   (a) How many samples of size 2 are possible when sampling from this population with replacement?

   (b) Show all possible samples of size 2 when sampling with replacement.

   (c) Compute the mean of each of these samples.

   (d) Compute the mean of these means.

   (e) Compute the variance of these means; remember, the means represent a population.

Answers

1. (a) $N^n = 3^2 = 9$

   (b) (1, 1), (1, 3), (1, 5), (3, 1), (3, 3), (3, 5), (5, 1), (5, 3), (5, 5).

   (c) 1, 2, 3, 2, 3, 4, 3, 4, 5.

   (d) $\mu_m = 3$.

   (e) $\sigma_m^2 = 12/9 = 1.333$.

Parameters and Shape of the Distribution of Means

The parameters of the distribution of means are related in very simple ways to the parameters of the distribution of scores from which the samples were drawn.

The Mean of the Distribution of Means

We saw earlier that our population of four scores has a mean of $\mu = 4$. The mean of the distribution of means is shown in Table 5.1 to be $\mu_m = 4$ as well; i.e.,

$$\mu_m = \mu. \tag{5.3}$$

What is true of our small population is true of all populations. That is, the **mean of the distribution of means** (μ_m) will always equal the mean of the population of scores (μ) no matter how many scores there are in the population (N), no matter how many scores in the samples (n), and no matter the shape of the distribution being sampled. This is kind of amazing and is very helpful knowledge in the context of estimation.

Statisticians often refer to the mean of a statistical distribution as the **expected value (E)** of the statistic. (This term is just useful shorthand.) The expected value of the sample mean is denoted $E(m)$. An individual sample mean may be greater than μ, or it may be less than μ because of sampling error (see Chapter 1). However, the average value of all possible sample means is μ. Because the mean of the distribution of means is equal to the mean of the population being sampled, we say that the sample mean is an **unbiased** estimator of its parameter. In other words, m is an unbiased estimator of μ because $E(m) = \mu$. Not all statistics are unbiased estimators of their parameters, as we will see later.

The Variance of the Distribution of Means

We just saw how the mean of the distribution of means relates to the mean of the population of scores. We will now ask how the **variance of the distribution of means (σ_m^2)** relates to the variance of the population of scores.

μ_m is the **mean of the distribution of means**. μ_m always equals μ, the mean of the distribution from which samples were drawn.

The **expected value** (E) of a statistic is the mean of its sampling distribution. We denote the expected value of m as $E(m)$.

A statistic is **unbiased** when its expected value equals the parameter it estimates.

σ_m^2 is the **variance of the distribution of means**. σ_m^2 always equals σ_x^2/n, the variance of the population of scores divided by sample size.

At the beginning of the chapter we noted that sample means cluster more tightly around μ than do scores. We can see evidence of this when we compare the variance of our small population and the variance of the distribution of means. We saw earlier that our population of four scores had a variance of $\sigma^2 = 5$. Table 5.1 shows that the variance of the distribution of means is $\sigma^2_m = 2.5$. There is a very simple relationship between these two variances:

$$\sigma^2_m = \frac{\sigma^2}{n}. \tag{5.4}$$

That is, the variance of the distribution of means (σ^2_m) equals the variance of the population (σ^2) of scores divided by sample size (n). As with the mean, the relation shown in equation 5.4 is true no matter how many scores there are in the population (N), no matter how many scores in the samples (n), and no matter the shape of the distribution being sampled.

The Standard Error of the Mean

In Chapter 3 we saw that the population standard deviation was simply the square root of the population variance. The square root of the variance of the distribution of means (σ^2_m) is called the **standard error of the mean** (or *SEM* for short):

$$\sigma_m = \sqrt{\sigma^2_m} = \sqrt{\frac{\sigma^2}{n}}. \tag{5.5a}$$

An equivalent expression is

$$\sigma_m = \frac{\sigma}{\sqrt{n}}. \tag{5.5b}$$

The standard error of the distribution of means (also called the **standard error of the mean**, or *SEM*) is denoted σ_m. The value of σ_m always equals $\sigma/\sqrt{n}$, which is the standard deviation of the population of scores divided by the square root of the sample size.

(Plugging $\sigma = 5$ and $n = 16$ into equations 5.5a and 5.5b confirms that they produce the same result; i.e., $\sqrt{25/16} = 1.25$ and $5/\sqrt{16} = 1.25$.)

The standard error of the mean is a critically important part of estimation. We've noted twice already that sample means cluster more tightly around μ than do scores. Furthermore, the larger the sample size, the more tightly sample means cluster around μ. Figure 5.1 illustrates this point by plotting σ_m as a function of sample size. In this example, $\sigma = 16$, and the solid black line shows σ_m for sample sizes of 1 to 128. The filled dots show σ_m for a subset of sample sizes (i.e., 1, 2, 4, 8, 16, 32, 64, and 128).

Figure 5.2 makes the same point as Figure 5.1 by showing how the distribution of means changes as a function of a sample size. In this example, the scores were drawn from normal distribution with μ = 65 and σ = 4. The lighter curve shows the normal distribution of scores being sampled. (This is the same as saying that sample size is $n = 1$.) The two other curves show the distribution of means for sample sizes of 4 and 16. Note how the spreads (standard errors) of the distributions change as sample size increases (as expected from equations 5.4 and 5.5).

Figures 5.1 and 5.2 make clear why the means of large samples are more precise estimates of μ than means of small samples. The larger the sample size, the closer sample means are to μ on average.

If the number of *Facebook friends* in a population of North American university students is normally distributed with a mean of μ = 300 and standard deviation of σ = 50, then the 34-14-2 approximation says that approximately 68% of *scores* fall in the interval μ ± σ = 300 ± 50 = [250, 350].

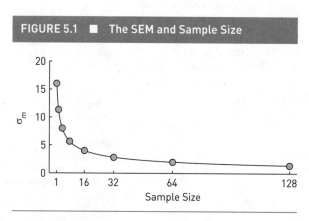

FIGURE 5.1 ■ The SEM and Sample Size

Sample Size

Now let's consider *means* computed from samples of size $n = 100$. In this case, the standard error of the mean (σ_m) is $\sigma/\sqrt{n} = 50/10 = 5$. Therefore, the 34-14-2 approximation says that approximately 68% of these *means* fall in the interval $\mu \pm \sigma/\sqrt{n} = 300 \pm 5 = [295, 305]$. Again, the larger the sample, the better the estimate.

Standard Error Versus Standard Deviation

You might ask why σ_m is called the standard error of the mean rather than the standard deviation of the mean. The reason is that σ_m is, roughly, the average *sampling error*. Recall from Chapter 1 that sampling error is the difference between a sample statistic and the parameter it estimates. If one were to compute the absolute value of the sampling error for all possible sample means, the average of these absolute values would be the *average sampling error*:

$$\text{Average Sampling Error} = \frac{\sum|m_i - \mu|}{N_m},$$

where N_m is the number of all possible sample means. The standard error of the mean (*SEM*), described in equations 5.5a and 5.5b, is closely related to the average sampling error. The standard error of the mean is defined as

$$\sigma_m = \sqrt{\sigma_m^2} = \sqrt{\frac{\sum(m - \mu)^2}{N_m}}.$$

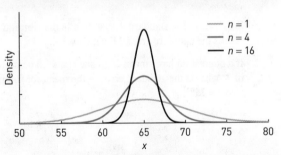

FIGURE 5.2 ■ **Three Distributions of Means**

The light blue line plots a population of scores having $\mu = 65$ and $\sigma = 4$. One can think of it as the distribution of means where sample size is $n = 1$. The dark blue line represents the distribution of means when sample size is 4. It has the same mean as the population (65) but its standard error is $\sigma/\sqrt{4} = 4/2 = 2$. The black line represents the distribution of means when sample size is 16. It also has the same mean as the population but its standard error is $\sigma/\sqrt{16} = 4/4 = 1$.

We can see similarities in these two definitions, but they yield slightly different values.

For mathematical reasons, and because $\sum|m_i - \mu|/N_m$ does not relate to normal distributions, the average sampling error is not much used in statistics, and we consider it no further. Because σ_m is similar to, but not exactly equal to, the average sampling error, we say that the standard error of the mean is *roughly* the average sampling error.

There is a standard error for every statistic, not just the mean. Therefore, a *standard error* is, roughly, the average absolute sampling error for a given statistic. Standard deviations relate to scores, and standard errors relate to statistics.

> ### AN IMPORTANT DISTINCTION: STANDARD DEVIATION VERSUS STANDARD ERROR
>
> The standard deviation relates to a distribution of scores; it is, roughly, the average distance between *scores* in the distribution and the mean of the distribution. The standard error relates to a distribution of statistics; it is, roughly, the average distance between the *statistics* in the distribution and the parameter they estimate.

Shape

Table 5.2 shows the distribution of means derived from Table 5.1. By referring back to Table 5.1, you can see that there is only one way to get a sample mean of 7 and there are two ways to get a sample mean of 6, three ways to get a sample mean of 5, and so on.

LEARNING CHECK 2

1. We can call the standard deviation of a distribution of scores the standard error. [True, False]

2. If a population has a mean of 10 and standard deviation of 2, what is the standard error of the mean for sample size $n = 25$?

3. If a population has a mean of 10 and standard deviation of 2, what is the variance of the distribution of means for sample size $n = 64$?

Answers

1. False. The standard error relates to the distribution of a statistic.

2. $2/5 = .4$.

3. $4/64 = .0625$.

Mean	f	p	P
7	1	.06	1.00
6	2	.13	.94
5	3	.19	.81
4	4	.25	.63
3	3	.19	.38
2	2	.13	.19
1	1	.06	.06

TABLE 5.2 ■ Frequency Table of Means

Note: A frequency distribution of sample means drawn from population $y = \{1, 3, 5, 7\}$ for samples of size $n = 2$.

The **central limit theorem** states (roughly) that the distribution of means will be a normal distribution if the population being sampled is normal. It also states that the distribution of sample means will converge on a normal distribution as sample size increases, irrespective of the shape of the distribution being sampled.

Figure 5.3a plots the distribution means from Table 5.2 as a histogram. Because the y-axis shows the proportion of times each possible mean occurs, it can be properly called a probability distribution. That is, it shows the probability of each possible sample mean occurring in an infinite series of identical sampling experiments. It is important to note that the distribution of means has a different shape than that of the population of scores from which the samples were drawn. The population of scores (shown in Figure 5.3b) is a flat distribution, because each score in the population (1, 3, 5, and 7) has the same probability (¼) of being chosen with repeated random sampling. The distribution of means, by contrast, is roughly normal.

Shape and the Central Limit Theorem

The example that we have been working with nicely illustrates how the parameters of the distribution of means relate to the parameters of the distribution of scores from which the samples were drawn. The example also shows that the distribution of means may have a shape that differs from the shape of the population from which the scores were drawn. However, this simple example does not illustrate one of the most important features of the distribution of means, which is that *the distribution of means is a normal distribution* when certain general conditions are met.

When samples of size n are drawn with replacement from a distribution of scores, the distribution of means will be a normal distribution if the population being sampled is normal, and it will *approach* a normal distribution as sample size (n) increases, irrespective of the shape of the distribution from which the samples were drawn. This statement is known as the **central limit theorem**. (Strictly speaking, the central limit theorem requires sampling with replacement, but we will assume it holds in large populations that are sampled without replacement.)

Figure 5.3 hints at the nature of the central limit theorem. The scores were drawn from a flat distribution, but the distribution of means has a roughly normal shape. If the sample size had been 30, which is possible only when sampling with replacement, then the distribution of means would look very normal.

Figure 5.4 is similar to Figure 5.3 except that the scores were drawn from a skewed distribution. The light blue curve shows the skewed distribution of scores being sampled. (This is the same as saying that sample size is $n = 1$.) The two other curves show the distribution of means for sample sizes of 4 and 16. As sample size increases, the distribution of means becomes narrower (as expected from equations 5.4 and 5.5) and increasingly normal (as stated by the central limit theorem). All distributions in Figure 5.4 have the same mean ($\mu = 9.0275$), and the variance of each distribution is equal to σ^2/n.

If the distribution of means tends toward the normal distribution as sample size increases, the immediate question is "How large must sample size be for the distribution of means to be normal?" The answer to this question depends on the nature of the distribution being sampled. For variables that diverge only slightly from a normal distribution, the distribution of means will approach the normal distribution very quickly as sample size increases. If variables differ radically from normality, then the distribution of means will approach the normal distribution more slowly as sample size increases. For the purposes of subsequent exercises, we will assume that the distribution of means is normal when sample size is 30 or greater. This is just a rule of thumb that we'll use for later exercises only.

The Central Limit Theorem and Populations of Scores

In statistics we often assume that populations of scores are normally distributed, and this assumption is very often true. Appendix 5.2 (available at study.sagepub.com/gurnsey) describes how the central limit theorem explains the normality of so many populations of scores. The short explanation is that each score is a consequence of many independent "factors," just like a sample mean is the result of many independent scores.

If a statistical method assumes that scores were drawn from a normal population, it would be useful to assess this assumption. Appendix 4.3 (available at study.sagepub.com/gurnsey) describes methods of assessing the normality assumption in Excel and SPSS.

Sampling Distributions

Every statistic has a sampling distribution. For example, there are sampling distributions for the mean, median, mode, range, variance, standard deviation, skew, kurtosis, and many other statistics. The concept is exactly the same as the sampling distribution of the mean. In a later section, we will consider the sampling distribution of variances. In Appendix 5.5 (available at study.sagepub.com/gurnsey), we will discuss the sampling distribution of proportions.

Appendix 5.3 (available at study.sagepub.com/gurnsey) describes a fantastic online demonstration of sampling distributions. You should absolutely check this out. You can use this demo to see how violations of normality affect the normality of the distribution of means.

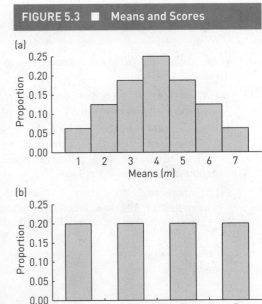

FIGURE 5.3 ■ Means and Scores

Distributions of means and scores. (a) A histogram of 16 sample means, based on samples of size $n = 2$ drawn from the population $y = \{1, 3\ 5, 7\}$. (b) The distribution of scores from which the samples were drawn.

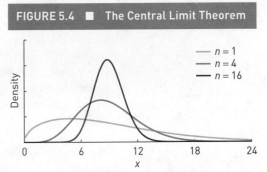

FIGURE 5.4 ■ The Central Limit Theorem

The light blue line plots a population of scores. The dark blue line represents the distribution of means when sample size is 4. It has the same mean as the population (9.0275) but its standard error is $\sigma/\sqrt{4}$. The black line represents the distribution of means when sample size is 16. It also has the same mean as the population but its standard error is $\sigma/\sqrt{16}$.

LEARNING CHECK 3

1. State whether the following statements are true or false.

   (a) Strictly speaking, the central limit theorem requires sampling with replacement.

   (b) If a normally distributed population has a mean of 10 and standard deviation of 100, then the sampling distribution of the mean will be a normal distribution when sample size is $n = 5$.

   (c) If a population has a mean of 10 and standard deviation of 100 and follows a rectangular distribution, then the sampling distribution of the mean will be a normal distribution when sample size is $n = 5$.

   (d) If a right-skewed distribution has a mean of 100 and standard deviation of 15, then the sampling distribution of the mean will be almost normal distribution when sample size is $n = 500$.

   (e) If a leptokurtic distribution has a mean of 88 and standard deviation of 0.15, then, for the purposes of this course, we will assume that the sampling distribution of the mean will be a normal distribution when sample size is $n = 25$.

Answers

1. (a) True.

   (b) True.

   (c) False. When the sample size is small (e.g., 5), the distribution of means is normal only if the population of scores is normal.

   (d) True. The sampling distribution of the mean approaches normality as sample size increases.

   (e) False. We will make the arbitrary assumption that the distribution of means is normal if $n = 30$.

AREA-UNDER-THE-CURVE QUESTIONS

We are now able to answer questions of the sort that opened this chapter. Here is an example:

> If we know that the number of *Facebook friends* is normally distributed with $\mu = 300$ and $\sigma = 50$, what is the exact probability that the mean number of *Facebook friends* of four individuals, randomly selected from the population, will be 350 or greater?

The question asks about drawing a mean from a distribution of means. There is enough information in the question to determine everything about this distribution of means. The mean of the distribution is $\mu_m = 300$, because the mean of the distribution of means is always equal to the mean of the population being sampled. The standard error of the mean is $\sigma_m = \sigma/\sqrt{n} = 50/2 = 25$, as defined in equation 5.5b. Finally, we are told that the variable (number of *Facebook friends*) is normally distributed. This means that the distribution of means will also be normal. With this information, we can convert the sample mean $m = 350$ to a *z*-score:

$$z = \frac{m - \mu_m}{\sigma_m}. \tag{5.6}$$

When we substitute in the numbers we've obtained, we find that

$$z = \frac{m - \mu_m}{\sigma_m} = \frac{350 - 300}{25} = 2.$$

Because we were asked about the exact probability of a randomly selected sample mean exceeding 350, we would consult the z-table (on page 91) to determine the exact proportion of the standard normal distribution lying above z. That probability is $p = .0228$.

We've just seen the power of the central limit theorem. We can apply all the methods used in Chapter 4 to solve area-under-the-curve problems for means drawn from a normal distribution of means. Let's work through problems similar to those we did in Chapter 4, making use of the following scenario.

Life at a university can be full of hassles and unexpected challenges. Kohn, Lafreniere, and Gurevich (1990) developed a questionnaire called the Inventory of College Students' Recent Life Experiences (ICSRLE) to identify students' exposure to negative life events in the past month. The ICSRLE is a 55-item questionnaire that asks students to rate on a 4-point scale (0 to 3) the degree to which they've had negative life events such as "conflicts with professor(s)" and "lower grades than hoped for." Scores on this scale can range from 0 to 165. We will imagine that this questionnaire has been administered to all members of a psychology department participant pool. The resulting distribution of ICSRLE scores is normal with a mean of 30 and standard deviation of 10. We will use these numbers to solve area-under-the-curve questions relating to sample means. There are four forms of these questions corresponding to the four rules introduced in Chapter 4.

Rule 1: Probability of Obtaining *m* or Less

■ If 25 scores are selected at random from a normal distribution with $\mu = 30$ and $\sigma = 10$, what is the exact probability that the mean of these 25 scores will be $m = 29$ or less?

Step 1. Choose the rule. Rule 1: $P(z)$.

Step 2. Compute the standard error of the mean. Because $\sigma = 10$ and $n = 25$, we know that

$$\sigma_m = \sigma/\sqrt{n} = 10/5 = 2.$$

Step 3. Convert m to a z-score. To do this, we must know m, μ, and σ_m. These values are $m = 29$, $\mu = 30$, and $\sigma_m = 2$. When we insert these values into the equation for a z-score, we obtain

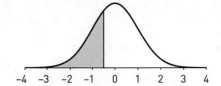

$$z = \frac{m - \mu_m}{\sigma_m} = \frac{29 - 30}{2} = \frac{-1}{2} = -0.5.$$

Step 4. Determine the proportion of the z-distribution below z. When we consult the z-table, we find that the proportion below $z = -0.5$ is $P(-0.5) = .3085$.

Step 5. Answer the question. The probability is $p = .3085$ that a random sample of 25 scores will have a mean of 29 or less when drawn from a normal distribution having $\mu = 30$ and $\sigma = 10$.

Rule 2: Probability of Obtaining *m* or Greater

■ If 25 scores are selected at random from a normal distribution with $\mu = 30$ and $\sigma = 10$, what is the exact probability that the mean of these 25 scores will be $m = 33.2$ or greater?

Step 1. Choose the rule. Rule 2: $1 - P(z)$.

Step 2. Compute the standard error of the mean. As before, $\sigma = 10$ and $n = 25$, so we know that

$$\sigma_m = \sigma / \sqrt{n} = 10/5 = 2.$$

Step 3. Convert m to a z-score. To do this, we must know m, μ, and σ_m. These values are $m = 33.2$, $\mu = 30$, and $\sigma_m = 2$. When we insert these values into the equation for a z-score, we obtain

$$z = \frac{m - \mu_m}{\sigma_m} = \frac{33.2 - 30}{2} = \frac{3.2}{2} = 1.6.$$

Step 4. Determine the proportion of the z-distribution above z. When we consult the z-table, we find that the proportion below $z = 1.6$ is $P(1.6) = .9452$. Therefore, the proportion above $z = 1.6$ is $1 - P(1.6) = 1 - .9452 = .0548$. [We could have solved this in fewer steps by determining $P(-1.6)$.]

Step 5. Answer the question. The probability is $p = .0548$ that a random sample of 25 scores will have a mean of 33.2 or greater when drawn from a normal distribution having $\mu = 30$ and $\sigma = 10$.

Rule 3: Probability of Obtaining m Within the Interval m_1 to m_2

■ If 25 scores are selected at random from a normal distribution with $\mu = 30$ and $\sigma = 10$, what is the probability that the mean of these 25 scores will be between $m_1 = 29$ and $m_2 = 33.2$?

Step 1. Choose the rule. Rule 3: $P(z_2) - P(z_1)$.

Step 2. Compute the standard error of the mean. As before, $\sigma = 10$ and $n = 25$, so we know that

$$\sigma_m = \sigma / \sqrt{n} = 10/5 = 2.$$

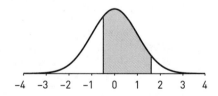

Step 3. Convert m_1 and m_2 to z-scores. To do this, we must know m_1, m_2, μ, and σ_m. These values are $m_1 = 29$, $m_2 = 33.2$, $\mu = 30$, and $\sigma_m = 2$. When we insert these values into the equation for a z-score, we obtain

$$z_1 = \frac{m_1 - \mu_m}{\sigma_m} = \frac{29 - 30}{2} = \frac{-1}{2} = -0.5 \text{ and}$$

$$z_2 = \frac{m_2 - \mu_m}{\sigma_m} = \frac{33.2 - 30}{2} = \frac{3.2}{2} = 1.6.$$

Step 4. Determine the proportion of the z-distribution between z_1 and z_2. When we consult the z-table, we find that the proportion below $z_1 = -0.5$ is $P(z_1) = .3085$, and the proportion below $z_2 = 1.6$ is $P(z_2) = .9452$. Therefore, the proportion of the z-distribution between z_1 and z_2 is $P(z_2) - P(z_1) = .9452 - .3085 = .6367$.

Step 5. Answer the question. The probability is $p = .6367$ that a random sample of 25 scores will have a mean between 29 and 33.2 when drawn from a normal distribution having $\mu = 30$ and $\sigma = 10$.

Rule 4: Probability of Obtaining m Outside the Interval m_1 to m_2

■ If 25 scores are selected at random from a distribution with $\mu = 30$ and $\sigma = 10$, what is the probability that the mean of these 25 scores will be outside the interval $m_1 = 29$ and $m_2 = 33.2$?

Step 1.   Choose the rule. Rule 4: $1 - [P(z_2) - P(z_1)]$.

Step 2.   Compute the standard error of the mean. As before, $\sigma = 10$ and $n = 25$, so we know that

$$\sigma_m = \sigma/\sqrt{n} = 10/5 = 2.$$

Step 3.   Convert m_1 and m_2 to z-scores. To do this, we must know m_1, m_2, μ, and σ_m. These values are $m_1 = 29$, $m_2 = 33.2$, $\mu = 30$, and $\sigma_m = 2$. When we insert these values into the equation for a z-score, we obtain

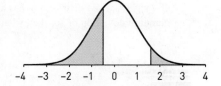

$$z_1 = \frac{m_1 - \mu_m}{\sigma_m} = \frac{29 - 30}{2} = \frac{-1}{2} = -0.5 \text{ and}$$

$$z_2 = \frac{m_2 - \mu_m}{\sigma_m} = \frac{33.2 - 30}{2} = \frac{3.2}{2} = 1.6.$$

Step 4.   Determine the proportion of the z-distribution outside of the interval z_1 to z_2. In the previous question, we saw that the proportion of the z-distribution in the interval z_1 to z_2 was $P(z_2) - P(z_1) = .6367$. Therefore, the proportion of the z-distribution outside the interval z_1 to z_2 is $1 - [P(z_2) - P(z_1)] = 1 - .6367 = .3633$. [We could have solved this in fewer steps by determining $P(z_1) + P(-1*z_2) = P(-.5) + P(-1.6) = .3085 + 0.0548 = 3633$.]

Step 5.   Answer the question. The probability is $p = .3633$ that a random sample of 25 scores will have a mean outside the interval 29 and 33.2 when drawn from a normal distribution having $\mu = 30$ and $\sigma = 10$.

Before moving on, it is important to reiterate that if the distribution of means is not normal, then we can't solve questions of the sort we've just done. For the distribution of means to be normal, either the distribution of scores must be normal or the sample size must be large. For the purposes of the exercises in this book, if the sample size is 30 or greater, then you can assume that the distribution of means is normal, even if the distribution of scores is not. If the distribution of scores is normal, then questions can be answered as above for any sample size.

CRITICAL VALUE PROBLEMS

Converting $P(m)$ to a Mean

In Chapter 4 we saw that we could determine the value of x when given μ, σ, and $P(x)$. Therefore, we can determine the value of m when given μ, σ, n, and $P(m)$, where $P(m)$ is the proportion of the distribution of means below m. The following question is of exactly this sort, and we show the steps necessary to solve it.

LEARNING CHECK 4

1. If a normally distributed population has a mean of 10 and standard deviation of 64, then what proportion of the distribution of means falls below −6 when sample size is $n = 16$?

2. If a normally distributed population has a mean of 10 and standard deviation of 64, then what proportion of the distribution of means falls below 11.2 when sample size is $n = 64$?

3. If a right-skewed distribution has a mean of 100 and standard deviation of 16, then what proportion of the distribution of means falls below 103.1 when sample size is $n = 64$?

4. If a leptokurtic distribution has a mean of 88 and standard deviation of 28, then what proportion of the distribution of means falls between 83 and 92 when sample size is $n = 49$?

5. If a left-skewed and platykurtic distribution has a mean of 42 and standard deviation of 18, then what proportion of the distribution of means falls between 43 and 45 when sample size is $n = 81$?

Answers

1. $z = (m - \mu_m)/\sigma_m = (-6 - 10)/16 = -1; P(-1) = .1587.$

2. $z = (m - \mu_m)/\sigma_m = (11.2 - 10)/8 = .15; P(.15) = .5596.$

3. $z = (m - \mu_m)/\sigma_m = (103.1 - 100)/2 = 1.55; P(1.55) = .9394.$

4. (i) $z_1 = (m - \mu_m)/\sigma_m = (83 - 88)/4 = -1.25; P(-1.25) = .1056;$

(ii) $z_2 = (m - \mu_m)/\sigma_m = (92 - 88)/4 = 1; P(1) = .8413;$

(iii) $P(1) - P(-1.25) = .8413 - .1056 = .7357.$

5. (i) $z_1 = (m - \mu_m)/\sigma_m = (43 - 42)/2 = .5; P(.5) = .6915;$

(ii) $z_2 = (m - \mu_m)/\sigma_m = (45 - 42)/2 = 1.5; P(1.5) = .9332;$

(iii) $P(1.5) - P(0.5) = .9332 - .6915 = .2417.$

■ If 25 scores are selected at random from a normal distribution with $\mu = 30$ and $\sigma = 10$, and we know that 58.71% of means fall below the sample mean, what is the sample mean?

Step 1. Find the proportion below m in the z-table. Because 58.71% corresponds to a proportion of .5871, we can use the z-table to determine the z-score to which this proportion corresponds. When we consult the z-table, we find that .5871 corresponds to $z = 0.22$.

Step 2. Compute the standard error of the mean. We were given that $\sigma = 10$ and $n = 25$, so we know that

$$\sigma_m = \sigma/\sqrt{n} = 10/5 = 2.$$

Step 3. Compute the value of m. We know $z = 0.22$, $\mu = 30$, and σ_m, so we can substitute these into equation 4.4 from Chapter 4 to find that

$$m = \mu_m + z (\sigma_m) = 30 + 0.22(2) = 30.44.$$

Step 4. Answer the question. 58.71% of a distribution of means falls below 30.44, when $\mu = 30$ and $\sigma = 2$.

Critical Values of m and $z_{\alpha/2}$

We now arrive at one of the most important computations associated with z-scores and the distribution of means. Let's return to our example involving *Facebook friends*. We will continue to assume that the population of scores is normally distributed with a mean of

$\mu = 300$ and standard deviation of $\sigma = 50$. We can now ask, what two sample means enclose the central 95% of the distribution of means when sample size is $n = 4$? As in Chapter 4, we can express this question in the following way: What two sample means enclose the central $(1-\alpha)100\%$ of the distribution of means when sample size is $n = 4$ and $\alpha = .05$?

- Assume that the population of *Facebook friends* for North American university students is normally distributed with a mean of $\mu = 300$ and standard deviation of $\sigma = 50$. What two sample means enclose the central $(1-\alpha)100\%$ of the distribution of means when sample size is $n = 4$ and $\alpha = .05$?

  Step 1. Calculate $z_{\alpha/2}$. $\alpha/2 = .05/2 = .025$; therefore, $z_{\alpha/2} = 1.96$. (You may need to review the steps for determining $z_{\alpha/2}$ from Chapter 4.)

  Step 2. Compute the standard error of the mean. We were given that $\sigma = 50$ and $n = 4$, so we know that

  $$\sigma_m = \sigma / \sqrt{n} = 50/2 = 25.$$

  Step 3. Compute the two values of m. We use $\mu \pm z_{\alpha/2}(\sigma_m)$ to determine our two values of m as follows:

  $$\mu_m \pm z_{\alpha/2}(\sigma_m) = 300 \pm 1.96(25) = 300 \pm 49 = [251, 349].$$

  Step 4. State the answer. The interval [251, 349] encloses the central 95% of a distribution of means with $\mu_m = 300$ and $\sigma_m = 25$.

The normality assumption is important in this example because the sample size is less than 30. If we hadn't been told that the distribution of scores is normal, we would not have been able to do the question. Let's consider another example.

- Assume that the population of *Facebook friends* for North American university students is normally distributed with a mean of $\mu = 300$ and standard deviation of $\sigma = 50$. What two sample means enclose the central $(1-\alpha)100\%$ of the distribution of means when sample size is $n = 64$ and $\alpha = .05$?

  Step 1. Calculate $z_{\alpha/2}$. $\alpha/2 = .05/2 = .025$; therefore, $z_{\alpha/2} = 1.96$.

  Step 2. Compute the standard error of the mean. We were given that $\sigma = 50$ and $n = 64$, so we know that

  $$\sigma_m = \sigma / i\sqrt{n} = 50/8 = 6.25.$$

  Step 3. Compute the two values of m. We use $\mu \pm z_{\alpha/2}(\sigma_m)$ to determine our two values of m as follows:

  $$\mu_m \pm z_{\alpha/2}(\sigma_m) = 300 \pm 1.96(6.25) = 300 \pm 12.25 = [287.75, 312.25].$$

  Step 4. State the answer. The interval [287.75, 312.25] encloses the central 95% of a distribution of means with $\mu_m = 300$ and $\sigma_m = 6.25$.

- Assume that the population of *Facebook friends* for North American university students is normally distributed with a mean of $\mu = 300$ and standard deviation of $\sigma = 50$. What two sample means enclose the central $(1-\alpha)100\%$ of the distribution of means when sample size is $n = 160,000$ and $\alpha = .05$?

Step 1.   Calculate $z_{\alpha/2}$. $\alpha/2 = .05/2 = .025$; therefore, $z_{\alpha/2} = 1.96$.

Step 2.   Compute the standard error of the mean. We were given that $\sigma = 50$ and $n = 160,000$, so we know that

$$\sigma_m = \sigma/\sqrt{n} = 50/400 = 0.125.$$

Step 3.   Compute the two values of m. We use $\mu \pm z_{\alpha/2}(\sigma_m)$ to determine our two values of m as follows:

$$\mu_m \pm z_{\alpha/2}(\sigma_m) = 300 \pm 1.96(0.125) = 300 \pm 0.245 = [299.755, 300.245].$$

Step 4.   State the answer. The interval $[299.76, 300.25]$ encloses the central 95% of a distribution of means with $\mu_m = 300$ and $\sigma_m = 0.125$.

When we put together everything that we've seen in this chapter, the obvious and main conclusion is that the larger the sample size, the closer sample means fall to the population mean (μ) on average. In the preceding three examples, we saw that

● when $n = 4$, 95% of sample means fall in the interval $[251, 349]$;

● when $n = 64$, 95% of sample means fall in the interval $[287.75, 312.25]$; and

● when $n = 160,000$, 95% of sample means fall in the interval $[299.76, 300.25]$.

The reason that these intervals become narrower with sample size is that σ_m, roughly the average sampling error, decreases with sample size:

● when $n = 4$, $\sigma_m = 49$;

● when $n = 64$, $\sigma_m = 6.25$; and

● when $n = 160,000$, $\sigma_m = 0.125$.

Therefore, as shown in Figures 5.1 and 5.2, statistics computed from large samples provide better estimates of population parameters than statistics computed from small samples.

In Chapter 6, this relationship between sample size and sampling error will be an important part of constructing *confidence intervals* around sample means. Knowing how far sample means fall from μ, on average, will allow us to compute an interval around a *sample mean* and state the confidence we have that the interval contains μ.

In Chapters 1 and 4 we briefly discussed confidence intervals. These are always constructed as *statistic ± margin of error* and are used to estimate a population parameter. In Chapter 6 we will see that confidence intervals around sample means are computed as *mean ± margin of error*. In Chapter 4 we said the margin of error has two parts: one is $z_{\alpha/2}$ and the other is the standard error of the mean. We can now say that when estimating the population mean (μ), the margin of error is $z_{\alpha/2}(\sigma_m)$. We will use this margin of error throughout Chapter 6.

THE DISTRIBUTION OF SAMPLE PROPORTIONS

Many important questions in psychology rely on proportions. For example, what proportion of children are autistic? What proportion of individuals living in a house with a gun die of a gunshot wound? What proportion of adults have a phobia? Appendix 5.5 (available at study.sagepub.com/gurnsey) explains that the distribution of proportions is really a special case of the distribution of means, and it follows naturally from the central limit theorem.

LEARNING CHECK 5

1. If a normally distributed population has a mean of 10 and standard deviation of 2, what sample mean has 95.05% of the distribution of means below it when sample size is 16?

2. If a normally distributed population has a mean of 10 and standard deviation of 2, what sample mean has 5.05% of the distribution of means above it when sample size is 16?

3. What is the value of $z_{\alpha/2}$ such that $\pm z_{\alpha/2}$ contains the central 89.9% of the z-distribution?

4. What is the value of $z_{\alpha/2}$ such that $\pm z_{\alpha/2}$ contains the central 99.02% of the z-distribution?

5. If a normally distributed population has a mean of 10 and standard deviation of 2, what two sample means contain the central 89.9% of the distribution of means when sample size is 16?

6. If a normally distributed population has a mean of 10 and standard deviation of 2, what two sample means contain the central 99.02% of the distribution of means when sample size is 16?

Answers

1. 95.05% corresponds to $P = .9505$, $z = 1.65$, $\sigma_m = 2/\sqrt{16} = 0.5$, $m = \mu + z(\sigma_m) = 10 + 1.65(0.5) = 10.825$.

2. 5.05% corresponds to $P = 1 - .0505$, $z = 1.64$, $\sigma_m = 2/\sqrt{16} = 0.5$, $m = \mu + z(\sigma_m) = 10 + 1.64(0.5) = 10.82$.

3. $\alpha/2 = (1 - .899)/2 = .0505$; therefore $z_{\alpha/2} = 1.64$.

4. $\alpha/2 = (1 - .9902)/2 = .0049$; therefore, $z_{\alpha/2} = 2.58$.

5. $z_{\alpha/2} = 1.64$, $\sigma_m = 2/\sqrt{16} = 0.5$; therefore, $\mu + z(\sigma_m) = 10 \pm 0.82 = [9.18, 10.82]$.

6. $z_{\alpha/2} = 2.58$, $\sigma_m = 2/\sqrt{16} = 0.5$; therefore, $\mu + z(\sigma_m) = 10 \pm 1.29 = [8.71, 11.29]$.

The appendix also shows Excel functions that can be used to solve area-under-the-curve problems and critical value problems.

THE DISTRIBUTION OF SAMPLE VARIANCES

The sample mean and the variance are the two statistics we use most often in this book. We've just discussed the sampling distribution of the mean in some detail, and now we will have a brief look at the sampling distribution of the variance. The distribution of sample variances is like any other statistical distribution. It is simply the distribution of all possible values of the sample variance computed from samples of the same size.

In Chapter 3 we described two formulas for computing the sample variance. We referred to these as

$$s_{\text{pop}}^2 = \frac{\sum(y-m)^2}{n}$$

and

$$s^2 = \frac{\sum(y-m)^2}{n-1}.$$

I promised in Chapter 3 that I'd explain why we divide the sum of squares by $n-1$ for the sample variance. The explanation comes down to the question of bias. s_{pop}^2 is a biased statistic and s^2 is an unbiased statistic. Earlier we saw that a statistic is unbiased when its expected value equals the parameter it estimates. (Remember, the expected value of a statistic is its average value, which is the same as saying the mean of its sampling distribution.) Therefore, a statistic is **biased** when its expected value does not equal the parameter it estimates. This means that the expected value of s^2 equals σ^2, whereas the expected value of s^2_{pop} does not.

Table 5.3 illustrates this point. The column labeled "Samples" in Table 5.3 shows all possible samples of size $n = 2$ drawn, with replacement, from our population of four scores, $y = \{1, 3, 5, 7\}$. The column labeled "s^2_{pop}" shows the sample variances computed with n in the denominator for each sample. The column labeled "s^2" shows the sample variance computed with $n-1$ in the denominator for each sample. The column headers show how s^2_{pop} and s^2 are computed for each sample. The subscripts on y_i and m_i in the numerators of s^2_{pop} and s^2 are to indicate that we are computing a variance for each sample.

These two columns (s^2_{pop} and s^2) show all possible values of s^2_{pop} and s^2 for samples of size 2 drawn from our population of four scores, $y = \{1, 3, 5, 7\}$. That is, these two columns are populations of sample statistics. Therefore, we can compute the mean for each column, which is shown at the bottom of the columns. The mean value of s^2_{pop} is $\mu_{s^2_{pop}} = 2.5$. and the mean value of s^2 is $\mu_{s^2} = 5$. We can also call these means expected values.

The point made in Table 5.3 is that s^2_{pop} is a biased statistic and s^2 is an unbiased statistic. In this case, s^2_{pop} and s^2 are both estimates of the population variance σ^2. If you think way back to the beginning of this chapter, you may remember that the variance of our population, $y = \{1, 3, 5, 7\}$, is 5. (You might take a moment to compute σ^2 to make sure you agree.)

Table 5.3 shows that the mean value of s^2_{pop} is $\mu_{s^2_{pop}} = 2.5$. Clearly, 2.5 does not equal 5 (i.e., σ^2); therefore, s^2_{pop} is a biased statistic. We say that s^2_{pop} *underestimates* σ^2 because its average value will always be less than σ^2, no matter the sample size or the nature of the distribution of scores from which our samples were drawn.

Table 5.3 also shows that the mean value of s^2 is $\mu_{s^2} = 5$. Obviously $\mu_{s^2} = \sigma^2 = 5$. Therefore, s^2 is an unbiased statistic. The average value of s^2 will always equal σ^2 no matter the sample size or the nature of the distribution of scores from which our samples were drawn.

Central to all statistical analyses is the estimation of population parameters from sample statistics. If we are using statistics to estimate parameters, it only makes sense that the average value of the statistic should equal the parameter it estimates. (We would not have much use for a trip-planning app whose average estimate of a trip's duration is less than the actual trip duration. If we used this app, we would be late for meetings more often than not.) Therefore, when estimating population parameters, we should only be interested in unbiased statistics. We've seen two examples of unbiased statistics so far: the sample mean (m) and the sample variance (s^2) are both unbiased statistics. In subsequent chapters we will encounter many other statistics, and our focus will be on those that are unbiased.

In this section we have discussed the mean of the sampling distribution of s^2. Of course, this sampling distribution also has a variance and a shape. Appendix 5.4 (available at study.sagepub.com/gurnsey) provides further information on these aspects of the distribution of variances. Appendix 5.4 also provides information about Excel functions that allow you to do area-under-the-curve problems related to sample variances. Other functions show how to solve critical value problems, such as finding the two values of s^2 that enclose the central 95% of the distribution of variances.

TABLE 5.3 ■ The Sampling Distribution of s_{pop}^2 and s^2

i	Sample	$s_{pop}^2 = \dfrac{\sum(y_i - m_i)^2}{n}$	$s^2 = \dfrac{\sum(y_i - m_i)^2}{n-1}$
1	$y_1 = [1, 1]$	0	0
2	$y_2 = [1, 3]$	1	2
3	$y_3 = [1, 5]$	4	8
4	$y_4 = [1, 7]$	9	18
5	$y_5 = [3, 1]$	1	2
6	$y_6 = [3, 3]$	0	0
7	$y_7 = [3, 5]$	1	2
8	$y_8 = [3, 7]$	4	8
9	$y_9 = [5, 1]$	4	8
10	$y_{10} = [5, 3]$	1	2
11	$y_{11} = [5, 5]$	0	0
12	$y_{12} = [5, 7]$	1	2
13	$y_{13} = [7, 1]$	9	18
14	$y_{14} = [7, 3]$	4	8
15	$y_{15} = [7, 5]$	1	2
16	$y_{16} = [7, 7]$	0	0
		$\begin{aligned} E\left(s_{pop}^2\right) &= \mu_{s_{pop}^2} \\ &= \dfrac{\sum s_{pop}^2}{N} \\ &= 2.5 \end{aligned}$	$\begin{aligned} E\left(s^2\right) &= \mu_{s^2} \\ &= \dfrac{\sum s^2}{N} \\ &= 5 \end{aligned}$

Note: The population is $y = \{1,3,5,7\}$, and sample size is $n = 2$.

SUMMARY

In this chapter we have laid the foundations of inferential statistics by introducing the notion of a sampling distribution. A *sampling distribution* is a probability distribution of all possible values of a sample statistic computed from samples of the same size. There is a sampling distribution for every possible statistic, and we often refer to sampling distributions as *statistical distributions*.

The sampling distribution of the mean (or the distribution of sample means) has a mean equal to the mean of the population being sampled (i.e., $\mu_m = \mu$) no matter how large the sample (n) is, no matter what the population standard deviation is (σ), and no matter what the shape of the distribution is. The variance of the distribution of means equals the variance of the distribution being sampled divided by n (i.e., $\sigma_m^2 = \sigma^2/n$). The square root of the variance is the *standard error of the mean* $\sigma_m = \sqrt{\sigma^2/n}$, or *SEM*.

The *central limit theorem* states that the sampling distribution of the mean will be normal if the population being sampled is normal. It also states that if the population being sampled is not normal, the sampling distribution of the mean will approach a normal distribution as sample size (n) increases. Knowing these facts allows us to compute the probability of a sample mean (m) falling in any given interval.

A statistic is *unbiased* when the mean of its sampling distribution (its *expected value*) equals the parameter it estimates. The sample mean is an unbiased statistic. When the mean of a statistic's distribution does not equal the parameter it estimates, the statistic is *biased*. The sample variance, computed as the average squared deviation from the mean, is a biased statistic. Its expected value is $\sigma^2(n-1)/n$, because it *underestimates* σ^2. To correct this bias, we could multiply the biased statistic by $n/(n-1)$, or we could simply compute the variance by dividing the sum of squares (ss) by $(n-1)$.

KEY TERMS

biased statistic   120
central limit theorem   110
distribution of means   104
expected value (E)   107
mean of the distribution of means (μ_m)   107

sampling distribution of the mean   104
sampling with replacement   104
sampling without replacement   104

standard error of the mean (σ_m)   108
unbiased statistic   107
variance of the distribution of means (σ_m^2)   107

EXERCISES

Please note that asterisks (*) indicate questions relating to material covered in the appendixes or requiring insight beyond what was covered in the chapter.

Definitions and Concepts

1. Put into words the meaning of the distribution of means.

2. Explain the difference between sampling with replacement and sampling without replacement.

3. How does μ_m relate to μ?

4. Define expected value.

5. What does it mean for a statistic to be unbiased? Give an example.

6. What does it mean for a statistic to be biased? Give an example.

7. How does σ_m^2 relate to σ^2?

8. What is the standard error of the mean?

9. When do we use the term "standard error" and when do we use "standard deviation"?

10. State the central limit theorem.

11. What is a sampling distribution?

True or False

State whether the following statements are true or false.

12. Some sampling distributions are not normal distributions.

13. The distribution of means is always normal if the population being sampled is normal.

14. $\sigma_m^2 = \sigma^2/n$ is the standard error of the mean.

15. If a normal distribution has a mean of 100 and standard deviation of 10, then the expected value of s^2 is 100 no matter what sample size is.

16. The expected value of the biased sample variance is $\sigma^2(n-1)/n$.

17. If a distribution has a mean of 100 and standard deviation of 10, then the standard error of the mean is 10.

18. *The sampling distribution of the range is a biased statistic.

19. *The distribution of sample variances is generally left skewed.

20. Strictly speaking, the central limit theorem requires sampling with replacement.

21. *The distribution of variances approaches a normal distribution as sample size increases.

22. The standard error of the mean is $\sqrt{s^2/n}$.

Calculations

23. If a normally distributed population has a mean of 50 and standard deviation of 12, then what proportion of the distribution of means falls below 56 when sample size is $n = 16$?

24. If a normally distributed population has a mean of 82 and standard deviation of 11.3, then what proportion of the distribution of means falls below 91.6 when sample size is $n = 4$?

25. If a right-skewed distribution has a mean of 200 and standard deviation of 32, then what proportion of the distribution of means falls below 206.2 when sample size is $n = 64$?

26. If a leptokurtic distribution has a mean of 1 and standard deviation of .35, then what proportion of the distribution of means falls between 0.98 and 1.11 when sample size is $n = 49$?

27. If a left-skewed and platykurtic distribution has a mean of 99 and standard deviation of 18, then what proportion of the distribution of means falls between 97 and 98 when sample size is $n = 81$?

28. If a normal distribution has a mean of 16 and standard deviation of 2, then what proportion of the distribution of means falls between 15 and 18 when sample size is $n = 2$?

29. If a normal distribution has a mean of 800 and standard deviation of 0.1, then what proportion of the distribution of means falls between 800.01 and 800.02 when sample size is $n = 25$?

Extra Practice

30. What is the approximate probability of drawing a random sample from a normal distribution with $\mu = 100$ and $\sigma = 10$ and finding that the sample mean is
    (a) 99 or less if sample size $n = 100$,
    (b) 98 or less if sample size $n = 25$,
    (c) 102 or less if sample size $n = 100$,
    (d) 102 or less if sample size $n = 25$, and
    (e) 104 or less if sample size $n = 25$?

31. What is the approximate probability of drawing a random sample from a normal distribution with $\mu = 100$ and $\sigma = 10$ and finding that the sample mean is
    (a) 99 or greater if sample size $n = 100$,
    (b) 98 or greater if sample size $n = 25$,
    (c) 102 or greater if sample size $n = 100$,
    (d) 102 or greater if sample size $n = 25$, and
    (e) 104 or greater if sample size $n = 25$?

32. What is the approximate probability of drawing a random sample from a normal distribution with $\mu = 100$ and $\sigma = 10$ and finding that the sample mean is between
    (a) 98 and 99 if sample size $n = 100$,
    (b) 99 and 101 if sample size $n = 100$,
    (c) 96 and 102 if sample size $n = 25$,
    (d) 96 and 104 if sample size $n = 25$, and
    (e) 100.2 and 100.4 if sample size $n = 2500$?

33. A sample of 2500 scores was drawn from a normal distribution with $\mu = 100$ and $\sigma = 10$. What was the sample mean if approximately
    (a) 2% of the distribution of means falls below it,
    (b) approximately 16% of the distribution of means falls below it,
    (c) approximately 84% of the distribution of means falls below it,
    (d) approximately 98% of the distribution of means falls above it,

(e) approximately 2% of the distribution of means falls above it, and

(f) approximately 16% of the distribution of means falls above it?

34. What is the exact probability of drawing a random sample from a normal distribution with $\mu = 102$ and $\sigma = 15$ and finding that the sample mean is

(a) 99 or less if sample size $n = 100$,

(b) 98 or less if sample size $n = 25$,

(c) 102 or less if sample size $n = 100$,

(d) 102 or less if sample size $n = 25$, and

(e) 104 or less if sample size $n = 25$?

35. What is the exact probability of drawing a random sample from a normal distribution with $\mu = 102$ and $\sigma = 15$ and finding that the sample mean is

(a) 99 or greater if sample size $n = 100$,

(b) 98 or greater if sample size $n = 25$,

(c) 102 or greater if sample size $n = 100$,

(d) 102 or greater if sample size $n = 25$, and

(e) 104 or greater if sample size $n = 25$?

36. A sample of 2500 scores was drawn from a normal distribution with $\mu = 100$ and $\sigma = 10$. What was the sample mean if exactly

(a) 2% of the distribution of means falls below it,

(b) 16% of the distribution of means falls below it,

(c) 84% of the distribution of means falls below it,

(d) 98% of the distribution of means falls above it,

(e) 2% of the distribution of means falls above it, and

(f) 16% of the distribution of means falls above it?

37. Assume that scores on a measure of narcissism are normally distributed with a mean of $\mu = 8$ and standard deviation of $\sigma = 4$. What two sample means enclose the central $(1-\alpha)100\%$ of the distribution of means when sample size is

(a) $n = 64$ and $\alpha = .05$,

(b) $n = 16$ and $\alpha = .0164$,

(c) $n = 256$ and $\alpha = .1212$,

(d) $n = 25$ and $\alpha = .025$, and

(e) $n = 64$ and $\alpha = .303$?

APPENDIX 5.1: STATISTICAL DISTRIBUTION FUNCTIONS IN EXCEL

In Chapter 4 we were introduced to the **NORM.DIST** family of functions in Microsoft Excel. We used these functions to solve area-under-the-curve questions relating to scores drawn from normal populations. We can use the same functions to solve area-under-the-curve questions relating means drawn from normal distributions of means.

NORM.DIST and the Distribution of Means

To solve area-under-the-curve questions for distributions of means, we use **NORM.DIST** as follows:

NORM.DIST(m,μ,σ_m,cumulative),

where **m** is a sample mean, **μ** is the population mean, σ_m is the standard error of the mean, and **cumulative** is 0 or 1, depending on whether you want **NORM.DIST** to return **p** (density) or **P** (proportion of the distribution below *m*). (Please note that the function **NORMDIST** will produce exactly the same results.)

Figure 5.A1.1 shows how to compute the proportion of a distribution of means:

- below m_1; i.e., $P(m_1)$ in cell **B10**;

- below m_2; i.e., $P(m_2)$ in cell **B11**;

- between m_1 and m_2; i.e., $P(m_2) - P(m_1)$, assuming $m_2 > m_1$ in cell **B12**; and

- the proportion outside the interval m_1 to m_2; i.e., $1 - [P(m_2) - P(m_1)]$ in cell **B13**.

The main new feature here is the computation of the standard error of the mean σ_m, which is computed in cell **B5** as $\sigma_m = \sigma / \sqrt{n}$.

Although it is not shown in Figure 5.A1.1, there is a function called **NORM.S.DIST** that is defined as follows:

NORM.S.DIST(z,cumulative).

It takes a *z*-score as an argument and returns a density if **cumulative** = 0 and the proportion of the standard normal distribution below **z** if **cumulative** = 1.

FIGURE 5.A1.1 ■ NORM.DIST and Means

	A	B	C
1	**Quantities**	**Values**	**Formulas**
2	μ	300	
3	σ	50	
4	n	64	
5	σ_m	6.25	=B3/sqrt(B4)
6			
7	m_1	314	
8	m_2	326	
9			
10	$P(m_1)$	0.9875	=NORM.DIST(B7,B2,B5,1)
11	$P(m_2)$	1.0000	=NORM.DIST(B8,B2,B5,1)
12	$P(m_2)$-$P(m_1)$	0.0125	=B11-B10
13	1-[$P(m_2)$-$P(m_1)$]	0.9875	=1-B12

FIGURE 5.A1.2 ■ NORM.S.INV and Means

	A	B	C
1	**Quantities**	**Values**	**Formulas**
2	μ	65	
3	σ	8	
4	n	16	
5	σ_m	2	=B3/sqrt(B4)
6			
7	α	0.1	
8	$(1-\alpha)100\%$	90	=(1-B7)*100
9	$\alpha/2$	0.05	=B7/2
10	$z_{\alpha/2}$	1.645	=ABS(NORM.S.INV(B9))
11			
12	$\mu-z_{\alpha/2}(\sigma_m)$	51.84	=B2-B10*B5
13	$\mu+z_{\alpha/2}(\sigma_m)$	78.16	=B2+B10*B5

NORM.S.INV

For many of the statistical analyses to come, inverse functions are very useful. We met **NORM .INV** and **NORM.S.INV** in Appendix 4.2. We used **NORM.S.INV** to compute $z_{\alpha/2}$ when we wanted to determine the two values of x that enclose the central $(1-\alpha)100\%$ of a normal distribution of scores. We can use **NORM.S.INV** in the same way to determine two values of m that enclose the central $(1-\alpha)100\%$ of a normal distribution of means.

In cell **B7** of Figure 5.A1.2, α has been set to .1. Therefore $(1-\alpha)100\% = 90\%$, as shown in cell **B8**. To determine the values of m that enclose the central $(1-\alpha)100\%$ of a normal distribution of means, we must compute $z_{\alpha/2}$. First, $\alpha/2$ is calculated in cell **B9**, and then $z_{\alpha/2}$ is computed in cell **B10**, using **NORM.S.INV**. Note that in cell **B10**, the **ABS** function has been used. **ABS** takes a real number as an argument and returns its absolute value. To determine our two values of m, we compute $\mu - z_{\alpha/2}(\sigma_m)$ in cell **B12** and $\mu + z_{\alpha/2}(\sigma_m)$ in cell **B13**.

ESTIMATION AND SIGNIFICANCE TESTS (ONE SAMPLE)

ESTIMATING THE POPULATION MEAN WHEN THE POPULATION STANDARD DEVIATION IS KNOWN

INTRODUCTION

When we glance at scenes such as those in Figure 6.1, we get their gist very quickly (Oliva, 2005). In this case, the gist is "people boarding a plane." Research has shown, however, that we notice far less detail in scenes than we think we do (e.g., Chabris & Simons, 2011). For example, there is an important difference in the two panels of Figure 6.1, but it takes a while to find it. Such displays can be frustrating, because although you know there is a difference, you just can't detect it. Once you do find the difference, however, you wonder how you could have missed it. Our inability to rapidly detect differences in stimuli such as these has been referred to as *change blindness*. You can find many fascinating examples of change blindness on YouTube.

So, one might ask, how long does it take to detect a change in stimuli such as those shown in Figure 6.1? When we ask this, we're typically not interested in how long it takes a particular individual to notice the difference. We're also typically not interested in the average time it takes a sample of participants from a participant pool to detect the change. Rather, we are more likely interested in the average (mean) time it takes members of a particular population (e.g., college students ages 18 to 24) to detect the change. Because it is usually impossible to answer such questions by taking measurements from all members of the population, we have to *estimate* the population mean (μ) from the mean of a sample (m) drawn from the population.

FIGURE 6.1 ■ Change Blindness

There is a difference between these images that is difficult to notice when the two images alternate on a computer screen (Rensink, O'Regan, & Clark, 1997).

iStock.com/brytta

Early experimental demonstrations of change blindness involved alternating images, such as those in Figure 6.1, at the rate of about one or two images per second, with a blank interval between each image. The time required to detect the difference was measured for each participant (Rensink, O'Regan, & Clark, 1997). Imagine that you've obtained a random sample of 144 college students (ages 18 to 24). For each participant, you've measured the time required to detect the difference between the images shown in Figure 6.1. Let's say that the mean detection time is $m = 12$ seconds, with a standard deviation of $s = 3$. We know that because of *sampling error*, m will not exactly equal μ. So, how precise is our estimate of μ?

We saw in Chapter 5 that the answer to this question involves the standard error of the mean (σ_m), which is, roughly, the average distance between sample means (m) and the population mean (μ). We've seen that σ_m depends on the population standard deviation (σ) and sample size (n). As sample size increases, the standard error decreases; thus, m becomes a more precise estimate of μ. To convey information about the precision of our estimate, it is typical to specify a range of values around m that reflects the standard error. In most situations, large intervals reflect imprecise estimates, and small intervals reflect precise estimates. Such intervals will be called *confidence intervals* when we can specify the confidence we have that they contain μ. In our change-blindness example, we might say that we have 95% confidence that the interval [11.51, 12.49] contains the mean detection time (μ) in the population of 18- to 24-year-old college students.

Confidence is an important new concept that you can think about in the following way. For any population, there is an enormous number of possible sample means. Because an interval can be placed around each sample mean, there is an equally large number of intervals. If we somehow know that 95% of all such intervals will capture μ, then we say they are 95% confidence intervals. This chapter explains *how* confidence intervals are computed and *why* we are able to specify our degree of confidence in them.

As we work through this chapter, keep in mind that measurement and estimation are two of the most fundamental issues facing scientists. Some examples from psychological research might include measuring aggression in schoolchildren, reaction times of 18-year-olds, or risk tolerance in alcoholics. Such measurements are made for theoretical or practical reasons, and researchers typically use the measures obtained from samples to make inferences about the populations from which the samples were drawn. In other words, if aggression, reaction times, and risk tolerance have been measured in *samples* of schoolchildren, 18-year-olds, and alcoholics, we would use these measurements to infer (i.e., estimate) something about the *populations* of schoolchildren, 18-year-olds, and alcoholics, respectively.

AN EXAMPLE

To develop our understanding of confidence intervals, we will start with an artificial example. This example will allow us to define a population whose properties are completely known. We will then estimate the mean of the population from the mean of a sample drawn from it. Our estimation procedure will require us to know the sample mean (m), the sample size (n), and the population standard deviation (σ). A moment's thought shows that knowing σ is implausible. When would we ever know the standard deviation σ of a population and yet not know its mean? In the real world, we would estimate μ when σ is not known. However, we will see in Chapter 10 that the real-world problem of estimating μ when σ *is not known* is structurally identical to the artificial problem of estimating μ when σ *is known*. Therefore, we start with an artificial example as a stepping-stone to understanding more realistic situations.

Our example population comes from students registered in a participant pool at a hypothetical university. At the beginning of each term, all members of the participant pool were invited to a vision research lab in the psychology department to participate in a change-blindness experiment. Each participant was shown two images like those in Figure 6.1. The two images alternated at the rate of 1 second each, with a blank interval (called a mask) presented after each image for 200 milliseconds. Participants were instructed to click a button to start the trial and then to use the computer mouse to click on the region that was changing once the change had been detected. The computer recorded the time in seconds it took each participant to detect the change.

Over the last 4 years, thousands of students have participated in this experiment and the records show that students take an average of 15 seconds to detect the change, with a standard deviation of 8 seconds. Therefore, our population of interest is a distribution of change detection times, which has a mean of $\mu = 15$ and standard deviation of $\sigma = 8$.

We will now go a little meta and think about a statistics instructor who wonders how much students in his advanced statistics course (e.g., you in the future) remember about their introductory statistics course (e.g., you at the end of this term). On the first day of classes, the instructor told students about the experiment and the distribution of change detection times. He then posed the following question:

> The population of change detection times has a standard deviation of $\sigma = 8$. A random sample of $n = 64$ scores has been drawn from this population. The mean of the sample is $m = 14$ seconds. Give me a *point* and *interval estimate* of the population mean, and tell me what *confidence* you have in the interval.

Note, the instructor did not tell students about the mean of the distribution (μ). This is what he asked them to estimate. One student gave the following answer:

APA Reporting

It is estimated that, on average, students require $M = 14$ seconds to detect the change, 95% CI [12.04, 15.96].

This answer got full marks for its accuracy and for the format in which it was presented. Our task in this chapter is to understand how the student arrived at this answer and what it means. We will address these questions in order.

POINT ESTIMATES VERSUS INTERVAL ESTIMATES

Point Estimates

A **point estimate** is a statistic, such as the sample mean, that is used to estimate the corresponding population parameter. For example, we use the mean change detection score (m) of a sample to estimate the mean change detection score (μ) of the population. In our example, the mean of the sample was $m = 14$, which is a point estimate of μ. When our hypothetical student estimated that "on average, students require $M = 14$ seconds to detect the change," she provided a point estimate of μ.

> A **point estimate** is a single number that estimates a population parameter. In other words, a point estimate is a statistic.

Interval Estimates

Point estimates are useful, of course, but they convey nothing about the precision of the estimate. For example, if you were told that the mean change detection score of the sample is 14 seconds, you would not know if this is a precise and useful estimate of μ or an imprecise estimate that you should treat skeptically. To convey some sense of the precision of the estimate, researchers make use of interval estimates.

An **interval estimate** specifies a range of values (i.e., an interval) believed to contain the population parameter. For example, if we estimate that μ is in the interval 14 ± 1.96 seconds, then we have provided an interval estimate. We estimate, in this case, that the mean change detection score of the population (i.e., μ) lies in the interval 12.04 to 15.96. Looking back to the student's answer, we can see that this is the interval she calculated.

> An **interval estimate** is a range of values believed to contain the population parameter.

95% CONFIDENCE INTERVALS

When we express the confidence we have that an interval contains the parameter, then the interval is called a **confidence interval**. Confidence is expressed as a percentage, and researchers most frequently report 95% confidence intervals. Therefore, when the student said "95% CI [12.04, 15.96]," she was saying that she has 95% confidence that the population mean is in this interval.

Having sketched out some of the general concepts underlying confidence intervals, we must now answer three important questions:

> An interval estimate is called a **confidence interval** when we can express the confidence we have that it contains the parameter. Confidence is expressed as a percentage.

- How are 95% confidence intervals calculated?

- What does it mean to say we have 95% confidence in an interval?

- Where does our confidence come from?

These questions are deeply connected, so we will deal with each in turn.

Constructing a 95% Confidence Interval

To calculate the 95% confidence interval, the student used the following formula:

$$m \pm 1.96(\sigma_m), \tag{6.1}$$

where m is the sample mean and σ_m is the standard error of the mean, $\sigma_m = \sigma/\sqrt{n}$. We know that $m = 14$, but before we can compute the confidence interval, we must determine σ_m. In our example, the statistics professor told students that $\sigma = 8$ and $n = 64$, which is the information necessary to compute σ_m. The following two steps show how to calculate a 95% confidence interval.

Step 1. Compute σ_m:

$$\sigma_m = \sigma/\sqrt{n} = 8/\sqrt{64} = 8/8 = 1.$$

Step 2. Compute $m \pm 1.96(\sigma_m)$. When we substitute m and σ_m into equation 6.1, we get the following:

$$CI = m \pm 1.96(\sigma_m) = 14 \pm 1.96(1) = 14 \pm 1.96 = [12.04, 15.96].$$

This is exactly what our hypothetical student reported. So, that is *how* the 95% confidence interval was calculated.

CALCULATION TIP

Confidence intervals are the first of many multi-step calculations that we will do in this book. For the most accurate results, you should not round until the last step in the calculation. For example, when you compute a 95% confidence interval when $\sigma = 2.6$ and $n = 9$, you should enter $1.96*2.6/\sqrt{9}$ into your calculator. This will produce $1.698\dot{6}$, where the dot over the 6 means 6 repeating. Don't round this quantity before adding it to the mean. Instead, round to two decimal places only after the addition or the subtraction. Answers in the learning checks and end-of-chapter exercises are obtained this way.

Example calculations will typically involve quantities having many digits after the decimal place. However, displaying such quantities to four, six, or more decimal places in our examples would be messy and somewhat difficult to read. Therefore, intermediate quantities in multi-step computations are generally *displayed* as rounded to two decimal places, even though the underlying quantities involve much higher decimal accuracy (*precision*). Here is an example in which $m = 12.36$, $\sigma = 2.6$, and $n = 9$:

$$m \pm 1.96(\sigma/\sqrt{n}) = 12.36 \pm 1.96(2.6/\sqrt{9}) = 12.36 \pm 1.96(0.87)$$
$$= 12.36 \pm (1.70) = [10.66, \ 14.06].$$

In this case, the intermediate quantities are *displayed* to two decimal places, but the actual results are based on much higher *precision*. If you try to confirm this calculation by typing $1.96*.87$ into your calculator (or Excel), you will find that the result, rounded to two decimal places, is 1.71, rather than 1.70. Remember, there is a difference between the precision of the numbers displayed and the precision of the numbers actually used in the underlying calculations.

A **95% confidence interval** is an interval around a statistic. We have 95% confidence in this interval when we know that 95% of all such intervals will capture the parameter estimated by the statistic.

The Meaning of 95% Confidence

We say that this is a **95% confidence interval** because 95% of all confidence intervals constructed this way will contain (or capture) the population mean, μ. In other words, if we were to select all possible samples of size $n = 64$ from this population and compute $m \pm 1.96(\sigma_m)$ for each sample, we would find that μ falls within exactly 95% of these intervals. You can memorize this fact, but it is far better to know *why* we're certain that 95% of all confidence intervals constructed this way will capture μ. The answer is that our confidence is built on our understanding of the distribution of sample means.

Sampling Distributions and Confidence

We can think of $\pm 1.96(\sigma_m)$ as two arms centered on either μ or m. We can make two observations about these arms:

If the arms capture m when centered on μ, then the same arms centered on m will capture μ. That is, if $\mu \pm 1.96(\sigma_m)$ captures m, then $m \pm 1.96(\sigma_m)$ will capture μ.

If the arms *do not* capture m when centered on μ, then the same arms centered on m will not capture μ. That is, if $\mu \pm 1.96(\sigma_m)$ does not capture m, then $m \pm 1.96(\sigma_m)$ will not capture μ.

If the arms around μ capture m, then the same arms around m will capture μ. Therefore, *whenever* the arms around μ capture m, the same arms around m will capture μ.

From Chapter 5 we know that 95% of all sample means fall within the interval $\mu \pm 1.96(\sigma_m)$. This means that for 95% of all sample means, the interval $m \pm 1.96(\sigma_m)$ will capture μ. The flip side of this is that for the 5% of sample means falling outside the interval $\mu \pm 1.96(\sigma_m)$, the interval $m \pm 1.96(\sigma_m)$ will not include μ. That is, for 5% of all sample means, the interval $m \pm 1.96(\sigma_m)$ will not capture μ.

The points above will be made concrete with an example. Let's now assume that in reality our population/distribution of change detection scores has a mean of $\mu = 15$, with $\sigma = 8$. If we were to compute m for every possible sample of size $n = 64$, drawn from the distribution of change detection scores, the result would be the sampling distribution of the mean. This sampling distribution would be normal with mean $\mu = 15$ and standard error $\sigma_m = 8/\sqrt{64} = 1$. This distribution is shown at the bottom of Figure 6.2. The vertical black line represents the mean of the distribution ($\mu = 15$), and the other vertical lines represent $\mu \pm 1.96(\sigma_m)$. We know from Chapter 5 that 95% of all sample means are in the interval $\mu \pm 1.96(\sigma_m) = [13.04, 16.96]$.

If the arms around μ do not capture m, then the same arms around m will not capture μ. Therefore, *whenever* the arms around μ do not capture m, the same arms around m will not capture μ. Adapted from iStock.com/HitToom

Each of the 32 dots in Figure 6.2 represents a mean that has been drawn at random from the distribution of means. (There is nothing special about 32; there just isn't enough space to show all possible means.) The white dots are means from the central 95% of the distribution; i.e., *within* the interval $\mu \pm 1.96(\sigma_m)$. The filled dots are means from outside the central 95% of the distribution; i.e., *outside* the interval $\mu \pm 1.96(\sigma_m)$. The gray dot corresponds to the mean from our example; i.e., $m = 14$. Each dot is the center of an interval having two arms, both of which are $1.96(\sigma_m)$ long; i.e., $m \pm 1.96(\sigma_m)$. Therefore, the width of the interval centered on μ [i.e., $\pm 1.96(\sigma_m)$] is the same as the width of the intervals centered on the sample means [i.e., $\pm 1.96(\sigma_m)$].

Figure 6.2 shows that the arms of the white dots (and the gray dot) *capture* μ, and those of the filled dots do not. Therefore, if a mean falls *within* the central 95% of the distribution, the confidence interval around it will capture μ; if a mean falls *outside* the central 95% of the distribution, the confidence interval around it will not capture μ. Because 95% of all sample means fall in the interval $\mu \pm 1.96(\sigma_m)$, the interval $m \pm 1.96(\sigma_m)$ will include μ for 95% of all sample means.

The example in Figure 6.2 is for a small set of 32 means and so, because of randomness and this small number, only 90.6% of the intervals capture μ. If we had a piece of paper long enough to show all possible means for samples of size $n = 64$, then exactly 95% of the intervals around the means would capture μ. (That piece of paper would have to be light-years long!)

I've said that 95% of intervals computed as $m \pm 1.96(\sigma_m)$ will capture μ. You might think that this is a special case that applies only to our example with $\mu = 15$, $\sigma = 8$, and $n = 64$. In fact, this is not a special case. The interval $m \pm 1.96(\sigma_m)$ will capture μ 95% of the time no matter what μ, σ, and n are, as long as the distribution of means is normal.

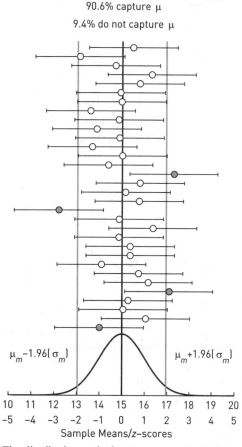

FIGURE 6.2 ■ 95% Confidence Intervals

90.6% capture μ

9.4% do not capture μ

$\mu_m - 1.96(\sigma_m)$ $\mu_m + 1.96(\sigma_m)$

Sample Means/z–scores

The distribution at the bottom represents the sampling distribution of the mean for μ = 15, σ = 8, and n = 64. The standard error of the mean is $\sigma_m = 1$. (This normal distribution can be transformed to the z-distribution, as shown.) The light blue lines $[\mu \pm 1.96(\sigma_m)]$ enclose the central 95% of the sampling distribution. Each dot represents a sample mean, and the arms around each dot represent the 95% confidence interval. White dots are the centers of confidence intervals that capture μ, and filled dots are the centers of confidence intervals that do not capture μ. The gray dot corresponds to the mean from our example (i.e., m = 14).

Let's have a final look at Figure 6.2. Each mean in Figure 6.2 can be transformed into a z-score through the following simple transformation:

$$z = \frac{m - \mu}{\sigma_m}.$$

Therefore, each of the means in Figure 6.2 can also be associated with a z-score. The x-axis of Figure 6.2 shows the z-score corresponding to the mean above it. We've seen many times that 95% of z-scores lie in the interval 0 ± 1.96, where 0 is the mean of the z-distribution. Therefore, for 95% of z-scores, 0 is no more than 1.96 standard units away. This means that for 95% of z-scores, z ± 1.96 will capture 0. Because any z-score can be transformed into a mean through the simple transformation

$$m = \mu \pm z(\sigma_m),$$

it is also true that 95% of all sample means lie in the interval $\mu \pm 1.96(\sigma_m)$, and so 95% of intervals computed as $m \pm 1.96(\sigma_m)$ will capture μ.

Notice that our conclusion said nothing about what μ, σ, and n are. Therefore, the argument is true for any values of μ, σ, and n. The only assumption we made here is that the distribution of means is normal, which it will be if the conditions of the central limit theorem are met. If you're super skeptical about this, there is a more formal demonstration in Appendix 6.2 (available at study .sagepub.com/gurnsey).

Our Example Revisited

Let's think back to the example that started the chapter. Students were told that a population has a standard deviation of σ = 8, and the mean of a sample of n = 64 scores was m = 14. From this the student computed the 95% confidence interval using the expression $m \pm 1.96(\sigma_m)$. Her confidence in the interval came from knowing that 95% of all intervals computed this way will capture μ. We've seen how she calculated this interval, we know what 95% confidence means, and we now know why we can have 95% confidence in the interval $m \pm 1.96(\sigma_m)$. We will next see that we can compute intervals for any possible level of confidence.

(1−α)100% CONFIDENCE INTERVALS

The preceding paragraphs worked through the logic of the 95% confidence interval. However, it is possible to construct intervals in which we have anywhere from 0% to 100% confidence. A 0% confidence interval doesn't sound particularly useful. In fact, a 0% confidence interval is an interval with 0 width, making it a point estimate. On the other hand, an interval in which you have 100% confidence sounds great, but it is not useful either because it would have to be infinitely wide. For example, I have 100% confidence that the average age of students in your class is between 0 and 200 years old, and that doesn't tell me much. So, we want to be confident,

LEARNING CHECK 1

1. What is a point estimate?

2. What is an interval estimate?

3. If a confidence interval is computed as $m \pm 1.96(\sigma_m)$,

   (a) What proportion of all such intervals will capture μ?

   (b) What proportion of all such intervals will not capture μ?

4. Answer the questions for the following statement: "It is estimated that, on average, students spend 27 hours per week surfing the Internet, 95% CI [25.04, 28.96]."

   (a) What is the point estimate?

   (b) What is the interval estimate?

   (c) Complete the following sentence: "I have 95% confidence in this interval because _____."

5. The mean of 25 scores drawn from a population with a standard deviation of 6 is 14. Fill in the blank for the following: Therefore, (a) $m =$ _____,

(b) $n =$ _____, (c) $\sigma =$ _____, and (d) $\sigma_m =$ _____. (e) The 95% confidence interval around m is [_____, _____].

6. The Beck Depression Inventory (BDI) is a questionnaire whose purpose is to assess the degree of depression present in an individual. Assume that in the population of North American university students, the distribution of BDI scores has a standard deviation of 8.5. A random sample of nine North American university students was selected, and a BDI score was obtained from each student.

   (a) The nine BDI scores were [8, 20, 4, 1, 4, 4, 9, 2, 2]. Compute the 95% confidence interval around the mean of these nine scores. (Round your answer to two decimal places.)

   (b) Given the confidence interval you've just computed, would you be surprised to learn that the mean BDI score in the population of North American university students is $\mu_{BDI} = 10$?

Answers

1. A point estimate is a single number that estimates a population parameter.

2. An interval estimate is a range of values believed to contain the population parameter.

3. (a) .95. (b) .05.

4. (a) 27.

   (b) [25.04, 28.96].

   (c) I have 95% confidence in this interval because 95% of all intervals computed this way will capture μ.

5. (a) 14.

(b) 25.

(c) 6.

(d) 1.2.

(e) $14 \pm 1.96(1.2) = [11.65, 16.35]$.

6. (a) Compute the 95% confidence interval around m as shown in the equations in the following table.

   (b) One should not be surprised that $\mu_{BDI} = 10$ because 10 is within the interval just computed. We know that 95% of all such intervals will capture the mean of the population from which the sample was drawn. Therefore, nothing unusual or surprising has occurred.

Compute m	Compute σ_m	Compute $m \pm 1.96(\sigma_m)$
$m = \sum y/n = \sum [8, 20, 4, 1, 4, 4, 9, 2, 2]/n$ $= 54/9 = 6.$	$\sigma_m = \sigma/\sqrt{n} = 8.5/\sqrt{9} = 2.83$	$CI = m \pm 1.96(\sigma_m) = 6 \pm 1.96(2.83) = 6 \pm 5.55$ $= [0.45, 11.55]$

but not too confident. Over the years, researchers have converged on the idea that 95% confidence generally satisfies this requirement. Therefore, the most commonly encountered confidence intervals are 95% confidence intervals.

Although 95% confidence intervals are most common in the literature, they are not the only ones found, so we should look at how to construct others. Furthermore, a proper understanding of confidence intervals requires us to think about the general case. In the general case, the level of confidence we want to achieve is expressed as $(1-\alpha)100\%$ confidence, where α is the proportion of all such intervals that *will not* capture μ. When $\alpha = .05$, the corresponding confidence interval would be the 95% confidence interval:

$$(1-\alpha)100\% = (1 - .05)100\% = (.95)100\% = 95\%.$$

When $\alpha = .1$, the corresponding confidence interval would be the 90% confidence interval:

$$(1-\alpha)100\% = (1 - .1)100\% = (.9)100\% = 90\%.$$

α can take any value from 0 to 1. If $\alpha = 1$, then our interval is a 0% confidence interval. If $\alpha = 0$, then our interval is a 100% confidence interval. The smaller α is, the wider the interval and the greater our confidence.

The general formula for a confidence interval is

$$m \pm z_{\alpha/2}(\sigma_m). \tag{6.2}$$

A $(1-\alpha)100\%$ **confidence interval** is an interval around a statistic. We have $(1-\alpha)100\%$ confidence in this interval when we know that $(1-\alpha)100\%$ of all such intervals will capture the parameter estimated by the statistic.

We call this the $(1-\alpha)100\%$ **confidence interval** because this is the percentage of all possible confidence intervals computed as $m \pm z_{\alpha/2}(\sigma_m)$ that will capture the population mean, μ. The only new term here is $z_{\alpha/2}$, which is familiar to us from Chapters 4 and 5. Remember that $\pm z_{\alpha/2}$ contains the central $(1-\alpha)100\%$ of the standard normal distribution.

To compute a $(1-\alpha)100\%$ confidence interval, we need to find the appropriate $z_{\alpha/2}$. This can be done in a few simple steps, some of which we've seen before. We will use a 68.26% confidence interval to illustrate the procedure.

Step 1. Express the percentage confidence as a proportion. In this example, $p = 68.26/100 = .6826$.

Step 2. Calculate α as $1-p$. In this case, $\alpha = 1 - p = 1 - .6826 = .3174$.

Step 3. Calculate $\alpha/2 = .3174/2 = .1587$.

Step 4. Find $\alpha/2$ in the z-table and note the z-score to which it corresponds. For this example, this means finding .1587 in the z-table. The z-score corresponding to $\alpha/2$ for this example is -1.

Step 5. $z_{\alpha/2}$ is the absolute value of the z-score we found. For this example, $z_{\alpha/2} = 1$.

Figure 6.3 is structured exactly like Figure 6.2, except that the central region is defined as $\mu \pm 1(\sigma_m)$ (i.e., $z_{\alpha/2} = 1$), which ranges from 14 to 16. All white dots represent means that fall within this range and have confidence intervals ($m \pm \sigma_m$) that capture $\mu = 15$. All of the filled dots represent means that fall outside the range 14 to 16 and have confidence intervals that do not capture $\mu = 15$. Note that the confidence intervals are narrower in Figure 6.3 than in Figure 6.2. This is because $\alpha = .3174$ and $z_{\alpha/2} = 1$ in Figure 6.3, whereas $\alpha = .05$ and $z_{\alpha/2} = 1.96$ in Figure 6.2. Therefore, the intervals in Figure 6.2 are almost twice as wide as those in Figure 6.3 because $z_{\alpha/2}$ is almost twice as large. Wide intervals will capture μ more often than narrow intervals.

Important Values of $z_{\alpha/2}$

We've just seen how to compute $(1-\alpha)100\%$ confidence intervals for the general case. However, it is rare to see a 68.26% confidence interval, and it is rarer still to see 57% or 83%

confidence intervals. The most common alternatives to the 95% confidence interval are the 90% and 99% confidence intervals. The $z_{\alpha/2}$ values for these are 1.645 and 2.575, respectively. These values are shown in Table 6.1 and are worth memorizing. (You can use steps 1 to 5 above to confirm that these values are correct.)

Factors Affecting the Width of a Confidence Interval

Four quantities are required to construct a confidence interval around a sample mean when σ is known. These are the sample mean (m), the sample size (n), the population standard deviation (σ), and the confidence we wish to have. As we just saw, the confidence level is determined by the proportion of all such intervals $(1-\alpha)$ that will capture μ.

The sample mean is always the center of a confidence interval and thus has no effect on the width of the interval. Sample size (n) and the population standard deviation (σ) affect the width of a confidence interval because both contribute to the calculation of $\sigma_m = \sigma/\sqrt{n}$. As σ increases (and n remains constant), σ_m will increase. Conversely, as n increases (and σ remains constant), σ_m will decrease.

- All things being equal, the width of the confidence interval increases as σ increases and decreases as n increases.

This leaves the question of how the confidence level affects our confidence interval. Recall that α is the proportion of all such intervals that *will not capture* μ. Therefore, interval width is *inversely* related to α, and so the proportion of intervals capturing μ is also inversely related to α. As α *increases*, our confidence *decreases* and the width of the confidence interval also *decreases*.

- All things being equal, the width of the confidence interval increases as the desired level of confidence [$(1-\alpha)100\%$] increases. This is equivalent to saying that the width of the confidence interval *increases* as α *decreases*.

CAUTIONS ABOUT INTERPRETATION

Estimation Implies Uncertainty

Several critical points need to be made about the proper interpretation of confidence intervals. First, our hypothetical student did not know that the mean of this distribution was μ = 15. Therefore, when she estimated "that, on average, students require $M = 14$ seconds to detect the change, 95% CI [12.04, 15.96]," this was all she knew. Her point estimate is 14 seconds, and she has 95% confidence that the actual μ lies in the interval [12.04, 15.96]. From her point of view, μ is just as likely to be 12.04 as 15.96. When confidence intervals are calculated in real research, researchers are in exactly the same position. They can't know exactly what the population mean is, but they can have $(1-\alpha)100\%$ confidence that it is contained within the interval they computed.

FIGURE 6.3 ■ 68.26% CIs

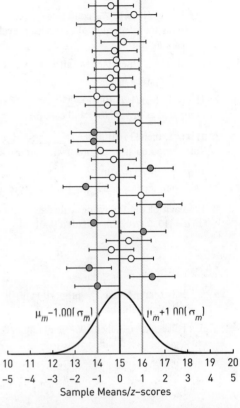

68.8% capture μ
31.2% do not capture μ

The distribution at the bottom represents the sampling distribution of the mean for μ = 15, σ = 8, and $n = 64$. The standard error of the mean is $\sigma_m = 1$. The light blue vertical lines [$\mu \pm (\sigma_m)$] enclose the central 68.26% of the sampling distribution. Each dot represents a sample mean, and the arms around each dot represent the 68.26% confidence interval. White dots are the centers of confidence intervals that capture μ, and filled dots are the centers of confidence intervals that do not capture μ. The gray dot corresponds to the mean from our example (i.e., $m = 14$).

TABLE 6.1 ■ Common Values of $z_{\alpha/2}$

% Confidence	$z_{\alpha/2}$
99	2.575
95	1.960
90	1.645

Note: Values of values of $z_{\alpha/2}$ for 99%, 95%, and 90% confidence intervals.

LEARNING CHECK 2

1. State whether the following statements are true or false.

   (a) For a $(1-\alpha)100\%$ confidence interval, α represents the proportion of all such intervals that will capture μ.

   (b) If $\alpha = .2$, then 90% of all intervals computed as $m \pm z_{\alpha/2}(\sigma_m)$ will capture μ.

2. If $\alpha = .1$, then $(1-\alpha)100\% =$ _____ of all intervals computed as $m \pm z_{\alpha/2}(\sigma_m)$ will capture μ.

3. If $\alpha = 1$, then $(1-\alpha)100\% =$ _____ of all intervals computed as $m \pm z_{\alpha/2}(\sigma_m)$ will capture μ.

4. If $\alpha = .5$, then $(1-\alpha)100\% =$ _____ of all intervals computed as $m \pm z_{\alpha/2}(\sigma_m)$ will capture μ.

5. For each of the following, calculate $z_{\alpha/2}$: (a) $\alpha = .05$, (b) $\alpha = .10$, (c) $\alpha = .20$, and (d) $\alpha = .01$.

6. The mean of 36 scores drawn from a population with a standard deviation of 18 is 88. Fill in the blanks for the following: Therefore, (a) $m =$ _____, (b) $n =$ _____, (c) $\sigma =$ _____, and (d) $\sigma_m =$ _____. (e) The 90% confidence interval around m is [_____, _____]. (Round your answer to two decimal places.)

7. The width of a confidence interval [increases, decreases] as α increases.

8. The width of a confidence interval [increases, decreases] as n increases.

9. The width of a confidence interval [increases, decreases] as σ increases.

Answers

1. (a) False. α represents the proportion of all such intervals that will *not* capture μ.

   (b) False. If $\alpha = .2$, then 80% of all intervals computed as $m \pm z_{\alpha/2}(\sigma_m)$ will capture μ.

2. 90%.

3. 0%.

4. 50%.

5. (a) 1.96. (b) 1.645. (c) 1.28. (d) 2.575.

6. (a) 88. (b) 36. (c) 18. (d) 3. (e) [83.07, 92.94].

7. Decreases.

8. Decreases.

9. Increases.

Confidence and Probability

Second, care must be taken not to confuse the term *confidence* with probability. If we have computed a 95% confidence interval (e.g., 12.04 to 15.96), it is *incorrect* to say that the probability is .95 that the interval contains μ. Once the interval is created, it either contains μ or it doesn't. So, the probability is either 1 or 0 that μ is in the interval. It is correct to say that 95% of all intervals computed as $m \pm 1.96(\sigma_m)$ will contain μ. However, this is a statement *about all possible intervals created this way*, not a statement about a specific interval. Probability in this case refers to the proportion of all possible intervals that capture μ. This is why we talk about the *confidence* we have that the interval contains μ rather than the probability that the interval contains μ.

What Not to Say About a Specific Confidence Interval

Finally, if you've computed a 95% confidence interval, be careful not to construe 95% confidence to mean that 95% of all sample means will fall in the interval $m \pm 1.96(\sigma_m)$. If you look back to Figure 6.2, you will see 32 different confidence intervals. Some of the intervals that capture μ have means that are rather far from μ. Clearly, such intervals do not enclose 95% of the distribution of means. Therefore, it is impossible to say, of any given confidence interval, that 95% of sample means will fall within it. There is *only* one interval for which 95% of sample means fall within it. This interval corresponds to the case when $m = \mu$ and, of course, you don't know if m equals μ.

LEARNING CHECK 3

1. State whether the following statements are true or false.

   (a) I have 95% confidence that μ lies in the interval [10.2, 13.6]; therefore, 95% of all sample means will fall in this interval.

   (b) I have 95% confidence that μ lies in the interval [63.1, 125.2]; therefore, the probability is .95 that μ is in this interval.

   (c) I have 95% confidence that μ lies in the interval $m \pm z_{\alpha/2}(\sigma_m) = [16.8, 28.4]$ because 95% of all possible intervals computed this way will capture μ.

   (d) $(1-\alpha)100\%$ intervals computed as $m \pm z_{\alpha/2}(\sigma_m)$ will capture m.

Answers

1. (a) False. This would only be true if $m = \mu$ (refer to Figure 6.2).

   (b) False. μ is either in the interval or not. Therefore, the probability that μ lies in the interval [63.1, 125.2] is either 0 or or 1.

   (c) True. This is the definition of confidence.

   (d) False. Because m is always at the center of the interval, 100% of intervals computed as $m \pm z_{\alpha/2}(\sigma_m)$ will capture m. Make sure you understand this point! In my experience, many students fall victim to endorsing statement (d) as true.

ESTIMATING μ WHEN SAMPLE SIZE IS LARGE

To this point we have illustrated the construction of confidence intervals in the artificial context of a statistics exercise. We can use this method whenever we know σ, but there are very few situations in which we would know σ but not μ. However, if our sample is very large, and certain assumptions are made, we can use the method we've been discussing to construct an *approximate* $(1-\alpha)100\%$ confidence interval around m.

If we don't know σ, we can use the sample standard deviation (s) to estimate it. Remember, as sample size increases, s becomes a better and better estimate of σ (see Appendix 5.4, available at study.sagepub.com/gurnsey). That is, as sample size increases, the distribution of sample standard deviations becomes narrower, meaning that individual standard deviations fall closer to σ, on average. Therefore, because we can use s to estimate σ, we can use s_m to estimate σ_m. The *estimated standard error of the mean* (s_m) is computed as follows:

$$s_m = \frac{s}{\sqrt{n}}.$$ (6.3)

This *estimated standard error* is just like the real standard error, except that it is computed with s rather than σ in the numerator. If σ is not known, but sample size is large, we can use the following formula to compute an approximate $(1-\alpha)100\%$ confidence interval around m:

$$m \pm z_{\alpha/2}(s_m).$$ (6.4)

Let's return to the change-blindness example that was used to introduce this chapter. Rather than thinking about a known population of change detection scores, we will think about running an actual experiment. The question is this: What is our best estimate of change detection times in a population of university students? You are the experimenter here and will address this question with a sample of volunteers from your department's participant pool.

The two images shown in Figure 6.4 are the stimuli. Each image will be presented for 1 second and followed by a blank interval (mask) for 200 milliseconds. The two images will alternate in this way until the participant clicks on the region that changes between the two images to indicate that the change has been detected. The computer will record the time in seconds it took each participant to detect the change.

Let's say you enlisted 100 participants from your department's participant pool, so you have gathered 100 change detection times. Let's also suppose that the mean of these 100 scores is 15.4, with a standard deviation of 8.4. Now, compute the approximate 95% confidence interval around the mean.

To compute this approximate confidence interval, we do the following things.

Step 1. Compute s_m.

$$s_m = s/\sqrt{n} = 8.4/\sqrt{100} = 0.84$$

Step 2. Find $z_{\alpha/2}$. Because this is the $(1-\alpha)100\% = 95\%$ confidence interval, $\alpha = .05$.

(a)   Therefore, $\alpha/2 = .025$.

(b)   When we use the z-table, we find that 2.5% of the z-distribution lies below -1.96.

(c)   Therefore, $z_{\alpha/2} = 1.96$.

Step 3. Compute the approximate confidence interval.

$$CI = m \pm z_{\alpha/2}(s_m) = 15.4 \pm 1.96(0.84) = [13.75, 17.05].$$

Therefore, we have *approximately* 95% confidence that the mean change detection time is between 13.75 and 17.05 seconds. Appendix 6.1 shows how to do these calculations in Excel.

The Meaning of Approximate

So far the term *approximate confidence interval* has been used very loosely. We can now explain the sense in which such intervals are approximate. When we use equation 6.4 to

FIGURE 6.4 ■ Another Change-Blindness Stimulus

compute an approximate confidence interval, the interval will be *narrower* than it should be. This means that if (theoretically) we computed all possible confidence intervals using $m \pm z_{\alpha/2}(s_m)$, then μ will be captured by less than $(1-\alpha)100\%$ of these. If α = .05, then $m \pm z_{\alpha/2}(s_m)$ will capture μ less than 95% of the time. This is illustrated in Figure 6.5.

In Figure 6.5, the *x*-axis represents sample size (*n*) and the *y* axis shows the percentage of all confidence intervals, computed using $m \pm 1.96(s_m)$, that will capture μ. If sample size were *n* = 2, for example, confidence intervals computed using equation 6.4 would capture μ only 70% of the time. This is the point where the light blue line touches the *x*-axis.

However, it is clear from Figure 6.5 that as *n* increases, the percentage of confidence intervals computed as $m \pm 1.96(s_m)$ increases toward 95%. That is, as sample size increases, our approximate confidence intervals get closer and closer to what we want. When *n* = 56, for example, 94.5% of all intervals computed using $m \pm 1.96(s_m)$ will capture μ. This is shown as the vertical line that goes from the *x*-axis up to 94.5. Therefore, when we compute $m \pm 1.96(s_m)$ with *n* = 56, we've actually computed a 94.5%

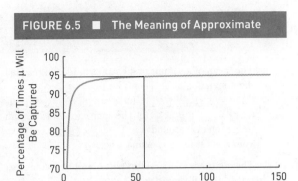

FIGURE 6.5   ■   The Meaning of Approximate

The actual percentage of times μ will be captured with intervals computed using the formula $m \pm 1.96(s_m)$. Our hope is to capture μ 95% of the time, but we would actually capture μ less than 95% of the time. The exact percentage of captures will depend on sample size (*n*). As sample size increases, the capture percentage approaches 95% asymptotically.

confidence interval rather than a 95% confidence interval. That's not too bad. To me, there is little practical difference between 94.5% confidence and 95% confidence. As sample size increases further, we approach 95% *asymptotically*, which means we get closer and closer to 95% without ever reaching it. In the case of *n* = 100, from our example, 94.72% of confidence intervals computed as $m \pm 1.96(s_m)$ will capture μ. So, this is what we mean by approximate, and this is why the approximate method works well for large samples.

The approximate confidence intervals just discussed hint at where we will be going in later chapters. The key point is that, using only statistics computed from samples, we can generate reasonably accurate confidence intervals for the population mean. I find this amazing, actually. Of course, these calculations are only approximately correct when sample size is large, and certain assumptions must be made about the population from which the scores were drawn. However, in Chapter 10 we will show how to compute accurate confidence intervals from sample statistics even when sample sizes are small. To do so, we will make use of the incredibly important work of the modest, brilliant, and intriguing William Sealy Gosset.

ASSUMPTIONS

Whenever we compute confidence intervals, there are certain assumptions that must be true in order for the intervals to be valid. In this section we will discuss these assumptions as they apply to the cases in which σ is known and unknown.

Normality

When σ Is Known

When σ is known, we must assume that *the distribution of means* is a normal distribution. This requirement is interesting because, as we saw in Chapter 3, many distributions (e.g., incomes, reaction times, and test scores) can be skewed because of floor effects. However, we saw in Chapter 5 that if sample size is large enough, the distribution of means will be normal even if the distribution of scores is not. Therefore, in our first example of change

LEARNING CHECK 4

1. The mean of a sample of 49 scores is 101, with a standard deviation of 21. Therefore, (a) $m = $ _____, (b) $n = $ _____, (c) $s = $ _____, and (d) $s_m = $ _____. (e) The approximate 95% confidence interval around m is [_____, _____]. (Round your answer to two decimal places.)

2. The confidence interval in part e will be [wider, narrower] than it *should be*.

3. Confidence intervals computed with s_m will capture μ more often than those computed with σ_m. [True, False]

4. Explain why we say $m \pm 1.96(s_m)$ is an approximate 95% confidence interval.

5. In Learning Check 1, we saw that a random sample of North American university students produced the following sample of BDI scores: [8, 20, 4, 1, 4, 4, 9, 2, 2].

   (a) Compute the approximate 95% confidence interval around the mean of these nine scores.

   (b) Should we have 95% confidence in this interval?

   (c) Is there any reason to think that the population of BDI scores is not normally distributed?

Answers

1. (a) 101. (b) 49. (c) 21. (d) 3. (e) [95.12, 106.88].

2. Narrower.

3. False.

4. It is approximate because such intervals won't capture μ 95% of the time. They will capture μ less than 95% of the time. The exact percentage of such intervals that will capture μ depends on sample size.

5. (a) The mean and sum of squares for the sample are $m = 6$, and $ss = 278$, respectively. The remaining calculations are shown in the following table.

   (b) We should not have 95% confidence in this approximate interval because when sample size is small, as in this case, far fewer than 95% of all intervals computed this way will capture the mean.

   (c) This small sample of scores seems to be right skewed. Six of the scores are between 1 and 4, and the rest are between 8 and 20. This suggests that the population from which the scores were drawn is also right skewed.

Compute s^2	Compute s_m	Compute approximate confidence interval
$s^2 = \dfrac{ss}{n-1} = \dfrac{278}{8} = 34.75$	$s_m = \sqrt{\dfrac{s^2}{n}} = \sqrt{\dfrac{34.75}{9}} = 1.96$	$CI = m \pm 1.96(s_m) = 6 \pm 1.96(1.96) = 6 \pm 3.85 = [2.15, 9.85]$

detection scores, we can safely assume that the distribution of means is normal because our sample was large (i.e., $n = 64$); we don't need to assume that the distribution of change detection times is normal.

When σ Is Not Known, but Sample Size Is Large

The second example involving change detection times presents an interesting case. When we compute an approximate confidence interval by using s_m rather than σ_m, we must assume that *the distribution of scores* is normal. The reason we must make this assumption is explained in detail in Chapter 10.

In our change detection example, there is reason to suspect that our distribution of scores is not normal. The scores in question are essentially reaction times, so there is a minimum detection time (zero seconds) but no maximum. Therefore, it is possible to have a floor effect that produces a positively skewed distribution of scores. We can use a QQ-plot (described in Appendix 4.3, available at study.sagepub.com/gurnsey) to look for violations of the normality assumption.

It is worth noting that the effects of violating the normality assumption tend to diminish as sample size increases. However, it is not possible to provide a specific sample size at which violations of the normality assumption are no longer a concern. This is because the sample size required to overcome a violation of the normality assumption depends on the nature of the violation (e.g., skew or kurtosis) and the severity of the violation.

Random Sampling

Finally, for the confidence interval to be valid, our sample must have been a random sample from the population of interest. Our first example of change detection scores was artificial and used for illustration only. In that example, of course it was possible to draw a random sample of scores from our population. The second example of change detection times is closer to a real-world case of estimation. In this scenario, we were trying to estimate the mean change detection time in the population of university students. If we look closely at the scenario, however, it stated that " . . . you enlisted 100 participants from your department's participant pool."

In Chapter 1 we called this a *convenience sample*, because it is conveniently available in your university. Although our sample was randomly drawn from the participant pool, it is not a random sample from all university students. As noted in Chapter 1, care must be taken when drawing inferences about the characteristics of populations from the characteristics of samples. We cannot draw valid conclusions about the population of university students from a convenience sample at a particular university.

PLANNING A STUDY

The two examples we've worked through involved samples of size 64 and 100, respectively. There was no particular reason for these choices, except that both are relatively large and have, conveniently, a square root that is a whole number. In real research, however, one should choose a sample size based on something a little bit more defensible. Because the standard error decreases as sample size increases, one can always obtain an extremely precise estimate by choosing an enormous sample. Practical considerations, such as the cost of obtaining the scores in the sample, might make it impossible to obtain a very large sample. On the other hand, small samples make for imprecise estimates. Therefore, these constraints must be balanced when planning a study.

In order to choose a sample size that will give a desired level of precision, we must first have a definition of precision. Typically, the precision of an estimate is called the margin of error, or *moe*. This is the quantity that we add and subtract from the mean when computing a confidence interval. Therefore, $moe = z_{\alpha/2}(\sigma_m)$. The precision we wish to achieve for an estimate is often expressed in terms of σ. For example, we might wish the *moe* for our estimate to be $0.5(\sigma)$, $0.25(\sigma)$, or $0.125(\sigma)$. In the general case, we could say we want the *moe* to be $f(\sigma)$, where f is some number less than 1 and greater than 0. In Appendix 6.3 (available at study.sagepub.com/gurnsey), we derive the following simple formula:

$$n = (z_{\alpha/2}/f)^2, \tag{6.5}$$

which shows the sample size required to produce an *moe* of $f(\sigma)$. For example, if you would like the *moe* for a 95% confidence interval to be $0.25(\sigma)$, then n would be 61.46, which we round up to 62. For further details and a worked example, you should consult Appendix 6.3.

LEARNING CHECK 5

1. A random sample of 14 incomes was drawn from a distribution of incomes having $\sigma = 100$. The confidence interval was computed as $m \pm z_{\alpha/2}(\sigma_m)$. Comment on the validity of this interval, stating any assumptions that may have been violated.

2. IQ is known to be normally distributed with a standard deviation of $\sigma = 15$. A random sample of 225 students was drawn from the participant pool in the Department of Psychology at Concordia University. The IQ of each was determined and a confidence interval was computed as $m \pm 1.96(\sigma_m)$.

   (a) Identify the population whose mean is estimated here.

   (b) If $m = 106$ and $\alpha = .08$, compute the $(1-\alpha)100\%$ around m.

3. To estimate the mean IQ of Brandeis University students, a random sample of 169 Brandeis students was selected and the IQ of each was computed. The mean was $m = 98$ and the standard deviation was 19.5. The confidence interval was computed as $m \pm z_{\alpha/2}(s_m)$.

   (a) If $\alpha = .05$, compute the $(1-\alpha)100\%$ around m.

   (b) Comment on the validity of this interval, stating any assumptions that may have been violated.

Answers

1. We know that the distribution of incomes is positively skewed. Because sample size is small ($n = 14$), there is no guarantee that the distribution of means is normal. Therefore, we have not met the normality assumption for a valid confidence interval.

2. (a) The population comprises all students in the participant pool in the Department of Psychology at Concordia University.

   (b) If $\alpha = .08$, then $z_{\alpha/2} = 1.751$ and $m \pm z_{\alpha/2}(\sigma_m) = 106 \pm 1.751(15/15) = [104.25, 107.75]$.

3. (a) If $\alpha = .05$, then $z_{\alpha/2} = 1.96$ and $m \pm z_{\alpha/2}(s_m) = 98 \pm 1.96(19.5/13) = [95.06, 100.94]$.

   (b) IQ is known to be normally distributed and sample size is large; therefore, the distributions of scores and means are normal, so we have met the normality assumption. Because we used an approximate confidence interval $[m \pm z_{\alpha/2}(s_m)]$, the interval will be slightly narrower than we want. But because sample size is relatively large, our approximate interval will be satisfactory, so we have met the sample size assumption for our confidence interval. All is well.

A WORD ABOUT JERZY NEYMAN

Four statisticians play an important role in the material covered in this book. These are William Sealy Gosset (1876–1937), Sir Ronald Fisher (1890–1962), Jerzy Neyman (1894–1981), and Jacob Cohen (1923–1998). We will say a few things about these important figures as we go along. We begin with Jerzy Neyman (Figure 6.6), who was the person most responsible for the material on confidence intervals covered in this chapter.

Neyman was born in Russia to Polish parents. His intellectual gifts emerged early, and he spoke five languages at the age of 10 (Reid, 1997). He attended the University of Kharkov (in present-day Ukraine), where he stayed on after graduation, working as a lecturer and tutor. He received his Ph.D. from the University of Warsaw in 1924 and taught there for a period. While there, he made visits to London, Paris, and Geneva.

In 1935, Neyman obtained a permanent position in the Biometrics Laboratory at University College London, where he stayed until 1938. The Biometrics Laboratory has an important place in the history of statistics. It was founded by Karl Pearson, whom David Salsburg (2001) credits with beginning the 20th century revolution in statistics. The Biometrics Laboratory was later jointly directed by Pearson's son Egon and Sir Ronald Fisher. Neyman and Egon Pearson collaborated on important work on statistical decision making, which will be covered in Chapter 8. Initially Neyman had a congenial relationship with Sir Ronald Fisher (one of the important figures mentioned above), but that degenerated when Fisher criticized Neyman's work. In fact, Fisher seems to have criticized Neyman at every possible opportunity. The Fisher-Neyman battles are legendary in statistics, and you can read Neyman's perspective on this (Neyman, 1961).

In 1938, Neyman was invited to take a position in the Mathematics Department at the University of California, Berkeley, following a series of very well received lectures given in the United States. He moved there in 1938 when he was 44 and stayed at Berkeley for the rest of his life. At Berkeley he built one of the strongest statistics departments in the United States.

Beyond Neyman's impressive list of research contributions, he is remembered as a generous and supportive mentor. The following quote from Lehmann (1994) captures the general feeling:

> On a personal level, the characteristic that perhaps remains above all in the minds of his friends and associates is his generosity—furthering the careers of his students, giving credit and doing more than his share in collaboration, and extending his help (including financial assistance out of his own pocket) to anyone who needed it. (p. 415)

FIGURE 6.6 ■ Jerzy Neyman

Courtesy of G. Paul Bishop Jr.

Neyman was a Polish statistician who developed the theory of confidence intervals covered in this chapter.

SUMMARY

In this chapter we saw how to compute a confidence interval around a sample mean when the population standard deviation (σ) is known. We compute the confidence interval as follows:

$$m \pm z_{\alpha/2}(\sigma_m),$$

where $z_{\alpha/2}$ is the z-score for which $(1-\alpha)100\%$ of the z-distribution lies within the interval $\pm z_{\alpha/2}$.

We have $(1-\alpha)100\%$ confidence in the interval computed as $m \pm z_{\alpha/2}(\sigma_m)$ because we know that $(1-\alpha)100\%$ of all such intervals will capture μ. For confidence intervals constructed this way to be valid, the sample must be a random sample from the population of interest, and the distribution of means must be normal.

An approximate confidence interval can be computed as follows:

$$m \pm z_{\alpha/2}(s_m),$$

where $s_m = s/\sqrt{n}$ is the *estimated* standard error of the mean. When sample size is small, the resulting confidence interval is narrower than it should be, and less than $(1-\alpha)100\%$ of all such intervals will capture μ. However, as sample size increases, the confidence intervals become increasingly similar to those computed with σ_m. For confidence intervals constructed this way to be valid, the sample must be a random sample from the population of interest, and the distribution of scores must be normal.

KEY TERMS

$(1-\alpha)100\%$ confidence interval   136

95% confidence interval   132

confidence interval   131

interval estimate   131

point estimate   130

EXERCISES

Please note that asterisks (*) indicate questions relating to material covered in the appendixes or requiring insight beyond what was covered in the chapter.

Definitions and Concepts

1. What is a point estimate?

2. What is an interval estimate?

3. What is a confidence interval?

4. What does it mean to say that we have 95% confidence in an interval?

5. What is $z_{\alpha/2}$?

6. What does it mean to say we have $(1-\alpha)100\%$ confidence in an interval?

7. What are the four quantities required to calculate a confidence interval when σ is known, and how does each of these affect the width of the confidence interval?

8. What assumptions must we make when computing a confidence interval when σ is known?

9. *Does knowing only the margin of error (e.g., $moe = 10$) tell you how much confidence you should have in an interval?

True or False

State whether the following statements are true or false.

10. 95% of all sample means will fall in the interval $m \pm 1.96(\sigma_m)$.

11. 90% of all intervals computed as $m \pm 1.96(\sigma_m)$ will contain μ.

12. $m \pm 1.96(\sigma_m) = [5.6, 9.3]$; therefore, the probability is .95 that this interval contains μ.

13. For $n = 64$, more than 94% of all $m \pm 1.96(s_m)$ will capture μ.

14. For $n = 1500$, more than 94% of all $m \pm 1.96(s_m)$ will capture μ.

15. As sample size increases, confidence intervals become narrower.

16. For $n = 15$, more than 5% of all $m \pm 1.96(s_m)$ will fail to capture μ.

17. For $n = 1500$, more than 5% of all $m \pm 1.96(\sigma_m)$ will fail to capture μ.

18. If $\alpha/2 = .025$, then $(1-\alpha)100\%$ describes 95% confidence.

Calculations

19. Which of the following most closely defines the 90% confidence interval:

    (a) $m \pm 1.64(\sigma_m)$,

    (b) $m \pm 1.96(\sigma_m)$,

    (c) $m \pm 2.58(\sigma_m)$, and

    (d) $m \pm 1.00(\sigma_m)$?

20. Determine $z_{\alpha/2}$ for each of the following values of α:

    (a) .484,

    (b) .242,

    (c) .126, and

    (d) .303.

21. For each of the following values of $z_{\alpha/2}$, determine the proportion of $(1-\alpha)100\%$ confidence intervals, $m \pm z_{\alpha/2}(\sigma_m)$, that will not capture μ:

    (a) 2.1,

    (b) 1.7,

    (c) 1.2, and

    (d) 2.5.

22. A population is known to be normally distributed with $\sigma = 10$. A sample of size $n = 36$ randomly selected from the population was found to have a mean of $m = 32$. Calculate the 90% and 95% confidence intervals about m.

23. A population is known to be normally distributed with $\sigma = 14$. A sample of size $n = 49$ randomly selected from the population was found to have a mean of $m = 86$. Calculate the 90% and 95% confidence intervals about m.

24. A population is known to be normally distributed with $\sigma = 14$. A sample of size $n = 4$ randomly selected from the population was found to have a mean of $m = 100$. Calculate the 60% and 70% confidence intervals about m.

25. A sample of size $n = 144$, randomly selected from a normally distributed population, was found to have a mean of $m = 100$ and standard deviation of $s = 18$. Calculate the approximate 90% and 95% confidence intervals about m.

Scenarios

26. An economist wanted to know the mean income of Americans. However, he was in a perplexing situation. A previous study had reported the mean and standard deviation of American incomes but, because of a filing problem, the mean of the distribution had gotten lost so he knew the only standard deviation. In order to estimate the mean income of Americans, the economist chose a random sample of 1600 Americans in the workforce and recorded their incomes. The mean of these 1600 incomes was $74,000 per year. Because the economist knew the standard deviation of incomes from the previous study ($\sigma = 12,000$), he could then calculate the 95% confidence interval around the sample mean $m = \$74,000$ using the formula $m \pm z_{\alpha/2}(\sigma_m)$. Please calculate this confidence interval.

27. Let's say that the standard deviation of the sample of 1600 scores in the preceding example was $s = 11,640.32$. Compute the approximate 95% confidence interval around m, and comment on the validity of this interval.

28. A cognitive psychologist has compiled a database of reaction times to detect a change in a change-blindness experiment. The standard deviation of these reaction times is $\sigma = .99$ seconds. A random sample of 36 of these reaction times has a mean of $m = 11.2$ seconds. Compute the 95% confidence interval around m. What assumptions must be correct for this confidence interval to be valid?

29. Let's say that the standard deviation of the sample of 36 scores in the preceding example was $s = 1.2$. Compute the approximate 95% confidence interval around m, and comment on the validity of this interval.

30. A recent study reported on the percentage of positive emotion words in 3 million Facebook posts. The standard deviation of these 3 million percentages was $\sigma = 3.92$. A sample of $n = 155,000$ posts was randomly selected from this population and found to have an average of $m = 5.24\%$ positive emotion words. Compute the 95% confidence interval around this mean.

    (a) *Is there any reason to believe that the distribution of these percentages is not normal? (*Hint:* Think about the sample mean and the population standard deviation.)

31. Compute the approximate 95% confidence interval for the mean of the set of scores below.

54	36	37	48	57
67	46	80	50	39
27	53	57	65	56
59	86	49	64	65
53	79	57	63	53

APPENDIX 6.1: COMPUTING CONFIDENCE INTERVALS IN EXCEL

This appendix shows how to set up an Excel workbook to compute confidence intervals around a sample mean when the population standard deviation is known or sample size is large. Cells **B2** to **B5** of Figure 6.A1.1 contain the four numbers required to compute the interval. These are m, σ, n, and α, respectively. To compute the confidence interval, we need to calculate σ_m and $z_{\alpha/2}$. σ_m is calculated in cell **B7**; it is simply $\sigma_m = \sigma/\sqrt{n}$. To compute $z_{\alpha/2}$, we first compute $\alpha/2$ in cell **B9**. In cell **B10** we use **NORM.S.INV** to determine the z-score with $(\alpha/2)$ *100% of the distribution below it. The absolute value (**ABS**) of the number is $z_{\alpha/2}$. Finally, the lower and upper

FIGURE 6.A1.1   ■   Computing CIs in Excel

	A	B	C
1	**Quantities**	**Values**	**Formulas**
2	m	14	
3	σ	8	
4	n	64	
5	α	0.05	
6			
7	σ_m	1	=B3/sqrt(B4)
8	(1-α)100%	95	=(1-B5)*100
9	α/2	0.025	=B5/2
10	$z_{\alpha/2}$	1.960	=ABS(NORM.S.INV(B9))
11			
12	m-$z_{\alpha/2}$(σ_m)	12.04	=B2-B10*B7
13	m+$z_{\alpha/2}$(σ_m)	15.96	=B2+B10*B7

Cells **B2** to **B5** contain the information required to compute the interval. From these numbers, we compute σ_m and $z_{\alpha/2}$ in cells **B7** and **B10**, respectively. The lower and upper limits of the interval are computed in cells **B12** and **B13**, respectively.

limits of the confidence interval are computed in cells **B9** and **B10**, respectively.

Once you have set up this workbook, you can use it to check the accuracy of any calculations you've done by hand. There is one word of caution, however. If you are doing calculations by hand, you must use the z-table to determine $z_{\alpha/2}$. The z-table only gives z-scores to two decimal places of accuracy, whereas **NORM.S.INV** will give much higher accuracy. Therefore, you can use the **ROUND** function in Excel to round the value returned by **NORM.S.INV** to two decimal places. To do this, you would enter

$$\text{=ROUND(ABS(NORM.S.INV(B9)), 2)}$$

into cell **B10**. This instructs Excel to round the value returned by **NORM.S.INV** to two decimal places. If the **2** at the end of the above **ROUND** function were changed to **4**, then the result would be rounded to four decimal places.

7 SIGNIFICANCE TESTS

INTRODUCTION

Sir Ronald Aylmer Fisher was born in 1890 and died in 1962. During his lifetime he made important contributions to many fields of scientific inquiry, including statistics, evolutionary biology, and genetics. Fisher is best known to psychologists for his contributions to statistics (Figure 7.1). He developed many statistical methods in the 1920s while employed at an agricultural research station, called Rothemsted, just north of London, England. Although these methods were originally applied to questions such as the determinants of yearly variations in crop yields, they were extremely general and have been applied routinely in fields as diverse as medicine, psychology, biology, economics, and political science. In 1925, Fisher wrote an influential textbook entitled *Statistical Methods for Research Workers*, which provided methods of data analysis that could be used by working scientists who did not have extensive statistical training. One of the main methods he promoted was the *significance test*. To this day, most researchers in psychology rely on statistical analyses that have their roots in Fisher's significance tests.

In 1935, Fisher wrote a second book entitled *Design of Experiments*, which went through eight editions and was also highly influential. Chapter 2 of that book opens with the following thought experiment to help develop the ideas underlying significance tests.

FIGURE 7.1 ■ Sir Ronald Aylmer Fisher

Fisher in 1913 at age 23 (*left*) and at 56 in 1946 (*right*). Fisher is arguably the most influential statistician of all time. The statistical methods he developed still dominate statistical analyses in most domains of psychology.

(*left*) Courtesy Barr Smith Library, University of Adelaide; (*right*) ©National Portrait Gallery, London.

A lady declares that by tasting a cup of tea made with milk she can discriminate whether the milk or the tea infusion was first added to the cup. (Fisher, 1935)

If we think about how to assess this claim, we are essentially thinking about how to assess any scientific claim or theory. So, thinking as scientists, how might one evaluate this claim about tea made with milk first versus tea first?

A taste test of some sort is the obvious answer. For example, the lady might be given two cups of tea, asked to take a sip from each, and then decide which of the two was made with milk first. If this taste test had been conducted and the lady made the correct decision, I think few people would be convinced that she could tell the difference. Most people would probably say that if she were simply *guessing*, she would have a 50% chance of being correct. That is, even if she hadn't tasted the two cups of tea, she would have a 50% chance of correctly guessing which cup contained the tea made with milk added first. Therefore, a correct response on a single trial is not convincing.

If one correct response is unconvincing, would we be convinced by two correct responses in two consecutive trials? This situation is like coin flipping. We saw in Chapter 2 (Appendix 2.3) that the probability of two successes in two flips is .5∗.5 = .25, or 25%. This means that there is a 25% chance of two correct responses in two consecutive trials by guessing alone. We wouldn't think that there is anything unusual about a flipped coin coming up heads twice in a row; therefore, getting two out of two correct in a taste test would convince few people that the lady's claim is true. However, intuition suggests that as the number of trials increases, the probability of getting a large number correct by guessing decreases.

When we step back and examine our approach to assessing the lady's claim, we can see that we're asking her to convince us that she is not just guessing. Our logic is that we'll be convinced if she correctly identifies an unusually large number of milk and tea mixtures. Fisher called the guessing hypothesis a *null hypothesis*. And he would say that an unusually large number of correct decisions is *statistically significant*. So, if the lady makes a statistically significant number of correct decisions, we would *reject the null hypothesis*. That is, we would reject the guessing hypothesis if the number of correct decisions has a low probability of occurring just by chance (guessing). So, what kind of experiment might we run to assess the lady's claim, and what accuracy would she have to achieve in this experiment to convince us that she really can tell the difference?

Fisher outlined the following efficient procedure to answer this question. According to Fisher, eight identical cups of tea should be prepared; four with milk first and four with tea first. These eight cups would be presented to the lady in a random order (see Figure 7.2). She must taste each and place the cups judged to have been made with milk first on a separate table. (She knows that four were milk first and four were tea first.) At the end of the experiment, we count the number of correctly placed cups on the milk-first table. We will see that there are 70 possible outcomes when a guessing strategy is used, and these can be grouped into five events corresponding to 0, 1, 2, 3, or 4 correctly placed cups. (You might review the distinction between events and outcomes in Chapter 2.) The importance of the null hypothesis is that it allows us to attach a probability to each event.

Figure 7.3 shows all outcomes corresponding to each of our five events. The numbers 1 to 4 (shown in blue type in the bars in Figure 7.3) represent milk-first cups, and the numbers 5 to 8 (shown in black) represent tea-first cups. We consider these numbers in groups of four to represent possible combinations of cups placed on the milk-first table. From Figure 7.3 we can see that

FIGURE 7.2 ■ Fisher's Taste Test

Eight cups of tea have been prepared. Cups 1 to 4 had milk added first and cups 5 to 8 had tea added first. These eight cups are arranged randomly on a table; of course, there weren't numbers hovering over the cups. The lady was to taste each cup and place those with milk added first on a separate table.

iStock.com/IndigoBetta

1. there is only one way to get no milk-first and four tea-first cups on the milk-first table,

2. there are 16 ways to get one milk-first and three tea-first cups on the milk-first table,

3. there are 36 ways to get two milk-first and two tea-first cups on the table,

4. there are 16 ways of getting three milk-first and one tea-first cup on the table, and

5. there is only one way to get four milk-first cups on the table.

In all, there are 1 + 16 + 36 + 16 + 1 = 70 possible outcomes.

The probability of getting all four milk-first mixtures on the milk-first table by chance alone (i.e., by guessing) is 1/70 = .014. If our lady achieved this level of accuracy, Fisher would say this event is statistically significant, because it would occur very infrequently if she were indeed guessing. When we obtain a statistically significant result in a significance test, we reject the null hypothesis and treat it *as though it is false*.

Would we consider three correctly placed cups to be statistically significant? Three correctly placed represents almost 25% of all possible outcomes (16/70 = .229, or 22.9% to be exact). As noted above, when two coins are flipped both will come up heads 25% of the time. Such a result seems completely unremarkable. Therefore, I doubt that three out of four correctly placed cups would lead you to seriously question the guessing hypothesis. That is, we would not treat this result as statistically significant.

The essence of a **significance test** is to specify a null hypothesis and then establish the probability of *each of the possible events* occurring by chance when the null hypothesis is true. If the event obtained in an experiment has a low probability of occurring by chance when the null hypothesis is true, we reject the null hypothesis and treat it as though it is false. We will elaborate on this concept in this chapter.

> A **significance test** is a statistical procedure for evaluating the plausibility of the null hypothesis.

According to Joan Fisher-Box (1978), Fisher's daughter and biographer, the lady-tasting-tea illustration arose from actual events that occurred on an afternoon at Rothemsted in the 1920s. The lady in question was Dr. B. Muriel Bristol, an algologist. (Google it.) Apparently, Dr. Bristol made a statistically significant number of correct choices and convinced the skeptics that she could tell when tea was made with milk first.

A SCENARIO: WHOLE LANGUAGE VERSUS PHONICS

We will now turn to a scenario to keep in mind as we introduce the basics of significance tests as typically used in psychology. Our scenario will share many features with the tea-tasting example that we just worked through. It will differ in that our judgment will be

---

FIGURE 7.3 ■ The Five Events

The five events, showing the 70 ways in which four of eight cups can be placed on a table, grouped into columns by number of milk-first cups (MF = milk added first).

0 MF — 1/70 = .014
5 6 7 8

1 MF — 16/70 = .229
1 5 6 7
1 5 6 8
1 5 7 8
1 6 7 8
2 5 6 7
2 5 6 8
2 5 7 8
2 6 7 8
3 5 6 7
3 5 6 8
3 5 7 8
3 6 7 8
4 5 6 7
4 5 6 8
4 5 7 8
4 6 7 8

2 MF — 36/70 = .514
1 2 5 6
1 2 5 7
1 2 5 8
1 2 6 7
1 2 6 8
1 2 7 8
1 3 5 6
1 3 5 7
1 3 5 8
1 3 6 7
1 3 6 8
1 3 7 8
1 4 5 6
1 4 5 7
1 4 5 8
1 4 6 7
1 4 6 8
1 4 7 8
2 3 5 6
2 3 5 7
2 3 5 8
2 3 6 7
2 3 6 8
2 3 7 8
2 4 5 6
2 4 5 7
2 4 5 8
2 4 6 7
2 4 6 8
2 4 7 8
3 4 5 6
3 4 5 7
3 4 5 8
3 4 6 7
3 4 6 8
3 4 7 8

3 MF — 16/70 = .229
1 2 3 5
1 2 3 6
1 2 3 7
1 2 3 8
1 2 4 5
1 2 4 6
1 2 4 7
1 2 4 8
1 3 4 5
1 3 4 6
1 3 4 7
1 3 4 8
2 3 4 5
2 3 4 6
2 3 4 7
2 3 4 8

4 MF — 1/70 = .014
1 2 3 4

The 70 ways in which four of eight cups can be placed on a table. These 70 outcomes correspond to five events. MF = milk added first.

based on the mean of a sample of scores, rather than the number of correct responses from a single individual in a sequence of trials.

Let's think about a school district in a city in the Northeastern United States that has a small number of elementary schools. An important part of the school curriculum is reading, which is taught in all grades. Until about 1980, reading was taught with an emphasis on *phonics*. That is, children were taught the correspondence between letter strings and speech sounds. For example, they were taught that [f], [ph], and [gh] can all correspond to the same sound. Once you know the rules that relate letters to sounds, you can *decode* new words.

About 35 years ago, the school district adopted the *whole language* approach to reading. Advocates of this approach believe we should think about the acquisition of reading skills in the same way we think about language acquisition. We're not taught how to speak and understand our first language. Rather, we figured out how language works through exposure and context. Therefore, the whole language approach places less emphasis on teaching decoding skills (i.e., phonics) and more emphasis on how to use context to figure out the meaning and sound of an unfamiliar word.

Let's say that over these 35 years, more than 2000 first-grade students in this school district have followed the whole language program. At the end of each year, all students are given a standardized test of reading skills. The mean of these reading scores is $\mu = 72$, with a standard deviation of $\sigma = 8$.

A new administrator within this school district read a report in a research journal arguing that students tend to do somewhat poorly in the whole language approach because they lack the decoding skills typically taught in phonics (Rayner et al., 2001). Therefore, the administrator hired a psychologist to conduct a year-long experiment to test the effect of adding 20 minutes of daily phonics instruction to the lesson plans of all 64 first-grade students in the school district. (We call this an *intervention*, or *treatment*, because we treat the members of one group differently than the members of another group.) The administrator strongly suspected that the additional phonics instruction would improve reading scores.

At the end of the school year, the reading skills of these 64 students were assessed using the same standardized test that had been used in the preceding 35 years. The results showed that the students with additional phonics instruction scored $m = 74.5$ on average on the standardized test. Now, 74.5 is greater than 72, and this supports the idea that additional phonics instruction improves reading skills. On the other hand, even if additional phonics instruction had *absolutely no effect* on reading skills, we wouldn't expect the sample mean to exactly equal 72. After all, there is always some difference between a statistic and the parameter it estimates. That is, there is always *sampling error*. If we compare this scenario to the tea-tasting scenario, we can ask whether a mean of 74.5 is like getting three out of four correct, which would be quite common if one is guessing, or like getting four out of four correct, which is quite unusual if one is guessing. In other words, we would like to know how to decide whether the difference between $\mu = 72$ and $m = 74.5$ is statistically significant, or whether it is about what we'd expect from sampling error.

LANGUAGE ALERT

The term *statistically significant* simply means unusual. An unusual event is one that has a low probability of occurring by chance. For example, it is unusual for a randomly selected American woman to be 6 feet, 2 inches or taller, because most American women are less than 6 feet, 2 inches tall. This means that there is a very low probability that a randomly selected American woman will be 6 feet, 2 inches or taller. Similarly, it would be unusual for a randomly selected Canadian to have an IQ of 145 or greater because most Canadians have IQs less than 145. This means that there is a low probability that a randomly selected Canadian will have an IQ of 145 or greater.

SIGNIFICANCE TESTS

The Logic of Significance Tests

The answer to the question we've posed is surprisingly simple. We can start with an analogue of the guessing hypothesis used in the tea-tasting example. The corresponding hypothesis in the reading scenario is that our treatment (additional phonics instruction) has *absolutely no effect on reading scores*. We call this hypothesis the **null hypothesis**. If the null hypothesis is true (i.e., if phonics instruction has absolutely no effect on reading scores), then our sample of 64 scores is just one of a huge number of possible samples of size $n = 64$, that could be drawn from the population of reading scores obtained over the previous 35 years. That distribution has $\mu = 72$ and $\sigma = 8$.

We use the sample mean (m) to test the effect of phonics instruction. To determine whether our sample mean of 74.5 is consistent with the null hypothesis, we would have to know something about the distribution of sample means *when the null hypothesis is true*. That is, as in the tea-tasting example, to assign a probability to our statistic, we have to know the distribution of all possible values of the statistic when the null hypothesis is true. *This distribution is the sampling distribution of the mean.*

We now have all the information necessary to determine the exact nature of the distribution of sample means when the null hypothesis is true. We know that this distribution will have $\mu = 72$, standard error $\sigma_m = \sigma/\sqrt{n}$, and because the sample size is relatively large, it's safe to assume that the distribution of means is normal. If we put all of this together, we can transform m into a z-score and then judge whether it is unusual under the null hypothesis. We first compute

$$\sigma_m = \frac{\sigma}{\sqrt{n}} = \frac{8}{\sqrt{64}} = \frac{8}{8} = 1.$$

Then we compute

$$z_{obs} = \frac{m - \mu}{\sigma_m} = \frac{74.5 - 72}{1} = 2.5. \tag{7.1}$$

We will call this z-score $z_{observed}$, or z_{obs} for short. This will help us distinguish it from z-scores in general. We call z_{obs} a **test statistic**.

Our hypothetical administrator predicted that the scores of the students receiving additional phonics instruction would be higher, on average, than those receiving no additional phonics instruction. To judge whether the phonics instruction had the predicted effect, the researcher relies on probability, in the same way Fisher did in the tea-tasting example. The researcher asks about the probability of obtaining $z_{obs} \geq 2.5$ when the null hypothesis is true (i.e., if phonics instruction had absolutely no effect).

By now we have strong intuitions about z-scores. We know that $z_{obs} = 2.5$ is very large (I'm sure you'd like your grade in this course to be 2.5 standard deviations above the class mean!), and therefore very unusual (it would be unusual that a randomly chosen student from your class would have a grade that corresponds to $z \geq 2.5$). If we consult the z-table on page 91, we find that 99.38% of the z-distribution lies *below* $z_{obs} = 2.5$. That is, $P(2.5) = .9938$. Stated the other way around, only 0.62% of the z-distribution lies *above* 2.5; that is, $1 - P(2.5) = .0062$. Therefore, m is very unusual, *under the null hypothesis* that $\mu = 72$. Figure 7.4 illustrates this point.

The **null hypothesis** is a hypothesis about the distribution of scores from which a sample was drawn.

A **test statistic** (such as z_{obs}) is computed from a sample statistic, such as the sample mean.

FIGURE 7.4 ■ Probability Under the Null

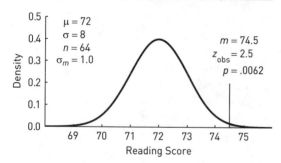

The probability of the test statistic, or one more extreme, under the null hypothesis. The normal distribution represents the sampling distribution of the mean under the null hypothesis. The sample mean (74.5) is indicated by the vertical line. The shaded region is the proportion of the distribution of means greater than 74.5. This shows that when we draw a sample of 64 scores from a distribution with a mean of 72 and standard deviation of 8, the probability that the sample mean will be 74.5 or greater is $p = .0062$.

p-Values

The *unusualness* of a statistic is typically expressed as a proportion, rather than a percentage. Such proportions are called **p-values**. In the reading scenario, we would say that $p = .0062$, because this is the proportion of the distribution of means, under the null, that lies above our statistic, $m = 74.5$, and the corresponding $z_{obs} = 2.5$. A statistic is said to be **statistically significant** if it has a *low probability of occurrence under the null hypothesis*. When a statistic is deemed statistically significant, we *reject the null hypothesis* because it seems implausible. When we reject the null hypothesis, we conclude that our prediction is supported.

In our example, the researcher would reject μ = 72 as a plausible hypothesis about the mean of the distribution from which her sample was drawn. She concludes that her result supports the hypothesis that additional phonics instruction will improve reading scores. When the researcher reports the results of her experiment in a research journal, she might say something like this:

A **p-value** expresses the probability of a statistic, or one more extreme, occurring by chance when the null hypothesis is true. (*"More extreme"* means further from the mean of the distribution.)

A statistic is said to be **statistically significant** when its associated *p*-value under the null is small.

APA Reporting

Additional phonics instruction was found to produce a statistically significant increase in reading scores, $z = 2.5$, $p = .01$.

Rejecting the Null Hypothesis

Rejecting the null hypothesis means that we treat it *as though* it is false, not that it *is* false. There is always some non-zero probability of obtaining our statistic, or one more extreme, by chance when the null hypothesis is true. In our reading example, the probability of obtaining our statistic, under the null, is $p = .0062$. So, even though it is unusual to obtain such a mean under the null, it is not impossible.

Retaining the Null Hypothesis

If rejecting the null hypothesis means that we treat it as though it is false, what should we say when we fail to reject the null hypothesis? Should we say that it is true? Should we treat it as though it is true? Actually, we should do neither of these things. When we fail to reject the null hypothesis, we say that we *retain* it. This just means that the mean specified by the null hypothesis remains a plausible hypothesis about the mean of the distribution from which our sample was drawn. However, this mean (72 in our example) is one of many plausible hypotheses about the mean of the distribution from which our sample was drawn. Therefore, retaining the null hypothesis does not mean that it is true (i.e., $\mu = 72$) or that we even treat it as though it is true. Rather, **retaining the null hypothesis** just means that the evidence (z_{obs}) is not sufficient for us to treat it as implausible.

What Do We Mean by Statistically Significant? $p < .05$

So far we have said that small p-values are statistically significant, but we haven't said how small they should be. Although Fisher was a relentless advocate for significance tests, he was inconsistent about how small a p-value must be for it to be considered statistically significant. Early in his career, he suggested that a statistic might be considered statistically significant if it is one that would occur less than 5% of the time when the null hypothesis is true (i.e., when $p < .05$).

> [I]t is convenient to take [the point for which $P = 0.05$, or 1 in 20] as a limit in judging whether a deviation is to be considered significant or not. (Fisher, 1925)

Later in his career, Fisher said that the experimenter is in the best position to make this judgment.

> . . . [N]o scientific worker has a fixed level of significance at which from year to year, and in all circumstances, he rejects hypotheses; he rather gives his mind to each particular case in the light of his evidence and his ideas. (Fisher, 1956, p. 42)

However, in spite of Fisher's later warnings, researchers have converged on the $p < .05$ criterion for statistical significance. When we adopt the .05 criterion, we say this is the **significance level** of the test. p-values less than .05 are statistically significant and p-values greater than .05 are not. Therefore, when you hear that some experiment produced a statistically significant result, it typically means that a result more extreme than the one observed would occur less than 5% of the time *when the null hypothesis is true*. For the remainder of this book, we will treat $p < .05$ as the definition of statistical significance.

COMPUTING EXACT p-VALUES: DIRECTIONAL AND NON-DIRECTIONAL TESTS

In the preceding example, it seemed completely reasonable that p should be the proportion of the z-distribution above z_{obs}. However, the way we compute a p-value depends on the prediction we're making, and we might make different predictions in different situations. For the kind of research question that we're considering, there are three possible predictions, and we will now step through each of these.

Retaining the null hypothesis means that it remains a plausible hypothesis about the distribution from which our sample was drawn. Retaining the null hypothesis does not mean that it is true or that we treat it as though it is true.

The **significance level** of a test defines the largest p-value that will be considered statistically significant; that is, p-values less than the significance level are statistically significant, and those that are greater are not. The significance level is typically set at .05.

Directional Tests: Predicting That $m - \mu > 0$

In the phonics scenario, the researcher predicted that the sample mean (m) would be greater than the mean of the null hypothesis distribution (μ). We can express this prediction as either

$$m > \mu,$$

or

$$m - \mu > 0.$$

The second form ($m - \mu > 0$) is preferable because $m-\mu$ is the numerator of our z_{obs} statistic given in equation 7.1. This makes it easier to connect the prediction to the test statistic. The prediction is simply that z_{obs} will be positive. We call this a *directional prediction*. When we predict that z_{obs} will be positive, then the p-value is the proportion of the z-distribution *above* z_{obs}, and this is exactly how the p-value was computed in the phonics example. Because the computed p-value represents an area in one tail of the distribution, **directional tests** are sometimes called **one-tailed tests**.

Directional tests make predictions about whether a statistic will be greater or less than a specified parameter. The p-value associated with a test statistic (such as z_{obs}) is the area in one tail of the distribution. Therefore, directional tests are also called **one-tailed tests**.

Directional Tests: Predicting That $m - \mu < 0$

The other possible directional prediction is that the sample (m) will be less than the mean of the null hypothesis distribution (μ). We can express this prediction as either

$$m < \mu,$$

or

$$m - \mu < 0.$$

Again, the second form ($m - \mu < 0$) is preferable. The prediction is simply that z_{obs} will be negative. When we predict that z_{obs} will be negative, the p-value is the proportion of the z-distribution *below* z_{obs}.

Let's revisit our phonics experiment from the point of view of a whole language enthusiast who predicts that additional phonics instruction will lead to *poorer* reading scores. In this case the directional prediction would be $m - \mu < 0$, meaning that the researcher predicts a negative z_{obs}. Now, if this researcher had found that $z_{obs} = -2.5$, contrary to our example, he would have reported $p = .0062$, because this is the proportion of the z-distribution falling *below* $z_{obs} = -2.5$. However, if this researcher had found $z_{obs} = 2.5$, as in our example, then he would have reported $p = .9938$, because this is the proportion of the z-distribution falling *below* $z_{obs} = 2.5$. A p-value of .9938 is not statistically significant. This situation is illustrated in Figure 7.5.

Non-directional Tests: Predicting That $m - \mu \neq 0$

We've just seen that directional tests can lead to peculiar conclusions. In the example we just considered, the whole language enthusiast predicted $m - \mu < 0$ but found that $z_{obs} = 2.5$, which is not statistically significant, $p = .9938$. This is a bit strange. Finding $z_{obs} = 2.5$ would be a very unusual result if the null hypothesis were true (i.e., if there was absolutely no effect of phonics instruction). This strange situation arises from the fact that the researcher's prediction was wrong. He predicted a negative z_{obs} but obtained a positive z_{obs}. The rules for computing p-values force him to calculate p as the proportion of the z-distribution below z_{obs}.

To avoid this potential problem, it is common for researchers to conduct **non-directional tests**. In this situation, the prediction is

$$m - \mu \neq 0.$$

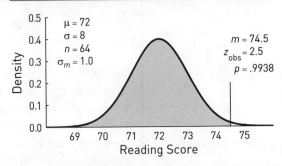

FIGURE 7.5 ■ Making the Wrong Prediction

The probability of the test statistic under the null hypothesis when the sign of z_{obs} is the opposite of that predicted. The normal distribution represents the sampling distribution of the mean under the null hypothesis. In this case, the prediction is $m - \mu < 0$, and so p is the proportion of the z-distribution below z_{obs}. The sample mean (74.5) is indicated by the vertical line. The shaded region is the proportion of the distribution of means below 74.5. When a sample of 64 scores is drawn from a distribution having a mean of 72 and standard deviation of 8, the probability that the sample mean will be 74.5 or less is .9938.

TABLE 7.1 ■ Rules for Computing p-Values

Prediction	Rule: p is the . . .
$m - \mu > 0$	proportion of the z-distribution *above* z_{obs}.
$m - \mu < 0$	proportion of the z-distribution *below* z_{obs}.
$m - \mu \neq 0$	proportion of the z-distribution *outside* $\pm |z_{obs}|$.

This means that we expect a large difference between m and μ, but we're not predicting the sign of the difference. In this case, we compute the p-value as the proportion of the z-distribution outside the interval $\pm |z_{obs}|$, that is, the proportion of the z-distribution above $|z_{obs}|$ plus the proportion of the z-distribution below $-|z_{obs}|$. Because the computed p-value in a non-directional test represents the area in two tails of the distribution, non-directional tests are sometimes called **two-tailed tests**.

In our example, the interval in question is ± 2.5, whether $z_{obs} = 2.5$ or $z_{obs} = -2.5$. We've seen that the proportion of the z-distribution above 2.5 is .0062, and the proportion below -2.5 is also .0062. Therefore, the proportion outside the interval ± 2.5 is $2*.0062 = .0124$. This is illustrated in Figure 7.6. Table 7.1 summarizes the rules for computing exact p-values.

Non-directional tests make no prediction about whether a statistic will be greater or less than a specified parameter. The p-value associated with a test statistic (such as z_{obs}) is the area in two tails of the distribution. Therefore, non-directional tests are also called **two-tailed tests**.

LANGUAGE ALERT

Previously we said that "a p-value expresses the probability of a statistic, or one *more extreme*, occurring by chance when the null hypothesis is true." Now that we've seen three different ways of computing p-values, we must clarify our notion of extreme, because it is a term we use quite often. The term *more extreme* means z-scores further from 0 than z_{obs} in the predicted direction, or directions. When predicting that $m - \mu > 0$, z-scores more extreme than z_{obs} are those that are greater than z_{obs}. When predicting that $m - \mu < 0$, z-scores more extreme than z_{obs} are those that are less than z_{obs}. When predicting that $m - \mu \neq 0$, there are two predicted directions. Therefore, scores more extreme than z_{obs} are those that are greater than $|z_{obs}|$ and those that are less than $-|z_{obs}|$.

FIGURE 7.6 ■ Two-Tailed Tests

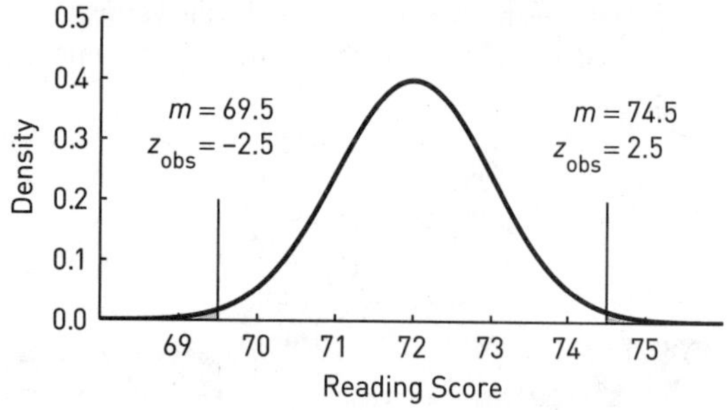

The normal distribution represents the sampling distribution of the mean under the null hypothesis. The shaded regions show the proportion of the distribution of means above 74.5 (corresponding to $z_{obs} = 2.5$) and below 69.5 (corresponding to $z_{obs} = -2.5$). The proportion of the z-distribution outside the interval ±2.5 is .0124.

Directional Versus Non-directional Tests: Why Is This a Thing?

There are two common responses to the problem of making the wrong prediction in a directional test: (i) why not use a non-directional test in all cases or (ii) why not wait and see whether z_{obs} is positive or negative and then compute the p-value accordingly? We will look at these two questions in turn.

Should We Always Use a Non-directional Test?

If we always compute p-values using a non-directional test, we will certainly avoid the problem of having a large z_{obs} that is not statistically significant because our prediction was wrong. But now things get a bit weird. As a rule, researchers prefer small p-values to larger p-values, and they uniformly prefer $p < .05$ to $p > .05$. Now, if one makes the correct prediction, then the directional p-value will always be half that of the non-directional p-value. This can make a difference if we have adopted the $p < .05$ criterion for statistical significance.

Let's say an experiment has been run and $z_{obs} = 1.8$. If the prediction had been $m - \mu > 0$, then $p = .0359$, because that's the proportion of the z-distribution lying above 1.8. In this case, the result is statistically significant. If we'd hedged our bets and not made a prediction—to avoid the awkwardness of an unusual result that is not statistically significant—then the p-value would be the proportion of the z-distribution lying outside the interval ±1.8; that is, $p = 2*.0359 = .0718$. In this case, the result is not statistically significant. Therefore, if the correct prediction has been made in a directional test, the null hypothesis can be rejected at the $p < .05$ level, with a smaller z_{obs} than would be needed to reject the null hypothesis in a non-directional test. However, as we've seen, the risk you take when making a directional prediction is that you may find an unusually large z_{obs} whose sign is the opposite of your prediction. In this case, you won't be able to declare your result statistically significant. For this reason, researchers do not always use non-directional tests.

Should We Wait for the Results Before Making a "Prediction"?

If we wait for the results before deciding which prediction to make, then we're not making a prediction at all. Let's think about a researcher who has started a new line of research testing the effect of a new drug on attentional focus. Because this is a new line of research, he is not sure what effect his intervention will have. After the researcher gathers his data, he finds that $z_{obs} = -1.8$, which indicates a negative effect on attention. If he computes the p-value as though his prediction had been $m - \mu < 0$, he would report $p = .0359$. However, taking this approach is fundamentally dishonest because he would have reported the same p-value if his result had been $z_{obs} = 1.8$. That is, he made no prediction but pretended as though he had. Because no prediction was made, he should report $p = .0718$. Sadly, this p-value does not meet the $p < .05$ criterion for statistical significance.

Critical Values of z

The issue we're considering arises from the $p < .05$ criterion for statistical significance, which is deeply embedded in statistical practice in psychology and related disciplines. When a researcher submits an article to a scientific journal, it is reviewed by other experts in the field of study. If the researcher claims that some difference was found (i.e., that there is a difference between m and μ), the reviewers will want to know if the difference is statistically significant. If the difference does not meet the $p < .05$ criterion for statistical significance, then it may be treated (incorrectly!) as no difference at all. In this case, the paper may not be accepted for publication. Therefore, researchers are highly motivated to report results that are statistically significant.

We will discuss the $p < .05$ criterion for statistical significance in greater detail in Chapter 9. For the moment, however, we need an answer to our question about directional and non-directional tests and how you should view reported p-values. The first answer is the most generous; assume honesty. If a researcher reports a one-tailed p-value, then we must accept that a prediction was honestly made. What else could we do?

A second observation can put our minds somewhat at ease. Up to this point, we have defined statistical significance in terms of p-values. However, statistical significance can also be related to the value that z_{obs} must exceed to be considered statistically significant. We call this value $z_{critical}$. For a one-tailed test that predicts $m - \mu > 0$, all values *above* $z_{critical} = 1.645$ will be statistically significant when $p < .05$ is the criterion for statistical significance. For a one-tailed test that predicts $m - \mu < 0$, all values *below* $z_{critical} = -1.645$ will be statistically significant. For a two-tailed test that predicts $m - \mu \neq 0$, all values *outside* the interval $z_{critical} = \pm 1.96$ will be statistically significant. Table 7.2 and Figure 7.7 show these *critical values*. From these observations, we note the following points:

TABLE 7.2 ■ Critical Values	
Prediction	$z_{critical}$
$m - \mu > 0$	1.645
$m - \mu < 0$	−1.645
$m - \mu \neq 0$	±1.96

Note: Critical values of z at the $p < .05$ criterion for statistical significance.

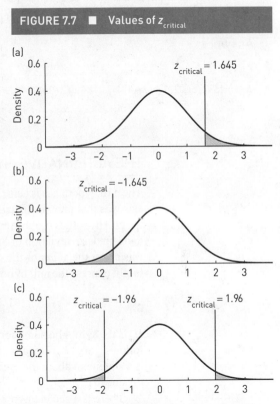

FIGURE 7.7 ■ Values of $z_{critical}$

(a)

(b)

(c)

Critical values of z are shown when the prediction is $m - \mu > 0$ (a), $m - \mu < 0$ (b), and $m - \mu \neq 0$ (c). The shaded regions represent all values of z_{obs} for which the null hypothesis will be rejected.

- Any z_{obs} *outside* the interval ±1.96 *will be* statistically significant whether we use a directional or non-directional test. For example, if $z_{obs} = 2$, then the p-value associated with a non-directional test is $p = .0456$, and the p-value associated with a directional test is $p = .0228$. Both are statistically significant.

- Any z_{obs} *within* the interval ±1.645 *will not be* statistically significant whether we use a directional or non-directional test. For example, if $z_{obs} = 1.6$, then the p-value associated with a non-directional test is $p = .1096$, and the p-value associated with a directional test is $p = .0548$. Neither of these is statistically significant.

These two points tell us that only when $|z_{obs}|$ is between 1.645 and 1.96 will we have suspicions about whether a directional test was reported honestly.

$z_{critical}$ is the z-score that z_{obs} must *exceed* for us to reject the null hypothesis. By *exceed* we mean further from 0 than the criterion. For example, 2.3 exceeds 1.645, and −2.3 exceeds −1.645.

LEARNING CHECK 2

1. Which of the following is a directional hypothesis: (a) $m - \mu > 0$, (b) $m - \mu \neq 0$, (c) $m > \mu$, and (d) $m < \mu$?

2. Use the z-table to determine the exact value of p for each of the following situations involving a prediction and a z_{obs}: (a) $m - \mu < 0$, $z_{obs} = -1.0$; (b) $m - \mu > 0$, $z_{obs} = 1.92$; (c) $m - \mu \neq 0$, $z_{obs} = 2.11$; and (d) $m - \mu \neq 0$, $z_{obs} = 1.6$.

3. Assuming the $p < .05$ criterion for statistical significance, what is the critical value of z for each of the following predictions: (a) $m - \mu > 0$, (b) $m - \mu \neq 0$, (c) $m > \mu$, and (d) $m < \mu$?

4. For each of the following predictions and values of z_{obs}, judge whether the statements about p are true or false: (a) $m - \mu > 0$, $z_{obs} = 1.3$, $p > .05$; (b) $m - \mu \neq 0$, $z_{obs} = 2$, $p > .05$; (c) $m > \mu$, $z_{obs} = -2$, $p < .05$; and (d) $m < \mu$, $z_{obs} = -2$, $p < .05$.

Answers

1. a, c, d.

2. (a) .1587. (b) .0274. (c) 2*.0174 = .0348. (d) 2*.0548 = .1096.

3. (a) 1.645. (b) ±1.96. (c) 1.645. (d) −1.645.

4. (a) True. (b) False. (c) False. (d) True.

THE ALTERNATIVE HYPOTHESIS

The null hypothesis is central to significance tests. In our phonics example, the null hypothesis was that phonics instruction has absolutely no effect on reading scores. According to the null hypothesis, our sample was drawn from a distribution of scores having a mean of 72 and standard deviation of 8. Statisticians usually think of the null hypothesis as a hypothesis about the *mean* of the distribution from which our sample was drawn. Therefore, we might express the null hypothesis as follows:

$$H_0: \mu_0 = 72. \tag{7.2}$$

H_0 is the symbolic name for the null hypothesis, and the colon (:) is a way of saying, "what follows." So, "H_0:" means "the null hypothesis is what follows." What follows in this case is a statement about the mean of the distribution from which a sample was drawn. We call this mean μ_0.

There is a second hypothesis lurking in the phonics example, but it wasn't treated as a hypothesis earlier. This second hypothesis is the prediction made by the researcher. The researcher had reason to believe that the scores of the students receiving additional phonics instruction would be higher, on average, than those receiving no additional phonics instruction. We call her hypothesis the **alternative hypothesis**, or the **research hypothesis**. (These terms are used interchangeably.) Implicit in this alternative hypothesis is the idea that if *all* children were to receive additional phonics instruction, then the distribution of year-end reading scores would have a mean greater than 72.

When we step back from the experiment, it is clear that the researcher ran the experiment because she wanted to know whether the population of all students receiving additional phonics instruction would have a mean higher than that of the regular population. She wasn't interested in simply *describing* the results in the sample. She wanted to *infer* something about the population from which the sample was drawn.

This second population is interesting because, in this example, it does not strictly exist. It might be said to exist only in our imaginations. If we can imagine a population of first-grade students who've all received an additional 20 minutes of phonics instruction daily,

The **alternative hypothesis** (also called the **research hypothesis**) is a hypothesis about the distribution of scores from which a sample was drawn. It is the alternative hypothesis that the researcher believes to be true.

then our (actual) sample is one of a very large number of samples that could have been drawn from this population. The research question is whether the mean reading score of this second population (phonics distribution) is the same as the mean reading score of the known population (whole language distribution). We can refer to the mean of this second (hypothetical) distribution as μ_1.

Because there are two populations to be considered, we can express the null hypothesis as a statement about their means as follows:

$$H_0: \mu_1 = \mu_0. \tag{7.3}$$

That is, according to the null hypothesis (H_0), our sample of scores was drawn from a distribution having a mean equivalent to the mean of the null hypothesis distribution. Therefore, we can also express H_0 as follows:

$$H_0: \mu_1 - \mu_0 = 0. \tag{7.4}$$

The alternative hypothesis (H_1) can be expressed in the same way. However, the nature of H_1 depends on the prediction made by the researcher. In the previous section, we saw that the three possible predictions were $m - \mu > 0$, $m - \mu < 0$, and $m - \mu \neq 0$. If we now express these predictions as statements about the relationship between μ_0 and μ_1, we have the following three possible alternative hypotheses:

$$H_1: \mu_1 - \mu_0 > 0, \tag{7.5}$$
$$H_1: \mu_1 - \mu_0 < 0, \text{ and} \tag{7.6}$$
$$H_1: \mu_1 - \mu_0 \neq 0 \tag{7.7}$$

An alternative hypothesis doesn't say much about this second distribution of scores. It says only where μ_1 lies on the x-axis relative to μ_0. Figure 7.8 illustrates the claims about the relationship between μ_1 and μ_0 expressed in equations 7.5, 7.6, and 7.7. Equation 7.5 says only that $\mu_1 > \mu_0$. There are infinitely many distributions for which this is true. Equation 7.6 says only that $\mu_1 < \mu_0$. Again, infinitely many distributions are consistent with this claim. Equation 7.7 says only that $\mu_1 \neq \mu_0$. Infinitely many distributions are consistent with this claim too.

The introduction of the alternative hypothesis has no effect on how we go about conducting significance tests. The null hypothesis is a known distribution of scores. Because this distribution is known, we can compute the probability of obtaining z_{obs}, or one more extreme, when H_0 is true. The alternative hypothesis is just a new way of expressing our predictions, and it determines how we compute a p-value, as shown in Table 7.3. It is important to keep in mind, however, that we know far less about the alternative hypothesis than the null hypothesis.

Fisher and the Alternative Hypothesis

My hands shake and my brow sweats as I write about the alternative hypothesis in a chapter about significance tests. I worry that zombie Fisher will come back and strike me dead for having polluted his pristine significance tests with the idea of an alternative hypothesis. Fisher viewed the alternative hypothesis as an abomination and rejected the idea that the alternative hypothesis is in any way relevant to significance tests.

One reason for Fisher's refusal to countenance the alternative hypothesis is that doing so would seem to give H_1 the same status as H_0, which it surely should not have. We can't compute the probability of some event under the alternative hypothesis as we can for the null hypothesis. In our reading example, we can say that the probability of obtaining a mean

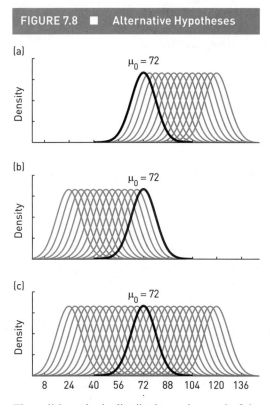

FIGURE 7.8 ■ Alternative Hypotheses

(a)

$\mu_0 = 72$

Density

(b)

$\mu_0 = 72$

Density

(c)

$\mu_0 = 72$

Density

8   24   40   56   72   88   104   120   136

The null hypothesis distribution and several of the infinitely many alternative hypothesis distributions. In (a) through (c), the distribution of scores under the null hypothesis is shown by the solid black line, and possible alternative distributions are shown as the blue lines. (a) The alternative hypothesis states that the mean of the alternative hypothesis distribution is greater than the mean of the null hypothesis distribution. (b) The alternative hypothesis states that the mean of the alternative hypothesis distribution is less than the mean of the null hypothesis distribution. (c) The alternative hypothesis states that the mean of the alternative hypothesis distribution is simply different from the mean of the null hypothesis distribution.

TABLE 7.3 ■ Rules for Computing p

Alternative Hypothesis	Rule: p is the . . .
$H_1: \mu_1 - \mu_0 > 0$	proportion of the z-distribution above z_{obs}.
$H_1: \mu_1 - \mu_0 < 0$	proportion of the z-distribution below z_{obs}.
$H_1: \mu_1 - \mu_0 \neq 0$	proportion of the z-distribution outside $\pm \lvert z_{obs} \rvert$.

of 74.5 or greater is .0062 under the null hypothesis. We cannot say anything about the probability of 74.5 under the alternative hypothesis. To do so, we'd have to know the mean and variance of the alternative hypothesis distribution, and we don't, so we can't. Because we cannot derive probabilities of events under the alternative hypothesis, Fisher would say we must avoid treating it as in any way comparable to the null hypothesis. For Fisher, this meant saying nothing about H_1.

A related reason for Fisher's rejection of the alternative hypothesis is that Jerzy Neyman (whom we met in Chapter 6) had an approach to statistical decision making that rested heavily on full knowledge of the alternative hypothesis—that is, knowing the mean and variance of the alternative hypothesis distribution. (Neyman had an enormous influence on the development of statistics in America in the middle of the 20th century.) Fisher rejected Neyman's model of decision making as in any way relevant to experimental sciences.

Fisher's refusal to acknowledge the importance of the alternative hypothesis puzzled many statisticians (Kruskal, 1980). As we've seen, we can't compute a p-value without making a prediction, which is essentially to specify an alternative hypothesis. This simple observation is reflected in Fisher's famous example of the lady tasting tea. In his example, the null hypothesis is that the lady cannot distinguish between milk-tea mixtures made with milk first or tea first. Although he didn't say so explicitly, Fisher's alternative hypothesis is a directional (one-tailed) hypothesis. Fisher says we'd reject the null hypothesis if the lady puts all four milk-first mixtures on the milk-first table. However, putting all *tea-first mixtures* on the milk-first table would be just as unusual ($p = 1/70 = .014$) as putting all milk-first mixtures on the milk-first table. If she put all tea-first mixtures on the milk-first table, this would strongly suggest an ability to discriminate the two mixtures, but an inability to identify which is which.

Clearly, Fisher's alternative hypothesis (forgive me Sir Ronald) *could* have been that the lady can distinguish the two mixtures but can't necessarily tell which is which. In this case, we would reject the null hypothesis if she puts all milk-first mixtures *or* all tea-first mixtures on the milk-first table. In other words, we would be conducting a two-tailed test. The probability of one of these two outcomes occurring is $2*1/70 = 2/70 = .029$, which is statistically significant by current convention. Not to beat a dead horse, but one can't compute a p-value without a prediction, which we can treat as an alternative hypothesis.

The strangeness of Fisher's refusal to consider an alternative hypothesis is well captured in the following quote from the distinguished statistician William Kruskal:

One of the puzzles in reading Fisher is his silence about the relevance of hypotheses alternative to the hypothesis under test, that is, to the hypothesis under which the sample significance level is computed or the formal test based. [. . .] During a visit by Fisher to the University of Chicago in the early 1950's [. . .] L. J. Savage and I, who had shared a puzzlement at Fisher's apparent obtuseness about alternative hypotheses, determined to try to understand him. We managed to arrange a conversation with the great man in the Mathematics Common Room of Eckhart Hall—just the three of us—and we talked for almost an hour. Savage and I were well aware of Fisher's animadversions about American mathematicians, and of the Fisher-Neyman battle, so we were especially tactful. For example, we did not use such expressions *power and critical region**; indeed we may even have avoided the word alternative. In the end, of course, Fisher agreed that, yes, naturally one had to think about distributions for the sample other than that of the hypothesis under test. And why were we making such a fuss about an elementary and trivial question?! (Kruskal, 1980, p. 1022)

(*Note: We will explain the terms *power* and *critical region* in Chapter 8.)

So, in spite of Fisher's "obtuseness" about alternative hypotheses, we will discuss them frequently.

LEARNING CHECK 3

1. Which of the following is a null hypothesis: (a) $\mu_1 - \mu_0 > 0$, (b) $\mu_1 - \mu_0 \neq 0$, (c) $\mu_1 - \mu_0 = 0$, and (d) $\mu_1 = \mu_0$?

2. Which of the following is a null hypothesis: (a) $\mu_1 = \mu_0$, (b) $\mu_1 \neq \mu_0$, (c) $\mu_1 > \mu_0$, or (d) $\mu_1 < \mu_0$?

3. Connect the formulas that express the same thing.

   (a) $\mu_1 = \mu_0$ (a) $\mu_1 - \mu_0 > 0$

   (b) $\mu_1 \neq \mu_0$ (b) $\mu_1 - \mu_0 < 0$

   (c) $\mu_1 > \mu_0$ (c) $\mu_1 - \mu_0 = 0$

   (d) $\mu_1 < \mu_0$ (d) $\mu_1 - \mu_0 \neq 0$

4. Does the researcher believe that the null hypothesis or the alternative hypothesis is true?

5. Does the researcher know the mean of the alternative hypothesis distribution?

6. Does the researcher know the mean of the null hypothesis distribution?

7. Which of the following is a directional hypothesis: (a) $\mu_1 - \mu_0 > 0$, (b) $\mu_1 - \mu_0 \neq 0$, (c) $\mu_1 > \mu_0$, and (d) $\mu_1 < \mu_0$?

8. Use the *z*-table to determine the exact value of *p* for each of the following situations: (a) H_1: $\mu_1 - \mu_0 < 0$, $z_{obs} = -1$; (b) H_1: $\mu_1 - \mu_0 > 0$, $z_{obs} = 1.92$; (c) H_1: $\mu_1 - \mu_0 \neq 0$, $z_{obs} = 2.11$; (d) H_1: $\mu_1 - \mu_0 \neq 0$, $z_{obs} = 1.6$.

Answers

1 c, d.

2. a.

3. (a, c), (b, d), (c, a), (d, b).

4. The alternative hypothesis.

5. No.

6. Yes.

7. a, c, d.

8. (a) .1587.

   (b) .0274.

   (c) 2 * .0174 = .0348.

   (d) 2 * .0548 = .1096.

p-Values Are Conditional Probabilities

Although *p*-values are at the heart of significance tests, it is very easy to get confused about how to interpret them. Some of our most distinguished researchers have been seduced into saying things about p-values that are simply false (Cohen, 1994). A p-value is a *conditional probability*. (Conditional probabilities were discussed in Appendix 2.3, available at study .sagepub.com/gurnsey.) A p-value is the probability of a statistic more extreme than the one observed, occurring by chance under the null hypothesis. The condition in this case is the truth of the null hypothesis (i.e., that the sample was drawn from the distribution of scores specified by the null hypothesis). We can write the meaning of a p-value as follows:

$$p = \text{Pr}(Statistic \mid Hypothesis).$$

We read this expression as

> p is the probability (Pr) of obtaining our *Statistic*, or one more extreme, by chance, given that (|) the *Hypothesis* is true.

In the phonics example, our statistic was $z_{obs} = 2.5$ and the hypothesis was H_0. We found that the proportion of the z-distribution exceeding $z_{obs} = 2.5$ was .0062. When we insert these things into the expression above, we have

$$.0062 = \text{Pr}(2.5 \mid H_0).$$

A frequent misinterpretation is to think that p-values represent the probability of H_0 being true. For example, one might think (incorrectly!) that $p = .0062$ is the probability of H_0 being true, and the probability it is false is therefore .9938. This interpretation is absolutely wrong because it reverses the conditional probability as follows:

$$p = \text{Pr}(Hypothesis \mid Statistic).$$

The **inverse probability error** is the mistaken belief that the p-value associated with our obtained statistic reflects the probability that H_0 is true.

This misinterpretation means something like p is the probability (Pr) of the *Hypothesis* being true given (|) *Statistic* we obtained. This error is so common that it has a name; it is called the **inverse probability error**. The error is simply mixing up the two terms in the conditional probability statement. If we fill in the quantities we've been working with, we would have

$$.0062 = \text{Pr}(H_0 \mid 2.5).$$

This is wrong, wrong, wrong.

Carver (1978) used the following colorful illustration of the inverse probability error. The expression below states the probability that someone is dead given that they've been hanged:

$$p = \text{Pr}(Dead \mid Hanged).$$

You'd expect this probability to be close to 1, given a reasonably competent executioner. However, if we switch the terms around, the following means something entirely different:

$$p = \text{Pr}(Hanged \mid Dead).$$

This expresses the probability that a person has been hanged given that he or she is dead. The probability that a dead person died from hanging is close to zero, because most people die from causes other than hanging. So, $p = \text{Pr}(Dead \mid Hanged)$ is not the same as $p = \text{Pr}(Hanged \mid Dead)$. In exactly the same way, $p = \text{Pr}(Statistic \mid Hypothesis)$ is not the same as $p = \text{Pr}(Hypothesis \mid Statistic)$. If you're ever confused about the meaning of a p-value, remember this example. A p-value is the conditional probability of the statistic *given the hypothesis*.

One more word on this point: a *p*-value is the proportion of the test statistic distribution under H_0 that *exceeds* the statistic that you've obtained. Depending on the alternative hypothesis, *p* is the proportion above z_{obs}, below z_{obs}, or outside the interval $\pm |z_{obs}|$. *The calculation of the* p*-value assumes that* H_0 *is true.* Therefore, a *p*-value can't be the probability that H_0 is true, because the calculation of a *p*-value is based on the assumption that H_0 *is* true.

I mentioned earlier that even the best statistical minds can slip up and say things about *p*-values that are wrong. *Jacob Cohen* (Figure 7.9) was an eminent psychologist who taught and conducted research for many years in the Psychology Department at New York University. He made many contributions to statistical analyses in psychology, some of which we'll see in subsequent chapters. One of his widely cited papers has the best title ever: "The Earth Is Round (*p* < .05)." (Google it.) In this paper, Cohen presents a partial list of eminent psychologists who have fallen victim to the inverse probability error.

> The inverse probability error in interpreting H_0 is not reserved for the great unwashed, but appears many times in statistical textbooks (although frequently together with the correct interpretation, whose authors apparently think they are interchangeable). Among the distinguished authors making this error are Guilford, Nunnally, Anastasi, Ferguson, and Lindquist. Many examples of this error are given by Robyn Dawes (1988, pp. 70–75); Falk and Greenbaum (1995); Gigerenzer (1993, pp. 316–329), who also nailed R. A. Fisher [for a related error]; and *Oakes (1986, pp. 17–20), who also nailed me for this error (p. 20).* Cohen (1994, p. 999; emphasis added)

If Fisher and Cohen can make this kind of slip, anybody can. So, be careful what you say (and think) about *p*-values.

FIGURE 7.9 ■ Jacob Cohen

Jacob Cohen (1923–1998). I am extremely grateful to Gideon Cohen for providing this photo.

LEARNING CHECK 4

1. Put this expression $p = \Pr(z_{obs}|H_0)$ into words.

2. Determine $p = \Pr(z_{obs}|H_0)$ given the following: H_0: $\mu_1 - \mu_0 = 0$, H_1: $\mu_1 - \mu_0 > 0$, and $z_{obs} = 2$.

3. Which of the following is the correct interpretation of $p = \Pr(z_{obs}|H_0)$:

   (a) *p* is the probability of H_0 being true given z_{obs},

   (b) *p* is the probability of H_0 being true,

   (c) *p* is the probability of H_1 being false,

   (d) *p* is the probability that H_1 is false given z_{obs},

   (e) *p* is the probability that H_1 is true given z_{obs}, or

   (f) *p* is the probability of obtaining a *z* more extreme than z_{obs} by chance given that H_0 is true?

Answers

1. $p = \Pr(z_{obs}|H_0)$ is the probability of obtaining z_{obs} (or one more extreme) by chance when H_0 is true.

2. $p = .0228$.

3. Only f is correct.

USING *s* TO ESTIMATE σ (AN APPROXIMATE *z*-TEST)

In Chapter 6 we saw that when sample size is large, we can use the sample standard deviation (*s*) to estimate the population standard deviation (σ) when computing confidence intervals about a sample mean. This method is slightly imprecise, but the imprecision diminishes as sample size increases. We can take the same approach when conducting significance tests. If our sample is large and σ is *unknown*, we can use *s* to conduct a significance test. So, if our scenario had been different and we knew only $\mu_0 = 72$ but not the standard deviation σ, we could have estimated σ using the standard deviation (*s*) of the 64 scores in our sample. Let's say the sample standard deviation was *s* = 7.8. In this case, we can compute the estimated standard error of the mean, which we denoted s_m in Chapter 6.

We compute s_m as follows:

$$s_m = \frac{s}{\sqrt{n}} = \frac{7.8}{\sqrt{64}} = \frac{7.8}{8} = .975.$$

If we then use s_m in our calculation of z_{obs}, we would obtain the following:

$$z_{obs} = \frac{m - \mu_0}{s_m} = \frac{74.5 - 72}{.975} = 2.56.$$

Because the sample standard deviation is slightly smaller than σ, the result is a slightly inflated z_{obs}. The one-tailed *p*-value associated with $z_{obs} = 2.56$ is *p* = .0052. We can report the results of this approximate test as follows:

<hr>

APA Reporting

An approximate *z*-test based on *n* = 64 scores showed that additional phonics instruction produced a statistically significant increase in reading scores, *z* = 2.56, *p* = .01.

<hr>

If *s* had been greater than σ (e.g., *s* = 8.2), we would have obtained a z_{obs} slightly smaller than 2.5; it would have been 2.44. The one-tailed *p*-value associated with $z_{obs} = 2.44$ is *p* = .0073. Therefore, when we use *s* to estimate σ, we can still conduct an approximate *z*-test. The resulting *p*-values will be somewhat imprecise, as noted above, but not seriously so if sample size is large. We will return to this question in Chapter 10; at that point, we will be able to say something about exactly how imprecise the test will be. For now, we will simply accept this as a useful approximation.

Which σ Does *s* Estimate?

I was not entirely forthcoming when I said that *s* can be used to estimate σ. In the discussion leading up to this section, σ was the known standard deviation of the null hypothesis distribution, which we may denote as σ_0. In our reading example, σ_0 was the standard deviation of the distribution of reading scores obtained over 35 years of whole language instruction. However, in this example, *s* was the standard deviation of the sample of reading scores obtained after 1 year of additional phonics instruction. Therefore, *s* estimates the standard deviation of the alternative hypothesis distribution, which we may denote as σ_1. Therefore, whether *s* estimates σ_1 or σ_0 seems to depend on whether the H_0 is true or false, which of course we don't know.

A way out of this paradox is to simply assume that $\sigma_1 = \sigma_0$. If this is true, then *s* is an estimate of both σ_1 and σ_0, so it doesn't matter whether H_0 is true or false. In fact, exactly this assumption is often made. However, if H_0 is false and $\sigma_1 \neq \sigma_0$, then *s* estimates σ_1, which is not what we described above. In Chapter 9 we will see that this complication does not arise when we conduct significance tests with confidence intervals. Therefore, we will sweep this problem under the rug for the moment.

LEARNING CHECK 5

1. If $\sigma_0 = \sigma_1$, what does s estimate?

2. If $\sigma_0 \neq \sigma_1$, what does s estimate?

3. A research team happens to know that male rats administered a placebo drug will spend 5 minutes per hour grooming. They believe that male rats administered a low dose of the street drug ecstasy will spend more than 5 minutes per hour grooming. They used an approximate significance test to evaluate their belief. They chose 100 male rats, administered a low dose of ecstasy to each, and measured the number of minutes per hour each rat spent grooming. The mean of this sample was $m = 6$, and the standard deviation was

$s = 5$. Use this information to answer the following questions and assume a significance level of .05:

(a) Express H_0 in symbols.

(b) Express H_1 in symbols.

(c) Calculate s_m.

(d) Calculate z_{obs}.

(e) Calculate p.

(f) Should you retain or reject H_0?

(g) When we are conducting this approximate test, what assumption are we making about σ_0 and σ_1?

Answers

1. s estimates both σ_0 and σ_1.

2. s estimates σ_1.

3. (a) H_0: $\mu_1 - \mu_0 = 0$.

   (b) H_0: $\mu_1 - \mu_0 > 0$.

   (c) $s_m = 5/\sqrt{100} = 0.5$.

(d) $z_{obs} = (6 - 5)/0.5 = 2$.

(e) $p \approx .0228$.

(f) We should reject H_0 because $p = .0228 < .05$.

(g) When we conducted this test, we assumed that $\sigma_0 = \sigma_1$.

STATISTICAL SIGNIFICANCE VERSUS PRACTICAL SIGNIFICANCE

Researchers often use the term *significant* (rather than *statistically significant*) to describe statistics that are unusual under the null hypothesis. This is an unfortunate choice of words. The problem is that the word *significant* is used differently in everyday life than in statistics. If you say that the result of an experiment is significant, to the average person this means that the result is important or meaningful. However, results can be statistically significant without being important or meaningful. We will consider two examples to illustrate this point.

Differences in IQ

First, let's say we know that the mean IQ of people born in Quebec, Canada, is $\mu_{Que} = 100$, with a standard deviation of $\sigma = 15$. Now let's say we measured the IQs of a huge random sample of $n = 250,000$ people born across the US border in Maine (I just checked; the population of Maine is greater than 250,000), and we find that their mean IQ is $m_{Maine} = 100.06$. We can perform a two-tailed test of the null hypothesis that

$$H_0:\ \mu_{Maine} - \mu_{Que} = 0$$

against the alternative hypothesis that

$$H_0:\ \mu_{Maine} - \mu_{Que} \neq 0.$$

We will declare our result statistically significant if $p < .05$.

The standard error in this case is

$$\sigma_m = \frac{\sigma}{\sqrt{n}} = \frac{15}{\sqrt{250,000}} = \frac{15}{500} = 0.03.$$

Therefore,

$$z_{obs} = \frac{m_{Maine} - \mu_{Que}}{\sigma_m} = \frac{100.06 - 100}{0.03} = \frac{0.06}{0.03} = 2.$$

If we consult the z-table, we find that the proportion of the standard normal distribution above $z_{obs} = 2$ is .0228, so the proportion below $z_{obs} = -2$ is also .0228. Therefore, the proportion of the z-distribution outside the interval $\pm\left| z_{obs}\right|$ is $p = 2(.0228) = .0456$. Because $p < .05$, we would say that this result is statistically significant.

If I were to say that the difference between μ_{Que} and m_{Maine} is *significant*, the average person, or even the average psychologist, might find this interesting and set about trying to explain this difference. However, an average difference of 0.06 IQ points is really, really small. Expressed as a proportion of the standard deviation ($\sigma = 15$), we find 0.06/15 = 0.004. That is, m_{Maine} is only 1/250 of a standard deviation above μ_{Que}. If I were to make a figure plotting two normal distributions having means of 100 and 100.06, with standard deviations of 15, you would not be able to tell the two distributions apart because the lines would overlap. This is not what most people would think of as a significant difference in the everyday sense of the word. Such a result is neither important nor meaningful, and we'd be considered unhinged if we were to spend time trying to explain it. That is, the result is statistically significant but is of no *practical significance*.

Bench-Pressing 5-Year-Olds

Let's say we know that the average 15-year-old European male can bench-press 125 pounds (56.7 kilograms). An honors student (at an unnamed European university) theorizes that the average 5-year-old can bench-press less than 125 pounds. Therefore, he would like to test the null hypothesis

$$H_0: \mu_{5YO} - \mu_{15YO} = 0$$

against the alternative hypothesis that

$$H_1: \mu_{5YO} - \mu_{15YO} < 0,$$

where μ_{5YO} is the unknown mean of all bench-pressing scores of 5-year-olds. To conduct the test, the student randomly selects one hundred 5-year-old European males and measures how much each can bench-press. He will compute z_{obs} to compare m_{5YO} and μ_{15YO} using the $p < .05$ criterion for statistical significance. The z_{obs} he will compute has $m_{5YO} - \mu_{15YO}$ in the numerator. His alternative hypothesis predicts that this difference will be negative.

After gathering his data, the student found that the mean of the 100 scores was $m_{5YO} = 10$ pounds, with a standard deviation of $s = 2$. Because he doesn't know the standard deviation of either the H_0 or H_1 distributions, he uses s to estimate σ. (Therefore, he assumes $\sigma_{5YO} = \sigma_{15YO}$.) He then carries out the calculations as follows. First he computes

$$s_m = \frac{s}{\sqrt{n}} = \frac{2}{\sqrt{100}} = \frac{2}{10} = 0.2.$$

Then he uses s_m in his calculation of z_{obs} and obtains the following:

$$z_{obs} = \frac{m_{5YO} - \mu_{15YO}}{s_m} = \frac{10 - 125}{0.2} = -575.$$

Our student researcher would like to compute the one-tailed p-value associated with $z_{obs} = -575$, but when he tries to compute it in Excel, the answer he gets is 0. This means that the p-value is so close to 0 that even Excel can't distinguish it from 0. Therefore, he reports, with a certain quiet satisfaction, in his honors thesis that there is, as he predicted, a statistically significant difference between the average bench-pressing ability of 15-year-olds and 5-year-olds.

Now, if you were assessing this thesis, I think you would worry about the sanity of the student conducting the research. There is nothing wrong with his calculations. And he is correct to conclude that there is a statistically significant difference between the average bench-pressing ability of 15-year-olds and 5-year-olds. But this difference is hardly significant in the everyday sense of the word. Any of the 5-year-olds tested could have told you that he can't bench-press as much as the average 15-year-old. Therefore, although this result is statistically significant, it is of no **practical significance**.

Practical Significance

Our examples showed two extreme cases in which differences are statistically significant but of no practical significance. In example 1, a tiny difference in mean IQ (0.06) was statistically significant. In this case, the sample size was enormous ($n = 250{,}000$), so the standard error was tiny (0.03). Dividing a tiny difference by an even tinier standard error produces a large and statistically significant z_{obs} ($z_{obs} = 0.06/0.03 = 2$).

In the second example, our sample size was not particularly large ($n = 100$) but the difference between the two means was enormous ($\mu_{15YO} = 125$ versus $m_{5YO} = 10$), relative to the estimated standard deviation ($s = 2$). Consequently, the resulting z_{obs} was absurdly large.

As we can see, both small and large differences can be statistically significant without having any practical significance. This brings us to the big question, when does a result have practical significance? In contrast to checking for statistical significance (is $p < .05$?), there is no universal answer to the question of practical significance. Whether a result has practical significance depends on the research question, and who is asking it.

Let's think back to our first-grade students who've had an additional 20 minutes of phonics instruction each day. We concluded that this treatment (or intervention) yielded a statistically significant improvement in reading skills. At first glance, this result seems significant in the everyday sense of the word, not just statistically significant. We all want children to be skilled readers, and added phonics instruction seems to improve reading skills. But let's think through different situations in which our result may or may not have practical significance.

We must first ask what a difference of 2.5 points (i.e., $74.5 - 72 = 2.5$) on the reading skill test *means*. If it means that at the end of first grade, students, on average, are reading at the same level as students midway through second grade, then this seems like an important difference. On the other hand, if a difference of 2.5 means that at the end of first grade, students, on average, are reading at the same level as students after the second week of second grade, then this seems like a much less important difference. So, judging the practical significance of a result requires knowing the implications of our measuring instrument.

Next, let's assume for a moment that an improvement of 2.5 points on the reading skill test means a half-grade improvement (i.e., at the end of first grade, students, on average,

Practical significance relates to the interpretation and implications of a statistical analysis. The practical significance of a result rests on how it matters.

are reading at the same level as students midway through second grade). I suggested that we might interpret this as an important difference. But are there considerations that affect our view of its practical significance? One question has to do with the difficulty one faces when changing the curriculum in a school district. In our scenario we considered a small-ish school district with only 64 first-grade students. Most school districts are far larger, and so one would have to consider the costs associated with changing the curriculum. These would include new materials, teacher training, scheduling changes, and possibly other things. Therefore, these considerations will affect our view of the practical significance of the result.

Perhaps, more seriously, it might also be that the additional 20 minutes of phonics instruction took time away from other subjects and thus had a detrimental effect on other parts of the children's education, such as writing, arithmetic, and physical skill development. These sound like terribly mundane concerns, but they are part of the reality of judging the practical significance of any result. One might have an excellent intervention that improves grades but it may be far too expensive or could have adverse consequences elsewhere and is thus impractical to implement on a large scale.

This discussion is not intended to induce a sense of pessimism. However, too often we treat statistical questions without any context. It might seem comforting to judge the results of a test to be statistically significant and consider the job done. However, research is always conducted in a specific context. Statistical analyses are an aid to human judgment, not a replacement for it. The use of context-independent, universal rules, such as $p < .05$, can obscure the meaning, or practical significance, of a result. A result cannot be judged as important or meaningful from its p-value alone.

REVIEW OF SIGNIFICANCE TESTS

Significance tests provide a way to draw inferences about the differences between two populations and, most often, inferences are drawn about the means of two populations. We can now review the steps involved in significance tests and put these steps in some order.

We started with a substantive question that asked about the effect of additional phonics instruction on reading. A researcher predicted that additional phonics instruction would improve reading skills. She decided to test this prediction by comparing reading performance on a standardized reading test for students who had received different amounts of phonics instruction. She posed her prediction in terms of two statistical hypotheses. The null hypothesis was that additional phonics instruction would have no effect on reading scores. The alternative hypothesis was that additional phonics instruction would lead to better reading. To test her prediction, the researcher proposed comparing the mean reading scores of two groups of participants. She knew that the mean reading score in a population having no added phonics instruction was $\mu_0 = 72$, with a standard deviation of $\sigma = 8$. She then gathered data from a sample of 64 students who had received additional phonics instruction, and she found that the mean of these 64 scores was $m = 74.5$. To assess whether the difference between the two means was greater than one would expect from sample-to-sample variation for samples drawn from the same population (i.e., sampling error), she computed a test statistic, $z_{obs} = 2.5$, which was consistent with her prediction and inconsistent with the null hypothesis of no difference. Our researcher consulted the z-table and found that the probability of obtaining a z-score greater than 2.5 by chance is $p = .0062$, which means such a score would occur very rarely when the null hypothesis is true. Therefore, she rejects the null hypothesis and concludes that the results support her prediction. She must now address the practical significance of these results. That is, she

LEARNING CHECK 6

1. A new treatment for depression involves a 1-month stay on a Caribbean island and daily therapy sessions with a team of eight professionals. A researcher knows that people with untreated depression miss an average of 4.3 days of work per month. It was found that a random sample of 121 depressed individuals who underwent the new treatment missed an average of $m = 4.1$ ($s = 1.1$) days of work per month following treatment. An approximate z-test showed that the reduction in days missed following treatment was statistically significant, $z_{obs} = -2$, $p = .0228$. Please assess the practical significance of this result from the point of view of (a) a depressed patient who has to pay for the treatment and (b) a large organization considering offering this treatment as part of its employee compensation package.

2. A new treatment for anxiety involves 10 minutes of guided meditation each morning when employees arrive at work. A researcher knows that people with untreated anxiety miss an average of 4.3 days of work per month. It was found that a random sample of 121 anxious individuals who underwent the new treatment missed an average of $m = 4.1$ ($s = 1.1$) days of work per month following treatment. An approximate z-test showed that the reduction in days missed following treatment was statistically significant, $z_{obs} = -2$, $p = .0228$. Please assess the practical significance of this result from the point of view of (a) an anxious patient who does not have to pay for the treatment, (b) a large organization considering offering this treatment as part of its employee compensation package, and (c) a clinical psychologist interested in treatments for anxiety.

Answers

1. This seems like a very expensive treatment, not to mention the money lost from individuals taking time off work. The reduction of 0.2 days per month seems rather modest, given the expense of the treatment. (a) This result seems to have low practical significance, given that the patient has to foot the bill. (b) The benefits of the treatment would not match the expense. Again, the practical significance is very low.

2. This treatment involves a very modest time commitment, but the benefits are also modest. (a) To an anxious patient, this result may or may not have practical significance. Even though on average the treatment produces 0.2 fewer days missed per month, any individual is not guaranteed to miss 0.2 fewer days per month. Furthermore, individuals may differ in how they feel about guided meditation. (b) One would have to compare the time lost to meditation with the reduction of days missed. An average reduction of 0.2 days per month corresponds to about 1.6 hours per month. Ten minutes a day corresponds to about 3.33 hours per month. So, this doesn't seem like a great deal because more hours per month are lost to meditation than to untreated anxiety. (c) The result may seem quite promising. The reduction in days missed suggests that the treatment had some effect, and it might be that the effects measured on different variables (such as happiness or well-being) would be more substantial.

must decide if the results suggest that a change to the school district's reading curriculum would be worthwhile. There are many practical issues to consider, and more extensive studies may be needed.

Table 7.4 summarizes the general steps in a significance test and shows how these steps were manifested in the phonics scenario that we have been considering. Of the eight steps, five (Steps 3 to 7) are statistical. Step 1 expresses the substantive question derived from the research literature. Step 2 relates to the question of operationalizing a psychological construct. In this example, the psychological construct is reading skill, and it is operationalized by the end-of-year reading test. Step 8 concerns practical significance. Assessing practical significance requires human judgment about what action or change in theory should follow from the results of the study.

TABLE 7.4 ■ The Steps Involved in a Significance Test	
General Method	**Phonics Scenario**
Step 1. Pose a substantive question related to the effect of some experimental manipulation or comparison between two populations, and make a prediction about the outcome.	Does adding 20 minutes of phonics instruction affect reading performance? We predict that it does.
Step 2. Choose a variable to assess that relates to the substantive question you're asking.	The variable is performance on a standardized, end-of-year reading test.
Step 3. Express the substantive question and prediction as two statistical hypotheses.	The null hypothesis is that the sample was drawn from a distribution with a mean of 72 and standard deviation of 8. The alternative hypothesis is that the sample was drawn from a distribution with a mean greater than 72. These hypotheses are expressed formally as follows: $H: \mu_1 - \mu_0 = 0$ and $H_1: \mu_1 - \mu_0 > 0$.
Step 4. Choose a sample statistic to assess the statistical hypotheses and specify its distribution under the null hypothesis.	The sample statistic is the sample mean. The distribution of this statistic is a normal distribution of sample means, having a mean of 72 and standard error of $\sigma = \sigma/\sqrt{n} = 8/8 = 1$.
Step 5. Gather data and compute the sample statistic.	The sample statistic is $m = 74.5$.
Step 6. Transform the sample statistic into a test statistic and compute its p-value under the null hypothesis.	The test statistic was $z_{obs} = (m - \mu_0)/\sigma_m = (74.5 - 72)/1 = 2.5$. Its p-value is $p = .0062$.
Step 7. Draw a statistical conclusion: If the p-value is less than .05, reject the null hypothesis and treat this as support for your alternative hypothesis; otherwise, retain the null hypothesis.	Because it would be very unusual to obtain $p = .0062$ when the null hypothesis is true, we say the result is statistically significant. Therefore, we reject the null hypothesis and interpret this as support for the alternative hypothesis.
Step 8. Draw an inference about the second population and assess the practical significance of the result.	We conclude that the sample was drawn from a distribution with a mean greater than 72. Interpreting the obtained result requires relating it to many practical concerns, such as what the measured difference implies about how much reading has improved, and whether the improvement justifies the costs associated with implementing a curriculum change.

SUMMARY

A significance test uses a sample statistic to assess which of two hypotheses is a more plausible account of the distribution of scores from which the corresponding sample was drawn. One hypothesis is the *null hypothesis* (H_0) and the other is the *alternative hypothesis* (H_1). The distribution of scores under H_0 is completely known and so the distribution of the statistic under H_0 is also known. Although the exact nature of the distribution of scores under H_1 is not known, hypotheses can be made about how its mean relates to the mean of the distribution of scores under H_0. Researchers test H_0 in hopes of finding it implausible and in doing so produce support for H_1. To test H_0, we determine the probability of obtaining our statistic, or one more extreme, by chance under H_0. We call this probability a p-*value*. We say that a small p-value is *statistically significant* because it indicates that our statistic has a low probability of occurring by chance under H_0. In this case, we reject H_0 as a plausible hypothesis about the distribution from which our sample was drawn, and conclude instead that H_1 is more plausible.

In this chapter, we used the sample mean (m) as a basis for judging the plausibility of the H_0 mean, μ_0. To do this, a test-statistic is computed as follows.

$$z_{obs} = \frac{m - \mu_0}{\sigma_m},$$

and then its probability under H_0 is determined by consulting the z-table. If there is a low probability of obtaining z-scores more extreme than z_{obs} by chance under H_0, we reject H_0.

If H_1 makes a prediction about whether the mean of the alternative distribution (μ_1) is greater or less than μ_0, then it is said to be a *directional hypothesis*. If H_1 simply predicts that μ_1 does not equal μ_0, then it is said to be a *non-directional hypothesis*. The p-value required to reject H_0 is called the *significance level* of the test. By convention, the significance level is set at .05. According to this convention, a statistically significant z_{obs} is one for which $p < .05$ under H_0.

The p-value associated with z_{obs} depends on H_1. If H_1 predicts $\mu_1 > \mu_0$, then p is the proportion of the z-distribution above z_{obs}. If H_1 predicts $\mu_1 < \mu_0$, then p is the proportion of the z-distribution below z_{obs}. If H_1 predicts $\mu_1 \neq \mu_0$, then p is the proportion of the z-distribution outside the interval $\pm|z_{obs}|$.

It is important to remember that p-values are *conditional probabilities* that express the probability of obtaining the observed statistic, or one more extreme, by chance, when H_0 is true. The *inverse probability error* is the mistaken belief that a p-value is the probability that H_0 is true.

Through most of the chapter it was assumed that the H_0 distribution is normal, with known μ_0 and σ_0. If we don't know σ_0, we can use the standard deviation of a sample (s) to compute an approximate z_{obs} as follows:

$$z_{obs} = \frac{m - \mu_0}{s_m},$$

where $s_m = s/\sqrt{n}$.

A difference between m and μ_0 may be statistically significant ($p < .05$) but of no *practical significance*. The practical significance of a result rests in how it matters and how it relates to the interpretation and implications of the result.

KEY TERMS

alternative hypothesis 160	practical significance 169	statistically significant 154
directional tests 156	p-value 154	test statistic 153
inverse probability error 164	research hypothesis 160	two-tailed tests 157
non-directional tests 157	retaining the null hypothesis 155	$z_{critical}$ 159
null hypothesis 153	significance level 155	
one-tailed tests 156	significance test 151	

EXERCISES

Definitions and Concepts

Define the following concepts.

1. The null hypothesis

2. The alternative hypothesis

3. Statistical significance

4. *p*

5. Significance test

6. Directional hypothesis

7. Non-directional hypothesis

8. Significance level

9. Retaining H_0

10. The inverse probability error

11. Practical significance

True or False

State whether the following statements are true or false.

12. Large p-values lead us to reject H_0.

13. Large p-values lead us to retain H_0.

14. If H_1: $\mu_1 - \mu_0 < 0$, then $z_{obs} = 1$ has a lower p-value than $z_{obs} = -1$.

15. If I reject the null hypothesis, this means that the null hypothesis is false.

16. $\mu_1 > \mu_0$ means that $\mu_1 - \mu_0 < 0$.

17. If H_1: $\mu_1 - \mu_0 < 0$ and $z_{obs} = -1$, then $p = .8413$.

18. $p = \Pr(Statistic \mid H_0)$ expresses the probability that H_0 is true.

19. Researchers typically believe that H_0 is true.

Calculations

20. If $\mu_0 = 32$, $\sigma_0 = 10$, $m = 27$, and $n = 25$, calculate (a) z_{obs} and (b) p for z_{obs}, assuming H_1: $\mu_1 - \mu_0 < 0$.

21. If $\mu_0 = 32$, $\sigma_0 = 10$, $m = 33$, and $n = 25$, calculate (a) z_{obs} and (b) p for z_{obs}, assuming H_1: $\mu_1 - \mu_0 > 0$.

22. If $\mu_0 = 54$, $s = 16$, $m = 58$, and $n = 64$, calculate (a) z_{obs} and (b) p for z_{obs}, assuming H_1: $\mu_1 - \mu_0 \neq 0$.

Scenarios

23. A perception researcher believes that the color of paper on which an exam is printed will influence the anxiety that students feel during an exam. Over the last 10 years, all of his tests were printed on white paper. Students in his classes scored 75% on average on his final exams, with a standard deviation of 3.5. This year he printed his final exam on pink paper. He predicted that this would have a soothing influence on students, which would lead to higher test scores. There were 49 students in his class. Please answer the following questions.

(a) What is H_0?

(b) What is H_1?

(c) The mean grade on his final exam was $m = 68$. Calculate z_{obs}.

(d) What p-value is associated with z_{obs}?

(e) If we adopt the $p < .05$ significance level, is z_{obs} statistically significant?

(f) Should you retain or reject H_0?

(g) Is the researcher's hypothesis supported?

(h) What do you conclude about the effect of pink paper on test scores?

(i) How would you explain these results?

24. The inferotemporal cortex (IT) is thought to play an important role in visual object recognition in human and non-human primates. A researcher knows that macaque monkeys can learn an object discrimination task in 10 trials, on average, with a standard deviation of $\sigma = 4$ trials. A surgical intervention makes it possible to deactivate IT temporarily, and the researcher believes that doing so will lead to longer learning times. A random sample of four adult male macaques underwent this procedure. Following recovery, the monkeys learned the object discrimination task while IT was deactivated. Please answer the following questions.

(a) What is H_0?

(b) What is H_1?

(c) The mean number of trials to learn the object discrimination task was $m = 14$ trials. Calculate z_{obs}.

(d) What p-value is associated with z_{obs}?

(e) If we adopt the $p < .05$ significance level, is z_{obs} statistically significant?

(f) Should you retain or reject H_0?

(g) Is the researcher's hypothesis supported?

(h) What do you conclude about the effect of deactivating IT on object discrimination learning?

25. College students read 450 words per minute, on average. A free, online training program claims to double reading speeds after 3 hours of training. A psychology student does not believe that reading speeds will double following 3 hours of training. That is, she thinks that reading speeds after training will be less than 900 words per minute, on average. Therefore, she would like to assess

the claim using an online reading speed assessment tool provided by Staples. (Please Google "Staples Reading Test.") The student chose a random sample of 121 students from her department's participant pool and had each participant follow the online training program for 3 hours. Afterward, each student took the Staples reading test. Please answer the following questions.

(a) Test the null hypothesis that $\mu_0 = 900$ using a significance level of $\alpha = .05$.

   (i) What is H_0?

   (ii) What is H_1? (Think carefully about whether this is a directional test.)

   (iii) The mean reading speed was $m = 650$ words per minute with a standard deviation of 165. Calculate the approximate z_{obs}.

   (iv) What p-value is associated with z_{obs}?

   (v) If we adopt the $p < .05$ significance level, is z_{obs} statistically significant?

   (vi) Should you retain or reject H_0?

   (vii) What do you conclude about the claim that the training method doubles reading speeds?

   (viii) Comment on the practical significance of this result.

   (ix) Would you recommend this training program to a friend? Explain your answer.

(b) Let's say that the student simply wondered if the training program had any effect at all on reading speed. Test the null hypothesis that $\mu_0 = 450$ using a significance level of $\alpha = .05$.

   (i) What is H_0?

   (ii) What is H_1? (Think carefully about whether this is a directional test.)

   (iii) The mean reading speed was $m = 650$ words per minute with a standard deviation of 165. Calculate the approximate z_{obs}.

   (iv) What p-value is associated with z_{obs}?

   (v) If we adopt the $p < .05$ significance level, is z_{obs} statistically significant?

   (vi) Should you retain or reject H_0?

   (vii) Comment on the practical significance of this result.

26. Professor Hebb ran an experiment with 100 participants. He calculated a z_{obs} that had a probability of $p = .048$. He concluded that the null hypothesis has only a 4.8% chance of being true. Please comment on Professor Hebb's conclusion.

APPENDIX 7.1: SIGNIFICANCE TESTS IN EXCEL

This appendix shows how to set up an Excel workbook to conduct a one-sample significance test when the population standard deviation is known or sample size is large (see Figure 7.A1.1). Cells **B2** to **B5** contain the four numbers required to conduct the test. These are **m**, $\mathbf{\mu_0}$, $\mathbf{\sigma}$, and **n**, respectively. We first calculate σ_m in cell **B8** and then z_{obs} in cell **B9**. The one- and two-tailed p-values are computed in cells **B10** and **B11**, respectively.

The one-tailed p-value has been computed in a very general way, but assumes that we've made the right prediction about the sign of z_{obs}. That is, the same p will be computed for positive and negative values of z_{obs}. To accomplish this, we take the absolute value of our z_{obs} to produce a positive number. We then make this number negative and determine the proportion of the z-distribution below it using **NORM.S.INV**. The two-tailed p-value is simply two times the one-tailed probability (cell **B11**).

FIGURE 7.A1.1 ■ One-Sample Significance Test

	A	B	C
1	**Quantities**	**Values**	**Formulas**
2	m	74.5	
3	μ_0	72	
4	σ	8	
5	n	64	
6			
7	m-μ_0	2.5	=B2-B3
8	σ_m	1	=B4/SQRT(B5)
9	z_{obs}	2.5	=B7/B8
10	$P_{(one-tailed)}$	0.0062	=NORM.S.DIST(-ABS(B9),1)
11	$P_{(two-tailed)}$	0.0124	=2*B10

Conducting a one-sample significance test in Excel. Cells **B2** to **B5** contain the information required to conduct the test. From these numbers we compute σ_m and z_{obs} in cells **B8** and **B9**, respectively. The one- and two-tailed p-values are computed in cells **B10** and **B11**, respectively.

8

DECISIONS, POWER, EFFECT SIZE, AND THE HYBRID MODEL

INTRODUCTION

In 2012, Amy Cuddy gave a compelling TED talk in which she argued that our body language affects not only others' perceptions of us but our self-perception and behavior as well. A portion of the talk was based on a research report by Carney, Cuddy, and Yap (2010), in which they claimed that assuming power poses (e.g., fists on hips, or arms raised in the air; Figure 8.1) for as little as 2 minutes can increase levels of testosterone, decrease levels of the stress hormone cortisol, and increase risk-taking behavior.

FIGURE 8.1 ■ A Power Pose

iStock.com/Maridav

Imagine that a final-year psychology student was inspired by the talk and was particularly interested in the behavioral changes associated with power poses. From the results of Carney et al. (2010), he theorized that the reduction in cortisol would reduce the stress that students feel in exams, which would lead to improved exam performance. The student brought this idea to a senior professor in his department who had been teaching a neuroscience course for the last 25 years. Together they devised a way to assess the student's prediction. The professor had taught thousands of students over the last 25 years, and his records showed that the distribution of final exam grades in his course had a mean of $\mu = 75$, with a standard deviation of $\sigma = 10$. The student and professor received ethical approval to have his current class of 64 students assume a power pose for 2 minutes before writing the final exam.

Before running the experiment, the student asked his professor whether 64 students would be a large enough sample. The professor assured him that a sample of 64 was huge, and certainly more than enough for this simple experiment. In fact, the professor said that most of his experiments with rats involve fewer than 10 subjects.

The student was not reassured by his professor, so he described the experiment to his stats teaching assistant (TA) and asked the same question about sample size. The TA's response was a question: "How much do you think power posing will raise grades?" The student hadn't thought about this before, but he ventured that a 2-percentage-point improvement would be interesting. "In that case," said the TA, "I think you'd need about 155 participants in your study."

In this chapter, we will see that statistical tests can vary in power, and we will use the concept of *power* to explain how the stats TA came up with her answer. The tools for controlling statistical power were introduced by Jerzy Neyman and Egon Pearson in the late 1920s and early 1930s in their model of statistical decision making (Neyman & Pearson, 1933). There are similarities between the Neyman-Pearson model of decision making and the significance tests of Sir

Ronald Fisher. However, the differences between the views of Fisher and Neyman-Pearson are important, and the debates between the two camps were fierce. By fierce I mean nasty. For example, Fisher (1955) described Neyman's work on confidence intervals as "childish."

Most psychologists of my generation were taught an approach to statistical decision making that combines the Fisherian and Neyman-Pearson methods and blurs the distinctions between the two. Many authors have noted that neither Fisher nor Neyman would have approved of this hybrid (e.g., Gigerenzer, 2004). These authors also believe that blurring the distinction between them can lead to serious misunderstandings. Although this is true, the hybrid model is not completely unreasonable, and explaining the model is the goal of this chapter.

STATISTICAL DECISIONS

Decisions and Decision Errors

Over the years, the $p < .05$ criterion for retaining or rejecting the null hypothesis became tightly bound with the idea that statistics is about making *decisions*. That is, we classify research outcomes as statistically significant or not statistically significant. Although Fisher later rejected the $p < .05$ criterion, it seems to have appealed to a basic human need to have clear rules to follow when making decisions and remains the default definition of statistical significance.

When we view significance tests as tools for making statistical decisions, there are two ways that our decisions can be wrong. The first type of decision error is to reject the null hypothesis when it is true. In our examples in Chapter 7, the null hypothesis was a hypothesis about the mean of a distribution of scores (μ_0) from which our sample was drawn. According to the $p < .05$ criterion, we reject this hypothesis if our result is one that would occur less than 5% of the time when the null hypothesis is true. Of course, 5% of the time does not mean never. Therefore, 5% of the time, a random sample from the null hypothesis distribution will produce a z_{obs} for which $p < .05$. Therefore, 5% of the time the null hypothesis will be rejected when it is true. We call this kind of decision error a **Type I error**.

The second kind of error occurs when we retain the null hypothesis when it is false. We call this kind of decision error a **Type II error**. Saying that the null hypothesis is false means that the sample of scores was not drawn from the null hypothesis distribution, but some other distribution. In Chapter 7 we called this other distribution the *alternative hypothesis distribution*.

Error Probabilities and Power

In the power-pose experiment, it is clear that the student believes (or predicts) that the population of all students holding a power pose has a mean higher than that of the regular population; i.e., he believes $\mu_1 - \mu_0 > 0$. He hopes to gain support for this hypothesis by rejecting the hypothesis that his sample of scores was drawn from the null hypothesis distribution with $\mu_0 = 75$. When the student initially designed the experiment with the neuroscience professor, he hadn't given much thought to what μ_1 might be, just that it was greater than μ_0. With the encouragement of his stats TA, the student concluded an increase of 2 percentage points on a final exam would be interesting. That is, he concluded that $\mu_1 = 77$ would be interesting.

If the student thinks $\mu_1 = 77$ would be interesting, his experiment is really designed to make a decision about which of two distributions his sample came from. One distribution has a mean of $\mu_0 = 75$ and the other has a mean of $\mu_1 = 77$. We said at the beginning of the

A **Type I error** occurs if the null hypothesis is rejected when it is true. This means that a sample was drawn from the distribution of scores specified by the null hypothesis, yet the statistic was statistically significant.

A **Type II error** occurs if the null hypothesis is retained when it is false. This means that a sample was not drawn from the distribution of scores specified by the null hypothesis, yet the null hypothesis was retained.

FIGURE 8.2 ■ Type I Errors

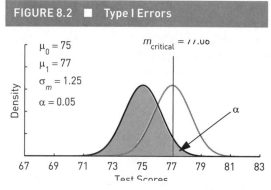

The distribution of means under the null and alternative hypotheses. The left distribution is the distribution of means under the null hypothesis, and that on the right is the distribution of means under the alternative hypothesis. All sample means above m_{critical} fall into the rejection region. The proportion of null hypothesis distribution of means above m_{critical} is referred to as α, which is the probability of a Type I error.

m_{critical} is the value that our sample mean must *exceed* in order for us to reject the null hypothesis.

chapter that the distribution of scores under the null hypothesis has a standard deviation of $\sigma = 10$. If we follow the advice of the student's professor, our sample size would be $n = 64$. With these numbers we will be able to work through the probability of Type I and Type II errors.

The Probability of a Type I Error: α

Figure 8.2 illustrates the state of affairs in our example. The distribution on the left (black curve) is the *distribution of means* under the null hypothesis. It has a mean of $\mu_0 = 75$ and standard error of $\sigma_m = \sigma/\sqrt{n} = 10/8 = 1.25$. The distribution on the right (light-gray curve) is the distribution of means under the alternative hypothesis. It has a mean of $\mu_1 = 77$ and standard error of $\sigma_m = 1.25$. (We will assume that both distributions of scores have the same standard deviation of $\sigma = 10$ and hence the same standard error.)

The focus of Figure 8.2 is the distribution of means under the null hypothesis, which is divided into two parts, separated by the line labeled m_{critical}. Any sample mean greater than m_{critical} would be statistically significant because m_{critical} is the sample mean that corresponds to z_{critical} (see Table 7.2). When the alternative hypothesis is $H_1: \mu_1 - \mu_0 > 0$, as in this example, then $z_{\text{critical}} = 1.645$. The sample mean corresponding to z_{critical} is m_{critical}, which is defined as follows:

$$m_{\text{critical}} = \mu_0 + z_{\text{critical}}(\sigma_m). \tag{8.1}$$

When numbers are inserted into equation 8.1, we find

$$m_{\text{critical}} = \mu_0 + z_{\text{critical}}(\sigma_m) = 75 + 1.645*(1.25) = 77.06.$$

Or, put the other way around,

$$z_{\text{crtical}} = \frac{m_{\text{crtitcal}} - \mu_0}{\sigma_m} = \frac{77.06 - 75}{1.25} = 1.645.$$

Any sample mean that exceeds m_{critical} is said to fall into the *rejection region*, because it will produce a z_{obs} that exceeds z_{critical} and thus a *p*-value less than .05. The proportion of the null hypothesis distribution of means above m_{critical} is .05 and is shown as the gray region in Figure 8.2. We refer to this proportion as α, which is the probability of a Type I error; i.e., the probability of rejecting the null hypothesis when it is true is $\alpha = .05$.

We use the symbol α to denote the probability of a Type I error.

The Probability of a Type II Error: β

Figure 8.3 switches focus to the distribution of means under the alternative hypothesis. This distribution too is divided into two parts, separated by the line labeled m_{critical}. Any sample mean *less than* m_{critical} would produce z_{obs} less than z_{critical}, which would not be statistically significant. If the sample were in fact drawn from the alternative hypothesis distribution (i.e., the alternative hypothesis is true), then means falling below m_{critical} would be Type II errors. The probability of a Type II error is the proportion of the alternative hypothesis distribution of means below m_{critical}. We refer to this proportion as β.

We use the symbol β to denote the probability of a Type II error.

The Probability of Correctly Rejecting the Null Hypothesis: Power

The proportion of the alternative hypothesis distribution *above* m_{critical} corresponds to the probability of correctly rejecting the null hypothesis. This means that the sample was drawn from the alternative hypothesis distribution and we correctly reject the null hypothesis. We refer to the proportion of the alternative hypothesis distribution above $m_{critical}$ as **power**. Because β and power are the proportions of the alternative hypothesis distribution of means on either side of $m_{critical}$, there is a very simple relationship between them:

$$\text{power} = 1 - \beta$$

and

$$\beta = 1 - \text{power}.$$

Even though we haven't seen how to compute β and power, Figure 8.3 shows that they are approximately equal in this example. (We'll see later that $\beta = .52$, and power = .48.) This is really terrible because it means that about 50% of the time when the alternative hypothesis is true, you will make a Type II error and retain the null hypothesis. Put the other way around, only about 50% of the time will the null hypothesis be correctly rejected. Researchers should want to run experiments with high power so that they won't waste time and resources. Therefore, our main concern will be how to increase power. Before we address this question, however, we will have a look at how to compute β and power and then say a few things about the historical context in which the concept of power emerged.

Calculating Power and β

Although power and β are new concepts, the tools we use to calculate them are very familiar. In Chapters 4 to 7 we did many *area-under-the-curve* problems. Given a normal distribution with known mean and standard deviation, we were able to use the z-table to determine the proportion of the distribution above or below any value of the variable in question. If we look at the H_1 distribution in Figure 8.3 this way, we can see that determining power and β are just area-under-the-curve problems relating to the alternative hypothesis distribution. There are five quantities needed to compute power and β, and once we have these, we're back on familiar ground.

The five quantities needed to compute power and β are μ_0, μ_1, σ, n, and α. In this example, these quantities are as follows: $\mu_0 = 75$, $\mu_1 = 77$, $\sigma = 10$, $n = 64$, and $\alpha = .05$. Power also depends on the nature of H_1. For the moment, we will consider only the case in which the alternative hypothesis states that $\mu_1 > \mu_0$ (or equivalently, $\mu_1 - \mu_0 > 0$). In this case, power is the proportion of the distribution of means under H_1 *above* $m_{critical}$ (Figure 8.3, blue region), and β is the proportion of the distribution of means under H_1 *below* $m_{critical}$ (Figure 8.3, gray region). To find power or β, we simply convert $m_{critical}$ to a z-score *on the distribution of means under H_1* and then consult the z-table on page 91. Below are the steps we follow.

Step 1. Find $m_{critical}$. If $\alpha = .05$ and $H_1: \mu_1 - \mu_0 > 0$, then $z_{critical} = 1.645$. (We've memorized this fact, but we could have also determined this from the z-table.) If $\sigma = 10$ and $n = 64$, then $\sigma_m = 1.25$. With these quantities, we can use equation 8.1 to determine

$$m_{critical} = \mu_0 + z_{critical}(\sigma_m) = 75 + 1.645(1.25) = 77.06.$$

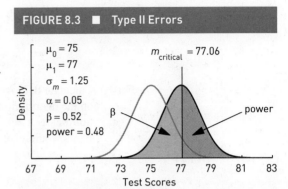

FIGURE 8.3 ■ Type II Errors

$\mu_0 = 75$
$\mu_1 = 77$
$\sigma_m = 1.25$
$\alpha = 0.05$
$\beta = 0.52$
power = 0.48
$m_{critical} = 77.06$

The distribution of means under the null and alternative hypotheses. The left distribution is the distribution of means under the null hypothesis, and that on the right is the distribution of means under the alternative hypothesis. All sample means above $m_{critical}$ fall into the rejection region. The proportion of alternative hypothesis distribution of means below $m_{critical}$ is referred to as β, which is the probability of a Type II error. The proportion above $m_{critical}$ is referred to as power, which is the probability of correctly rejecting the null hypothesis.

Power is the probability of rejecting the null hypothesis when it is false. That is, power is the probability of correctly rejecting the null hypothesis.

Step 2. Convert $m_{critical}$ to a z-score on distribution of means under H_1. This is done in the usual way:

$$z = (m_{critical} - \mu_1)/\sigma_m = (77.06 - 77)/1.25 = .0476.$$

Step 3. Calculate β. Now that we know that $m_{critical}$ corresponds to a z-score of 0.0476 on the distribution of means under H, we consult the z-table to find the proportion below $z = 0.0476$. (To use the z-table, we round 0.0476 to 0.05. Using Excel would give a more precise answer.) This gives

$$\beta = P(z) = P(0.05) = .5199.$$

Step 4. Calculate power. Power is just $1 - \beta$. Therefore, we find that

$$power = 1 - \beta = 1 - .5199 = .4801.$$

These calculations show that power = .48 and β = .52. This means that if we draw a random sample of size $n = 64$ from the distribution of scores specified by H_1, then the probability is only .4801 (power) that the sample mean (m) will exceed $m_{critical}$ and .52 (β) that it will not. If our student was only interested in an H_1 distribution that has a mean of 77 or greater, then the experiment he planned to run had only about a 48% chance of rejecting H_0 when the sample was drawn from this distribution (i.e., when $\mu_1 = 77$). Obviously, this is not a good situation. So the obvious question is what can be done to increase power? Before we examine this question, we will review a little history to understand how β and power became part of researchers' lexicon.

LEARNING CHECK 1

1. If $\mu_0 = 30$ and $\sigma = 16$, calculate $m_{critical}$ for each of the following cases, when $H_1: \mu_1 > \mu_0$: (a) $n = 16$ and $\alpha = .05$, (b) $n = 16$ and $\alpha = .01$, (c) $n = 64$ and $\alpha = .05$, and (d) $n = 64$ and $\alpha = .01$.

2. What symbol is used to denote the probability of a Type I error?

3. If $\alpha = .05$, what is the probability of retaining H_0 when it is true?

4. If $\alpha = .1$, what is the probability of retaining H_0 when it is true?

5. If the probability of retaining H_0 when it is true is .3, what is α?

6. If the probability of retaining H_0 when it is true is .9, what is α?

7. Compute power and β, when $H_1: \mu_1 > \mu_0$, $\mu_0 = 45$, $\mu_1 = 48$, $\sigma = 15$, $n = 25$, and $\alpha = .05$.

8. Compute power and β, when $H_1: \mu_1 > \mu_0$, $\mu_0 = 45$, $\mu_1 = 46.8$, $\sigma = 15$, $n = 36$, and $\alpha = .05$.

9. If I run a significance test, do I want power to be high or low?

Answers

1. Values for $m_{critical} = \mu_0 + z_{critical}(\sigma_m)$ equal the following:

   (a) $30 + 1.645(4) = 36.58$.

   (b) $30 + 2.33(4) = 39.32$.

   (c) $30 + 1.645(2) = 33.29$.

   (d) $30 + 2.33(2) = 34.66$.

2. α.

3. .95.

4. .9.

5. .7.

6. .1.

7. $\beta = .74$ and power = .26.

8. $\beta = .82$ and power = .18.

9. High! There is no point in running an experiment with low power.

NEYMAN AND PEARSON

Decision errors, α, β, and power are concepts that come from a model of statistical decision making developed by Jerzy Neyman (whom we met in Chapter 6) and Egon Pearson (see Figure 8.4). The core ideas in the Neyman-Pearson model grew out of correspondence between Egon Pearson and William Sealy Gosset, whom we will meet in Chapter 10. Pearson had written to Gosset and posed a question about the justification for using a statistic that Gosset had developed called the *t*-statistic. (In Chapter 10, we will see that the *t*-statistic is central to estimation and significance tests for small samples.) In the following passage, Pearson describes their communication and the role that Gosset played in the development of the Neyman-Pearson model:

> At that time I had been trying to discover some principle beyond that of practical expediency which would justify the use of [the *t*-statistic] in testing the hypothesis that the mean of the sampled population was [μ]. Gosset's reply had a tremendous influence on the direction of my subsequent work, for the first paragraph contains the germ of that idea which has formed the basis of all the later joint researches of Neyman and myself. *It is the simple suggestion that the only valid reason for rejecting a statistical hypothesis is that some alternative hypothesis explains the observed events with a greater degree of probability.* (Pearson, 1939, p. 242, *emphasis added*)

The Neyman-Pearson model, then, addresses the question of which of two distributions best explains an observed result. Of course the details of their analysis get into some pretty hairy mathematics, so we won't pursue them. However, we can say that researchers saw how Neyman-Pearson notions of decision errors and power apply to research questions in psychology and related disciplines.

We will see that power is now recognized as an important consideration when designing experiments and has been grafted onto Fisher's significance tests. Most components of the Neyman-Pearson model were described above. However, to distinguish it from Fisher's significance tests, we will think about an application of the model that even Sir Ronald would find acceptable (Fisher, 1955).

The example that we will consider relates to a quality control operation called an *acceptance procedure*. We will apply this notion to problems faced by a hypothetical company that manufactures potato chips. In this example, we will assume that an extensive (and expensive) analysis shows that a properly manufactured bag of potato chips weighs 55 grams on average. However, not every properly manufactured bag of chips weighs exactly 55 grams, so there is some variation in their weights. This variation is characterized by a standard deviation of 1 gram. When a specific component of the chip-making machine (called the knurled piston*) breaks, the chips are cut too thickly and thus quality is not what it should be. (Consumers can be very sensitive to these things.) Because the chips are cut too thickly, bags of chips will weigh more than they should. The same

---

*Not really.

FIGURE 8.4 ■ Neyman and Pearson

Neyman and Pearson in 1935. Jerzy Neyman (*left*) collaborated with Egon Pearson (*right*) at the Biometrics Laboratory at University College London. At that time, the Biometrics lab was co-directed by Egon Pearson and Sir Ronald Fisher, and early on Neyman had a congenial relationship with Sir Ronald. Their relationship was fractured irreparably when Fisher harshly attacked Neyman following a presentation Neyman gave to the Royal Statistical Society in 1935; somewhat different accounts of this event are given in Fisher-Box (1978) and Reid (1982). According to most accounts, Fisher opened his remarks by saying, in essence, that Neyman didn't know what he was talking about. Things went downhill from there and never recovered.

From Reid (1998). Reproduced with permission.

extensive analysis shows that when the knurled piston breaks, a bag of chips weighs 57 grams on average, with a standard deviation of 1.

Because the weight of a bag of chips provides a clue to whether the knurled piston is functioning properly, this information can be used in the future to check its state. Instead of checking every bag of chips produced by the machine, which might be costly and time-consuming, one can take a single bag of chips off the conveyor belt from time to time and weigh it. The weight of the chip will affect one's judgment about whether the knurled piston is broken. We know everything we need to know about the distribution of weights when the knurled piston is working properly (let's call this the H_A distribution, $\mu_A = 55$, $\sigma = 1$) and when it's not (let's call this the H_R distribution, $\mu_R = 57$, $\sigma = 1$). The subscripts A and R stand for *accept* (the knurled piston is working just fine) and *reject* (the knurled piston is broken).

So, how do we use the weight of the bag of chips we've selected to make a decision about the knurled piston? Because these two distributions overlap, the only thing we can do is set a *criterion*, or threshold. We will decide that something has gone wrong with the process if the bag's weight exceeds the criterion, and we will judge that all is well if the bag's weight does not exceed the criterion. If we judge that something has gone wrong, we must then shut down the manufacturing process to fix the knurled piston. A Type I error in this situation is deciding that the knurled piston is broken when in fact it is not, and a Type II error is deciding that the knurled piston is okay when in fact it is broken.

Trade-offs Between α and β

Where we set the criterion clearly affects the probability of Type I and Type II errors, as illustrated in Figure 8.5. The proportion of the *accept* distribution above the criterion is the probability of a Type I error (blue regions in Figure 8.5), and the proportion of the *reject* distribution below the criterion is the probability of a Type II error (gray regions in Figure 8.5). As the criterion increases, the probability of a Type I error (α) decreases and the probability of a Type II error (β) increases.

Only the people manufacturing the potato chips can say whether a Type I error or a Type II error is more serious, because there are costs associated with both. These costs may depend on where they are sending their chips. If one of the customers is extremely picky, they might drop you as a supplier if they receive even a small number of complaints about chips that are too thick. In this case, the manufacturer would want to avoid Type II errors (deciding that all is well when something is wrong) and set the criterion to a low value. Of course, this will lead to an increase in Type I errors (deciding that something is wrong when all is well). Frequently shutting down the manufacturing process to look for a nonexistent problem will slow production and increase costs. Clearly, many considerations like these will determine how the criterion is set. In the Neyman-Pearson model, no single criterion can apply to all situations. Setting $\alpha = .05$ in all cases would make no sense.

FIGURE 8.5	■ Trade-offs Between Errors

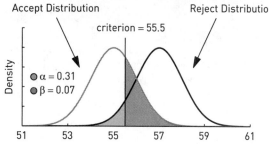

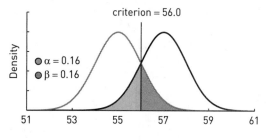

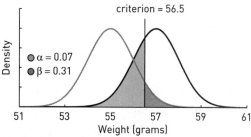

The trade-off between Type I and Type II errors. The distribution of weights under H_A (left) and the distribution of weights under H_R (right). As the criterion increases, α decreases (blue region under H_A) and β increases (region under H_R in gray).

Whatever probability the company specifies as tolerable for one type of error will immediately determine the criterion, and hence the probability of the other type of error (see Figure 8.5). Once the acceptance procedure has been set up, chip bags will be tested periodically. Over time many thousands of measurements will be made. Because the knurled piston is either broken or not, each bag of chips tested can be thought of as a sample from one of the two corresponding distributions. With *repeated sampling* from these two distributions, the process will achieve the specified balance between Type I and Type II errors.

From the Neyman-Pearson perspective, there is no such thing as statistical significance. Their model was about making decisions. They did not compute *p*-values because *p*-values are irrelevant to the decision-making process. The question is whether a score (measurement) exceeds the criterion, not by how much it exceeds the criterion.

Fisher Versus Neyman

It is an understatement to say that the Neyman-Pearson model infuriated Sir Ronald. He was convinced that Neyman had tried to improve on his significance tests, but in doing so had completely distorted their purpose. Acceptance procedures, Fisher (1955) argued, have no place in experimental research. Fisher was a thorn in Neyman's side until the day he died (Salsburg, 2001; see Figure 8.6). Their disagreements have been referred to as the "widest-cleft" in statistics (Savage, 1961), pitting followers of Fisher and Neyman against each other.

Fisher blasted Neyman-Pearson on two points. The first is the idea of repeated sampling. Although Fisher would grant that repeated sampling is reasonable and necessary in an acceptance procedure, he was adamant that repeated sampling is completely irrelevant to scientists trying to reason about experimental data. Fisher argued that scientists do not repeatedly sample from known distributions to achieve a specified balance between Type I and Type II errors. Rather, scientists use significance tests to learn about the way nature operates (e.g., whether one fertilizer provides higher crop yields than another).

The main issue that separated Fisher and Neyman-Pearson was the question of two distributions. As we saw in Chapter 7, for Fisher there is only one distribution that we know anything about, and that is the null hypothesis distribution. Although we might predict whether a sample mean will be above or below μ_0, in reality we don't have enough information about the alternative hypothesis distribution to compute the probability of a Type II error (see Figure 7.8). Fisher referred to Type II errors as errors of the second kind, and in his writings it is easy to picture the revulsion he felt for this concept. However, we've seen that psychologists acknowledge Type II errors and power, so we must have a look at how the Neyman-Pearson decision model was grafted onto Fisher's significance tests.

The Alternative Hypothesis Distribution in Empirical Research

In the Neyman-Pearson model, there are two distributions. We called these "A" and "R" in the chip manufacturing example, but they could have been called anything. Neither distribution has greater status than the other because we know everything we need to know about both. Psychologists recognized that, when applied to significance tests, one of these distributions could be thought of as the null hypothesis distribution (e.g., "A") and the other as the alternative hypothesis distribution (e.g., "R"). However, in psychological research, the null and alternative hypothesis distributions do not have the same status. As noted

FIGURE 8.6 ■ Fisher and Neyman

Sir Ronald Fisher (*left*) and Jerzy Neyman (*right*), with Prasanta Chandra Mahalanobis (*center*). I imagine that this was an uncomfortable moment for everyone.

Author: Konrad Jacobs, Source: Archives of the Mathematisches Forschungsinstitut Oberwolfach

several times, we know everything there is to know about the null hypothesis distribution and very little about the alternative hypothesis distribution.

Although it is true that scientists don't know exactly what the alternative distribution is, it is actually this distribution that interests them. In our power-posing example, the null hypothesis distribution was the distribution of final exam grades in the population of students who *had not* assumed a power pose for 2 minutes before the exam. Although it doesn't exist, we can *imagine* an entire population of final exam grades of students who *had* assumed a power pose for 2 minutes before the exam. Such a population would have a mean and standard deviation. It is this distribution that the student really wanted to know about. In fact, the student's hypothesis is that this is the distribution from which his sample will be drawn. In other words, the student wanted to run the experiment to infer how the population of exam grades would change if *all students* (not just those in the class he studied) assumed a power pose for 2 minutes before the exam. So, the alternative distribution is there, in our minds at least, and we should care about it.

Researchers hope to gain support for their alternative hypothesis by rejecting the null hypothesis. So, if you're hoping to reject the null hypothesis, you want to be sure that there is a high probability of rejecting it when it is false. However, complete knowledge of the alternative distribution is required to determine this probability. If we have complete knowledge of the alternative distribution, we wouldn't need to run the experiment. Therefore, we have a paradox: we want to run an experiment that has a high power so that we can learn something about the alternative hypothesis distribution, but to compute power requires complete knowledge of the alternative distribution.

The way out of this paradox is to think carefully about what alternative distributions would be interesting to us as experimenters. If we stick with our power-posing example, the student researcher, with prodding from his stats TA, said that he would be interested in an alternative distribution that has a mean of 77 or greater. *Now we're getting somewhere!* If you can specify what difference between μ_1 and μ_0 would be meaningful to you, you can take steps to control the probability of correctly rejecting the null hypothesis in a significance test.

We can see that this is exactly what was done in Figure 8.3. The alternative hypothesis distribution was assumed to have $\mu_1 = 77$ and $\sigma = 10$. When we assume sample size of $n = 64$, we can compute β and power. Unfortunately, Figure 8.3 showed that under these assumptions, power was very low. We can now ask what affects power and how we can control it.

THE DETERMINANTS OF POWER

There are three factors that affect power: α, effect size, and sample size. We will step through these in turn. Of these factors, effect size is the only one we haven't yet discussed. Even though sample size is the most important consideration in controlling power, most attention will be devoted to effect size, because effect size is more important in research in general.

α and Power

The first and most obvious determinant of power is α. As we saw in Figure 8.5, α and β trade-off. All things being equal, as α increases, $m_{critical}$ decreases (i.e., as the criterion moves closer to μ_0), and therefore power increases.

Figure 8.7 shows two distributions of *means* based on samples of size 16. Figures 8.7a through 8.7c show that as the probability of a Type I error (α) increases from .025 to .25, the probability of a Type II error (β) decreases and power increases. In Figures 8.7a through 8.7c, μ_0, μ_1, σ, and n remain constant.

Increasing power by increasing α is not particularly relevant in research because the criterion for statistical significance is almost always fixed at $p < .05$, which is equivalent

LEARNING CHECK 2

1. An acceptance procedure has been established in which $\mu_0 = 20$, $\mu_1 = 30$, and criterion = 22. If μ_1 is the mean of the distribution associated with a manufacturing error, which error, Type I or Type II, is considered more costly in this procedure? (It may help to draw these two distributions and the criterion.)

2. An acceptance procedure has been established in which $\mu_0 = 40$, $\mu_1 = 55$, and criterion = 50. If μ_1 is the mean of the distribution associated with a manufacturing error, which error, Type I or Type II, is considered more costly in this procedure?

3. An acceptance procedure has been established in which $\mu_0 = 20$ and $\mu_1 = 60$. If μ_1 is the mean of the distribution associated with a manufacturing error, which criterion, 30 or 40, would make for a more powerful test?

4. State whether the following statements are true or false.

   (a) Fisher advocated the importance of Type II errors.

   (b) Power is an unimportant concept in experimental psychology.

5. If I reject the null hypothesis when it is true, what kind of error have I committed?

Answers

1. In this case, $\beta < \alpha$ because the criterion is closer to μ_0 than μ_1, so the probability of a Type II error is lower than the probability of a Type I error. Therefore, Type II errors are considered more serious, and efforts are made to avoid them.

2. In this case, $\alpha < \beta$ because the criterion is closer to μ_1 than μ_0, so the probability of a Type I error is lower than the probability of a Type II error. Therefore, Type I errors are considered more serious, and efforts are made to avoid them.

3. The closer the criterion is to μ_0, the more powerful the test. Therefore, a criterion of 30 would make for a more powerful test.

4. (a) False. (b) False.

5. Type I error.

---

to setting α to .05. However, it is worth noting that a one-tailed (directional) test of the null hypothesis is more powerful than a two-tailed test, but only *if you've made the right prediction* about whether μ_1 is greater than or less than μ_0. Comparing Figures 8.7a and 8.7b where $\alpha = .025$ and .05 can help us understand this point.

When we conduct a one-tailed test that predicts $\mu_1 > \mu_0$ and $\alpha = .05$, $z_{critical} = 1.645$. For a two-tailed test with $\alpha = .05$, $z_{critical} = \pm1.96$, so there are two values of $m_{critical}$. For the example in Figure 8.7, the two values of $m_{critical}$ would be 70.01 and 79.9. Power in this case is the proportion of the alternative hypothesis distribution of means above 79.9 *plus* the proportion below 70.01.

Figure 8.7a shows the proportion above 79.9. The short, blue vertical line in Figure 8.7a shows 70.01. It is clear that very little of the alternative hypothesis distribution of means falls below 70.01. This means that the proportion of the alternative hypothesis distribution of means above 79.9 *plus* the proportion below 70.01 is essentially equal to the proportion above 79.9. Therefore, in this case, power is mainly determined by the proportion of the alternative hypothesis distribution of means above 70.09. More generally, the power of a two-tailed test with $\alpha = .05$ is very close to the power of a one-tailed test with $\alpha = .025$.

In Chapter 7 we noted that researchers often use one-tailed tests because they can declare statistical significance with a smaller z_{obs} in a one-tailed test than a two-tailed test. In the terminology of this chapter, we can say that one-tailed tests are more powerful than two-tailed tests. But, once again, this is true only if you've made the right prediction about whether μ_1 is greater than or less than μ_0.

FIGURE 8.7 ■ α and Power

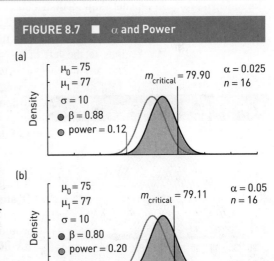

(a)

$\mu_0 = 75$
$\mu_1 = 77$
$\sigma = 10$
● $\beta = 0.88$
● power = 0.12

$m_{critical} = 79.90$ $\alpha = 0.025$
$n = 16$

(b)

$\mu_0 = 75$
$\mu_1 = 77$
$\sigma = 10$
● $\beta = 0.80$
● power = 0.20

$m_{critical} = 79.11$ $\alpha = 0.05$
$n = 16$

(c)

$\mu_0 = 75$
$\mu_1 = 77$
$\sigma = 10$
● $\beta = 0.45$
● power = 0.55

$m_{critical} = 76.69$ $\alpha = 0.25$
$n = 16$

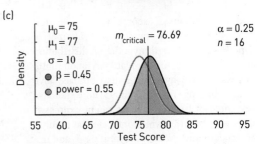

Test Score

(a through c) The association between α and power; as α increases, power increases.

Effect Size and Power

Effect size is a measure of the difference between the means of two distributions in units of standard deviation. In this discussion we will assume that the distributions of scores under H_0 and H_1 are normal, have the same standard deviations (σ), and differ only in their means. Using symbols, we can provide the following definition of effect size:

$$\delta = \frac{\mu_1 - \mu_0}{\sigma}. \tag{8.2}$$

Cohen's δ is an effect size. It is a standardized measure of difference between two population means in units of standard deviation. Cohen's δ is essentially a z-score.

δ is the Greek letter d and was chosen to denote difference. We will refer to this parameter as **Cohen's δ**, which is named for Jacob Cohen, who is well known for emphasizing the importance of effect size and power in psychological research (Cohen, 1988). (We met Cohen in Chapter 7 when we discussed the inverse probability error.) Although δ can be positive or negative, its magnitude is more important than its sign, because the sign simply reflects whether δ was computed with $\mu_1 - \mu_0$ or $\mu_0 - \mu_1$ in the numerator.

Equation 8.2 should look familiar because it describes a z-score. As with z-scores, δ is a unit-free measure of distance. The black curve in Figure 8.8 shows the distribution of *scores* under the null hypothesis that we've been discussing ($\mu_0 = 75$, $\sigma = 10$). The blue lines show a small number of possible alternative hypothesis distributions of scores with different means. Below the means on the x-axis are the corresponding effect sizes. For example, 85 is one standard deviation above 75, so it corresponds to an effect size of $\delta = (\mu_1 - \mu_0)/\sigma = 1$. Effect sizes defined in this way provide a convenient characterization of the difference between two population means without having to specify μ_0, μ_1, and σ.

Figure 8.9 shows that power increases as effect size increases ($\delta = .2$, $.5$, and $.8$). Each panel shows a distribution of *means* based on samples of size $n = 16$. In all cases, $\alpha = .05$ and $\sigma_m = \sigma/\sqrt{n} = 10/4 = 2.5$. Therefore, $m_{critical} = 75 + 1.645(2.5) = 79.11$ in Figures 8.9a through 8.9c. As before, the regions to the right of $m_{critical}$ represent power, and the gray regions to the left of $m_{critical}$ represent β. Clearly, the probability of correctly rejecting H_0 (power) increases as effect size increases from .2 to .8.

It would be a worthwhile exercise to use the numbers in Figure 8.9 to compute β and power in each panel. This simply involves expressing $m_{critical}$ as a z-score on the alternative hypothesis distribution of means and then using the z-table to find $P(z)$ and $1 - P(z)$. You will find that these three z-scores are 0.85, −0.36, and −1.56.

Although effect size is an important determinant of power, it is generally not completely under the control of the researcher. (I am making a simplification here by ignoring the contribution of measurement error to σ.) Rather, researchers specify the δ that would be meaningful, and they then arrange the experiment so that it will have a high probability of producing a statistically significant result if the true effect size is δ or greater. Earlier we considered our hypothetical student researcher who thought that an alternative distribution with a mean of 77 or greater would be interesting. When $\mu_1 = 77$ is expressed as an effect size, we have

$$\delta = \frac{\mu_1 - \mu_0}{\sigma} = \frac{77 - 75}{10} = 0.2.$$

Therefore, he could rephrase his previous statement to say that he would be interested in an alternative distribution for which the effect size is $\delta = .2$. We now have to figure out how to detect an effect of this size in a significance test.

FIGURE 8.8 ■ Effect Size

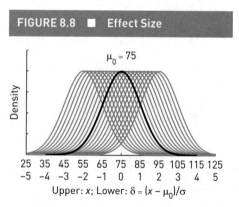

| 25 | 35 | 45 | 55 | 65 | 75 | 85 | 95 | 105 | 115 | 125 |
| −5 | −4 | −3 | −2 | −1 | 0 | 1 | 2 | 3 | 4 | 5 |

Upper: x; Lower: $\delta = (x - \mu_0)/\sigma$

The distribution of scores under the null hypothesis, and a number of alternative hypotheses. The null hypothesis distribution has a mean of 75 and standard deviation of 10. The alternative hypothesis distributions have means ranging from 55 to 95 and standard deviations of 10. These means can be expressed as effect sizes (δ), which indicate the number of standard deviations (σ) separating μ_0 and μ_1.

Sample Size and Power

The third determinant of power is sample size. If δ is fixed at a specific level, then power increases as n increases. At the same time, β decreases as n increases. In Figure 8.10, δ is fixed at .2 and sample sizes are $n = 32$, 64, and 155. Therefore, Figure 8.10 complements Figure 8.9, in which n was fixed and δ varied.

Because the distributions of means in each panel of Figure 8.10 are based on different sample sizes, $\sigma_m = \sigma/\sqrt{n}$ decreases from Figure 8.10a to Figure 8.10c. Although the distribution of means under H_0 and H_1 becomes narrower, their means don't change. Furthermore, as sample size increases, $m_{critical}$ moves closer to μ_0 because σ_m decreases; remember, $m_{critical} = \mu_0 + z_{critical}(\sigma_m)$. For these two reasons, increasing sample size increases power. Because sample size is under the control of the experimenter, this gives the experimenter control over power.

FIGURE 8.9 ■ Effect Size and Power	FIGURE 8.10 ■ Sample Size and Power

(a)

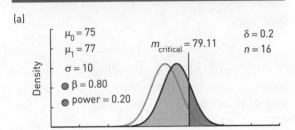

(a)

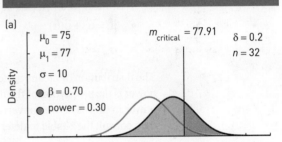

(b)

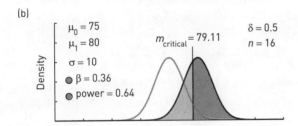

(b)

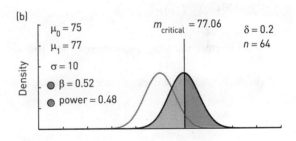

(c)

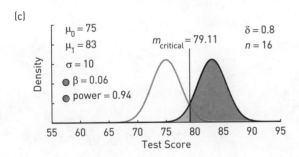

(c)

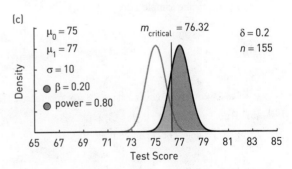

The influence of effect size (δ) on power and β. In (a) through (c), $\sigma_m = \sigma/\sqrt{n} = 10/4 = 2.5$ and $\alpha = .05$. Therefore, $z_{critical} = 1.645$ and $m_{critical} = \mu_0 + z_{critical}(\sigma_m) = 79.11$. As δ increases, the distribution of means under H_1 shifts to the right. As this distribution shifts to the right, power increases and β decreases.

The effect of sample size (n) on power and β. In (a) through (c), effect size is constant, $\delta = .2$, but sample size n increases from 32 to 155. In all panels, $\alpha = .05$, so $z_{critical} = 1.645$. Because n increases from (a) to (c), $\sigma_m = \sigma/\sqrt{n}$ decreases from (a) to (c). As n increases, the distribution of means under both H_0 and H_1 become narrower. As these distributions become narrower, $m_{critical}$ shifts to the left, power increases, and β decreases.

Figure 8.10b is the same as Figure 8.3. It shows that when sample size is 64, as suggested by the student's neuroscience professor, power is .48. Figure 8.10c shows that when sample size is 155, as suggested by the student's stats TA, power is .8. It seems that the stats TA was aiming for power of .8 and was somehow able to determine the sample size required to achieve this power.

In summary, power is the probability of rejecting H_0 when it is false, which is exactly the same as saying that power is the probability of rejecting H_0 when H_1 is true. When α is fixed at .05, only effect size and sample size affect power.

- For a given sample size, power increases as effect size increases.

- For a given effect size, power increases as sample size increases.

We can compute power when we know μ_0, δ, σ, n, α, and H_1. From these quantities, we can determine the following:

- $m_{critical} = \mu_0 + z_{critical}(\sigma_m)$
- $\mu_1 = \mu_0 + \delta(\sigma)$
- $\sigma_m = \sigma/\sqrt{n}$

The distribution of means under H_1 has mean μ_1 and standard error σ_m. When $H_1: \mu_1 - \mu_0 > 0$, power is the proportion of the distribution of means under H_1 lying above $m_{critical}$. To compute power in this case, we convert $m_{critical}$ to a z-score on the distribution of means under H_1. We then consult the z-table to determine the proportion of the z-distribution lying above z.

LEARNING CHECK 3

1. State whether the following statements are true or false.

   (a) Both effect size and sample size affect power.

   (b) If effect size is fixed at .5, power can be increased by decreasing sample size.

   (c) If effect size is fixed at .8, power can be increased by increasing sample size.

   (d) $\delta = \dfrac{(\mu_1 - \mu_2)}{\sigma\sqrt{n}}$.

2. If $\mu_0 = 15$, $\mu_1 = 15.6$, and $\sigma = 2$, what is δ?

3. If $\mu_0 = 15$, $\sigma = 3$, and $\delta = .4$, what is μ_1?

4. If $\mu_0 = 15$, $\sigma = 3$, and $\delta = .4$, what is power if $n = 36$, $\alpha = .05$, and $H_1: \mu_1 > \mu_0$?

Answers

1. (a) True.

   (b) False. Power increases as sample size increases.

   (c) True.

   (d) False. δ does not depend on sample size.

2. $\delta = (\mu_1 - \mu_2)/\sigma$. $\delta = (15.6 - 15)/2 = .3$.

3. $\mu_1 = 15 + .4(3) = 16.2$.

4. $\mu_1 = 16.2$, $m_{critical} = 15.8225$; $z = -0.75$, $\beta = .2266$, power $= .7734$.

PROSPECTIVE POWER ANALYSIS: PLANNING EXPERIMENTS

In this section, we will see that if a researcher knows what effect size would be meaningful in his or her research, it is relatively easy to determine the sample size required to achieve any level of statistical power. We call this *prospective power analysis* because it is conducted before the experiment is run.

The hardest part of prospective power analysis is determining the effect size that would be meaningful in the specific research setting. There is no rule to follow here. If a researcher thinks that an effect size of $\delta = .5$ would be meaningful, he or she must justify this decision. In some research settings, this might be considered an extraordinarily large effect; in others, it might be extraordinarily small.

Once one has decided on the effect size of interest and desired level of power, it is necessary to use programs like Excel to compute the required sample size. (Unfortunately, there is no simple formula to compute the required sample size for a given δ and desired level of power.) Appendix 8.1 (available at study.sagepub.com/gurnsey) provides a description of how to use Excel to compute the sample size required to achieve a desired power for a given δ.

When we are planning experiments, it is common to aim for power = .8. This is somewhat arbitrary, but it means that the probability of correctly rejecting H_0 is four times greater than incorrectly retaining H_0. That is, if power = .8, then power/β = .8/.2 = 4. Another common choice is to aim for power = .95. This means that if we adopt the $p < .05$ criterion, then $\beta = \alpha = .05$.

Table 8.1 allows us to see the connection between sample size, effect size, and power without having to do the calculations in Excel. Each row shows the sample sizes required to achieve a specified level of power (top row) for a given effect size (δ, left column) for *one-tailed* tests with $\alpha = .05$. The number of scores required depends on the magnitude of δ and not its sign, because the number of scores required to achieve a specific power would be the same whether δ is positive or negative.

The first row of Table 8.1 shows the sample sizes required to achieve power ranging from .1 to .95 when $\delta = .1$. An effect size of .1 may seem very small, but there are situations in which an intervention producing $\delta = .1$ would be worthwhile. However, to achieve power = .8, for example, requires a sample size of 619, which is a large number of participants. If a researcher could only test ⅓ of this number of participants (i.e., about 200), then he or she could only achieve power of about .4. This would mean that the probability is only .4 that the null hypothesis will be rejected when the alternative hypothesis ($\delta = .1$) is true. In this case, the thoughtful experimenter should probably not conduct a significance test. Sometimes we have to face facts. There may be an experiment we want to run, but we can't afford to run it because achieving reasonable power requires too many participants. Choosing not to run an underpowered study makes more sense to me than choosing to run it.

The fifth row of Table 8.1 shows the sample sizes required to achieve power ranging from .1 to .95 when $\delta = .5$. A researcher who is considering a clinical intervention for depression can achieve high power with relatively small samples if the effect size of interest is $\delta = .5$. With samples of size $n = 25$ or $n = 35$, she achieves power of .8 and .9, respectively. So, even if there is some expense associated with collecting data, the small number of participants required makes the experiment seem worthwhile.

Table 8.2 follows the same conventions as Table 8.1 but provides the sample sizes required to achieve a specified level of power (top row) for a range of effect sizes (left column) for *two-tailed* tests with $\alpha = .05$. Comparing Table 8.1 and Table 8.2 shows that two-tailed tests require larger samples to achieve power equivalent to the corresponding one-tailed test.

Using Published Effect Sizes to Guide Prospective Power Analysis

The hypothetical student in our example thought that a mean increase of 2 percentage points on the final exam would be interesting, and we saw that this corresponds to an effect size of $\delta = .2$. This judgment was based on not much more than intuition, and perhaps common sense; would such a minimal intervention really be expected to produce more than a 2-point increase in grades? A better guide to what would constitute an

TABLE 8.1 ■ Sample Size, Effect Size, and Power (one-tailed tests)											
					Power						
δ	.10	.20	.30	.40	.50	.60	.70	.80	.85	.90	.95
.1	14	65	126	194	271	361	471	619	720	857	1083
.2	4	17	32	49	68	91	118	155	180	215	271
.3	2	8	14	22	31	41	53	69	80	96	121
.4	1	5	8	13	17	23	30	39	45	54	68
.5	1	3	6	8	11	15	19	25	29	35	44
.6	1	2	4	6	8	11	14	18	20	24	31
.7	1	2	3	4	6	8	10	13	15	18	23
.8	1	2	2	4	5	6	8	10	12	14	17
.9	1	1	2	3	4	5	6	8	9	11	14
1.0	1	1	2	2	3	4	5	7	8	9	11

Note: Directional tests assuming $\alpha = .05$.

TABLE 8.2 ■ Sample Size, Effect Size, and Power (two-tailed tests)											
					Power						
δ	.10	.20	.30	.40	.50	.60	.70	.80	.85	.90	.95
.1	47	126	207	292	385	490	618	785	898	1051	1300
.2	12	32	52	73	97	123	155	197	225	263	325
.3	6	14	23	33	43	55	69	88	100	117	145
.4	3	8	13	19	25	31	39	50	57	66	82
.5	2	6	9	12	16	20	25	32	36	43	52
.6	2	4	6	9	11	14	18	22	25	30	37
.7	1	3	5	6	8	10	13	17	19	22	27
.8	1	2	4	5	7	8	10	13	15	17	21
.9	1	2	3	4	5	7	8	10	12	13	17
1.0	1	2	3	3	4	5	7	8	9	11	13

Note: Non-directional tests assuming $\alpha = .05$.

interesting or meaningful effect size is the published literature relating to the research question. Therefore, it would have been better for the student to base his expectation about effect size on the results reported by Carney et al. (2010) that Amy Cuddy mentioned in her TED talk.

Carney et al. (2010) compared a group of participants who held high-power poses (also called expansive poses) and a group of participants who held low-power poses (also called contractive poses). They looked at the effects of high- and low-power poses on levels of testosterone and cortisol, as well as risk-taking behaviors. For all three dependent measures, the differences between means of participants who held high-power poses and low-power poses corresponded to estimated effect sizes of approximately .6. (We won't explain how these estimates were calculated because this would take us beyond the scope of this chapter.)

If δ were approximately .6, the student would have predicted that $\mu_1 = \mu_0 + \delta(\sigma) = 75 + 0.6(10) = 81$. Table 8.1 shows that achieving power = .8 for $\delta = .6$ would require only 22 participants, and achieving power = .95 would require only 37 participants. However, Carney et al. (2010) compared participants who held both high- and low-power poses. If they had compared high-power poses to neutral poses, the estimated effect size might have been only half of what was found; i.e., estimated $\delta = .3$. In this case, achieving power = .8 would require 69 participants and achieving power = .95 would require 121 participants.

Reasonable arguments might be made for why 22, 37, 69, or 121 participants in the student's study would be appropriate. These arguments would have to (i) refer to the effect sizes estimated from the study by Carney et al. (2010), (ii) connect these effect sizes to the experiment to be run, and (iii) justify the choice of power that the study aims to achieve. Although any of these justifications might be questioned by other researchers, such justifications are far superior to simply going with the size of a sample that is conveniently available.

There is (for some) a disappointing coda to the power-pose study. A recent study by Ranehill, Dreber, Johannesson, Leiberg, Sul, and Weber (2015) attempted to replicate the original results of Carney et al. (2010) using sample sizes that would achieve power greater than .98, assuming that δ is approximately .6. Ranehill et al. found no statistically significant effects of the pose taken on testosterone or cortisol levels, nor was there a statistically significant effect on risk-taking behavior. Ranehill et al. did find a statistically significant difference in how powerful people reported *feeling*. The results of Ranehill et al. strongly suggest that assuming a power pose has no effect on hormones or behavior. In fact, Dana Carney, the lead author on the power pose paper, stated in a position paper posted on her website that she does not believe that "power pose" effects are real. To find the paper that explains her position on power poses, you can Google "Dana Carney power pose."

Sample Size Reveals, Implicitly, the Effect Size of Interest

Even if a researcher has not performed a prospective power analysis, the choice of sample size implicitly reveals what effect size the study was designed to detect. First, let's assume that a sensible researcher would aim to have power of at least .8. So, if the goal is to achieve power = .8, then the sample size chosen determines the effect size that yields this power. Imagine that a neuroscientist wants to test whether sleep deprivation increases the time it takes rats to learn a water maze task. In this task, a rat is put in a pool of water and swims until it finds a "safe" platform just below the surface of the water; the platform allows the rat to "escape" from the water. Because the platform is below the surface, the rat can't find it visually. Instead, he (rats used in this type of experiment are almost always male) must learn the location of the platform through trial-and-error exploration of the pool, and then rely on his spatial memory and visible landmarks to get back to the platform on later trials.

Assume that our neuroscientist knows that normal rats learn this task in 15 trials, on average, with a standard deviation of 5. Now let's say she chooses $n = 8$ rats, deprives them of sleep, then measures how long it takes them to learn the task. Her alternative hypothesis is $\mu_1 - \mu_0 > 0$. That is, she thinks it will take sleep-deprived rats longer to learn the task. So, what does her sample size imply about the effect size she hopes to detect? To answer this question, we consult Table 8.1; we find the column in which power = .8, and we then look down the column until we find 8. When we look across to the effect size corresponding to this choice of power and sample size, we find that $\delta = .9$. This means that the experiment was designed to detect an effect size of at least $\delta = .9$. (The researcher may not have recognized that she was predicting an effect of this size if she hadn't planned the study using a prospective power analysis.)

We may now ask what an effect size of $\delta = .9$ implies about the mean of the alternative hypothesis distribution. When we know δ, σ, and μ_0, we can compute μ_1 as follows:

$$\mu_1 = \mu_0 + \delta(\sigma). \tag{8.3}$$

When we plug what we know about the experiment into equation 8.3, we find

$$\mu_1 = \mu_0 + \delta(\sigma) = 15 + .9(5) = 19.5.$$

Therefore, the sample size suggests that the neuroscientist implicitly suspects that sleep deprivation yields a 30% increase in learning time; i.e., $(19.5 - 15)/15*100 = (4.5/15)*100 = 30\%$. So, even though a power analysis may not have been conducted, sample size implies the size of the effect the study is able to detect.

LEARNING CHECK 4

1. I plan to run a one-tailed test with $\alpha = .05$. I would only be interested in an effect size of .4. How large should my sample be if I wish to have power = .8?

2. I plan to run a one-tailed test with $\alpha = .05$. How large should my sample be if I'm interested in $\delta = .2$ and wish to have power = .9?

3. I plan to run a one-tailed test with $\alpha = .05$. I'm interested in $\delta = .2$ and I have 17 scores in my sample. What is my power?

4. I plan to run a one-tailed test with $\alpha = .05$. I have 18 scores in my sample. What effect size could I detect 90% of the time when H_0 is false?

5. I will run a one-tailed test of the null hypothesis with $\alpha = .05$ and $n = 25$. Assuming I would like to have power = .8, what effect size do I seem to be interested in?

Answers

1. $n = 39$.

2. $n = 215$.

3. Power = .2.

4. $\delta = .7$.

5. $\delta = .5$.

INTERPRETING EFFECT SIZE

We've just seen that before we conduct research, we should have some idea of what effect size would be meaningful. But how do we decide this? What is a big effect size? What does a given effect size imply about the two populations involved? What is the practical significance associated with a given effect size? These are really difficult questions, and they bump up against the limits of what statistics can answer for a researcher.

Statistical tests provide researchers with information to guide their theoretical reasoning or to help them choose the best course of action in clinical or applied settings. Statistics, however, cannot do our thinking for us. To conduct useful research requires, above all, a firm grasp of a research area. You must know what studies have been conducted, as well as what ambiguities or contradictions exist, so that you can properly interpret the statistical results of a given study or know whether a new study is even worth pursuing. Such questions concern the *content* of a research area; they are not questions about statistics. However, in this section, we review some of the contributions that effect size measures can make to reasoning about research.

Cohen's Classifications

Jacob Cohen is the person who brought power into the consciousness of psychologists (Cohen, 1988). By forcing us to consider power, Cohen also forced us to consider effect size, δ. In his writings, he gave a tentative answer to the question of what should be considered small, medium, and large effect sizes. His tentative answer is given in Table 8.3. An effect size of $\delta = .2$ is small, $\delta = .5$ is medium, and $\delta = .8$ is large. These are *very rough guidelines* with limited applicability. Cohen did not want to replace a rigid rule for what constitutes statistical significance with rigid rules for interpreting effect size. Just as statistical significance ($p < .05$) does not imply practical significance, saying that "$\delta = .8$ is a large effect size" is equally uninformative.

Cohen's so-called T-shirt model of effect sizes (they come in small, medium, and large) is not universally applicable, nor was it derived from serious statistical analysis. Cohen did not intend these classifications to be universally applied. There is no reason that $\delta = .5$ should be considered a medium effect size in all research domains. For example, $\delta = .5$ might be considered small in neuroscience or large in certain clinical settings. Therefore, one cannot make a judgment about the size of an effect independent of the research context. Rather, Cohen (1988) suggested that these guidelines could be employed *if no better guidance were available.*

TABLE 8.3 ■ Cohen's Guidelines for Effect Size (δ)	
Small	.2
Medium	.5
Large	.8

Note: These guidelines can be used as a last resort.

> The terms "small," "medium," and "large" are relative, not only to each other, but to the area of behavioral science or even more particularly to the specific content and research method being employed in any given investigation. . . . In the face of this relativity, there is a certain risk inherent in offering conventional operational definitions for these terms for use in power analysis in as diverse a field of inquiry as behavioral science. This risk is nevertheless accepted in the belief that more is to be gained than lost by supplying a common conventional frame of reference, which is recommended for use *only when no better basis for estimating the ES index is available.* (Cohen, 1988, p. 25; emphasis added)

In spite of the tentative nature of Cohen's guidelines, researchers commonly appeal to them as though they were well-established and valid rules. It would be a bad thing if $\delta = .8$ were to become the next $p < .05$.

Cohen's U_3

Cohen's δ can provide useful information about the relationship between two populations of scores. In particular, we can use Cohen's δ to determine the proportion of one distribution that is below the mean of another distribution. For example, if two normal distributions have the same standard deviation (σ) but one has a mean of μ_0 and the other has a mean of $\mu_1 = \mu_0 + \sigma$, then the 34-14-2 approximation tells us that approximately 84% of distribution 0 falls below the mean of distribution 1. This measure is captured by a quantity called Cohen's U_3, which is simply

U_3 is the proportion of one distribution of scores that falls below (or above) the mean of a second distribution.

$$U_3 = P(|\delta|), \tag{8.4}$$

where P denotes the cumulative distribution function of the standard normal (z) distribution; see Figure 4.6. (Remember, δ is just a z-score.)

Cohen's U_3 can be interpreted in two ways. To illustrate this point, we will consider the situation in which $\mu_1 > \mu_0$. Because of symmetry, U_3 is both

- the proportion of the H_0 distribution falling below μ_1 (see Figure 8.11), and
- the proportion of the H_1 distribution lying above μ_0 (see Figure 8.12).

When $\delta = 0$, the two distributions are identical and the proportion lying above and below the mean they share is .5. When $\delta = .2, .5,$ and $.8$, the corresponding U_3 values are .5793, .6915, and .7881, respectively. As shown in the corresponding panels of Figures 8.11 and 8.12, the proportion of the H_0 distribution below μ_1 is the same as the proportion of the H_1 distribution above μ_0.

To get some sense of what U_3 conveys, think of an intervention that increases IQ. Prior to this intervention, the distribution of IQs for individuals in the population of interest has $\mu_0 = 100$, and $\sigma = 15$. This distribution is shown as the black distribution in all panels of Figures 8.11 and 8.12. If the intervention produces an effect size of $\delta = .8$ (blue lines in Figures 8.11c and 8.12c), then the distribution following the intervention has $\mu_1 = 112$, and $\sigma = 15$. The U_3 associated with $\delta = .8$ is .7881. Therefore, 78.81% of the population before the intervention falls below the mean of the population after intervention (Figure 8.11c). Equivalently, 78.81% of the population after intervention lies above the mean of the population before intervention (Figure 8.12c).

For this example, the second interpretation is more straightforward. In this case, we can say that 50% of the population had an IQ above 100 before the intervention, and 78.81% of the population had an IQ above 100 after the intervention (see Figure 8.12). That's an increase of 28.81% of the population scoring above 100.

We noted in Chapter 7 that statistical significance is not synonymous with practical significance. Similarly, effect size is not directly related to practical significance. In the current example, an increase of 28.81% of the population having an IQ above 100 sounds impressive, and it would be if the intervention in question involved modest resources (e.g., reading the *New York Times* for half an hour a day). However, if the intervention in question involved 40 hours a week of intensive tutoring by Nobel Laureates, then the practical significance would be considerably less.

Practical Significance

There is growing concern that psychology, as a discipline, has become too reliant on p-values as arbiters of what is important and what is not. In Chapter 7, we saw an example of a statistically significant difference in IQs between Quebec residents and Maine residents. In that example, μ_{Que} was 100, and m_{Maine} was 100.06. Let's assume for the moment that 100.06 is the mean of the population of Maine (i.e., $\mu_{Maine} = 100.06$), rather than a sample mean, and that both populations have $\sigma = 15$. I noted in Chapter 7 that the difference between these two means was 1/250 of a standard deviation. Therefore, without using the term, I was referring to δ:

$$\delta = \frac{\mu_{Maine} - \mu_{Que}}{\sigma} = \frac{100.06 - 100}{15} = \frac{.06}{15} = 0.004.$$

$\delta = .004$ seems to be a very, very small effect size. Those concerned with the overreliance on p-values point to examples like this and say that estimates of δ should be a routine part of how statistical results are reported in the literature. It is impossible to disagree with this point.

We have to be careful, however, not to confuse effect sizes, large or small, with practical significance. We can't say, for example, that a small effect size ($\delta = .2$)

FIGURE 8.11   ■   Cohen's U_3

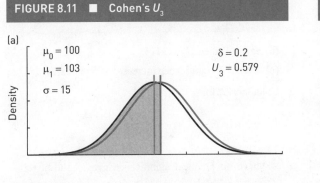

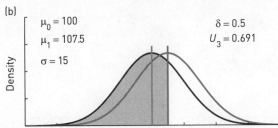

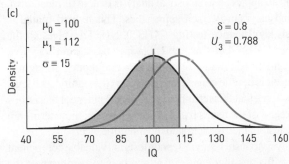

An illustration of U_3. Each panel shows the distribution of IQs under H_0 (black distribution) and H_1 (blue distribution). The three values of μ_1 are associated with δ = .2 (a), .5 (b), and .8 (c). U_3 is the proportion of the H_0 distribution below μ_1. U_3 is simply $P(|\delta|)$.

FIGURE 8.12   ■   Cohen's U_3

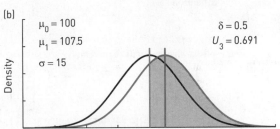

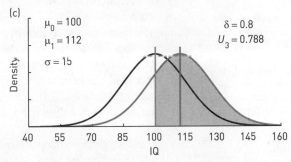

An illustration of U_3. Each panel shows the distribution of IQs under H_0 (black distribution) and H_1 (blue distribution). The three values of μ_1 are associated with δ = .2 (a), .5 (b), and .8 (c). U_3 is the proportion of the H_1 distribution above μ_0. U_3 is simply $P(|\delta|)$.

implies low practical significance, and a large effect size (δ = .8) implies high practical significance. Thinking back to Chapter 7 again, remember that a sample of 5-year-old European males can bench press 10 pounds, on average, and 15-year-old European males can bench-press 125 pounds, on average. In that example, we used the sample standard deviation of 2 to compute the estimated standard error in the denominator of our z_{obs} statistic. Let's use σ = 2 as the denominator of an effect size calculation. In this case, we would find

$$\delta = \frac{\mu_{5YO} - \mu_{15YO}}{\sigma} = \frac{10 - 125}{2} = -57.5.$$

Now, if effect size were the same as practical significance, this result would be headline news on cable channels around the globe. However, in this example, it is not news, because effect size is not synonymous with practical significance.

LEARNING CHECK 5

1. State whether the following statements are true or false.

   (a) Cohen's classification scheme treats an effect size of .5 as large.

   (b) If $\delta = 1$, then $U_3 = .3415$.

   (c) Practical significance increases as δ increases.

2. Calculate U_3 for $\delta = .7$.

Answers

1. (a) False. .5 is a medium effect size in Cohen's classification scheme.

   (b) False. $U_3 = .8415$.

   (c) False. A result may have practical significance if the effect size is small. Conversely, a result may have no practical significance even if the effect size is large.

2. .7580.

THE HYBRID MODEL: NULL HYPOTHESIS SIGNIFICANCE TESTING

The **hybrid model** combines elements of Fisher's significance tests and the Neyman-Pearson decision-making model. The hybrid model is referred to as **null hypothesis significance testing** (NHST).

Most researchers in psychology employ an approach to statistical analysis that, in its ideal form, incorporates elements of the Fisher and Neyman-Pearson approaches. This **hybrid model** is typically referred to as **null hypothesis significance testing** (NHST). NHST involves all the steps in a significance test (as shown in Table 7.4 of Chapter 7) but includes terminology and power analysis drawn from the Neyman-Pearson model. Within the hybrid model, the significance level is referred to as α. This by itself is of little consequence because the significance level is almost invariably .05. However, we will see that this terminology holds the potential for serious confusion. Psychologists will also make use of the terms β and *power*. However, these terms are most often used when explaining why a result failed to yield statistical significance.

Although the addition of power analysis to Fisher's significance tests is a valuable component of NHST, there are some potential confusions that can lead researchers to misinterpret the results of a significance test. We will highlight two examples of these confusions.

Confusion About p and α

One danger in the hybrid model is failing to appreciate the difference between p and α. For Fisher, a p-value is a *characteristic of the data* and a measure of the evidence against the null hypothesis.

> The actual value of P obtainable from the table . . . indicates the strength of the evidence against the hypothesis. (Fisher, 1925)

Therefore, according to Fisher, the smaller the p-value, the greater the evidence against the null hypothesis. This in itself is a somewhat dangerous perspective that has some truth but can lead to the inverse probability error. Nevertheless, for Fisher, a p-value is a characteristic of the data.

By contrast, in the Neyman-Pearson model, α is a *characteristic of the test*, not a characteristic of the data. That is, α is the probability that the test will reject the null hypothesis incorrectly. For Neyman and Pearson, this means that with repeated sampling from the same two distributions, α is the proportion of times the null hypothesis will be rejected incorrectly. Failing to distinguish between p and α can lead us to believe the p is the probability that the null hypothesis has been rejected incorrectly. If for some z-test we find $p = .0062$, *it is an error to believe* that the probability is $p = .0062$ that H_0 has been rejected in error; i.e., that a Type I error has been made. This error arises from thinking that p is a characteristic of the test. Therefore, it is important to keep p and α separate:

- p is the probability of a statistic, or one more extreme, occurring by chance when the null hypothesis is true. p is computed after the data have been collected and is therefore a property of the data. This probability is associated with z_{obs}.

- α is the probability that, in an infinite series of tests, a statistic will exceed the criterion when the null hypothesis is true. α is specified before data collection and is therefore a property of the test. This probability is associated with $z_{critical}$.

Confusion About Retaining the Null Hypothesis

A second confusion attributable to the hybrid model is the mistaken belief that retaining the null hypothesis means that the null hypothesis is true. This confusion can easily arise from the terminology of acceptance procedures built on the Neyman-Pearson model. A decision in the Neyman-Pearson framework amounts to accepting which one of two distributions produced the sample. In the potato chip example, this makes some sense; we may accept the hypothesis that all is well or we may accept the hypothesis that something has gone wrong. We then act according to our decision by doing nothing (all is well) or shutting down the process (something has gone wrong).

However, when the hybrid model is applied to psychological research, we must avoid the notion of accepting the null hypothesis. Accepting the null hypothesis implies that we've concluded that our sample was drawn from a population having a mean equal to μ_0. This is an incorrect interpretation, although one that is frequently made. To avoid this misinterpretation, we say we retain the null hypothesis when we fail to reject it. This simply means that it might be true or it might be false, but we haven't met the criterion for treating it *as though it is false*.

Fisher Versus Neyman: Who Had Greater Influence?

We may now ask, "Which tradition has had the greater influence on *statistical practice* in psychology?" When I use the term *statistical practice*, I'm referring to what researchers do, not what they think or know they should do. In my view, Fisher has had greater influence on statistical practice. The main reason for this view has to do with statistical power, which clearly derives from the Neyman-Pearson model. Our discussion makes clear that power *should* be a critical part of research planning. However, prospective power analysis is absent in the vast majority of papers published in research journals.

Lack of Power in Psychological Research

Many years ago, Jacob Cohen (1962) examined all studies published in the *Journal of Abnormal and Social Psychology* in 1960. He was interested in the power of the studies to detect small, medium, and large effect sizes. Specifically, he looked at the sample sizes used in the studies and then determined the power they would have to detect small, medium, and large effect sizes. Cohen found that median power was .17 for small effect sizes, .46 for medium effect sizes, and .89 for large effect sizes. This means that the studies only had sufficient power, on average, to detect large effect sizes at greater than chance levels. These results suggest that researchers either (i) assumed that only large effect sizes are of interest or (ii) hadn't considered the effect sizes that would be of interest, and they thus chose their sample sizes arbitrarily.

Twenty-seven years later, Sedlmeier and Gigerenzer (1989) repeated Cohen's (1962) analysis for all studies published in the *Journal of Abnormal Psychology* in 1984. They found that median power was .12 for small effect sizes, .37 for medium effect sizes, and .86 for large effect sizes. Sedlmeier and Gigerenzer concluded that researchers had not changed the designs of their studies in light of Cohen's original report, which highlighted the low power of many studies in psychology.

A recent report showed that in a large sample of neuroscience reports, the median power was a jaw-dropping .21 (Button, Ioannidis, Mokrysz, Nosek, Flint, Robinson, & Munafo, 2013a, 2013b, 2013c). Therefore, low-powered experiments are not unique to abnormal and social psychology, what some call soft psychology, but are also characteristic of so-called hard psychology.

Fixed Criterion for Statistical Significance and Exact p-Values

In spite of having adopted the $p < .05$ criterion for statistical significance, researchers feel some conflict between this and reporting the exact p-value associated with an obtained statistic. Researchers resolve these two impulses by using both. For example, in many of the notes on APA Reporting in Chapter 7, we said things such as "The results were statistically significant, $z_{obs} = 2.5$, $p = .0062$." There is no need to specify what the criterion for statistical significance is because everyone assumes it to be $p < .05$.

One might argue that adherence to the $p < .05$ criterion for statistical significance reflects the influence of the Neyman-Pearson model of statistical decision making. However, the criterion in an acceptance procedure, for example, is determined by the desired balance between Type I and Type II errors, which might differ depending on the distributor to whom you're sending your products. Therefore, Neyman-Pearson did not specify what α should be and certainly never said it should always be .05.

The practice of using the $p < .05$ criterion for statistical significance and simultaneously reporting exact p-values, as measures of the evidence against the null hypothesis, seems better explained by adherence to the contradictory advice of early and late Fisher. Therefore, in my view, the practice of statistics in psychology owes more to Fisher than to Neyman-Pearson. This is not a good thing. Significance tests without power considerations lead inevitably to much wasted effort.

LEARNING CHECK 6

1. State whether the following statements are true or false.

   (a) Fisher was a strong advocate of power analysis.

   (b) Neyman and Pearson advocated $\alpha = .5$ as a universal criterion for statistical significance.

   (c) Although they differed on many issues, p-values were important for both Fisher and Neyman.

   (d) Jerzy is a cooler name than Ronald.

   (e) p-values were irrelevant in the Neyman-Pearson model.

   (f) p is the probability that you've made a Type I error.

   (g) p is the probability that H_0 is true.

   (h) p is a measure of effect size.

   (i) p is the probability of a statistic, or one more extreme, under the null.

   (j) Because of Fisher's influence, psychologists view H_1 as an unimportant concept.

   (k) Fisher viewed p as a measure of the evidence favoring H_0.

   (l) $p < .05$ implies a large effect size.

   (m) δ can be converted to p without any further information.

Answers

1. (a) False.

   (b) False.

   (c) False.

   (d) True.

   (e) True.

   (f) False.

   (g) False.

   (h) False.

   (i) True.

   (j) False.

   (k) False.

   (l) False.

   (m) False.

SUMMARY

In this chapter, we introduced the decision-making framework of Neyman and Pearson. In this framework there are two known distributions, and our task is to decide from which of the two a sample was drawn. When this framework is applied in psychology, one of these distributions is the H_0 distribution and the other is the H_1 distribution. To make a decision, we set a criterion, $m_{critical}$, that a sample mean must exceed in order to reject H_0. Rejecting H_0 when it is true is a *Type I error*. The probability of a Type I error is the proportion of the H_0 distribution of means above $m_{critical}$. This probability is referred to as α.

Retaining H_0 when it is false (i.e., when H_1 is true) is a *Type II error*. Computing the probability of a Type II error requires that we know everything about the distribution of means under H_0 and H_1. If we know this, then the probability of committing a Type II error is given by the proportion of the distribution of means under H_1 that do not exceed $m_{critical}$. This probability is referred to as β. Power is the proportion of the distribution of means under H_1 that exceeds $m_{critical}$. Power is the probability of correctly rejecting H_0.

Power is affected by *effect size* and *sample size*. Effect size (δ) is the difference between two population means relative to their standard deviation, $\delta = (\mu_1 - \mu_0)/\sigma$. For a fixed sample size, power increases as δ increases. We have no control over δ, but it is the quantity most interesting to an experimenter. For any fixed effect size, power increases as sample size increases. This is because sample size does not affect the difference between the two population means, but causes the associated standard errors (σ_m) to decrease. Sample size is under the control of researchers.

If researchers wish to run experiments with high power, they must specify what δ would be meaningful to them. Once this is specified, one can determine the sample size to reject H_0 (power)100% of the time when it is false.

The importance of power draws attention to the importance of δ. Cohen suggested that $\delta = .2$ might be considered small, $\delta = .5$ might be considered medium, and $\delta = .8$ might be considered large. This classification scheme is widely used but should not be applied uniformly to all research areas. Effect sizes that are considered large in one research area might be considered small in another, and vice versa. Knowledge of δ can tell us the proportion of the H_0 distribution of scores that falls below the mean of the H_1 distribution of scores. This proportion was called U_3 by Cohen. Neither δ nor U_3 determines *practical significance* in a context-independent way. Knowing whether specific values of either of these quantities have practical significance requires deep familiarity with the research area in question.

KEY TERMS

α 178
β 178
Cohen's δ 186
hybrid model   196

$m_{critical}$ 178
null hypothesis significance
  testing   196
power   179

Type I error   177
Type II error   177
U_3 193

EXERCISES

Definitions and Concepts

Define the following concepts.

1. Type II error

2. Power

3. α and β

4. Effect size (δ)

5. Cohen's effect size classifications

6. U_3

True or False

State whether the following statements are true or false.

7. Power $= 1 - \beta$.

8. $\beta = (\mu_1 - \mu_0)/\sigma$.

9. $\delta = (\mu_1 - \mu_0)/\sigma_m$.

10. If we assume that those charged with crimes are assumed innocent until proven guilty, then a Type II error is analogous to convicting an innocent person.

11. Power is the probability of rejecting a true null hypothesis.

12. U_3 is the probability that a randomly chosen score from the H_0 distribution will exceed a randomly chosen score from the H_1 distribution.

13. If $\delta = 1$, the difference between μ_1 and μ_0 is meaningful.

14. Fisher agreed that β is a useful concept for researchers.

15. $\beta = 1 - \alpha$.

16. All things being equal, power increases as α increases.

Calculations

17. $\delta = .5$, $\mu_0 = 10$, $\sigma = 2$, and H_1: $\mu_1 - \mu_0 > 0$. Calculate β and power for each of the following situations:

    (a) $\alpha = .05$, $n = 16$;

    (b) $\alpha = .1$, $n = 16$;

    (c) $\alpha = .05$, $n = 25$; and

    (d) $\alpha = .025$, $n = 25$.

18. $\delta = .2$, $\mu_0 = 10$, $\sigma = 2$, and H_1: $\mu_1 - \mu_0 > 0$. Calculate β and power for each of the following situations:

    (a) $\alpha = .05$, $n = 16$;

    (b) $\alpha = .1$, $n = 16$;

    (c) $\alpha = .05$, $n = 64$; and

    (d) $\alpha = .05$, $n = 256$.

19. $\mu_1 = 11$, $\mu_0 = 10$. Calculate U_3 for each of the following situations:

    (a) $\sigma = 4$,

    (b) $\sigma = 2$,

    (c) $\sigma = 1$, and

    (d) $\sigma = .5$.

20. Use Table 8.1 to determine the sample size required for each of the following combinations of power and effect size:

    (a) $\delta = .8$, power $= .8$;

    (b) $\delta = .2$, power $= .8$;

    (c) $\delta = .8$, power $= .2$; and

    (d) $\delta = .2$, power $= .2$.

Scenarios

21. In Chapter 7, we described an experiment by a perception researcher who believes that the color of paper on which an exam is printed will influence the anxiety that students feel during an exam. Over the last 10 years, all of his tests were printed on white paper. Students in his classes scored 75% on average on his final exams, with a standard deviation of 3.5. This year, he printed his final exam on pink paper. He predicted that this would have a soothing influence on students, which would lead to higher test scores.

    (a) If the researcher was only interested in detecting effect sizes of .5 or greater, how many scores would he need in his sample to have power $= .8$? Assume a one-tailed test with $\alpha = .05$.

    (b) He uses this sample size to conduct the experiment and finds $m = 76.147$. Is this result statistically significant?

    (c) We can estimate δ by replacing μ_1 with the sample mean, m; i.e., using the formula $(m - \mu_0)/\sigma$. Use the information provided above to estimate δ.

    (d) Compute U_3 for this estimate.

    (e) How would his conclusion in part (b) have changed if $m = 76.147$ had been based on the sample size required to achieve power $= .9$?

22. A cognitive scientist believes that playing video games will lead to elevated alertness for several hours after game playing has ended. She quantifies alertness as the number of correctly detected targets in a 40-second stream of letters presented at random positions on a computer screen at the rate of about two letters per second. Previous studies have shown that following 2 hours of television watching, young adults detect 50 out of 75 targets during the 40-second period, with a standard deviation of 10. The researcher plans to obtain a random sample of young adults and have them play a first-person shooter video game for 1 hour. The participants will then spend 2 hours watching television. Afterward, they will do the 40-second target detection task.

    (a) If the researcher was only interested in detecting effect sizes of .3 or greater, how many participants would she need in her sample to have power $= .7$? Assume a one-tailed test with $\alpha = .05$.

    (b) She uses this sample size to conduct the experiment and finds $m = 52.489$. Is this result statistically significant?

(c) We can estimate δ by replacing μ_1 with the sample mean, m; i.e., using the formula $(m-\mu_0)/\sigma$. Use the information provided above to estimate δ.

(d) Compute U_3 for this estimate.

(e) How would the conclusion in part (b) have changed if $m = 52.489$ had been based on the sample size required to achieve power $= .6$?

23. A clinical psychologist specializes in treating seasonal affective disorder (SAD), which is a form of depression that occurs in northern countries during winter. One symptom of SAD is a loss of interest in activities that previously brought pleasure. In a long-term study of SAD, the clinician and her colleagues had patients keep a log of the hours spent on hobbies during the 3-month period from December 1 to March 1. So far, she has collected reports on 1024 patients and found that they spend an average of $\mu = 45$ minutes a day ($\sigma = 15$) engaged in hobbies during this period. She would like to examine the effects of light-box therapy on a sample of patients with SAD during the same 3-month period. (Light-box therapy involves 30 minutes per day of exposure to high-intensity lights that mimic the sun's rays.) Patients in her sample will keep the same diary as patients in the larger population. The researcher predicts that light-box therapy will increase the time spent on hobbies.

(a) The clinician is only interested in a very large effect size; i.e., $\delta = 1$. How many participants would she need in her sample to have power $= .8$? Assume a one-tailed test with $\alpha = .05$.

(b) She uses this sample size to conduct the experiment and the mean number $m = 60.564$. Is this result statistically significant? Calculate the p-value associated with this statistic.

(c) We can estimate δ by replacing μ_1 with the sample mean, m; i.e., using the formula $(m-\mu_0)/\sigma$. Use the information provided above to estimate δ.

(d) Compute U_3 for this estimate.

(e) How would her conclusion in part (b) have changed if $m = 60.564$ had been based on the sample size required to achieve power $= .6$?

SIGNIFICANCE TESTS

Problems and Alternatives

INTRODUCTION

You may be surprised to learn that statistics is currently seething with controversy. People do not disagree about basic things like sampling distributions. Rather, the controversy centers on the use of significance tests, which are by far the most widely used data analysis methods in psychology. People can get quite worked up about these issues (Bakan, 1966; Carver, 1978; Cohen, 1994; Lambdin, 2012; Rozeboom, 1960), so it can be very entertaining to read these debates.

When psychologists approach a research question, we reflexively form our questions in terms like "Does this intervention or experimental manipulation *work*?" When we state things this way, we really mean "Is there a statistically significant effect of our experimental manipulation?" Significance tests seem to provide an elegant way to make decisions about our questions, so what could be the problem? And, if there is a problem, what might be a better approach to data analysis?

In this chapter, we'll summarize some of the most frequently aired concerns about significance tests and then show how the routine use of estimation can go a long way toward addressing them. Rather than asking whether an intervention works (yes or no), it might be better to ask *how well* an intervention works.

SIGNIFICANCE TESTS UNDER FIRE

Given the prevalence of significance tests in psychology, you might think that all researchers endorse them. This is not so. Consider the following quote from Gerd Gigerenzer (2004), who has long criticized the use of significance tests in psychology:

> You would not have caught Jean Piaget [conducting a significance test]. The seminal contributions by Frederick Bartlett, Wolfgang Köhler, and the Noble laureate I. P. Pavlov did not rely on *p*-values. Stanley S. Stevens, a founder of modern psychophysics, together with Edwin Boring, known as the "dean" of the history of psychology, blamed Fisher for a "meaningless ordeal of pedantic computations" (Stevens, 1960, p. 276). The clinical psychologist Paul Meehl (1978, p. 817) called routine null hypothesis testing "one of the worst things that ever happened in the history of psychology," and the behaviorist B. F. Skinner blamed Fisher and his followers for having "taught statistics in lieu of scientific method" (Skinner, 1972, p. 319). The mathematical psychologist R. Duncan Luce (1988, p. 582) called null hypothesis testing a "wrongheaded view about what constituted scientific progress" and the Nobel laureate Herbert A. Simon (1992, p. 159) simply stated that for his

research, the "familiar tests of statistical significance are inappropriate." (Gigerenzer, 2004, pp. 591–592)

There are some really strong words in this quote from people you've probably read about. So, let's try to understand where these criticisms come from. Because many criticisms of significance tests are connected to the publication process, we will have to say a few things about this before moving on to the criticisms.

The Publication Process

I mentioned in Chapter 1 that most of your professors think of themselves as researchers. A routine part of research is publishing our research results in academic journals so that others can discuss them. Publishing is generally fun and exciting because it is one of the most important ways of engaging in a public conversation about research questions that are interesting to us. However, publishing is not an optional part of the job for your professors. They are expected to publish regularly, and their job performance is based on the number (and quality) of the papers they publish. Researchers who don't, or can't, publish their research results will not succeed, and they may lose their jobs. Because research productivity determines how they are viewed by their universities and professional colleagues, there is tremendous pressure on professors to publish. To understand some of the concerns about significance tests, we need to think about the process that researchers go through to get the results of their research published in academic journals.

Figure 9.1 illustrates the publication process. A professor typically has a laboratory housed in her university. In collaboration with other professors, graduate students, and research assistants, she runs experiments and collects data. When she thinks the results of her experiments answer her research question, she writes a paper describing the experiments, the results, and what the results mean.

When the paper is finished, the professor sends it to a research journal, where it is assigned to an editor. It is the editor's responsibility to ensure that the journal publishes high-quality research. Therefore, the editor sends the paper to two or three experts in the relevant field and asks them to read the paper to make sure that the experiments were properly run, that the statistical analyses are sound, and that the conclusions make sense.

Because the experts reviewing the paper work in the same field as the author, they probably know her from her previous publications or from meeting her at conferences. However, the reviews are typically anonymous so that the reviewers can feel free to express any concerns they have about the quality of the research. The reviewers want to be thorough but fair. Their job is to provide useful comments in a report to the editor that will help him decide whether to accept the paper.

When the editor receives the reports from the reviewers, he may be able to make a decision to accept or reject the paper right away. Very often, however, the reviewers will find the paper interesting but needing improvement. For example, the author may have failed to acknowledge relevant research from another researcher. Or the reviewers may find the conclusions unconvincing and ask for more experiments to be run or more analyses to be conducted. In such cases, the editor may ask the author to do additional work and then submit a revised version of the paper. The revised paper may be sent back to the same reviewers to see if their concerns have been answered. There can be several rounds of revisions and reviews before the editor makes a final decision to accept or reject the paper.

If the paper is accepted, it will be published in the journal and other scientists will be able to read about the research. If the paper is rejected, it will not be published in that journal, and the author will have to either look for another journal to publish it or give up and store the paper away in a filing cabinet.

FIGURE 9.1 ■ The Publication Process

A researcher runs an experiment and collects and analyzes data. She then writes a paper (manuscript) describing the experiment, the results, and what the results mean. She sends the paper to a journal, where it is assigned to an editor. The editor sends the paper to experts in the field and asks for their opinions on the merits of the paper. When the editor receives the reports from the experts, he makes a decision about whether to accept and then publish the paper or to reject it.

Figure courtesy of Danielle Sauvé.

Publication and Statistical Significance

At the heart of many research papers are claims such as A causes B. For example, we might claim that assuming a power pose for 2 minutes causes an increase in final exam grades, or that an additional 20 minutes of phonics instruction improves the reading scores of first-grade students. In the simplest case, such claims involve comparing two means (e.g., m and μ_0), computing a test statistic (e.g., z_{obs}), and determining its p-value under the null hypothesis. If $p < .05$, the result is considered statistically significant and the claim may be supported, assuming there are no problems that undermine the interpretation. Journal editors and reviewers often rely heavily on significance tests to judge whether claims are supported. In this way, statistical significance acts as a kind of filter that determines whether a paper is published and thus made available for other researchers to discuss. Unfortunately, many problems arise from the requirement for statistical significance.

CRITICISMS OF SIGNIFICANCE TESTS

The File-Drawer Problem

A major problem in psychology is the reluctance of journals and journal editors to publish papers in which statistically significant results have not been found. This form of **publication bias** means that many interesting results won't make it into the literature because the results were not supported by statistical significance. Such results may be filed away and thus not shared with other researchers, creating what we call the **file-drawer problem**.

The file-drawer problem means that results in the published literature are not representative of all results obtained from studies addressing the same question. Imagine that 16 studies independently addressed the effectiveness of a particular treatment for attention-deficit/hyperactivity disorder (ADHD). Let's say that a quarter of the studies found a statistically significant reduction in ADHD symptoms, and the other three-quarters found reductions that weren't statistically significant. If only the statistically significant results are published, they will not represent the effectiveness of the treatment.

Later in this chapter we will see that the population effect size [$\delta = (\mu_1 - \mu_0)/\sigma$] can be estimated from the sample mean [$d = (m - \mu_0)/\sigma$]. If we average the estimated effect sizes obtained in the four published studies, the resulting mean will overestimate the size of the effect in question. That is, the average of the four published effect sizes will be greater than the average of all 16 studies (including both published and unpublished). Averaging only effect sizes that made it through the $p < .05$ filter is like computing the class average on a statistics test from only those people whose grades exceeded a threshold of 75%.

Publishing only statistically significant results clearly distorts the literature and results in a misleading representation of the full body of evidence relating to a given question. This is a dangerous situation if the studies relate to health outcomes, such as the effectiveness of pharmacological treatments for depression (Turner, Matthews, Linardatos, Tell, & Rosenthal, 2008).

Proliferation of Type I Errors

A publication bias favoring statistically significant results leads to the strong possibility that many if not most published results are Type I errors (Ioannidis, 2005). Let's do a thought experiment and consider a theory predicting that a daily dose of 1000 mg of vitamin C increases IQ. I doubt this theory is true because I just made it up. If several research groups (possibly funded by the vitamin C industry) studied this theory, then most studies would fail to find a statistically significant effect because the null hypothesis is true. However, it is inevitable that some studies will find statistically significant results; i.e., Type I errors. In a world in which publication bias favors statistically significant findings, these Type I errors would have a higher probability of being published than those that failed to reject the null hypothesis. If these Type I errors are published, then anybody reviewing the literature pertaining to this theory about vitamin C and IQ would conclude that it has been supported because they would not know about the many studies, hidden away in filing cabinets, that correctly retained the null hypothesis.

Publication bias "occurs whenever the research that appears in the published literature is systematically unrepresentative of the population of completed studies" (Rothstein, Sutton, & Borenstein, 2006). One form of publication bias occurs when journals, editors, reviewers, and even authors favor publication of results that achieve statistical significance.

The **file-drawer problem** refers to the large number of papers filed away in cabinets (or hard drives) because they were unpublished. As a consequence, many valid and worthwhile results are not available to guide and inform other researchers.

A second form of publication bias favors novelty. If I predicted, long before Carney et al. (2010), that holding a power pose for 2 minutes would increase final exam grades, I think most people would have found this prediction implausible. Therefore, if I ran the experiment and found no such increase, people would be unsurprised and it would probably be very hard to get the paper published. However, if the same experiment produced a statistically significant increase in grades, this would be viewed as an exciting new finding and the chances of being published would be much greater.

A publication bias favoring novelty is compounded by the fact that it is the policy of some journals not to publish replications of previously reported, statistically significant results. A notorious example of this happened recently when Bem (2011) published a paper in the *Journal of Personality and Social Psychology* titled "Feeling the Future: Experimental Evidence for Anomalous Retroactive Influences on Cognition and Affect." Here is how one of the experiments was described to the participants:

> [T]his is an experiment that tests for ESP. It takes about 20 minutes and is run completely by computer. First you will answer a couple of brief questions. Then, on each trial of the experiment, pictures of two curtains will appear on the screen side by side. One of them has a picture behind it; the other has a blank wall behind it. Your task is to click on the curtain that you feel has the picture behind it. The curtain will then open, permitting you to see if you selected the correct curtain. There will be 36 trials in all.
>
> Several of the pictures contain explicit erotic images (e.g., couples engaged in nonviolent but explicit consensual sexual acts). If you object to seeing such images, you should not participate in this experiment. (Bem, 2011, p. 409)

The novel twist in the experiment was that the window showing the picture was chosen at random by a computer *after* the participant had made his or her choice. Therefore, the choice is (arguably) about the future state of the world.* The null hypothesis in this case is that participants would have a 50% chance of guessing which of the two curtains hid the erotic image. However, it was found that participants guessed correctly 53% of the time, *on average*, and this turned out to be statistically significant. Bem concluded that these results constitute positive evidence that people can see or sense future events.

If this study truly demonstrated that people can "feel the future," it would be the most important experiment ever reported in human history, and every domain of science would have to be revised fundamentally in view of it. If any experiment calls out for replication to ensure that it is not a Type I error, it is this one. However, when a paper reporting an unsuccessful attempt to find the same results was submitted to the same journal, the editor rejected it, saying that it was the journal's policy to publish only original studies and not replications. The editor in question was quoted in the *New Scientist* as saying, "This journal does not publish replication studies, whether successful or unsuccessful" (Aldhous, 2011).

This episode illustrates the unfortunate fact that Type I errors are far easier to get into the literature than to remove from the literature. Science is supposed to be self-correcting, but when journals devalue replication, errors become difficult to correct.**

---

*I'm not sure why it wasn't taken as evidence for participants reading the current state of the random number generator in the computer through extrasensory perception.

**A bit of hopeful news here is that the public outcry over this event caused the *Journal of Personality and Social Psychology* to accept attempted replications of the Bem experiments (Galak, LeBoeuf, Nelson, and Simmons, 2012). Not surprisingly, Galak et al. did not find any evidence that people can "feel the future."

p-Hacking: The Quest for Statistical Significance

Because there is a publication bias favoring results that are statistically significant, researchers naturally feel pressured to find statistically significant results. This pressure does not necessarily result in dishonest behavior, but it can result in a multitude of questionable practices in which researchers make undisclosed adjustments to their data analysis procedure in efforts to produce statistically significant results. These practices are collectively known as *p*-**hacking**. Uri Simonsohn defined *p*-hacking as "trying multiple things until you get the desired result" (Nuzzo, 2014), where the desired result is statistical significance.

We encountered a simple example of *p*-hacking in Chapter 7 when we noted that a researcher may decide, after the experiment has been run, to test for statistical significance using a one-tailed test rather than a two-tailed test as originally planned. Rex Kline (2013) refers to this as an instance of hypothesizing after the results are known, or HARKing.

A very common type of *p*-hacking involves running many experiments, or conducting multiple analyses on the same data, and reporting only the results that are statistically significant or those that are consistent with the predictions of the researcher. Such practices very often create a misleading picture of reality and increase the likelihood that Type I errors will make their way into the literature. This type of *p*-hacking is brilliantly illustrated by an xkcd cartoon that can be found at xkcd.com/882. The cartoon shows twenty significance tests, each one assessing the link between acne and jelly beans of a specific color. In one case (green jelly beans) a statistically significant result was found. If only this result is reported, as in the cartoon, then a reader has no idea about the other 19 analyses that did not produce statistically significant results. That is, the reader has no way of knowing that the result is almost certainly a Type I error. (If you ever need a break from studying statistics, visit xkcd.com for hours of top-notch entertainment. Some comics are NSFW.)

Another form of *p*-hacking is when researchers increase their sample size in hopes of eventually getting a statistically significant result. Let's think back to my theory about vitamin C and IQ. We will assume that IQ is normally distributed in college students, with a mean of 115 and standard deviation of 15; this is my null hypothesis distribution of scores. I then take a random sample of 25 students and put them on a high vitamin C diet for a month, after which I measure their IQs. I find that their mean IQ is 118.6, and from this, I compute $z_{obs} = (118.6 - 115/3 = 1.2, p = .12$. This result is not statistically significant, but the difference is in the predicted direction and the *p*-value is sort of small. I might conclude that my test lacked power because the sample size was too small. Therefore, I continue to add participants to my sample and test for statistical significance after each additional participant is added. Finally, I end the experiment when I find a statistically significant result.

We could call this the "run-until-statistically-significant" (RUSS) approach. Many researchers believe this is a perfectly legitimate strategy; I know I did for a long time. However, it is definitely not okay to do this. The problem with the RUSS approach is that even when the null hypothesis is true, there will be random variations in the statistic (and its associated *p*-value) as we add more participants to our sample. At some point, the statistic (e.g., z_{obs}) may become statistically significant because of these chance fluctuations.

The random fluctuations in *p*-values are illustrated in Figure 9.2. Starting with a sample of size 1, *drawn from the null hypothesis distribution*, we compute z_{obs} and record its *p*-value. We then add another score to the sample and test for statistical significance. We repeat this process until sample size is 50. Note that our 50 samples are not independent samples of different sizes. Rather, a sample of size *n* contains all the scores from the sample of size *n*−1, plus one more.

In Figure 9.2, sample size is plotted on the *x*-axis and *p*-values on the *y*-axis. As you can see, the *p*-values "wander around" as sample size increases. In this case, the *p*-value

The term *p*-hacking refers to a multitude of questionable practices in which researchers make undisclosed adjustments to their data analysis procedure in efforts to produce statistically significant results.

FIGURE 9.2 ■ *p*-Hacking: RUSS

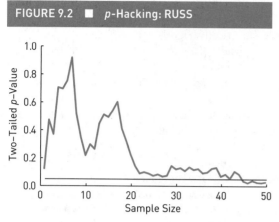

An illustration of how *p*-values can change as sample size increases. The sample is built up over time by drawing scores from the null hypothesis distribution. We start with a sample size of 1, and we add another randomly selected score on each of the remaining 49 trials. On each trial, z_{obs} and its two-tailed *p*-value are computed. The *p*-values change over time. In this example, the *p*-value drops below the $p < .05$ criterion for statistical significance (indicated by the black horizontal line) at trial 45. The RUSS strategy ends the experiment when *p* drops below the $p < .05$ criterion.

eventually drops below the $p < .05$ criterion for statistical significance, indicated by the black horizontal line. If data collection stops at this point, we might be inclined report that the obtained z_{obs} is statistically significant, $p < .05$. As we will see, however, the actual *p*-value associated with z_{obs} is much greater than .05.

Imagine running the process shown in Figure 9.2 10,000 times. The process stops when either (i) sample size is 50 or (ii) the sample produces a two-tailed z_{obs} for which $p < .05$. In this situation, about 30% of these 10,000 trials will stop because the $p < .05$ criterion has been obtained. The trials comprising this 30% are Type I errors because the scores were drawn from the null hypothesis distribution! The *p*-value associated with our z_{obs} statistic is clearly much greater than .05. Therefore, widespread use of the RUSS approach will inflate the number of Type I errors in the literature. As we've seen, Type I errors have an increased chance of being published, and once published it is difficult to publish replication studies to correct the error.

Binary Thinking

The quest for statistical significance produces a tendency to think of statistically significant results as important and meaningful, and results that are not statistically significant as unimportant and meaningless. As we will see at the end of the chapter, it is possible for two very similar results to differ in their statistical significance (e.g., $p = .04$ versus $p = .06$). It is silly to treat these two results as different in any meaningful way, but a reliance on statistical significance can seduce us into doing just that. Worse, psychologists in the past were instructed to think in exactly this way by the *APA Publication Manual*. The effects of this guidance are still with us.

Statistical Significance Versus Practical Significance

We've already discussed the important difference between statistical significance and practical significance. One effect of the $p < .05$ filter is that achieving statistical significance can become the goal of the research, rather than an aid to reasoning. Consequently, one might consider the job done once the null hypothesis has been rejected. We've seen, however, that statistical significance does not mean that the result is important or meaningful. Unfortunately, when the focus is on statistical significance, we may spend less time than we should considering the practical significance of a result. This can lead to very shallow research in which a large number statistically significant results "don't amount to a hill of beans," as Humphrey Bogart said to Ingrid Bergman in *Casablanca*. (You might have to Google this.)

It is a useful thought experiment to consider what would happen if researchers were not allowed to use significance tests (Harlow, Mulaik, & Steiger, 2016). In this case, a result would not carry weight just because it is statistically significant. Instead, researchers would have to explain things such as (i) why the magnitude of the difference is important, (ii) how it changes our view of a particular theory, or (iii) what it implies about the effect of some treatment on some population. I don't mean to imply that researchers don't do this at all. Rather, it is a question of balance. In my view, a result having low practical or theoretical significance can gain unwarranted stature if it is statistically significant.

Misunderstanding *p*-Values

A frequent complaint about significance tests is that many researchers don't know what statistical significance means. We saw in Chapter 7 that a *p*-value is a conditional probability. It is the probability of the obtained statistic, or one more extreme, occurring by chance when the null hypothesis is true. However, as noted by Cohen more than 20 years ago, many people who should know better are prone to committing the inverse probability error, which is the belief that the *p*-value represents the probability that the null hypothesis is true. As Cohen (1994) argues, the inverse probability error is really a case of wishful thinking. Researchers really want to know the probability of the null hypothesis being true, and they let their wishes lead them to believe that this is what *p* tells them, when it does nothing of the sort.

There are other misinterpretations of *p*-values documented by Oakes (1986), Haller and Kraus (2002), Carver (1978), and Kline (2004, Chapter 3), among many others. Among these misinterpretations are the following:

1. The belief that a *p*-value is the probability that you've made a Type I error. We saw in Chapter 8 that this is a confusion between *p* and α, arising from the hybrid Fisher-Neyman model. The correct interpretation is that α is the probability that *the test* will produce a Type I error, and *p* is the conditional probability of the statistic (derived from *the data*) under the null hypothesis. Unfortunately, someone believing that $p = .0062$ is the probability that a Type I error has been committed would have little reason to think that the study should be replicated.

2. The belief that $1-p$ is the probability that the same experiment will produce a statistically significant result if it is repeated. An innocent-sounding instance of this occurs when researchers say that small *p*-values indicate *reliable* (presumably replicable) results. It is true that there is a relationship between *p* and the probability of replication, but the relationship is not a simple one. The probability that a result will replicate is certainly not $1-p$. The consequences of this mistaken belief can be quite bad. If you think that $1-p$ is the probability that the same experiment will produce a statistically significant result if it is repeated, then when $p = .0062$ you think the probability is .9938 that your experiment will replicate. This misunderstanding also might lead one to think that doing an actual replication would be a waste of time.

3. The belief that $1-p$ is the probability that the alternative hypothesis is true. The Fisher-Neyman hybrid model probably explains this confusion. In an acceptance procedure, one of two hypotheses is accepted. When this is merged with a significance test, then rejecting H_0 is equivalent to accepting H_1. So, what do we do with a *p*-value in this case, given that it plays no role in an acceptance procedure? Well, it seems just a short step to see the *p*-value as a measure of the probability that H_0 is true (the inverse probability error) and $1-p$ as the probability that H_1 is true. This is impossible; *p*-values are derived from the assumption that H_0 is true, with no consideration of H_1.

Misunderstandings about *p*-values are so widespread that the American Statistical Association, in a highly unusual move, published a statement about the meaning of *p*-values (Wasserstein & Lazar, 2016). It is not a good thing when a professional does not understand the most commonly used tool in his or her toolbox. We'd be concerned about a surgeon who uses the wrong end of a scalpel.

In summary, statistical significance is a kind of filter that determines which papers make it into the publicly available research literature. This means that the published literature may seriously misrepresent the size of some effect. We would have a much better sense of any given effect if we knew about all relevant results, not just those that passed through the $p < .05$ filter. Furthermore, the criterion for statistical significance ensures that if some effect *does not exist*, it is more likely that a paper reporting that the effect does exist (a Type I error) will enter the literature than a paper that says it doesn't exist (correctly retains the null hypothesis). In a world in which replication is not valued, Type I errors will linger in the literature for years, misleading many other researchers and thus wasting their time and slowing the progress of science. This problem is compounded by the p-hacking that occurs when researchers (who are under pressure to publish) selectively report data in order to present a story that ends with a rejected null hypothesis that supports their claims. An emphasis on statistical significance also leads to a black-and-white worldview in which only results that are statistically significant are treated as interesting and meaningful. Equating statistical significance with significance can lead to very shallow thinking. If we add to these points reports that many researchers misunderstand the meaning of a p-value, then I'd say we have a problem.

CONFIDENCE INTERVALS

I have seen speakers deliver eloquent summaries of problems like those listed above to rooms full of researchers who routinely use significance tests as part of their professional activities. The really interesting thing is that it is rare for researchers to rise and defend significance tests against these criticisms. The most common response is, "True, true, but what's the alternative?" This dynamic also exists in the statistics literature in general. Kline (2013) notes that there are hundreds of papers criticizing significance tests, and there are only a handful of defenses. So, what is the alternative?

For many, estimation with confidence intervals provides the healthiest alternative to significance tests. Significance tests and confidence intervals rest on the same theoretical foundations (i.e., sampling distributions) but differ in their objectives. Significance tests are designed *to decide* what a parameter *is not*, whereas confidence intervals are designed *to estimate* what a parameter *is*. When you reject the null hypothesis, you are rejecting a specific hypothesis about the mean of the distribution from which a sample was drawn. When you construct a confidence interval, you are providing your best estimate about the mean of the distribution from which a sample was drawn. When we move attention away from decision making based on the $p < .05$ criterion, we can focus on the size of an effect and its practical significance.

> ### DECISIONS VERSUS ESTIMATION
>
> Significance tests are designed to decide what a parameter *is not*, whereas confidence intervals are designed to estimate what a parameter *is*.

The Advantages of Confidence Intervals

In my view, (almost) any steps that eliminate the $p < .05$ filter from the publication process in psychology and related disciplines will produce improvements over the current state of affairs. Without the $p < .05$ filter, there would be no need for p-hacking or binary thinking, less of a file-drawer problem, and fewer Type I errors lodged in the literature. Furthermore, without the $p < .05$ crutch, there would be more emphasis on replication and greater effort to explain the practical significance of a result.

LEARNING CHECK 1

1. With regard to the proper interpretation of p-values, explain why the following are wrong:

   (a) Somebody tells you that $1-p$ is the probability that an experiment will replicate.

   (b) Somebody tells you that $1-p$ is the probability that H_1 is true.

   (c) Somebody tells you that p is the probability that you've made a Type I error.

2. Provide an example in which a statistically significant result is of no practical significance. (Be creative.)

3. A researcher tells you that he plans to run an experiment to test the effect of THC (the psychoactive agent in marijuana) on a problem-solving task known to have a mean of $\mu_0 = 25$ in the general population. He says he thinks 18 participants should be sufficient to conduct a two-tailed test of the null hypothesis $\mu_1 = \mu_0$. What questions might you ask him that would help him run a better experiment?

4. A researcher wondered whether verbal reasoning might improve following 15 minutes of listening to a Mozart sonata. After running the experiment with 21 participants, he found an improvement of 6 points on a verbal reasoning task, $z_{obs} = 1$, $p = .159$. He thought this was encouraging, but concluded that his sample was not large enough. Therefore, he continued adding participants until he reached 36 participants, at which point he again found a 6-point improvement, $z_{obs} = 1.7$, $p = .045$. Would you have anything to say to this experimenter about the legitimacy of what he's done?

5. A researcher wondered about the effects of alcohol consumption on driving ability. She chose a random sample of 21 college students and measured the number of driving errors they made (on a controlled driving track) once their blood alcohol level was elevated to .05 g/dL, which is below the legal limit for intoxication in most jurisdictions. She found an elevation in driving errors corresponding to $z_{obs} = 1.4$, $p = .081$. Because this increase was not statistically significant according to the $p < .05$ criterion for statistical significance, she concluded that a blood alcohol level of .05 g/dL posed no increased risk of driving accidents. Please comment on this conclusion.

Answers

1. (a) $1-p$ is the probability of obtaining a statistic less extreme than the one you obtained *when the null hypothesis is true*. $1-p$ is not derived from any consideration of the alternative hypothesis and so can't possibly be the probability that the result will replicate.

   (b) For the same reason given in (a), $1-p$ cannot be the probability that H_1 is true.

   (c) This claim confuses p with α. α specifies the probability of a Type I error. p is the probability of the obtained data (or data more extreme) occurring by chance when the null hypothesis is true. α is a property of the test, whereas p is a property of the data.

2. You can choose your own example because infinitely many silly tests can be imagined. If I take a random sample of 121 psychology students and measure the number of push-ups that each can do, I'm pretty sure that the difference between the mean of these 121 scores and $\mu_0 = 200$ would be statistically significant. I can't attach any meaning to this result.

3. I'd ask why he chose $n = 18$ participants. If he were to say, "Well, that seems like enough," you could then ask him to think about what effect size would be interesting to him, and then tell him a little bit about power analysis. On the other hand, he might have said he is only interested in an effect size of $\delta = .6$ or greater, and 18 participants gives him power = .8. In this case, he gets a pat on the head and a big gold star.

4. I would say: "Running the experiment until you find a statistically significant result increases the probability of obtaining statistical significance when the null hypothesis is true. You should have done a prospective power analysis, chosen your sample size, and stuck with it."

5. Failing to reject the null hypothesis does not mean that the null hypothesis is true. In this case, the data suggest that an increased blood alcohol level increases driving errors and in the real world would increase the probability of driving accidents, which often have devastating consequences.

The routine use of confidence intervals would go a long way to realizing these benefits. With confidence intervals, the focus is on the precision of a parameter estimate for some variable of interest. The actual size of a measured difference would be central to our thinking, rather than the probability of the difference under the null hypothesis. Because the focus would be on the size of a measured difference, we would not be drawn into binary thinking. Furthermore, any measured difference would be seen as a single estimate with unavoidable imprecision. Therefore, replication with an eye to combining measures from different studies would be far more common.

Significance Tests With Confidence Intervals

At a purely technical level, estimation with confidence intervals is a far more general method of data analysis than significance tests. In this section we'll see that once you have computed a confidence interval, you can test any null hypothesis of interest simply by asking whether μ_0 falls within the interval. For this reason, confidence intervals are the method recommended by the most recent versions of the *APA Publication Manual*. We will now revisit examples of significance tests from previous chapters from the perspective of computing confidence intervals.

Please note that *I am absolutely not recommending the routine use of confidence intervals to test null hypotheses*! This would be an egregious misuse of estimation. Rather, the main point is to emphasize that significance tests represent a minor feature of what can be done with confidence intervals. Understanding the secondary role of significance tests helps us recognize the poverty of the information they yield.

Whole Language Versus Phonics: Revisited

In the phonics example that opened Chapter 7, we were told that the mean of the null hypothesis distribution was $\mu_0 = 72$. We were also given the sample mean ($m = 74.5$), the population standard deviation ($\sigma = 8$), and sample size ($n = 64$). Using σ and sample size, we were able to compute $\sigma_m = \sigma/\sqrt{n} = 8/\sqrt{64} = 1$. This information is exactly what is needed to compute a confidence interval. The 95% confidence interval around m is as follows:

$$\text{CI} = m \pm z_{\alpha/2}(\sigma_m) = 74.5 \pm 1.96(1) = [72.54, 76.46].$$

Therefore, we have 95% confidence that the mean of the distribution from which the sample was drawn is in the interval [72.54, 76.46]. This confidence interval allows us to reject $\mu_0 = 72$ as a plausible hypothesis about the mean of the distribution from which the phonics-enriched students were drawn, *because 72 does not fall within it*.

Figure 9.3 illustrates the connection between confidence intervals and significance tests. The normal distribution in Figure 9.3 is the sampling distribution of the mean under the null hypothesis. Figure 9.3 compares two intervals: $\mu_0 \pm 1.96(\sigma_m)$ (the light blue vertical lines) and $m \pm 1.96(\sigma_m)$ (the confidence interval around the dot, representing the sample mean). The arms of both intervals are the same. Therefore, if m is outside the interval $\mu_0 \pm 1.96(\sigma_m)$, the confidence interval $m \pm 1.96(\sigma_m)$ will not capture μ_0. If m is within the interval $\mu_0 \pm 1.96(\sigma_m)$, the confidence interval $m \pm 1.96(\sigma_m)$ will capture μ_0. (Think back to the arms in the illustrations from Chapter 6 in the section on sampling distributions and confidence.)

FIGURE 9.3 ■ Estimation and Significance Tests

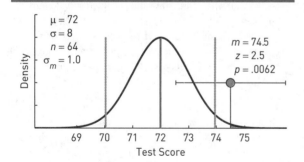

The connection between confidence intervals and two-tailed significance tests. The normal distribution is a distribution of sample means for size $n = 64$, drawn from a population of reading scores having a mean of $\mu_0 = 72$ and standard deviation of $\sigma = 8$. The mean of the sample is $m = 74.5$ and $\sigma_m = 1$. The dot plots the sample mean. The arms around the dot represent the 95% confidence interval. Notice that the interval does not capture μ_0, which is the mean of the distribution, according to the null hypothesis. Therefore, we can reject a two-tailed test of the null hypothesis at the $p < .05$ level.

Another way to see this relationship is to note that when a 95% confidence interval does not capture μ_0, then $z_{obs} = (m - \mu_0)/\sigma_m$ will be outside the interval ±1.96. If the 95% confidence interval captures μ_0, then $z_{obs} = (m - \mu_0)/\sigma_m$ will be within the interval ±1.96. Therefore, any time a 95% confidence interval does not capture μ_0, we can reject the null hypothesis at the $p < .05$ criterion for statistical significance.

If the 95% confidence interval around the sample mean does not capture μ_0 specified by H_0, we are faced with the same two possibilities we face in any significance test. Either H_0 is true, and this is one of those rare times that the sample mean falls a long way from the mean of the distribution (μ_0), or the confidence interval does not capture μ_0 because H_0 is false, and so μ_0 is not the mean of the distribution from which the sample was drawn. Of course, the second interpretation means we reject the null hypothesis. Therefore, in the phonics example, we could reject a two-tailed test of H_0 at the $p < .05$ level because the 95% confidence interval around the sample mean ($m = 74.5$) did not capture the mean specified by H_0 ($\mu_0 = 72$). More importantly, the confidence interval provides evidence of what μ_1 (i.e., $\mu_{phonics}$) *is*, not just what it *is not*.

Differences in IQ: Revisited

In the IQ example from Chapter 7, we were told that the mean IQ of Quebec residents was $\mu_{Que} = 100$ with a standard deviation of $\sigma = 15$. We were also told that the mean IQ of $n = 250,000$ people from Maine was $m = 100.06$. Using σ and sample size ($n = 250,000$), we were able to compute $\sigma_m = 0.03$. When we compute the 95% confidence interval around m, we find the following:

$$CI = m \pm z_{\alpha/2}(\sigma_m) = 100.06 \pm 1.96(0.03) = [100.0012, 100.1188].$$

As in the reading example, our confidence interval does not capture the known population mean, $\mu_{Que} = 100$. Therefore, we can reject a two-tailed test of H_0 at the $p < .05$ level. Besides revealing that 100 is an implausible hypothesis about the mean IQ of Maine residents, the confidence interval provides the best estimate of what μ_{Maine} is (i.e., 100.06). It also provides and interval around this mean, and we have 95% confidence that μ_{Maine} is in this interval (95% CI [100.0012, 100.1188]). As with the reading example, confidence intervals provide us with information about what the parameter in question *is*, in addition to what it *is not*.

Bench-Pressing 5-Year-Olds: Revisited

In the bench-pressing example from Chapter 7, we were told that the mean bench-pressing weight for 15-year-old European males was $\mu_{15YO} = 125$ pounds. We were also told that the mean bench-pressing weight for a sample of one hundred 5-year-old European males was $m = 10$ pounds, with sample standard deviation, $s = 2$. Using s and sample size, we were able to compute $s_m = s/\sqrt{n} = 2/10 = 0.2$. When we compute the approximate 95% confidence interval around m, we find the following:

$$CI = m \pm z_{\alpha/2}(s_m) = 10 \pm 1.96(0.2) = [9.61, 10.39].$$

As in the previous two examples, our confidence interval does not capture the known population mean, $\mu_{15YO} = 125$. Therefore, we can reject H_0 at the $p < .05$ level. In addition to revealing that 125 is an implausible hypothesis about μ_{5YO}, the confidence interval provides the best estimate of μ_{5YO} (i.e., 10). It also provides an interval around this estimated mean, and we have 95% confidence that μ_{5YO} is in this interval (95% CI [9.61, 10.39]). Again, this confidence interval provides us with information about what the parameter in question *is*, in addition to what it *is not*.

As a final point, note that in this example we used the sample standard deviation s in the calculation of s_m. The question of what s estimates does not arise in this situation. The sample standard deviation, s, clearly estimates σ_{5YO}. We didn't have to assume that σ_{15YO} and σ_{5YO} are the same, as we did in Chapter 7, so this makes things a little cleaner.

Retaining the Null Hypothesis

We noted in Chapters 7 and 8 that retaining H_0 does not mean that it is true. This is particularly clear when we test H_0 with a confidence interval. Think of the following null hypothesis in a case where $m = 26$, $\sigma = 8$, and $n = 16$:

$$H_0: \mu_0 = 25.$$

The 95% confidence interval around $m = 26$ is computed as follows:

$$\text{CI} = m \pm z_{\alpha/2}(\sigma_m) = 26 \pm 1.96(2) = [22.08, 29.92].$$

This confidence interval contains $\mu_0 = 25$, so we would retain H_0. In this example, it can be seen that many population means (μ) are consistent with this interval. That is, 25 is just one of many plausible hypotheses about the mean of the population from which the sample was drawn. Therefore, this example makes it very clear that retaining $H_0: \mu_0 = 25$ *does not mean* that 25 *is* the population mean. In fact, the most plausible estimate is that $\mu = m$.

*One- and Two-Tailed Tests (Optional Material)

In this section, we will see how to conduct one-tailed tests with confidence intervals. This material is presented for completeness only, and it can be skipped without doing damage to your understanding of the connection between confidence intervals and significance tests.

In the original phonics scenario, the researcher predicted that phonics instruction would improve reading scores. Therefore, her significance test was a one-tailed test in which she predicted $\mu_1 - \mu_0 > 0$. This raises the question of how to conduct a one-tailed significance test using confidence intervals. The easiest way to do this (when $\alpha = .05$) is to compute the 90% confidence interval around the mean, and then reject H_0 *only if* (i) the interval does not capture μ_0 and (ii) the sample mean is on the predicted side of μ_0. We will illustrate this procedure while referring to Figure 9.4.

First, in the phonics example, the 90% CI would be

$$\text{CI} = m \pm z_{\alpha/2}(\sigma_m) = 74.5 \pm 1.645(1) = [72.86, 76.15].$$

This confidence interval is shown in Figure 9.4. The interval does not capture μ_0, so that is the first requirement for concluding that there is a statistically significant difference between m and μ_0. The second requirement is that the sample mean must be on the side of μ_0 predicted by the alternative hypothesis. In this case, the alternative hypothesis predicts m will be greater than μ_0, so this requirement is satisfied as well. Therefore, we can reject the null hypothesis at the $p < .05$ level of statistical significance. As in the case of significance tests with z_{obs}, if the alternative hypothesis had been $\mu_1 - \mu_0 < 0$, we would have retained the null hypothesis.

Having shown that one can do significance tests with confidence intervals, I must reiterate that I don't recommend doing so. If you feel that you must do a significance test, then simply use a standard z-test; i.e., compute z_{obs}. That is, if you find yourself living in the high-contrast world of statistical significance, then use a test statistic such as z_{obs}, which is the coin of the realm.

FIGURE 9.4 ■ Estimation and Directional Tests

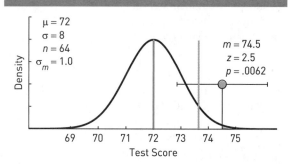

The connection between confidence intervals and one-tailed significance tests. This is a modified version of Figure 9.3. The distribution of sample means of size $n = 64$ was drawn from a population of reading scores having a mean of $\mu_0 = 72$ and standard deviation of $\sigma = 8$. The mean of the sample is $m = 74.5$ and $\sigma_m = 1$. The dot plots the sample mean. The arms around the dot represent the 90% confidence interval. Notice that the interval does not capture the μ_0, which is the mean of the distribution, according to the null hypothesis. Only if the alternative hypothesis predicts $\mu_1 - \mu_0 > 0$ could we reject a one-tailed test of the null hypothesis at the $p < .05$ level. If the alternative hypothesis predicts $\mu_1 - \mu_0 < 0$, we would have to retain the null hypothesis.

LEARNING CHECK 2

1. A researcher conducts a two-tailed test of H_0: $\mu_1 - \mu_0$ = 0, where $\mu_0 = 25$. His alternative hypothesis is H_1: $\mu_1 - \mu_0 \neq 0$. In which of the following cases can he reject H_0 at the $p < .05$ significance level: (a) 95% CI [23, 26], (b) 95% CI [23, 24], (c) 95% CI [26, 27], and (d) 95% CI [24, 26]?

2. *A researcher conducts a one-tailed test of H_0: $\mu_1 - \mu_0 = 0$, where $\mu_0 = 25$. His alternative hypothesis is H_1: $\mu_1 - \mu_0 > 0$. In which of the following cases can

he reject H_0 at the $p < .0$ significance level: (a) 90% CI [22, 27], (b) 95% CI [23, 24], (c) 90% CI [26, 27], and (d) 90% CI [18, 24]?

3. *A researcher conducts a one-tailed test of H_0: $\mu_1 - \mu_0$ = 0, where $\mu_0 = 25$. His alternative hypothesis is H_1: $\mu_1 - \mu_0 < 0$. In which of the following cases can he reject H_0 at the $p < .05$ significance level: (a) 90% CI [22, 27], (b) 95% CI [23, 24], (c) 90% CI [26, 27], and (d) 90% CI [18, 24]?

Answers

1. (a) No. The interval contains $\mu_0 = 25$.
   (b) Yes. The interval does not contain $\mu_0 = 25$.
   (c) Yes. The interval does not contain $\mu_0 = 25$.
   (d) No. The interval contains $\mu_0 = 25$.

2. (a) No. The interval contains $\mu_0 = 25$.
   (b) No. The interval does not contain $\mu_0 = 25$, but it is on the wrong side of μ_0.
   (c) Yes. The interval does not contain $\mu_0 = 25$, and it is on the predicted side of μ_0.

   (d) No. The interval does not contain $\mu_0 = 25$, but it is on the wrong side of μ_0.

3. (a) No. The interval contains $\mu_0 = 25$.
   (b) Yes. The interval does not contain $\mu_0 = 25$, and it is on the predicted side of μ.
   (c) No. The interval does not contain $\mu_0 = 25$, but it is on the wrong side of μ_0.
   (d) Yes, the interval does not contain $\mu_0 = 25$, and it is on the predicted side of μ_0.

ESTIMATING $\mu_1 - \mu_0$

There is an important point about significance tests that questions whether H_0 could ever be true. Consider pairs of populations, such as the IQs of men and women, the IQs of US citizens living east and west of the Mississippi River, or the heights of 20-year-old men born on odd and even days of the year. In none of these situations would we have any reason to expect a big difference in the means of the two populations. On the other hand, would we ever expect the difference between the two population means to be exactly 0? Not 0.1, 0.0032, or 0.0089, but exactly 0. It seems highly unlikely that the difference between any two populations, no matter how similar, will be exactly 0. If this is the case, there is a good argument that it makes little sense to test a null hypothesis that there is exactly zero difference.

If the null hypothesis that $\mu_1 = \mu_0$ has very little chance of ever being true, then a better approach to data analysis is to estimate the difference between μ_1 and μ_0. Therefore, to judge "how wrong" the null hypothesis is, we can estimate $\mu_1 - \mu_0$. Estimating $\mu_1 - \mu_0$ involves a minor modification to the method covered in the previous section.

In the phonics scenario, the known distribution of reading scores with $\mu_0 = 72$ and $\sigma = 8$ was the null hypothesis distribution. The distribution of reading scores under the alternative hypothesis has a mean (μ_1) that is unknown. However, it is assumed that the standard deviation (σ) of this distribution is equal to the standard deviation of the known (null hypothesis) distribution. From these observations it follows that the sampling distribution of the mean under the alternative hypothesis would have mean μ_1 and a standard error of $\sigma_m = \sigma/\sqrt{n}$.

To estimate $\mu_1 - \mu_0$, we must consider subtracting μ_0 from each mean in this distribution of means under the alternative hypothesis. Doing so will produce a distribution of the statistic $m - \mu_0$. The sampling distribution of $m - \mu_0$ will have mean $\mu_1 - \mu_0$ and standard error $\sigma/\sqrt{n}$. This standard error is just the standard error of the mean that we've been using all along. However, we will refer to it in a way that makes clear that it is associated with the statistic $m - \mu_0$. Therefore, the standard error of $m - \mu_0$ will be called $\sigma_{m-\mu_0}$, and its value is

$$\sigma_{m-\mu_0} = \sigma/\sqrt{n}. \tag{9.1}$$

A confidence interval around $m - \mu_0$ is computed as follows:

$$\text{CI} = (m - \mu_0) \pm z_{\alpha/2}\left(\sigma_{m-\mu_0}\right). \tag{9.2}$$

This is exactly like all confidence intervals we've computed before, except that now we're estimating $\mu_1 - \mu_0$ rather than μ_1.

Using the numbers from the phonics scenario, the 95% confidence interval around $m - \mu_0$ is computed as follows:

Step 1. Compute $m - \mu_0$.

$$m - \mu_0 = 74.5 - 72 = 2.5.$$

Step 2. Compute $\sigma_{m-\mu_0}$.

$$\sigma_{m-\mu_0} = \sigma/\sqrt{n} = 8/\sqrt{64} = 1.$$

Step 3. Find $z_{\alpha/2}$. Because this is the 95% confidence interval, $\alpha = .05$. Therefore, $\alpha/2 = .025$. Using the z-table, we find that 2.5% of the z-distribution falls below -1.96. Therefore, $z_{\alpha/2} = 1.96$.

Step 4. Compute $(m - \mu_0) \pm z_{\alpha/2}\left(\sigma_{m-\mu_0}\right)$.

$$\text{CI} = (m - \mu_0) \pm z_{\alpha/2}\left(\sigma_{m-\mu_0}\right) = (74.5 - 72) \pm 1.96(1) = [0.54, \ 4.46].$$

We have 95% confidence that the true difference between μ_1 and μ_0 is in the interval [0.54, 4.46]. Our confidence comes from knowing that 95% of all intervals computed this way will capture $\mu_1 - \mu_0$. This means that our best estimate of the difference between the means of the whole language and phonics distributions is 2.5. However, a difference of 0.54 is just as likely as a difference of 4.46. The practical significance of this difference depends on all the factors discussed in Chapter 7.

To see that nothing mysterious has been done here, we can look at the situation slightly differently. The confidence interval around m was [72.54, 76.46] and the confidence interval around $m - \mu_0$ is [0.54, 4.46]. Therefore, the confidence interval around $m - \mu_0$ is simply equal to the confidence interval around m minus μ_0. That is, [72.54, 76.46] − 72 = [0.54, 4.46]; μ_0 has been subtracted from the upper and lower limits of the 95% confidence interval around m.

According to the null hypothesis, $H_0: \mu_1 - \mu_0 = 0$. Testing this null hypothesis is just a matter of asking whether 0 is in this interval $(m - \mu_0) \pm z_{\alpha/2}(\sigma_{m-\mu_0})$. Because it is not, a two-tailed test of the null hypothesis can be rejected at the $p < .05$ level of significance. However, this confidence interval also provides our best estimate of what $\mu_1 - \mu_0$ *is* (95% CI [0.54, 4.46]) in addition to what it is *not* (0). So, if the null hypothesis is never really true, a confidence interval around $m - \mu_0$ provides a simple method of estimating just how wrong it is.

LEARNING CHECK 3

1. A researcher with a long-standing interest in visual memory has collected data on the ability of university students to recall the details of five 30-second video clips. All participants were asked 100 questions about the videos; for example, How many people were in the video about the car theft? What color was the cottage in the video about the summer camp? What was the name on the store in the video about the hockey team? The average score on this test was $\mu_0 = 38$ with $\sigma = 10$. A random sample of 25 university students was shown the same five videos and asked the same 100 questions. Unlike all previous participants, these 25 participants were asked to answer the questions with their eyes closed (Nash, Nash, Morris, & Smith, 2015). The mean score for the sample of 25 students was $m = 42$. Use this information to compute the 95% confidence interval about an estimate of $\mu_1 - \mu_0$. Use this estimate to perform a two-tailed test of the null hypothesis in which $H_0: \mu_1 - \mu_0 \neq 0$.

Answers

1. The estimate of $\mu_1 - \mu_0$ is $m - \mu_0 = 42 - 38 = 4$. The standard error is $\sigma_{m-\mu_0} = \sigma/\sqrt{n} = 10/\sqrt{25} = 2$. Therefore, the 95% confidence interval is

$$\text{CI} = m \pm z_{\alpha/2}\left(\sigma_{m-\mu_0}\right) = 4 \pm 1.96(2) = [0.08, \ 7.92].$$

Because this confidence interval does not capture 0, we can reject the null hypothesis that $H_0: \mu_1 - \mu_0 = 0$ at the $p < .05$ level.

ESTIMATING $\delta = (\mu_1 - \mu_0)/\sigma$

Let's now return to the topic of power and effect size. In Chapter 8, the following points were made:

- Significance tests are used to decide whether two population means are different.

- If you have insufficient power to detect a difference, you may be wasting your time.

- Therefore, you should think about what effect size would be meaningful to you and then choose a sample size that provides sufficient power to reject the null hypothesis when it is false.

These points strongly imply that δ should be a primary concern. Therefore, rather than using δ as part of a power analysis conducted before running a significance test, it would seem more direct to estimate δ from the sample data. In fact, it is a simple matter to estimate δ, and a confidence interval around the estimate can be constructed if we know its standard error.

In Chapter 8, the population effect size was defined as follows:

$$\delta = \frac{\mu_1 - \mu_0}{\sigma}.$$

When the population standard deviation is known, δ is estimated with the statistic d as follows:

$$d = \frac{m - \mu_0}{\sigma}. \tag{9.3}$$

In equation 9.3, μ_1 has been replaced by m. As with any statistic, d is subject to sampling error. When σ is known, the standard error of d is the following:

$$\sigma_d = \frac{1}{\sqrt{n}}. \tag{9.4}$$

(There is an explanation of why this is the standard error of d in a later section.) To compute the $(1-\alpha)100\%$ confidence interval around d, we would use our familiar formula:

$$d \pm z_{\alpha/2}(\sigma_d). \tag{9.5}$$

To illustrate the construction of a confidence interval around d, we will return to the power-pose example. In Chapter 8 we were told that the mean of a population of final exam grades was $\mu_0 = 75$ with a standard deviation of $\sigma = 10$. Let's assume that a sample of $n = 64$ students held a power pose for 2 minutes before their final exam and that the average grade on the exam was $m = 76$. With this information we can compute an estimate of δ and a confidence interval around the estimate as follows:

Step 1. Compute d.

$$d = \frac{76-75}{10} = \frac{1}{10} = 0.1.$$

Step 2. Compute σ_d

$$\sigma_d = 1/\sqrt{n} = 1/\sqrt{64} = 0.125.$$

Step 3. Find $z_{\alpha/2}$. Because this is the 95% confidence interval, $\alpha = .05$. Therefore, $\alpha/2 = .025$. Using the z-table, we find that 2.5% of the z-distribution falls below -1.96. Therefore, $z_{\alpha/2} = 1.96$.

Step 4. Compute $d \pm z_{\alpha/2}(\sigma_d)$.

$$\text{CI} = d \pm 1.96\,(\sigma_d) = 0.1 \pm 1.96(0.125) = [-0.145, 0.345].$$

We have 95% confidence that δ lies in the interval $[-0.15, 0.35]$. Our confidence comes from knowing that 95% of all intervals computed this way will capture δ. This means that our best estimate is that the mean of the power-pose distribution is 0.1 standard deviations above the mean of the regular population. According to Cohen's classification scheme, this is a very small effect size. The confidence interval shows that a difference of -0.15 standard deviations, which means a drop in the average final exam grade, is just as likely as a difference of 0.35.

One way to approach the practical significance of this result is to ask what effect this intervention would have if applied to all members of the population. This question can be addressed with Cohen's U_3. Our best estimate of δ is 0.1. The z-table shows that $U_3 = P(0.1) = 0.5793$, which means that we estimate the proportion of the H_0 (power-pose) distribution above $\mu_0 = 75$ (the mean of the H_0 distribution) to be 0.5398. Therefore, it is estimated that adding 2 minutes of power posing before the final exam produces about a 4% increase $[(0.5398 - 0.5)*100 = 3.98\%]$ in the number of students scoring above the previous mean of 75. This seems like an interesting result given how minimal the intervention was.

However, we should also consider the limits of the confidence interval. U_3 for the lower limit (-0.15) is 0.4424. This means about a 6% *decrease* $[(0.4424 - 0.5)*100 = -5.76\%]$ in the number of students scoring above the previous mean of 75. U_3 for the upper limit (0.35) is 0.6350. This means about a 13% *increase* [(0.6350 − 0.5)*100 = 13.5%] in the number of students scoring above the previous mean of 75.

All things considered, these results seem rather imprecise and not very compelling. Our best estimate is that a 2-minute intervention will increase by 4% the number of students scoring above the previous mean. The lower limit of the confidence interval suggests that power posing could reduce by 6% the number of students scoring above the previous mean. This seems like a pretty big risk. The upper limit suggests that power posing could increase by 13% the number of students scoring above the previous mean. Although this suggests a large potential benefit, it must be weighed against the risk of a large cost. If one believes these results to be interesting, a study could be designed to achieve a far more precise estimate of δ using the "precision planning" method described in Appendix 6.3 (available at study.sagepub.com/gurnsey).

From the hypothesis testing point of view, the null hypothesis is that power posing has no effect on exam grades. We can state this as

$$H_0: \delta = 0.$$

If the null hypothesis were true, we would expect our interval to capture 0. The interval ([−0.15, 0.35]) does capture 0. Therefore, we retain the null hypothesis and say that the result is not statistically significant.

*A Detail

Let's return to the question of why $\sigma_d = 1/\sqrt{n}$. The following expression needs no explanation at this point:

$$\text{CI} = m \pm z_{\alpha/2}\left(\sigma_m\right) = m \pm z_{\alpha/2}\left(\sigma/\sqrt{n}\right).$$

We also know from equation 9.3 that

$$d = \frac{m - \mu_0}{\sigma}.$$

We say that we've *standardized* the difference between m and μ_0 by dividing the difference by σ. (Remember, d is a kind of z-score, and z-scores are called standard scores.) We can standardize the limits of the 95% confidence interval about m in the same way. That is,

$$\frac{\left[m \pm z_{\alpha/2}\left(\sigma/\sqrt{n}\right)\right] - \mu_0}{\sigma}.$$

With a little manipulation, we can derive the following:

$$\frac{\left[m \pm z_{\alpha/2}\left(\sigma/\sqrt{n}\right)\right] - \mu_0}{\sigma} = \frac{\left(m - \mu_0\right) \pm z_{\alpha/2}\left(\sigma/\sqrt{n}\right)}{\sigma}$$

$$= \frac{m - \mu_0}{\sigma} \pm \frac{z_{\alpha/2}\left(\sigma/\sqrt{n}\right)}{\sigma} = \frac{m - \mu_0}{\sigma} \pm z_{\alpha/2}\left(1/\sqrt{n}\right) = d \pm z_{\alpha/2}\left(\frac{1}{\sqrt{n}}\right).$$

Or, more simply,

$$\sigma_d = \frac{\sigma_m}{\sigma} = \frac{\sigma/\sqrt{n}}{\sigma} = 1/\sqrt{n} = \frac{1}{\sqrt{n}}.$$

The Connection Between z_{obs} and d

As a final note, it is important to recognize the connection between d and z_{obs}. We define z_{obs} as follows:

$$z_{obs} = \frac{m - \mu}{\sigma/\sqrt{n}},$$

but this is equivalent to

$$z_{obs} = \frac{m - \mu}{\sigma} * \sqrt{n}.$$

(You should plug numbers into examples like this to convince yourself that the statement is true.) The first term in this expression [i.e., $(m - \mu_0)/\sigma$] is the definition of d given in equation 9.2. Therefore,

$$z_{obs} = d * \sqrt{n}, \tag{9.6}$$

and

$$d = z_{obs}/\sqrt{n}. \tag{9.7}$$

Because $z_{obs} = d * \sqrt{n}$ we can see that no matter how small d is, as long as it is not exactly 0 it will become statistically significant if n is large enough. The value of equation 9.7 is that it allows one to determine an estimated effect size from a published report, even when a researcher has not reported it. The importance of this will be seen in the third part of this book when meta-analysis is discussed.

LEARNING CHECK 4

1. Let's say that the mean IQ for American adults living west of the Mississippi River is 100, with a standard deviation of 15. A random sample of 105,625 American adults living east of the Mississippi River is found to have a mean of 100.1.

   (a) Compute the 95% confidence interval around an estimate of δ.

   (b) Would this confidence interval lead you to reject a two-tailed test of the null hypothesis at the $p < .05$ level of significance? Why or why not?

   (c) Show how to convert the estimate of δ into z_{obs}.

Answers

1. (a) The estimate of δ is $d = (100.1 - 100)/15 = .006667$. The standard error of d is $1/\sqrt{n} = 1/\sqrt{105,625} = 0.0031$. Therefore, the 95% confidence interval is

   $$CI = d \pm Z_{\alpha/2}(\sigma_d) = .006667 \pm 1.96(0.0031)$$
   $$= [0.0006, 0.0127].$$

   (b) Because this confidence interval does not capture 0, we can reject a two-tailed test of the null hypothesis that $H_0: \delta = 0$ at the $p < .05$ level.

   (c) $z_{obs} = d * \sqrt{n} = (.1/15)(325) = 2.167$.

ESTIMATION VERSUS SIGNIFICANCE TESTING

Let's start with a few kind words for significance tests. Significance tests are used to make *decisions* about the null hypothesis. If we adopt the $p < .05$ criterion for statistical significance, then we seem to have a simple rule that makes research decisions very easy. A universal criterion for statistical significance could be seen as a referee in scientific debates. Without such a referee, judgments about a particular result may be determined by who can shout the loudest. A senior researcher might claim that a difference reported by a junior researcher is not important, while claiming that the same difference is important when she reports it. The apparent impartiality of $p < .05$ explains its role in the publication process.

Unfortunately, as discussed earlier in this chapter, many problems arise from adopting $p < .05$ as a universal criterion for statistical significance, and these far outweigh the potential value of $p < .05$ as an impartial referee. One of the major problems is the binary thinking that leads one to believe that when a statistic has an associated $p < .05$, the result is important and meaningful, and that when a statistic has an associated $p > .05$, the result is unimportant and meaningless. Binary thinking is put into stark relief when we consider the following question: What if H_1: $\mu_1 \neq \mu_0$, and $z_{obs} = 1.96$?

What if H_1: $\mu_1 \neq \mu_0$, and $z_{obs} = 1.96$?

If we adopt the $p < .05$ criterion for statistical significance, then $z_{critical} = \pm 1.96$ for a non-directional test. So, what do we do if z_{obs} is exactly equal to 1.96, or exactly equal to –1.96? That is, should we treat 1.96 like 1.95 (not statistically significant) or 1.97 (statistically significant)? Of course, these two z-scores (1.95 and 1.97) are almost identical; if we hadn't heard about the $p < .05$ criterion for statistical significance, we wouldn't have thought for a moment that we should treat them differently. Unfortunately, psychologists of my generation were instructed to do exactly this. The 1974 edition of the *APA Publication Manual* provided the following guidance about the interpretation of p-values:

> Caution: Do not infer trends [*read as statistical significance*] from data that fail by a small margin to meet the usual levels of significance. Such results are best interpreted as caused by chance and are best reported as such. (p. 19)

This is an explicit instruction to treat $z_{obs} = 1.97$ and $z_{obs} = 1.95$ differently. In fairness, this (mis)advice may derive from Fisher (1926), who said:

> Personally, the writer prefers to set a . . . standard of significance at the 5 per cent point, *and ignore entirely all results which fail to reach this level.* (p. 504, emphasis added)

Imagine an experiment that tested the effects of regular exercise on the IQs of men and women. In the general population, the IQs of both men and women have $\mu = 100$, and $\sigma = 15$. Let's say a random sample of 100 men and a random sample of 100 women were put on the same exercise routine for 3 months. At the end of 3 months, the mean IQ for men was found to be $m_{men} = 102.955$, and the mean IQ for women was found to be $m_{women} = 102.925$. If you do the calculations, you'll find that $z_{obs} = 1.97$ for men, and $z_{obs} = 1.95$ for women. The guidance from the 1974 *APA Publication Manual* says we should report this result as follows:

> The effect of exercise on IQ was significant for men ($z = 1.97$, $p < .05$) but not for women ($z = 1.95$, $p > .05$).

This kind of binary thinking has been passed from one generation of psychologists to the next, so it is no wonder that some engage in *p*-hacking to try to get statistically significant results.

Fortunately, things have changed, and more recent editions of the APA manual provide better guidance. The sixth edition (published in 2010) stated the following:

> Because confidence intervals combine information on location [*read as point-estimates*] and precision and can often be directly used to infer significance levels [*read as can be used to test H_0*], they are, in general, the best reporting strategies. The use of confidence intervals is therefore strongly recommended. (p. 34)

According to this guidance, we could report the results of our experiment in the following way:

APA Reporting

Following the 3-month exercise regime, the mean IQ for men was $M = 102.96$ (95% CI [100.02, 105.90]) and the mean IQ for women was $M = 102.93$ (95% CI [99.99, 105.87]). The mean IQ in the general populations of men and women is 100. Therefore, both after-exercising sample means are associated with increases in IQ. The effect of exercise on IQ yielded an effect size of $d = .197$ for men (95% CI [.001, .393]) and $d = .195$ for women (95% CI [−.001, .391]). Therefore, exercise leads to very similar improvements for both men and women.

This report gives a much clearer sense of the results of the experiment. In this situation, it doesn't seem to be particularly important that one confidence interval contains $\mu_0 = 100$, and the other doesn't.

What if H_1: $\mu_1 < \mu_0$, and $z_{obs} = 2$?

Another problem we've already encountered within the Fisher-Neyman hybrid model of significance testing is the question of what to do with a large z_{obs} that is inconsistent with our directional alternative hypothesis. For example, let's say that a review of the literature suggested to a researcher that extensive video gaming would impair the ability of college students to solve crossword puzzles. (The speculation is that video gaming would make a person more impulsive and this would lead to fewer correct words on the crossword puzzle.) The researcher happens to know that college students in general complete $\mu_0 = 24$ ($\sigma = 4.8$) words in 20 minutes on a standardized crossword puzzle. The researcher selects a random sample of $n = 36$ college students and has them play a video game for 30 minutes, after which they are given 20 minutes to complete the standardized crossword puzzle. The variable of interest is the number of correct responses on the crossword puzzle.

In this situation, the null hypothesis is that H_0: $\mu_1 - \mu_0 = 0$ and a directional alternative hypothesis that H_1: $\mu_1 - \mu_0 < 0$ has been chosen to increase the power of the experiment. (As noted in Chapter 8, one-tailed tests are more powerful than two-tailed tests, if your prediction is correct.) The alternative hypothesis states that following video game playing, the participants are expected to get fewer words correct on the crossword puzzle. Because this is a directional test that predicts $m - \mu_0 < 0$, $z_{critical} = -1.645$, with $\alpha = .05$; i.e., the $p < .05$ criterion for statistical significance. After running the experiment, the researcher finds that the average number of words completed is $m = 25.6$, which corresponds to $z_{obs} = 2$. Now she faces a conundrum; z_{obs} is large, but she is only allowed to reject H_0 if $z_{obs} < -1.645$. Should she say there is no statistically significant effect of video game playing on crossword performance? This would be an odd conclusion because $z_{obs} = 2$ would be unusual if H_0 were true.

This conundrum occurs only because the Fisher-Neyman hybrid model forces us to make a *decision* based on the statistic. The simplest way to deal with these data is to avoid the decision-making language of significance tests and say something like the following:

APA Reporting

Following 30 minutes of video gaming, the mean number of correctly completed crossword items was $M = 25.6$, 95% CI [24.03, 27.17]. This is an improvement over the population mean of $\mu_0 = 24$ and corresponds to an effect size of $d = .33$, 95% CI [.01, .66]. The magnitude of effect size is in line with past research, but its sign is opposite of what was expected. Therefore, further analysis is required to determine what features of this experiment may have differed from those of past experiments.

Decisions Versus Measurements

In this chapter and in Chapters 6 to 8, we have considered two approaches to inferential statistics: estimation and significance testing. Even though these approaches rest on exactly the same foundations, they differ fundamentally in emphasis. Estimation emphasizes what a parameter *is*. The hybrid model of significance testing emphasizes what a parameter *is not*. The estimation approach is the more general of the two, because once we have an estimate of what the parameter is (e.g., $m_{men} = 102.96$, 95% CI [100.02, 105.90]), we can, if we wish, make a statistical judgment about what it is not (e.g., we can reject the null hypothesis that $\mu_{men} = 100$).

The questions raised in the two preceding sections ("What if H_1: $\mu_1 - \mu_0 \neq 0$, and $z_{obs} = 1.96$?" and "What if H_1: $\mu_1 - \mu_0 < 0$, and $z_{obs} = 2$?") arise from a focus on *decision making*. When we establish a criterion for statistical significance (α, $z_{critical}$, or $m_{critical}$), it leads us to binary thinking: either H_0 is true or it is false; exercise either affects IQ or it does not; either the earth is round or it is not. That is, the answers are either yes or no, black or white. From the estimation point of view, the questions are more along the lines of "How wrong is H_0?," "How big is the effect of exercise on IQ?," and "How round is the earth?" Estimation does not lure us into binary thinking.

Addressing any research question should be thought of as part of an ongoing effort to understand a phenomenon of interest to researchers, or to society in general. Therefore, no single study should be seen as providing a verdict on such questions. Ideally, evidence accumulates as more studies are run, and researchers eventually form a consensus about the size of some effect. The hybrid model of significance testing, with its focus on decision making, can obscure the fact that no single study is definitive. When making a decision about H_0, one can be led to feel that the decision is final, particularly if we misunderstand p-values. With estimation, however, our view is more like that of a pollster, who finds that on August 10, 2016, 48% of decided voters prefer Hillary Clinton and 40% prefer Donald Trump. This result is seen as a single, more or less imprecise estimate. A better estimate would come from combining many such imprecise estimates. In statistics, we refer to the combination of many imprecise estimates as *meta-analysis*. We will discuss meta-analysis in Part III of this book. There we will see that it is easier to combine parameter estimates in a meta-analysis than the results of significance tests. For a meta-analysis to be valid, however, it is important to combine all relevant results, not just those that have made it into the literature through the $p < .05$ filter.

SUMMARY

Throughout their existence, significance tests have been criticized by many distinguished psychologists. The criticisms include the following points:

- The requirement for statistical significance for publication leads to publication bias and the file-drawer problem, causing the published literature to misrepresent the true effect of an intervention or the true difference between two populations.

- The requirement for statistical significance for publication also makes it more likely that Type I errors will be published.

- The requirement for statistical significance encourages p-hacking, which also distorts the literature and increases the probability that Type I errors will be published.

- A focus on obtaining small p-values is like the tail wagging the dog and distracts from the important questions of practical significance.

- The quest for small p-values promotes a tendency toward binary thinking, in which only statistically significant results are viewed as important and meaningful.

- Many researchers misunderstand p-values and thus misinterpret their results.

Many view confidence intervals as the healthiest alternative to significance tests because the emphasis is taken off decision making and placed on estimation. In a world in which estimation is standard practice, estimates would enter the literature without having to pass through the $p < .05$ filter of statistical significance. Furthermore, confidence intervals can be used to conduct significance tests. If μ_0 does not fall in the interval

$$m \pm z_{\alpha/2}(\sigma_m),$$

then we can reject H_0 at the $p < \alpha$ level of significance.

We can also estimate the difference between two population means $(\mu_1 - \mu_0)$ using the statistic $m - \mu_0$. This statistic has a standard error of $\sigma_{m-\mu_0} = \sigma/\sqrt{n}$. A confidence interval around $m - \mu_0$ is calculated as

$$\left(m - \mu_0\right) \pm z_{\alpha/2}\left(\sigma_{m-\mu_0}\right).$$

We can not only estimate $\mu_1 - \mu_0$ but also test the null hypothesis that $\mu_1 - \mu_0 = 0$ by asking whether 0 falls in the confidence interval.

When we think of the hybrid model of significance testing that involves power analysis, we become aware of the central role that δ should play in our thinking about research questions. δ is estimated by

$$d = \left(m - \mu_0\right)/\sigma.$$

Of course, d is subject to sampling error, so a confidence interval can be computed around d. The standard error of d is $\sigma_d = 1/\sqrt{n}$, so the $(1-\alpha)100\%$ confidence interval around d can be computed as

$$d \pm z_{\alpha/2}(\sigma_d).$$

Many researchers feel that estimating, μ, $\mu_1 - \mu_0$, or δ is healthier and more informative than testing a null hypothesis about μ_0. Because confidence intervals for μ, $\mu_1 - \mu_0$, or δ can be used to test hypotheses about μ_0, $\mu_1 - \mu_0$, or δ, they represent a more general approach to data analysis. Estimation avoids the binary (black-and-white) thinking that we may fall into with significance tests and lends itself more readily to meta-analysis, which we'll cover in Part III.

KEY TERMS

file-drawer problem   205

p-hacking   207

publication bias   205

EXERCISES

Definitions and Concepts

1. What is the definition of a p-value?

2. Why does the file-drawer problem happen?

3. Why should one not use the run-until-statistically-significant procedure?

4. How will the literature be biased if only statistically significant results ($p < .05$) are published?

5. Why is it important for replications to be published?

6. Give three examples of p-hacking.

7. Explain why failing to reject the null hypothesis does not mean that the null hypothesis is true.

True or False

State whether the following statements are true or false.

8. If I retain H_0, then it is true.

9. If I reject H_0, then H_1 is true.

10. I ran an experiment in which H_1: $\mu_1 - \mu_0 < 0$. I obtained $z_{obs} = -1.43$. Therefore, my study is not worth submitting for publication.

11. $\mu_0 = 22$ and my 95% confidence interval around m is [21, 28]. Therefore, I should retain a two-tailed test of H_0, assuming $\alpha = .05$.

12. $\mu_0 = 5$ and my 95% confidence interval around $m - \mu_0$ is [–1, 6]. Therefore, I should retain a two-tailed test of H_0, assuming $\alpha = .05$.

13. $\mu_0 = 22$ and my 95% confidence interval around m is [18, 21]. Therefore, I should retain a two-tailed test of H_0, assuming $\alpha = .05$.

14. $\mu_0 = 5$ and my 95% confidence interval around $m - \mu_0$ is [1, 6]. Therefore, I should reject a two-tailed test of H_0, assuming $\alpha = .05$.

15. If $\mu_0 = 5$ and my 95% confidence interval around $m - \mu_0$ is [1, 7], then $m = 4$.

16. If $\mu_0 = 5$ and my 95% confidence interval around $m - \mu_0$ is [1, 7], then $m = 9$.

Calculations

17. If $\mu_0 = 32$, $\sigma_0 = 10$, $m = 27$, and $n = 25$, answer the following:

    (a) Calculate the 95% confidence interval around m.

    (b) Calculate the 95% confidence interval around $m - \mu_0$.

    (c) Calculate the 95% confidence interval around d.

    (d) Assuming the $p < .05$ criterion for statistical significance, explain why these intervals would or would not allow you to reject the null hypothesis when H_1: $\mu_1 - \mu_0 \neq 0$.

18. If $\mu_0 = 16.8$, $\sigma_0 = 2.4$, $m = 18.3$, and $n = 36$, answer the following:

    (a) Calculate the 95% confidence interval around m.

    (b) Calculate the 95% confidence interval around $m - \mu_0$.

    (c) Calculate the 95% confidence interval around d.

    (d) Assuming the $p < .05$ criterion for statistical significance, explain why these intervals would or would not allow you to reject the null hypothesis when H_1: $\mu_1 - \mu_0 \neq 0$.

19. If $\mu_0 = 100$, $\sigma_0 = 15$, $m = 101$, and $n = 225$, answer the following:

    (a) Calculate the 95% confidence interval around m.

    (b) Calculate the 95% confidence interval around $m - \mu_0$.

    (c) Calculate the 95% confidence interval around d.

    (d) Assuming the $p < .05$ criterion for statistical significance, explain why these intervals would or would not allow you to reject the null hypothesis when H_1: $\mu_1 - \mu_0 \neq 0$.

20. If $\mu_0 = 100$, $\sigma_0 = 15$, $m = 101$, and $n = 225$, answer the following:

    (a) Calculate the 90% confidence interval around m.

    (b) Calculate the 90% confidence interval around $m - \mu_0$.

    (c) Calculate the 90% confidence interval around d.

    (d) Assuming the $p < .05$ criterion for statistical significance, explain why these intervals would or would not allow you to reject the null hypothesis when H_1: $\mu_1 - \mu_0 > 0$.

21. If $\mu_0 = 100$, $\sigma_0 = 15$, $m = 102$, and $n = 225$, answer the following:

    (a) Calculate the 90% confidence interval around m.

    (b) Calculate the 90% confidence interval around $m - \mu_0$.

    (c) Calculate the 90% confidence interval around d.

    (d) Assuming the $p < .05$ criterion for statistical significance, explain why these intervals would or would not allow you to reject the null hypothesis when H_1: $\mu_1 - \mu_0 > 0$.

    (e) Assuming the $p < .05$ criterion for statistical significance, explain why these intervals would or would not allow you to reject the null hypothesis when H_1: $\mu_1 - \mu_0 < 0$.

22. If $\mu_0 = 75$, $\sigma_0 = 15$, $m = 80$, and $n = 25$, answer the following:

    (a) Calculate the 90% confidence interval around m.

    (b) Calculate the 90% confidence interval around $m - \mu_0$.

    (c) Calculate the 90% confidence interval around d.

    (d) Assuming the $p < .05$ criterion for statistical significance, explain why these intervals would or would not allow you to reject the null hypothesis when H_1: $\mu_1 - \mu_0 > 0$.

    (e) Assuming the $p < .05$ criterion for statistical significance, explain why these intervals would or would not allow you to reject the null hypothesis when H_1: $\mu_1 - \mu_0 < 0$.

Scenarios

23. A difference threshold is the smallest difference between two levels of a stimulus that can be discriminated with a given level of accuracy. Let's say that on average, university students require a 3% increase in weight to discriminate one weight from another. For example, if a standard stimulus weighs 100 grams, then a comparison stimulus would need to be 3 grams heavier (i.e., 103 grams) for the difference to be noticeable. A researcher wondered how much difference thresholds could be changed by hypnotic suggestion. Each member of a random sample of 100 university students was given a suggestion under hypnosis that his or her sensory sensitivity had increased. Following hypnosis, difference thresholds for weights were measured in all participants. The standard stimulus was 100 grams and the mean difference threshold after hypnosis was $m = 2.5$ grams, with an estimated standard deviation of $s = 2$. Compute an approximate 95% confidence interval around $m - \mu_0$ and use this interval to determine whether a two-tailed test of H_0 is statistically significant at the $p < .05$ level.

24. Cell phone usage is a fact of life for many people these days, and there is some evidence of separation anxiety when individuals cannot access their phones. Let's say that the mean systolic blood pressure for cell phone users in possession of their phones is 110 mmHg with a standard deviation of 18. We wonder if the anxiety associated with cell phone separation will lead to increased blood pressure. A random selection of 36 cell phone users was contacted and asked to complete a survey about cell phone use. They were asked to leave their phones in a locked cabinet in another room while they completed the survey. At the end of the survey, their blood pressure was taken and the mean was found to be 128 mmHg. Compute the 95% confidence interval around $m - \mu_0$ and state whether a two-tailed test of H_0 is statistically significant at the $p < .05$ level.

25. People often feel pressure to conform to the attitudes and behaviors of groups. There are experiments showing that people will even deny the evidence of their own senses to conform to group behaviors (Asch, 1951). Imagine that 16 students in a psychology department participant pool volunteered for a study about perceptual judgments. They were asked to show up to a large classroom to take the test. On the day of the test, there were about 64 students in the classroom altogether. All 64 people were shown a large number of stimuli and asked which of three alternatives (A, B, C) matched a target line (e.g., the line on the left). Twenty-one different stimuli were shown. For each stimulus, the experimenter asked for a show of hands in response to the questions "How many think A is the match?" "How many think B is the match?" and "How many think C is the match?"

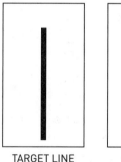 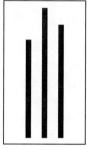

TARGET LINE   A B C

Unbeknownst to the 16 volunteers, the majority of people in the classroom were confederates of the researcher. Whenever the obvious match was C (as in this example), all of the confederates raised their hands in response to a non-match (e.g., B). The experimenter (and colleagues) noted how many times each of the 16 volunteers raised their hands along with the majority, and thus gave an obviously wrong answer. They found that on average the volunteers gave the wrong answer on $m = 5.1$ times out of 7 opportunities, with a standard deviation $s = 1.2$. Compute the approximate 95% confidence interval around m. If you were to compute an approximate confidence interval around $m - \mu_0$, what would μ_0 be? What do you think this result says about conformity? Would more or less than 95% of all such intervals be expected to capture $\mu_1 - \mu_0$?

26. It is known that the way a question is posed can affect the answer given (Loftus & Palmer, 1974). This is particularly important in court trials. Imagine that a large population of American adults had seen a movie of a traffic accident in which a car ran a stop sign and struck another. After viewing the movie, they were asked to judge how fast the car (that ran the stop sign) was going when it contacted the other car. On average, they judged that it was going $\mu = 32$ mph, with a standard deviation $\sigma = 6$. In a subsequent study, a random sample of 25 participants from the same population were shown the movie and asked how fast the car was going when it smashed into the other car. On average, they judged that it was going 42 mph (51.5 kph). Compute the 95% confidence interval around d. What value of z_{obs} does d correspond to? Do you think that it would be easier for a non-statistician to understand a confidence interval around d or a confidence interval around $m - \mu_0$?

27. Does memory for line drawings change with age (Harwood & Naylor, 1969)? A standardized test shows that when young adults (mean age 24 years) have learned to recognize 20 line drawings of common objects, they recognize about 75% ($\sigma = 5$) of these in a surprise test 4 weeks later. The same experiment examined the recognition performance of 36 older adults (mean age 71.2 years). The mean percentage of recalled items in the older participants was found to be lower than the mean percentage recalled in the younger participants. A one-tailed test ($\mu_0 = 75$, $\sigma = 5$) showed that this difference was statistically significant, $z_{obs} = -2.4$, $p = .0082$. Compute the 95% confidence interval around d. Use the information given, as well as the d-statistic you just computed, to determine the mean recognition rate of the older adults. (*Hint:* Remember that d represents change

in standardized units.) Do you judge the decrease in recall ability to be severe?

28. Early work by Hetherington and Ranson (1940) suggested a role for the ventromedial hypothalamus in the regulation of food intake. It is known that Wistar rats weigh 275 grams on average, with a standard deviation of 12. A random sample of four Wistar rats was selected and subjected to a surgery that lesioned (destroyed) their ventromedial hypothalamus. After a 2-month period, during which these brain-damaged rats had free access to unlimited food, each was weighed. The mean weight of these brain-damaged rats was greater than that of the average Wistar rat, $z_{obs} = 91.7$, $p < .05$. Compute the 95% confidence interval around d. What was the mean weight of the four brain-damaged rats? Do you think these results are of any practical significance?

29. Here is a strange passage of text taken from an interesting study published in the early 1970s (Dooling & Lachman, 1971).

> With hocked gems financing him, our hero bravely defied all scornful laughter that tried to prevent his scheme. "Your eyes deceive," he had said. "An egg, not a table, correctly typifies this unexplored planet." Now three sturdy sisters sought proof. Forging along, sometimes through calm vastness, yet more often over turbulent peaks and valleys, days became weeks as many doubters spread fearful rumors about the edge. At last from nowhere welcome winged creatures appeared, signifying momentous success.

Imagine that a large population of psychology students had been read this passage and were asked to recall as many words as they could. The mean number of correctly recalled words was 13.23, with a standard deviation of $\sigma = 4.2$. A random sample of 16 psychology students was read the same passage but, unlike for the larger group, the passage was preceded by the title "Christopher Columbus." The mean number of words recalled by the 16 participants was $m = 15.67$. Compute the 95% confidence interval around $m - \mu_0$. What do you make of these results? Can you think of any situations in which these results would be seen as having practical significance?

10 ESTIMATING THE POPULATION MEAN WHEN THE STANDARD DEVIATION IS UNKNOWN

INTRODUCTION

If you are asked to name the actor who played Harry Potter, you would probably have no trouble answering, Daniel Radcliffe. You retrieve this fact from the large store of facts in your *long-term memory*. On the other hand, when your friend gives you his phone number, you must hold it in your *short-term memory* until you can enter it into your phone or write it down. If you are distracted by an incoming text, the phone number might decay from your short-term memory before you can save it. In the early days of memory research, short-term memory was viewed as a temporary storage bin with a limited number of slots into which words, numbers, and facts could be placed. When these items are not transferred to long-term memory, they decay and are lost forever.

Our understanding of long- and short-term memory has advanced a great deal in the last 60 years, but let's place ourselves back in 1950 or so and think about how we might estimate the *mean* capacity (mean number of slots) in short-term memory. A possible measure of short-term memory capacity is *digit span*. To measure digit span, an experimenter presents a participant with a list of digits (at a fixed number of digits per second) that the participant must repeat back to the experimenter. The task starts with a list of one or two digits and ends when the participant makes his or her first mistake. Therefore, digit span is the maximum number of digits that can be reported back without error.

So imagine we're back in the 1950s and a student working in the lab of the great G. A. Miller has been asked to use the digit-span task to estimate the mean short-term memory capacity of undergraduate psychology students at Harvard University. To do so, the student selects a random sample of 25 participants from the department's participant pool and measures the digit span of each one. As part of the written report for Professor Miller, the results might have been stated as follows:

APA Reporting

The mean digit span for the 25 participants was $M = 7.20$, with a standard deviation of $s = 1.41$, 95% CI [6.62, 7.78]

Confidence intervals were reported in exactly this way in Chapter 6. The mean of our sample, $m = 7.2$, is a point estimate of the population mean, μ. The intended population is the population of undergraduate psychology students at Harvard University. From this study, we can say that we have 95% confidence that the true mean digit span is in the interval [6.62, 7.78]. As before, our confidence comes from knowing that 95% of all such intervals will capture μ. Notice that in this example, the population standard deviation (σ) was not given. Therefore, this raises the question of how the interval was constructed.

The groundwork for this question was laid in Chapter 6, where we saw how an approximate confidence interval could be placed around a sample mean using s_m rather than σ_m. However, the approximation only works well for large samples. In this chapter we'll see how to compute confidence intervals using s_m for small samples. This will involve the critical new concept of a t-statistic. With an understanding of t-statistics and t-distributions, we will see how to compute a confidence interval around a sample mean when sample size is small and σ is not known.

We will also see that t-distributions play an important role in hypothesis tests when sample size is small and σ is not known. Rather than computing z_{obs}, as we did in Chapters 7 and 8, in this chapter we will compute t_{obs} to test H_0. Finally, in Chapter 9 we saw how to place a confidence interval around an estimated effect size when σ is known. We will revisit this question for the typical case in which σ is not known.

t-SCORES: σ_m VERSUS s_m

In Chapter 6, we used the following formula to construct an approximate $(1-\alpha)100\%$ confidence interval around a sample mean when σ was not known:

$$m \pm z_{\alpha/2}(s_m). \tag{10.1}$$

In this formula, s_m replaced σ_m. We noted that equation 10.1 is a valid approximation only for large samples because as sample size increases, the sample standard deviation (s) is a better and better estimator of the population standard deviation (σ). This means that as sample size increases, the sampling error of s decreases (see Figure 5.A4.2 in Appendix 5.4, available at study.sagepub.com/gurnsey). When sample size is very large, s will differ little from sample to sample and will always be very close to σ. Therefore, when sample size is large, the estimated standard error of the mean

$$s_m = \frac{s}{\sqrt{n}} \tag{10.2}$$

will differ very little from the real thing, σ_m, making equation 10.1 a reasonable approximation.

For small samples, however, the confidence intervals computed with equation 10.1 will be too narrow, thus capturing μ less than $(1-\alpha)100\%$ of the time (see Chapter 6, Figure 6.5). We can solve this problem by using the following, slightly modified version of equation 10.1:

$$m \pm t_{\alpha/2}(s_m). \tag{10.3}$$

The only difference between equation 10.3 and equation 10.1 is that s_m is multiplied by $z_{\alpha/2}$ in equation 10.1 and by $t_{\alpha/2}$ in equation 10.3. Our task in the first part of this chapter is to understand what $t_{\alpha/2}$ is and how it differs from $z_{\alpha/2}$. To accomplish this, we will first define t-scores and then t-distributions. We will encounter t-distributions in almost all subsequent chapters because they are among the most important distributions in statistics.

z-Scores Versus t-Scores

Remember from Chapters 6 to 8 that any distribution of means can be transformed to the z-distribution by computing

$$z = \frac{m - \mu}{\sigma_m} \tag{10.4}$$

for all sample means. When s_m replaces σ_m in equation 10.4, the result is a *t*-statistic defined as follows:

$$t = \frac{m - \mu}{s_m}. \tag{10.5}$$

In general, a distribution of *t*-scores is quite different from a distribution of *z*-scores. To understand why this is so, we will take a closer look at *t*-scores and *z*-scores, which means considering the consequences of dividing $m - \mu$ by s_m rather than σ_m.

The denominator of equation 10.4 is based on the *population* standard deviation and sample size (i.e., $\sigma_m = \sigma/\sqrt{n}$) and so *z does not* depend on the sample standard deviation. To make this clear, think of two samples that have the same means; e.g., $m_1 = m_2 = 7$. Let's say that m_1 is based on the scores $y_1 = [5, 7, 9]$ and m_2 is based on the scores $y_2 = [1, 7, 13]$. Despite the big difference in variability within these two samples, when m_1 and m_2 are transformed into *z*-scores they produce $z = (7 - \mu)/\sigma_m$ in both cases. For illustration, if $\mu = 5$ and $\sigma_m = 4$, then whenever *m* happens to exactly equal 7, equation 10.4 will produce a *z*-score of $(7 - 5)/4 = 0.5$, no matter what three numbers produced this mean of 7.

The denominator of equation 10.5 is based on the *sample* standard deviation and sample size (i.e., $s_m = s/\sqrt{n}$), and so a *t*-score depends on the sample standard deviation. Again, think about m_1 based on the scores $y_1 = [5, 7, 9]$, and m_2 based on the scores $y_2 = [1, 7, 13]$. The standard deviation associated with m_1 is $s_1 = 2$, and the standard deviation associated with m_2 is $s_2 = 6$. Therefore, the corresponding *estimated* standard errors are $s_{m_1} = 2/\sqrt{3} = 1.155$, and $s_{m_2} = 6/\sqrt{3} = 3.464$. If $\mu = 5$, then the *t*-scores based on these two samples would be different, despite having the same numerator, as shown below:

$$t_1 = (7 - 5)/s_{m_1} = (7 - 5)/1.155 = 1.73$$
$$t_2 = (7 - 5)/s_{m_2} = (7 - 5)/3.464 = .58$$

Because *t*-scores depend on the sample standard deviation and *z*-scores do not, there is no reason to think that the distribution of *t*-scores should be the same as the distribution of *z*-scores. As noted already, the distribution of *t*-scores is (usually) quite different from the distribution of *z*-scores, and this is why we cannot make use of $z_{\alpha/2}$ when placing a confidence interval around a sample mean when σ is not known and sample size is small. To put this concretely, we know that ±1.96 encloses the central 95% of the *z*-distribution, but we don't (yet) know what values of *t* enclose the central 95% of a *t*-distribution. We will now look at this point more closely.

t-DISTRIBUTIONS

A **t-distribution** is a distribution of *t*-scores [$t = (m - \mu)/s_m$] computed from all possible samples of size *n*, drawn from a normal distribution. If the distribution from which the samples were drawn is not normal, we do not call the resulting distribution of *t*-scores a *t*-distribution.

A **t-distribution** is a distribution of *t*-scores computed from all possible samples* of size *n* *drawn from a normal distribution.* Even though it is possible to compute $t = (m - \mu)/s_m$ for all samples of size *n* drawn from a non-normal distribution of scores, the resulting distribution of *t*-scores is not called a *t*-distribution. We saw something very similar in Chapter 4 when the *z*-distribution was introduced. Any normal distribution can be transformed into the standard normal (*z*) distribution. However, if a distribution of scores is not normal, then transforming all scores in the distribution to *z*-scores would not yield the standard normal (*z*) distribution.

---

*Please keep in mind that we do not literally compute the statistic for all possible samples. Rather, mathematicians are able to determine what the distribution of statistics would be *if one were to compute* the statistic for all possible samples.

LEARNING CHECK 1

1. Imagine that the population of digit spans for students at the University of Toronto has a mean of 7.5 and standard deviation of 1.25. Let's say that four random samples, each with 25 scores, were drawn from this population. Strangely, each of the four samples has a mean of 8, but their standard deviations were (a) 0.8, (b) 1.5, (c) 0.95, and (d) 1.25.

   For each of these four samples, compute $z = (m - \mu)/\sigma_m$ and $t = (m - \mu)/s_m$.

2. What proportion of the z-distribution lies in the interval ±1.96?

3. What proportion of the z-distribution lies in the interval ±1.67?

Answers

1. (a) $z = (8 - 7.5)/(1.25/5) = 2, t = (8 - 7.5)/(.8/5) = 3.125$;

   (b) $z = (8 - 7.5)/(1.25/5) = 2, t = (8 - 7.5)/(1.5/5) = 1.67$;

   (c) $z = (8 - 7.5)/(1.25/5) = 2, t = (8 - 7.5)/(.95/5) = 2.63$;

   (d) $z = (8 - 7.5)/(1.25/5) = 2, t = (8 - 7.5)/(1.25/5) = 2$.

2. .95.

3. .905.

Unlike the z-distribution, the *shape* of a t-distribution depends on sample size. This means that there is not just one t-distribution but a t-distribution for each sample size.

Degrees of Freedom

Although there is a different t-distribution for each sample size, statistical tables and statistical software associate t-distributions with **degrees of freedom (df)** rather than sample size (n). When estimating the mean of a population, $n-1$ is said to be the number of *degrees of freedom* associated with the estimate.

There is a reason for this seemingly arbitrary convention. Remember, $n-1$ is the denominator in the definition of the unbiased sample variance

$$s^2 = \frac{\sum(y-m)^2}{n-1},$$

which could therefore be written as

$$s^2 = \frac{\sum(y-m)^2}{df}.$$

> **Degrees of freedom (df)** refers to the number of independent variables required to account for a given source of variance (treatment, variable, or error) in a regression (or ANOVA) analysis.

The unbiased sample variance cannot be computed when sample size is 1, because the denominator goes to 0 (i.e., $n-1 = 1 - 1 = 0$). Therefore, the smallest sample size for which the unbiased sample variance can be computed is 2, and so the smallest sample for which $t = (m - \mu)/s_m$ can be calculated is also 2. A sample of size $n = 2$ is associated with 1 df.

If degrees of freedom were always $n-1$, then it would seem like an unnecessary convention. However, in later chapters we'll see that different statistics are associated with $n-2$, $n-3$, or even fewer degrees of freedom. Therefore, the same number of total scores can be associated with different degrees of freedom.

How t-Distributions Change With df

Now that we know what t-scores and degrees of freedom are, we can have a look at t-distributions. Figure 10.1 provides examples of t-distributions for 1, 5, and ∞ degrees of

FIGURE 10.1 ■ Three *t*-Distributions

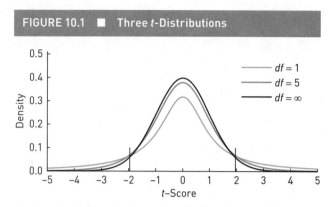

Three *t*-distributions having 1, 5, and infinite degrees of freedom. The two vertical lines represent $t = \pm1.96$.

freedom. Don't worry about the impossibility of ∞ degrees of freedom. This is a theoretical concept. Just read ∞ degrees of freedom to mean a very, very large sample. Although we'll see that the *t*- and *z*-distributions are *generally* different, the *t*-distribution becomes increasingly similar to the *z*-distribution as sample size increases (i.e., as degrees of freedom increase toward infinity). When $df = ∞$ (theoretically), the *t*-distribution is identical to the *z*-distribution. This case is shown by the black line in Figure 10.1.

The fact that the *t*-distribution converges to the *z*-distribution makes perfect sense. As noted at the beginning of this section, when sample size is very large, all sample standard deviations (*s*) will be very similar to the population standard deviation (σ). Consequently, all $s_m = s/\sqrt{n}$ will be similar to $\sigma_m = \sigma/\sqrt{n}$. Therefore, dividing by s_m is essentially the same as dividing by σ_m, and so *t*-scores $[t = (m - \mu)/s_m]$ become *z*-scores $[z = (m - \mu)/\sigma_m]$.

At the opposite extreme ($df = 1$), the *t*-distribution diverges dramatically from the *z*-distribution, as shown by the light blue line in Figure 10.1. (Comparing these two curves in Figure 10.1 might lead you to think that you and I have different definitions of the word *dramatic*. Let's see if I can change your mind.) When $df = 1$, we can see that *t*-scores are less densely packed around 0 than are *z*-scores and the distribution is leptokurtic (see Chapter 2). As well, the density of scores in the tails of the *t*-distribution is greater than the density of scores in the tails of the *z*-distribution. We say that the tails of *t*-distributions are *heavier* than the tails of the *z*-distribution. For $df = 5$ (dark blue line), the *t*-distribution looks more like the *z*-distribution but still has heavier tails and lower density around the mean of 0.

We noted earlier that 95% of the *z*-distribution lies in the interval ±1.96. We might now ask what proportion of a *t*-distribution lies in the interval ±1.96. The answer depends on sample size. The vertical lines on the left and right sides of Figure 10.1 are at $t = \pm1.96$. When sample size = 2 ($df = 1$, light blue line), only 70% of the *t*-distribution lies in the interval 0±1.96. (I haven't told you how I came up with 70%; just trust me on this one.) If 70% of the distribution lies in the interval ±1.96, then 30% lies in the tails outside the interval ±1.96. This is what we mean when we say that *t*-distributions have heavier tails than *z*-distributions. For *any* value of *t*, a larger proportion of the *t*-distribution than of the *z*-distribution will lie outside the interval ±*t*.

So now the big question is, what values of *t* contain the central 95% of a *t*-distribution? More generally, what is $t_{\alpha/2}$? Knowing the answer to this question will allow us to compute a confidence interval around *m* when σ is not known, using the formula $m \pm t_{\alpha/2}(s_m)$.

The *t*-Table

The *z*-table introduced in Chapter 4 shows the proportion of the *z*-distribution below each of a large number of *z*-scores; i.e., $P(z)$. We defined $z_{\alpha/2}$ as the absolute value of the *z*-score having $(\alpha/2)100\%$ of the *z*-distribution below it. Tables like this would be unwieldy for *t*-distributions, because there would have to be a *t*-table for each sample size. To simplify matters, Table 10.1 provides a selection of *t*-scores. The left column of Table 10.1 lists degrees of freedom for a small number of sample sizes. All *t*-scores in the table are values of $t_{\alpha/2}$.

There are two column headers in Table 10.1. The first is labeled "Area in two tails." The proportions under this header are .5, .2, .1, .05, .02, and .01. The second header is "Area in one tail." The proportions under this header are .25, .1, .05, .025, .01, and .005. In the context of confidence intervals, the column headers in "Area in one tail" correspond to the proportion of the *t*-distribution above $t_{\alpha/2}$, and the column headers in "Area in two tails"

TABLE 10.1 ■ The *t*-Table: *t*-Values Having a Proportion of .5 to .01 Outside the Interval ±*t*, for Selected *df*

df	Area in two tails (α)					
	.50	**.20**	**.10**	**.05**	**.02**	**.01**
	Area in one tail (α/2)					
	.250	**.100**	**.050**	**.025**	**.010**	**.005**
1	1.000	3.078	6.314	12.706	31.821	63.657
2	0.816	1.886	2.920	4.303	6.965	9.925
3	0.765	1.638	2.353	3.182	4.541	5.841
4	0.741	1.533	2.132	2.776	3.747	4.604
5	0.727	1.476	2.015	2.571	3.365	4.032
6	0.718	1.440	1.943	2.447	3.143	3.707
7	0.711	1.415	1.895	2.365	2.998	3.499
8	0.706	1.397	1.860	2.306	2.896	3.355
9	0.703	1.383	1.833	2.262	2.821	3.250
10	0.700	1.372	1.812	2.228	2.764	3.169
11	0.697	1.363	1.796	2.201	2.718	3.106
12	0.695	1.356	1.782	2.179	2.681	3.055
13	0.694	1.350	1.771	2.160	2.650	3.012
14	0.692	1.345	1.761	2.145	2.624	2.977
15	0.691	1.341	1.753	2.131	2.602	2.947
16	0.690	1.337	1.746	2.120	2.583	2.921
17	0.689	1.333	1.740	2.110	2.567	2.898
18	0.688	1.330	1.734	2.101	2.552	2.878
19	0.688	1.328	1.729	2.093	2.539	2.861
20	0.687	1.325	1.725	2.086	2.528	2.845
21	0.686	1.323	1.721	2.080	2.518	2.831
22	0.686	1.321	1.717	2.074	2.508	2.819
23	0.685	1.319	1.714	2.069	2.500	2.807
24	0.685	1.318	1.711	2.064	2.492	2.797
25	0.684	1.316	1.708	2.060	2.485	2.787
26	0.684	1.315	1.706	2.056	2.479	2.779
27	0.684	1.314	1.703	2.052	2.473	2.771
28	0.683	1.313	1.701	2.048	2.467	2.763
29	0.683	1.311	1.699	2.045	2.462	2.756
30	0.683	1.310	1.697	2.042	2.457	2.750
31	0.682	1.309	1.696	2.040	2.453	2.744
32	0.682	1.309	1.694	2.037	2.449	2.738
33	0.682	1.308	1.692	2.035	2.445	2.733
34	0.682	1.307	1.691	2.032	2.441	2.728
35	0.682	1.306	1.690	2.030	2.438	2.724
36	0.681	1.306	1.688	2.028	2.434	2.719
37	0.681	1.305	1.687	2.026	2.431	2.715
38	0.681	1.304	1.686	2.024	2.429	2.712
39	0.681	1.304	1.685	2.023	2.426	2.708
40	0.681	1.303	1.684	2.021	2.423	2.704
50	0.679	1.299	1.676	2.009	2.403	2.678
60	0.679	1.296	1.671	2.000	2.390	2.660
70	0.678	1.294	1.667	1.994	2.381	2.648
80	0.678	1.292	1.664	1.990	2.374	2.639
90	0.677	1.291	1.662	1.987	2.368	2.632
100	0.677	1.290	1.660	1.984	2.364	2.626
200	0.676	1.286	1.653	1.972	2.345	2.601
500	0.675	1.283	1.648	1.965	2.334	2.586
1000	0.675	1.282	1.646	1.962	2.330	2.581
10,000	0.675	1.282	1.645	1.960	2.327	2.576

Note: All values calculated by the author in Excel.

correspond to the proportion of the *t*-distribution outside the interval $\pm t_{\alpha/2}$. We will see that this table can be used for both confidence intervals and significance tests.

To orient ourselves to the table, we will look at the last row, which shows so-called *critical values* of *t* for 10,000 *df.* The distribution of *t*-scores for 10,000 *df* is almost identical to the distribution of *z*-scores. Therefore, in the column corresponding to area in two tails = .05 (area in one tail = .025), we find a *t*-score of 1.96. This means that for 10,000 *df*, the two *t*-scores having 5% of the distribution outside the interval $\pm t$ are ± 1.96. (This should sound very familiar!) Put in more familiar terms, when $\alpha = .05$ and $df = 10,000$, $t_{\alpha/2} = 1.96$.

If we remain with the same column and look from *bottom* to *top*, we see that the *t*-scores increase. When we reach the top row, corresponding to 1 *df*, we find that 2.5% of the distribution lies above 12.706, and so 5% of the distribution lies outside the range ± 12.706. Therefore, $t = \pm 12.706$ encloses the central 95% of the *t*-distribution, when $n = 2$. That is, $t_{\alpha/2} = 12.706$ when $df = 1$. I think you'll agree that 12.706 is *dramatically* different from 1.96!

The remaining numbers in the fourth column of Table 10.1 show how $t_{\alpha/2}$ changes as a function of sample size when $\alpha = .05$. Because $t_{\alpha/2}$ *increases* as sample size *decreases*, it is clear that large *t*-scores are far more common for small sample sizes than for large sample sizes. This is true no matter what value of α we consider. The remaining columns in Table 10.1 show $t_{\alpha/2}$ for $\alpha = .5, .2, .1, .02,$ and .01.

We can illustrate how to use the *t*-table by asking what *t*-score has 5% of the distribution above it when sample size is 6. A sample size of 6 corresponds to 5 *df*, so that's the row we use to answer our question. Because we're asking about the *t*-score having 5% of the distribution above it, we look in the third column (proportion of the distribution in one tail = .05). We find that the *t*-score in question is 2.015. If we had wanted to know what *t*-score has 5% of the distribution *below* it when sample size is 6, the answer would be −2.015, because the *t*-distribution is symmetrical about its mean of 0. Therefore, the proportion of the *t*-distribution, with 5 *df*, *outside* the interval ± 2.015 is $.05 + .05 = .1$. So, if $\alpha = .1$, then $t_{\alpha/2} = 2.015$.

Table 10.1 shows *t*-scores for a limited number of degrees of freedom and proportions. In real-world statistical analyses, more precise values of *t* can be obtained using software such as Excel. Appendix 10.1 describes the **T.INV** and **T.INV.2T** functions in Excel that can be used for cases not covered by the *t*-table. For example, you can use **T.INV.2T** to determine $t_{\alpha/2}$, when $\alpha = .036$ and $df = 108$. If you do this, you will find that the answer is 2.123.

FIGURE 10.2 ■ Student

William Sealy Gosset (Student) (1876 –1937). This picture was taken in 1908 (age 32).

Scanned from Gosset's obituary in *Annals of Eugenics*

A Word About William Sealy Gosset

It is possible to think that statistical distributions have come down to us through the ages from the ancient Greeks, just like Pythagoras's theorem. This is not so. The initial work on *t*-distributions was done quite recently by William Sealy Gosset (Figure 10.2). Gosset was trained as a chemist and was the head chemist at the Guinness Brewery in Dublin, Ireland, for many years. In 1908, he published a paper titled "The Probable Error of the Mean" in the journal *Biometrika*, which was run by Karl Pearson, who was Egon Pearson's father (see Chapter 8, Figure 8.4). For reasons that are not altogether clear, Guinness did not allow their employees to publish the results of their research. However, Gosset was given special permission to publish his results, but to conceal his identity and that of his employer, he published under the pseudonym "Student." Student was interested in the sampling distribution of $t = (m - \mu)/s_m$ for small samples. His paper laid the groundwork for what we know today as Student's *t*-distribution. Sir Ronald Fisher was the first to recognize the immense importance of Student's work and was responsible for a rigorous formal characterization of the distribution; Fisher was able to show some things about *t*-distributions that Student could not, and it was Fisher who introduced the concept of degrees of freedom. Although *t*-distributions are among the most important distributions in statistics, Student's original paper attracted very little attention from statisticians of the day because they focused on large samples, whereas Student's work focused on small samples. Largely due to Fisher's use of Student's work, "The Probable Error of the Mean" holds a special place in the history of statistics (Zabell, 2008).

It is worth noting that Gosset was able to maintain very amicable relationships with the two Pearsons, Neyman, *and* R. A. Fisher, even though Fisher seemed constantly at war with the Pearsons and Neyman (Salsburg, 2001). Furthermore, it was Gosset whom Egon Pearson (1939) credits with "the germ of the idea" that is at the core of the Neyman-Pearson model of statistical decision making.

In a tribute to Student, Fisher (1939) wrote the following: "His life was one full of fruitful scientific ideas and his versatility extended beyond his interests in research. In spite of his many activities it is the student of Student's test of significance who has won, and deserved to win, a unique place in the history of scientific method."

LEARNING CHECK 2

1. A distribution of scores on a very easy, nationwide test has a mean of 18 and standard deviation of 3. If we compute $t = (m - \mu)/s_m$ for all possible samples of size 11, the result will be a t-distribution with 10 df. [True, False]

2. What is the smallest sample size for which the unbiased standard deviation can be computed?

3. If $n = 25$, then $df =$ _____.

4. If $df = 36$, then $n =$ _____.

5. Which t-distribution will be more similar to the z-distribution, one with $df = 100$ or one with $df = 105$?

6. Which t-distribution will have heavier tails, one with $df = 5$ or one with $df = 105$?

7. Which t-distribution will have a greater proportion lying above $t = 2$, one with $df = 1$ or one with $df = 105$?

8. What proportion of a t-distribution with 25 df lies outside the interval ±2.060?

9. If $\alpha = .05$ and $df = 15$, what is $t_{\alpha/2}$?

10. What t-score has 1% of the distribution above it if $df = 50$?

Answers

1. False. The distribution of scores will be negatively skewed. t-distributions only result when $t = (m - \mu)/s_m$ has been computed from scores drawn from normal distributions.

2. $n = 2$.

3. $df = 24$.

4. $n = 37$.

5. $df = 105$; the t-distribution becomes more similar to the z-distribution as sample size increases.

6. $df = 5$. As sample size decreases, the tails of the t-distribution become heavier.

7. $df = 1$. t-distributions with small df have heavier tails than those with large df.

8. .05.

9. 2.131.

10. 2.403.

CONFIDENCE INTERVALS: ESTIMATING μ

Now that we are familiar with t-distributions, we can return to the question of how to compute a confidence interval around a sample mean when σ is not known. Once again, the formula for the confidence interval is

$$m \pm t_{\alpha/2}(s_m).$$

FIGURE 10.3 ■ 95% Confidence Intervals

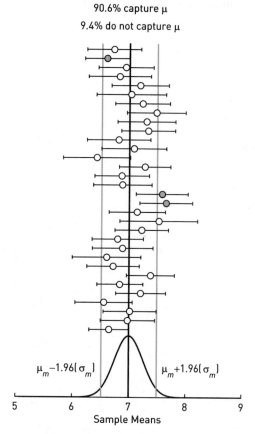

90.6% capture μ

9.4% do not capture μ

$\mu_m - 1.96(\sigma_m)$ $\mu_m + 1.96(\sigma_m)$

Sample Means

95% confidence intervals for μ when σ is estimated by s. The distribution at the bottom represents the sampling distribution of the mean for μ = 7, σ = 1, and n = 16. The standard error of the mean is σ_m = 0.25. The light blue vertical lines [μ_m ± 1.96(σ_m)] enclose the central 95% of the sampling distribution; these are for reference only. Each dot represents a sample mean, and the arms around each dot represent the 95% confidence interval. White dots are the centers of confidence intervals that capture μ, and the filled dots are the centers of confidence intervals that do not capture μ.

Figure 10.3 illustrates many important points about confidence intervals. Let's consider digit spans and assume that in some population the mean digit span is μ = 7 with standard deviation σ = 1. The normal distribution at the bottom of Figure 10.3 depicts the distribution of sample means for samples of size n = 16, drawn from this normal population. In this situation, the standard error of the mean is σ_m = 0.25. The vertical black line indicates the mean of the distribution, and the two light blue vertical lines correspond to μ − 1.96(σ_m) = 6.51 (*left*), and μ + 1.96(σ_m) = 7.49 (*right*). These light blue lines enclose the central 95% of the distribution of means, but *they are shown for reference only*, as we'll see. As before, white dots are sample means having 95% confidence intervals that capture μ, and the filled dots are sample means having 95% confidence intervals that do not capture μ.

There is an interesting difference between Figure 10.3 and Figures 6.2 and 6.3 from Chapter 6. The widths of the confidence intervals in Figure 10.3 differ from sample to sample (compare the top two intervals), whereas all confidence intervals were the same width in Chapter 6. The sample-to-sample variation in interval width is because the sample standard deviation (s) is different for each sample. Because s is different for each sample, s_m will also be different for each sample. Because s_m is different for each sample, ±$t_{\alpha/2}(s_m)$ (i.e., interval width) is different for each sample.

This sample-to-sample variation in interval width means that a 95% confidence interval *may not* capture μ even if its mean comes from the central 95% of the distribution of means. For example, one of the filled dots in Figure 10.3 is within the central 95% of the distribution of means. Conversely, a 95% confidence interval *may* capture μ even if its mean comes from outside the central 95% of the distribution of means. There is one example of this in Figure 10.3.

As sample size increases, the sample-to-sample variation in interval width decreases; look back to Figure 5.A4.2 (available at study.sagepub.com/gurnsey). As a consequence, means from the central 95% of the distribution of means will almost always produce confidence intervals that capture μ. Whether sample size is large or small, however, exactly (1−α)100% of all confidence intervals computed as m ± $t_{\alpha/2}(s_m)$ will capture μ. Please see Appendix 10.5 (available at study.sagepub.com/gurnsey) if you would like a demonstration that this is true.

What Affects the Width of a Confidence Interval?

There are four quantities required to construct a confidence interval when σ is estimated by s. These are the sample mean (m), the sample size (n), the sample standard deviation (s), and the level of confidence [(1−α)100%] we wish to have. The level of confidence is determined by the probability (α) that an interval *won't* capture μ. The sample mean is always the center of a confidence interval and thus has no effect on its width. The remaining three quantities (n, s, α) have effects similar to the corresponding quantities used when σ is known:

- interval width *decreases* as *n* increases; $s_m = s/\sqrt{n}$
- interval width *decreases* as *s* decreases; $s_m = s/\sqrt{n}$
- interval width *increases* as α *decreases*; α is the probability that an interval won't capture μ, so as α gets smaller, the interval gets wider.

LEARNING CHECK 3

1. What is the formula for the $(1-\alpha)100\%$ confidence interval when σ is not known?

2. What does it mean to say we have 95% confidence in an interval?

3. Imagine that we have computed $m \pm t_{\alpha/2}(s_m)$ for all possible means drawn from a normal population with μ = 26, when α = .05 and *n* = 16. Based on this scenario, state whether the following statements are true or false.

   (a) 95% of all such intervals will capture μ.

   (b) 90% of all such intervals will capture μ.

   (c) All intervals will be the same width.

   (d) Only means within $\mu \pm 1.96(\sigma_m)$ will have 95% confidence intervals that capture μ.

   (e) Only means outside $\mu \pm 1.96(\sigma_m)$ will have 95% confidence intervals that do not capture μ.

4. As sample size increases, sample means fall [closer to, further from] μ on average.

5. As sample size increases, interval widths get [narrower, wider] on average.

6. As sample size increases, interval widths become [more, less] variable.

7. As α increases, interval widths [increase, decrease].

8. As σ increases, interval widths [increase, decrease].

Answers

1. $m \pm t_{\alpha/2}(s_m)$.

2. We have 95% confidence in an interval when we know that 95% of all intervals computed as $m \pm t_{\alpha/2}(s_m)$ will capture μ.

3. (a) True.

   (b) False.

   (c) False. Because of sample-to-sample variations in *s*, interval widths will differ.

   (d) False. Because of sample-to-sample variation, it is possible for *s* to be very large, *m* to be outside the central 95% of the distribution, and the confidence interval captures μ.

   (e) False. Because of sample-to-sample variation, it is possible for *s* to be very small, *m* to be inside the central 95% of the distribution, and the confidence interval fails to capture μ.

4. Closer to.

5. Narrower.

6. Less.

7. Decrease.

8. Increase.

AN EXAMPLE

Let's now return to the example that we considered at the beginning of the chapter. Table 10.2 shows the 25 digit spans collected for Professor Miller by the student researcher. The mean digit span is *m* = 7.20 and *ss* = 48. From *ss* and *df* = 24, we can calculate the unbiased variance as $s^2 = ss/df = 48/24 = 2$, and we calculate the sample standard deviation as $s = \sqrt{2} = 1.41$.

Recall from the introduction that the report to Professor Miller stated the following:

The mean digit span for the 25 participants was *M* = 7.20, with a standard deviation of *s* = 1.41, 95% CI [6.62, 7.78].

TABLE 10.2 ■ Twenty-Five Digit Spans				
9	11	6	5	9
9	7	7	7	6
6	7	9	7	6
7	7	7	9	6
7	6	6	6	8
$m = 7.20$				
$ss = 48$				

This brief statement provides enough information for us to compute the 95% confidence interval ourselves. Remember that the confidence interval is defined as $m \pm t_{\alpha/2}(s_m)$. We know the sample mean $m = 7.20$, but we must figure out s_m and $t_{\alpha/2}$. Let's determine each of these in turn and then compute the 95% confidence interval.

Step 1. Compute s_m:

$$s_m = s/\sqrt{n} = 1.414/\sqrt{25} = 0.28$$

Step 2. Find $t_{\alpha/2}$. Because this is the 95% confidence interval, $\alpha = .05$. Therefore, $\alpha/2 = .025$. When we use the t-table for $n-1 = 24$ df, we find that 2.5% of the t-distribution lies above 2.064. Therefore, $t_{\alpha/2} = 2.064$.

Step 3. Compute $m \pm t_{\alpha/2}(s_m)$.

$$CI = m \pm t_{\alpha/2}(s_m) = 7.2 \pm 2.064(0.28) = 7.2 \pm 0.584 = [6.62, 7.78]$$

So this is how our (hypothetical) student estimated the mean digit span of undergraduate psychology students at Harvard University. We have 95% confidence in this interval because we know that 95% of all intervals computed as $m \pm t_{\alpha/2}(s_m)$ will capture μ.

In this example, the student chose a sample of 25 students. This was convenient because the square root of 25 is a whole number, making the calculation of $s_m = s/\sqrt{n}$ very simple. Of course, in real research, some thought should be given to sample size because it determines the precision (margin of error) of the estimate. Appendix 10.4 (available at study.sagepub.com/gurnsey) reviews precision planning for estimating μ when σ is not known.

Assumptions

I have always found it remarkable and beautiful that we can have $(1-\alpha)100\%$ confidence in the interval $m \pm t_{\alpha/2}(s_m)$, which has been computed using only the statistics of the sample. (Thank you William Sealy Gosset, Ronald Fisher, and Jerzy Neyman!) I encourage you to look at the demonstration in Appendix 10.5 (available at study.sagepub.com/gurnsey), which shows that $m \pm t_{\alpha/2}(s_m)$ will capture μ $(1-\alpha)100\%$ of the time, no matter what μ, σ, and n are. However, for this estimate to be valid, certain things must be true:

1. The population we sampled must be normal.

2. The sample is a random sample from the population of interest.

Normality

Note that requirement 1 is different from the corresponding assumption when σ is known. When σ is known, and the conditions of the central limit theorem are met, the distribution of *means* will be normal. Consequently, when σ is known, we can compute a valid confidence interval around a sample mean even if the scores in the population are not normally distributed. However, t-distributions characterize the distribution of t-scores computed from samples drawn from *normal populations*. If the scores in the population we've sampled are not normally distributed, then the t-values shown in Table 10.1 will not be appropriate. Therefore, if we had evidence that the population of digit spans is not normally distributed, our confidence interval would be suspect.

It's natural to ask how violations of the normality assumption affect the validity of confidence intervals. The answer is sometimes a little, sometimes a lot. It really depends on the kind of non-normality, the degree of non-normality, and sample size. In general, the effects of violating the normality assumption diminish as sample size increases, but no rule can be given that applies in all circumstances. Therefore, it is best to examine your data for violations of the normality assumption, using the QQ and PP plots described in Appendix 4.3 (available at study.sagepub.com/gurnsey).

Random Sampling

As we saw in Chapter 6, a sample must be a random sample from the population of interest in order for the confidence interval to be valid. The population of interest in our example comprises the digit spans of Harvard psychology students. The scenario stated that the sample was drawn from a participant pool, but nothing was said about who was in the participant pool. If those in the participant pool were volunteers, then it is very likely that not all members of the target population (Harvard psychology students) had an equal chance of being selected in the sample, because some students may not have volunteered. If the digit spans of those who volunteer to be in the pool differ in some way from those who don't volunteer, then the confidence interval will be invalid. The scenario stated that the confidence interval was computed in about 1950, which was a time when departments felt fewer ethical qualms about requiring students to register in participant pools. So, maybe this interval would have been valid.

Interpretation

What Population Are We Interested In?

Even though we've estimated the mean digit span in undergraduates at Harvard, Professor Miller's interest may extend beyond this small population. For example, he might be interested in the digit spans of all university students, or all young adults between 18 and 24, or all people on earth, for that matter. The sample of 25 Harvard undergraduates is a sample of each of these populations, but it is not a random sample from these populations. Because this is not a random sample from these populations, its mean and variance may not be valid estimates of the means and variances of these larger populations.

The validity of inferences from samples to populations is referred to as **external validity**. There will be high external validity if a sample is a simple random sample from the population of interest. There will be lower external validity if the sample is not a random sample from the population of interest. Of course, there are degrees of validity depending on the sample and intended population. There would be higher external validity when generalizing the digit span results from Harvard psychology students to psychology students in general than there would be in generalizing the digit span results from Harvard psychology students to humans in general.

This question of generalizing from samples to populations is particularly important because a great deal of behavioral science research is conducted on Western-educated students from industrialized, rich democracies. Because researchers typically wish to *generalize* beyond the population of undergraduate participant pool volunteers, they may not be inclined to draw attention to this kind of limitation in published accounts of their research. This does not mean that you should ignore such limitations. When you read research papers, you must be alert to the question of whether the authors are making legitimate inferences from their samples. That is, you should pay attention to issues of external validity.

> **External validity** refers to the validity of an inference from a sample to a population. External validity is high when a sample is a random sample from the population of interest, and lower when it is not.

Connecting Statistics to Psychological Mechanisms

In our scenario, we *assumed* that the digit-span task could be used to measure the capacity of short-term memory. Might there be any problems with this assumption? Let's think of some questions that could be asked. The digits in the study were presented at the same fixed rate to all participants. Would digit span change if presentation rate were faster or slower? If so, what does this tell us about the capacity of short-term memory? Does the capacity of short-term memory depend on the voice used to present the digits? Does the capacity of short-term memory depend on the kinds of items participants are asked to remember? For example, would the results change if words had been used rather than digits? All of these are important and interesting questions, and they have been pursued in the years since 1950. I raise these considerations not to suggest that progress in psychological research is impossible, but to remind us about the assumptions that connect our measurements to hypothetical psychological mechanisms. Furthermore, a single result, by itself, rarely answers any question of general interest. It is only in the accumulation of many results, involving the manipulation of many possible important variables, that a full picture of any psychological function can emerge.

LEARNING CHECK 4

1. If $m = 10$ and $s = 2$, calculate the $(1 - \alpha)100\%$ confidence interval around m for the following situations: (a) $\alpha = .05$, $n = 16$; (b) $\alpha = .05$, $n = 25$; (c) $\alpha = .01$, $n = 16$; (d) $\alpha = .10$, $n = 25$.

2. A 95% confidence interval was computed around the mean of a random sample of 64 Canadian household incomes using the formula $m \pm t_{\alpha/2}(s_m)$. Is this is a valid confidence interval? [Yes, No]

3. A researcher wanted to estimate the stress levels of American college students, so he recruited volunteers for his studies by posting advertisements around the computer science department at Tulane University. He used $m \pm t_{\alpha/2}(s_m)$ to compute a 95% CI, which was [22.63, 34.98]. Comment on the external validity of this confidence interval.

4. A researcher wanted to estimate the mean intelligence of fast food restaurant workers in North America. She was able to obtain a truly random sample of 128 individuals. Her measure of intelligence was the number of correct responses on a widely known trivia game. Her 95% CI [computed as $m \pm t_{\alpha/2}(s_m)$] on this measure was [111.32, 142.11]. Please comment on the validity of this interval.

5. Compute the 95% confidence interval around the mean of the following scores: $y = \{6, 7, 9, 7, 6\}$

Answers

1. $m \pm t_{\alpha/2}(s_m)$ equals the following:

   (a) $10 \pm 2.131(.5) = [8.93, 11.07]$.

   (b) $10 \pm 2.064(.4) = [9.17, 10.83]$.

   (c) $10 \pm 2.947(.5) = [8.53, 11.47]$.

   (d) $10 \pm 1.711(.4) = [9.32, 10.68]$.

2. Maybe not. The distribution of incomes is positively skewed, which violates the assumption that the scores were drawn from a normal population.

3. External validity is probably low because the researcher has used a convenience sample whose characteristics many not represent the characteristics of the intended population.

4. From a sampling point of view, there seems to be no problem. However, one can question the *validity* of the trivia game as a measure of intelligence. (See Chapter 1.) Something has been measured, but we're not exactly sure what.

5. The following table shows the calculation of the sample mean and sum of squares. From ss and sample size, we find $s^2 = 6/4 = 1.5$; therefore, $s_m = 0.548$. There are 4 degrees of freedom, so $t_{\alpha/2} = 2.776$. Therefore, $m \pm t_{\alpha/2}(s_m) = 7 \pm 2.776(0.548) = [5.48, 8.52]$.

Scores (y)	$y - m_y$	$(y - m_y)^2$
6	−1	1
7	0	0
9	2	4
7	0	0
6	−1	1
$m = 7$		$ss = 6$

ESTIMATING THE DIFFERENCE
BETWEEN TWO POPULATION MEANS

Estimating the mean of a single population has important uses. Often, however, researchers are interested in the difference between two population means. In Chapter 9, we showed that it is possible to estimate the difference between two population means $(\mu_1 - \mu_0)$ if one of those means is known (μ_0) and the other has been estimated from a sample mean (m). In that case, it was assumed that standard deviations of the two populations were known and equal (i.e., $\sigma_1 = \sigma_2$). In more realistic situations, neither σ is known but they are still assumed to be equal. Let's continue with a modified version of Professor Miller's digit-span study to see how $\mu_1 - \mu_0$ is estimated when σ is unknown.

Imagine that Professor Miller believes linguistic factors affect digit span. In particular, he suspects that digit span is affected by the number of syllables in number-words. His theory is that the more syllables in number-words, the harder it will be to maintain digits in memory. He recently read a report showing that the mean digit span of psychology students at the University of Helsinki is 6.6. He found this interesting because there are more syllables in Finnish number-words than there are in English number-words (e.g., the Finnish word for nine is yhdeksän). Therefore, he wanted to estimate the difference in the mean digit spans of Harvard and Helsinki undergraduates. That is, he wanted to estimate $\mu_{\text{Harvard}} - \mu_{\text{Helsinki}}$.

Professor Miller has the following information available: the mean digit span of $n = 25$ Harvard undergraduates ($m = 7.2$), the standard deviation of this sample ($s = 1.41$), and the mean digit span of undergraduates at the University of Helsinki ($\mu_{\text{Helsinki}} = 6.6$). The statistic $m - \mu_{\text{Helsinki}}$ estimates $\mu_{\text{Harvard}} - \mu_{\text{Helsinki}}$. This point estimate, $m - \mu_{\text{Helsinki}} = 7.2 - 6.6 = 0.6$, supports his theory that the more syllables in number-words, the shorter the mean digit span. However, he would like to construct a 95% confidence interval around this estimate to convey something about its precision.

In Chapter 9, we showed that the sampling distribution of $m - \mu_0$ has a mean of $\mu_1 - \mu_0$ (in this case, $\mu_{\text{Harvard}} - \mu_{\text{Helsinki}}$), and standard error

$$\sigma_{m-\mu_0} = \sigma/\sqrt{n}.$$

The standard deviation of the sample of digit spans (s) estimates σ. Therefore, $\sigma_{m-\mu_0}$ is estimated by

$$s_{m-\mu_0} = s/\sqrt{n}.$$

With this in mind, the $(1-\alpha)100\%$ confidence interval around $m - \mu_0$ is computed as follows:

$$(m - \mu_0) \pm t_{\alpha/2}\left(s_{m-\mu_0}\right). \tag{10.6}$$

To compute the 95% confidence interval around $m - \mu_0$, we do the following things:

Step 1. Compute $m - \mu_0$.

$$m - \mu_0 = 7.2 - 6.6 = 0.6$$

Step 2. Compute $s_{m-\mu_0}$.

$$s_{m-\mu_0} = s/\sqrt{n} = 1.41/\sqrt{25} = 0.28$$

Step 3. Find $t_{\alpha/2}$. Because this is the 95% confidence interval, $\alpha = .05$. Therefore, $\alpha/2 = .025$. When we use the t-table for $n-1 = 24$ df, we find that 2.5% of the t-distribution lies above 2.064. Therefore, $t_{\alpha/2} = 2.064$.

Step 4. Compute $m \pm t_{\alpha/2}(s_m)$.

$$CI = (m - \mu_0) \pm t_{\alpha/2}\left(s_{m-\mu_0}\right) = (7.2 - 6.6) \pm 2.064(0.28) = [0.02,\ 1.18]$$

Therefore, we have 95% confidence that $\mu_1 - \mu_0$ lies in the interval [0.02, 1.18]. Our confidence comes from knowing that 95% of all intervals computed as $(m - \mu_0) \pm t_{\alpha/2}(s_{m-\mu_0})$ will capture $\mu_1 - \mu_0$.

As in Chapter 9, this confidence interval is a simple modification of the 95% confidence interval around m, which was [6.62, 7.78]. When $\mu_0 = 6.6$ is subtracted from this interval, we obtain [6.62, 7.78] − 6.6 = [0.02, 1.18], which is the interval that we just computed. Professor Miller's best estimate of the difference between the two population means is $m - \mu_0 = 0.6$. As with all estimates, however, this estimate is subject to sampling error, and this is characterized by the confidence interval. Therefore, the true difference between population means ($\mu_{\text{Harvard}} - \mu_{\text{Helsinki}}$) is just as likely to be 0.02 as 1.18.

The practical significance of this hypothetical result is that it provides insight into the mechanisms of memory. It suggests that the short-term memory does not store abstract quantities (e.g., the abstract notion of digits). If it did, digit span should be independent of language. Rather, the information that must be maintained in short-term memory seems to involve a linguistic representation of these abstract quantities as well. In some sense, there is more information in Finnish number-words to maintain in short-term memory than in English number-words. Maybe short-term memory should not be thought of as a number of slots, but as something more complex.

LEARNING CHECK 5

1. If $m = 22$, $s^2 = 14$, and $\mu_0 = 20$, calculate the $(1-\alpha)100\%$ confidence interval around $m - \mu_0$ for the following situations: (a) $\alpha = .05$, $n = 16$; (b) $\alpha = .05$, $n = 40$; (c) $\alpha = .01$, $n = 25$; and (d) $\alpha = .10$, $n = 25$.

2. If $\mu_0 = 15$, compute the 95% confidence interval around $m - \mu_0$, given the following scores: $y = \{18, 12, 24, 30, 24, 18\}$.

Answers

1. $(m - \mu_0) \pm t_{\alpha/2}(s_{m-\mu_0})$ equals the following:

   (a) $2 \pm 2.131(0.94) = [0.01, 3.99]$.

   (b) $2 \pm 2.023(0.59) = [0.80, 3.20]$.

   (c) $2 \pm 2.797(0.75) = [-0.09, 4.09]$.

   (d) $2 \pm 1.711(0.75) = [0.72, 3.28]$.

2. The following table shows the calculation of the sample mean and sum of squares. From ss and 5 degrees of freedom, we find $s^2 = 198/5 = 39.6$; therefore, $s_{m-\mu_0} = 2.569$. There are 5 degrees of freedom, so $t_{\alpha/2} = 2.571$. Therefore, $(m - \mu_0) \pm t_{\alpha/2}(s_{m-\mu_0}) = (21 - 15) \pm 2.571(2.57) = [-0.61, 12.61]$.

Scores (y)	$y - m_y$	$(y - m_y)^2$
18	−3	9
12	−9	81
24	3	9
30	9	81
24	3	9
18	−3	9
$m = 21$		$ss = 198$

ESTIMATING δ

In Chapter 8 Cohen's effect size (δ) was defined as the difference between two population means in units of standard deviation [$\delta = (\mu_1 - \mu_2)/\sigma$]. Although δ was discussed in the context of power calculations, it is useful in its own right. That is, one might wish to estimate δ to provide some sense of the difference between two population means, without formulating the question as a significance test.

An Approximate Confidence Interval for *d*

In Chapter 9 it was shown that when σ is *known*, δ is estimated by

$$d = \frac{m - \mu_0}{\sigma}.$$

We rarely know σ, so it typically must be estimated by the sample standard deviation, *s*. In this case, δ is estimated by

$$d = \frac{m - \mu_0}{s}. \tag{10.7}$$

That is, σ is replaced with the sample standard deviation, *s*. The logic of estimation is the same in both cases.

As with any statistic, confidence intervals are used to characterize the precision of an estimate. A widely used *approximate* confidence interval for *d* has a familiar form:

$$CI - d \pm z_{\alpha/2}(s_d). \tag{10.8}$$

Because of the nature of the approximation, s_d is multiplied by $z_{\alpha/2}$ rather than $t_{\alpha/2}$. The only part of the equation that we haven't seen yet is s_d, which is defined as

$$s_d = \sqrt{\frac{d^2}{2*df} + \frac{1}{n}}. \tag{10.9}$$

Equation 10.9 bears a strong resemblance to the formula for $\sigma_d = 1/\sqrt{n}$ that was given in Chapter 9. The difference is that equation 10.9 adds the term $d^2/(2*df)$ under the radical, which has the effect of inflating s_d somewhat relative to σ_d.

This approximate method is very convenient for hand (and Excel) calculations and is generally quite accurate. The approximation is needed because the sampling distribution of *d* is a skewed distribution and its shape depends on both δ and sample size.

To construct an approximate confidence interval around *d*, we need to know *m*, *s*, *n*, and μ_0. We will continue the scenario involving the digit spans of Harvard and Helsinki undergraduates. Therefore, $m = 7.2, s = 1.41, n = 25$, and $\mu_0 = 6.6$. To compute the approximate 95% confidence interval around *d* from the digit-span example, we do the following:

Step 1. Compute *d*.

$$d = \frac{m - \mu}{s} = \frac{7.2 - 6.6}{1.41} = 0.424$$

Step 2. Compute s_d.

$$s_d = \sqrt{\frac{d^2}{2*df} + \frac{1}{n}} = \sqrt{\frac{0.424^2}{2*24} + \frac{1}{25}} = .2093$$

Step 3. Determine $z_{\alpha/2}$. Because this is the $(1-\alpha)*100\% = 95\%$ confidence interval, $\alpha = .05$. Therefore, $\alpha/2 = .025$. When we use the z-table, we find that 2.5% of the z-distribution lies below -1.96. Therefore, $z_{\alpha/2} = 1.96$.

Step 4. Compute $CI = d \pm z_{\alpha/2}(s_d)$.

$$CI = d \pm z_{\alpha/2}(s_d) = 0.424 \pm 1.96(0.2092) = [0.01, 0.83].$$

The value of d computed in this example is between a small and medium effect size, according to Cohen's classification scheme, which of course should be taken with a grain of salt. Our best estimate is that $\mu_{Harvard}$ lies 0.424 (approximately $\frac{2}{5}$) standard deviations above $\mu_{Helsinki}$. Thus, we have approximately 95% confidence that δ lies in the interval [0.01, 0.83], because approximately 95% of all such intervals will capture δ.

Why Do We Use $z_{\alpha/2}$?

In a chapter largely devoted to t-distributions it may seem odd to find a confidence interval for a statistic that involves $z_{\alpha/2}$ rather than $t_{\alpha/2}$. To understand this oddity, let's think about the effect of $z_{\alpha/2}$ versus $t_{\alpha/2}$ on the width of a confidence interval, and the effect of σ_d versus s_d on the width of a confidence interval. A confidence interval involving $t_{\alpha/2}$ will be wider than a confidence interval involving $z_{\alpha/2}$, because $t_{\alpha/2}$ is larger than $z_{\alpha/2}$. Similarly, a confidence interval involving s_d will be wider than a confidence interval involving σ_d, because s_d is greater than σ_d. Therefore, in this approximation, the required increase in confidence interval width is achieved by using s_d rather than $t_{\alpha/2}$.

An Exact Confidence Interval

An exact confidence interval for d cannot be computed easily in Excel or SPSS. However, an exact confidence interval is very easily computed in **R**. Appendix 1.3 showed you how to download and install **R** on your computer, and then we showed you how to do some simple calculations. Appendix 10.3 shows you how to compute an exact confidence interval for d in **R**. You will see that the approximation described above produces a solution that is very similar to the exact solution.

LEARNING CHECK 6

1. If $m = 91$, $ss = 96$, and $\mu_0 = 88$, calculate the approximate $(1-\alpha)100\%$ confidence interval around d for the following situations: (a) $\alpha = .1$, $n = 25$; (b) $\alpha = .05$, $n = 49$; (c) $\alpha = .05$, $n = 21$; and (d) $\alpha = .05$, $n = 6$.

2. If $\mu_0 = 15$, compute the approximate 95% confidence interval around d given the following scores: $y = \{18, 12, 24, 30, 24, 18\}$.

Answers

1. Note that ss and sample size have been given, so you must compute $s = \sqrt{ss/df}$ before you can calculate $d = (m - \mu_0)/s$. $d \pm z_{\alpha/2}(s_d)$ equals the following:

   (a) $1.500 \pm 1.645(0.295) = [1.015, 1.985]$.

   (b) $2.121 \pm 1.96(0.259) = [1.613, 2.630]$.

   (c) $1.369 \pm 1.96(0.307) = [0.767, 1.972]$.

   (d) $0.685 \pm 1.96(0.462) = [-0.221, 1.591]$.

2. The table in Learning Check 5 shows that for these data, $m = 21$ and $ss = 198$. From ss and 5 degrees of freedom, we find $s = 6.29$. $d = (21 - 15)/6.29 = 0.953$. $s_d = \sqrt{d^2/(2df) + 1/n} = .508$. Because this is the 95% confidence interval, $z_{\alpha/2} = 1.96$. Therefore, $d \pm z_{\alpha/2}(s_d) = 0.953 \pm 1.96(0.508) = [-0.041, 1.948]$.

SIGNIFICANCE TESTS

We've seen that significance tests are used to make decisions. A measured difference is judged statistically significant if it has a low probability of occurring by chance when the null hypothesis is true. In Chapter 8, significance tests were conducted by computing z_{obs} and its associated *p*-value. In Chapter 9, significance tests were conducted with confidence intervals. In both cases, it was assumed that the standard deviation (σ) of one population was known. Because it is rare for σ to be known, it is usually estimated from the sample standard deviation, *s*. In this case a significance test is conducted by computing t_{obs} rather than z_{obs}, or by constructing one of the three types of confidence intervals described in the previous sections. In all cases, the null hypothesis is that two population means are identical; i.e., $H_0: \mu_1 = \mu_0$.

Significance Tests With Confidence Intervals

In the illustration that opened this chapter, a confidence interval was used to estimate the mean digit span of Harvard psychology undergraduates (μ_{Harvard}). Later, the difference between two population means ($\mu_{\text{Harvard}} - \mu_{\text{Helsinki}}$) was estimated. Professor Miller suspected that $\mu_{\text{Harvard}} \neq \mu_{\text{Helsinki}}$, so he could have structured his question as a significance test. In this case, the null and alternative hypotheses would be as follows:

$$H_0: \mu_{\text{Harvard}} = \mu_{\text{Helsinki}}$$
$$H_1: \mu_{\text{Harvard}} \neq \mu_{\text{Helsinki}}$$

Stated more generally, the null and alternative hypotheses would be as follows:

$$H_0: \mu_1 = \mu_0$$
$$H_1: \mu_1 \neq \mu_0$$

For this illustration, $\mu_{\text{Harvard}} = \mu_1$ and $\mu_{\text{Helsinki}} = \mu_0$. The known population mean is μ_{Helsinki}, which means that $\mu_0 = 6.6$. The null hypothesis states that the sample of Harvard digit spans was drawn from a population with a mean of 6.6, and the alternative hypothesis says that it was not. Miller's best estimate of μ_{Harvard} is $m = 7.2$, 95% CI [6.62, 7.78]. To test the null hypothesis that his sample was drawn from a distribution with mean $\mu_0 = 6.6$, he simply asks if 6.6 is within the 95% confidence interval. Because 6.6 is not within the confidence interval, he can reject a two-tailed test of the null hypothesis at the $p < .05$ level. That is, because 95% of all intervals computed as $m \pm t_{\alpha/2}(s_m)$ will capture the mean of the population from which the sample was drawn, $\mu_0 = 6.6$ can be rejected as a plausible hypothesis about this population mean because it doesn't fall within the 95% confidence interval.

The same logic applies when using the confidence interval around $m - \mu_0$. If the null hypothesis is true, then $\mu_{\text{Harvard}} - \mu_{\text{Helsinki}} = 0$. Our best estimate of $\mu_{\text{Harvard}} - \mu_{\text{Helsinki}}$ is $m - \mu_{\text{Helsinki}} = 0.6$, 95% CI [0.02, 1.18]. Because 0 does not fall in this interval, we can reject a two-tailed test of the null hypothesis at the $p < .05$ level.

Finally, the approximate 95% confidence interval around *d* was [0.01, 0.83]. If the null hypothesis were true, then $\delta = (\mu_{\text{Harvard}} - \mu_{\text{Helsinki}})/\sigma = 0$. Because 0 does not fall in this interval, we can reject a two-tailed test of the null hypothesis that $\delta = 0$ at the $p < .05$ level.

Significance Tests With a *t*-Statistic

When σ is not known, hypothesis tests can be conducted using t_{obs} as our *test statistic*, which is computed as follows:

$$t_{obs} = \frac{m - \mu_0}{s_m}. \tag{10.10}$$

FIGURE 10.4 ■ Values of $t_{critical}$

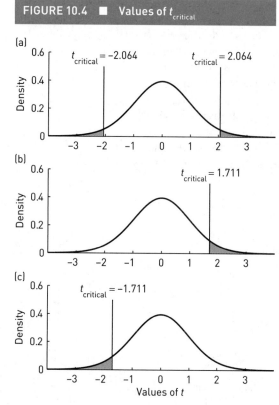

Values of $t_{critical}$ are shown for α = .05 and df = 24. (a) When H_1 predicts that $\mu_1 \neq \mu_0$, $t_{critical} = \pm2.064$ because 5% of the distribution is outside this interval. (b) When H_1 predicts that $\mu_1 > \mu_0$, $t_{critical} = 1.711$ because 5% of the distribution is above 1.711. (c) When H_1 predicts that $\mu_1 < \mu_0$, $t_{critical} = -1.711$ because 5% of the distribution is below −1.711.

Determining $t_{critical}$

To determine whether t_{obs} is statistically significant, it can be compared with a critical value of t ($t_{critical}$) in the same way z_{obs} was compared with $z_{critical}$, or m was compared with $m_{critical}$. As always, H_0: $\mu_1 = \mu_0$ or, equivalently H_0: $\mu_1 - \mu_0 = 0$. There are three possible alternative hypotheses depending on what prediction is made:

$$H_1: \mu_1 - \mu_0 \neq 0$$
$$H_1: \mu_1 - \mu_0 > 0$$
$$H_1: \mu_1 - \mu_0 < 0$$

When no prediction is made about the direction of the difference, the two values of $t_{critical}$ are those that enclose the central $(1-\alpha)100\%$ of the t-distribution. Therefore, $t_{critical}$ is identical to $t_{\alpha/2}$. In the current example, $n = 25$, $df = 24$, and α = .05. The t-table shows that $t_{critical} = \pm2.064$ because, as shown in Figure 10.4a, ±2.064 encloses the central 95% of the t-distribution with 24 degrees of freedom. Equivalently, $(\alpha)100\% = 5\%$ of the t-distribution falls outside the interval ±2.064.

When μ_1 is predicted to be greater than μ_0, $t_{critical}$ is the t-score with $(\alpha)100\%$ of the t-distribution above it. In the current example, $n = 25$, $df = 24$, and α = .05. From the t-table, we find that $t_{critical} = 1.711$ because, as shown in Figure 10.4b, $(\alpha)100\% = 5\%$ of the t-distribution lies above it.

When μ_1 is predicted to be less than μ_0, $t_{critical}$ is the t-score with $(\alpha)100\%$ of the t-distribution below it. In the current example, $n = 25$, $df = 24$, and α = .05. From the t-table, we find that $t_{critical} = -1.711$ because, as shown in Figure 10.4c, $(\alpha)100\% = 5\%$ of the t-distribution lies below −1.711.

Computing t_{obs} and Making a Decision

We now have only to compute t_{obs}. In previous sections we saw that $m = 7.2$, $s_m = 0.28$, and $\mu_0 = 6.6$. With this information, t_{obs} is computed as follows:

$$t_{obs} = \frac{m - \mu_0}{s_m} = \frac{7.2 - 6.6}{0.28} = 2.12.$$

Our calculation shows that t_{obs} exceeds the two-tailed value of $t_{critical} = 2.064$, so the null hypothesis that $\mu_{Harvard} = \mu_{Helsinki}$ can be rejected at the $p < .05$ level of significance. The exact p-value associated with t_{obs} cannot be determined from the t-table. Rather, one would have to use software to do this. The **T.DIST.2T** function in Excel (see Appendix 10.1), for example, shows that the exact p-value associated with $t_{obs} = 2.12$ for a two-tailed test with 24 df is $p = .045$. The results of this experiment may be reported as follows:

APA Reporting

The difference between the mean digit spans of Harvard and Helsinki students was statistically significant, $t(24) = 2.12$, $p = .045$.

In this report, the number 24 indicates the number of degrees of freedom associated with t_{obs}. For a one-tailed test, the exact p-value would be half the two-tailed value, or $p = .022$.

Approximate p-Values

When software is not available, the t-table can be used to *categorize* the p-value associated with t_{obs}. To do this, we report that the p-value is that associated with the largest t-score in the table that t_{obs} exceeds; e.g., $p < .01$. This categorization depends on df and whether the test is one- or two-tailed.

In the current example, $t_{obs} = 2.12$ and $df = 24$. Therefore, we look in Table 10.1 (a fragment of which is shown in Table 10.3) in the row with 24 df and then scan across until we find two t-scores that enclose t_{obs}. We find that $t_{obs} = 2.12$ exceeds $t = 2.064$ but not 2.492, which means that 2.064 and 2.492 enclose 2.12. For a two-tailed test, the p-values associated with 2.064 and 2.492 are $p = .05$ and $p = .02$, respectively. Therefore, for a two-tailed test, the exact p-value for t_{obs} is between .05 and .02, so we report $p < .05$. That is, we report that the p-value is less than that associated with the largest t-score in the table that t_{obs} exceeds.

For a one-tailed test, the p-values associated with 2.064 and 2.492 are $p = .025$ and $p = .01$, respectively. Therefore, for a one-tailed test, the exact p-value for t_{obs} is between .025 and .01, so we report $p < .025$. That is, we report that the p-value is less than that associated with the largest t-score in the table that t_{obs} exceeds.

TABLE 10.3 ■ A Fragment of the t-Table Used to Illustrate Approximate p-Values						
	Area in two tails (α)					
	.50	.20	.10	.05	.02	.01
	Area in one tail ($\alpha/2$)					
df	.250	.100	.050	.025	.010	.005
21	.686	1.323	1.721	2.080	2.518	2.831
22	.686	1.321	1.717	2.074	2.508	2.819
23	.685	1.319	1.714	2.069	2.500	2.807
24	.685	1.318	1.711	2.064	2.492	2.797
25	.684	1.316	1.708	2.060	2.485	2.787
26	.684	1.315	1.706	2.056	2.479	2.779

Planning Significance Tests: Power

We saw in Chapter 8 that before running a significance test one should consider (i) the effect size (δ) that would be meaningful given the research question and (ii) the power one would like the test to have. The values of δ and power combine to determine the sample size (n) required for your experiment. Table 10.4 provides the sample sizes required to achieve a specified level of power for a range of effect sizes for one-tailed tests with $\alpha = .05$. Table 10.5 provides the sample sizes required to achieve a specified level of power for a range of effect sizes for two-tailed tests with $\alpha = .05$.

Appendix 8.1 (available at study.sagepub.com/gurnsey) showed that it is relatively easy to determine the sample size required to achieve a specified level of power in Excel when σ is known. Things are not as simple when σ is not known because Excel does not have the required functions, which relate to non-central t-distributions. However, for those interested in calculating power for situations not covered in Tables 10.4 and 10.5, there is an excellent, freely available program called *G*Power* that you can download and use. (A description of how to use G*Power is given in Appendix 10.6 (available at study .sagepub.com/gurnsey). On the other hand, if you compare Table 10.5 with Table 8.1 from

TABLE 10.4   ■   Sample Size, Effect Size, and Power (one-tailed tests)

δ	Power										
	.10	.20	.30	.40	.50	.60	.70	.80	.85	.90	.95
.1	15	66	127	195	272	362	472	620	721	858	1084
.2	5	18	33	50	70	92	120	156	182	216	272
.3	3	9	16	23	32	42	54	71	82	97	122
.4	3	6	10	14	19	24	31	41	47	55	70
.5	2	5	7	10	13	16	21	27	31	36	45
.6	2	4	6	7	9	12	15	19	22	26	32
.7	2	3	5	6	8	9	12	15	17	19	24
.8	2	3	4	5	6	8	9	12	13	15	19
.9	2	3	4	4	5	7	8	10	11	13	15
1.0	2	3	3	4	5	6	7	8	9	11	13

Note: Directional tests assuming $\alpha = .05$.

TABLE 10.5   ■   Sample Size, Effect Size, and Power (two-tailed tests)

δ	Power										
	.10	.20	.30	.40	.50	.60	.70	.80	.85	.90	.95
.1	45	127	208	294	387	492	620	787	900	1053	1302
.2	13	33	54	75	98	125	157	199	227	265	327
.3	7	16	25	35	45	57	71	90	102	119	147
.4	5	10	15	21	26	33	41	52	59	68	84
.5	4	7	11	14	18	22	27	34	38	44	54
.6	4	6	8	11	13	16	20	24	27	32	39
.7	3	5	7	8	10	13	15	19	21	24	29
.8	3	4	6	7	9	10	12	15	17	19	23
.9	3	4	5	6	7	9	10	12	14	16	19
1.0	3	4	5	5	6	7	9	10	12	13	16

Note: Non-directional tests assuming $\alpha = .05$.

Chapter 8, you will see that the sample sizes are only slightly larger when σ is not known. Therefore, if you use the Excel method described in Appendix 8.1, you will achieve a good approximation to the required sample size when σ is not known. Just add one or two participants to this result and you should be fine.

Relationship Between d and t_{obs}

We saw in Chapter 9 that d and z_{obs} were related in a simple way. When the population variance is estimated, d and t_{obs} are also related in same way. When we compare the following definitions of d and t_{obs},

$$d = \frac{m - \mu_0}{s} \text{ and}$$

$$t_{obs} = \frac{m - \mu_0}{s/\sqrt{n}} = \frac{m - \mu_0}{s} * \sqrt{n},$$

it is clear that the relationship between them is simply

$$t_{obs} = d * \sqrt{n} \text{ or}$$

$$d = t_{obs}/\sqrt{n}.$$

This means that if sample size (n) is very large, t_{obs} can be very large and statistically significant even if d is very small. Therefore, before accepting a statistically significant t_{obs} as meaningful, we should at least check to see what estimated effect size it corresponds to.

LEARNING CHECK 7

1. A researcher believes that a fish-rich diet will increase cognitive processing speed among elderly men aged 75–79 years. The measure of cognitive processing speed is the time it takes to complete a trail-making task, in which participants are asked to connect a sequence of numbers (e.g., 1, 2, 3, . . . , 24) in a figure like the one below. In this situation, short completion times represent faster cognitive processing. Let's say that in the normal healthy population of men aged 70–79, the average time to complete the task is 50 seconds. Therefore, we can say $\mu_{normal} = 50$. A random sample of 25 men, aged 75–79, was selected to participate in this study. All 25 were put on a high-fish diet for 6 weeks, after which they all completed the trail-making task. The mean completion time was $m = 44$ seconds with a standard deviation of 10 seconds. Use this information to answer the following questions.

   (a) Should a one- or two-tailed test be conducted?
   (b) Use the symbols μ_{normal} and $\mu_{FishDiet}$ to express H_0.
   (c) Use the symbols μ_{normal} and $\mu_{FishDiet}$ to express H_1.
   (d) Show the formula for the t-statistic that you would use to test H_0.
   (e) Do you expect the difference $m - \mu_{normal}$ to be greater or less than 0?
   (f) If $\alpha = .05$, what is $t_{critical}$?
   (g) Calculate t_{obs}.
   (h) Should you retain or reject H_0?
   (i) Compute the 95% confidence interval around m.
   (j) Should this confidence interval lead you to reject $\mu_{normal} = 50$?

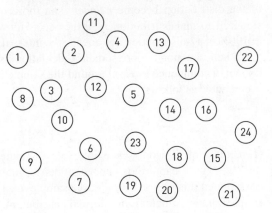

2. I plan to run a one-tailed significance test with $\alpha = .05$. I would only be interested in an effect size of .4. How large should my sample be if I wish to have power = .8?

3. I plan to run a one-tailed significance test with $\alpha = .05$. How large should my sample be if I'm interested in $\delta = .2$ and wish to have power = .9?

4. I plan to run a one-tailed significance test with $\alpha = .05$. I'm interested in $\delta = .2$, and I have 18 scores in my sample. What is my power?

5. I plan to run a one-tailed significance test with $\alpha = .05$. I have 19 scores in my sample. What effect size could I detect 90% of the time when H_0 is false?

6. If $m = 10$, $\mu_0 = 9$, $s = 4$, and $n = 25$, answer the following:

   (a) Calculate t_{obs}.

   (b) Calculate d.

   (c) Construct a 95% CI around d.

   (d) Does this interval contain 0?

7. If $m = 100.4$, $\mu_0 = 100$, $s = 4$, and $n = 400$,

   (a) Calculate t_{obs}.

   (b) Calculate d.

   (c) Construct a 95% CI around d.

   (d) Does this interval contain 0?

8. If $d = .5$ and $n = 16$, calculate t_{obs}.

9. If $d = .05$ and $n = 1600$, calculate t_{obs}.

Answers

1. (a) A one-tailed test because the researcher is making a prediction about the outcome.

   (b) H_0: $\mu_{FishDiet} - \mu_{normal} = 0$.

   (c) H_1: $\mu_{FishDiet} - \mu_{normal} < 0$.

   (d) $t_{obs} = (m - \mu_{normal})/s_m$.

   (e) Less than zero because the fish diet is expected to reduce completion times, so m is predicted to be less than μ_{normal}.

   (f) Given that there are $25 - 1 = 24$ degrees of freedom, $t_{critical}$ is -1.711.

   (g) $t_{obs} = (m - \mu_{normal})/s_m = (44 - 50)/(10/\sqrt{25}) = -3$.

   (h) Reject because $t_{obs} < t_{critical}$.

   (i) $CI = m \pm z_{\alpha/2}(s_m) = 44 \pm 2.064(2) = [39.87, 48.13]$.

   (j) Yes, 50 is not in the confidence interval so we should reject it as a plausible hypothesis about the mean of the distribution from which the 25 scores were drawn.

2. 41.

3. 216.

4. .2.

5. .7.

6. (a) $t_{obs} = (m - \mu_0)/s_m = (10 - 9)/0.8 = 1.25$.

   (b) $d = (m - \mu_0)/s = (10 - 9)/4 = .25$.

   (c) $CI = d \pm z_{\alpha/2}(s_d) = .25 \pm 1.96(0.2032) = [-.148, 648]$.

   (d) Yes, this interval contains 0.

7. (a) $t_{obs} = (m - \mu_0)/s_m = (100.4 - 100)/0.2 = 2$.

   (b) $d = (m - \mu_0)/s = (100.4 - 100)/4 = .1$.

   (c) $CI = d \pm z_{\alpha/2}(s_d) = .1 \pm 1.96(0.0501) = [.002, .198]$.

   (d) No, this interval does not contain 0.

8. $t_{obs} = d*\sqrt{n} = .5*4 = 2$.

9. $t_{obs} = d*\sqrt{n} = 0.5*40 = 2$.

SUMMARY

In this chapter we introduced t-*scores*, which are computed like *z*-scores except that the difference between a sample mean and a population mean is divided by the *estimated standard error of the mean* ($s_m = s/\sqrt{n}$) rather than the standard error of the mean ($\sigma_m = \sigma/\sqrt{n}$),

$$t = \frac{m - \mu}{s_m}.$$

Distributions of *t*-scores are called t-*distributions* only if samples are drawn from a normal population. There is a *t*-distribution for every possible sample size. Each *t*-distribution is associated with $n-1$ *degrees of freedom* (*df*), where *n* is sample size. When samples are very large ($df \to \infty$), the *t*-distribution approaches the standard normal (*z*) distribution. As sample size decreases ($df \to 1$), *t*-scores become less densely packed around the mean (0) of the distribution, and the tails of the distribution become *heavier* than those of the standard normal distribution. $t_{\alpha/2}$ is the *t*-score for which (α)100% of a *t*-distribution lies outside the interval $\pm t_{\alpha/2}$.

When the population standard deviation is not known, a confidence interval around the sample mean is computed as follows:

$$m \pm t_{\alpha/2}(s_m).$$

$(1-\alpha)$100% of all confidence intervals computed in this way will capture μ. For confidence intervals constructed this way to be valid, the sample must be a simple random sample from a normal distribution.

Confidence intervals can be used to test any null hypothesis about μ_0. If μ_0 is not in the interval, we can reject the null hypothesis that H_0: $\mu_1 = \mu_0$.

A confidence interval around $m - \mu_0$ estimates the difference between μ_1 and μ_0 (i.e., $\mu_1 - \mu_0$). The estimated standard error of $m - \mu_0$ is

$$s_{m-\mu_0} = \frac{s}{\sqrt{n}},$$

which means that $s_{m-\mu_0} = s_m$. The confidence interval around $m - \mu_0$ is

$$(m - \mu_0) \pm t_{\alpha/2}(s_{m-\mu_0}).$$

In this case, $df = n-1$, just as in the case of the confidence interval around m. If 0 is not in the interval, we can reject the null hypothesis that H_0: $\mu_1 = \mu_0$.

We also revisited the important concept of effect size. The parameter δ is estimated by

$$d = \frac{m - \mu_0}{s}.$$

An approximate confidence interval around d is computed as

$$CI = d \pm z_{\alpha/2}(s_d),$$

where

$$s_d = \sqrt{\frac{d^2}{2df} + \frac{1}{n}}.$$

Approximately $(1-\alpha)100\%$ of all intervals computed this way will capture δ. If 0 is not in the interval, we can reject the null hypothesis that H_0: $\delta = 0$.

We can also use t-scores to test hypotheses about population means. If our observed t-score,

$$t_{obs} = \frac{m - \mu_0}{s_m},$$

exceeds a critical value of t ($t_{critical}$), we reject H_0. $t_{critical}$ is determined by our significance level (α), degrees of freedom (df), and the nature of H_1 (directional or non-directional). There is a simple relationship between d and t_{obs}: $d = t_{obs}/\sqrt{n}$ and $t_{obs} = d*\sqrt{n}$.

KEY TERMS

EXERCISES

Please note that asterisks (*) indicate questions relating to material covered in the appendixes or requiring insight beyond what was covered in the chapter.

Definitions and Concepts

1. What is a t-score?

2. What is a t-distribution?

3. Define degrees of freedom.

4. Explain what d estimates.

5. What proportion of a t-distribution falls above $t_{\alpha/2}$?

6. What is the smallest sample size for which the unbiased sample variance can be computed?

7. A population has a mean of $\mu = 100$ and standard deviation of $\sigma = 16$. I've drawn a sample of size 16 and computed the 95% confidence interval around the sample mean using the formula $m \pm$ $t_{\alpha/2}(s_m)$. The sample mean equals 98 and the confidence interval does not capture the population mean. Is this possible? If it is, explain why. If it is not, explain why not.

8. A novice statistician wanted to place a 95% confidence interval around a sample mean (m). There were $n = 15$ scores in the sample. The sample had a mean of 7 and a standard deviation of 2. He computed $s_m = 2/\sqrt{15}$ and used the following equation to compute the confidence interval: $m \pm 1.96(s_m)$. Explain the mistake that has been made and calculate the correct 95% confidence interval.

9. What assumptions must be correct for $m \pm t_{\alpha/2}(s_m)$ to be valid?

10. What are the four quantities required to calculate a confidence interval when σ is not known, and how does each of these affect the width of the confidence interval?

True or False

State whether the following statements are true or false.

11. All distributions of t-scores are t-distributions.

12. All 95% confidence intervals computed as $m \pm t_{\alpha/2}(s_m)$ will capture μ.

13. The centers of all 95% confidence intervals computed as $m \pm t_{\alpha/2}(s_m)$ will be within $1.96(\sigma_m)$ of μ.

14. As sample size decreases, the t-distribution becomes more like the z-distribution.

15. If $\alpha = .05$ and $n = 16$, then $t_{\alpha/2} = 2.12$.

16. If $\alpha = .1$ and $n = 56$, then $m \pm 1.96(s_m)$ will capture μ more than 90% of the time.

17. A t-distribution with 4 df is leptokurtic.

18. If $\alpha = .05$, $n = 16$, and $H_1: \mu_1 - \mu_0 > 0$, then $t_{obs} = 2.12$ is statistically significant.

Calculations

19. For sample sizes (a) 10, (b) 20, (c) 36, (d) 14, (e) 88*, (f) 92*, and (g) 1089*, determine $t_{\alpha/2}$ when $\alpha = .05, .01$, and .1.

20. If I draw a sample of size $n = 15$ from a normal distribution with a mean of $\mu = 10$, what is the probability that $t = (m - \mu)/s_m$ will fall in each of the following intervals: (a) ±1.761, (b) ±1.345, (c) ±2.624, (d) ±2.145, (e) ±0.692, (f) ±2.415*, (g) ±1.887*, and (h) ±1.791*?

21. Which of the following defines the 95% confidence interval if sample size is $n = 25$: (a) $m \pm 1.711(s_m)$, (b) $m \pm 2.064(s_m)$, (c) $m \pm 2.060(s_m)$, and (d) $m \pm 1.708(s_m)$?

22. Assume you have drawn a sample of size 51 from a normal distribution with an unknown standard deviation and an unknown mean. Suppose your sample mean (m) was 14 and the unbiased sample standard deviation (s) was 7. Calculate the upper and lower limits of the 95% confidence interval about m.

23. Assume you have drawn a sample of size 46 from a normal distribution with an unknown standard deviation and an unknown mean. Suppose your sample mean was 22 and the unbiased sample standard deviation was 2.63. Calculate the upper and lower limits of the 99% confidence interval about m.

24. Assume you have drawn a sample of size 12 from a normal distribution with an unknown standard deviation and an unknown mean. Suppose your sample mean was 8 and the unbiased sample standard deviation was 1.33. Calculate the upper and lower limits of the 90% confidence interval about m.

25. Compute the 95% confidence interval for each of the following samples:

    (a) $y = [13, 14, 5, 14, 9, 8, 9, 8]$,

    (b) $y = [5, 11, 11, 11, 9, 6, 9, 11, 9, 10, 9, 7, 9]$,

    (c) $y = [8, 3, 10, -1, 4, 7, 10, 12, 4, -3, 0, -1, 16, 1]$,

    (d) $y = [19, 24, 19, 21, 14, 14, 21, 24, 33]$.

26. For each part of question 25, test the non-directional null hypothesis that $\mu = 13$. State the null and alternative hypotheses in symbols. (Assume $\alpha = .05$.) State whether you reject or retain the null hypothesis.

27. For each part of question 25, assuming $\mu = 13$, compute the best estimate of δ and construct an approximate 95% confidence interval around this estimate.

28. *Assume a population has an unknown standard deviation. For each of the following situations, determine the approximate sample size (n) required to achieve the the following:

    (a) $moe = \sigma$ for the 95% confidence interval,

    (b) $moe = \sigma$ for the 99% confidence interval,

    (c) $moe = \sigma$ for the 90% confidence interval,

    (d) $moe = .05*\sigma$ for the 99% confidence interval,

    (e) $moe = .36*\sigma$ for the 85% confidence interval.

Scenarios

29. Do Facebook posts affect your emotions? In a recent study (Kramer, Guillory, & Hancock, 2014), the news feeds of a random sample of 155,000 Facebook users were manipulated to reduce the number of news posts containing negative emotion words by 10%. That is, each of these 155,000 Facebook users saw 10% fewer negative posts than usual. Following this, researchers measured the number of positive emotion words in posts made by each of these 155,000 people. The mean percentage of positive emotion words in these posts was $m = 5.29\%$, with a standard deviation of

$s = 4.12$. This mean is slightly higher than the mean percentage of positive emotion words in Facebook posts when the newsfeed is not manipulated ($\mu = 5.24\%$).

(a) Calculate the 95% confidence interval around m.

(b) Using the confidence interval computed above, determine whether the difference between $m = 5.29\%$ and $\mu = 5.24\%$ is statistically significant.

(c) Use a t-statistic to conduct a two-tailed significance test to determine whether the difference between $m = 5.29\%$ and $\mu = 5.24\%$ is statistically significant.

(d) *Use Excel to determine the exact p-value associated with this t-statistic.

(e) Use the information given above to estimate δ.

(f) Compute the approximate 95% confidence interval around this estimate of δ.

(g) Put into words how reducing the number of negative posts one receives affects the emotional content of subsequent posts.

30. Do people enjoy being alone with their thoughts? A recent study (Wilson, Reinhard, Westgate, Gilbert, Ellerbeck, Hahn, Brown, & Shaked, 2014) measured how people feel about being alone with their thoughts, with nothing else to do. One study involved 74 participants, each of whom was brought to a featureless room, where they were left alone for 15 minutes (without personal belongings such as phones, books, or writing materials) and asked to entertain themselves with their thoughts, without moving from their chair or falling asleep. At the end of the 15 minutes they were asked to rate their experience. One question was "How enjoyable was this part of the experiment?" Participants responded on a 9-point scale, where 1 means not enjoyable. The mean response was $m = 4.67$ with a standard deviation of $s = 1.8$. Calculate the 95% confidence interval around m.

(a) A score of 5 on this 9-point scale means neither enjoyable or unenjoyable. Using the confidence interval computed above, determine whether the difference between $m = 4.67$ and $\mu_0 = 5$ is statistically significant.

(b) Use a t-statistic to conduct a two-tailed significance test to determine whether the difference between $m = 4.67$ and $\mu_0 = 5$ is statistically significant.

(c) *Use Excel to determine the exact p-value associated with this t-statistic.

(d) Use the information given above to estimate δ.

(e) Compute the approximate 95% confidence interval around this estimate of δ.

(f) Put into words what this study demonstrates about how much people enjoy being alone with their thoughts.

31. A second study was conducted in the same way, but participants were given the opportunity to self-administer an electric shock (that they had previously experienced) during the free-thinking period. The authors of the study report the following:

> Many participants elected to receive negative stimulation over no stimulation—especially men: 67% of men (12 of 18) gave themselves at least one shock during the thinking period [range = 0 to 4 shocks, mean (M) = 1.47, $SD = 1.46$, not including one outlier who administered 190 shocks to himself], compared to 25% of women (6 of 24; range = 0 to 9 shocks, $M = 1.00$, $SD = 2.32$). Note that these results only include participants who had reported that they would pay to avoid being shocked again. (Wilson et al. 2014, *Science*, 345, 75, p. 76).

(a) Calculate the 95% confidence interval around the mean number of self-administered shocks for men.

(b) Calculate the 95% confidence interval around the mean number of self-administered shocks for women.

(c) Please comment on whether there is any reason to think these confidence intervals are invalid.

(d) Put into words what this study demonstrates about how much people enjoy being alone with their thoughts.

APPENDIX 10.1: CONFIDENCE INTERVALS AND SIGNIFICANCE TESTS IN EXCEL

Four functions in Excel allow us to solve area-under-the-curve questions related to t-distributions. These four functions are **T.DIST**, **T.DIST.2T**, **T.INV**, and **T.INV.2T**. We will discuss how each of these is used to construct confidence intervals and to conduct significance tests. First we will consider a confidence interval around m and $m-\mu_0$ when given s, n, and α. Then we will conduct a significance test for $m-\mu_0$. Finally, we will compute a confidence interval around the estimated effect size, d.

Cells **B2**, **B3**, **B4**, and **B6** of Figure 10.A1.1 contain the quantities **m**, **s**, **n**, and **α**, which are typically given when computing a confidence interval around m. **df** is calculated in cell **B5**, and s_m has been calculated in cell **B8**. These are straightforward calculations that need little explanation. The percent confidence we will have in this interval, **(1−α)100%**, is calculated in cell **B9** just as it was in Chapter 6. Finally, **α/2** is calculated in cell **B10**.

T.INV and T.INV.2T

To compute the $(1-\alpha)100\%$ confidence interval around m, we need to determine $t_{\alpha/2}$. To do this, we can use either **T.INV** or **T.INV.2T**. The **T.INV** function takes the following arguments:

$$\text{T.INV(P, df).}$$

The first argument is a proportion (**P**) and the second is the degrees of freedom (**df**). **T.INV** returns the t-score having the specified proportion (**P**) of the distribution *below* it.

There are two ways that we can use **T.INV** to determine $t_{\alpha/2}$. The first is shown in cell **B11** of Figure 10.A1.1. In this case, we ask **T.INV** to return the t-score with $(\alpha/2)100\%$ of the distribution below it:

$$\text{=T.INV(α/2, df).}$$

Then we take the absolute value of the result, using the **ABS** function, to make the negative number positive. (We could also have multiplied the negative number by −1 to achieve the same result.) The second way to use **T.INV** to determine $t_{\alpha/2}$ is shown in cell **B12**. In this case, we have provided **T.INV** with $1-\alpha/2$:

$$\text{=T.INV(1−α/2, df).}$$

In effect, we are asking for the t-score having $(\alpha/2)100\%$ of the distribution *above* it. Of course, the same thing is returned in both cases.

Finally, we can use the **T.INV.2T** function to do the same thing. The **T.INV.2T** function takes the following arguments,

$$\text{T.INV.2T(α, df),}$$

and returns the t-score for which the proportion outside the interval $\pm t$ is α. That is, it takes α and returns $t_{\alpha/2}$. Notice that **T.INV.2T** takes α as an argument (cell **B13**), whereas to achieve the same result using **T.INV** we make use of $\alpha/2$ (cell **B11**).

Cells **B15** and **B16** compute the lower and upper limits of the $(1-\alpha)100\%$ confidence interval around the sample mean, making use of $t_{\alpha/2}$ computed in one of the three ways shown in cells **B11** to **B13**. As discussed in the chapter, we can also compute a confidence interval around $m-\mu_0$ in much the same way as the confidence interval around m. The statistic in this case is $m-\mu_0$, computed in cell **B19**, and the standard error is $s_{m-\mu_0}$. As noted in the chapter, $s_{m-\mu_0}$ (computed in cell **B20**) has exactly the same value

---

FIGURE 10.A1.1	■	T.INV and T.INV.2T

	A	B	C
1	**Quantities**	**Values**	**Formulas**
2	m	7.2	
3	s	1.4142	
4	n	25	
5	df	24	=B4-1
6	α	0.05	
7			
8	s_m	0.2828	=B3/sqrt(B4)
9	(1−α)100%	95	=(1-B6)*100
10	α/2	0.025	=B6/2
11	$t_{α/2}$	2.064	=ABS(T.INV(B10,B5))
12	$t_{α/2}$	2.064	=T.INV(1-B10,B5)
13	$t_{α/2}$	2.064	=T.INV.2T(B6,B5)
14			
15	$m-t_{α/2}(s_m)$	6.62	=B2-B10*B7
16	$m+t_{α/2}(s_m)$	7.78	=B2+B10*B7
17			
18	μ_0	6.60	
19	$m-\mu_0$	0.6	=B2-B18
20	$s_{m-\mu_0}$	0.2828	=B3/sqrt(B4)
21	$(m-\mu_0)-t_{α/2}(s_{m-\mu_0})$	0.02	=B19-B11*B20
22	$(m-\mu_0)+t_{α/2}(s_{m-\mu_0})$	1.18	=B19+B11*B20

The **T.INV** (t inverse) function in Excel. **T.INV** is given a proportion and returns a t-score. When one provides a proportion (**α/2**) and degrees of freedom (**df**), **T.INV** returns the t-score for which $(\alpha/2)100\%$ of the t-distribution lies below it.

as s_m (computed in cell **B8**). The lower and upper limits around $m-\mu_0$ are calculated in cells **B21** and **B22**.

T.DIST and T.DIST.2T

Significance tests can be conducted by computing t_{obs}. Once t_{obs} has been computed, we can determine its p-value under the null hypothesis. In Figure 10.A1.2, **m**, **μ_0**, **s**, and **n** are given in cells **B2** to **B5**, and **df** is computed in cell **B6**. The difference between m and μ_0 is computed in cell **B8**, the estimated standard error of m (**s_m**) is computed in cell **B9**, and t_{obs} in cell **B10**.

We use the **T.DIST** function to compute the p-value associated with t_{obs}. **T.DIST** takes the following arguments:

T.DIST(t, df, cumulative).

When **cumulative** is set to 1, **T.DIST** returns the proportion of the t-distribution below the t-score given. When cumulative is set to 0, **T.DIST** returns the density of the t-distribution at t.

In cell **B12** of Figure 10.A1.2, **T.DIST** has been used to compute the one-tailed p-value associated with t_{obs}. To make this work for both positive and negative values of t_{obs}, we take the absolute value of t_{obs} (making it positive) and then put a negative sign in front of it. The proportion of the distribution below $-t_{obs}$ is the same as the proportion above t_{obs}. Cell **B13** shows that the same thing can be accomplished by subtracting the proportion below the absolute value of t from 1. The two-tailed p-value is simply twice the one-tailed value, as shown in cell **B14**.

The **T.DIST.2T** function calculates the proportion of the t-distribution outside the interval $\pm t$:

T.DIST.2T(t, df).

T.DIST.2T takes a t-score (**t**) and the degrees of freedom (**df**) as does **T.DIST**. Unfortunately, **T.DIST.2T** does not work with negative t-scores, so we must take the absolute value of t_{obs} before submitting it to **T.DIST.2T** (as shown in cell **B16**).

An Approximate Confidence Interval for d

For completeness, Figure 10.A1.3 shows how to compute an approximate confidence interval for the estimated effect size, d. The data in this example (cells **B2** to **B7**) are the same as in Figures 10.A1.1 and 10.A1.2. The main calculations are in cells **B9** and **B10**. Cell **B9** shows the calculation of d (**d**), and **B10** shows the approximate estimated standard error. This approximation uses $z_{\alpha/2}$

FIGURE 10.A1.2 ■ T.DIST and T.DIST.2T

	A	B	C
1	Quantities	Values	Formulas
2	m	7.2	
3	μ_0	6.6	
4	s	1.4142	
5	n	25	
6	df	24	=B5-1
7			
8	m-μ_0	0.6	=B2-B3
9	s_m	0.2828	=B4/SQRT(B5)
10	t_{obs}	2.1213	=B8/B9
11			
12	$p_{(one-tailed)}$	0.0222	=T.DIST(-ABS(B10),B6,1)
13	$p_{(one-tailed)}$	0.0222	=1-T.DIST(ABS(B10),B6,1)
14	$p_{(two-tailed)}$	0.0444	=2*B13
15			
16	$p_{(two-tailed)}$	0.0444	=T.DIST.2T(ABS(B10),B6)

The **T.DIST** and **T.DIST.2T** functions in Excel. **T.DIST** is given a t-score and returns the proportion of the t-distribution below it (**P**) or the density of the distribution at that point. **T.DIST.2T** takes a positive t-score and returns the proportion of the t-distribution outside the interval $\pm t$.

FIGURE 10.A1.3 ■ Estimated Effect Size

	A	B	C
1	Quantities	Values	Formulas
2	m	7.2	
3	μ_0	6.60	
4	s	1.4142	
5	n	25	
6	df	24	
7	α	0.05	
8			
9	d	0.4243	=(B2-B3)/B4
10	s_d	0.2092	=SQRT(B9^2/(2*B6)+1/B5)
11			
12	$(1-\alpha)100\%$	95	=(1-B7)*100
13	$\alpha/2$	0.025	=B7/2
14	$z_{\alpha/2}$	1.960	=ABS(NORM.S.INV(B13))
15			
16	$d-z_{\alpha/2}(s_d)$	0.01	=B9-B14*B10
17	$d+z_{\alpha/2}(s_d)$	0.83	=B9+B14*B10

An approximate confidence interval for the estimated effect size, d (**d**).

rather than $t_{\alpha/2}$, so $z_{\alpha/2}$ is computed in cell **B14**. The lower and upper limits of the approximate confidence interval are computed in cells **B16** and **B17**.

APPENDIX 10.2: CONFIDENCE INTERVALS AND SIGNIFICANCE TESTS IN SPSS

A Confidence Interval for the Mean

There are two ways to compute confidence intervals related to sample means in SPSS. The first is through the Analyze→Descriptive Statistics→Explore . . . menu. For this example, the 25 digit spans from Table 10.2 have been arranged in a single column (Digit_Span) in an SPSS data file. In the Explore . . . dialog, Digit_Span has been moved to the Dependent List: region (Figure 10.A2.1). Clicking the ⬛ OK ⬛ button produces the output shown in Figure 10.A2.2. The mean, estimated standard error, and 95% confidence interval are shown in the top two rows of the SPSS output.

A number of other statistics are shown, only two of which we've not seen before. The 5% Trimmed Mean is the mean obtained when the highest and lowest 5% of scores are removed. That is, SPSS eliminates potential outliers before computing the mean. If the data are heavily skewed, the mean and the 5% trimmed mean can be quite different.

The second new quantity is the Interquartile Range. When scores are sorted from highest to lowest, three numbers divide the scores into four groups of equal size. These groups are called quartiles. For example, the median divides the complete set into two equal sized groups and each of these two groups has a median of its own. Therefore, the medians of these two groups contain the central 50% of scores. When we compute the range of the central 50% of scores, we have computed the so-called interquartile range. In this example, the interquartile range is 3.

FIGURE 10.A2.2 ■ Explore . . . Dialog Output

Descriptives

			Statistic	Std. Error
Digit_Span	Mean		7.20	.283
	95% Confidence Interval for Mean	Lower Bound	6.62	
		Upper Bound	7.78	
	5% Trimmed Mean		7.12	
	Median		7.00	
	Variance		2.000	
	Std. Deviation		1.414	
	Minimum		5	
	Maximum		11	
	Range		6	
	Interquartile Range		3	
	Skewness		.961	.464
	Kurtosis		.619	.902

The descriptive statistics produced through the Explore . . . dialog.

A Confidence Interval and Significance Test for $m - \mu_0$

To compute a confidence interval and significance test for a single sample of scores, we choose the Analyze→Compare Mean→One-Sample T Test . . . menu. Figure 10.A2.3 shows this dialog. The variable Digit_Span has been moved into the Test Variable(s): region. The quantity that we call μ_0 is entered in the text box titled Test Value. Clicking ⬛ OK ⬛ produces the output shown in Figure 10.A2.4. All quantities shown were calculated exactly as described in the body of the chapter and exactly the same results were obtained. Note that when Test Value is set to 0, the confidence interval in Figure 10.A2.4 will be exactly the same as in Figure 10.A2.2.

FIGURE 10.A2.1 ■ The Explore . . . Dialog

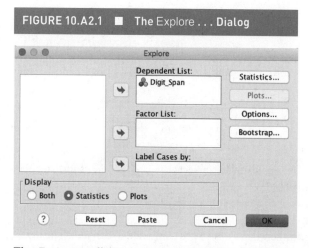

The Explore . . . dialog is used to compute a confidence interval around a sample mean in SPSS.

FIGURE 10.A2.3 ■ The One-Sample T Test

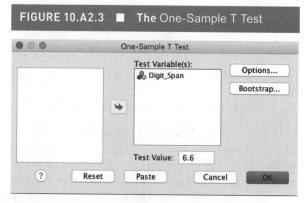

The One-Sample T Test dialog used to compute a confidence interval around a sample mean in SPSS.

FIGURE 10.A2.4 ■ The One-Sample T Test **Output**

One-Sample Test

	Test Value = 6.6					
					95% Confidence Interval of the Difference	
	t	df	Sig. (2–tailed)	Mean Difference	Lower	Upper
Digit_Span	2.121	24	.044	.600	.02	1.18

The t statistic, df, p-value and 95% confidence interval computed using the One-Sample T Test dialog.

APPENDIX 10.3: EXACT CONFIDENCE INTERVALS FOR *d* USING MBESS IN R

Appendix 1.3 introduced **R**. There we saw how to download and install **R**, as well as how to use **R** to do simple computations such as computing GPA. We can use **R** to compute exact confidence intervals around an estimated effect size (*d*). However, to do so, we have to add some functionality. Therefore, this appendix describes how to add this functionality. The discussion that follows assumes that you've read Appendix 1.3 and have **R** installed on your computer.

MBESS

Professor Ken Kelley at Notre Dame University has written a package called MBESS, which originally stood for "Methods for the Behavioral, Educational, and Social Sciences" (Kelley, 2007). However, as Professor Kelley describes on his website, MBESS has methods that are applicable to a far wider range of fields than educational and social sciences. Nevertheless, the name of the package remains unchanged.

We saw that when **R** is launched, you are presented with a window that looks like the one shown in Figure 10.A3.1. We've seen that commands can be typed into the **R** prompt (>) at the bottom (see the bottom of Figure 10.A3.1) and **R** will execute the command and return the result. To use the MBESS package, it must be first downloaded from the Internet. In Figure 10.A3.1, I have clicked on the Packages & Data menu item. We will first choose Package Installer from this menu. When this is chosen, the window shown in Figure 10.A3.2 will appear.

R packages are stored in digital repositories around the world. To install the MBESS package, you must choose a repository from which to download it. One of these is shown in Figure 10.A3.2. In the top left corner of Figure 10.A3.2, you will see a drop-down

menu that is currently set to Other Repository. The default repository that shows up is R.research.att .com/, as shown to the right of Other Repository in Figure 10.A3.2. Under the top left drop-down list is a button saying Get List. When this is clicked, an enormous list of packages will appear in the window below. These packages are listed alphabetically so you can scroll down to find MBESS, as shown in the highlighted line in Figure 10.A3.2. To install this package, you can click the radio button that says At System Level (in R framework), then the check box that says Install Dependencies; dependencies are other packages used by MBESS. Once this is done, click the Install Selected button to install the package.

Before MBESS can be used in **R** you must use the R Package Manager to load it into the **R** workspace. If you look back to Figure 10.A3.1, you will see that Package Manager is one of the options in the Packages & Data menu. When this is chosen, the window shown in Figure 10.A3.3 will appear. This window has a list of all **R** packages available on your computer. Because MBESS has been downloaded, it is one of the items on the list. When you check the box beside it (✓) MBESS becomes available to use in **R**.

Notice that at the bottom of Figure 10.A3.3, there is a section labeled Documentation for package 'MBESS' version 3.3.3. If you continue to scroll down, you will find a list of all MBESS functions, some of which will be discussed in this text. In this chapter, we will use the function `ci.sm`, which stands for *confidence interval for a standardized mean*. You can get to the documentation for this function by clicking on the underlined blue c, then scrolling until you get to `ci.sm`. When you click on `ci.sm`, the documentation will appear. We will not look at this documentation because the routine is very general and can be used in many ways that are beyond our current interest.

FIGURE 10.A3.1　■　The R Window

The front page of the **R** Project for Statistical Computing.

FIGURE 10.A3.2　■　The R Package Installer

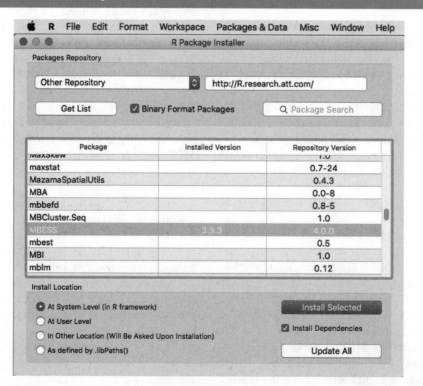

Using the **Package Installer** to install MBESS (and the packages it uses) on your computer.

FIGURE 10.A3.3 ■ The Package Manager **in R**

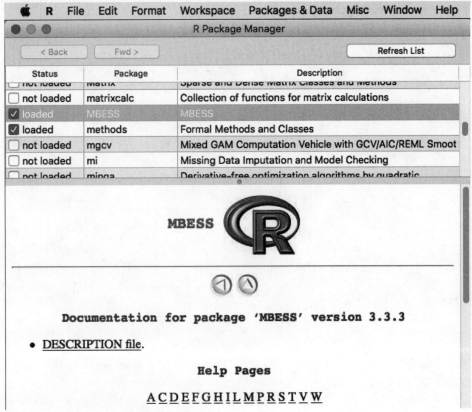

Using the **Package Manager** to load MBESS into **R**.

We will illustrate the `ci.sm` function using the example from the body of the chapter. In this case, our standardized mean (*d*) was 0.424 and sample size was 25. The first line of text entered at the **R** prompt below will produce the 95% confidence interval around *d* = 0.424.

```
> ci.sm(sm=0.424, N=25,conf.level=0.95)
```
```
[1] "The 0.95 confidence limits for the
standardized mean are given as:"
$Lower.Conf.Limit.Standardized.Mean
[1] 0.01022244
$Standardized.Mean
[1] 0.424
```

```
$Upper.Conf.Limit.Standardized.Mean
[1] 0.8297169
```

The function name `ci.sm` is typed first and the arguments to the function are in parentheses. The first argument is the standardized mean; i.e., `sm` = 0.424. The second argument is sample size; i.e., `N` = 25. The third argument is the desired confidence level; i.e., `conf.level` = .95. When we hit return, `ci.sm` returns the lower limit of the interval (0.1022244), the center of the interval (0.424), and the upper limit of the interval (0.8297169). When rounded to two decimal places, these numbers are exactly what we calculated in the body of the chapter using the approximate method.

ESTIMATION AND SIGNIFICANCE TESTS (TWO SAMPLES)

ESTIMATING THE DIFFERENCE BETWEEN THE MEANS OF INDEPENDENT POPULATIONS

INTRODUCTION

In Part I, we established the importance of statistical distributions. In Part II, these distributions were shown to underlie the construction of confidence intervals around point estimates of parameters. Estimating a parameter of a single distribution has important uses. For example, in the early stages of research, one may wish to estimate mean digit span, mean number of depressive symptoms, or mean response times in specific populations. However, as research progresses, our interest inevitably shifts to how such parameters differ between populations.

In Chapters 9 and 10 we saw examples of estimating $\mu_1 - \mu_0$ using $m - \mu_0$. In some situations, it is plausible to assume we know μ_0. For example, assuming $\mu_{IQ} = 100$ may be plausible for a given population. Also, when a population is completely known, such as the distribution of reading scores obtained over 30 years, it seems quite reasonable to use the mean of this population as μ_0. In other cases, however, it is a wild stretch to assume we know μ_0. For example, it seems highly implausible to assume, as we did earlier, "that college students in general complete 24 words in 20 minutes ($\sigma = 4.8$) on a standardized crossword puzzle."

When the goal is to estimate the difference between two population means (μ_1 and μ_2), it is almost always best to obtain samples from each of the two populations and then use the difference between the two sample means, $m_1 - m_2$, to estimate $\mu_1 - \mu_2$. When we estimate $\mu_1 - \mu_2$, we say that the two populations (e.g., male versus female participants, or control versus experimental groups) represent two levels of the **independent variable** and the scores measured represent the **dependent variable**. In Chapters 11 through 15, the dependent variables we consider will be scale variables.

An **independent variable** is a qualitative or quantitative variable whose values define the two (or more) groups of interest.

A **dependent variable** is the variable for which we obtain scores.

THE TWO-INDEPENDENT-GROUPS DESIGN

Variables that have just two values are called *dichotomous variables*. In this chapter, our independent variable will be dichotomous. The left column of Table 11.1 shows examples of five different qualitative dichotomous variables and their values. The right column shows examples of five scale variables and their values.

Table 11.1 provides 25 different questions that can be asked. Five such questions can be formed by combining the dichotomous variable *sex* with the five scale variables. For example, how much do males and females differ on the following variables: *depression, fitness, intelligence, math ability,* and *belief in extrasensory perception* (ESP)? The same five questions can be asked for each of the four remaining dichotomous variables. For any scale variable (e.g., *fitness*), it is almost certain that the means of the two populations differ,

TABLE 11.1 ■ Five Independent Variables and Five Dependent Variables	
Dichotomous Independent Variable	**Scale Dependent Variable**
Smoking: smokers versus nonsmokers	*Depression:* scores on a clinical test
Sex: males versus females	*Fitness:* scores on a fitness test
Political views: conservatives versus liberals	*Intelligence:* IQ scores
Religious views: believers versus atheists	*Math ability:* scores on a math test
Experimental condition: treatment versus control	*Belief in ESP:* scores on an ESP questionnaire

as we saw in Chapter 9. Therefore, our goal is to estimate the size of such differences and place confidence intervals around them.

Experimental Versus Quasi-Experimental Studies

There are two ways that dichotomous groups can be formed. Some groups form naturally (e.g., smokers versus nonsmokers, age younger than 40 versus older than 40, Christian versus Muslim, depressed versus not depressed). In other cases, groups are formed by an experimenter (e.g., treatment versus control conditions). Studies involving experimenter-created groups are called **experimental** studies, and those involving naturally occurring groups are called **quasi-experimental** studies.

The critical difference between experimental and quasi-experimental studies is random assignment. One can't randomly assign individuals to levels of the first four dichotomous variables in Table 11.1 because they were formed naturally. However, one can randomly assign individuals to different conditions in an experiment. Random assignment permits inferences about causality that are much more straightforward than without it. For example, the mean score on a measure of cardiovascular health may be lower for smokers than for nonsmokers. However, we can't say that smoking caused this difference, because some other factor may have disposed people to both smoking and poor cardiovascular health. On the other hand, imagine selecting 1000 rats, dividing them randomly into two groups, and then raising one group in a smoky environment and the other in a smoke-free environment. Any measured differences in cardiovascular health following these two rearing methods may be reasonably attributed to the presence or absence of smoke.

Real Versus Hypothetical Populations

There is an interesting (and provocative) difference between the populations considered in experimental and quasi-experimental studies that we touched on in Chapter 7 (see the section on the alternative hypothesis). The populations in quasi-experimental studies are easy to imagine, such as men versus women, smokers versus nonsmokers, Americans versus Canadians, or Sprague-Dawley rats versus Wistar rats, to name just a few randomly chosen examples. We can easily imagine drawing samples from each of these populations and computing the difference between the means of the samples.

Now let's think about a new treatment for alcohol abuse. We could choose a random sample of alcohol abusers, divide them at random into two groups, and administer the new treatment to one group but not the other. The untreated sample represents a random sample of alcohol abusers. What about the treated group? You may think it odd, but we will

An **experiment** involves random assignment of individuals to treatment conditions. In an experiment, it is plausible to conclude that group differences on the dependent variable are caused by the different treatment conditions.

A **quasi-experiment** compares two naturally occurring groups (i.e., groups not formed as a result of random assignment). In a quasi-experiment, it is not always plausible to conclude that group differences on the dependent variable are caused by group membership.

consider this group to be a sample from the population of alcohol abusers administered the new treatment. But, you might say, this group represents the entire population because no other individuals exist that have been administered this treatment. However, if things had been different and all alcohol abusers had been given this treatment, then our sample is one of the many random samples that could have been drawn from this population. Therefore, we can use the characteristics of this sample to infer properties of the **hypothetical population** of alcohol abusers who have been given this new treatment.

Confounding Variables

When we compare two samples of scores, it is important that there be no **confounding variables**. A confounding variable is one that affects the scores in our samples differently but is not the independent variable of interest. For example, if rats raised in the smoke-free environment were allowed more exercise than those raised in the smoke-filled environment, we would say that the effect of smoke is *confounded* with the effect of *exercise*. In this case, *exercise* is a confounding variable. Another example would be assigning the first 20 volunteers to the treatment condition of an experiment and the next 20 to the control condition. There may be something different about those who volunteer early and those who volunteer late that confounds our ability to assess the effect of the independent variable. In this case, *time to enroll* is the confounding variable.

We make all efforts to control for (i.e., eliminate) the effects of confounding variables in experiments, typically through random sampling from a population (e.g., alcoholics) and random assignment to experimental conditions (e.g., treatment and control). However, random sampling and random assignment do not guarantee that all confounds are eliminated. Consider a random sample from a psychology department participant pool. Because there are typically more females than males in such pools, a random sample will typically contain more women than men. When these participants are randomly assigned to the two conditions of an experiment, it is possible for all the men to be in one of the two conditions. No matter what the experimental manipulation is, the result will be confounded because one group is all women and the other is a mixture of men and women. In the long run, random assignment will cancel out confounds such as these, but any given sample may not.

Independent Versus Dependent Samples

In this chapter we will consider **independent samples**. This simply means that the scores in the two samples are not related in any systematic way. For example, if I choose 100 people and divide them randomly into two groups of 50, then the scores obtained from the individuals in the two samples are *independent* of each other. This means that knowing the score (on the dependent variable) of an individual in one sample tells you nothing about the score of any individual in the other sample.

However, the same 100 people could be *paired* based on their similarity on variables such as height, IQ, or extroversion. If each member of each pair is then assigned to one of the two groups, we would have individuals matched for specific characteristics in the two groups. In this case, we say there is a dependency between the members of the two groups, and the scores obtained from the individuals in the samples are dependent. This means that knowing the score (on the dependent variable) of an individual in one sample does tell you something about the score of the corresponding individual in the other sample.

Another case of **dependent samples** is when two measurements are taken from each individual. For example, depression scores could be measured in individuals before and after treatment. These samples of before and after scores are dependent samples. When two or more scores are obtained from each individual, we say we have a *repeated-measures design*.

A **hypothetical population** is one that does not exist, but which could exist. Individuals in an experimental group that undergo a novel treatment can be considered a sample from a hypothetical population of all individuals that undergo the same treatment.

A **confounding variable** is an uncontrolled variable that affects the scores in our samples differently.

Independent samples comprise scores that are completely unrelated to each other. This means that knowing the score of an individual in one sample tells you nothing about the score of any individual in the other sample.

Dependent samples comprise scores that are related to each other in some way; either pairs of scores come from the same individual (repeated measures) or pairs of scores come from individuals matched on some characteristic (matched samples). This means that knowing the score of an individual in one sample tells you something about the score of the corresponding individual in the other sample.

Two Populations, Two Distributions

Figure 11.1 illustrates the state of affairs examined in this chapter. Each level of a dichotomous independent variable is associated with a distribution of scores on the scale dependent variable. One distribution in Figure 11.1 has a mean (μ_1) of 14 and the other has a mean (μ_2) of 10. Both distributions have a variance of 16. One of these distributions might correspond to females and the other to males. Or one distribution might correspond to a control group and the other to an experimental group. The scores on the dependent variable may correspond to cardiovascular health, belief in supernatural phenomena, anxiety, digit span, or anything else we choose to measure. Whatever the two distributions correspond to, our objective will be to estimate the difference between the two population/distribution means (i.e., $\mu_1 - \mu_2$).

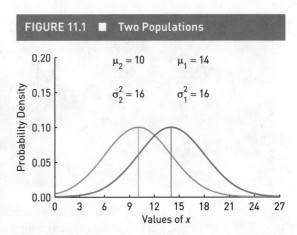

FIGURE 11.1 ■ Two Populations

Two normal populations with the same variance but different means.

LEARNING CHECK 1

1. What is the primary difference between an experimental study and a quasi-experimental study?

2. Explain why it is possible to study a hypothetical population. Give an example.

3. What is the difference between an independent-groups design and a dependent-groups design?

4. For the purposes of this chapter, is it the independent or dependent variable that is dichotomous?

5. Can you imagine a two-independent-groups design in which both variables are dichotomous? If so, give an example.

Answers

1. Random assignment. We have random assignment in experiments but not in quasi-experiments.

2. In an experimental study, our treatment group may be the only group of individuals who've received the treatment in question. Nevertheless, for the purposes of estimation, we may think of this group as a sample from a population of individuals who've all received the same treatment. For example, we could estimate the mean heart rate of chimpanzees in space. If all chimps on earth had been launched into space, then we would have a population of heart rates from spacefaring chimps. If instead we choose a random sample of chimps, launch them into space, and measure their heart rates while they are in space, we would have a sample from the population of interest even though that population doesn't (currently) exist.

3. In an independent-groups design, knowing the score of an individual in one sample tells you absolutely nothing about any score in the other sample. In the dependent-groups design, there is an association between pairs of individuals in the two groups. The clearest example is the repeated-measures design, in which two scores are obtained from the same individuals at different times.

4. The independent variable is dichotomous.

5. Sure. The independent variable could be smoking status (smoker, nonsmoker) and the dependent variable could be cancer status (has cancer, doesn't have cancer). (This is not a situation that is covered in this chapter.)

AN EXAMPLE

Susan Cain (2012), a former Wall Street lawyer, wrote a well-received book titled *Quiet: The Power of Introverts in a World That Can't Stop Talking*, in which she argued that

Western society, and the United States in particular, values the traits of extroverts more than those of introverts. She argues that introverts thrive in quiet settings where they are free to pursue their thoughts and nurture their creativity. One of Cain's strongest claims is that the group work favored in schools and industrial settings often works against the natural inclinations of introverts. In other words, introverts are not able to flourish when forced to work in groups at school or in open offices in the workplace.

Let's say that an industrial psychologist at an American university shares the view that open offices are detrimental to activities requiring insight and creativity, but she believes this is true for both introverts and extroverts. She would like to assess this idea using what she believes is a novel methodology. To assess problem solving and creativity, she will have participants solve riddles of the following sort:

"Yesterday I went to the zoo and saw the giraffes and ostriches. Altogether they had 30 eyes and 44 legs. How many animals were there?"*

"Marsha and Marjorie were born on the same day of the same month of the same year to the same mother and the same father, yet they are not twins. How is that possible?"**

Solving riddles requires insight and creative thinking; therefore, the researcher chose the number of riddles correctly solved in a 30-minute period as her dependent variable.

Our researcher's hypothesis is that more riddles will be solved in quiet settings than in noisy settings. In her experiment, all participants will be tested in a university classroom where students are learning statistics. One group will be tested when the class is writing an exam (quiet condition) and the second group will be tested while the students are engaged in group work (noisy condition).

Because our hypothetical researcher believes that the nature of the task itself requires quiet, she expects both introverts and extroverts to benefit from quiet. Therefore, when she selects participants, the researcher does not assess their positions on the introversion-extroversion continuum.

Twenty-two individuals were randomly selected from the department's participant pool and then divided at random into two groups of 11 participants each. One group was assigned to the quiet condition and the other group was assigned to the noisy condition. Each participant took the riddle test, and the number of correct solutions was counted. The results of the experiment are shown in Table 11.2 and summarized in Figure 11.2. The mean number of riddles solved in the quiet condition was 12 with a standard deviation of 4.94, and the mean number of riddles solved in the noisy condition was 9 with a standard deviation of 3.95.

The goal of the riddle-solving study was to estimate the difference between two population means (i.e., $\mu_1 - \mu_2$). The difference between the two sample means (i.e., $m_1 - m_2$) is the best point estimate of $\mu_1 - \mu_2$. The confidence interval around the difference between two sample means is computed as follows:

$$(m_1 - m_2) \pm t_{\alpha/2}(s_{m_1 - m_2}).$$ (11.1)

---

*Answer: Each animal has two eyes, so there were 30/2 = 15 animals.

**Answer: They are triplets.

This confidence interval, like those we saw in Chapters 6, 9, and 10, involves a point estimate (i.e., a statistic) and a margin of error. The statistic in this case is the difference between the two sample means $(m_1 - m_2)$, and the margin of error is $t_{\alpha/2}(s_{m_1 - m_2})$.

We were reminded in Chapter 10 that the smallest sample size for which the sample variance can be computed is two. In the two-independent-groups design, we require two groups with at least two scores in each, because computing a confidence interval will require computing the variance for both samples. This means that $t_{\alpha/2}$ is based on the sum of the degrees of freedom *within* the two groups; i.e., $df_{within} = df_1 + df_2$ degrees of freedom. We can also express this as $df_{within} = n_1 + n_2 - 2$. (There is an even more interesting reason why we associate $t_{\alpha/2}$ with $df_1 + df_2$ degrees of freedom, but we will have to wait until Chapter 16 for this explanation.)

Knowing the point estimate and $t_{\alpha/2}$ leaves only the quantity $s_{m_1 - m_2}$ to be explained. This quantity is the *estimated standard error* of the statistic $m_1 - m_2$. In the special case in which (i) sample sizes are the same and (ii) we can assume that the two populations have the same variance (i.e., $\sigma_1^2 = \sigma_2^2$), we compute the estimated standard error as follows:

$$s_{m_1 - m_2} = \sqrt{\frac{s_1^2}{n_1} + \frac{s_2^2}{n_2}}, \tag{11.2}$$

where s_1^2 and s_2^2 are the two sample variances, and n_1 and n_2 are the two sample sizes.

With these definitions (and the assumptions that $n_1 = n_2$ and $\sigma_1^2 = \sigma_2^2$), we can compute a 95% confidence interval around the difference between the two sample means in Table 11.2. To make things clearer, we will use the subscripts q (quiet) and n (noisy) on sample means, variances, and sample sizes.

Step 1. Calculate $m_n - m_q$. $m_q = 12$ and $m_n = 9$, so $m_n - m_q = 12 - 9 = 3$.

Step 2. Calculate $s_{m_1 - m_2}$ using equation 11.2:

$$s_{m_1 - m_2} = \sqrt{\frac{s_q^2}{n_q} + \frac{s_n^2}{n_n}} = \sqrt{\frac{24.4}{11} + \frac{15.6}{11}} = 1.91.$$

Step 3. Determine $t_{\alpha/2}$. There are $11 + 11 - 2 = 20$ degrees of freedom. Because this is a 95% confidence interval, $\alpha/2 = .025$. When we consult the *t*-table, we find that $t_{\alpha/2} = 2.086$. [The same thing can be accomplished using **T.INV.2T** in Excel by typing '=**T.INV.2T(.05, 20)**' into a cell in a spreadsheet.]

Step 4. Calculate the 95% confidence interval around the difference between the two means as follows:

$$\text{CI} = \left(m_q - m_n\right) \pm t_{\alpha/2}\left(s_{m_1 - m_2}\right) = (12 - 9) \pm 2.086(1.91) = [-0.98, 6.98].$$

TABLE 11.2 ■ Raw Data		
	Quiet	Noisy
	12	5
	4	7
	13	11
	11	3
	17	14
	21	10
	11	9
	5	8
	15	7
	14	8
	9	17
n	11	11
m	12	9
SS	244	156
s^2	24.4	15.6
s	4.94	3.95

FIGURE 11.2 ■ Raw Data and Statistics

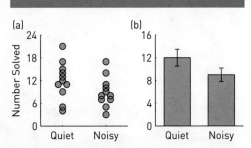

(a) Raw scores for participants in the quiet and noisy conditions. (b) The means of the two groups. Error bars represent $\pm SEM$.

The results of this study might be reported as follows:

APA Reporting

The mean number of riddles solved in the quiet condition was $M_q = 12$ with a standard deviation of $s = 4.94$, and the mean number of riddles solved in the noisy condition was $M_n = 9$ with a standard deviation of $s = 3.94$. The difference between the two means was 3, 95% CI [−0.98, 6.98]. Although the results are somewhat imprecise, they support the idea that quiet conditions are more conducive to problem solving than noisy conditions.

The general form of this conclusion is familiar. The best point estimate of the difference in the number of riddles solved in the quiet and noisy populations is $12 - 9 = 3$. In this example, we have 95% confidence that the true difference between the population means is in the interval [−0.98, 6.98]. Our confidence comes from knowing that 95% of intervals computed this way will capture $\mu_q - \mu_n$.

LEARNING CHECK 2

1. Bargh, Chen, and Burrows (1996) thought that exposure to words associated with age and fragility would have a subconscious effect on people and cause them to walk more slowly than individuals exposed to neutral words. In their experiment, Bargh et al. brought participants to a research lab and asked them to solve word problems. One group of 15 participants solved problems involving words such as *old*, *bingo*, and *Florida*. The other group of 15 participants solved problems involving words such as *thirsty*, *clean*, and *private*. Without their knowledge, the researchers measured participants' walking speeds when they left the research lab. Bargh et al. found that those exposed to words suggesting age and fragility walked at 2.63 mph, on average, with a variance of about 0.1933. This was slower than the speed of participants exposed to neutral words, who walked at 2.99 mph, on average, with a variance of about 0.2233. Compute the 95% confidence interval around the difference between these two means.

2. Do the results of this study support the idea that exposure to words associated with age and fragility has a subconscious effect on people and causes them to walk more slowly than individuals exposed to neutral words?

Answers

1. $m_1 = 2.63$, $s_1^2 = 0.1933$, $n_1 = 15$, $m_2 = 2.99$, $s_2^2 = 0.2233$, $n_2 = 15$. We first calculate $m_1 - m_2 = 2.63 - 2.99 = -0.36$. We then note that there are $15 + 15 - 2 = 28$ degrees of freedom. Then, because this is a 95% confidence interval, we find that $t_{\alpha/2} = 2.048$ when we consult the t-table. The calculations for $s_{m_1 - m_2}$ and CI $= (m_1 - m_2) \pm t_{\alpha/2}(s_{m_1 - m_2})$ are shown in the table below.

2. Yes, the results support the proposal. Those exposed to words associated with age and fragility walked $m_1 - m_2 = 2.63 - 2.99 = -0.36$ mph slower than those exposed to neutral words, 95% CI [−0.70, −0.02]. This result has become quite controversial in the last few years, and it's worth Googling.

Calculate $s_{m_1-m_2}$	Calculate CI $= (m_1 - m_2) \pm t_{\alpha/2}(s_{m_1-m_2})$
$s_{m_1-m_2} = \sqrt{\dfrac{s_1^2}{n_1} + \dfrac{s_2^2}{n_2}}$ $= \sqrt{\dfrac{0.1933}{15} + \dfrac{0.2233}{15}}$ $= 0.17$	CI $= (m_1 - m_2) \pm t_{\alpha/2}(s_{m_1-m_2})$ $= -0.36 \pm 2.048(0.17)$ $= -0.36 \pm 0.34$ $= [-0.70, -.02]$

THEORETICAL FOUNDATIONS FOR THE $(1-\alpha)100\%$ CONFIDENCE INTERVAL FOR $\mu_1 - \mu_2$

In the scenario discussed above, our hypothetical researcher was interested in the difference between two population means. One population comprises riddle scores obtained in quiet conditions and the other comprises riddle scores obtained in noisy conditions. Our researcher had in mind a situation like the one shown in Figure 11.1. She assumed that the two distributions have different means but she didn't know how different they are.

We will use the two distributions in Figure 11.1 when we need concrete illustrations in this section. Keep in mind that Population 1 has a mean of 14 and a variance of 16, and Population 2 has a mean of 10 and a variance of 16. Of course, these are things that we know but our hypothetical researcher did not.

The Sampling Distribution of $m_1 - m_2$

As in previous chapters, understanding a confidence interval requires understanding the sampling distribution of the statistic in question. In this case, the statistic is $m_1 - m_2$. To understand the distribution of this statistic, we must consider the means of *all possible* samples of size n_1 drawn from Population 1 and the means of *all possible* samples of size n_2 drawn from Population 2. Given these two (enormous) distributions of means, we can now imagine computing the difference between all possible pairings of the means from Populations 1 and 2. The resulting distribution is called the **sampling distribution of the difference between two means**, or the sampling distribution of $m_1 - m_2$ for short.

The parameters of the sampling distribution of $m_1 - m_2$ are directly related to the parameters of the distributions (populations) from which the samples were drawn. The mean of the distribution of $m_1 - m_2$ is

$$\mu_{m_1-m_2} = \mu_1 - \mu_2. \tag{11.3}$$

Because the sample mean is an unbiased statistic, the mean of all sample means drawn from Population 1 will be μ_1, and the mean of all sample means drawn from Population 2 will be μ_2. Therefore, on average, the difference between two sample means is equal to the difference between the two population means. This is illustrated in Figure 11.3, which shows that the mean of the distribution of $m_1 - m_2$ is $\mu_1 - \mu_2 = 14 - 10 = 4$.

The variance of the distribution of $m_1 - m_2$ is

$$\sigma^2_{m_1-m_2} = \frac{\sigma_1^2}{n_1} + \frac{\sigma_2^2}{n_2}. \tag{11.4}$$

This formula contains familiar components. If you think back to our discussion of the distribution of means in Chapter 5, you will remember that its variance is $\sigma^2_m = \sigma^2/n$. Equation 11.4 shows that the variance of the distribution of $m_1 - m_2$ is simply the variance of the sampling distribution of m_1 (i.e., σ_1^2/n_1) plus the variance of the sampling distribution of m_2 (i.e., σ_2^2/n_1).

FIGURE 11.3 ■ The Distribution of $m_1 - m_2$

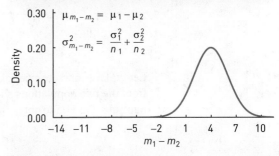

The sampling distribution of the difference between two independent means. The two distributions in question are those shown in Figure 11.1. Population 1 has a mean of 14 and variance of 16, and Population 2 has a mean of 10 and variance of 16. If we consider all possible samples of size n_1 drawn from Population 1 and all possible samples of size n_2 drawn from population 2, then we can compute the difference between all possible pairs of these sample means (i.e., $m_1 - m_2$). The resulting distribution is shown. In this example, $n_1 = n_2 = 11$; therefore, $\sigma^2_{m_1-m_2} = 2.91$. and $\sigma_{m_1-m_2} = 1.71$.

The **sampling distribution of the difference between two means** is a probability distribution of all possible differences between two sample means, m_1 and m_2, of size n_1 and n_2, respectively, drawn at random from two independent populations. In other words, it is the sampling distribution of all possible values of $m_1 - m_2$.

Knowing the variance of the distribution of $m_1 - m_2$ allows us compute the standard error of $m_1 - m_2$ as follows:

$$\sigma_{m_1 - m_2} = \sqrt{\frac{\sigma_1^2}{n_1} + \frac{\sigma_2^2}{n_2}}. \tag{11.5}$$

Let's take a concrete example and think about the distribution of $m_1 - m_2$ for samples drawn from the two populations shown in Figure 11.1. If $n_1 = n_2 = 11$, we can substitute numbers into equation 11.5 to obtain the following:

$$\sigma_{m_1 - m_2} = \sqrt{\frac{\sigma_1^2}{n_1} + \frac{\sigma_2^2}{n_2}} = \sqrt{\frac{16}{11} + \frac{16}{11}} = 1.71.$$

This is the standard error of the distribution shown in Figure 11.3.

Looking back to our opening example, we computed a confidence interval around $m_1 - m_2$ using the usual definition $(m_1 - m_2) \pm t_{\alpha/2}(s_{m_1 - m_2})$, with the estimated standard error computed as in equation 11.2. A comparison of equations 11.2 and 11.5 shows that they are structurally identical, with σ_1^2 and σ_2^2 in equation 11.5 replacing s_1^2 and s_2^2 from equation 11.2. Therefore, $s_{m_1 - m_2}$ is our best estimate of $\sigma_{m_1 - m_2}$.

We've noted that using $s_{m_1 - m_2}$ as defined in equation 11.2 is valid only when (i) the two sample sizes are the same, and (ii) it reasonable to assume that $\sigma_1^2 = \sigma_2^2$. Creating confidence intervals for $m_1 - m_2$ becomes slightly more complex when these conditions do not hold, but we don't need to worry about these additional complexities for the moment.

LEARNING CHECK 3

1. Calculate $\sigma_{m_1 - m_2}^2$ for $\sigma_1 = 6$, $n_1 = 16$, $\sigma_2 = 5$, $n_2 = 10$.

2. Calculate $\sigma_{m_1 - m_2}^2$ for $\sigma_1 = 20$, $n_1 = 16$, $\sigma_2 = 5$, $n_2 = 100$.

3. Calculate $\sigma_{m_1 - m_2}^2$ for $\sigma_1 = 12$, $n_1 = 8$, $\sigma_2 = 9$, $n_2 = 2$.

Answers

1. $36/16 + 25/10 = 4.75$.

2. $\sqrt{400/16 + 25/100} = 5.02$.

3. $144/8 + 81/2 = 58.5$.

The Logic of a Confidence Interval for $\mu_1 - \mu_2$

We will now return to the familiar logic underlying confidence intervals. In Chapter 10 it was shown that when the sampling distribution of the mean is normal, it can be transformed to a t-distribution by applying the following transformation to each sample mean:

$$t = \frac{m - \mu}{s_m}. \tag{11.6}$$

Because $(1-\alpha)100\%$ of all t-scores fall in the interval $\pm t_{\alpha/2}$, we also know that $(1-\alpha)100\%$ of all possible $t \pm t_{\alpha/2}$ intervals will contain 0, which is the mean of the t-distribution. From this, we were able to show that $(1-\alpha)100\%$ of all possible intervals computed as

$$m \pm t_{\alpha/2}(s_m) \tag{11.7}$$

will capture the population mean, μ. There was a formal demonstration of this in Appendix 10.4 (available at study.sagepub.com/gurnsey).

We can apply the same logic to define the (1–α)100% confidence interval for $\mu_1 - \mu_2$. As with the distribution of means, we can transform the distribution of $m_1 - m_2$ into a t-distribution as follows:

$$t = \frac{(m_1 - m_2) - (\mu_1 - \mu_2)}{s_{m_1 - m_2}}.$$

(11.8)

If you compare equation 11.8 with equation 11.6, you will see that they are structurally identical. In equation 11.8, $m_1 - m_2$ replaces m, $\mu_1 - \mu_2$ replaces μ, and $s_{m_1 - m_2}$ replaces s_m. The result is a t-distribution with $df_{within} = df_1 + df_2$ degrees of freedom.

As before, because (1–α)100% of all t-scores fall in the interval $\pm t_{\alpha/2}$, we also know that (1–α)100% of all possible intervals $\pm t_{\alpha/2}$ will contain 0. From this, it follows that (1–α)100% of all possible intervals computed as

$$(m_1 - m_2) \pm t_{\alpha/2} \left(s_{m_1 - m_2} \right)$$

(11.9)

will capture $\mu_1 - \mu_2$. A formal demonstration of this is provided in Appendix 11.3 (available at study.sagepub.com/gurnsey).

Figure 11.4 shows that the logic underlying confidence intervals for $\mu_1 - \mu_2$ is exactly the same as the logic underlying confidence intervals for μ in Chapter 10. The sampling distribution of $m_1 - m_2$ at the bottom of Figure 11.4 was previously shown in Figure 11.3. It is a normal distribution with a mean of $\mu_1 - \mu_2 = 4$ and a standard error of $\sigma_{m_1 - m_2} = 1.71$. Each dot above the distribution represents the difference between two sample means ($m_1 - m_2$) and the arms around each dot define the 95% confidence interval. White dots are at the centers of intervals that capture $\mu_1 - \mu_2$, and filled dots are at the centers of intervals that do not capture $\mu_1 - \mu_2$. If we were to compute $(m_1 - m_2) \pm t_{\alpha/2} \left(s_{m_1 - m_2} \right)$ for all possible pairs of sample means, then exactly 95% of these intervals would capture $\mu_1 - \mu_2$.

Estimating the Standard Error of the Difference Between Two Means

We saw earlier that computing a confidence interval around the difference between two means requires estimating the standard error of the difference between two means. The formula we use to do this depends on our sample sizes and assumptions we make about the two population variances. In this section, we will describe two ways to compute the estimated standard error of $m_1 - m_2$ when we can assume that the two populations have the same variance. When we aren't able to make this assumption, computing confidence intervals becomes a bit more complicated, and this method is described in Appendix 11.4 (available at study.sagepub.com/gurnsey).

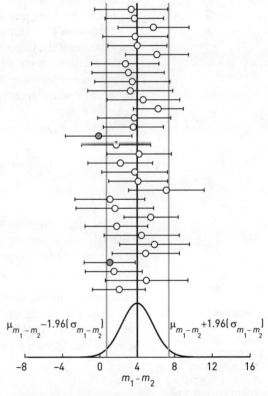

FIGURE 11.4 ■ 95% Confidence Intervals

93.8% capture $\mu_1 - \mu_2$
6.2% do not capture $\mu_1 - \mu_2$

95% confidence intervals for $\mu_1 - \mu_2$ when the population standard deviations are unknown. The distribution at the bottom represents the sampling distribution of $m_1 - m_2$ for $\mu_1 = 14$, $\mu_2 = 10$, $\sigma_1 = \sigma_2 = 4$, and $n_1 = n_2 = 11$. The standard error of the distribution of $m_1 - m_2$ is $\sigma_{m_1 - m_2} = 1.71$. The light blue lines enclose the central 95% of the sampling distribution. Each dot represents the difference between two sample means ($m_1 - m_2$), and the arms around each dot represent the 95% confidence interval. White dots are the centers of confidence intervals that capture $\mu_1 - \mu_2$ and filled dots are the centers of confidence intervals that do not capture $\mu_1 - \mu_2$.

Case 1: Equal Sample Sizes and Equal Population Variances

In the example that we worked through earlier, $\sigma_{m_1-m_2}$ was estimated using equation 11.2:

$$s_{m_1-m_2} = \sqrt{\frac{s_1^2}{n_1} + \frac{s_2^2}{n_2}}.$$

As noted twice before, this way of estimating $\sigma_{m_1-m_2}$ is only appropriate when it is reasonable to assume that $\sigma_1^2 = \sigma_2^2$ and when sample sizes are the same ($n_1 = n_2$).

Case 2: Unequal Sample Sizes and Equal Population Variances

If we assume that $\sigma_1^2 = \sigma_2^2$ but *sample sizes are different*, then estimating $\sigma_{m_1-m_2}$ becomes more interesting and involves an intermediate step to estimate the variance common to the two distributions, which we denote σ^2. That is, subscripts on the population variances are unnecessary because they are assumed to be identical. The two sample variances, s_1^2 and s_2^2, are therefore independent estimates of σ^2.

In Chapter 5, we saw that parameter estimates based on large samples are more precise than estimates based on small samples. Therefore, when two samples are different sizes, the variance of the larger sample provides a more precise estimate of σ^2 than the variance of the smaller sample. However, the variance of the smaller sample can't be ignored. Therefore, to estimate σ^2, we combine s_1^2 and s_2^2 in a way that gives greater weight to the variance from the larger sample. This estimate of σ^2 is called s_{pooled}^2, because we *pool* two sample variances.

When s_{pooled}^2 has been computed, we use it to estimate $\sigma_{m_1-m_2}$ as follows:

$$s_{m_1-m_2} = \sqrt{\frac{s_{\text{pooled}}^2}{n_1} + \frac{s_{\text{pooled}}^2}{n_2}}. \tag{11.10}$$

That is, s_{pooled}^2 replaces both σ_1^2 and σ_2^2 from equation 11.5. This leaves us with the relatively minor problem of computing s_{pooled}^2. There are two ways to do this. The first is more conceptual in nature and the second produces exactly the same result but is easier to compute by hand when you've been given the sums of squares for the two samples.

Estimating the Population Variances by "Pooling" the Sample Variances. Sample variances are pooled by making use of a so-called **weighted sum**. (We used weighted sums to compute GPA in Appendixes 1.1 through 1.3.) This means that we multiply s_1^2 by a weight and s_2^2 by a different weight and then add these products as follows:

$$s_{\text{pooled}}^2 = w_1\left(s_1^2\right) + w_2\left(s_2^2\right).$$

> A **weighted sum** is a way of computing the mean of two or more statistics by multiplying each statistic by a weight related to sample size and then summing the products.

In the case of s_{pooled}^2, w_1 and w_2 are defined as

$$w_1 = \frac{df_1}{df_{\text{within}}}, \text{ and } w_2 = \frac{df_2}{df_{\text{within}}},$$

> s_{pooled}^2 is the **pooled variance**. It is computed as a weighted sum of two separate estimates of σ^2.

where $df_{\text{within}} = df_1 + df_2$ as before. Therefore, these two weights always sum to 1; i.e., $w_1 + w_2 = 1$. The **pooled variance** (s_{pooled}^2) can be computed as a weighted sum as follows:

$$s_{\text{pooled}}^2 = \frac{df_1}{df_{\text{within}}}\left(s_1^2\right) + \frac{df_2}{df_{\text{within}}}\left(s_2^2\right). \tag{11.11a}$$

To illustrate the value of weighted sums, we will consider an extreme example. Imagine that we've drawn samples from Populations 1 and 2 from Figure 11.1. (Remember, for both distributions $\sigma^2 = 16$). Let's say $s_1^2 = 15$ and $s_2^2 = 25$. However, these two samples have very different sizes. There are $n_1 = 99$ scores in the sample drawn from Population 1 and $n_2 = 3$ scores in the sample drawn from Population 2. This means that $df_1 = 98$ and $df_2 = 2$. In this case,

$$w_1 = \frac{df_1}{df_{within}} = \frac{98}{100} = .98$$

and

$$w_2 = \frac{df_2}{df_{within}} = \frac{2}{100} = .02.$$

When we compute the weighted sum of the two sample variances, we have

$$s_{pooled}^2 = \frac{df_1}{df_{within}}\left(s_1^2\right) + \frac{df_2}{df_{within}}\left(s_2^2\right) = .98(15) + .02(25) = 15.2.$$

If we had computed the simple mean of s_1^2 and s_2^2, it would be $(15 + 25)/2 = 20$. Another way to express this average would be

$$.5\left(s_1^2\right) + .5\left(s_2^2\right) = .5(15) + .5(25) = 20.$$

So, when we give equal weight to s_1^2 and s_2^2, we obtain a poor estimate of σ^2, which in this example is 16. Therefore, when sample sizes are unequal, we compute the pooled variance as a weighted sum that gives greater weight to the variance from the larger sample.

Estimating the Population Variances by "Pooling" Sums of Squares. There is a second way to compute s_{pooled}^2 that is more useful for hand calculations. Remember that the sample variance (s^2) is the sum of squares divided by degrees of freedom ($s^2 = ss/df$). Therefore, the sum of squares is the sample variance multiplied by degrees of freedom ($ss = s^2*df$). With this in mind, we can show a simpler version of equation 11.11a:

$$s_{pooled}^2 = \frac{ss_1 + ss_2}{df_{within}}. \tag{11.11b}$$

The following sequence shows the equivalence of 11.11a and 11.11b:

$$s_{pooled}^2 = \frac{df_1}{df_{within}}\left(s_1^2\right) + \frac{df_2}{df_{within}}\left(s_2^2\right) = \frac{df_1\left(s_1^2\right)}{df_{within}} + \frac{df_2\left(s_2^2\right)}{df_{within}} = \frac{ss_1}{df_{within}} + \frac{ss_2}{df_{within}} = \frac{ss_1 + ss_2}{df_{within}}.$$

Therefore, if we've been given ss_1 and ss_2, we can compute s_{pooled}^2 in fewer steps using equation 11.11b.

Computing the Estimated Standard Error Using the Pooled Variance. Let's return to our opening example that estimated the effect of noise on riddle solving. Steps 1, 3, and 4 in the calculation of the confidence interval are exactly as before (so they are not recalculated). However, the calculation of $s_{m_1-m_2}$ is broken into two parts (steps 2a and 2b).

Step 2a. Compute s^2_{pooled} using equation 11.11b.

$$s^2_{\text{pooled}} = \frac{ss_q + ss_n}{df_{\text{within}}} = \frac{244 + 156}{10 + 10} = 20.$$

Step 2b. Calculate $s_{m_1-m_2}$ using equation 11.10.

$$s_{m_1-m_2} = \sqrt{\frac{s^2_{\text{pooled}}}{n_q} + \frac{s^2_{\text{pooled}}}{n_n}} = \sqrt{\frac{20}{11} + \frac{20}{11}} = 1.91.$$

If you compare $s_{m_1-m_2}$ calculated in Step 2b, you will see that it is exactly the same as $s_{m_1-m_2}$ computed in our original example. Therefore, the confidence interval computed in Step 4 will be the same in both cases.

Which Formula to Use? We've just seen that when sample sizes are the same, then equations 11.5 and 11.10 will produce exactly the same result, so *either one* can be used. However, when sample sizes are different, these two methods will produce different results, and only equation 11.10 will provide a valid estimate of $\sigma_{m_1-m_2}$. Therefore, when sample sizes are unequal, you *must* use s^2_{pooled} in the calculation of $s_{m_1-m_2}$.

Case 3: Unequal Population Variances

It is not always plausible to assume that $\sigma_1 = \sigma_2$. Sometimes interventions can change the variance in a distribution. For example, a weight loss treatment might reduce weight by the same percentage for each individual in a population. The distribution of weights would be smaller after the weight loss treatment than before the treatment. (See the discussion in Chapter 3 about the effect of multiplying the variance by a constant.)

If we cannot assume that $\sigma_1 = \sigma_2$, then confidence intervals are computed differently. First, the estimated standard error is computed using equation 11.2. However, the result will underestimate $\sigma_{m_1-m_2}$. To compensate for this underestimation, our second step is to reduce the degrees freedom to widen the confidence interval. That is, reducing the degrees of freedom increases $t_{\alpha/2}$. The amount by which the degrees of freedom are reduced depends on how different s_1^2 and s_2^2 are. Appendix 11.4 (available at study.sagepub.com/gurnsey) provides a detailed explanation of how to adjust the degrees of freedom.

Assumptions

For a confidence interval to be valid, a number of assumptions must be made.

1. The two populations we sampled are normal.
2. The two populations have the same variance.
3. The samples are random samples from the populations of interest.
4. The two samples are *independent*.

Normality

The assumption of normal populations is a common one in inferential statistics. We make this assumption because our margin of error was computed using $t_{\alpha/2}$, and the *t*-distribution assumes that the scores in our samples were drawn from normal distributions. Obvious violations of the normality assumption can be detected using QQ or PP plots or other methods discussed in Appendix 4.3 (available at study.sagepub.com/gurnsey).

Equal Variances

When we compute confidence intervals for the difference between two sample means, we often assume that the two samples were drawn from distributions having the same variance. We made this assumption in the riddle example above when we computed s^2_{pooled}. If the two populations in question were to have different variances, our confidence interval will generally be too narrow. The Welch-Satterthwaite correction factor (described in Appendix 11.4 available at study.sagepub.com/gurnsey) provides a way around this problem. Most statistical packages that compute confidence intervals around the difference between two means will provide both versions. (We will see this in Appendix 11.2.)

Random Sampling

Of course, we always have to assume random sampling from the populations of interest. The scenario at the beginning of the chapter stated that a random sample of 22 participants was drawn from the participant pool. These 22 individuals were then assigned to the two conditions (quiet and noisy) at random. Because of random selection and random assignment to conditions, there is no reason to suspect a systematic difference between the two groups; hence, there is no reason to suspect sampling bias.

The populations of interest were populations of scores (number of riddles solved) of individuals who performed the task in quiet and noisy conditions. We treat our samples as random samples from hypothetical populations that would exist if all members of the participant pool had solved riddles in a quiet environment, or if all members of the participant pool had solved riddles in a noisy environment.

Independence

Finally, we assume that there is no *association* between the scores in the two groups. In other words, the two samples are independent. In our riddle-solving example, the sample of 22 participants was divided into two equal groups at random. That is, pairs of participants were not matched on any task-relevant variables before being placed in different groups. For this reason, the two groups are independent.

EFFECT SIZE δ

In Chapters 9 and 10 we showed how to estimate Cohen's δ, which is the difference between two population means divided by their shared standard deviation. It is typically the magnitude or absolute value of δ that interests us (i.e., $|\delta|$) and not so much its sign. Nevertheless, we can compute δ in either of the following ways:

$$\delta = \frac{\mu_1 - \mu_2}{\sigma} \tag{11.12a}$$

or

$$\delta = \frac{\mu_2 - \mu_1}{\sigma}. \tag{11.12b}$$

Therefore, we can estimate δ in either of the following ways:

$$d = \frac{m_1 - m_2}{s_{pooled}} \tag{11.13a}$$

LEARNING CHECK 4

1. If $m_1 = 10$, $m_2 = 9$, $s_{m_2-m_1} = 1.6$, and $t_{\alpha/2} = 2$, calculate $(m_1 - m_2) \pm t_{\alpha/2}(s_{m_1-m_2})$.

2. Calculate $\sigma^2_{m_1-m_2}$ for $\sigma_1 = 6$, $n_1 = 16$, $\sigma_2 = 5$, $n_2 = 10$.

3. Calculate s^2_{pooled} for $s_1 = 6$, $n_1 = 16$, $s_2 = 5$, $n_2 = 16$.

4. Calculate s^2_{pooled} for $s_1 = 6$, $n_1 = 16$, $s_2 = 5$, $n_2 = 11$.

5. If $\alpha = .05$, $n_1 = 16$, and $n_2 = 10$, what is $t_{\alpha/2}$?

6. The 95% confidence interval is computed as $(m_1 - m_2) \pm 2.086(s_{m_1-m_2})$ What is the total number of scores in the two samples?

7. If $n_1 = 11$ and $n_2 = 13$, what proportion of all intervals computed as $(m_1 - m_2) \pm 1.717(s_{m_1-m_2})$ will capture $\mu_1 - \mu_2$?

8. Calculate the 95% confidence interval for $m_1 = 100$, $m_2 = 90$, $s_1 = 6$, $s_2 = 7$, $n_1 = 11$, $n_2 = 16$.

9. What four assumptions must be correct for $(m_1 - m_2) \pm t_{\alpha/2}(s_{m_1-m_2})$ to be a valid confidence interval?

Answers

1. $(10 - 9) \pm 2(1.6) = [-2.2, 4.2]$.

2. $36/16 + 25/10 = 4.75$.

3. $(36 + 25)/2 = 30.5$.

4. $15/25*36 + 10/25*25 = 31.6$.

5. 2.064 for $df_{within} = 16 + 10 - 2 = 24$.

6. When we look at the t-table, we find that 2.086 is associated with $df_{within} = 20$ when $\alpha = .05$; therefore, $n_1 + n_2 = df_{within} + 2 = 22$.

7. $df_{within} = 22$; therefore, $t_{\alpha/2} = 1.717$ corresponds to $\alpha = .1$, so this is the 90% confidence interval. Therefore, the proportion of all such intervals that will capture $\mu_1 - \mu_2$ is .9.

8. $(100 - 90) \pm 2.060(2.592) = 10 \pm 5.340 = [4.66, 15.34]$.

9. The two populations we sampled are normal, the two populations have the same variance, the samples are random samples from the populations of interest, and the two samples are independent.

or

$$d = \frac{m_2 - m_1}{s_{pooled}}.$$ (11.13b)

In equations 11.13a and 11.13b, s_{pooled} is simply the square root of s^2_{pooled}, as defined in equations 11.11a and 11.11b. That is, s_{pooled} estimates σ, which is the standard deviation common to the two distributions.

An Approximate Confidence Interval Around an Estimate of δ

As with any statistic, a point estimate is of limited use if we don't know the precision of the estimate. Therefore, we next show how to construct an approximate $(1-\alpha)100\%$ confidence interval around d. Of course, computing a confidence interval requires knowing the sampling distribution of the statistic. Unfortunately, the sampling distribution of d is not usually normal, as illustrated in Figures 11.5 and 11.6. Figure 11.5 illustrates sampling distributions of d, computed as in equation 11.13a, for samples of size $n = 5$, drawn from two populations whose means are separated by $\delta = 0$ to 2. When $\delta = 0$, the sampling distribution of d resembles a t-distribution. As δ increases, the distributions shift to the right, because the average value of d increases as δ increases. More important is the fact that the distributions become increasingly skewed as δ increases.

The skew in the distributions of d is more apparent when it is computed from small samples than when it is computed from large samples. Figure 11.6 illustrates sampling distributions of d for $\delta = 0$ to 2 when both sample sizes are 50. As in Figure 11.5, the distributions in Figure 11.6 shift to the right as δ increases. In addition, the distributions in Figure 11.6 are narrower than those in Figure 11.5 because the samples are larger, and thus the estimates of δ are less variable. Finally, the right skew of the distributions (particularly $\delta = 2$) is much less pronounced in Figure 11.6 than in Figure 11.5.

We can compute an approximate confidence interval around d using a simple extension of the method used in Chapter 10 as follows:

$$d \pm z_{\alpha/2}(s_d). \tag{11.14}$$

The only part of equation 11.14 that we haven't yet explained is s_d, which is the approximate standard error of d. In Chapter 9, when we had one sample and σ was known, the standard error of d was

$$\sigma_d = \sqrt{\frac{1}{n}}.$$

In Chapter 10, when we had one sample and σ was unknown, the approximate estimated standard error of d was

$$s_d = \sqrt{\frac{d^2}{2df} + \frac{1}{n}}.$$

In the present case, we have two samples and σ is unknown. The approximate estimated standard error of d is

$$s_d = \sqrt{\frac{d^2}{2df_{within}} + \frac{1}{n_1} + \frac{1}{n_2}}. \tag{11.15}$$

In equation 11.15, d is our estimate of δ, computed from our samples as described in equation 11.13a. As before, $df_{within} = df_1 + df_2$. Because of the nature of this approximation, we use $z_{\alpha/2}$ rather than $t_{\alpha/2}$. Confidence intervals computed using this approximation will always be slightly different from the exact confidence interval. For small values of d, the confidence intervals will be slightly narrower; for large values of d, the confidence interval will be slightly wider. Fortunately, these differences are often negligible, particularly as sample sizes increase.

Our Example (Continued)

We will now step through the calculation of an approximate confidence interval for an estimated effect size. For quick reference, Table 11.3 shows the data from the riddle study for which we computed the confidence interval around $m_1 - m_2$. We will use the same data to compute d and the 95% confidence interval around it.

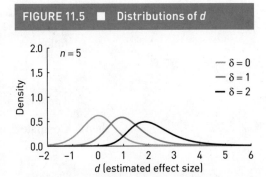

FIGURE 11.5 ■ Distributions of d

Sampling distributions for d when $\delta = 0$, 1, and 2 when both sample sizes are 5. As δ increases, the distributions shift to the right and become wider and increasingly skewed.

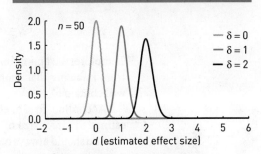

FIGURE 11.6 ■ Distributions of d

Sampling distributions for d when $\delta = 0$, 1, and 2 when both sample sizes are 50. As δ increases, the distributions shift to the right and become wider, but they are markedly less skewed than in Figure 11.5.

TABLE 11.3 ■ Data From Riddle Study

	Quiet (q)	Noisy (n)
N	11	11
M	12	9
SS	244	156

Step 1. Compute s_{pooled}. Table 11.3 shows that $ss_q = 244$ and $ss_n = 156$. Because $n_1 = n_2 = 11$, $df_{within} = 11 + 11 - 2 = 20$. With this information, we can compute s_{pooled} as follows:

$$s_{pooled} = \sqrt{\frac{ss_q + ss_n}{df_{within}}} = \sqrt{\frac{244 + 156}{10 + 10}} = 4.47.$$

Step 2. Compute d, using equation 11.13.

$$d = \frac{m_q - m_n}{s_{pooled}} = \frac{12 - 9}{4.47} = 0.67.$$

Step 3. Compute s_d, using equation 11.15.

$$s_d = \sqrt{\frac{d^2}{2 * df_{within}} + \frac{1}{n_1} + \frac{1}{n_2}} = \sqrt{\frac{0.67^2}{2 * 20} + \frac{1}{11} + \frac{1}{11}} = 0.4394.$$

Step 4. Compute the approximate 95% confidence interval around d using equation 11.14.

$$\text{CI} = d \pm z_{\alpha/2}(s_d) = 0.67 \pm 1.96(0.4394) = [-0.19, 1.53].$$

Therefore, the approximate 95% CI is [−0.19, 1.53]. Our confidence in this interval comes from knowing that approximately 95% of all confidence intervals computed in this way will contain δ.

According to Cohen's classification scheme, described in Chapters 9 and 10, our estimated effect size of 0.67 is between medium and large. Of course, this classification scheme should always be treated with some skepticism. In a later section we will reconsider a more quantitative approach to giving meaning to our estimate of δ.

An Exact Confidence Interval Around an Estimate of δ

In Appendix 10.3 we described how to compute an exact confidence interval using MBESS `ci.sm` in **R** for an estimated effect size (d) based on a sample mean and a known population mean. We can also compute an exact confidence interval for d when computed as in equations 11.13a and 11.13b using `ci.smd` from MBESS. The text in the code fragment below shows the arguments provided to `ci.smd`: smd is the standardized mean difference (i.e., d), n.1 and n.2 are the sample sizes, and conf.level is the confidence level. With values assigned to each of these arguments, we press return and `ci.smd` returns the lower and upper limits of the interval, [−0.1986694, 1.522964] as well as the center of the interval. When it is rounded to two decimal places, the 95% confidence interval [−0.20, 1.53] is very similar to the approximate interval computed in the previous section.

```
> ci.smd (smd =.67, n.1 = 11, n.2 = 11, conf.level =.95)
$Lower.Conf.Limit.smd
[1] −0.1986694
$smd
[1] 0.67
$Upper.Conf.Limit.smd
[1] 1.522964
```

U_3

In Chapters 8 through 10, we discussed Cohen's U_3 as a measure of overlap between two populations of scores. This allowed us to think in concrete terms about the effect of some treatment or intervention at the level of populations. We can use U_3 in the context of the two-independent-groups design just as we did for the one-sample design in earlier chapters.

Figure 11.7 illustrates our best estimate of the relationship between the two populations. Both distributions are assumed to be normal, and our sample statistics are the best estimates of the population parameters. Our best estimate of μ_q is $m_q = 12$, our best estimate of μ_n is $m_n = 9$, and our best estimate of σ is $s_{pooled} = 4.47$. Therefore, our best estimate of δ is $d = 0.67$. U_3 is the proportion of the noise distribution below the mean of the quiet distribution (shown in gray in Figure 11.7) or, equivalently, the proportion of the quiet distribution above the mean of the noise distribution. When estimating δ from two independent samples, U_3 is calculated exactly as before:

$$U_3 = P(|d|),$$

where $P(|d|)$ is the proportion of the standard normal (z) distribution below the absolute value of d. In our example, our best estimate is that $U_3 = .7486$.

Let's think about the population of scores for the quiet condition. The distribution is assumed to be normal, with 50% of scores lying above and below its mean, μ_q. Our results suggest that if all individuals from this population had been tested under noisy conditions, the mean would have dropped by 0.67(σ). This means that 74.86% of scores would now fall below μ_q, as shown in the shaded gray region in Figure 11.7. Therefore, we estimate that testing in the noisy versus quiet conditions leads to an increase of $74.86 - 50 = 24.86\%$ of scores falling below μ_q. Conversely, testing in quiet conditions would result in 24.86% of scores falling above μ_n. This seems like a substantial effect on performance. But, like any statistical result, the practical significance of U_3 depends on one's perspective, as we'll see in a later section.

FIGURE 11.7 ■ Cohen's U_3

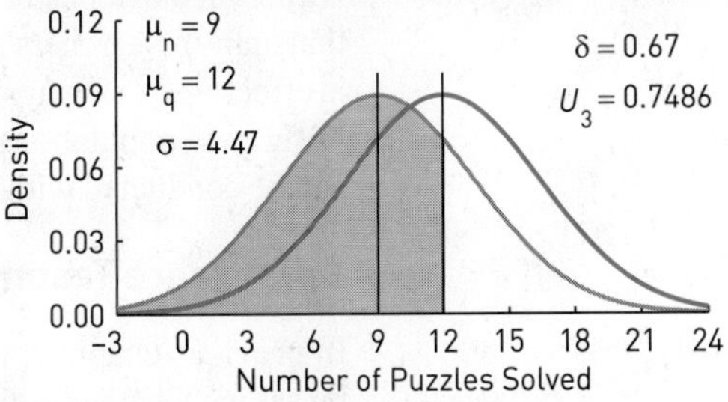

An illustration of our best estimate of the relationship between quiet and noisy populations. Because $d = .67$, we estimate that 74.86% of the noisy distribution lies below the mean of the quiet distribution.

LEARNING CHECK 5

1. What does s_{pooled} estimate?

2. What does $d = (m_1 - m_2)/s_{pooled}$ estimate?

3. For fixed sample sizes, how does the shape of the sampling distribution of d change as δ gets further from 0?

4. Compute the approximate 95% confidence interval for d for $m_1 = 100$, $m_2 = 90$, $s_1 = 8$, $s_2 = 8$, $n_1 = 20$, $n_2 = 20$.

Answers

1. σ, the standard deviation assumed to be common to the two populations in question.

2. $\delta = (\mu_1 - \mu_2)/\sigma$.

3. The sampling distribution of d becomes increasingly skewed as δ gets further from 0.

4. $d = (100 - 90)/8 = 1.25$.

$$s_d = \sqrt{d^2/(2df_{within}) + 1/n_1 + 1/n_2}$$

$$= \sqrt{.0206 + .1} = .3472.$$

$$\text{CI} = d \pm z_{\alpha/2}(s_d) = 1.25 \pm 1.96(.3472) = [0.57, 1.93].$$

SIGNIFICANCE TESTING

Although our focus has been on estimating $\mu_1 - \mu_2$ and δ, we've seen in previous chapters that many researchers follow the custom of significance testing to judge whether a treatment is effective. As always, the null hypothesis is that there is no difference between the means of the two populations under consideration. We saw in Chapter 10 that significance tests can be conducted with confidence intervals and t-statistics, so we will review these in turn.

Significance Testing With Confidence Intervals

In previous chapters, two-tailed tests of the null hypothesis with $\alpha = .05$ were conducted by asking whether μ_0 falls in the 95% confidence interval around m. In the present case, we are estimating the mean of the distribution of $m_1 - m_2$. According to the null hypothesis, our two populations have the same means ($\mu_1 = \mu_2$), so we can state it as

$$H_0: \mu_1 - \mu_2 = 0.$$

For a two-tailed test of H_0, our alternative hypothesis is simply

$$H_1: \mu_1 - \mu_2 \neq 0.$$

To conduct a two-tailed test of our null hypothesis, we ask if 0 falls in our confidence interval defined as $(m_1 - m_2) \pm t_{\alpha/2}(s_{m_1 - m_2})$. If it does, we retain H_0; if not, we reject H_0. In the riddle-solving example that we worked through above, the 95% confidence interval around $m_1 - m_2$ was [−0.98, 6.98]. Because this interval contains 0, we retain the null hypothesis. (Remember that retaining the null hypothesis does not mean that it is true.)

Hypothesis Testing With t-Statistics

Traditionally, significance tests are conducted with t-statistics rather than confidence intervals. However, the logic of the hypothesis test is the same. If the researcher predicts that quiet conditions will lead to more correct solutions than noisy conditions, then the null and alternative hypotheses would be as follows:

$$H_0: \mu_q - \mu_n = 0$$

$$H_1: \mu_q - \mu_n > 0$$

If the researcher makes no prediction about which condition will lead to more correct solutions, then the null and alternative hypotheses will be as follows:

$$H_0: \mu_q - \mu_n = 0$$

$$H_1: \mu_q - \mu_n \neq 0$$

In either case, t_{obs} is defined as follows:

$$t_{obs} = \frac{m_1 - m_2}{s_{m_1 - m_2}}. \tag{11.16}$$

When we fill in the quantities defined earlier, we find that

$$t_{obs} = \frac{m_q - m_n}{s_{m_1 - m_2}} = \frac{12 - 9}{1.91} = 1.57.$$

To make a decision about t_{obs}, we would have to compare it with $t_{critical}$. As we saw in Chapter 10, $t_{critical}$ depends on α, df, and H_1. Let's make the conventional assumption that $\alpha = .05$. The degrees of freedom are $df_{within} = n_q + n_n - 2 = 20$, exactly as in the 95% confidence interval we computed. Finally, assuming the directional alternative hypothesis, H_1 predicts that $\mu_q > \mu_n$, so we expect a positive value of t_{obs}. With this information we can consult the t-table for the t-score having $\alpha(100)\% = 5\%$ of the distribution above it when $df_{within} = 20$. In doing so, we find $t_{critical} = 1.725$. Because $t_{obs} = 1.57$ fails to exceed $t_{critical}$, we retain H_0.

Using the **T.DIST** function in Excel, we can determine that the exact proportion of a t-distribution (with 20 df) above 1.57 is $p = .0657$. Again, because $p = .0657$ is greater than $\alpha = .05$, we retain H_0. (For a two-tailed test, $p = 2*.0657 = .1314$.)

APA Reporting

The mean number of riddles solved for 11 participants in the quiet condition was $M_q = 12$, and the mean number of riddles solved for the 11 participants in the noisy condition was $M_n = 9$. The difference between the two means was 3, 95% CI [−0.98, 6.98]. A one-tailed test of the difference between the two means showed that the difference was not statistically significant, $t(20) = 1.57$, $p = .07$.

The Connection Between d and t_{obs}

The vast majority of research in psychology over the past 75 years or so has relied on significance tests to draw inferences about the effects of experimental manipulations. Until recently, very little emphasis has been placed on measures of effect size. As a consequence, you may read an older paper in which the authors report t_{obs} but don't mention the corresponding effect size. If you are in this situation, you can take heart in the following, very simple connection between d and t_{obs}:

$$d = t_{obs}\sqrt{\frac{1}{n_1} + \frac{1}{n_2}}. \tag{11.17}$$

Therefore, if an author provides t_{obs} and the sample sizes, one can easily recover d and then put a confidence interval around it.

The APA reporting section above provides the information required by equation 11.17 to recover d. We do this as follows:

$$d = t_{obs}\sqrt{\frac{1}{n_1} + \frac{1}{n_2}} = 1.57\sqrt{\frac{1}{11} + \frac{1}{11}} = 0.67.$$

One of the unfortunate things about the past emphasis on significance tests is that t_{obs} may not be reported when it is not statistically significant. This means that we can't recover information about the effect size associated with the difference in question. Of course, this produces a very unbalanced representation of effect sizes in the literature.

INTERPRETATION OF OUR RIDDLE STUDY

The researcher in the riddle-solving experiment predicted that more riddles would be solved in a quiet setting than in a noisy setting. This prediction is supported by the data. Participants in the quiet condition solved three more riddles, on average, than those in

LEARNING CHECK 6

1. A researcher at the University of Dallas wonders what effect challenging cognitive activities has on outcomes on the Montreal Cognitive Assessment (MoCA) test of cognitive functioning. Thirty-six elderly men (ages 65 to 75) were selected at random and divided, at random, into a treatment group and a control group; there were 18 individuals in each group. The treatment group participated in a digital photography course for 3 hours a day for 4 weeks. The control group spent 3 hours a day watching television in the common room of a local senior home. At the end of the 4-week period, all 36 men were administered the MoCA. The results of the experiment were as follows: $m_T = 29.6$, $m_C = 27.2$, $s_T = 2.2$, and $s_C = 3.1$.

(a) State the null and alternative hypotheses in symbols.

(b) Assuming $\alpha = .05$, compute a confidence interval to test the null hypothesis.

(c) If you were to conduct the hypothesis test using a t-test, what would be the value of $t_{critical}$?

(d) Calculate t_{obs}.

(e) Based on your calculations, should you retain or reject the null hypothesis?

(f) Show how to compute an estimate of δ from t_{obs}.

Answers

1. (a) H_0: $\mu_{m_T - m_C} = 0$ and H_1: $\mu_{m_T - m_C} \neq 0$; no prediction was made.

(b) $t_{\alpha/2} = 2.032$. $(m_1 - m_2) \pm t_\alpha (s_{m_1 - m_2}) = (29.6 - 27.2$ $2.032(.896) = [0.58, 4.22]$.

(c) $t_{critical} = \pm 2.032$.

(d) $t_{obs} = (29.6 - 27.2)/.896 = 2.679$.

(e) $t_{obs} = 2.679$, $p = .011$. Reject H_0 (a p-value computed in Excel, $p < .02$, is also acceptable).

(f) $d = t_{obs}\sqrt{1/n_1 + 1/n_2} = 2.679\sqrt{2/18} = .89$.

the noisy condition, 95% CI [−0.29, 6.29]. This represents a 33% improvement (12/9 = 1.33) and corresponds to an estimated effect size of 0.67, 95% CI [−0.19, 1.53]. From the estimated effect size of 0.67 we estimate that in the two populations, 74.86% of the noisy distribution lies below the mean of the quiet distribution.

Although the researcher's prediction is supported, the estimate of the difference between the two population means is somewhat imprecise. In fact, the 95% confidence interval contains 0. So, although 3 is our best estimate of the difference between the two population means, 0 remains one of many plausible values for $\mu_q - \mu_n$. Some would say that the difference is not statistically significant, but this should not be taken to mean that there is no evidence of a difference between the two population means.

This hypothetical experiment provided an analog of the real-world situation of working in a quiet or a noisy setting. The results suggest that for some types of work, a quiet setting may produce better outcomes. One could imagine many real-world contexts in which work quality would suffer in a noisy and active open office. Computer programming and architectural design might be examples of jobs that are done more effectively in quiet environments. Such jobs, like riddle solving, seem to require extended periods of focus, during which many choices are considered and the implications of each choice must be weighed. Interrupting these thought processes might lead to poor work.

On the other hand, people are different. It may be that introverts gravitate to quieter environments and perform better there, whereas extroverts seek out more dynamic environments. Therefore, it might be interesting to see whether both introverts and extroverts benefit from quiet environments on the riddle task, or whether introverts show greater benefit. The methods to address such questions go beyond the two-groups design and will be considered in Part IV of this book.

One mustn't forget, however, that there are jobs for which a dynamic open environment is essential. Think of comedy writers. So, although we have a (hypothetical) experimental example of quiet conditions yielding better outcomes, different results might be obtained for different populations of individuals and different types of tasks.

A Note on Sample Size

As with most examples, we have used small samples so that we will have manageable numbers to illustrate the calculations involved. However, it is important to remember that for both estimation and significance testing, it is possible to make rational and informed choices about sample sizes.

From the estimation point of view, sample size determines the margin of error. For the example we've been working with in this chapter, the margin of error associated with the confidence interval around $m_1 - m_2$ was shown above to be $t_{\alpha/2}(s_{m_1-m_2}) = 2.086 * 1.9069 = 3.98$. Appendix 11.5 (available at study.sagepub.com/ gurnsey) shows that this margin of error is approximately $0.88(\sigma)$, or almost a full standard deviation wide. In some circumstances, this lack of precision might be acceptable. For example, in the present case we were simply asking if there is any evidence that quiet rooms are more conducive to riddle solving than noisy rooms. The results support the idea that there is an advantage to working in a quiet setting. However, because the samples are small, we are unable to state how big the effect is with much precision. Appendix 11.5 continues the discussion of how to choose an appropriate sample size to achieve a desired level of precision.

From the significance testing perspective, we use prospective power analysis to choose sample sizes. We can use G*Power to determine sample sizes required to achieve a specified power, given an assumed effect size (δ). However, G*Power can also be used to determine the power of the experiment *post hoc* (i.e., after the experiment has been run). The experiment described in our example has very low power. The G*Power application shows that if the population effect size were $\delta = 0.67$ (as estimated in our hypothetical study), then using two samples of size 11 would yield power of only .45. Furthermore, we can use G*Power to determine what effect size can be detected with a specified power, for a given α and sample size. We call this the *sensitivity* of the experiment. For example, assuming a one-tailed test with $\alpha = .05$ and power = .8, the choice of 11 participants per group suggests that the researcher was interested in an effect size of $\delta = 1.1$. That is, the experiment has the sensitivity to detect an effect size of $\delta = 1.1$ power* 100% of the time when sample sizes are 11 and $\alpha = .05$. Appendix 11.6 (available at study.sagepub.com/ gurnsey) illustrates how to use G*Power to obtain both post hoc power and sensitivity.

PARTITIONING VARIANCE

Overview

In this section we will discuss the fascinating interconnectedness of statistics. This interconnectedness is a major theme that we will see often in the remainder of this book. We've already seen that d and t_{obs} are connected in a relatively straightforward way (equation 11.17). We'll now see that things go deeper than this.

The interconnectedness of statistics derives from the notion of partitioning variance. To see where we're going, let's think about merging the 11 scores in each of the two groups of our riddle study into a single group of 22 scores. Partitioning variance means that we can decompose the variability in this merged set of scores into two components. One component is related to the variability within each of the two subgroups of 11 scores. The other component is related to the difference between the two sample means. From this decomposition we will discover a new statistic that we will call r^2, which represents the proportion of variance in the merged group of scores that is attributable to the difference between the two group means. Although r^2 is derived quite differently from Cohen's d, we will see that they are two expressions of the same thing. We will also see how other statistics are connected to each other through r^2 and Cohen's d.

There will be quite a few formulas in this section, but introducing them is not an intellectual exercise to illustrate the interconnectedness of statistics. Rather, the research literature abounds with different statistics, and when we understand the connections between them, we are in a much better position to relate studies to each other and to see the order in what might otherwise seem a chaotic set of results. Seeing the unity in statistics puts us in a better position to understand an important statistical technique called meta-analysis, which will be introduced in the last section of this chapter.

Between- and Within-Group Variability

We could continue with the data from the riddle-solving example to illustrate the notion of partitioning variance. However, an example with fewer scores will be easier to work with. Therefore, we will use a concrete example involving cats and dogs.

Assume that we have a collection of three cats and three dogs and we've measured the weight of each animal. Figure 11.8a shows these weights. Cats are denoted by circles and dogs are denoted by squares. It should be clear from Figure 11.8 that cats weigh less than dogs on average. More importantly, the variability within the group of cats, and the variability within the group of dogs, is less than the variability in the two groups combined. It is this difference that we will focus on. To simplify the discussion, we will use sums of squares as a measure of variability.

In the discussion that follows, we will refer to weights as scores and denote the scores for cats and dogs as y_{cats} and y_{dogs}, respectively. The scores for the cats, $y_{cats} = \{6, 9, 12\}$, have a mean of $m_{cats} = 9$. The scores for the dogs, $y_{dogs} = \{19, 25, 31\}$, have a mean of $m_{dogs} = 25$. To compute the sums of squares within each group, we subtract the group mean from each score and then square and sum these deviation scores. The deviation scores for cats are $\{-3, 0, 3\}$, producing $ss_{cats} = 18$. The deviation scores for dogs are $\{-6, 0, 6\}$, producing $ss_{dogs} = 72$.

iStock.com/GlobalP

The sums of squares for cats and dogs reflect **within-group variability**. If we add the sums of squares for the two groups, we can denote the result as ss_{within}. Therefore, the total within-group variability is $ss_{within} = 18 + 72 = 90$.

Now let's merge our two groups of three scores into one group of six scores. The resulting merged set will be $y_{total} = \{6, 9, 12, 19, 25, 31\}$. This merged set of scores has a mean and a sum of squares, which we'll call m_{total} and ss_{total}, respectively. As shown in Figure 11.8a, $m_{total} = 17$. Subtracting 17 from each score in y_{total} produces deviation scores, $y_{total} - m_{total} = \{-11, -8, -5, 2, 8, 14\}$. Squaring and summing these deviation scores produces $ss_{total} = 474$.

We now have two sums of squares, ss_{within} and ss_{total}. The within-group variability makes up part of ss_{total} but not all of it. The difference between ss_{total} and ss_{within} is $ss_{total} - ss_{within} = 474 - 90 = 384$. We refer to this difference as **between-group variability** and denote it with $ss_{between}$. We will next see that $ss_{between}$ is derived from the mean scores of cats and dogs.

Within-group variability is the variability about the mean of a sample or population.

Between-group variability reflects the variability in a collection of scores resulting from scores having been drawn from two or more populations.

To see more clearly what $ss_{between}$ represents, we replace each score in y_{total} with the mean of the sample from which it came. This will produce the following collection of numbers: y_{means} = {9, 9, 9, 25, 25, 25}. The mean of these six numbers is m_{means} = 17; i.e., the mean of the means equals m_{total}. When we subtract m_{means} from each of the means, we obtain the following deviation scores: $y_{means} - m_{means}$ = {−8, −8, −8, 8, 8, 8}. Squaring and summing these deviation scores produces ss_{means} = $6*8^2$ = 384. (Notice, we've seen 384 before.) We will refer to ss_{means} as $ss_{between}$ because it represents variability associated with the difference between group means.

From this example, we can see the following relationship:

$$ss_{total} = ss_{between} + ss_{within}$$

$$474 = 384 + 90.$$

That is, we have decomposed the total variability in our six scores (ss_{total}) into a part associated with variability within groups (ss_{within}) and a part associated with variability between groups ($ss_{between}$). This is what we mean by partitioning variance.

The numbers in Table 11.4 summarize the calculations we've just seen and relate to Figures 11.8a through 11.8c. The first column of Table 11.4 identifies the groups in question (i.e., cats and dogs). The second column (labeled "Subject") provides a number that we can use to refer to each individual. Each of the six scores shown in the third column comes from a specific individual (see Figure 11.8a). The fourth column shows the means of the corresponding groups; i.e., m_{cats} = 9 and m_{dogs} = 25 (see Figure 11.8b). The fifth column shows the deviation of each subject's score from its group mean (see Figure 11.8c). The deviations are denoted with the letter e because deviation scores are sometimes called error scores, and we don't want confusion with Cohen's d.

Table 11.4 shows that each individual's score can be broken down into two parts. One part is the group mean and the other is its deviation from the group mean. For example,

FIGURE 11.8 ■ Partitioning Variability

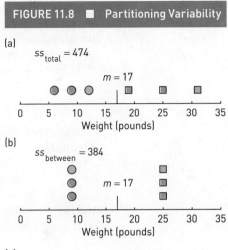

(a) ss_{total} = 474

(b) $ss_{between}$ = 384

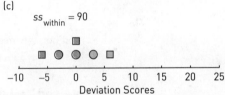

(c) ss_{within} = 90

A graphical illustration of partitioning variability. (a) The six scores from Table 11.4 are shown; cats are circles and dogs are squares. (b) Each of these six scores is replaced with the group mean (9 or 25). (c) Each of the six scores is replaced with its deviation from its group mean.

TABLE 11.4 ■ Illustrating the Concept of r^2				
Group	Subject	Scores (y)	Means (m)	$e = y - m$
Cats	1	6	9	−3
Cats	2	9	9	0
Cats	3	12	9	3
Dogs	4	19	25	−6
Dogs	5	25	25	0
Dogs	6	31	25	6
	Mean	17	17	0
	ss	474	384	90
	r^2		0.8101	

subject 4 (a dog) has a score of 19, which equals the group mean of 25, plus the score's deviation from the mean; i.e., −6. Therefore, we can say that $y_4 = m_4 + e_4 = 25 − 6 = 19$. Breaking down (or decomposing) each score into these two parts is the essential feature of partitioning the total variability into within-group and between-group variability.

In summary, the total variability in our scores (column y in Table 11.4) is defined as

$$ss_{total} = \Sigma(y − m_y)^2.$$

The variability attributable to the difference between group means (column m in Table 11.4) is defined as

$$ss_{between} = \Sigma(m − m_y)^2.$$

And the variability not attributable to the difference between group means (column e in Table 11.4) is defined as

$$ss_{within} = \Sigma e^2.$$

Remember, the mean of the e scores is 0. The row labeled ss in Table 11.4 shows that $ss_{total} = 474$, $ss_{between} = 384$, and $ss_{within} = 90$.

Explained Variance as an Effect Size

We can now ask, what *proportion* of ss_{total} is attributable to group means? Well, that's easy. We define the proportion of ss_{total} attributable to group means as

$$r^2 = \frac{ss_{between}}{ss_{total}}.$$

In this case,

$$r^2 = \frac{384}{474} = .8101.$$

That is, 81.01% of the variability in our set of six scores is *attributable* to (or explained by) group means. Put differently, 81.01% of the variability in our set of six scores comes from between-group variability.

Furthermore, the variability not explained by means is

$$1 − r^2 = \frac{ss_{within}}{ss_{total}},$$

which in this case is

$$1 − r^2 = \frac{90}{474} = .1899.$$

That is, 18.99% of the variability in our set of six scores is *not attributable* to the differences in the group means. Put differently, 18.99% of the variability in our set of six scores comes from within-group variability.

The calculations we just worked through are cumbersome, and we did them to establish the general notion of partitioning variance, and r^2 in particular. It turns out that there is a very simple relationship between t_{obs} and r^2. It is just this:

$$r^2 = \frac{t_{obs}^2}{t_{obs}^2 + df_{within}}. \tag{11.18}$$

Although we haven't computed t_{obs}, the data in Table 11.4 give us enough information to do so. We saw that $m_{cats} = 9$ and $m_{dogs} = 25$. The deviation scores in column 5 of Table 11.4 allow us to compute the corresponding variances. That is, $s_{cats}^2 = \Sigma\{-3,\ 0,\ 3\}^2/2 = 9$ and $s_{dogs}^2 = \Sigma\{-6,\ 0,\ 6\}^2/2 = 36$. Therefore,

$$t_{obs} = \frac{m_{dogs} - m_{cats}}{s_{m_1-m_2}} = \frac{m_{dogs} - m_{cats}}{\sqrt{\dfrac{s_{dogs}^2}{n_{dogs}} + \dfrac{s_{cats}^2}{n_{cats}}}} = \frac{25-9}{\sqrt{\dfrac{9}{3} + \dfrac{36}{3}}} = 4.1312.$$

If we put t_{obs} into equation 11.18, we find the following

$$r^2 = \frac{t_{obs}^2}{t_{obs}^2 + df_{within}} = \frac{4.1312^2}{4.1312^2 + 4} = .8101,$$

which is exactly what was computed previously as $r^2 = ss_{between}/ss_{total}$. Therefore, if we know t_{obs} and df_{within}, we can easily determine the proportion of variability in our scores explained by group means.

We can now think back to the riddle example that we've worked with throughout this chapter. Equation 11.18 provides a very simple way to determine the proportion of variability in our 22 scores explained by the difference between group means. Remember that t_{obs} for the riddle example was 1.57 and df_{within} was 20. When we put these numbers into equation 11.18 we find the following:

$$r^2 = \frac{t_{obs}^2}{t_{obs}^2 + df_{within}} = \frac{1.57^2}{1.57^2 + 20} = .1097.$$

This means that about 11% of the total variability in our 22 scores is explained by the difference between the group means, and the remaining 89% is explained by within-group variability; i.e., the vast majority of variability in our set of 22 scores represents within-group variability.

As noted at the beginning of this section, r^2 is an effect size. However, the interpretation of r^2 (or its square root r) as an effect size presents the same complications as the interpretation of d. An r^2 that is considered large in one field of study (e.g., social psychology) may be considered small in another (e.g., neuroscience). Therefore, there are no universal guidelines that allow us to say what are small, medium, and large effects when r is the measure of effect size. Well, there are almost no universal guidelines. As with d, Jacob Cohen provided guidelines that he found useful in his field of study. These are summarized in Table 11.5.

TABLE 11.5 ■ Cohen's Guidelines		
Classification	r	r^2
Small	.10	.01
Medium	.25	.0625
Large	.40	.16

Note: Cohen's guidelines for effect size (r and r^2) are to be used as a last resort!

The Connection Between r^2 and d

Equation 11.17 showed that t_{obs} and d are directly related, and equation 11.18 shows that t_{obs} and r^2 are directly related. This suggests that r^2 and d should be directly related, and indeed they are. The relationship is this:

$$d = \sqrt{df_{within}\left(\frac{r^2}{1-r^2}\right)\left(\frac{1}{n_1}+\frac{1}{n_2}\right)}. \tag{11.19}$$

In the section on effect size (Cohen's d), we determined that our estimate of δ in the riddle study was $d = .67$. Therefore, if we substitute $r^2 = .1097$ into equation 11.19, we should obtain $d = .67$, which we do:

$$d = \sqrt{df_{within}\left(\frac{r^2}{1-r^2}\right)\left(\frac{1}{n_1}+\frac{1}{n_2}\right)} = \sqrt{20\left(\frac{.1097}{1-.1097}\right)\left(\frac{1}{11}+\frac{1}{11}\right)} = 0.67.$$

You should work through these numbers yourself to confirm that the calculations are correct.

What we've just shown is really elegant. The standardized difference between two means (d) is directly related to the proportion of total variance explained by the difference between the two means (r^2). These are two units-free measures of how different the means of two distributions are. This result is not only elegant but also very useful.

The F-Statistic

Before drawing this discussion to a close, there is one last statistic to be thrown into the mix. This is the F-statistic (in honor of Sir Ronald Fisher) and it is widely used in the advanced analyses that are covered in Part IV of this book. However, F can be computed for the independent-groups design that we are currently considering. For the two-groups design, the F-statistic is defined as

$$F = \frac{r^2}{1-r^2}df_{within}. \tag{11.20}$$

For our riddle example,

$$F = \frac{r^2}{1-r^2}df_{within} = \frac{.1097}{1-.1097}20 = 2.46.$$

Interestingly, F is related to t_{obs} as follows:

$$F = t_{obs}^2 \text{ or} \tag{11.21}$$

$$t_{obs} = \sqrt{F}. \tag{11.22}$$

In the riddle example, $t_{obs} = 1.57$ and when squared (equation 11.21), we find $F = t_{obs}^2 = 1.57^2 = 2.46$, exactly as we found above using equation 11.20.

The Interconnectedness of Statistics

We've just worked through a lot of formulas that connect statistics to each other. Interconnectedness is one of the major themes that emerge from the study of statistics. This is fascinating from a purely intellectual point of view. In fact, this kind of interconnectedness is what makes mathematics beautiful. We can enjoy this on a small scale.

However, beyond aesthetic reasons, there are important practical reasons to understand these connections. Once understood, these interconnections allow us to compare the results of statistical analyses that have used different statistics. For example, different researchers may have run essentially the same experiment but reported the results using different statistics, including d, r^2, t_{obs}, and F. When we know how to translate between these four statistics, we can more easily compare the results of the experiments. Furthermore, we can put all results on a common scale and combine the results. In the next section we will see how this works.

LEARNING CHECK 7

1. A researcher at the University of Arizona was curious about how much male and female psychology students know about the local football team. (The Arizona Cardinals were formerly the St. Louis Cardinals, who, in their glory years, featured Jim Hart and Mel Gray as the core of an incredibly productive passing offense.) The researcher obtained three female and three male volunteers from the Psychology Department participant pool and asked each to name as many current Cardinals players as possible. The results of this small study showed that $y_{males} = \{8, 9, 10\}$ and $y_{females} = \{4, 5, 6\}$.

   (a) What proportion of the variability in these two sets of scores is explained by sex?

   (b) What proportion of the variability in these two sets of scores is not explained by sex?

   (c) Convert r^2 to d.

   (d) Compute the approximate 95% confidence interval around d.

   (e) What does this confidence interval say about the null hypothesis that $m_{males} = m_{females}$?

Answers

1. (a) $r^2 = .8571$.

   (b) $1 - r^2 = .1429$.

   (c) $d = \sqrt{4\left(\dfrac{.8571}{.1429}\right)\left(\dfrac{2}{3}\right)} = 4$.

   (d) $s_d = \sqrt{\dfrac{16}{8} + \dfrac{1}{3} + \dfrac{1}{3}} = 1.633$. CI $= 4 \pm 1.96(1.633)$
   $= [.80, 7.20]$.

   (e) If the null hypothesis were true, we would expect 0 to fall within this interval. Because it does not, we can reject the null hypothesis.

META-ANALYSIS

All the examples we've worked through to this point have dealt with the results of single studies. This reflects how research is typically done. A researcher has a question that he or she would like answered. The study is conducted, the analysis is completed, and a paper is written describing the experiment, results, and conclusions. The paper is then submitted to a professional journal in the hope that it will be judged suitable for publication.

The examples in this and preceding chapters show that the results of individual studies can be rather imprecise. That is, the confidence intervals around our point estimates can be quite wide. The appendixes for Chapters 6 and 10 as well as Appendix 11.5 (available at study.sagepub.com/gurnsey) show that with proper planning, we can have some control over our margin of error. However, in practical terms there may be limits to how precise a single study can be. In any case, it is rare for a question of any importance to be settled by the results of a single study. Rather, we rely on replication and the cumulative weight of evidence arising from many related studies to draw our conclusions.

One method used to combine the results of several studies is **meta-analysis** (Ellis, 2010; Hedges & Olkin, 1995; Hunter & Schmidt, 1990; Smith & Glass, 1977). The logic of the method is extremely simple: if the results of several individual studies are somewhat imprecise, then the results of these individual studies can be averaged to yield a more precise result. For example, if several studies had addressed the effects of a new treatment for depression, then the results of all such studies can be averaged to yield a more precise estimate of the benefits of the treatment. The logic is exactly the same as all estimation procedures we've done to this point. In single studies, we combine *measures taken from individuals* to estimate a population parameter. In meta-analysis, we combine the *results of studies* to get a more precise estimate of a population parameter.

Throughout this chapter we have worked with the example of the effect of noise (or quiet) on riddle solving. We found that the estimated effect size was $d = .67$. Let's say a second study of the same sort had been conducted, with 25 participants in each condition (quiet and noisy), and a statistically significant increase in the number of riddles solved was found, $t(48) = 3.7$, $p = .0006$. Perhaps a third study used the same method with 20 participants in each group. In this study too there was an increase in the number of riddles solved, which corresponded $r^2 = .25$. That is, the difference between sample means accounted for 25% of the variability in the scores. Finally, a fourth study was conducted with two groups of 16 participants and a statistically significant increase in the number of riddles solved was found, $F = 6.5$, $p = .016$.

Now we have four studies addressing the same question, but each has presented the results in a different way. However, we now know that all three measures can be converted to an estimate of δ. That is, all three results can be expressed as d. This is shown in Table 11.6. Using equation 11.17, we find that $t(48) = 3.7$ corresponds to $d = 1.05$. Using equation 11.22, we find that $r^2 = .25$ corresponds to $d = 1.13$. And, using equation 11.17, we find that $F = 6.5$ corresponds to $d = 0.90$; in this case, $t_{obs} = \sqrt{F}$.

Meta-analysis is a quantitative method of data analysis that combines the results of many individual studies to obtain a more precise estimate of a population parameter.

TABLE 11.6 ■ Converting t_{obs} and r^2 to d for a Meta-Analysis				
Study	Statistic	Value	Expressed as d	Transformation
Study 1	d	0.67	0.67	
Study 2	t_{obs}	3.70	1.05	$d = t_{obs} * \sqrt{\dfrac{1}{n_1} + \dfrac{1}{n_2}}$
Study 3	r^2	0.25	1.13	$d = \sqrt{df_{within}\left(\dfrac{r^2}{1-r^2}\right)\left(\dfrac{1}{n_1} + \dfrac{1}{n_2}\right)}$
Study 4	F	6.50	0.90	$d = \sqrt{F} * \sqrt{\dfrac{1}{n_1} + \dfrac{1}{n_2}}$
Mean			0.94	
Standard deviation			0.20	

Now that all statistics have been converted to the same units (d), we can compute the mean and standard deviation of these four values of d. At the bottom of Table 11.6, we can see that the mean of the four d values is 0.94 with a standard deviation of 0.20. This is a meta-analysis. We have combined the results of four studies to obtain a more precise estimate of the true population effect size. Although we won't do it here, we could go on to compute a confidence interval around this average value of d. (We could also test it for statistical significance if we thought this would be useful, or if a journal editor forced us to.)

It is important to recognize a second important role of d in meta-analysis. Remember, d, like z, is a unitless measure. Therefore, the responses on many different dependent variables can be converted to d and then averaged. For example, in the four studies we just considered, the dependent variable was the number of riddles solved. However, different dependent variables could have been used. For example, a researcher might have measured the time required to solve five riddles, rather than the number solved in 30 minutes. Response times are supposed to reflect the same underlying psychological construct measured by the number of riddles solved. Therefore, participants who solve more riddles would be expected to solve a fixed number of riddles in a shorter time. So, on average, participants in the quiet condition would be expected to solve riddles faster than those in the noisy condition. The difference between these two *time-to-solve* means can be converted to d and combined in a meta-analysis with d derived from two *number solved* means.

Meta-analysis is a powerful tool for combining research results, and estimates of δ play an important role. First, many different *statistics* can be converted to d so that they are put on a common scale and averaged. Second, many different dependent variables, all reflecting the same underlying psychological construct, can be converted to d and averaged.

When you read about individual research results that seem interesting, you should always be thinking meta-analytically (Kline, 2013), which means

- thinking about how the dependent variable in one study relates to the dependent variables used in related studies, and

- what the weight of evidence suggests about the question being addressed in these related studies.

Thinking meta-analytically keeps us from thinking that the results of a single study are definitive, as we might be led to believe from a misinterpretation of p-values.

Because meta-analysis is a very general but simple analysis, we will wait until we've seen a few more statistical methods before returning to the question of how results obtained using these different methods can be combined.

SUMMARY

In this chapter we saw how to compute a confidence interval around the difference between two independent sample means when the population standard deviations are unknown. The confidence interval is computed as follows:

$$(m_1 - m_2) \pm t_{\alpha/2}(s_{m_1 - m_2}),$$

where $m_1 - m_2$ is the difference between two means drawn from two independent populations. Population 1 has mean μ_1 and variance σ_1^2, and population 2 has mean μ_2 and variance σ_2^2. One can (theoretically) compute all possible values of m_1 from samples of size n_1 from population 1, and all possible values of m_2 from samples of size n_2 from population 2. If $m_1 - m_2$ is computed for all possible combinations of m_1 and m_2, the result will be *the sampling distribution of the difference between two means*. This distribution will have a mean of $\mu_1 - \mu_2$ and a variance of $\sigma_1^2/n_1 + \sigma_2^2/n_2$. If sample sizes are equal and it is assumed that $\sigma_1^2 = \sigma_2^2$, then $\sigma_1^2/n_1 + \sigma_2^2/n_2$

can be estimated with $s_1^2/n_1 + s_2^2/n_2$. If sample sizes are unequal but we can assume that $\sigma_1^2 = \sigma_2^2$, then we can compute

$$s_{\text{pooled}}^2 = \frac{df_1}{df_{\text{within}}}(s_1^2) + \frac{df_2}{df_{\text{within}}}(s_2^2)$$

to estimate the variance (σ^2) common to the two populations. s_{pooled}^2 can then be substituted into

$$\sigma_{m_1-m_2} = \sqrt{\frac{\sigma_1^2}{n_1} + \frac{\sigma_2^2}{n_2}}$$

as follows to yield the estimated standard error of the difference between two means, $s_{m_1-m_2}$:

$$s_{m_1-m_2} = \sqrt{\frac{s_{\text{pooled}}^2}{n_1} + \frac{s_{\text{pooled}}^2}{n_2}}.$$

We use $s_{m_1-m_2}$ to compute our confidence interval. If σ_1^2 and σ_2^2 cannot be assumed to be identical, then we can use the Welch-Satterthwaite correction (described in Appendix 11.4 available at study.sagepub.com/gurnsey) to increase $t_{\alpha/2}$ appropriately.

The information used to compute a confidence interval around $m_1 - m_2$ can also be used to estimate δ and to place an approximate confidence interval around this estimate. Specifically,

$$d = \frac{m_1 - m_2}{s_{\text{pooled}}}$$

estimates δ. An approximate $(1-\alpha)100\%$ confidence interval around d is computed as follows:

$$d \pm z_{\alpha/2}(s_d)$$

where

$$s_d = \sqrt{\frac{d^2}{2df_{\text{within}}} + \frac{1}{n_1} + \frac{1}{n_2}}.$$

The estimated effect size d is related to t_{obs} as follows:

$$d = t_{\text{obs}}\sqrt{\frac{1}{n_1} + \frac{1}{n_2}}.$$

This is a very useful connection because many research studies report only t_{obs}. So, if you know t_{obs} and the size of the two samples, you can recover d and place an approximate confidence interval around it.

Confidence intervals, $(m_1 - m_2) \pm t_{\alpha/2}(s_{m_1-m_2})$, can be used to test the null hypothesis that $H_0: \mu_1 - \mu_2 = 0$. If 0 does not fall within the interval, we can reject a two-tailed test of the null hypothesis with a significance level of α. The traditional method of testing the null hypothesis is to compute

$$t_{\text{obs}} = \frac{m_1 - m_2}{s_{m_1-m_2}}$$

and reject H_0 if t_{obs} exceeds the t_{critical} value that is established based on α, df_{within}, and H_1.

The concept of partitioning variance is important for two reasons. The first is that r^2 is an alternative estimate of a population effect size, which we will call ρ^2 in later chapters. r^2 is the proportion of variability in our two samples, treated as a single set of scores, that is explained by the difference between the two group means. r^2 is related to t_{obs} in a very simple way:

$$r^2 = \frac{t_{\text{obs}}^2}{t_{\text{obs}}^2 + df_{\text{within}}}.$$

r^2 is also related in a simple way to d. Specifically,

$$d = \sqrt{df_{\text{within}}\left(\frac{r^2}{1-r^2}\right)\left(\frac{1}{n_1} + \frac{1}{n_2}\right)}.$$

In statistics, everything is related. Because t_{obs} and r^2 can be converted to d, we can combine the results of studies using different statistics and different dependent variables in a meta-analysis.

KEY TERMS

between-group variability 284
confounding variable 264
dependent samples 264
dependent variable 262
experiment 263

hypothetical population 264
independent samples 264
independent variable 262
meta-analysis 290
pooled variance (s_{pooled}^2) 272

quasi-experiment 263
sampling distribution of the difference between two means 269
weighted sum 272
within-group variability 284

Definitions and Concepts

1. What is the difference between an experiment and a quasi-experiment?

2. Please explain the concept of a hypothetical population.

3. What is the difference between independent and dependent samples?

True or False

State whether the following statements are true or false.

4. For the two-groups design, the independent variable is dichotomous and the dependent variable is quantitative.

5. The mean heart rate of a sample of 32 Wistar rats living on the international space station is an estimate of the mean heart rate of Wistar rats living on the international space station.

6. It is impossible to estimate the mean of a hypothetical population.

7. Statistical significance implies high practical significance.

8. $s_{m_1-m_2}$ is the standard error of the difference between two means.

9. $\sigma^2_{m_1-m_2} = \dfrac{\sigma^2_1}{n_1} + \dfrac{\sigma^2_2}{n_2}$.

10. If $n_1 = n_2 = 5$, and $s_1 = s_2 = 15$, then $\sigma_{m_1-m_2} = 6$.

11. If $n_1 = n_2 = 5$, and $s_1 = s_2 = 15$, then $s_{pooled} = 15$.

12. If $n_1 = n_2 = 11$, and $t_{obs} = 4$, then $r^2 = .6154$.

13. If $n_1 = n_2 = 11$, and $t_{obs} = 4$, then $d = 1.706$.

Calculations

14. For each of the following parameters, compute the mean and variance of the sampling distribution of the difference between two means.

    (a) $\mu_1 = 10$, $\mu_2 = 12$, $\sigma_1 = 6$, $\sigma_2 = 5$, $n_1 = 15$, $n_2 = 15$

    (b) $\mu_1 = 10$, $\mu_2 = 12$, $\sigma_1 = 6$, $\sigma_2 = 5$, $n_1 = 22$, $n_2 = 15$

    (c) $\mu_1 = 20$, $\mu_2 = 6$, $\sigma_1 = 6$, $\sigma_2 = 25$, $n_1 = 15$, $n_2 = 4$

    (d) $\mu_1 = 18$, $\mu_2 = 30$, $\sigma_1 = 28$, $\sigma_2 = 5$, $n_1 = 4$, $n_2 = 15$

    (e) $\mu_1 = 4$, $\mu_2 = 8$, $\sigma_1 = 16$, $\sigma_2 = 8$, $n_1 = 32$, $n_2 = 64$

15. For each of the following scenarios, (i) calculate the 95% confidence interval around $m_1 - m_2$, (ii) compute the approximate 95% confidence interval around d, (iii) compute t_{obs}, and (iv) compute r^2.

    (a) $m_1 = 10$, $m_2 = 12$, $s_1 = 6$, $s_2 = 5$, $n_1 = 15$, $n_2 = 15$

    (b) $m_1 = 9$, $m_2 = 11$, $s_1 = 5$, $s_2 = 5$, $n_1 = 10$, $n_2 = 15$

    (c) $m_1 = 8$, $m_2 = 7$, $s_1 = 4$, $s_2 = 3$, $n_1 = 9$, $n_2 = 15$

    (d) $m_1 = 7$, $m_2 = 8$, $s_1 = 3$, $s_2 = 4$, $n_1 = 8$, $n_2 = 15$

Scenarios

16. Twenty university students (chosen at random from students enrolled at the University of Vermont) took part in an experiment testing the effect of watching FOX News versus CNN on knowledge of world events. The 20 participants were divided into two equal groups of 10 participants each. One group was assigned to watch FOX and the other was assigned to watch CNN. Participants then watched their assigned station (FOX or CNN) for 3 hours a night for 4 weeks. At the end of the 4-week period they were administered a test of world knowledge. The results were as follows: $m_{FOX} = 110$, $m_{CNN} = 99$, $s_{FOX} = 10$, $s_{CNN} = 6$.

    (a) Compute the 95% confidence interval around $m_{FOX} - m_{CNN}$.

    (b) What does it mean to have 95% confidence in this interval?

    (c) Use the 95% confidence interval to test the null hypothesis H_0: $\mu_{FOX} - \mu_{CNN} = 0$ against the alternative hypothesis H_1: $\mu_{FOX} - \mu_{CNN} \neq 0$.

    (d) What do you conclude from these data?

17. This font is called Comic Sans MS. An honors student at Concordia University theorized that reading comprehension is reduced when

text is presented in Comic Sans MS rather than the more commonly used Helvetica. She chose to test her theory in a population of university students using a widely available test of reading comprehension. The student drew a random sample of 42 university students in Canada, which she then divided, at random, into two groups of 21. One group was given the reading test formatted using Comic Sans MS, and the other was given the test formatted in Helvetica. After she collected the results from the 42 students, she found the following: $m_{CSMS} = 110$, $m_{Helvetica} = 122$, $s_{CSMS} = 16$, $s_{Helvetica} = 20$.

(a) What proportion of the variability is explained by the difference between the two means?

(b) Calculate the approximate 95% confidence interval around d.

(c) If H_1: $\mu_{CSMS} - \mu_{Helvetica} < 0$, can we reject the null hypothesis at the $\alpha = .05$ significance level? Explain your answer.

18. Researchers at Concordia University have a theory that drinking beer while studying will reduce test anxiety and result in improved test grades. To test their theory, the researchers chose 16 students at random from those currently enrolled in an introductory statistics course at Concordia. Half of the students were asked to consume two bottles of beer during their normal studying periods, and the other half were asked to refrain from alcohol consumption while studying. The students then wrote the test in their regular class period. The results were as follows: $m_{Beer} = $

75, $m_{NoAlc} = 70$, $s_{Beer} = 8$, $s_{NoAlc} = 6$. Conduct a test to assess the researchers' theory about the relationship between beer drinking and test performance.

(a) Assuming $\alpha = .05$, is there a statistically significant difference between the means of the two groups? Explain your answer.

(b) What proportion of the variability in the test scores is explained by between-groups variability?

(c) Would Cohen consider this a small, medium, or large effect size?

19. A professor of linguistics was dismayed by the prevalence of the word "like" in contemporary language. Thinking this was a generational difference, he measured the number of times a random selection of young and old subway passengers used the word "like" in a 5-minute conversation. (This study did not have the ethical approval of his university's ethics committee and verges on the creepy.) The results were as follows: $m_{young} = 60$, $m_{old} = 75$, $s_{young} = 15$, $s_{old} = 17$. There were 25 individuals in the group of older subway passengers and 36 in the group of younger subway passengers.

(a) Compute the 95% confidence interval around $m_{old} - m_{young}$.

(b) If his alternative hypothesis was H_1: $\mu_{old} - \mu_{young} < 0$, would he be able to reject the null hypothesis based on this confidence interval?

(c) Compute the approximate 95% confidence interval around his best estimate of δ.

APPENDIX 11.1: ESTIMATION AND SIGNIFICANCE TESTS IN EXCEL

Confidence Intervals and t_{obs} for $m_1 - m_2$

In previous appendixes we introduced all the Excel functions needed to conduct the analyses described in the body of this chapter. Figure 11.A1.1 shows how to compute a confidence interval around $m_1 - m_2$ and how to calculate t_{obs}. In cells **B2** to **B7**, we are provided with values for m_1, m_2, s_1, s_2, n_1, and n_2. We are also provided with the value for α in cell **B9**. The within-groups degrees of freedom (df_{within}) are computed in cell **B8** as $n_1 + n_2 - 2$.

From the numbers given, $m_1 - m_2$ is calculated in cell **B11**, and

$$s^2_{pooled} = \frac{df_1}{df_{within}}(s_1^2) + \frac{df_2}{df_{within}}(s_2^2)$$

is calculated in cell **B12**. s^2_{pooled} is used in the calculation of

$$s_{m_1-m_2} = \sqrt{\frac{s^2_{pooled}}{n_1} + \frac{s^2_{pooled}}{n_2}},$$

in cell **B13**. To compute a confidence interval around $m_1 - m_2$ requires $t_{\alpha/2}$, which is computed in cell **B14** using the **T.INV** function. The lower and upper bounds of the confidence interval are computed in cells **B16** and **B17**.

A two-tailed test of the null hypothesis can be conducted by asking whether 0 falls in the 95% confidence interval around $m_1 - m_2$. One can also compute

$$t_{obs} = \frac{m_1 - m_2}{s_{m_1 - m_2}}$$

as shown in cell **B19**. The **T.DIST** function is used in cell **B20** to compute the one-tailed p-value associated with t_{obs}. The two-tailed p-value is computed in cell **C21**.

Confidence Intervals for d

Cells **B2** to **B12** of Figure 11.A1.2 contain exactly the same quantities as cells **B2** to **B12** in Figure 11.A1.1. In cell **B13**, s_{pooled} is computed simply as the square root of s_{pooled}^2. In cell **B14**,

$$d = \frac{m_1 - m_2}{s_{pooled}}$$

is computed, and the estimated standard error of d is computed in cell **B15** as

FIGURE 11.A1.1 ■ Confidence Intervals and t_{obs}

	A	B	C
1	**Quantities**	**Values**	**Formulas**
2	m_1	12	
3	m_2	9	
4	s_1	4.94	=SQRT(244/10)
5	s_2	3.95	=SQRT(156/10)
6	n_1	11	
7	n_2	11	
8	df_{within}	20	=B6+B7-2
9	α	0.05	
10			
11	$m_1 - m_2$	3	=B2-B3
12	s_{pooled}^2	20	=(B6-1)/B8*B4^2 + (B7-1)/B8*B5^2
13	s_{m1-m2}	1.907	=SQRT(B12/B6 + B12/B7)
14	$t_{\alpha/2}$	2.086	=T.INV(1-B9/2,B8)
15			
16	$(m_1 - m_2) - t_{\alpha/2}(s_{m1-m2})$	-0.978	=B11-B14*B13
17	$(m_1 - m_2) + t_{\alpha/2}(s_{m1-m2})$	6.978	=B11+B14*B13
18			
19	t_{obs}	1.573	=B11/B13
20	$p_{one-tailed}$	0.066	=T.DIST(-ABS(B19),B8,1)
21	$p_{two-tailed}$	0.131	=2*B20

Confidence intervals and t-tests for the two-independent-groups design.

FIGURE 11.A1.2 ■ Confidence Intervals for d

	A	B	C
1	**Quantities**	**Values**	**Formulas**
2	m_1	12	
3	m_2	9	
4	s_1	4.94	=SQRT(244/10)
5	s_2	3.95	=SQRT(156/10)
6	n_1	11	
7	n_2	11	
8	df_{within}	20	=B6+B7-2
9	α	0.05	
10			
11	$m_1 - m_2$	3	=B2-B3
12	s_{pooled}^2	20	=(B6-1)/B8*B4^2 + (B7-1)/B8*B5^2
13	s_{pooled}	4.472	=SQRT(B12)
14	d	0.671	=B11/B13
15	s_d	0.439	=SQRT(B14^2/(2*B8) + (1/B6 + 1/B7))
16	$z_{\alpha/2}$	1.960	=NORM.S.INV(1-B9/2)
17			
18	$d - z_{\alpha/2}(s_d)$	-0.190	=B14-B16*B15
19	$d + z_{\alpha/2}(s_d)$	1.532	=B14+B16*B15

Approximate confidence intervals for d.

$$s_d = \sqrt{\frac{d^2}{2df_{within}} + \frac{1}{n_1} + \frac{1}{n_2}}.$$

NORM.S.INV is used in cell **B16** to compute $z_{\alpha/2}$. The lower and upper limits of the confidence interval around d [i.e., $d \pm z_{\alpha/2}(s_d)$] are computed in cells **B18** and **B19**.

Partitioning Variance

Figure 11.A1.3 illustrates how variability in the dependent variable is partitioned into within- and between-groups sources. Cells **B2** to **B23** are the raw data from Table 11.2. The mean of each condition is computed in cells **C2** to **C23** using the **AVERAGE** function. (The '**$**' signs shown to the right in column **D** have to do with absolute referencing, which was introduced in Appendix 2.1.) Cells **E2** to **E23** show the deviations of scores from their group mean. For example, cell **B13** shows a score of 5 in the noisy condition, which has a mean of 9. Therefore, the deviation score is $5 - 9 = -4$, as shown in cell **E13**.

The **DEVSQ** function has been used to compute ss_{total} (cell **B26**), $ss_{between}$ (cell **B27**), and ss_{within} (cell **B28**). Note that $ss_{total} = ss_{between} + ss_{within} = 49.5 + 400 = 449.5$. Cell **B30** shows the calculation of r^2 as

$$r^2 = \frac{ss_{between}}{ss_{total}}.$$

FIGURE 11.A1.3 ■ Partitioning Variance

	A	B	C	D	E	F
1	Condition	Scores	Mean	Formulas	Errors	Formulas
2	Quiet	12	12	=AVERAGE($B$2:$B$12)	0	=B2-C2
3	Quiet	4	12	=AVERAGE($B$2:$B$12)	-8	=B3-C3
4	Quiet	13	12	=AVERAGE($B$2:$B$12)	1	=B4-C4
5	Quiet	11	12	=AVERAGE($B$2:$B$12)	-1	=B5-C5
6	Quiet	17	12	=AVERAGE($B$2:$B$12)	5	=B6-C6
7	Quiet	21	12	=AVERAGE($B$2:$B$12)	9	=B7-C7
8	Quiet	11	12	=AVERAGE($B$2:$B$12)	-1	=B8-C8
9	Quiet	5	12	=AVERAGE($B$2:$B$12)	-7	=B9-C9
10	Quiet	15	12	=AVERAGE($B$2:$B$12)	3	=B10-C10
11	Quiet	14	12	=AVERAGE($B$2:$B$12)	2	=B11-C11
12	Quiet	9	12	=AVERAGE($B$2:$B$12)	-3	=B12-C12
13	Noisy	5	9	=AVERAGE($B$13:$B$23)	-4	=B13-C13
14	Noisy	7	9	=AVERAGE($B$13:$B$23)	-2	=B14-C14
15	Noisy	11	9	=AVERAGE($B$13:$B$23)	2	=B15-C15
16	Noisy	3	9	=AVERAGE($B$13:$B$23)	-6	=B16-C16
17	Noisy	14	9	=AVERAGE($B$13:$B$23)	5	=B17-C17
18	Noisy	10	9	=AVERAGE($B$13:$B$23)	1	=B18-C18
19	Noisy	9	9	=AVERAGE($B$13:$B$23)	0	=B19-C19
20	Noisy	8	9	=AVERAGE($B$13:$B$23)	-1	=B20-C20
21	Noisy	7	9	=AVERAGE($B$13:$B$23)	-2	=B21-C21
22	Noisy	8	9	=AVERAGE($B$13:$B$23)	-1	=B22-C22
23	Noisy	17	9	=AVERAGE($B$13:$B$23)	8	=B23-C23
24						
25	Quantities	Values	Formulas			
26	SS_{Total}	449.5	=DEVSQ(B2:B23)			
27	$SS_{Between}$	49.5	=DEVSQ(C2:C23)			
28	SS_{Within}	400	=DEVSQ(E2:E23)			
29	df	20	=COUNT(B2:B23)-2			
30	r^2	0.110	=B27/B26			
31	F	2.475	=B30/(1-B30) * C33B29			
32	t_{obs}	1.573	=SQRT(B31)			
33	r^2	0.110	=B32^2/(B32^2+B29)			

Partitioning the variance in scores from two independent samples.

Cell **B31** shows the calculation of the F-statistic as

$$F = \frac{r^2}{1-r^2} df.$$

And as noted in this chapter, t is the square root of F, as shown in cell **B32**. (Note that 1.573 is the value of t_{obs} calculated in the chapter.) Finally, r^2 is calculated a second time in cell **B33** using the formula

$$r^2 = \frac{t_{obs}^2}{t_{obs}^2 + df}.$$

APPENDIX 11.2: ESTIMATION AND SIGNIFICANCE TESTS IN SPSS

To compute a confidence interval around the difference between independent sample means in SPSS, we first enter our data as a single column of numbers. In Figure 11.A2.1a the column labeled N_Solved contains the scores from Table 11.2 arranged in a single column. To the right, in the column labeled Group, are numbers used to identify the groups. In this example, 1s corresponds to the quiet condition and 2s correspond to the noisy condition.

To compute a confidence interval and significance test for these two groups of scores, we choose the Analyze→Compare Mean→Independent-Samples T Tests . . . menu. When this has been chosen, the Independent-Samples T Tests dialog appears, as shown in Figure 11.A2.1b. The variable N_Solved has been moved into the Test Variable(s): region and Group has been moved into the Grouping Variable: region. Before we can proceed, we are required to identify the groups (the numbers in the Group variable) in the analysis. To do this, we click on the Define Groups . . . button, and the Define Groups dialog appears. As shown in Figure 11.A2.1c, 1s are associated with Group 1 and 2s are associated with Group 2. (These are values that I entered.) With this done, we click Continue to return to the Independent-Samples T Tests dialog, and there we click OK to proceed with the analysis.

The output of the analysis is shown in Figure 11.A2.2. There are two rows of numbers. The top row shows the analysis when the population variances are assumed to be equal (i.e., $\sigma_1 = \sigma_2$), and the second row shows the analysis when population variances are not assumed to be equal. The first two columns show the results of Levene's test for equal variances. We won't discuss this analysis, but it is in essence a kind of significance test. If the p-value (Sig.) associated with

the computed statistic (F) is small, then a statistically significant difference in the two sample variances exists.

The next column (t) shows t_{obs} computed exactly as described in the body of the chapter. The degrees of freedom are $n_1 + n_2 - 2$ when equal variances are assumed. When equal variances are not assumed, the degrees of freedom are adjusted downward using the Welch-Satterthwaite procedure described in Appendix 11.4 (available at study.sagepub.com/gurnsey). In this case, there has been very little adjustment because the sample variances are very similar. The two p-values [Sig (2-tailed)] are slightly different because they are based on the same t_{obs} (1.573) but slightly different degrees of freedom.

The difference between the two means (Mean Difference) and the estimated standard error (Std. Error Difference) used to compute t_{obs} are shown in the next two columns. These two quantities do not depend on whether the population variances are assumed to be equal. The last two columns show the 95% confidence intervals for the difference between two means. The interval computed in

FIGURE 11.A2.1   Independent-Groups Analysis

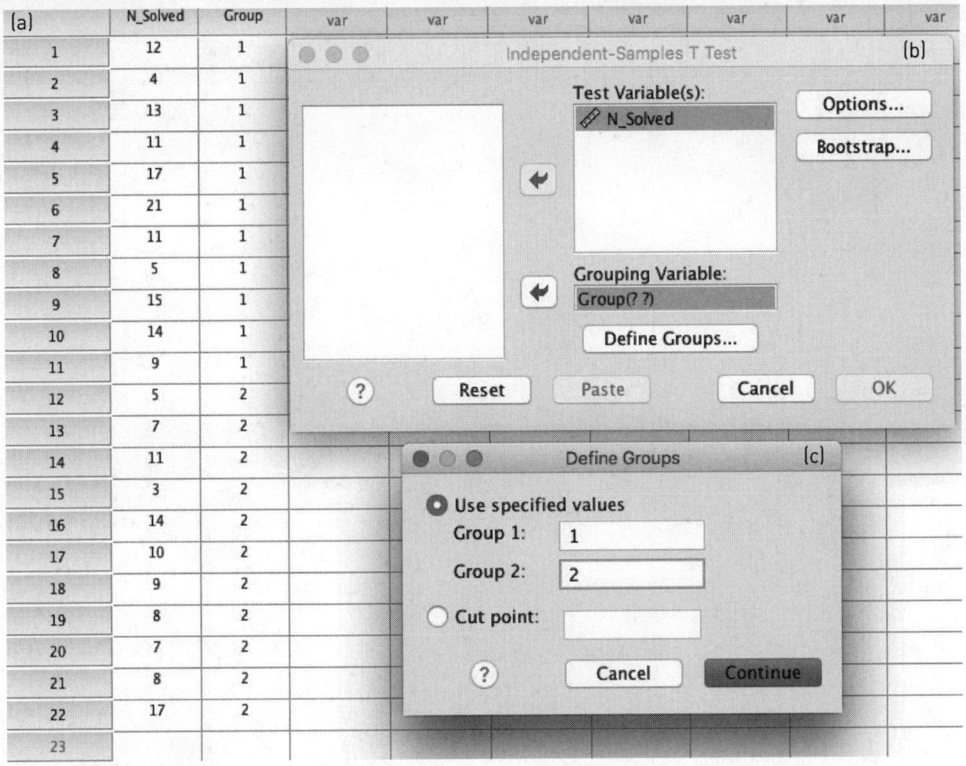

Conducting an independent-groups analysis in SPSS.

FIGURE 11.A2.2 ■ Independent-Groups Output

Independent Samples Test

| | | Levene's Test for Equality of Variances | | t–test for Equality of Means | | | | | | |
| | | F | Sig. | t | df | Sig. (2–tailed) | Mean Difference | Std. Error Difference | 95% Confidence Interval of the Difference | |
									Lower	Upper
N_Solved	Equal variances assumed	.360	.555	1.573	20	.131	3.000	1.907	−.978	6.978
	Equal variances not assumed			1.573	19.077	.132	3.000	1.907	−.990	6.990

Conducting an independent-groups analysis in SPSS.

the first row is exactly the same as the one computed in the chapter. The second row shows the Welch-Satterthwaite corrected confidence interval. In this case, $t_{\alpha/2}$ is based on 19.077 degrees of freedom rather than 20, which means $t_{\alpha/2}$ is a little larger than in the first row and hence the confidence interval is a little wider.

ESTIMATING THE DIFFERENCE BETWEEN THE MEANS OF DEPENDENT POPULATIONS

INTRODUCTION

We will begin this chapter with an odd but useful thought experiment. Let's consider a random sample of Concordia University psychology students whose weights are measured before and after lunch at Cosmos restaurant in Montreal. We obtained these before- and after-lunch weights because we would like to estimate the average weight gained at lunch in this population. Table 12.1 summarizes our results, along with the means (m_B and m_A) and variances (s_B^2 and s_A^2) of the two groups.

Let's approach this estimation problem just as we did in the last chapter. We first estimate the standard error of $m_1 - m_2$ as follows:

iStock.com/bhofack2

$$s_{m_1 - m_2} = \sqrt{\frac{s_B^2}{n_1} + \frac{s_A^2}{n_2}} = \sqrt{\frac{260.7}{5} + \frac{285.3}{5}} = 10.45.$$

We then compute our confidence interval as

$$\text{CI} = (m_A - m_B) \pm t_{\alpha/2}(s_{m_1 - m_2}) = (146.4 - 144.2) \pm 2.306(10.45) = [-21.898, 26.298].$$

Therefore, we have 95% confidence that $\mu_1 - \mu_2$ lies in the interval $[-21.898, 26.298]$. Our confidence comes from knowing that . . . Wait! What?! −21.898 to 26.298? That can't be right! How could $\mu_1 - \mu_2$ possibly be as low as −21.898 pounds or as high as 26.298 pounds? These numbers clearly make no sense. What's going on? Well, that's the question we address in this chapter.

DEPENDENT VERSUS INDEPENDENT POPULATIONS

In Chapter 11 we distinguished between independent and dependent samples and populations, and we then considered estimates of $\mu_1 - \mu_2$ for independent populations. Remember that two populations are independent when there is no systematic association between

TABLE 12.1 ■ Sample Data		
Participant	Before	After
1	151	154
2	136	138
3	137	139
4	128	129
5	169	172
Mean	144.2	146.4
Variance	260.7	285.3

Note: Before- and after-lunch weights (in pounds) for five individuals.

their scores. For example, the populations of Americans born east and west of the Mississippi River are independent populations. On the other hand, when there is an association between the scores in two populations, we say that we have *dependent populations*. The clearest example of dependent populations is when two scores come from the same individual, as in the lunch example we just considered. More generally, we think of dependent populations as comprising pairs of related scores. The members of the pairs are dependent, or related, in the sense that knowing one of the scores in a pair allows us to make a reasonable prediction about the other score. For example, knowing that a before-lunch weight is 128 pounds allows you to predict that an after-lunch weight of 130 is far more likely than an after-lunch weight of 190. Knowing that a before-lunch weight is 200 allows you to predict that an after-lunch weight of 202 is far more likely than an after-lunch weight of 150.

Difference Scores

The relatedness of scores in dependent populations and dependent samples allows us to eliminate sources of variability that made the confidence interval computed above absurdly wide. If you look back to Table 12.1, you will notice that there is a great deal of variability among the before- and after-lunch weights. The variances of these two samples are $s_B^2 = 260.7$ and $s_A^2 = 285.3$, respectively. These variances reflect the fact that there are big differences in the weights of the individuals we've chosen to measure. When we computed the confidence interval on the previous page, these variances were part of the estimated standard error

$$s_{m_1 - m_2} = \sqrt{\frac{s_B^2}{n_1} + \frac{s_A^2}{n_2}},$$

which in turn contributed to the margin of error, $moe = t_{\alpha/2}(s_{m_1 - m_2})$. As a consequence, the variability among the weights of the diners overwhelms the weights gained at lunch, which is the thing we wanted to estimate.

Table 12.2 reproduces the data from Table 12.1, with an added column to show the difference (D) in the before- and after-lunch weights, computed as $D = after - before$. We call D a difference score, and column D shows that each individual gained between 1 and 3 pounds at lunch. The mean of these difference scores is $m_D = 2.2$, which is exactly the same as $m_A - m_B = 146.4 - 144.2 = 2.2$. The mean of the difference scores (m_D) will always equal the difference between the two means ($m_1 - m_2$).

TABLE 12.2 ■ Difference Scores			
Participant	Before	After	D
1	151	154	3
2	136	138	2
3	137	139	2
4	128	129	1
5	169	172	3
Mean	144.2	146.4	2.2
Variance	260.7	285.3	0.7

At the bottom of column D in Table 12.2, you will see that the variance of the difference scores (s_D^2) is 0.7. That's much smaller than variances within the two samples; i.e., $s_B^2 = 260.7$ and $s_A^2 = 285.3$. We will see that by computing difference scores, we reduce the margin of error.

Statisticians refer to the individuals who contributed scores to a sample or a population as "subjects." Subjects could be people, rats, maple trees, or worms. Therefore, we refer to the variability *within* the before and after samples as **subject variability**. Table 12.2 shows that there is far more subject variability in the before- and after-lunch weights than there is in the difference scores. The question we asked in the opening example was about weight gained at lunch, not about how much the individual weights differ. Therefore, computing

Subject variability is the variability of individuals' scores around the group mean.

difference scores provides a way of separating variability in subjects' average *weights* from variability in subjects' *weight gains*. Separating these two sources of variability is the key to computing more sensible confidence intervals in the case of dependent samples.

Computing a Confidence Interval Around m_D

Because $m_1 - m_2$ and m_D both estimate $\mu_1 - \mu_2$, a confidence interval around m_D provides an estimate of $\mu_1 - \mu_2$. To compute a confidence interval around m_D, we use the following formula:

$$m_D \pm t_{\alpha/2}(s_{m_D}). \tag{12.1}$$

In this formula, $t_{\alpha/2}$ is based on $n-1$ degrees of freedom, where n is the number of difference scores and s_{m_D} is the estimated standard error of m_D, which is computed as

$$s_{m_D} = \sqrt{\frac{s_D^2}{n}}, \tag{12.2}$$

where s_D^2 is the variance of the difference scores. From Table 12.2, we can see that $s_D^2 = 0.7$. Therefore, we have everything necessary to compute a 95% confidence interval around m_D, which we'll assume has been computed.

Step 1. Compute s_{m_D}. When we substitute the values of s_D^2 and n into equation 12.2, we obtain the following:

$$s_{m_D} = \sqrt{\frac{s_D^2}{n}} = \sqrt{\frac{0.7}{5}} = 0.374.$$

Step 2. Compute $t_{\alpha/2}$. There are $n-1 = 4$ degrees of freedom. Because this is the 95% confidence interval, $\alpha = .05$. When we consult the t-table, we find that $t_{\alpha/2} = 2.776$.

Step 3. Compute the 95% confidence interval around m_D.

$$CI = m_D \pm t_{\alpha/2}(s_{mD}) = 2.2 \pm 2.776(0.374) = [1.16, 3.24].$$

We have 95% confidence that $\mu_D = \mu_A - \mu_B$ lies in the interval [1.16, 3.24]. Our confidence comes from knowing that 95% of intervals computed this way will capture μ_D. This is a far more precise (and sensible) confidence interval than the one we computed earlier.

THE DISTRIBUTIONS OF D AND m_D

To understand why we computed the confidence interval around m_D as we did, we need to think about the distribution of difference scores (D) and then the sampling distribution of the mean of difference scores (m_D). Figure 12.1 illustrates two dependent populations of scores representing before- and after-lunch weights. These distributions have means of $\mu_B = 149$ and $\mu_A = 151$, respectively, and both distributions have the same standard deviation of $\sigma_B = \sigma_A = 15$. Five individuals

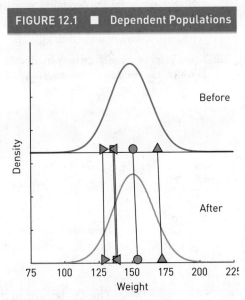

FIGURE 12.1 ■ Dependent Populations

The before and after distributions are both normal and have $\sigma_B = \sigma_A = 15$. The mean of the before distribution (μ_B) is 149 and the mean of the after distribution (μ_A) is 151. This means that, on average, people weigh 2 pounds more after lunch than before lunch. The individual symbols show the scores of five individuals from these dependent populations. Participant 1 = ●, participant 2 = ◀, participant 3 = ■, participant 4 = ▶, and participant 5 = ▲.

LEARNING CHECK 1

1. Which of the following are dependent samples of scores:

   (a) maze-solving times for 50 rats randomly assigned to a control condition and an experimental condition,

   (b) IQ scores from 34 monozygotic (identical) twins,

   (c) IQ scores from 34 dizygotic (non-identical) twins,

   (d) heart rates for 83 individuals before and after playing *Grand Theft Auto* for 25 minutes,

   (e) math scores from two random samples of eighth-grade students from Ontario, or

   (f) scores on a measure of social conservatism for pairs of adult siblings who have lived in two different municipalities for at least 10 years?

2. State whether the following statements are true or false.

   (a) I have two dependent samples of scores for which $m_1 - m_2 = 3.1415$; therefore, $m_D = 3.1415$.

   (b) A 95% confidence interval will generally be wider when computed for dependent samples than for independent samples.

   (c) s_D will always be larger than s_{m_D}.

3. Seven pairs of identical twins participated in a study in which they were asked to complete a Sudoku puzzle while traveling on the subway. Both members of each pair were seated next to a confederate (a person who, unknown to the participant, was part of the experiment). One confederate talked on her cell phone throughout the trip, and the other confederate listened to very loud death metal music throughout the trip. The following table shows the numbers of correctly solved items. Compute the following:

   (a) s_D^2 (compute the differences as *metal music – conversation*),

   (b) s_{m_D},

   (c) the 95% confidence interval around m_D.

Conversation	Metal Music
12	16
10	10
7	10
8	10
6	7
9	10
8	11

Answers

1. b, c, d, f.

2. (a) True. (b) False. (c) True.

3. (a) 2. (b) .5345.

   (c) $m_D \pm t_{\alpha/2}(s_{m_D}) = 2 \pm 2.447(0.5345) = [.69, 3.31]$.

have been selected at random, which means we have five (dependent) pairs of scores drawn from these dependent distributions. Each pair of scores is shown as a different symbol in Figure 12.1.

The Distribution of Difference Scores

The density function in Figure 12.2 shows the distribution of difference scores (*D*) for all individuals in the dependent populations shown in Figure 12.1. This distribution is much narrower than the distributions of weights shown in Figure 12.1; compare the *x*-axes in Figures 12.1 and 12.2. The variability among the difference scores represents only the variability in weight gained at lunch. The difference scores for the five individuals shown in Figure 12.1 are also shown in Figure 12.2 for reference.

The mean of the population of difference scores is

$$\mu_D = \frac{\sum D}{N},$$

(12.3)

which we noted before is equivalent to $\mu_1 - \mu_2$. We've already seen that μ_D is estimated by

$$m_D = \frac{\sum D}{n}, \tag{12.4}$$

which is the mean of the difference scores in a sample. The variance of the population of difference scores is

$$\sigma_D^2 = \frac{\sum(D - \mu_D)^2}{N}. \tag{12.5}$$

This is estimated by

$$s_D^2 = \frac{\sum(D - m_D)^2}{n-1}, \tag{12.6}$$

which is the variance of the difference scores in a sample.

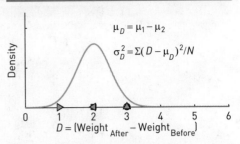

FIGURE 12.2   ■   Distribution of D Scores

The differences between the before- and after-lunch weights have been computed for each individual. We call this difference score D. The mean of this distribution is equal to $\mu_A - \mu_B$ and the variance is $\sigma_D^2 = \sum(D - \mu_D)^2/N$. Participant 1 = ●, participant 2 = ◀, participant 3 = ■, participant 4 = ▶, and participant 5 = ▲.

The Sampling Distribution of m_D

The sample of difference scores that we've been considering (i.e., those in Table 12.1) is one of an infinite number of samples (with $n = 5$) that could be drawn from the distribution of differences (D). Therefore, the mean of our sample, m_D, has been drawn from the distribution of all possible m_D values. The sampling distribution of m_D has a mean

$$\mu_{m_D} = \mu_D = \mu_1 - \mu_2, \tag{12.7}$$

a variance

$$\sigma_{m_D}^2 = \frac{\sigma_D^2}{n}, \tag{12.8}$$

and a standard error

$$\sigma_{m_D} = \sqrt{\frac{\sigma_D^2}{n}}. \tag{12.9}$$

Therefore, when we constructed our confidence interval, the quantity computed in equation 12.2 (i.e., s_{m_D}) was an estimate of σ_{m_D}.

We can now see that constructing a confidence interval around m_D follows exactly the same logic as constructing a confidence interval around m, as described in Chapter 10. In the case of m, our sample was a random sample of scores from a normal distribution of scores. In the case of m_D, our sample was a random sample from a normal distribution of difference scores.

A Second Look at the Variance of the Distribution of Difference Scores

Let's have a final look at the distribution of difference scores and return to the question of subject variability unrelated to weight gain. Table 12.3 reproduces the data in Table 12.2 but adds a final column labeled "Mean." The numbers in this column are the means of the before- and after-lunch weights ([*before* + *after*]/2). For example, participant 3 has before- and after-lunch weights of 137 and 139, respectively. Therefore, the mean weight for participant 3 is (137 + 139)/2 = 138.

TABLE 12.3 ■ Subject Means				
Participant	Before	After	D	Mean
1	151	154	3	152.5
2	136	138	2	137.0
3	137	139	2	138.0
4	128	129	1	128.5
5	169	172	3	170.5
Mean	144.2	146.4	2.2	145.3
Variance	260.7	285.3	0.70	272.825

At the bottom of the "Mean" column are the mean and variance of the mean weights. Notice that the mean of the mean weights is equal to the mean of the before- and after-lunch means; i.e., $(144.2 + 146.4)/2 = 145.3$. More interesting to us is how the variance of the mean weights relates to s_D^2, the variance of the sample of difference scores.

First, let's call the variance of the mean weights s_S^2, where the subscript "S" denotes *subjects*. Now, here is the interesting thing: the variance of the sample of difference scores is related to s_A^2, s_B^2, and s_S^2 as follows:

$$s_D^2 = 2\left(\left(s_B^2 - s_S^2\right) + \left(s_A^2 - s_S^2\right)\right). \tag{12.10a}$$

The important point is that variance associated with the mean weights is *subtracted* from the sample variances. We can rewrite equation 12.10a in a simpler form as follows:

$$s_D^2 = 2\left(s_B^2 + s_A^2 - 2s_S^2\right). \tag{12.10b}$$

If we insert the numbers from Table 12.3 into equation 12.10b, we can verify that it produces the result we obtained before:

$$s_D^2 = 2\left(s_B^2 + s_A^2 - 2s_S^2\right) = 2(260.7 + 285.3 - 2*272.825) = 0.7.$$

That is, $s_D^2 = 0.7$ as we saw before. Therefore, equation 12.10b shows that the variance of the distribution of difference scores eliminates variability unrelated to the weight gain s_S^2.

Although equation 12.10b deals with the variance of a sample of difference scores, the same relationship holds for the population of difference scores. Specifically,

$$\sigma_D^2 = 2\left(\sigma_B^2 + \sigma_A^2 - 2\sigma_S^2\right). \tag{12.11}$$

We can now see exactly why estimating $\mu_1 - \mu_2$ with dependent samples provides much more precise estimates than do estimates derived from independent samples. This is generally a good thing, but it does come with some risks, as we'll discuss in the next section.

LEARNING CHECK 2

1. State whether the following statements are true or false.

   (a) A distribution of difference scores will generally be wider than either of the populations from which dependent samples were drawn.

   (b) All things being equal, σ_D^2 will decrease as σ_S^2 increases.

   (c) $\sigma_D^2 = \sum(D - m_D)^2/(n-1)$.

   (d) $s_D^2 = 2(\sigma_1^2 + \sigma_2^2 - 2\sigma_S^2)$.

2. Calculate s_D^2 for each of the following situations:

   (a) $s_1 = 10, s_2 = 10, s_S = 2$;

   (b) $s_1 = 10, s_2 = 10, s_S = 9$;

   (c) $s_1 = 11, s_2 = 11, s_S = 10$.

Answers

1. (a) False.

   (b) True (see equation 12.11).

   (c) False. The right side of the equation defines s_D^2.

   (d) False. The right side of the equation defines σ_D^2.

2. $s_D^2 = 2\left(s_B^2 + s_A^2 - 2s_S^2\right)$ equals the following:

   (a) $2(100 + 100 - 8) = 384$.

   (b) $2(100 + 100 - 162) = 76$.

   (c) $2(121 + 121 - 200) = 84$.

REPEATED MEASURES AND MATCHED SAMPLES

There are different ways that populations can be dependent. The example we've been discussing is the most obvious case of dependent samples. We saw in Chapter 11 that when two measures are taken from an individual, we have what is called a **repeated-measures design**. As we've just seen, repeated-measures designs permit very precise estimates of differences by factoring out irrelevant subject variability.

Repeated-measures designs have some obvious potential problems, however. For example, imagine that some wonder drug is supposed to improve intelligence immediately. To assess the drug, you will give a randomly chosen group of participants a puzzle-solving task twice: once before taking the drug and once after taking the drug. Your goal is to measure the change in puzzle-solving speed after taking the drug. If you find participants solve puzzles faster after taking the drug, it might be that the drug really worked. That would be great! On the other hand, and perhaps more realistically, participants may solve puzzles faster because they are seeing them for the second time. This kind of problem is sometimes called a **carryover effect**.

The wonder drug example is a glaring case of a carryover effect. Others might be more subtle. For example, a rat receiving a reward on one trial may be less motivated to behave for a reward on a subsequent trial; a rat can eat just so much kibble before it becomes sated. One must always be on guard for this kind of issue when using a repeated-measures design.

An alternative way to obtain dependent samples without the dangers of carryover effects is the **matched-samples design**. In this design, we compute difference scores (D) from pairs of individuals. Pairs of participants in the matched-samples design are chosen to be similar on task-relevant variables. So, continuing with the wonder drug example, let's say you have a large pool of potential participants available. For each participant chosen, you match that participant with another that is similar on variables of interest. If a given participant is a 23-year-old woman who weighs 130 pounds and has an IQ of 122, you would pair, or match, this participant to another that has similar scores on these variables. IQ is probably the most important feature to match because that is the dependent variable

A **repeated-measures design** involves scores obtained from participants on more than one occasion. That is, we've taken repeated measures from the same participants.

A **carryover effect** occurs when obtaining measurements on previous trials affect the measurements made on subsequent trials.

The **matched-samples design** creates dependent samples without measuring individuals twice. Individuals in two samples are matched on features that might be associated with the dependent variable, such as age, sex, socioeconomic status, IQ, health status, income, height, weight, or empathy, to name just a few that might be important in psychological research.

in this study. Weight could be important as well because the concentration of the drug in the bloodstream will depend on weight. Sex is probably important, and age maybe less so; however, one probably shouldn't match a 23-year-old woman to a 71-year-old woman because a big age difference in itself could produce a big performance difference, thus defeating the purpose of matching. Once you have a sample of matched participants, you give one member of each pair the drug and the other a placebo. The puzzle-solving task is then given to all participants, and difference scores are computed for each matched pair.

The matched-samples design capitalizes on the dependencies between pairs of scores but avoids the problem of carryover effects. However, one needs twice as many participants in the matched-samples procedure. Furthermore, the matched-samples procedure typically reduces subject variability less than the repeated-measures procedure, because there is no better match to an individual than the individual himself. Therefore, estimates are likely to be more precise in the repeated-measures design. However, increased precision has to be balanced against the possible confounding consequences of carryover effects. Nevertheless, both dependent-samples designs allow us to make more precise estimates of $\mu_1 - \mu_2$ than the independent-samples design.

LEARNING CHECK 3

1. Do all studies using dependent samples involve repeated measures? Explain your answer.

2. What is a carryover effect? Use the following example to illustrate your answer. Let's say you think that negative imagery (reliving past negative experiences) will reduce physical strength. To test your theory, you measure the maximum number of pushups that a participant can do before and after the negative imagery.

3. What are the advantages and disadvantages of repeated-measures designs versus matched-samples designs?

Answers

1. No. A matched-samples design does not involve repeated measures from the same individuals. For example, a matched-samples design might measure the effect of a pharmacological intervention on participants matched for age, sex, and weight. Each participant is measured only once.

2. A carryover effect occurs in repeated-measures designs when taking one measurement affects a subsequent measurement. If the number of pushups is reduced after the negative imagery, this may be a consequence of the exertion before the negative imagery, rather than the negative imagery itself.

3. The repeated-measures design will produce more precise estimates with fewer participants than the matched-samples design. However, the repeated-measures design is subject to carryover effects, whereas the matched-samples design is not.

ESTIMATING δ FOR DEPENDENT POPULATIONS

As with the independent-samples design, we can express the difference between scores on the two variables using a standardized effect size. The definition of the effect size for dependent populations is

$$\delta = \frac{\mu_1 - \mu_2}{\sigma_D} = \frac{\mu_D}{\sigma_D},$$

(12.12)

where μ_D is the mean of the distribution of difference scores as before, and σ_D is the standard deviation of the distribution. Of course, an estimate of δ is easily derived from sample statistics, as follows:

$$d = \frac{m_D}{s_D}. \tag{12.13}$$

An Approximate Confidence Interval for δ

To place an approximate confidence interval around an estimated effect size, we use the following formula:

$$d \pm z_{\alpha/2}(s_d). \tag{12.14}$$

For an approximate 95% confidence interval $z_{\alpha/2}$ would be 1.96. The estimated standard error is the same as the estimated standard error for $d = (m - \mu)/s$ that we saw in Chapter 10. The formula is as follows:

$$s_d = \sqrt{\frac{d^2}{2df} + \frac{1}{n}}. \tag{12.15}$$

In this case, $df = n-1$, where n is the number of pairs.

Equations 12.13, 12.14, and 12.15 can be used to compute an approximate 95% confidence interval around d computed from the lunch data in Table 12.2. To compute an approximate 95% confidence interval around d, we would do the following:

Step 1. Compute s_D. We previously found that $s_D^2 = 0.7$; therefore, s_D is simply $\sqrt{0.7} = .837$.

Step 2. Compute d, using equation 12.13.

$$d = \frac{m_D}{s_D} = \frac{2.2}{0.837} = 2.628.$$

Step 3. Compute s_d, using equation 12.15. From Step 2, we know that $d = 2.628$. We also know that $n = 5$ and $df = n-1 = 4$. Therefore,

$$s_d = \sqrt{\frac{d^2}{2df} + \frac{1}{n}} = \sqrt{\frac{2.628^2}{2*4} + \frac{1}{5}} = 1.031.$$

Step 4. Compute the approximate confidence interval.

$$CI = m_D \pm z_{\alpha/2}(s_d) = 2.628 \pm 1.96(1.031) = [.61, 4.65].$$

Therefore, we have approximately 95% confidence that δ lies in the interval $[.61, 4.65]$. Our confidence comes from knowing that approximately 95% of all intervals computed this way will capture δ.

An Exact Confidence Interval for δ

An exact confidence interval around d is computed in **R** exactly like the confidence interval around a standardized mean as described in Appendix 10.3. At the **R** prompt, we type the first line shown in the following code fragment. (Make sure that the MBESS package has been loaded in the Package Manager.) In this case, the standardized mean (i.e., *d*) is $sm = 2.628$, sample size is $N = 5$, and `conf.level` = .95. The output shows that the exact 95% confidence interval is [.6610099, 4.565259], which is not far off from our approximate confidence interval. However, it is worth noting that approximate intervals are least accurate when *d* is large and *n* is small.

```
> ci.sm(sm=2.628, N=5, conf.level=.95)
[1] "The 0.95 confidence limits for the standardized mean are given as:"
$Lower.Conf.Limit.Standardized.Mean
[1] 0.6610099
$Standardized.Mean
[1] 2.628
$Upper.Conf.Limit.Standardized.Mean
[1] 4.565259
```

Estimating δ for Dependent Versus Independent Populations

It is important to note that the effect size estimated for dependent samples is different from that estimated for independent samples; therefore, caution must be exercised when comparing them. The difference arises from the nature of the distributions in question. The effect size is $\delta_{ind} = (\mu_1 - \mu_2)/\sigma$ for independent populations and $\delta_{dep} = (\mu_1 - \mu_2)/\sigma_D$ for dependent populations. Because σ_D is generally smaller than σ, δ_{dep} will generally be larger than δ_{ind}.

Let's think through an example of comparing estimated effect sizes derived from dependent- and independent-samples designs. Many researchers have wondered what it is that makes a human face attractive. One argument is that faces are judged to be more attractive when they are closer to the average face. Support for this view comes from results

---

FIGURE 12.3 ■ Average Faces

(a)

(b)

(a) An average non-celebrity created and made available by Lisa Debruine and Benedict Jones (2017) (b) An average of celebrity faces derived from materials available at faceresearch.org/demos/famous using the Webmorph.org application provided by Lisa DeBruine.

(a) DeBruine, Lisa (2016). Young adult composite faces. figshare. (b) Based on images from DeBruine (2016) using faceresearch.org and webmorph.org.

showing that participants typically judge the faces of individuals to be less attractive than faces that are the average of several individuals (Langlois & Roggman, 1990). Figure 12.3 shows two average faces. Figure 12.3a shows the average of non-celebrity, young female faces (Debruine & Jones, 2017) and Figure 12.3b shows the average of celebrity female faces, derived from materials available at faceresearch.org/demos/famous. Typically, average faces are seen as more attractive than any of the individual faces contributing to the average.

So let's imagine two experiments in which participants rate the attractiveness of faces on a 10-point scale, with 10 representing maximum attractiveness. In both experiments, the stimuli are either (i) individual faces or (ii) averaged faces (see Figure 12.3). Faces of female celebrities were shown in Study 1, and faces of female non-celebrities were shown in Study 2. Study 1 was run with independent groups and Study 2 was run with dependent groups.

Study 1. Participants in one group rated individual celebrity faces, and participants in the other group rated faces that were averages of 32 celebrity faces. The results showed that $m_{average} = 7.2$, $m_{individual} = 6$, $s_{pooled} = 3$. Therefore, $d_{ind} = .4$.

Study 2. Participants rated both individual non-celebrity faces and averages of 32 non-celebrity faces. The results showed that $m_{average} = 6.2$, $m_{individual} = 5$, $s_D = .8$. Therefore, $d_{dep} = 1.5$.

From these two results, one might be tempted to conclude that the difference between the average and individual faces is greater for non-celebrity faces ($d_{dep} = 1.5$) than for celebrity faces ($d_{ind} = .4$). Although this is tempting to consider, we must keep in mind that the difference between $m_{average}$ and $m_{individual}$ was 1.2 in both cases. The two effect sizes differ because d_{dep} is based on a denominator (s_D) that has eliminated irrelevant subject variability, whereas d_{dep} is based on a denominator (s_{pooled}) that has not. Therefore, great care should be taken when comparing these effect sizes. Effect sizes computed from independent and dependent samples should not be combined in a meta-analysis.

SIGNIFICANCE TESTING

Our main goal in this chapter has been to estimate μ_D when our samples come from dependent populations. We've noted, however, that many researchers ask the narrower question of whether some measured difference is different from zero. To answer this question, they employ significance tests. For our lunch example, the null hypothesis is simply that there is no difference between the before- and after-lunch weights. The null hypothesis is

$$H_0: \mu_D = 0,$$

and the alternative hypothesis could be

$$H_1: \mu_D \neq 0.$$

Significance Testing With Confidence Intervals

We can test the null hypothesis that $\mu_D = 0$ by asking whether the confidence interval around m_D includes 0. We saw previously that the 95% confidence interval around m_D was [1.16, 3.24]. Because this interval does not include 0, we can *reject* H_0. This means that our computed confidence interval would be very unusual if H_0 were true, so we treat H_0 *as though* it is false. Of course, in our lunch example, there is no chance that H_0 is true, so significance testing makes little sense in this case. Furthermore, many would argue that the null hypothesis is never true, so rejecting it is simply a matter of choosing sufficiently large samples.

Significance Testing With *t*-Statistics

We can also test the null hypothesis by computing t_{obs}. As you might expect, we compute t_{obs} as follows:

$$t_{obs} = \frac{m_D}{s_{m_D}}. \tag{12.16}$$

Therefore, if we were to hypothesize that there is no difference in the before- and after-lunch weights, we could compute t_{obs} as follows:

$$t_{obs} = \frac{m_D}{s_{m_D}} = \frac{2.2}{0.374} = 5.882.$$

If we consult Excel using the **T.DIST** function, we find that for $t(4) = 5.882$, $p = .0021$ for a one-tailed test, and $t(4) = 5.882$, $p = .0042$ for a two-tailed test.

LEARNING CHECK 4

1. Estimates of δ will generally be larger for dependent than independent samples. [True, False]

2. A dependent-samples design showed the following: $m_D = 5$, $s_1^2 = 10$, $s_2^2 = 10$, $s_S^2 = 9$, and $n = 10$. Compute the 95% confidence interval around the estimate of δ. (*Hint:* See equation 12.10.)

3. Use the data from question 2 to compute a 95% confidence interval around m_D. If H_0: $\mu_D = 0$, should we retain or reject H_0?

4. Use the data from question 2 to compute t_{obs}.

Answers

1. True.

2. CI = 2.5 ± 1.96(0.669) = [1.19, 3.81].

3. CI = 5 + 2.262(0.632) = [3.57, 6.43]. Because this confidence interval does not contain 0, we should reject H_0.

4. $t_{obs} = m_D/s_{m_D} = 5/0.632 = 7.91$. (Gadzooks, that's big!)

PARTITIONING VARIANCE

In Chapter 11 we saw how to partition the variability in two samples of scores into between-group and within-group variance. We can take this analysis a step further in the dependent-groups design. Rather than partitioning variance into just two sources, we can partition it into three sources: (i) between-group variability attributable to the treatment (independent variable), (ii) variability attributable to subjects, and then (iii) the variability remaining when these two sources have been accounted for.

The column labeled "Scores" in Table 12.4 rearranges the data from our five lunchers into a single column. If you compare these data with those in Table 12.1, you will see that the before and after scores are the same in both tables. For example, in both tables, the weight of subject 1 is 151 pounds before lunch and 154 pounds after lunch. At the bottom of this column, you will see the sum of squared deviations from the overall mean. We call this ss_{total}, and it is this that we wish to partition. As you can see, $ss_{total} = 2196.1$.

TABLE 12.4 ■ Partitioning Variance

Condition	Subject	Scores	Mean (C)	Mean (S)
Before	1	151	144.2	152.5
Before	2	136	144.2	137.0
Before	3	137	144.2	138.0
Before	4	128	144.2	128.5
Before	5	169	144.2	170.5
After	1	154	146.4	152.5
After	2	138	146.4	137.0
After	3	139	146.4	138.0
After	4	129	146.4	128.5
After	5	172	146.4	170.5
	Mean	145.3	145.3	145.3
	ss	2196.1	12.1	2182.6

In Chapter 11 we determined the proportion of ss_{total} attributable to group means by replacing each score with the mean of its group and then computing the sum of squares. Exactly the same thing has been done in the column labeled "Mean (C)" in Table 12.4. In this case, C stands for condition. Therefore, each subject's before-lunch weight has been replaced with the mean of the before-lunch weights (144.2) and each subject's after-lunch weight has been replaced with the mean of the after-lunch weights (146.4). At the bottom of this column is the sum of squared deviations from the overall mean, which we call ss_C. As you can see, $ss_C = 12.1$, which represents a very small proportion of the total variability in the 10 scores. The exact proportion of ss_{total} explained by the difference between the means of the two conditions is

$$r_C^2 = ss_C/ss_{total} = 12.1/2196.1 = .0055.$$

We can now ask what proportion of ss_{total} is accounted for by variability in the average subject weights. In the column labeled "Mean (S)" of Table 12.4, each subject's weight in the two conditions (before and after lunch) has been replaced with the subject's mean weight ([*before* + *after*]/2). For example, subject 1 weighs 151 pounds before lunch and 154 after lunch. Therefore, the average of these two weights is $(151 + 154)/2 = 152.5$. This value has been placed in the two rows of Table 12.4 corresponding to subject 1. At the bottom of this column, you will see the sum of squared deviations from the overall mean, which we call ss_S. As you can see, $ss_S = 2182.6$, which represents a very large proportion of the total variability in the 10 scores. The exact proportion of ss_{total} explained by the variability among subjects is

$$r_S^2 = ss_S/ss_{total} = 2182.6/2196.1 = .9939.$$

Now the question is whether there is any variability left over after we've accounted for ss_C and ss_S. The answer is yes, and we can calculate the residual (or error) variability as

$$ss_{error} = ss_{total} - (ss_C + ss_S) = 2196.1 - (12.1 + 2182.6) = 1.4.$$

Therefore, we see that ss_{total} can be divided into three components:

$$ss_{total} = ss_C + ss_S + ss_{error}.$$

For our example,

$$ss_{total} = ss_C + ss_S + ss_{error} = 12.1 + 2182.6 + 1.4 = 2196.1.$$

The F-Statistic for Dependent Samples

We saw in Chapter 11 that the F-statistic is derived from the quantities obtained through partitioning variance. The F-statistic for dependent samples is computed as follows:

$$F = \frac{ss_C}{ss_{error}}(df_{error}). \tag{12.17}$$

In this case, $df_{error} = n-1$, or 4. If we use the numbers obtained above for ss_C and ss_{error}, we obtain the following:

$$F = \frac{ss_C}{ss_{error}}(df_{error}) = \frac{12.1}{1.4}(4) = 34.5714.$$

In Chapter 11 we also noted that F is the square of the t-statistic; so by taking the square root of F, we obtain t_{obs} as follows:

$$t_{obs} = \sqrt{F} = \sqrt{34.5714} = 5.88.$$

If you look back to the end of the previous section, you will see that our computed t_{obs} was 5.88. We will say much more about the F-statistic in Part IV of this book. For the moment, we will just note that the question of partitioning variance relates directly to many tests of statistical significance.

The F-Statistic for Independent Samples (Review)

As a final note, we will calculate F for the independent-groups design. In this case, we do not remove subjects as a source of variance. Therefore, the formula we use to compute F is as follows:

$$F = \frac{ss_C}{ss_{error}}(df_{within}). \tag{12.18}$$

So, let's pretend for a moment that the 10 scores we've been working with are two independent samples of five scores each. In this case, ss_C is again 12.1, but

$$ss_{error} = ss_{total} - ss_C = 2196.1 - 12.1 = 2184$$

That is, ss_{error} in this case includes subject variability, ss_S. In the two-independent-groups design, we have $n_1 + n_2 - 2 = 5 + 5 - 2 = 8$ degrees of freedom, so $df_{within} = 8$. Therefore,

$$F = \frac{ss_C}{ss_{error}}(df_{within}) = \frac{12.1}{2184}8 = 0.04432.$$

Once again, this F-statistic is the square of the corresponding t-statistic, so that

$$t_{obs} = \sqrt{F} = \sqrt{0.04432} = 0.2105.$$

We can check that this answer is correct by referring to the first calculation we did in this chapter. When we computed the 95% confidence interval around the difference between two means (treated as independent), we found that $s_{m_1 - m_2} = 10.45$. If we divide the difference between the two means by this standard error, we get

$$t_{obs} = (m_1 - m_2)/s_{m_1 - m_2} = 2.2/10.45 = 0.2105.$$

This shows, as before, that t_{obs} is directly related to explained variance.

LEARNING CHECK 5

1. Explain why ss_C/ss_{error} will be larger for a dependent-samples design than when the same data are treated as an independent-samples design.

2. Would $F = 16$ be statistically significant at $\alpha = .05$ for a dependent-samples design if $n = 10$?

Answers

1. For the dependent-samples design, $ss_{error} = ss_{total} - (ss_C + ss_S)$. For the independent-samples design, $ss_{error} = ss_{total} - ss_C$. Therefore, ss_C/ss_{error} will be greater for the dependent-samples design because ss_{error} is smaller.

2. $F = 16$ corresponds to $t_{obs} = 4$. If $n = 10$, we have 9 degrees of freedom, and thus $t_{critical} = 2.262$ for a

two-tailed test. Therefore, $t_{obs} = 4$ is statistically significant for a two-tailed test, and it is statistically significant for a one-tailed test that predicted a positive difference.

SUMMARY

In this chapter we saw how to compute a confidence interval around the difference between two dependent means when the population standard deviations are unknown. To eliminate variability unrelated to the difference of interest, we compute difference scores $D = x_1 - x_2$, where x_1 and x_2 represent a dependent pair of scores. The population of difference scores has mean μ_D, which is equivalent to the difference between the means of the dependent populations, $\mu_1 - \mu_2$. The variance of the distribution of difference scores is

$$\sigma_D^2 = \frac{\Sigma(D - \mu_D)^2}{N},$$

where N is the number of difference scores in the population. Except in unusual circumstances, σ_D^2 will be smaller than the variances, σ_1^2 and σ_2^2, of the two populations treated separately. We can express σ_D^2 as follows:

$$\sigma_D^2 = 2(\sigma_1^2 + \sigma_2^2 - 2\sigma_S^2),$$

where σ_S^2 is the variance of the mean scores for subjects. In this way, we can see that σ_D^2 eliminates the variance that is unrelated to the difference of interest.

The average difference score in a sample is

$$m_D = \Sigma D/n,$$

where n is the number of dependent pairs. The mean difference (m_D) is the same as the difference between the two sample means ($m_1 - m_2$). Both m_D and $m_1 - m_2$ estimate $\mu_1 - \mu_2$, which is the difference between the two dependent population means. The variance of the sample difference scores is

$$s_D^2 = \frac{\Sigma(D - m_D)^2}{n-1},$$

which estimates σ_D^2. We can also write the variance of the sample difference scores as

$$s_D^2 = 2(s_1^2 + s_2^2 - 2s_S^2).$$

The sampling distribution of m_D has mean $\mu_1 - \mu_2$, and standard error

$$\sigma_{m_D} = \sqrt{\frac{\sigma_D^2}{n}}.$$

Therefore, the estimated standard error of m_D is

$$s_{m_D} = \sqrt{\frac{s_D^2}{n}}.$$

The $(1-\alpha)100\%$ confidence interval is computed as follows:

$$m_D \pm t_{\alpha/2}(s_{m_D}).$$

Our confidence in this interval comes from knowing that $(1-\alpha)100\%$ of all intervals computed this way will capture $\mu_1 - \mu_2$.

The effect size for dependent populations is

$$\delta = \frac{\mu_1 - \mu_2}{\sigma_D} = \frac{\mu_D}{\sigma_D},$$

which is estimated by

$$d = \frac{m_D}{s_D}.$$

An approximate confidence interval for δ is computed as

$$d \pm z_{\alpha/2}(s_d),$$

where

$$s_d = \sqrt{\frac{d^2}{2df} + \frac{1}{n}}.$$

In this case, $df = n-1$, where n is the number of dependent pairs. The estimated effect size (d) for

dependent samples will typically be larger than d for independent samples because s_D is typically smaller than s_{pooled}. Therefore, we must be cautious when comparing or combining estimated effect sizes derived from different studies.

We can use a confidence interval around m_D to conduct a significance test by simply checking to see if 0 falls in the interval. If not, we can reject H_0. We can also compute a t-statistic as follows:

$$t_{\text{obs}} = \frac{m_D}{s_{m_D}}.$$

In this case, t_{critical} is based on $df = n-1$, α, and H_1.

Finally, we revisited the question of partitioning variance. In the dependent-groups design, there are three sources of variability. These are (i) variability attributable to the treatment conditions (ss_C), (ii) variability attributable to subjects (ss_S), and (iii) error variability (ss_{error}). Therefore, the total variability (ss_{total}) in n pairs of scores can be described as

$$ss_{\text{total}} = ss_C + ss_S + ss_{\text{error}}.$$

The proportion of ss_{total} explained by treatment conditions is

$$r_C^2 = \frac{ss_C}{ss_{\text{total}}},$$

and the proportion of ss_{total} explained by subjects is

$$r_S^2 = \frac{ss_S}{ss_{\text{total}}}.$$

The F-statistic for dependent samples is defined as

$$F = \frac{ss_C}{ss_{\text{error}}}(df_{\text{error}}) = \frac{ss_C}{ss_{\text{total}} - ss_C - ss_S}(df_{\text{error}}),$$

where $df_{\text{error}} = n-1$. There is a simple relationship between F and t_{obs}:

$$t_{\text{obs}} = \sqrt{F}.$$

KEY TERMS

EXERCISES

Definitions and Concepts

1. What is the difference between independent and dependent populations?

2. What is meant by subject variability?

3. What is a repeated-measures design?

4. What is a matched-samples design?

5. What is a carryover effect?

6. What are the three components of ss_{total} in a dependent-samples design?

7. What is the difference between σ_{m_D} and s_{m_D}?

True or False

State whether the following statements are true or false.

8. s_{m_D} is the standard error of the sampling distribution of mean differences.

9. $\mu_D = \mu_1 - \mu_2$.

10. $\sigma_D^2 = 2(\sigma_1^2 + \sigma_2^2 - 2\sigma_s^2)$.

11. $\sigma_{m_D}^2 = \sigma_D^2/n$.

12. $\sigma_{m_D}^2 = \sigma_D^2/n = 2(\sigma_1^2 + \sigma_2^2 - 2\sigma_s^2)/n$.

13. $\sigma_{m_D}^2$ is always greater than σ_1^2.

14. $F = \sqrt{t}$.

15. $F = t^2$.

16. $d = \mu_D / \sigma_D$.

17. $\delta = m_D / s_D$.

18. $d = \dfrac{m_D}{2(s_1^2 + s_2^2 - 2s_s^2)}$.

19. A matched-samples design will generally yield narrower confidence intervals than a repeated-measures design.

20. A matched-samples design is not susceptible to carryover effects.

21. $(\mu_1 - \mu_2)/\sigma_D = (\mu_1 - \mu_2)/\sigma_{m_1 - m_2}$

22. We have 95% confidence in $m_D \pm t_{\alpha/2}(s_{m_D})$ because we know that 95% of all such intervals will capture m_D.

23. We have 95% confidence in $d \pm z_{\alpha 2}(s_d)$ because we know that approximately 95% of all such intervals will capture δ.

Calculations

24. For each of the following parameters, compute the mean and variance of the sampling distribution of the difference between two dependent means.

    (a) $\mu_1 = 10$, $\mu_2 = 12$, $\sigma_D = 6$, $n = 15$
    (b) $\mu_1 = 15$, $\mu_2 = 8$, $\sigma_D = 16$, $n = 4$
    (c) $\mu_1 = 25.6$, $\mu_2 = 21.9$, $\sigma_D = 8.8$, $n = 121$
    (d) $\mu_1 = 43$, $\mu_2 = 42$, $\sigma_D = 31$, $n = 97$
    (e) $\mu_1 = 102$, $\mu_2 = 101$, $\sigma_D = 15$, $n = 225$
    (f) $\mu_1 = 68$, $\mu_2 = 63$, $\sigma_D = 17$, $n = 12$

25. For each of the following scenarios, (i) calculate the 95% confidence interval around $m_1 - m_2$, (ii) compute the approximate 95% confidence interval around d, (iii) compute t_{obs}, and (iv) say whether the difference between m_1 and m_2 is statistically significant at the $p < .05$ level, assuming a nondirectional test.

    (a) $m_1 = 18$, $m_2 = 15$, $s_D = 18$, $n = 9$
    (b) $m_1 = 69$, $m_2 = 72$, $s_D = 24$, $n = 16$
    (c) $m_1 = 88$, $m_2 = 101$, $s_D = 48$, $n = 64$
    (d) $m_1 = 17$, $m_2 = 14.2$, $s_D = 6$, $n = 25$
    (e) $m_1 = 99$, $m_2 = 98$, $s_D = 1$, $n = 25$
    (f) $m_1 = 1000$, $m_2 = 1001$, $s_D = 20$, $n = 1600$

26. For each of the following parameters, compute the mean and variance of the sampling distribution of the difference between two dependent means.

    (a) $\mu_1 = 10$, $\mu_2 = 12$, $\sigma_1 = 6$, $\sigma_2 = 6$, $\sigma_S = 4$, $n = 25$
    (b) $\mu_1 = 12$, $\mu_2 = 11$, $\sigma_1 = 6$, $\sigma_2 = 6$, $\sigma_S = 5$, $n = 16$
    (c) $\mu_1 = 88$, $\mu_2 = 91$, $\sigma_1 = 16$, $\sigma_2 = 16$, $\sigma_S = 10$, $n = 36$
    (d) $\mu_1 = 100$, $\mu_2 = 88$, $\sigma_1 = 40$, $\sigma_2 = 40$, $\sigma_S = 36$, $n = 64$
    (e) $\mu_1 = 67$, $\mu_2 = 65$, $\sigma_1 = 10$, $\sigma_2 = 10$, $\sigma_S = 5$, $n = 225$
    (f) $\mu_1 = 20$, $\mu_2 = 19$, $\sigma_1 = 18$, $\sigma_2 = 18$, $\sigma_S = 26$, $n = 25$ (There is something unusual about this question. Can you find a rule that explains what has happened here?)

27. For each of the following scenarios, (i) calculate the 95% confidence interval around $m_1 - m_2$, (ii) compute the approximate 95% confidence

interval around d, (iii) compute t_{obs}, and (iv) state whether the difference between m_1 and m_2 is statistically significant at the $p < .05$ level assuming a nondirectional test.

(a) $m_1 = 10$, $m_2 = 12$, $s_1 = 6$, $s_2 = 6$, $s_S = 4$, $n = 25$

(b) $m_1 = 82$, $m_2 = 79$, $s_1 = 36$, $s_2 = 36$, $s_S = 25$, $n = 15$

(c) $m_1 = 41$, $m_2 = 43$, $s_1 = 24$, $s_2 = 24$, $s_S = 21$, $n = 144$

(d) $m_1 = 88$, $m_2 = 84$, $s_1 = 50$, $s_2 = 50$, $s_S = 40$, $n = 384$

(e) $m_1 = 1234$, $m_2 = 999$, $s_1 = 300$, $s_2 = 300$, $s_S = 290$, $n = 225$

(f) $m_1 = 74.5$, $m_2 = 72$, $s_1 = 8$, $s_2 = 8$, $s_S = 7$, $n = 64$

Scenarios

28. The following data show the results of a matched-samples design. Half of the participants were in a control condition (C) and the other half were in a treatment condition (T). Participants on the same row were matched for IQ.

Participant (C)	Control	Participant (T)	Treatment
1	151	6	154
2	136	7	138
3	137	8	139
4	128	9	129
5	169	10	172
Mean	144.2		146.4
Variance	260.7		285.3

(a) Compute the 95% confidence interval around m_D.

(b) Compute the approximate 95% confidence interval around d.

(c) Compute t_{obs} and the two-tailed p-value associated with t_{obs}.

(d) State whether m_D is statistically significant at the $p < .05$ level.

29. The following data show the results of a repeated-measures design. Each participant was in both the control condition and treatment condition.

Participant	Control	Treatment
1	95	80
2	93	71
3	95	91
4	87	91
5	90	60
6	92	81
7	98	78
8	102	88
Mean	94.0	80.0
Variance	21.7	113.1

(a) Compute the 95% confidence interval around m_D.

(b) Compute the approximate 95% confidence interval around d.

(c) Compute t_{obs} and the two-tailed p-value associated with t_{obs}.

(d) State whether m_D is statistically significant at the $p < .05$ level.

30. The data from question 29 have been rearranged in the following table.

Participant	Control
1	95
2	93
3	95
4	87
5	90
6	92
7	98
8	102

Participant	Treatment
1	80
2	71
3	91
4	91
5	60
6	81
7	78
8	88
Mean	87.0

(a) Compute ss_{total}, ss_C, ss_S, and ss_{error}.

(b) Compute r_C^2.

(c) Compute F.

(d) Compute t_{obs}.

(e) What is $t_{critical}$ for a two-tailed test of H_0 with $\alpha = .05$?

(f) Is the difference between the control and treatment conditions statistically significant?

31. The motion aftereffect (MAE) is a well-documented perceptual phenomenon that you have probably experienced many times in the past. A classic example of the MAE results from watching a waterfall steadily for 1 or 2 minutes. When you look away from the waterfall to a stationary part of the scene (e.g., a cliff beside the waterfall), it will appear to move upward (i.e., in the direction opposite the waterfall motion to which you adapted).

An interesting question is whether the strength of the MAE depends on whether you're paying attention to the motion. Avi Chaudhuri (1990) ran exactly this experiment using a stimulus like the one shown. The random dots moved downward for about a minute while the participant gazed steadily at the center of the screen, where a sequence of letters and digits was presented at the rate of about four per second. In one condition, participants were asked to simply fixate on the letters; in a second condition, the same participants were asked to press a button whenever a digit appeared in the sequence. So, participants could ignore the letter sequence in one condition, but they had to pay attention to the letter sequence in the other condition. Attending to the letter sequence removes attention from the motion. When the downward motion stopped, the participant experienced an upward MAE. The participant pressed a key to indicate when the upward MAE started and another to indicate when it ended. The question was whether the duration of the MAE depended on whether the participant was attending to the letter sequence. The following table shows hypothetical data from this experiment. Using these data, answer the following questions.

Results of the Motion Aftereffect Experiment		
Participant	Ignore Letters	Attend Letters
1	2	2
2	10	3
3	8	2
4	6	3
5	6	4
Mean	6.4	2.8
Variance	8.8	0.7

(a) What is the independent variable?

(b) What is the dependent variable?

(c) Is this a repeated-measures or matched-samples experiment?

(d) How does attention affect the duration of the MAE?

(e) Compute the 95% confidence interval around m_D.

(f) Compute the approximate 95% confidence interval around d.

(g) Compute t_{obs} and the two-tailed p-value associated with t_{obs}.

(h) State whether m_D is statistically significant at the $p < .05$ level.

(i) Do you think that these results are meaningful?

32. Different sentences can have the same meaning. For example, "He sent a letter about it to Galileo" is in the active voice and "A letter about it was sent

to Galileo" is in the passive voice. These two sentences differ in their surface structures (i.e., word selection and order) but both convey the same meaning. Jacqueline Sachs (1967) wondered how the passage of time affects our ability to detect changes in surface structure versus changes in meaning.

Sachs had participants listen to passages of text and later presented a sentence that was either identical to one of those in the passage they had just heard or changed in some way. When there was a change, it could be either a change in surface structure or a change in meaning. The participant was to respond "identical" or "changed." The question was how time affects our ability to notice these two types of change. Therefore, in one condition, participants were prompted to make their judgment immediately following the sentence in question (0 syllables later). In a

second condition, they were prompted to make their judgment 180 syllables after the sentence in question. Therefore, participants were tested for their ability to detect a surface structure change 0 or 180 syllables following the sentence, as well as a meaning change 0 or 180 syllables following the sentence.

The dependent variable was the percentage of correct responses. If participants were simply guessing, they would have a 50% chance of being correct, on average. Hypothetical results for eight participants are shown in the following table. We assume that all participants made judgments about surface and meaning changes at 0 and 180 syllables after the test sentence. We also assume that each participant was presented with many different sentences. Use the data in the table to answer the following questions.

Accuracy (% Correct) for Detecting a Surface or Meaning Change				
Participant	Surface Change (0)	Surface Change (180)	Meaning Change (0)	Meaning Change (180)
1	83	51	87	72
2	89	68	91	69
3	92	70	87	83
4	93	56	84	88
5	91	45	84	54
6	98	62	91	80
7	86	58	95	75
8	96	62	93	79
Mean	91	59	89	75
ss	172	490	118	760

(a) What is the difference in accuracy for the two times (0 and 180) in the surface change condition?

(b) What is the difference in accuracy for the two times (0 and 180) in the meaning change condition?

(c) In which of these two conditions does accuracy decrease more?

(d) Compute the 95% confidence interval around m_D (i.e., $m_0 - m_{180}$) for both the surface change and meaning change conditions.

(e) Compute the approximate 95% confidence interval around d (i.e., $m_0 - m_{180}$) for both the surface change and meaning change conditions.

(f) Compute t_{obs} and the two-tailed p-value associated with t_{obs} for both the surface change and meaning change conditions.

(g) State whether one or both of these m_D values is statistically significant at the $p < .05$ level.

(h) Explain what these results mean.

APPENDIX 12.1: ESTIMATION AND SIGNIFICANCE TESTS IN EXCEL

Confidence Interval for m_D and d, and Computing t_{obs}

The calculations shown in Figure 12.A1.1 mirror those shown in Tables 12.2 and 12.3. The before- and after-lunch weights are shown under **Before** and **After**, the difference scores are shown in column **A - B**, and the mean weights are shown in column **(A + B)/2**. The variance of the difference scores is computed two ways. In cell **B19**, it is simply the variance of the difference scores in cells **C2** to **C6**. In cell **B20**, it is computed as $2(s_B^2 + s_A^2 - 2s_S^2)$.

Confidence Intervals for m_D and d

Figure 12.A1.2 shows the computation of an approximate $(1-\alpha)100\%$ confidence interval around d. The sample statistics are provided in cells **B2** to **B6**, and the confidence interval is computed from these using formulas provided in the chapter.

FIGURE 12.A1.1 ■ Confidence Interval for m_D

	A	B	C	D	E
1	**Before**	**After**	**A - B**	**Formulas**	**(A+B)/2**
2	151	154	3	=B2-A2	152.5
3	136	138	2	=B3-A3	137.0
4	137	139	2	=B4-A4	138.0
5	128	129	1	=B5-A5	128.5
6	169	172	3	=B6-A6	170.5
7					
8	**Quantities**	**Values**	**Formulas**		
9	m_B	144.20	=AVERAGE(A2:A6)		
10	m_A	146.40	=AVERAGE(B2:B6)		
11	s_B^2	260.70	=VAR.S(A2:A6)		
12	s_A^2	285.30	=VAR.S(B2:B6)		
13	s_S^2	272.65	=VAR.S(E2:E6)		
14	n	5			
15	df	4	=B14-1		
16	α	0.05			
17					
18	m_D	2.200	=B10-B9		
19	s_D^2	0.700	=VAR.S(C2:C6)		
20	s_D^2	0.700	=2*(B11+B12-2*B13)		
21	s_{mD}	0.374	=SQRT(B20/B14)		
22	$t_{\alpha/2}$	2.776	=T.INV.2T(B16)		
23	moe	1.039	=B22*B21		
24					
25	$m_D - t_{\alpha/2}(s_{mD})$	1.161	=B18-B23		
26	$m_D + t_{\alpha/2}(s_{mD})$	3.239	=B18+B23		

Computing a confidence interval around m_D.

FIGURE 12.A1.2 ■ Confidence Interval for d

	A	B	C
1	**Quantities**	**Values**	**Formulas**
2	m_1	146.4	
3	m_2	144.2	
4	s_1^2	285.3	
5	s_2^2	260.7	
6	s_S^2	272.825	
7	n	5	
8	df	4	=B5-1
9	α	0.05	
10			
11	m_D	2.2	=B2-B3
12	s_D^2	0.7	=2*(B4 + B2 -2*B6)
13	s_D	0.837	=SQRT(B12)
14			
15	d	2.630	=B11/B13
16	sd	1.032	=SQRT(B15^2/(2*B8) + 1/B7)
17	$z_{\alpha/2}$	1.960	=NORM.S.INV(1-B9/2)
18	moe	2.022	=B17*B16
19			
20	$d - z_{\alpha/2}(s_d)$	0.608	=B15-B18
21	$d + z_{\alpha/2}(s_d)$	4.651	=B15+B18

Computing an approximate confidence interval around d.

Partitioning Variance in Dependent Samples

Figure 12.A1.3 shows how to use Excel to partition the total variability (ss_{total}) in two dependent samples into ss_C, ss_S, and ss_{error}. The data used here are the before- and after-lunch weights for five individuals. These are the same data shown in Table 12.1. Cells **C2** to **C6** are the before-lunch weights and cells **C8** to **C12** are the after-lunch weights. In cells **D2** to **D6**, the mean of the before-lunch weights has been computed. The formulas in **E2** to **E6** use absolute referencing, as shown in Appendix 2.1. In cells **D8** to **D12**, the means of the after-lunch weights have been computed. In column **F**, the mean weight for each subject has been computed. For example, for subject 1, the mean weight is the average of the values in cells **C2** and **C8**. This weight appears in cells **F2** and **F8**, because both of these correspond to subject 1.

Once the condition means (column **D**) and subject means (column **F**) have been computed, it is a simple matter to compute the three sources of variability in our two dependent samples. First, the total variability (ss_{total})

FIGURE 12.A1.3 ■ Partitioning Variance in the Dependent Samples Design

	A	B	C	D	E	F	G
1	Condition	Subject	Scores	Mean(C)	Formulas	Mean (S)	Formulas
2	Before	1	151	144.2	=AVERAGE($C$2:$C$6)	152.5	=AVERAGE(C2,C8)
3	Before	2	136	144.2	=AVERAGE($C$2:$C$6)	137.0	=AVERAGE(C3,C9)
4	Before	3	137	144.2	=AVERAGE($C$2:$C$6)	138.0	=AVERAGE(C4,C10)
5	Before	4	128	144.2	=AVERAGE($C$2:$C$6)	128.5	=AVERAGE(C5,C11)
6	Before	5	169	144.2	=AVERAGE($C$2:$C$6)	170.5	=AVERAGE(C6,C12)
7							
8	After	1	154	146.4	=AVERAGE($C$8:$C$12)	152.5	=F2
9	After	2	138	146.4	=AVERAGE($C$8:$C$12)	137.0	=F3
10	After	3	139	146.4	=AVERAGE($C$8:$C$12)	138.0	=F4
11	After	4	129	146.4	=AVERAGE($C$8:$C$12)	128.5	=F5
12	After	5	172	146.4	=AVERAGE($C$8:$C$12)	170.5	=F6
13							
14	Quantities	Values		Formulas			
15	ss_{Total}	2196.1		=DEVSQ($C$2:$C$12)			
16	$ss_{Condition}$	12.1		=DEVSQ($D$2:$D$12)			
17	$ss_{Subjects}$	2182.6		=DEVSQ($F$2:$F$12)			
18	ss_{Error}	1.4		=B15-B16-B17			
19							
20	df	4		=COUNT(B2:B6)-1			
21	r^2_C	0.0055		=B16/B15			
22	F	34.5714		=B16/B18*B20			
23	t_{obs}	5.8797		=SQRT(B22)			

Decomposing the variability in the dependent variable (ss_{total}) into ss_C, ss_S, and ss_{error}.

---

is computed in cell **B15** using the **DEVSQ** function, which computes the sum of squares in a collection of numbers. We also use **DEVSQ** to compute ss_C in cell **B16** and ss_S in cell **B17**. In this example, conditions and subjects explain most of the variability in ss_{total}. That is, $ss_{total} = 2196.1$ and $ss_C + ss_S = 12.1 + 2182.6 = 2194.7$. Therefore, $(ss_C + ss_S)/ss_{total} = 2194.7/2196.1 = .9994$. The remaining variability is ss_{error}, which is computed in cell **B18**.

As we noted in the chapter, ss_C represents a very small proportion of the total variability. This is shown as r^2_C in cell **B21**. Expressed as a percentage, $r^2_C = .0055$ is 0.55%, or about ½ of 1 percent. As seen in the chapter, the F statistic is $F = ss_C/ss_{error}*df$. This calculation is shown in cell **B22**. As also noted, $t_{obs} = \sqrt{F}$, which is computed in cell **B23**. This is exactly the same t_{obs} computed for these data in the chapter itself. Much of statistical analysis comes down to partitioning variance.

APPENDIX 12.2: ESTIMATION AND SIGNIFICANCE TESTS IN SPSS

To compute a confidence interval around the difference between dependent-sample means in SPSS, we first enter our data as two columns of numbers. In Figure 12.A2.1, the columns labeled Before and After contain the scores from Table 12.1. To compute a confidence interval and significance test for the mean difference, we choose the Analyze→Compare Mean→Paired-Samples T Tests . . . menu. When this has been chosen, the Paired-Samples T Tests dialog appears. The variables Before and After have been moved into the

Paired Variables: region. Within SPSS, the After scores will be treated as Variable 1 and the Before scores will be treated as Variable 2. SPSS computes the difference as Variable 1 − Variable 2, so the way that variables are assigned determines the sign of the difference. In this case, the After − Before differences will be positive. Once the variables have been assigned, we click OK to proceed with the analysis. The output of the analysis is shown in Figure 12.A2.2. All of the quantities shown were computed in this chapter.

FIGURE 12.A2.1 ■ Dependent-Groups Analysis

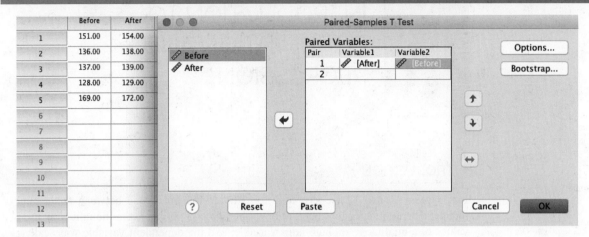

The dependent-groups analysis in SPSS.

FIGURE 12.A2.2 ■ Dependent-Groups Output

Paired Samples Test

| | Paired Differences | | | | | | | |
| | | Std. Deviation | Std. Error Mean | 95% Confidence Interval of the Difference | | t | df | Sig. (2-tailed) |
	Mean			Lower	Upper			
Pair 1 After – Before	2.200	.837	.374	1.161	3.239	5.880	4	.004

Output of the dependent-groups analysis in SPSS.

INTRODUCTION TO CORRELATION AND REGRESSION

INTRODUCTION

In Chapters 11 and 12 we saw how to estimate the difference between the means of independent and dependent populations. The two populations were thought of as the two values of a qualitative dichotomous variable, and the dependent variable was scale (i.e., quantitative). In many situations, researchers are interested in the *association* between *two scale variables*. Association means that the scores of one variable vary systematically but imperfectly with scores of the other variable. For example, we might expect there to be an association between the number of hours one spends studying for a test and the score one gets on a test. We would expect test scores to increase as the number of hours spent studying increases. Conversely, we would expect the number of errors on a test to decrease as the number of hours spent studying increases. However, we wouldn't expect two people who both studied for 6 hours to get exactly the same grade or make exactly the same number of errors. Therefore, saying that there is an association between two variables means that knowing an individual's score on one variable allows us to *predict*, with some degree of precision, his or her score on the other variable.

Other examples of such associations include associations between the characteristics of parents and their children. For example, parents' heights predict their children's heights. We might expect a parent's anxiety score to predict, to some degree, a child's anxiety score. The same might be true for depression, intelligence, extroversion, or any other psychological characteristic of interest. The time spent waiting to be served in a restaurant is probably associated with one's satisfaction with the restaurant. The number of days on a calorie-restricted diet is associated with weight. Age is associated with visual acuity, and the number of hours of sleep deprivation presumably predicts performance on an attention-demanding task.

In all of these examples, both variables are scale variables. Typically we treat one variable as the *predictor* variable and the other as the *outcome* variable. In some cases, this assignment of labels can seem arbitrary, but generally it is easy to think of one variable as predicting the other. For example, it is natural to think of a parent's anxiety as predicting a child's anxiety, but not vice versa.

In many of the above examples, the associations under study cannot be experimentally manipulated. For example, we can't randomly assign an individual to have a specific level of anxiety. Rather, anxiety is a characteristic of the individual and we can ask only if that characteristic varies systematically with other characteristics of the individual. In other cases, however, the associations under study can be experimentally manipulated. For example, an experiment may involve depriving participants of sleep for 8, 6, 4, 2, or 0 hours before they perform an attention-demanding task. We can then assess the association between number of hours of sleep deprivation and performance on the task.

There are two related questions that we can ask about the associations between scale variables. The first is "How strong is the association between the predictor variable and the outcome variable?" The second is "How do we predict scores on the outcome variable from scores on the predictor variable?" We use a quantity call the *correlation coefficient* to answer the first question, and we use the *regression equation* to answer the second question. In this chapter, we will show how to calculate the correlation coefficient and the regression equation. In Chapters 14 and 15, we will discuss inferential techniques for regression and correlation.

ASSOCIATIONS BETWEEN TWO SCALE VARIABLES

Linear Functions

If we can perfectly predict y from x, we say that y is a **function** of x. This means that there is exactly one y value for every x value. The most important function in this chapter is the linear function. You may recall from high school math that a straight line is defined as

$$y = a + b(x).$$

> If there is exactly one value of y for each value of x, we say that y is a **function** of x. When y is a function of x, we can perfectly predict y from x.

This equation describes a function because for every value of x, there is exactly one value of y. In this equation, we call a the intercept and b the slope. The intercept (a) is the value of y when x equals 0. The slope of the equation (b) is the change in y produced by a given change in x. So, if $y_1 = a + b(x_1)$ and $y_2 = a + b(x_2)$, then

$$b = \frac{y_2 - y_1}{x_2 - x_1}.$$

To illustrate this, let's think about a linear function that relates x to y, when $a = 3$ and $b = 2$. If $x_1 = 7$ and $x_2 = 9$, then

$$y_1 = 3 + 2(7) = 17 \text{ and}$$

$$y_2 = 3 + 2(9) = 21.$$

Therefore,

$$b = \frac{y_2 - y_1}{x_2 - x_1} = \frac{21 - 17}{9 - 7} = 2.$$

We would have obtained the same result ($b = 2$) for any two values of x_1 and x_2. Once again, the slope (b) is the change in y produced by a given change in x.

If the slope of the equation is positive, then values of y *increase* as values of x *increase*. If the slope of the equation is negative, then values of y *decrease* as values of x *increase*. Figure 13.1 illustrates several linear functions having different intercepts and slopes.

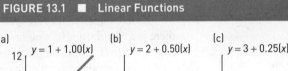

FIGURE 13.1 ■ Linear Functions

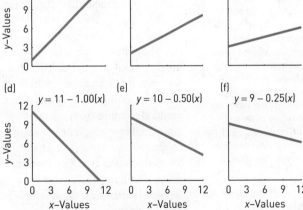

(a through f) Each graph shows a different linear function. Each line is defined by a different intercept and slope. The equation for each line is shown at the top of each graph.

We encounter many linear functions in our daily lives. For example, if you buy food at the bulk store, you will find it is priced by the pound. If demerara sugar is $3 per pound, then buying 2 pounds will cost you $3 * 2 = $6. If you buy ½ pound, it will cost you $3 * 0.5 = $1.50. As the amount you buy increases, so does the amount you pay. The relation between x (pounds of demerara sugar) and y (cost in dollars) is described as

$$y = 0 + \$3(x).$$

In this case, the intercept is 0 (if you don't buy any sugar you pay $0) and the slope is $3.

Let's think of another example relating time to money. If you need your computer, phone, or tablet repaired, you may have to pay a flat fee to bring it to the repair shop and you may then pay an amount per hour for the technician to work on it. For example, if your computer is broken, you may find that the cost to fix it is a flat fee of $50 and then $80 for each hour the technician works on it. The amount you would pay (before taxes) in this case is

$$y = \$50 + \$80(x).$$

The intercept is $50 and the slope is $80. Because the intercept is not 0, this means that you'll pay a flat fee of $50, even if you change your mind about the repair before the technician starts working on your computer.

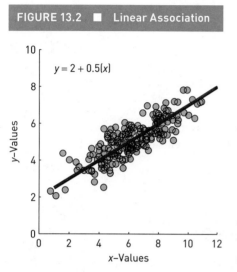

FIGURE 13.2 ■ Linear Association

$y = 2 + 0.5(x)$

The circles show a linear association between x and y. That is, for each x value, there is a distribution of y values. The linear association between x and y is well captured by the linear function $y = 2 + .5(x)$.

Linear Associations

In many areas of psychology, we study **associations** between scale variables. This means that for each value of x, there may be many values of y. Take the standard example of height and weight. There is an association between height and weight because taller people tend to weigh more than shorter people. However, this association is not perfect. In any population, there is a distribution of weights for any given height. We sometimes describe associations as *probabilistic*. For example, if we consider all individuals who are 5 feet, 7 inches tall, we can attach a probability to each possible weight. That is, there is a higher probability that a person who is 5 feet, 7 inches tall will weigh 135 pounds than 235 pounds.

Figure 13.2 depicts the association between 200 *xy* pairs in a **scatterplot**. Scatterplots can be created in statistical programs such as Excel, SPSS, **R**, or Matlab, to name a few that are commonly used. Each dot in a scatterplot corresponds to an x score and a y score. If the scores come from the same individual, then x might represent height and y weight, as we've been discussing. Or x could be time spent studying for an exam and y the test grade. Scores could also come from *pairs of individuals*, like parents and children or younger and older siblings. For example, x values might represent a parent's depression score and the y values might represent a child's depression score.

Association between two variables can take many forms, but a very common one is linear. In Figure 13.2, there is a general linear increase in y with increases in x. Therefore, a straight line through the scatter of points provides a good characterization of this association. The *best-fitting straight line* through the scatterplot in Figure 13.2 is $y = 2 + .5(x)$.

Non-linear Functions

Figure 13.3 depicts a non-linear function and a non-linear association. Non-linear simply means *not a straight line*. The black line in Figure 13.3 plots the function $y = 4 + (x - 5)^2$; that is, 4 plus the squared distance of x from 5. This is a function because there is a single value of y for each value of x. Of course, there are infinitely many non-linear functions; I've shown just one for illustration.

Non-linear Associations

Associations may be non-linear. A classic example of a non-linear association is the effect of arousal on test performance. When you are too aroused (anxious), you may make many errors because your arousal interferes with clear thinking. When you're not sufficiently aroused, you may make many errors because you are not motivated to excel. However, when your arousal is just right, you'll make the fewest errors. If we were able to measure both arousal and test performance in a population of individuals, we would expect the facts we've just mentioned to reveal themselves as a non-linear association.

The plot symbols in Figure 13.3 illustrate this example. In this case, each dot represents an individual. The number of errors made (y-axis) is lowest in the middle range of arousal (x-axis), but the error rate increases as arousal moves above or below this optimal level. The association between arousal and error rate is probabilistic because for any given level of arousal, there is a distribution of error rates. In this example, the probabilistic association between x and y is best described by the equation $y = 4 + (x - 5)^2$.

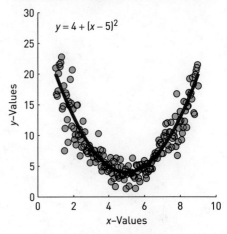

FIGURE 13.3   ■   Non-linear Association

The circles show a non-linear association between x and y. That is, for each x value, there is a distribution of y values. The non-linear association between x and y is well captured by the non-linear function $y = 4 + (x - 5)^2$.

LEARNING CHECK 1

1. State whether the following statements are true or false.

   (a) $y = x^2$ is a linear function.

   (b) $y = 4 + 5(x)$ is a linear function.

   (c) $y = 4 + 5(\sqrt{x})$ is a linear function.

   (d) $y = 4 + x/5$ is a linear function.

   (e) $y = \sin(x)$ is a non-linear function.

   (f) The association between height and weight is probabilistic.

   (g) The association between height in inches and height in centimeters is probabilistic.

Answers

1. (a) False.

   (b) True.

   (c) False.

   (d) True. Dividing x by 5 is the same as multiplying by 1/5.

   (e) False. If you don't understand why, you should Google "sinusoidal functions."

   (f) True.

   (g) False. Height in centimeters = 2.54 times height in inches. There is a one-to-one association between these two measures, meaning that one is a *function* of the other.

CORRELATION AND REGRESSION

Many associations between variables of interest to psychologists are linear. Figure 13.4 shows six examples of linear associations; in each plot, the values of y tend to change

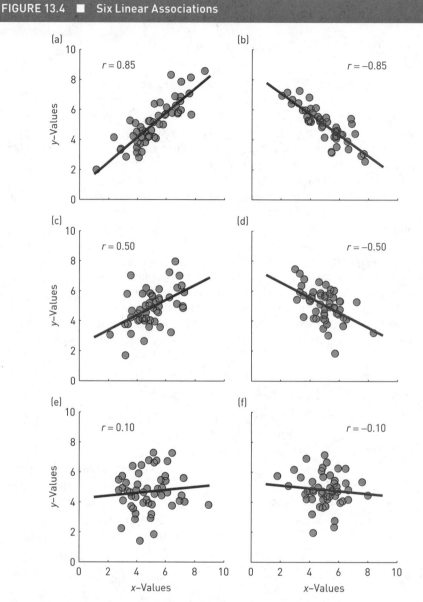

FIGURE 13.4 ■ Six Linear Associations

(a through f) The association is positive if the average value of y increases as x increases and negative if the average value of y decreases as x increases. The degree of linear association between x and y is denoted with the symbol r. When points are tightly clustered around the regression line, values of r are close to 1 or –1. When points are loosely clustered around the regression line, values of r approach 0.

linearly with x, just as shown in Figure 13.2. Several important features of linear associations are illustrated in Figure 13.4.

Linear Associations Between x and y Can Be Positive or Negative

Figure 13.4 shows that linear associations can be positive or negative, just as the slopes of straight lines (Figure 13.1) can be positive or negative. For example, you would expect the number of correct answers on a test to increase as the number of hours spent studying for the test increases. This would be a positive linear association. Conversely, you would expect the number of errors made on a test to decrease as the number of hours spent studying for the test increases. This would be a negative linear association. Figures 13.4a, 13.4c, and 13.4e show positive linear associations, and Figures 13.4b, 13.4d, and 13.4f show negative linear associations.

The Regression Line Is the Line of Best Fit Through the Scatterplot

The linear trends in the scatterplots are conveyed by the black lines. Even without knowing how these lines were computed, you would probably agree than they capture the association between the x and y values. For example, you would probably draw something similar if asked to draw the straight line that best captures the association between x and y. The lines drawn in these scatterplots are called regression lines. We will see in subsequent sections how to compute the regression line for any collection of xy pairs in a scatterplot. For now, we will note that the **linear regression line** is the best-fitting line through a scatter of points. That is, many straight lines could be drawn through any scatter of points, but there is one line (the regression line) that provides the best fit to the data. Of course, you may wonder how we define "best fit." If so, that's good. We will return to this point later.

The **linear regression line** is the best-fitting straight line through a scatter of xy points.

The Correlation Coefficient (*r*) Is a Measure of Fit

The scatterplots in Figure 13.4 differ in how tightly the dots cluster around the regression line. If the x- and y-axes of a scatterplot are properly scaled—as they are in Figure 13.4—then the more tightly points cluster around the regression line, the stronger the association between x and y.* It is possible to quantify the strength of the association between x and y using a quantity called the **correlation coefficient**, which is denoted with the symbol r. The correlation coefficient can only take on values from -1 to 1. When the x- and y-axes of a scatterplot are properly scaled, the closer r is to 1 or -1, the more tightly packed points are about the regression line. When r approaches 0, the points become increasingly scattered about the regression line. Figure 13.4 illustrates this point. Whether r is positive or negative, points are tightly clustered about the regression line when the magnitude (absolute value) of r approaches 1.

The **correlation coefficient** is a quantitative measure of the precise degree of linear association between two variables.

The Slope of the Regression Line Is Related to *r*

Finally, Figure 13.4 shows that the slope of the line of best fit is related to the correlation coefficient. When the x- and y-axes of a scatterplot are properly scaled, the closer r is to 1 or -1, the steeper the slope of the regression line. As r approaches 0, the slope of the regression line decreases. When r is exactly equal to 0, the regression line is parallel to the x-axis. We will return to each of these points as we work our way through this chapter.

THE CORRELATION COEFFICIENT

The correlation coefficient computed from a sample is denoted with the letter r. This statistic estimates the population parameter ρ (pronounced rho), which is the Greek letter r. In this section, we will see how to calculate r; in Chapter 15, we will show how to estimate

---

*By "properly scaled" I mean that the x- and y-axes cover the same number of standard deviations of the x and y variables. This is true of all graphs in Figure 13.4. Later in the chapter, we will see why proper scaling of the x- and y-axes is important.

LEARNING CHECK 2

1. Let's say we have measured the time it takes to be served at a restaurant and satisfaction with the restaurant. Would we expect r to be closer to .5 or −.5? Explain.

2. Let's say we have measured the number of alcoholic drinks consumed in the last 2 hours and the number of errors on a driving test. Would we expect r to be closer to .5 or −.5? Explain.

3. Let's say we've recorded the number of hours spent studying for a statistics test and the resulting grades on the test. Does is seem plausible to you that the correlation between these two variables would be $r = .99$?

4. The correlation between x and y is $r = .3$. Explain why the regression line in this case cannot be $3 + (−4x)$.

Answers

1. We would expect satisfaction with the restaurant to *decrease* as the time to be served *increases*. Therefore, we would expect a negative correlation, and we could expect r to be closer to −.5 than to .5.

2. We would expect the number of driving errors to *increase* as the number of alcoholic drinks consumed *increases*. Therefore, we would expect a positive correlation, and we could expect r to be closer to .5 than to −.5.

3. A correlation of $r = .99$ is an almost perfect correlation, which means that the association is very close to a function. This means that almost everyone who studied for 5 hours got almost exactly the same grade. Correlations in psychology are rarely if ever this strong.

4. If $r = .3$, then y tends to increase as x increases. However, the slope of this line is negative (−4), which is impossible if $r = .3$.

FIGURE 13.5 ■ Example Data

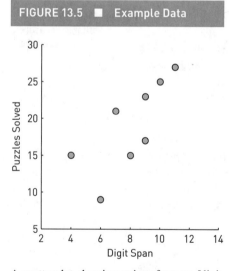

A scatterplot showing pairs of scores [digit spans, puzzles solved] for eight individuals.

ρ (the population parameter) from r (the sample statistic). Because r and ρ are calculated in essentially the same way, we will focus on calculations for samples in this chapter.

So that we have an example to work with, let's say that your stats instructor asked you to collect a small data set comprising two sets of scores that might be positively correlated. You decide to measure digit span and the number of simple verbal puzzles solved in a 10-minute period for eight of your friends. After you collect the data, you create the scatterplot shown in Figure 13.5, which seems to show a positive correlation between digit span and the number of simple verbal puzzles solved. We will use these numbers to illustrate how we compute the correlation coefficient.

Computing the Correlation Coefficient: The Covariance

To help us understand the correlation coefficient, we will introduce a new quantity called the **covariance**, which is defined as follows:

$$cov_{xy} = \frac{\sum(x-m_x)(y-m_y)}{n-1}. \tag{13.1}$$

The covariance is closely related to the sample variance. Remember, the sample variance is defined as

$$s^2 = \frac{\sum(y-m_y)^2}{n-1}.$$

But this could also be written as

$$s^2 = \frac{\sum(y - m_y)(y - m_y)}{n-1}.$$

In this form, you can see that the deviation scores in y are multiplied by themselves. In the covariance, however, each deviation score in y is multiplied by the corresponding deviation score in x. In both cases, the sum of products is divided by $n-1$. Unlike the variance, the covariance can be positive or negative.

If y values tend to increase as x values *increase* (Figures 13.4a, 13.4c, and 13.4e), the covariance will be positive. This is because deviation scores for both x and y will tend to have the same signs, so their products (in the numerator of equation 13.1) will be positive. For example, if a given $x - m_x$ is positive, then the corresponding $y - m_x$ is more likely to be positive than negative. And, if a given $x - m_x$ is negative, then the corresponding $y - m_x$ is more likely to be negative than positive. Therefore, there will be more positive than negative products, so the sum of these products (the numerator of the covariance) will be positive.

On the other hand, if y values tend to *decrease* as x values increase (Figures 13.4b, 13.4d, and 13.4f), the covariance will be negative. This is because pairs of deviation scores will tend to have opposite signs; hence, there will be more negative than positive products. Therefore, the sum of these products will be negative.

We will see later that the covariance by itself is not a good measure of the association between two variables. However, it plays an important role in the definition of the correlation coefficient. The correlation coefficient is defined as

$$r = \frac{cov_{xy}}{\sqrt{s_x^2 * s_y^2}} \tag{13.2}$$

where cov_{xy} is covariance between x and y, s_x^2 is the variance of the x scores, and s_y^2 is the variance of the y scores. There is really no complexity here, because r is defined in terms of quantities we've seen before. We will next illustrate the calculation of r.

Table 13.1 shows the eight pairs of digit spans (x) and the number of puzzles solved (y) that were plotted in Figure 13.5. The sample means, sums of squares, and variances are shown below the x and y scores. The next two columns show the x and y deviation scores, and the last column shows the product of the deviation scores. We denote the sum of these products as sp_{xy}, which is the numerator of the covariance. Dividing sp_{xy} by $n-1$ yields the covariance. In this example, $cov_{xy} = sp_{xy}/(n-1) = 72/7 = 10.29$. When we use these quantities in equation 13.2, we find that the correlation coefficient is

$$r = \frac{cov_{xy}}{\sqrt{s_x^2 * s_y^2}} = \frac{10.29}{\sqrt{5.14 * 36.57}} = .75.$$

A Simplification

Although computing the correlation coefficient is not complicated, it can be simplified even further. To compute s_x^2, s_y^2, and cov_{xy}, we divided ss_x, ss_y, and sp_{xy} by $n-1$. As a consequence, we can bypass this division, which is common to these three quantities, and define the correlation coefficient as follows:

$$r = \frac{sp_{xy}}{\sqrt{ss_x * ss_y}}. \tag{13.3}$$

TABLE 13.1 ■ Example Data				
x	y	$x - m_x$	$y - m_y$	$(x - m_x)(y - m_y)$
4	15	−4	−4	16
6	9	−2	−10	20
7	21	−1	2	−2
8	15	0	−4	0
9	17	1	−2	−2
9	23	1	4	4
10	25	2	6	12
11	27	3	8	24
m_x	m_y			
8	19			
ss_x	ss_y			sp_{xy}
36	256			72
s_x^2	s_y^2			cov_{xy}
5.14	36.57			10.29

Note: Quantities required to compute the correlation coefficient for eight pairs of digit spans (x) and number of puzzles solved (y).

Therefore, using ss_x and ss_y, and sp_{xy} from Table 13.1, we obtain exactly the same result in our example:

$$r = \frac{sp_{xy}}{\sqrt{ss_x * ss_y}} = \frac{72}{\sqrt{36 * 256}} = .75.$$

Should you ever be asked to compute the correlation coefficient on an exam, this simplification means fewer steps in the calculation and hence fewer opportunities for rounding and computation errors.

r Is Invariant Under Linear Transformations of x and y

The correlation coefficient has an important advantage over the covariance as a measure of association between two variables. Specifically, r is unaffected by the "scales" on which x and y are measured, whereas cov_{xy} changes when the scales of the variables change. For example, the correlation between height and weight will remain the same whether height is measured in inches, feet, or miles or weight is measured in ounces, pounds, or tons. These conversions between units (e.g., inches to centimeters) are examples of **linear transformations**. Because r does not change when either x, y, or both undergo a linear transformation, we say that r is *invariant under linear transformations*.

A **linear transformation** of a set of scores (x) is any combination of additions or multiplications that can be reduced to $a + b(x)$.

Linear transformations are simple and familiar. Any time we add a constant (positive or negative) to a collection of numbers, we have performed a linear transformation. Any time we multiply or divide a set of numbers by a constant, we have performed a linear transformation. If we multiply (or divide) a set of numbers by a constant and then add (or subtract) another constant, we have also performed a linear transformation.

Converting miles per hour (mph) to kilometers per hour (kph) is a linear transformation:

$$mph = 0.6214(kph),$$

and vice versa,

$$kph = \frac{1}{0.6214}(mph).$$

Converting degrees Fahrenheit (F°) to degrees Celsius (C°) is a linear transformation:

$$F° = 32 + 1.8(C°),$$

and vice versa,

$$C° = \frac{F° - 32}{1.8} = -\frac{32}{1.8} + \frac{1}{1.8}(F°).$$

Converting x to z_x is also a linear transformation:

$$z_x = \frac{x - m_x}{s_x} = -\frac{m_x}{s_x} + \frac{1}{s_x}(x),$$

(in this case, $a = -m_x/s_x$, $b = 1/s_x$) and vice versa

$$x = m_x + z_x(s_x).$$

These examples should look familiar because they involve the arithmetic operations that define a line, $y = a + b(x)$, and hence the term *linear* transformation. Therefore, if x_1 and x_2 are related by $x_2 = a + b(x_1)$, we say that x_2 is a linear transformation of x_1. This means that we have multiplied x_1 by any number (except 0) and added another number (possibly 0) to the result.

To illustrate the invariance of r under a linear transformation, we will *multiply* each x score in Table 13.1 by 10. These new values of x are shown in the first column of Table 13.2. The third column shows that the resulting deviation scores for x increase by a factor of 10. Therefore, sp_{xy} and cov_{xy}, which are based on the products of x and y deviation scores, also increase by a factor of 10; previously, cov_{xy} was 10.286 and now it is 102.86. So, even though the basic association between x and y has not changed, the covariance has, which is why the covariance by itself is not a useful measure of association.

Multiplying all the x scores by 10 causes ss_x and s_x^2 to increase by a factor of 100 (see Chapter 3 for a reminder of how multiplication affects the variance). Table 13.2 illustrates this as well. Previously, ss_x was 36 and now it is 3600. However, when we compute the correlation coefficient from these new quantities, we obtain exactly the same result as before:

$$r = \frac{sp_{xy}}{\sqrt{ss_x * ss_y}} = \frac{720}{\sqrt{3600 * 256}} = \frac{720}{\sqrt{100}\sqrt{36 * 256}} = \frac{72}{\sqrt{36 * 256}} = .75.$$

The third step in this little derivation relies on the simple mathematical fact that $\sqrt{a * b} = \sqrt{a}\sqrt{b}$. In our example, $\sqrt{3600} = \sqrt{36 * 100} = \sqrt{36}\sqrt{100} = 60$. The important thing

TABLE 13.2 ■ Invariance Under Scale Change				
x	y	$x - m_x$	$y - m_y$	$(x - m_x)(y - m_y)$
40	15	−40	−4	160
60	9	−20	−10	200
70	21	−10	2	−20
80	15	0	−4	0
90	17	10	−2	−20
90	23	10	4	40
100	25	20	6	120
110	27	30	8	240
m_x	m_y			
80	19			
ss_x	ss_y			sp_{xy}
3600	256			720
s_x^2	s_y^2			cov_{xy}
514.29	36.57			102.86

Note: Values of the predictor variable (x) from Table 13.1 have been multiplied by 10.

to remember is that r is *invariant under linear transformations* of x and y. This is just a fancy way of saying that the correlation coefficient does not change when we change the units of measurement for one or both of our variables.

The Correlation Coefficient Ranges From −1 to 1

There is a **perfect correlation** between x and y when $y = a + b(x)$. In this case, r will be −1 or 1 depending on the sign of b, the slope of the linear transformation.

Two variables are **perfectly correlated** when they are linear transformations of each other. This means that one can perfectly predict y from x; i.e., y is a linear function of x. Therefore, the correlation coefficient will be either 1 or −1, depending on whether the slope of the linear transformation that relates y to x is positive or negative. Table 13.3 provides an illustration of this point. In this example, the x values are 1, 2, 3, 4, and 5, and the y values are $y = 10 - 2(x)$, as shown in the second column. If you were to plot y against x, it would be a perfectly straight line with a slope of −2.

When we compute the correlation coefficient for x and y, we find that it is $r = -1$. To see why this is so, remember that equation 13.2 states that $r = cov_{xy} / \sqrt{s_x^2 * s_y^2}$. Table 13.3 shows that $cov_{xy} = -5$, $s_x^2 = 2.5$, and $s_y^2 = 10$. Therefore, $s_x^2 * s_y^2 = 25$, which has a square root of 5. This means that the numerator and the denominator of the formula for r have the same magnitude (absolute value) but different signs. Therefore, the correlation coefficient is $r = -1$.

The example in Table 13.3 illustrates the general rule that if y is a linear transformation of x, then

$$\left| cov_{xy} \right| = \sqrt{s_x^2 * s_y^2}.$$

TABLE 13.3 ■ "Perfect" Correlation

x	$y = 10 - 2(x)$	$x - m_x$	$y - m_y$	$(x - m_x)(y - m_y)$
1	8	−2	4	−8
2	6	−1	2	−2
3	4	0	0	0
4	2	1	−2	−2
5	0	2	−4	−8
m_x	m_y			
3	4			
ss_x	ss_y			sp_{xy}
10	40			−20
s_x^2	s_y^2			cov_{xy}
2.50	10.00			−5.00

Note: Two variables are perfectly correlated when one is a linear transformation of the other.

For this reason, r can never be less than −1 or greater than 1 because these represent cases of perfect negative and positive correlation, respectively.

Converting to z-Scores Is a Linear Transformation

Figure 13.6 brings together important points about scale change and correlation, and it also previews some points about regression that we'll cover in the next section. Figures 13.6a, 13.6c, and 13.6e show scatterplots of x and y scores. The only difference between the scatterplots is that the scales of the variables have been changed.

- In Figure 13.6a, $s_x = s_y = 5$.
- In Figure 13.6c, the x scores from Figure 13.6a have been scaled to have $s_x = 1$, while s_y remains 5.
- In Figure 13.6e, the y scores from Figure 13.6a have been scaled to have $s_y = 1$, while s_x remains 5.

When the scales (standard deviations) of the xy values change, the regression equations change, as shown in Figures 13.6a, 13.6c, and 13.6e. However, the correlation coefficient is $r = .6$ for Figures 13.6a through 13.6f, because, as we've seen, the correlation coefficient is invariant to linear transformations of one or both variables.

To see that the scale changes do not affect the correlation coefficient, consider Figures 13.6b, 13.6d, and 13.6f. These scatterplots show the scores in Figures 13.6a, 13.6c, and 13.6e converted to z-scores through the following linear transformations: $z_x = (x - m_x)/s_x$ and $z_y = (y - m_y)/s_y$. Notice that Figures 13.6b, 13.6d, and 13.6f are identical. This confirms that the xy pairs on the left are just differently scaled versions of the $z_x z_y$ pairs on the right. Therefore, the correlation coefficients are exactly the same in all panels of Figure 13.6; i.e., $r = .6$.

FIGURE 13.6 ■ **Invariance Under Linear Transformation**

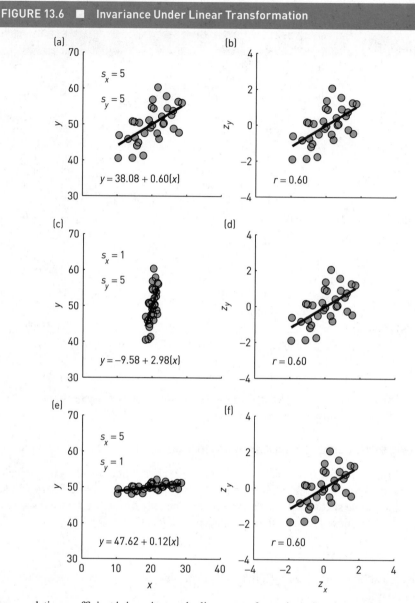

The correlation coefficient is invariant under linear transformations. See the text for an explanation.

Finally, Figures 13.6a, 13.6c, and 13.6e illustrate a practical point about the importance of paying very close attention to the axes of scatterplots. The scatterplots in Figures 13.6c and 13.6e seem to show points tightly clustered around the regression line, suggesting a high correlation. However, the seemingly tight clustering actually reflects inappropriate scaling of the x- and y-axes. Note that the x-axes cover the range 0 to 40 in Figures 13.6a, 13.6c, and 13.6e, and the y-axes cover the range 30 to 70. For Figure 13.6a, this is appropriate because the range of values on the x- and y-axes is about $m \pm 4(s)$.

However, the ranges of the x and y values on the axes of Figures 13.6c and 13.6e have not been adapted to the distributions of the x and y scores. For Figure 13.6c, the range of the x-axis is about $m_x \pm 20(s_x)$, because the x-axis covers the range 0 to 40, whereas $s_x = 1$.

For Figure 13.6e, the range of the y-axis is about $m_y \pm 20(s_y)$, because the y-axis covers the range 30 to 70, whereas $s_y = 1$.

When we construct a scatterplot, the range of x and y values must relate to the standard deviations of the x and y scores. Therefore, for Figure 13.6c, the range of values on the x-axis should have about $m_x \pm 4(s_x)$, which would have produced a range of about 16 to 24 on the x-axis. For Figure 13.6e, the range of the y-axis should have about $m_y \pm 4(s_y)$, which would have produced a range of about 46 to 54 on the y-axis. Had the axes in Figures 13.6a, 13.6c, and 13.6e been scaled in these ways, all three scatterplots on the left would have been identical and indistinguishable from those on the right. Of course, one way to ensure that the axes are properly scaled is to convert both x and y to z-scores, as in Figure 13.6b, 13.6d, and 13.6f, and then make sure the range of z_x and z_y values is the same on the two axes.

Remember that when I discussed Figure 13.4, I made a point of saying r relates to how tightly points cluster around the regression line *when the x- and y-axes are properly scaled*. We now see that points can *appear* to cluster tightly around the regression line if the axes have been improperly scaled. This can create the illusion that there is a strong correlation when there is not.

LEARNING CHECK 3

1. If $x = [1, 3, 5, 7, 9]$ and $y = [9, 7, 5, 1, 3]$, compute r.

2. If $r = [1, 2, 3, 4, 5, 6, 7, 8, 9]$ and $y = 5 + 3(r)$, determine r without calculating it.

3. If $x = [1, 2, 3, 4, 5, 6, 7, 8, 9]$ and $y = 5 - 4(x)$, determine r without calculating it.

4. Convert 10°C to degrees Fahrenheit.

5. State whether the following statements are true or false.

   (a) The correlation between the same reaction times measured in milliseconds and measured in hours is $r = 1$.

   (b) If $x = [1, 2, 3]$ and $y = [-2, -4, -6]$, then $r = -1$.

Answers

1. $ss_x = 40$, $ss_y = 40$, and $sp_{xy} = -36$; therefore, $r = -.9$.

2. $r = 1$ because $y = 5 + 3(x)$ is a linear transformation of x. y is a positive linear function of x.

3. $r = -1$ because $y = 5 - 4(x)$ is a linear transformation of x. y is a negative linear function of x.

4. $F° = 32 + 1.8(10) = 32 + 18 = 50$.

5. (a) True. These numbers are linear transformations of each other, so they are perfectly correlated.

   (b) True. $y = -2(x)$ is a negative linear transformation of x. Or you could compute the correlation coefficient.

THE REGRESSION EQUATION

The lines drawn through the points in the scatterplots in Figures 13.2 and 13.4 are called regression lines, or lines of best fit. So now we have two tasks ahead of us. The first is to find out how to compute a regression line through a scatter of points. The second is to explain what it means for the regression line to be the *best-fitting* line through a scatter of points. The main theme of this section is the deep connection between the correlation coefficient and the regression equation. Therefore, the correlation coefficient is central to both questions addressed here.

Prediction

We describe the regression line as a prediction line and define it as follows:

$$\hat{y} = a + b(x).$$

The hat (^) over y in the left side of the equation means "predicted," so $\hat{y}$ should be read as "y-predicted." This terminology reflects the fact that the regression equation can't specify exactly what y is for a given value of x because there may be many values of y for any value of x. Therefore, $\hat{y}$ is the best prediction of y from x. (In Chapter 14, we will see that the regression equation relating x and y in a population predicts the average value of y for a given value of x.)

The regression equation allows us to predict $\hat{y}$ from x for values of x that do not appear in our sample. For example, no individual in the sample (shown in Figure 13.5) had a digit span of 5. However, once we compute the regression equation from the scores in Table 13.1, we can make a prediction about the number of puzzles solved for a person with a digit span of 5 by computing $\hat{y}$ for $x = 5$; i.e., $\hat{y} = a + b(5)$.

The Correlation Coefficient and Prediction

There is an intimate connection between the correlation coefficient and the regression equation. To see this connection, we will first think about x and y scores that have been transformed into z-scores; i.e., $z_x = (x - m_x)/s_x$ and $z_y = (x - m_y)/s_y$. We saw earlier that the correlation coefficient can be computed from x and y scores as follows:

$$r = \frac{cov_{xy}}{\sqrt{s_x^2 * s_y^2}}.$$

This equation implies something really interesting about computing the correlation coefficient from z-scores.

In Chapter 4 we saw that a sample (or population) of z-scores has a mean of 0 and a standard deviation of 1. Of course, if the standard deviation is 1, then so too is the variance; i.e., $1^2 = 1$. So, if we're computing the correlation coefficient from two sets of z-scores, both of which have a variance of 1, then the denominator of the correlation coefficient formula is also 1. This means that the correlation coefficient computed from z-scores is simply the covariance of the z-scores. Therefore, we can write the correlation coefficient equation for z-scores as

$$r = \frac{\sum z_x z_y}{n-1}. \tag{13.4}$$

This is a really interesting observation and yields some insight into the nature of the correlation coefficient. However, it is probably not the equation you would use to calculate r by hand, if you were ever asked to do so on a test.

Now let's address the connection between the correlation coefficient and the regression equation by having a look at Figure 13.7. Each scatterplot in Figure 13.7 shows 64 pairs of z_x and z_y scores. From Figure 13.7a to 13.7f, r decreases from 1 down to 0. The thick black line in each plot is the regression line through the values of z_x and z_y. Figure 13.7 shows two important points made earlier in Figure 13.4 about the regression line when two samples have the same variance:

- The closer r is to 1, the more tightly points cluster around the regression line; and the closer r is to 0, the more dispersed points are around the regression line.

FIGURE 13.7   ■   Correlation and Slope

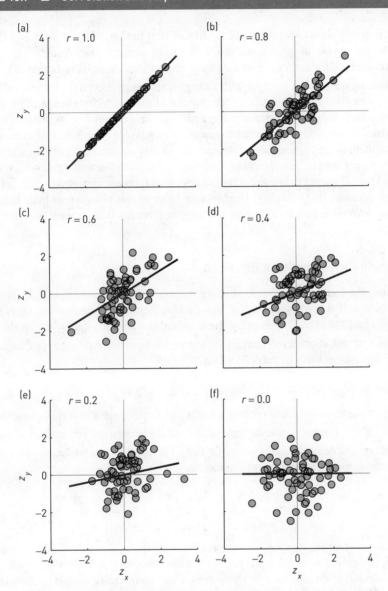

(a through f) The correlation coefficient is the slope of the regression equation that relates z_y to z_x; i.e., $\hat{z}_y = rz_x$. (a) When $r = 1$, our prediction is perfect. As r moves from 1 toward 0, the precision of the prediction decreases. (f) When $r = 0$ (bottom right), x provides no information about y. Therefore, the only prediction we can make is that y equals the mean value of y, which corresponds to $z_y = 0$; i.e., $\hat{z}_y = rz_x = 0_{z_x} = 0$.

- Simultaneously, the closer r is to 1, the steeper the regression line; and the closer r is to 0, the flatter the regression line.

The really interesting thing is that the slope of the regression line in each plot is simply r. The intercept of this regression line is 0, and so the regression equation for z-scores is simply

$$\hat{z}_y = r(z_x). \tag{13.5}$$

Each of the regression lines in Figure 13.7 was computed this way. The hat over z_y means "predicted" exactly as discussed earlier. Therefore, $\hat{z}_y$ should be read as z_y-predicted.

r is the slope of the regression line predicting z_y from z_x.

Our example shows that r defines the line of best fit that relates z_x to z_y and therefore relates to the notion of prediction. So, r not only conveys the strength of the linear association between x and y (or z_x and z_y) but also provides the best prediction of z_y from z_x. We are finding out that r is a very useful little guy, and there is even more to come!

We saw earlier in this chapter that the slope of a line tells us the change in y that results from a one-unit change in x. To help understand what this means for z-scores, let's take the standard deviation as our unit of measurement. In this case, $r = 1$ means that a one standard deviation increase in z_x produces a one standard deviation increase in $\hat{z}_y$. Similarly, $r = .5$ means that a one standard deviation increase in z_x produces a ½ standard deviation increase in $\hat{z}_y$. And $r = -.5$ means that a one standard deviation increase in z_x produces a ½ standard deviation decrease in $\hat{z}_y$. Because the absolute value of r is always less than 1, it tells us what *proportion of a standard deviation* change in $\hat{z}_y$ results from a one standard deviation change in z_x.

Computing the Regression Equation

We can use the fact that r is the slope of the regression line relating z_y to z_x to explain the regression equation relating y to x; i.e., $\hat{y} = a + b(x)$. Remember, converting x and y scores to z-scores had the effect of equating the standard deviations $s_{z_x} = s_{z_y} = 1$ of the scores. Therefore, our task is simply to convert the regression equation relating z_y to z_x back into an equation that relates x to y in the original units.

The Slope

The slope of the regression line relating x to y is just a scaled version of r. Because the original x and y scores typically have different standard deviations, the actual change in $\hat{y}$ with a one s_x change in x depends on the ratio of s_y to s_x. Therefore, when we know s_x, s_y, and r, we compute the slope of the regression equation as

$$b = r\frac{s_y}{s_x}. \tag{13.6}$$

The ratio s_y/s_x compensates for the different standard deviations of the x and y scores. For example, imagine that x and y are perfectly correlated, so $r = 1$. This means that a one standard deviation change in x (s_x) produces a one standard deviation change in y (s_y). Equation 13.6 tells us how much this change would be in units of y. So, if $s_x = 2$ and $s_y = 10$, then $s_y/s_x = 10/2 = 5$. Therefore, when $r = 1$, a one s_x change in x would produce an $r*5*s_x = 10$ unit change in y; i.e., a one s_y change in y. When $r = .5$, a one s_x change in x would produce a $r*5*s_x = 5$ unit change in y; i.e., a $.5*s_y$ change in y.

The Intercept

The definition of z-scores means that m_x and m_y correspond to $z_x = 0$ and $z_y = 0$, respectively. The fine vertical and horizontal lines in Figure 13.7 show that the regression line for z-scores [$r(z_x)$] always passes through the point [0, 0]. Consequently, the raw score regression line passes through the point [m_x, m_y]. Therefore, in the original units, m_y is the value of the regression equation at m_x. This means that

$$m_y = a + b(m_x).$$

With a simple rearrangement of this equation, we would have

$$m_y - b(m_x) = a, \text{ or}$$

$$a = m_y - b(m_x). \tag{13.7}$$

So, if we know m_x, m_y, and b, we can solve for a using equation 13.7.

The Regression Equation for Our Example

Let's put equations 13.6 and 13.7 to work on the data from Table 13.1, which have been reproduced in Table 13.4 on the next page. Table 13.4 shows that $m_x = 8$, $s_x = 2.268$, $m_y = 19$, and $s_y = 6.047$. Furthermore, we determined earlier that $r = .75$. With these numbers, we can compute the slope of the regression equation as follows:

$$b = r\frac{s_y}{s_x} = .75\frac{6.047}{2.268} = 2,$$

and then the intercept as

$$a = m_y - b(m_x) = 19 - 2(8) = 19 - 16 = 3.$$

Therefore, the regression equation is $\hat{y} = a + b(x) = 3 + 2(x)$.

Applying the Regression Equation

This regression equation allows us to make a prediction about the y value corresponding to any value of x. The third column of Table 13.4 shows $\hat{y}$ for each value of x. We will look at $\hat{y}$ for two values of x. Let's first look at $\hat{y}$ when $x = m_x$. When we put $x = m_x = 8$ into the regression equation, we find that

$$\hat{y} = a + b(m_x) = 3 + 2(8) = 19.$$

Therefore, as stated above, our regression equation predicts $m_y = 19$ when $x = m_x$.

Now let's think about $x = 5$. When we put $x = 5$ into the regression equation, we find that

$$\hat{y} = a + b(x) = 3 + 2(5) = 13.$$

Even though $x = 5$ was not part of the data set from which the regression equation was computed, we are able to use the regression equation to make a reasonable prediction about what the corresponding y would be.

The regression line is useful because it conveys information about the association between y and x, and it allows us to predict the value of the outcome variable from the predictor variable, even if the xy pair in question does not exist in our data set. Of course, the quality of our prediction will depend on the strength of the correlation between x and y. That is, our prediction of y will be more precise as r approaches 1 or -1 and less precise as r approaches 0.

The Line of Best Fit: Least Squares

Let's now turn to the question of what it means for the regression line to be the line of best fit through a scatter of points. We will refer to Table 13.4 and Figure 13.8 as we go. The first two columns in Table 13.4 are the xy pairs from Table 13.1. We just saw that the regression line for these data is

$$\hat{y} = 3 + 2(x).$$

TABLE 13.4 ■ The Least-Squares Fit			
x	y	$\hat{y}$	$y - \hat{y}$
4	15	11	4
6	9	15	−6
7	21	17	4
8	15	19	−4
9	17	21	−4
9	23	21	2
10	25	23	2
11	27	25	2
m_x	m_y	$m_{\hat{y}}$	$m_{y-\hat{y}}$
8	19	19	0
ss_x	ss_y (ss_{total})	$ss_{\hat{y}}$ $(ss_{\text{regression}})$	$ss_{y-\hat{y}}$ (ss_{error})
36	256	144	112
s_x	s_y		
2.268	6.047		

Note: Partitioning variance and the least-squares fit: $\hat{y} = 3 + 2(x)$.

FIGURE 13.8 ■ Partitioning Variance

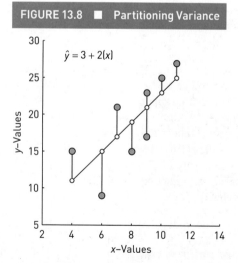

The filled dots represent the xy pairs. The small white dots represent $\hat{y}$ for each value of x. The vertical gray lines connecting the white dots ($\hat{y}$) to the filled dots (y) represent errors of prediction ($y - \hat{y}$).

Figure 13.8 shows the scatterplot of these xy pairs (the filled dots replicate Figure 13.5) along with the regression line. The means and sums of squares (ss) for x and y are shown at the bottom of Table 13.4; $m_x = 8$, $ss_x = 36$, $m_y = 19$, and $ss_y = 256$.

Partitioning Variance

In Chapter 11 we partitioned the variability in a set of scores by decomposing each score into two parts. One part was the group mean and the other was the score's deviation from the group mean. We can do the same thing in the case of regression. In this case, we will decompose each y into $\hat{y}$ and its deviation from $\hat{y}$; i.e., $y - \hat{y}$. The third and fourth columns of Table 13.4 show the values of $\hat{y}$ and $y - \hat{y}$.

We will refer to the total variability in y as ss_{total}. It is this quantity that we want to partition, or decompose. We compute ss_{total} as follows:

$$ss_{\text{total}} = \Sigma(y - m_y)^2. \tag{13.8}$$

Therefore, ss_{total} is just a new name for the sum of squares for the y scores shown in the second column of Table 13.4 and depicted as filled dots in Figure 13.8. Table 13.4 shows that $ss_{\text{total}} = 256$.

We will now compute the variability in y associated with $\hat{y}$ and $y - \hat{y}$. We say that $\hat{y}$ is that part of y that *can be predicted* from x, or that part of y that is *explained* by x. The predicted value of y (i.e., $\hat{y}$) plays the same role that group means played in Chapter 11. We will refer to $ss_{\hat{y}}$ as $ss_{\text{regression}}$, as shown at the bottom of the third column of Table 13.4. The small white dots in Figure 13.8 correspond to these $\hat{y}$ values. Therefore, we compute $ss_{\text{regression}}$ as follows:

$$ss_{\text{regression}} = \Sigma(\hat{y} - m_{\hat{y}})^2. \tag{13.9}$$

Table 13.4 shows that $ss_{\text{regression}} = 144$.

The differences between y and $\hat{y}$ (i.e., $y - \hat{y}$) represent that part of y that *cannot be predicted* from or *explained by* x. These differences are shown as vertical gray lines in Figure 13.8 connecting y and $\hat{y}$. We call the $y - \hat{y}$ values the *errors of prediction*. Therefore, we will refer to $ss_{y-\hat{y}}$ as ss_{error}, as shown at the bottom of the fourth column of Table 13.4. We compute ss_{error} as follows:

$$ss_{\text{error}} = \Sigma(y - \hat{y})^2. \tag{13.10}$$

Table 13.4 shows that $ss_{\text{error}} = 112$.

Explained Variance, r²

We can now ask how much of the variability in y can be *explained* by regression. As in Chapter 11, we answer this question by forming the following ratio:

$$r^2 = \frac{ss_{\hat{y}}}{ss_y}. \tag{13.11a}$$

Or, using the terminology of regression, we can say that

$$r^2 = \frac{ss_{\text{regression}}}{ss_{\text{total}}}. \tag{13.11b}$$

In our example, we find that

$$r^2 = \frac{ss_{\text{regression}}}{ss_{\text{total}}} = \frac{144}{256} = .5625.$$

Therefore, 56.25% of the variability in y is explained by its association with x. So, if 56.25% of the variability in y is explained by regression, then the remaining variability in y represents prediction errors.

The three quantities we've just discussed (ss_{total}, $ss_{\text{regression}}$, and ss_{error}) are related in a simple way. That is,

$$ss_{\text{total}} = ss_{\text{regression}} + ss_{\text{error}}. \tag{13.12}$$

In other words, the total variability in y (ss_{total}) can be decomposed into that part associated with x ($ss_{\text{regression}}$) and that part that is not associated with x (ss_{error}). Therefore, the proportion of variability in y not explained by regression is

$$1 - r^2 = \frac{ss_{\text{error}}}{ss_{\text{total}}} \quad \text{or} \tag{13.13a}$$

$$1 - r^2 = \frac{ss_{\text{total}} - ss_{\text{regression}}}{ss_{\text{total}}}. \tag{13.13b}$$

We can now state that the regression line is the best-fitting line through a scatter of xy pairs because *no other line would produce a smaller* ss_{error}. This fact is often referred to as the **least-squares fit** because the regression line is the one that produces the smallest (*least*) sum of squared prediction errors. Another way to say this is that no other line would produce a larger r^2.

Finally, r^2 is the square of the correlation coefficient. In our example, we found that $r^2 = .5625$. The square root of .5625 is .75. That is, $r = \sqrt{r^2} = \sqrt{.5625} = .75$!

The **least-squares fit** means the fit (line) through a scatter of points that produces the smallest sum of squared deviations of data points from the line.

Correlation and Regression

The correlation coefficient is an amazing thing. We've seen that it expresses the exact degree of linear association between two variables. We've also seen that r^2 is the proportion of variability in y explained by the regression equation. This means that when we compute the correlation coefficient, we simply have to square it to determine the proportion of variability in y explained by the regression equation. In addition, r defines the regression equation when x and y have been transformed to z_x and z_y. Because the transformations from x and y to z_x and z_y are linear transformations, it is no surprise that the slope of the regression equation (b) for raw scores is a simple linear transformation of r; i.e., $b = r*s_y/s_x$. As we'll see in later chapters, almost all statistics can be reduced to questions of correlation.

LEARNING CHECK 4

1. State whether the following statements are true or false.

   (a) $r^2 = ss_{\hat{y}}/ss_y$.

   (b) $r^2 = ss_x/ss_y$.

   (c) $r^2 = ss_{total}/ss_{error}$.

   (d) $r^2 = ss_{regression}/ss_{error}$.

   (e) $r^2 = ss_{regression}/ss_{total}$.

   (f) If $x = [1, 2, 3]$ and $y = [-2, -4, -6]$, then $r^2 = 1$.

   (g) If $x = [1, 2, 3]$ and $y = [-2, -4, -6]$, then $r = 1$.

   (h) $ss_{regression} = \Sigma(y - \hat{y})^2$.

   (i) $ss_{error} = \Sigma(y - \hat{y})^2$.

   (j) $ss_{total} = \Sigma(y - m_y)^2$.

   (k) $ss_{total} = ss_{regression} + ss_{error}$.

2. For $x = [1, 2, 3, 6]$ and $y = [10, 3, 2, 5]$, compute (a) r and (b) the regression equation.

Answers

1. (a) True. (b) False. (c) False.

   (d) False. (e) True. (f) True.

   (g) False. $r = -1$. Make a scatterplot from x and y, or notice that $y = -2*x$.

   (h) False. The expression defines ss_{error}.

   (i) True. (j) True. (k) True.

2. From the data in the table below, we determine that $m_x = 3$, $m_y = 5$, $s_x = 2.16$, $s_y = 3.56$. (a) r −.35 (b) $b = -.35*3.56/2.16 = -.57$, $a = 5 + .57(3) = 6.71$. (These answers were computed without rounding intermediate quantities.)

x	y	$x - m_x$	$y - m_y$	$(x - m_x)(y - m_y)$
1	10	−2	5	−10
2	3	−1	−2	2
3	2	0	−3	0
6	5	3	0	0
m_x	m_y	$ss_x = \Sigma(x - m_x)^2$	$ss_y = \Sigma(y - m_y)^2$	$sp_{xy} = \Sigma(x - m_x)(y - m_y)$
3	5	14	38	−8

MANY BIVARIATE DISTRIBUTIONS HAVE THE SAME STATISTICS

In this chapter we have been considering **bivariate distributions**, which are so named because they are distributions of *pairs* of scores drawn from dependent populations. So far, we've mainly depicted bivariate distributions in scatterplots; in all cases, the scatter of *xy* pairs has formed an oriented ellipse. That is, if you were to draw an outline around the scatter of points in Figures 13.2, 13.4, 13.6, and 13.7, the result would be an ellipse whose long axis is described by the regression line. Not all bivariate distributions have this shape. In this section, we'll see that the correlation coefficient itself tells us nothing about the shape of the bivariate distribution of scores from which it was computed.

Anscombe (1973) developed an excellent illustration of how samples with very similar statistics can differ dramatically in their shapes. Table 13.5 shows four small data sets, comprising *xy* pairs. All *x* variables have the same mean and standard deviation (9 and 11, respectively), and all *y* variables have the same mean and standard deviation (7.5 and 4.1, respectively). Furthermore, all samples produce the same correlation coefficient ($r = .82$). However, when the data are plotted (see Figure 13.9), we can see that they are distributed in radically different ways.

The first pair (x_1, y_1) looks like a subset of Figure 13.2. The second pair (x_2, y_2), however, is clearly non-linear. In this case, *y* scores increase over much of the range of *x* but then decrease at the largest values of *x*. So, one could say that there is a general increase in *y* with increases in *x*, but a straight line is a very poor characterization of these data. The third pair (x_3, y_3) is almost perfectly linear with one *outlier* that is out of line with the pattern of the data. Finally, without the outlier in the fourth pair (x_4, y_4), the correlation between *x* and *y* would be zero. That is, for all pairs except the outlier, the *x* value is 8; so for 10 of 11 pairs, *y* does not vary with *x*.

In Chapters 14 and 15 we will discuss inference about the population regression slope (β) and the population correlation coefficient (ρ). For these inferences to be valid, we

Bivariate distributions are distributions of pairs of scores.

TABLE 13.5 ■ Anscombe's Quartet								
	x_1	y_1	x_2	y_2	x_3	y_3	x_4	y_4
	10.0	8.0	10.0	9.1	10.0	7.5	8.0	6.6
	8.0	7.0	8.0	8.1	8.0	6.8	8.0	5.8
	13.0	7.6	13.0	8.7	13.0	12.7	8.0	7.7
	9.0	8.8	9.0	8.8	9.0	7.1	8.0	8.8
	11.0	8.3	11.0	9.3	11.0	7.8	8.0	8.5
	14.0	10.0	14.0	8.1	14.0	8.8	8.0	7.0
	6.0	7.2	6.0	6.1	6.0	6.1	8.0	5.3
	4.0	4.3	4.0	3.1	4.0	5.4	19.0	12.5
	12.0	10.8	12.0	9.1	12.0	8.2	8.0	5.6
	7.0	4.8	7.0	7.3	7.0	6.4	8.0	7.9
	5.0	5.7	5.0	4.7	5.0	5.7	8.0	6.9
Mean	9.0	7.5	9.0	7.5	9.0	7.5	9.0	7.5
Variance	11.0	4.1	11.0	4.1	11.0	4.1	11.0	4.1
r	$r_{x_1 y_1}$	0.82	$r_{x_2 y_2}$	0.82	$r_{x_4 y_4}$	0.82	$r_{x_4 y_4}$	0.82

FIGURE 13.9 ■ The Anscombe Quartet

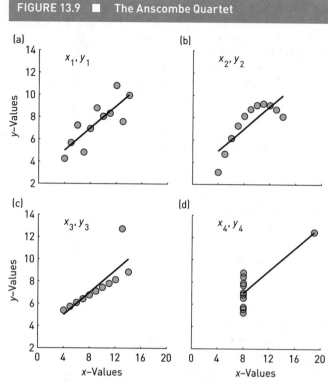

(a through d) Each plots y as a function of x. The correlation between x and y is $r = .82$ in all plots. It is clear, however, that the four bivariate distributions are very different. The lesson here is to always check your data to make sure that the distributional assumptions of your statistical procedure have been met.

will have to make some assumptions about the population from which the sample was drawn. The lesson here is that it's critical to check our data to make sure they conform to the assumptions made about our statistical procedure.

RANDOM VARIABLES, EXPERIMENTS, AND CAUSATION

You may be familiar with the caution that *correlation does not imply causation*. This statement can be misleading if it is not properly understood. In one sense, correlation simply refers to the correlation coefficient. However, when people caution against inferring causation from correlation, they are really referring to the nature of the two variables from which the correlation coefficient has been computed. We can infer a causal association between x and y under appropriate circumstances.

We met essentially the same issue of causation and correlation when we discussed the distinction between quasi-experiments and experiments in Chapter 11. There we saw that a difference between two naturally occurring groups (e.g., smokers and nonsmokers) on some variable of interest (e.g., cardiovascular health) does not mean that the difference was *caused* by the behaviors that distinguish the groups. For example, if smokers have poorer cardiovascular health than nonsmokers, we cannot conclude that poor cardiovascular health in smokers was caused by smoking. However, when there is random assignment to groups, inferences about causation are more easily made. Even though the difference between experiments and quasi-experiments was shown to have important implications for inferences about causation, the mechanics of computing confidence intervals around $m_1 - m_2$ were the same in both cases. The same is true in the case of correlation. That is, whether causation can be inferred from a correlation coefficient depends on whether there is random assignment.

Random Variables

When the score on some variable (e.g., hours of study, or midterm test grade) results from random selection, statisticians refer to the variable as a **random variable**. This terminology is new, but it simply provides a name for something we've dealt with throughout this book. If we draw a random selection of first-year statistics students and measure their hours of study and midterm test grades, then we have dependent samples of scores for two random variables. We can compute the correlation coefficient between these two variables to assess the precise degree of linear association between them. The caution about inferring causation from correlation applies to cases like this, in which

A variable is said to be a **random variable** when scores on the variable arise from random selection.

the correlation coefficient is computed from two random variables drawn from dependent populations.

Correlation and Causation: Random Variables

Let's now consider two other cases in which the caution about correlation and causation applies. Imagine a random selection of North American cities. We can count the number of street vendors and the number of cases of food poisoning for each city. These two variables (number of street vendors and number of cases of food poisoning) are random variables. Imagine that the correlation between these two variables is $r = .85$ (see Figure 13.4a). How should we interpret this correlation? Our minds often look for causal connections between correlated events. It's easy to imagine (rightly or wrongly) that street vendors sell food of questionable quality, and so we can quickly jump to the conclusion that this explains the correlation. That is, more people are getting food poisoning because more people are eating sketchy food. We might even think that the regulations governing street vendors ought to be tightened.

Even though we can readily generate this causal explanation, is there any reason to think that something else might explain the association between the number of street vendors and the number of cases of food poisoning? Well, one factor that comes to mind is city size. The more people in a city, the more street vendors, and the more people in a city, then, inevitably, the more cases of food poisoning. Therefore, the association between these two random variables might be strong because both variables are strongly associated with a third variable. In this case, the third variable is city size. In this case, we may say that the correlation between the number of street vendors and the number of cases of food poisoning is *spurious*.

Here's a related example. Many years ago, people started to notice an association between smoking and lung disease. People who smoked seemed to have more lung-related problems. From our vantage point, we know that smoking causes cancer, but this connection wasn't always known. The early evidence about the association between smoking and lung disease was based on the associations between random variables, so a third variable may have been responsible for both the tendency to smoke and the tendency to suffer lung problems. For example, a genetic factor might predispose people to smoke and to develop lung cancer.

As we saw in Chapter 7, Sir Ronald Fisher was arguably the most important statistician in history (Figure 13.10). Fisher argued vehemently that there was no evidence of a causal link between smoking and lung cancer because all available data were based on correlations between random variables. From today's perspective this seems silly, but we have the benefit of many years of experimental study that have clearly established a causal link between smoking and lung cancer. With only correlations between random variables, however, it is impossible to make a causal link. So, Fisher's argument was actually very logical at the time. It should be noted that most people think Fisher, in spite of his brilliance, did not take sufficient account of a growing body of experimental work that strongly suggested a causal link (see Stolley, 1991).

These two illustrations raise questions about the role of correlation in research. Just because correlation does not imply causation does not mean that correlations can never reflect causal associations. There was a correlation between smoking

FIGURE 13.10 ■ R.A. Fisher: Smoker

Sir Ronald Fisher argued for years that the evidence relating smoking to lung cancer involved correlations between random variables, and hence the link was not necessarily causal.

(Photo courtesy of The Barr Smith Library, The University of Adelaide)

and cancer and we now know that it is a causal association. What about street vendors and food poisoning? The association might or might not be causal, but with only a correlation between two random variables, we can't tell which. There are ways to approach, these questions. One approach is statistical. When a third variable (e.g., population) is thought to explain the correlation between two others (e.g., street vendors and food poisoning), it is possible to "control for" the third variable and then see if the correlation still exists. This is a common and useful approach, and we will learn more about it when we discuss multiple regression in Chapter 17.

The best solution to ambiguous correlations is to run experiments. One way to address the question of whether food from street vendors causes digestive problems would be to choose a random sample of individuals and divide them into two equal groups. Half would eat their usual diet and the other half might eat one meal a day from a street vendor. If we measure the number of instances of food poisoning for each participant over the course of a month, then we can estimate the difference in digestive problems between the two groups. With large enough groups we can precisely estimate the difference and then decide if it is big enough to warrant actions that might affect the livelihoods of street vendors.

Given the ambiguities associated with correlation, it is best to view associations between random variables as contributing to the preliminary stages of research. Most research in science tries to explain *why* something occurs. Simple correlation between two random variables can convey the strength of an association but not why it exists.

Correlation, Causation, and Random Assignment

Figure 13.11a shows the association between hours of sleep the previous night (*x*) and time required to solve a simple puzzle (*y*) for a sample drawn from a hypothetical population of university students. This sample was drawn during the final exam period, a time during which many students sleep less than usual. In this case, there is a negative correlation between the two sets of scores: the more sleep one had the previous night, the less time it takes to solve the puzzle. As with any correlation between two random variables, the question of causation arises. It seems quite plausible that sleep deprivation leads to slower cognitive functioning, but this can't be inferred from correlation alone.

A correlational approach can be used in an *experimental* context to produce stronger evidence for a causal association between sleep deprivation and speed of cognitive functioning. Such an experiment could involve randomly assigning participants to have between 0 and 8 hours of sleep the night before being asked to complete a number of simple puzzles. In this case, the predictor variable (*number of hours of sleep*) *is not a random variable*. Rather, the levels of the variable are fixed by the experimenter. We say that *number of hours of sleep* is the independent variable, and study participants are randomly assigned to one of its levels. The dependent variable (*puzzle-solving time*) is a random variable. That is, at each level of the independent variable, the scores represent a random selection of puzzle-solving times. Figures 13.11b through 13.11f show fixed levels of the independent variable (*number of hours of sleep*) that have been determined by the experimenter. At each of these fixed levels, 12 puzzle-solving times have been obtained from 12 participants.

No matter how many levels of the independent variable there are, the correlation between *x* (*number of hours of sleep*) and *y* (*puzzle-solving time*) is computed in exactly the same way as when the two variables are random variables. Figure 13.11 shows the

FIGURE 13.11 ■ Fixed Versus Random Variables

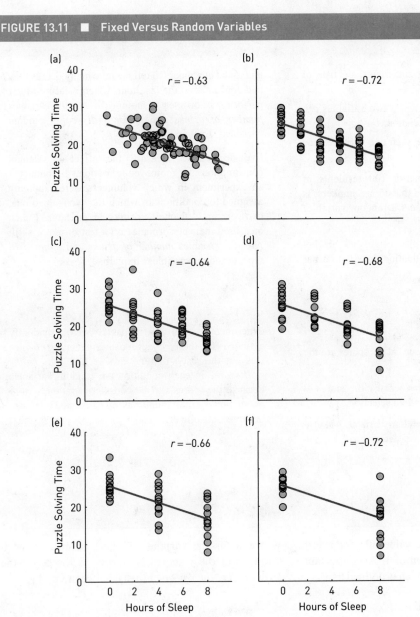

(a) Both x and y are random variables. (b through f) The x values are fixed by the experimenter and y is a random variable. The correlation coefficient is shown in each graph, along with the regression line through the points.

correlation between x and y, along with the regression line. In these cases, there is less ambiguity about the association between the two variables. Because there was random assignment, we can say with more confidence that the decrease in puzzle-solving time was caused by increased sleep. Put the other way around, the increase in puzzle-solving time was caused by decreased sleep.

LEARNING CHECK 5

1. State whether the following statements are true or false.

    (a) A positive correlation between two variables can never imply a causal association.

    (b) A negative correlation between two variables can never imply a causal association.

2. I've taken a random sample of 100 residents of Kingston, Jamaica, and found that 15 are smokers. Is the variable *number of smokers* a random variable? Explain.

3. I've decided to measure the effect of alcohol consumption on motor coordination. I went to a bar and asked customers to tell me how many drinks they had consumed in the last hour. I then measured their performance on a test of motor skill. Are the variables *number of drinks* and *motor skill scores* random variables? Explain.

4. I've decided to measure the effect of alcohol consumption on motor coordination. I conducted an experiment in which volunteers were randomly assigned to conditions in which they consumed one, two, or three alcoholic drinks in an hour. I then measured their performance on a test of motor skill. Are the variables *number of drinks* and *motor skill scores* random variables? Explain.

Answers

1. (a) False. (b) False.

2. Yes. It is a random variable because the scores are the result of random selection.

3. Yes. Both are random variables because the scores in both cases are the result of random selection.

4. *Number of drinks* is not a random variable because it has fixed levels that were determined by the experimenter and participants were randomly assigned to these conditions. *Number of drinks* was not the result of random selection. The *motor skill scores* in each condition are random variables because the scores can be considered samples from populations of individuals having consumed one, two, or three alcoholic drinks.

SUMMARY

When there is one and only one value of y for every value of x, we say y is a *function* of x. An important function is the linear function, which is defined as

$$y = a + b(x).$$

When there are many values of y for each value of x, we say that there is an *association* between x and y. When a straight line best describes the association between x and y, we say that the association is a *linear association*.

The correlation coefficient characterizes the precise degree of linear association between two scale variables. The correlation coefficient can be defined in three equivalent ways:

$$r = \frac{cov_{xy}}{\sqrt{s_x^2 s_y^2}},$$

where s_x^2 is the variance of the x scores, s_y^2 is the variance of the y scores, and cov_{xy} is the covariance $[cov_{xy} = \Sigma(x - m_x)(y - m_y)/(n-1)]$ between x and y;

$$r = \frac{sp_{xy}}{\sqrt{ss_x ss_y}},$$

where $ss_x = \Sigma(x - m_x)^2$, $ss_y = \Sigma(y - m_y)^2$, and $sp_{xy} = \Sigma(x - m_x)(y - m_y)$; and

$$r = \frac{\sum z_x z_y}{n-1},$$

where $z_x = (x - m_x)/s_x$, $z_y = (y - m_y)/s_y$, and n is the number of xy pairs.

The regression equation is defined as

$$\hat{y} = a + b(x),$$

where a and b are the intercept and slope of the regression line, respectively. $\hat{y}$ denotes the best prediction of y from x. The regression equation can be computed from the statistics of x and y as follows:

$$b = r\frac{s_y}{s_x} \text{ and}$$

$$a = m_y - b(m_x),$$

where r is the correlation coefficient.

We say the regression line is the *line of best fit in the least squares sense*, because no other straight line would produce a smaller sum of squared deviations of data points from the line; i.e.,

$$ss_{y-\hat{y}} = \sum(y-\hat{y})^2.$$

$ss_{y-\hat{y}}$ is often called ss_{error} because it is the sum of *squared prediction errors*. The total sum of squares in y,

$$ss_{total} = \sum(y-m_y)^2,$$

can be partitioned into two components. One component is ss_{error}, and the other is $ss_{regression}$:

$$ss_{regression} = \sum(\hat{y}-m_y)^2,$$

which is also defined as

$$ss_{regression} = ss_{total} - ss_{error}.$$

The proportion of variability in y that is explained by y's association with x is

$$r^2 = \frac{ss_{regression}}{ss_{total}}.$$

r^2 is also the square of the correlation coefficient. Therefore, correlation and regression are two faces of the same underlying association.

The correlation coefficient tells us nothing about the shape of the distribution of scores from which it was computed. This point is well illustrated by the Anscombe quartet, which comprises four sets of xy pairs yielding dramatically different scatterplots while having identical statistics (m_x, s_x, m_y, s_y, and r).

The caution that *correlation does not imply causation* applies to situations in which x and y are *random variables* from dependent populations. The term "random variable" simply means that the scores of the variable result from random selection. It is possible to choose random samples of y scores for values of x that have been fixed by an experimenter. In this case, it is possible to ascribe a causal association between x and y.

KEY TERMS

association 324	function 323	perfect correlation 332
bivariate distributions 343	least-squares fit 342	r 338
correlation coefficient 327	linear regression line 327	random variable 344
covariance 329	linear transformation 330	scatterplot 324

EXERCISES

Please note that asterisks (*) indicate questions relating to material covered in the appendixes or requiring insight beyond what was covered in the chapter.

Definitions and Concepts

1. What does it mean for y to be a function of x?

2. What does it mean for the association between x and y to be probabilistic?

3. What does it mean for the association between x and y to be linear?

4. What does it mean for the association between x and y to be non-linear?

5. What kind of association between two variables will we be considering in this section of the book?

6. What is a scatterplot?

7. If I say that grades increase with the number of hours studied, which variable is the predictor variable and which is the outcome variable? If I were to plot this association on a scatterplot, which would go on the x-axis and which on the y-axis?

8. Give examples of pairs of variables that you would expect are (a) positively correlated, (b) negatively correlated, and (c) uncorrelated.

9. What range of values can the correlation coefficient take on?

10. Show the "computational" equation for the correlation coefficient.

11. How would the correlation between $x = [1, 2, 3]$ and $y = [4, 2, 3]$ change if all the values in y were multiplied by 12.543?

True or False

State whether the following statements are true or false.

12. A correlation coefficient can never imply a causal association between x and y.

13. $r^2 = ss_{\hat{y}}/ss_{total}$.

14. $r^2 = ss_{regression}/ss_y$.

15. $r^2 = ss_{total}/ss_y$.

16. $r^2 = ss_{y-\hat{y}}/ss_{error}$.

17. $r^2 = (ss_{total} - ss_{error})/ss_{total}$.

18. $\sum z_x z_y/(n-1) = sp_{xy} * \sqrt{ss_x ss_y}$.

19. If $x = [1, 2, 3, 4, 5, 6, 7, 8, -9]$ and $y = 5 + 3(x^2)$, $r = 1$.

20. If $x = [1, 2, 3, 4, 5, 6, 7, 8, -9]$ and $y = 5^2 + 0.312(x)$, $r = 1$.

21. The association between height in miles and weight in grams is probabilistic.

Calculations

22. Compute the correlation coefficient for each of the following xy pairs:
    (a) $x = [1, 2, 3], y = [1, 2, 3]$;
    (b) $x = [1, 2, 3], y = [3, 2, 1]$;
    (c) $x = [1, 2, 3], y = [1, 3, 2]$; and
    (d) $x = [1, 2, 3, 4, 5, 6, 7], y = [1, 3, 2, 4, 7, 6, 5]$.

23. Calculate the regression equation for each of the following situations:
    (a) $r = -.34, m_x = 12, s_x = 1.2, m_y = 12, s_y = 1.2$
    (b) $r = -.80, m_x = 22, s_x = 12, m_y = 12, s_y = 1.2$
    (c) $r = .50, m_x = 22, s_x = 6, m_y = 12, s_y = 4$
    (d) $r = 1.3, m_x = 12, s_x = 2, m_y = 12, s_y = .2$

24. What is the relationship between $ss_{error}, ss_{regression}$, and ss_{total}?

25. If $r = .50$, what proportion of the total variability in y is explained by y's association with x?

26. *If $r = .50, n = 25$, and $s_y = 10$, answer the following questions:
    (a) What is ss_{total}?
    (b) What is $ss_{regression}$?
    (c) What is ss_{error}?
    (d) What proportion of the total variability in y is explained by y's association with x?
    (e) What proportion of the total variability in y is not explained by y's association with x?

Scenarios

27. For the following data set, compute the following: (a) sp_{xy}, (b) r, (c) s_x, (d) s_y, (e) b, (f) a, and (g) for $x = 13$, compute $\hat{y}$.

x	y
9	24
16	28
11	22
10	25
11	24
10	18
10	21
11	16
12	24
11	14
10	15
m_x	m_y
11	21
ss_x	ss_y
34	212

28. After a recent statistics midterm, several students were disappointed in their grades. Six of these students sought tutoring. The following table shows the number of hours of tutoring that each of these six students received, as well as the change in their grades between the first and second midterm. That

is, the column labeled "Change" shows *midterm2* – *midterm1* for each student.

Hours	Change
16	2
6	5
24	2
8	2
11	–3
25	16
m_{hours}	m_{change}
15	4
ss_{hours}	ss_{change}
328	206

(a) Compute the correlation coefficient between *hours* and *change*.

(b) What proportion of variability in *change* is explained by *hours*?

(c) If this is treated as a regression problem, which variable is best described as the predictor and which is best described as the outcome?

(d) Compute the regression equation.

(e) Compute $\hat{y}$ for *hours* = 10.

(f) Which pair of scores in the table is unusual? (It may help to plot these data to determine which pair is unusual.)

(g) Compute the correlation coefficient with this unusual pair of scores removed.

APPENDIX 13.1: CORRELATION AND REGRESSION IN EXCEL

This appendix introduces four new functions relating to correlation and regression. The first two columns in Figure 13.A1.1 show values for **x** and **y** and are the first pair from the Anscombe quartet shown in Figure 13.7. In cells **D2** to **D5**, the means and variances of x and y have been computed in the usual way.

Cell D6 introduces the **CORREL** function, which computes the correlation coefficient for us. In this case, $r = .819$. To compute **r²**, we could simply square **r**. Or we can use the **RSQ** function provided by Excel and shown in cell **D7**. Excel also provides functions to compute the slope (**b**) and intercept (**a**) of the regression equation. These functions, called **SLOPE** and **INTERCEPT**, are shown in cells **D8** and **D9**, respectively. Notice that when using both **SLOPE** and **INTERCEPT**, the first argument (**B2:B12**) refers to scores on the outcome variable (y) and the second argument (**A2:A12**) refers to scores on the predictor variable (x). Of course, we can always compute the regression equation from the statistics of the samples as shown in cells **D11** and **D12**. In **D11**, we have computed

FIGURE 13.A1.1 ■ **Correlation and Regression**

	A	B	C	D	E
1	**x**	**y**	**Quantity**	**Values**	**Formulas**
2	10.0	8.0	m_x	9.000	=AVERAGE(A2:A12)
3	8.0	7.0	s^2_x	11.000	=VAR.S(A2:A12)
4	13.0	7.6	m_y	7.500	=AVERAGE(B2:B12)
5	9.0	8.8	s^2_y	4.084	=VAR.S(B2:B12)
6	11.0	8.3	r	0.819	=CORREL(A2:A12,B2:B12)
7	14.0	10.0	r^2	0.671	=RSQ(A2:A12,B2:B12)
8	6.0	7.2	b	0.499	=SLOPE(B2:B12,A2:A12)
9	4.0	4.3	a	3.008	=INTERCEPT(B2:B12,A2:A12)
10	12.0	10.8			
11	7.0	4.8	b	0.499	=D6*SQRT(D5)/SQRT(D3)
12	5.0	5.7	a	3.008	=D4-D11*D2

$$b = r\frac{s_y}{s_x}.$$

In cell **D12**, we have computed

$$a = m_y - b(m_x).$$

14 INFERENTIAL STATISTICS FOR SIMPLE LINEAR REGRESSION

INTRODUCTION

In Chapter 13, we saw that the regression equation computed from a sample of scores is $\hat{y} = a + b(x)$. Because it was computed from a sample, the slope (b) and intercept (a) are statistics. In this chapter, we will see that b estimates β, which is the slope of the regression equation in a population, and a estimates α, which is the intercept of the regression equation in a population. Of these two parameters, β is of more interest because it tells us how much change in y results from a unit change in x.

We also saw in Chapter 13 that regression analysis can be applied when x is fixed by the researcher and when x is a random variable. We will work through examples involving both of these situations, and we will comment on the assumed sampling distributions associated with each. Before we discuss the statistical model underlying inferential statistics for regression, we will introduce two examples to work with.

Example 1: Regression When x Values Are Fixed

We previously theorized that the time to perform cognitive operations *increases* as sleep deprivation *increases*. To assess this theory, let's say we choose 20 individuals (at random) from our department's participant pool and we randomly assign participants to one of four conditions in which they are deprived of 0, 2, 4, or 6 hours of sleep. Sleep deprivation is achieved by having the participants go to bed 0, 2, 4, or 6 hours later than usual. (In this example, our independent variable is *hours of sleep deprivation*, rather than hours of sleep as it was in Chapter 13.) At 9:00 the following morning, we measure the time it takes each participant to complete four simple puzzles. In this case, we have used *puzzle-solving time* as an operational definition

iStock.com/Rawpixel

of cognitive processing speed. After we collect our data, we find that the correlation between *sleep deprivation* and *puzzle-solving time* is $r = .67$.

Although this is useful, the correlation coefficient doesn't tell us directly *how much puzzle-solving time* increases as *hours of sleep deprivation* increases. The answer to this question comes from the slope of the regression equation. Let's say that the regression equation has slope $b = 1.1$ and intercept $a = 16.7$. Therefore, our best estimate is that each additional hour of *sleep deprivation* increases *puzzle-solving time* by 1.1 minutes. It is in this way that the regression equation answers the "how much" question.

Not surprisingly, we compute a confidence interval around b using the following equation:

$$CI = b \pm t_{\alpha/2}(s_b).$$

The only thing new in this equation is s_b. In the next section, we will discuss how s_b is calculated and what it estimates. When we work through this example in detail, we will see that the 95% confidence interval around b is [0.49, 1.71]. This means that we have 95% confidence that the true slope of the regression equation (β) relating *hours of sleep deprivation* to *puzzle-solving time* is between 0.49 minutes per hour and 1.71 minutes per hour.

APA Reporting

The slope of the regression equation relating *hours of sleep deprivation* to *puzzle-solving time* is estimated to be $b = 1.1$, 95% CI [0.49, 1.71].

Example 2: Regression When *x* Is a Random Variable

Our second example illustrates how simple linear regression is used in the early stages of research to explore associations between two random variables. For example, we may wonder about the association between television exposure and cognitive development (Tomopoulos, Dreyer, Berkule, Fierman, Brockmeyer, & Mendelsohn, 2010). To address this question, we will imagine a sample of 200 mother-child dyads selected at random from the population of a large urban center. (A dyad is a group of two people.) This hypothetical study was conducted in two stages. In the first stage, researchers visited the homes of infants aged 6 months and obtained estimates from the mothers of how much *non-child-directed* television their infant was exposed to on the previous day. Eight months following the first visit, the researchers returned to the homes (when the children were 14 months old) and assessed the infants' cognitive development using the third edition of the Bayley Scales of Infant and Toddler Development (Bayley III; Bayley, 2006). The question is "What is the nature of the association between exposure to television at 6 months and cognitive development at 14 months?" In this study, *television exposure* (hours of television viewing) is the predictor variable (x) and the *cognitive development score* is the outcome variable (y).

The results of the (hypothetical) study showed that the mean television viewing was $m_x = 2.5$ hours ($s_x = 2$) and the mean cognitive development score was $m_y = 95$ ($s_y = 12$). The correlation between the two variables was $r = -.2$. From these data, we can compute the regression equation as

$$b = r\frac{s_y}{s_x} = -0.2\frac{12}{2} = -1.2, \text{ and}$$

$$a = m_y - b(m_x) = 95 + 1.2(2.5) = 98.$$

The regression equation shows that each additional hour of television viewing at 6 months is associated with a decrease of 1.2 units on the Bayley III at 14 months. The 95% confidence interval around b is [−2.03, −0.37]. So, we have 95% confidence that the true regression slope relating television viewing to cognitive development falls in the interval [−2.03, −0.37]. This finding may be reported in the following way:

iStock.com/Ghislain & Marie David de Lossy

The slope of the regression equation relating cognitive development scores at 14 months to non-child-directed television exposure at 6 months is estimated to be $b = -1.2$, 95% CI [−2.03, −0.37].

This confidence interval is calculated exactly as in the previous example:

$$\text{CI} = b \pm t_{\alpha/2}(s_b).$$

Therefore, the mechanics of computing confidence intervals in Examples 1 and 2 are the same, even though we'll see that the underlying statistical models are somewhat different.

REGRESSION WHEN VALUES OF *x* ARE FIXED: THEORY

We will first describe the assumed bivariate distribution of scores underlying inferences about the regression coefficient (*b*) when values of *x* are fixed. We will then describe the sampling distribution of *b*.

The Assumed Bivariate Population

It will help to use the concrete example of sleep deprivation from Example 1 to illustrate the nature of the bivariate population of scores underlying the regression model. Therefore, we will think about the population from which the sample of *xy* pairs in our example was drawn. In this example, the *x* variable, *sleep deprivation*, has *fixed* values of 0, 2, 4, and 6. Although one could be sleep deprived for 5 hours or more than 6 hours, such values play no role in the model. The only values relevant to the model are those fixed by the researcher.

Conditional Distributions and Expected Values

> In regression analysis, a **conditional distribution** is the distribution of *y* scores at a given value of *x*.

Figure 14.1 illustrates the nature of the bivariate population underlying the regression model. The *x*-axis in Figure 14.1a shows the four levels of sleep deprivation in our illustration. For each value of *x*, there is a normal distribution of *y* scores, illustrated by the gray dots. We call these **conditional distributions** because the mean of each distribution depends, or is *conditional*, on *x*. In Chapter 5, we called the mean of a distribution its *expected value* and denoted it as $E(y)$. When the expected value of *y* is conditional on *x*, we denote it as $E(y|x)$.

Figure 14.1b shows the conditional distributions as light blue lines rotated 90° from their usual orientation with their density on the *x*-axis and values of the outcome variable on the *y*-axis. Although it is a bit unusual, this rotation makes it easier to see the connection between both Figure 14.1a and 14.1b, and the normality of the conditional distributions. The means of the conditional distributions in Figure 14.1b are aligned with the means of the conditional distributions in Figure 14.1a. The dark-gray distribution in Figure 14.1b is called a *marginal distribution*, which we will discuss later.

The Regression Equation and Expected Values

The regression model assumes that the means of the conditional distributions change linearly with *x*. Therefore, we can express the association between *x* and $E(y|x)$ as follows:

$$E(y|x) = \alpha + \beta(x).$$

The parameters of this equation, α and β, are the intercept and slope of a line, respectively. In Figure 14.1, $\alpha = 20$ and $\beta = 1.5$. Therefore, $E(y|x) = 20 + 1.5(x)$, which means that $E(y|x)$ increases by 1.5 minutes for each additional hour of sleep deprivation.

The white circles in Figure 14.1a show $E(y|x)$ for each conditional distribution.

Because each individual in a bivariate population corresponds to an xy pair, there is a distribution of x scores and a distribution of y scores. We call these two distributions **marginal distributions**. These marginal distributions are not assumed to be normal. The mean and standard deviation of the marginal distribution of x scores are denoted μ_x and σ_x, and the mean and standard deviation of the marginal distribution of y scores are denoted μ_y and σ_y.

The marginal distribution of y is shown as the dark-gray line in Figure 14.1b. This is just the distribution of all y scores treated as a single population rather than four "subpopulations." We encountered the same idea in Figure 3.5 when we discussed multimodal distributions arising from subpopulations having different means. In this example, the marginal distribution of y values has a standard deviation of $\sigma_y = \sqrt{17.5} = 4.18$, as shown in Figure 14.1b.

Because the population variance of the x-values [0, 2, 4, 6] is 5, the marginal distribution of x values has a standard deviation of $\sigma_x = \sqrt{5} = 2.24$. (We will return to this point below.)

The correlation coefficient describing the strength of the linear association between all xy pairs in a bivariate population is denoted ρ (the Greek letter r, pronounced "rho"), as mentioned in Chapter 13. The population correlation coefficient can be computed using any of the three formulas described in Chapter 13. For example, we can use the simple computational formula to compute

$$\rho = \frac{sp_{xy}}{\sqrt{ss_x ss_y}}. \tag{14.1}$$

For the bivariate population in our example, $\rho = .8018$, as shown in Figure 14.1a.

In parallel to what we saw in Chapter 13, the slope of the population regression equation can be derived from ρ, σ_x, and σ_y as follows:

$$\beta = \rho \frac{\sigma_y}{\sigma_x}. \tag{14.2}$$

Furthermore, the intercept of the regression equation can be derived from β, μ_x, and μ_y as follows:

$$\alpha = \mu_y - \beta(\mu_x). \tag{14.3}$$

Therefore, for any bivariate population of xy pairs, we can determine the regression equation from μ_x, μ_y, σ_x, σ_y, and ρ. This regression equation is the best-fitting line in the least squares sense because no other straight line would produce a smaller sum of squared deviations. Equivalently, no other line would produce a larger ρ^2, which is a measure of explained variance.

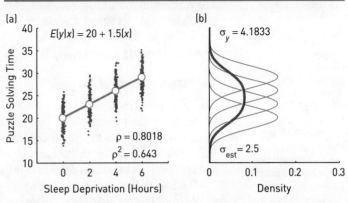

FIGURE 14.1 ■ Conditional and Marginal Distributions

Illustrations of the distributions underlying the regression model. (a) The graph shows four conditional distributions whose means are $E(y|x) = \alpha + \beta(x)$. (b) The four conditional distributions are normal, and the graph shows their density functions, rotated 90° from their usual orientation to make the connection between (a) and (b) clearer. The dark-gray line in (b) is the marginal distribution of y values.

A **marginal distribution** is a distribution of the x scores from all xy pairs or the y score from all xy pairs.

Homoscedasticity

The regression model assumes that all conditional distributions are normal and have the same standard deviation. Conditional distributions with the same standard deviations are said to be *homoscedastic*. Therefore, inferential statistics for regression requires the **homoscedasticity** assumption to be true.

The Standard Error of Estimate

We say that conditional distributions are homoscedastic when they all have the same standard deviation. Inferential statistics for regression requires the **homoscedasticity** assumption.

The variability in a conditional distribution reflects scores' deviations from the mean of the conditional distribution; i.e., deviations from $E(y|x)$. This is equivalent to saying that the variability in the conditional distributions reflects deviations from the regression line. The standard deviation that is common to the conditional distributions is called the **standard error of estimate** and is denoted with the symbol σ_{est}. We will now see how σ_{est} is computed.

If we were to compute $y - E(y|x)$ for every individual in the bivariate population, we would have a single distribution of prediction errors, which are sometimes called *residuals*. Because the conditional distributions are assumed to be homoscedastic, the variance of the distribution of prediction errors (residuals) is the same as the variance of all conditional distributions. This variance is based on the sum of squared prediction errors, which is computed as follows:

The **standard error of estimate** (σ_{est}) is the square root of the average squared deviation from the regression line in a population. σ_{est} is the standard deviation common to all conditional distributions.

$$ss_{error} = \sum (y - E(y|x))^2. \tag{14.4}$$

When ss_{error} is divided by the number of xy pairs in the population, the result is a quantity called the *variance of estimate*:

$$\sigma^2_{est} = \frac{ss_{error}}{N}. \tag{14.5}$$

The square root of σ^2_{est} is the standard error of estimate (σ_{est}), which is simply

$$\sigma_{est} = \sqrt{\frac{ss_{error}}{N}}. \tag{14.6}$$

As mentioned above, the standard error of estimate (σ_{est}) is the standard deviation of the y scores about the regression line, which is the same as saying that σ_{est} is the standard deviation common to all conditional distributions. In our example in Figure 14.1, $\sigma_{est} = 2.5$.

The variance of estimate can be defined more simply in terms of σ^2_y and ρ^2 as follows:

$$\sigma^2_{est} = (1 - \rho^2)\sigma^2_y.$$

Likewise, the standard error of estimate can be defined as

$$\sigma_{est} = \sqrt{(1 - \rho^2)\sigma^2_y}. \tag{14.7}$$

(Appendix 14.3, available at study.sagepub.com/gurnsey, provides a simple explanation for why equations 14.6 and 14.7 are equivalent. However, it can be skipped without losing the thread of this discussion.) Earlier we noted that, in our example, the marginal distribution of y values has a variance of $\sigma^2_y = 17.5$ and the correlation between x and y is $\rho = .8018$. When we insert these numbers into equation 14.7, we find that

$$\sigma_{est} = \sqrt{(1 - \rho^2)\sigma^2_y} = \sqrt{(1 - .8018^2)*17.5} = 2.5.$$

Equation 14.7 shows that σ_{est} is related to the variability in the marginal distribution of y scores that is *not* explained by regression. That is, the variance of the marginal distribution of y scores is σ_y^2, and the proportion of this variance not explained by regression is $(1 - \rho^2)$. Therefore, σ_{est} is the square root of the unexplained variance $(1 - \rho^2)\sigma_y^2$. For the purposes of what follows, equation 14.7 will be more convenient than equation 14.6, but they are two different ways of computing the same thing.

The Sampling Distribution of b

To construct a confidence interval around b, we have to know about its sampling distribution. We've said many times that a sampling distribution is the probability distribution for all possible values of a sample statistic for samples of size n. Therefore, the **sampling distribution of b** is a probability distribution for all possible values of b, for samples of size n.

Figure 14.2 provides the intuitions behind the sampling distribution of b. The white dots in Figure 14.2a represent the means [i.e., $E(y|x)$] of the four homoscedastic conditional distributions from our example. The black line is the *population regression line*, $E(y|x) = \alpha + \beta(x) = 20 + 1.5(x)$. There are 32 blue lines in Figure 14.2a, and each is the *sample regression line* [i.e., $\hat{y} = a + b(x)$] derived from 20 xy pairs made up of five y scores drawn at random from each of the four conditional distributions. Each sample regression line has a different slope (b).

When $\hat{y} = a + b(x)$ is computed for all possible samples of $4*5 = 20$ xy pairs, we obtain the sampling distribution of b as shown in Figure 14.2b. It is a normal distribution with mean

> The **sampling distribution of b** is a probability distribution for all possible values of b, for samples of size n, drawn from a homoscedastic bivariate distribution with fixed values of x.

$$\mu_b = \beta \tag{14.8}$$

and standard error

$$\sigma_b = \frac{\sigma_{est}}{\sqrt{ss_x}}. \tag{14.9}$$

Equation 14.8 shows that b is an unbiased estimator of β. Equation 14.9 shows that σ_b decreases as sample size increases, as is always the case. Remember, ss_x is the sum of squared deviations, so the larger the sample, the more squared deviations there are and thus the larger the denominator of equation 14.9 will be.

We have already seen that $\sigma_{est} = 2.5$; so to compute σ_b, we need only to think about ss_x. The four x values are 0, 2, 4, and 6, which have $\mu_x = 3$. Therefore, with five scores at each of these four levels, $ss_x = 5\Sigma(x - \mu_x)^2 = 100$. That is, $ss_x = 5(9 + 1 + 1 + 9) = 100$. (And so the variance is $s_x^2 = 100/20 = 5$.) From these numbers, we can determine that the standard error of b in this example is

$$\sigma_b = \frac{\sigma_{est}}{\sqrt{ss_x}} = \frac{2.5}{\sqrt{100}} = 0.25.$$

This is the value of σ_b for the distribution shown in Figure 14.2b. Ninety-five percent of this distribution lies in the interval $\mu_b \pm 1.96(\sigma_b) = 1.5 \pm 1.96(0.25)$, or [1.01, 1.99].

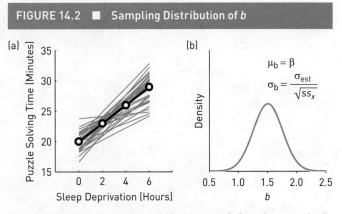

FIGURE 14.2 ■ Sampling Distribution of b

(a) The white dots represent the means of four homoscedastic conditional distributions; i.e., $E(y|x)$. The black line is the population regression line, $E(y|x) = \alpha + \beta(x) = 20 + 1.5(x)$. Each blue line is a sample regression line [$\hat{y} = a + b(x)$] derived from 20 xy pairs. Each sample regression line has a different slope (b). (b) The sampling distribution of the regression coefficient (b) has a mean equal to β and a standard error of $\sigma_b = \sigma_{est}/\sqrt{ss_x}$. The shape of the sampling distribution is normal if the samples are drawn from a homoscedastic, bivariate distribution with fixed values of x.

A Confidence Interval Around b

To compute a confidence interval around b, we must first estimate the standard error of its sampling distribution (σ_b), which will require an estimate of σ_{est}. The simplest estimator of σ_{est} is a slightly modified version of equation 14.7:

$$s_{est} = \sqrt{s_y^2(1-r^2)\frac{n-1}{n-2}}. \tag{14.10}$$

In equation 14.10, s_y^2 replaces σ_y^2 and r^2 replaces ρ^2. However, the term $s_y^2(1-r^2)$ is a biased estimate of $\sigma_y^2(1-r^2)$; therefore, $s_y^2(1-r^2)$ is multiplied by $(n-1)/(n-2)$ to increase it slightly and remove the bias.

The sum of squares for x, ss_x, is computed from all x values in the sample, exactly as in the population. With estimates of σ_{est} and ss_x, we can now modify equation 14.9 to compute an estimate of σ_b as follows:

$$s_b = \frac{s_{est}}{\sqrt{ss_x}}. \tag{14.11}$$

As shown in the opening paragraphs of this chapter, a confidence interval for b relies on the t-distribution. Because the sampling distribution of b is normal, it can be transformed into a t-distribution by computing

$$t = \frac{b-\beta}{s_b}$$

for all possible samples of size n, where n is the number of xy pairs. By definition, we know that $(1-\alpha)100\%$ of all t-scores fall in the interval $0 \pm t_{\alpha/2}$. As a consequence, $(1-\alpha)100\%$ of all intervals defined as $t \pm t_{\alpha/2}$ will capture 0, which is the mean of the t-distribution. Therefore, $(1-\alpha)100\%$ of confidence intervals computed as

$$CI = b \pm t_{\alpha/2}(s_b) \tag{14.12}$$

will capture β, which is the parameter we're estimating. In this case, $t_{\alpha/2}$ is based on $n-2$ degrees of freedom, where n is the number of xy pairs.

REGRESSION WHEN x VALUES ARE FIXED: AN EXAMPLE

The data in Table 14.1 are from our hypothetical sleep deprivation experiment. There were four conditions in which participants were deprived of 0, 2, 4, or 6 hours of sleep on the night before being given puzzles to solve. There were five participants in each group, and the numbers in the table show the time (in minutes) required to complete the four puzzles for each participant. The sample means, sums of squares, variances, and standard deviations are shown at the bottom of the table. Figure 14.3 displays the data from Table 14.1 as a scatterplot along with the regression line.

Computing the Regression Equation

We can now rearrange the data from Table 14.1 to form columns of x and y values and then compute the regression equation

TABLE 14.1 ■ Sample Data

	0	2	4	6
	23	21	26	25
	17	20	25	24
	16	18	22	23
	15	16	21	21
	13	15	19	20
m	16.8	18.0	22.6	22.6
ss	56.8	26.0	33.2	17.2
s^2	14.2	6.5	8.3	4.3
s	3.77	2.55	2.88	2.07

Note: Data from a hypothetical experiment assessing the association between sleep deprivation and puzzle-solving time.

LEARNING CHECK 1

1. State whether the following statements are true or false.

   (a) σ_b is the variance of the sampling distribution of b.

   (b) $s_y^2(1-r^2)$ is a biased estimator of $\sigma_y^2(1-\rho^2)$.

   (c) If $E(y|x) = 30 - 1.25(x)$, then $E(y|20) = 5$.

   (d) If $E(y|x) = 30 - 1.25(x)$, then $E(y|30) = -7.5$.

   (e) The regression model assumes that the conditional distributions are normal.

   (f) The regression model assumes that marginal distributions are normal.

2. If x is the number of hours slept (2, 4, 6, or 8) and y is the time taken to solve four puzzles, what can we say about our regression equation, $E(y|x) = 30 - 1.25(x)$, when $x = 24$?

3. Provide an example of distributions whose means increase with x, but whose variances are unlikely to be homoscedastic.

4. Is b a biased statistic? Explain why or why not.

Answers

1. (a) False. σ_b is the square root of the variance.

   (b) True.

   (c) True. $30 - 1.25(20) = 30 - 25 = 5$.

   (d) True. $30 - 1.25(30) = 30 - 37.5 = -7.5$.

   (e) True.

   (f) False. (See Figure 14.1b, for example.)

2. This tells us that $E(y|x) = 30 - 1.25(24) = 30 - 30 = 0$. This makes no sense because we can't solve four puzzles in 0 minutes. The problem here is that our regression equation was derived from the x values of 2, 4, 6, and 8 hours. Therefore, the equation cannot be applied to values outside this range. We don't know anything about how long it would take to solve the puzzles if we had more than 8 hours sleep or less than 2 hours.

3. Any distribution for which the variance increases with the mean would be an example of a heteroscedastic distribution. For example, think of the relationship between age and bench-pressing ability. As age increases, the mean bench-pressing ability would be expected to increase. However, the variability about the mean might be 10 pounds for 5-year-olds, but it may be more like 50 pounds for 20-year-olds.

4. b is an unbiased statistic because its expected value is β, which is the parameter it estimates.

---

exactly as we did in Chapter 13. These rearranged data are shown in Table 14.2. From these data, we first calculate r and then the regression equation. To calculate r, we compute deviation scores ($x - m_x$ and $y - m_y$) as shown in the third and fourth columns of Table 14.2. The products of these deviation scores [$(x - m_x)(y - m_y)$] are shown in the last column, and the sum of products (sp_{xy}) is shown below the products.

From the information given in Table 14.2, we can compute the correlation coefficient as follows:

$$r = \frac{sp_{xy}}{\sqrt{ss_x * ss_y}} = \frac{110}{\sqrt{100 * 272}} = .67.$$

Table 14.2 shows that the mean hours of sleep deprivation was $m_x = 3$ hours ($s_x = 2.29$) and the mean time to solve puzzles was $m_y = 20$ ($s_y = 3.78$). The correlation between the two variables is $r = .67$. From these data, we can compute the regression equation as

$$b = r\frac{s_y}{s_x} = .67\frac{3.74}{2.29} = 1.1, \text{ and}$$

$$a = m_y - b(m_x) = 20 - 1.1(3) = 16.7.$$

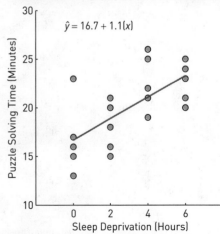

FIGURE 14.3 ■ Sample Data

$\hat{y} = 16.7 + 1.1(x)$

Puzzle Solving Time (Minutes) vs. Sleep Deprivation (Hours)

Scatterplot of the scores of each of five participants at each of the four levels of the independent variable. The independent variable was the number of hours of sleep deprivation (0, 2, 4, 6) on the night before performing a puzzle-solving task. The dependent variable was the time to complete the puzzles, in minutes.

TABLE 14.2 ■ Rearranged Data				
Sleep Deprivation (x)	Time to Solve (y)	$x - m_x$	$y - m_y$	$(x - m_x)(y - m_y)$
0	23	−3	3	−9
0	17	−3	−3	9
0	16	−3	−4	12
0	15	−3	−5	15
0	13	−3	−7	21
2	21	−1	1	−1
2	20	−1	0	0
2	18	−1	−2	2
2	16	−1	−4	4
2	15	−1	−5	5
4	26	1	6	6
4	25	1	5	5
4	22	1	2	2
4	21	1	1	1
4	19	1	−1	−1
6	25	3	5	15
6	24	3	4	12
6	23	3	3	9
6	21	3	1	3
6	20	3	0	0
m_x	m_y	ss_x	ss_y	$sp_{xy} = \sum (x - m_x)(y - m_y)$
3	20	100	272	110
s_x	s_y			$r = \dfrac{sp_{xy}}{\sqrt{ss_x * ss_y}}$
2.29	3.78			.667

Note: Data from a hypothetical sleep deprivation experiment rearranged for a regression analysis.

Computing the Confidence Interval

We now have everything in place to compute a confidence interval around b, and we have enough theoretical development to understand why it is computed the way that it is. To compute a confidence interval around b, we must compute s_b as in equation 14.11. However, to compute s_b, we first need to compute s_{est}. Let's now put this all together. We'll compute s_{est}, then s_b, and then the 95% confidence interval.

Step 1. Compute s_{est}. To compute s_{est}, we need to know n, s_y, and r. These were given earlier: $n = 20$, $s_y = 3.7836$, and $r = .667$. We can put these numbers into equation 14.10 and determine s_{est} as follows:

$$s_{est} = \sqrt{s_y^2(1-r^2)\frac{n-1}{n-2}} = \sqrt{3.78^2(1-.67^2)\frac{19}{18}} = 2.89.$$

Step 2. Compute s_b. To compute s_b, we need to know s_{est}, and ss_x. We just found that $s_{est} = 2.89$. Table 14.2 shows that $ss_x = 100$. Therefore, we can calculate s_b as follows:

$$s_b = \frac{s_{est}}{\sqrt{ss_x}} = \frac{2.89}{10} = 0.29.$$

Step 3. Compute $t_{\alpha/2}$. The degrees of freedom in this case are $n-2$, or 18. Because this is the 95% confidence interval, $\alpha = .05$. When we consult the t-table, we find that $t_{\alpha/2} = 2.101$.

Step 4. Compute the 95% confidence interval. Having determined s_b and $t_{\alpha/2}$, we can now do what we set out to do:

$$CI = b \pm t_{\alpha/2}(s_b) = 1.1 \pm 2.101(0.29) = [0.49, 1.71].$$

Remember that b estimates β, which is the slope of the regression equation in the population from which the sample was drawn. Therefore, we have 95% confidence that β falls in the interval $[0.49, 1.71]$. Our confidence comes from knowing that 95% of all intervals computed this way will capture β.

Significance Tests

The confidence interval we just computed, 95% CI [0.49, 1.71], tells us that 0 is an implausible hypothesis about β, because 0 is not in the interval. The traditional way to test the null hypothesis that $\beta = 0$ is to compute a t-statistic. t-statistics are always computed by dividing a statistic by its estimated standard error. In this case, we compute t_{obs} as

$$t_{obs} = \frac{b}{s_b}. \tag{14.13}$$

When we substitute b and s_b into equation 14.13, we obtain the following:

$$t_{obs} = \frac{b}{s_b} = \frac{1.1}{0.29} = 3.80.$$

This statistic is tested on $n-2$ degrees of freedom. By now, our understanding of t-statistics tells us right away that it is statistically significant at the $p < .05$ level of significance, whether $H_1: \beta \neq 0$, or $H_1: \beta > 0$. The exact p-value can be computed in Excel.

Interpretation

The results of this regression analysis support the idea that lack of sleep slows cognitive processing speed. Our best estimate is that each hour of sleep deprivation increases *puzzle-solving time* by 1.1 minutes, 95% CI [0.49, 1.71]. Because participants were randomly assigned to conditions, the data strongly suggest that lack of sleep caused slower *puzzle-solving times*.

However, *puzzle-solving time* per se may not be our principal interest. Rather, *puzzle-solving time* is a stand-in for more general cognitive processing speed. Therefore, we probably expect sleep deprivation to cause many other cognitive functions to slow in a similar way. For example, we might expect that reduced sleep will yield

- slower responses to events such as road signs, light changes, and approaching cars in a driving simulator or actual road conditions;

- poorer performance on a statistics exam; or

- slowed responses to client disclosures in a therapy session.

In each of these cases, it may be possible to attach meaning to the association between sleep loss and task performance. For example, we might be able to say that each hour of sleep deprivation leads to (i) a given percentage increase in average number of driving errors, (ii) a given percentage decrease in average statistics grade, or (iii) a given average number of missed opportunities for progress in a therapy session. So, the practical significance of the association between *sleep deprivation* and *puzzle-solving time* rests in the possible generality of the result. There are many follow-up studies that could assess the exact association between sleep deprivation and performance on tasks that have real-world significance.

In this kind of study, we expect that random assignment should average out systematic variations that might otherwise influence our results. Without random assignment, we run the risk of finding an association resulting from the influence of a third variable. For example, if *sleep deprivation* were a random variable, then it might be that a third variable (e.g., baseline rate of depression) is associated with both *sleep deprivation* and number of errors in a driving simulator.

On the other hand, random assignment (again, when x is not a random variable) might disguise some subtleties. For example, professional drivers (truck drivers and cabbies) might be less affected by sleep deprivation than are casual drivers (you and me). A sample that includes both professional and casual drivers might then underestimate the effect of sleep deprivation on casual drivers and overestimate its effect on professional drivers. Similar cautions can be made in the case of students' exam performance and therapists' effectiveness.

Assumptions for the Regression Model

For a confidence interval around b (or a t-test) to be valid, the following assumptions must be true.

- In the population there is a linear association between the specific fixed values of x and $E(y|x)$.

- The conditional distributions of the bivariate distribution are homoscedastic.

- The conditional distributions are normal.

- There was random sampling.

LEARNING CHECK 2

1. A researcher was interested in how well students retain information about a course after the final exam is written. Therefore, he arranged to conduct a study involving 100 statistics students, who were divided, at random, into five groups of 20. One of these five groups completed a statistics quiz immediately following their final exam. The other groups completed the quiz 1, 2, 3, or 4 months after the final. The dependent variable was the score on the statistics quiz. In this case, $m_x = 2$ and $ss_x = 200$. The mean score on the quiz was $m_y = 75$ with a standard deviation of $s_y = 10$. The correlation between x and y was $r = -.6$. Compute the 95% confidence interval around the slope of the regression line.

2. What assumptions must be true for this interval to be valid?

Answers

1. Take the following steps to compute the answer:

   (i) $b = r*s_y/s_x = -.6*10/1.42 = -4.22$.

   (ii) $s_{est} = \sqrt{s_y^2(1-r^2)*(n-1)/(n-2)}$

   $= \sqrt{100(.64)*99/98} = 8.04$.

   (iii) $s_b = s_{est}/\sqrt{ss_x} = 8.02/14.14 = 0.57$.

   (iv) $CI = b \pm t_{\alpha/2}(s_b) = -4.22 \pm 1.987(0.57)$
   $= [-5.35, -3.09]$. (*Note:* $t_{\alpha/2}$ was taken from the t-table using 90 df, because 98 df is not in the table.)

2. The four assumptions are as follows: in the population there is a linear relationship between the specific fixed values of x and $E(y|x)$, the conditional distributions are normal and homoscedastic, and there was random sampling.

REGRESSION WHEN x IS A RANDOM VARIABLE

The second example at the beginning of this chapter concerned the association between hours of non-child-directed television viewing at age 6 months and cognitive development scores at 14 months, both of which are random variables. In this case, we can compute the regression coefficient exactly as we did above and in Chapter 13. However, inferential statistics in this case are also based on the assumption that the x values are fixed, which is not exactly plausible, but this implausibility turns out to be of little practical consequence.

Bivariate Normal Distributions

Figure 14.4 illustrates the kind of distribution that may have been sampled in the *television viewing* and *cognitive development* example. Although there are many kinds of distributions in which x and y are random variables, the one shown in Figure 14.4 is a **bivariate normal distribution**. In this case, as in all bivariate normal distributions, both x and y are normally distributed. That is, the marginal distributions of x and y are normal. These marginal distributions are shown below and to the right of the scatterplot in Figure 14.4a.

In a bivariate normal distribution, there is a conditional distribution of y scores for every possible value of x. Because x is continuous, there are infinitely many values of x and thus infinitely many conditional distributions. Seven of these conditional distributions are shown as sideways normal distributions in Figure 14.4a. These seven conditional distributions are also shown in Figure 14.4b in their usual orientations. Each conditional distribution has a mean and standard deviation. As in the case of fixed values of x, the conditional distributions are homoscedastic, meaning that they have the same standard deviation (σ_{est}). Finally, the expected values of y change linearly with x according to the regression equation, $E(y|x) = \alpha + \beta(x)$.

In a **bivariate normal distribution**, x and y are normally distributed random variables and the conditional distributions are normal and homoscedastic, with expected values defined by the regression line.

FIGURE 14.4   ■   A Bivariate Normal Population

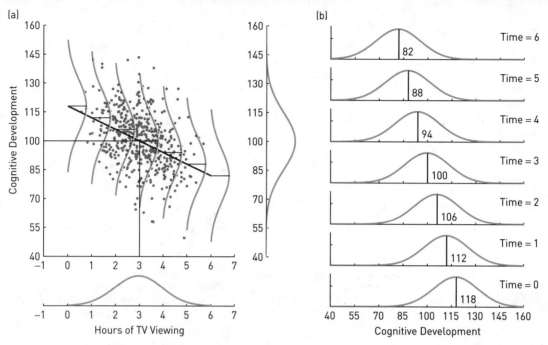

(a) The variables x and y are normally distributed. Variable x has a mean of $\mu_x = 3$ and a standard deviation of $\sigma_x = 1$. Variable y has a mean of $\mu_y = 100$ and a standard deviation of $\sigma_y = 15$. There is a negative linear association between x and y, and the line of best fit is plotted through the data points. The correlation between x and y is $\rho = -.4$. Each y value on the line of best fit represents the mean of the conditional distribution for the corresponding value of x. All conditional distributions have the same standard deviation of 13.75. (b) The seven conditional distributions illustrated as blue lines in (a) are shown in their more usual orientation.

In the bivariate normal population shown in Figure 14.4, the x variable has a mean of $\mu_x = 3$ and a standard deviation of $\sigma_x = 1$. The y variable has a mean of $\mu_y = 100$ and a standard deviation of $\sigma_y = 15$. The correlation between x and y is $\rho = -.4$. Therefore, there is a negative linear association between x and y, as shown by the regression line plotted through the data points. Each y value on the regression line represents the mean of the conditional distribution $[E(y|x)]$ for the corresponding value of x. The slope of the regression line is

$$\beta = \rho \frac{\sigma_y}{\sigma_x} = -.4 \frac{15}{1} = -6.$$

The intercept of the regression line is

$$\alpha = \mu_y - \beta(\mu_x) = 100 - -6(3) = 118.$$

The standard deviation of all conditional distributions is

$$\sigma_{est} = \sqrt{(1 - \rho^2)\sigma_y^2} = \sqrt{(1 - .4^2)15^2} = 13.75.$$

To summarize, for a distribution to be bivariate normal, the following five things must be true:

- Both the x and y distributions of scores (*marginal distributions*) are normal.

- For all values of x there is a distribution of y values (a *conditional distribution*).

- All conditional distributions are normal.

- All conditional distributions have the same standard deviation (*homoscedasticity*).

- The means of the conditional distributions $[E(y|x)]$ of y values change *linearly* with x; i.e., $E(y|x) = \alpha + \beta(x)$.

Unlike the case of fixed values of x, the marginal distribution of x is normal in the case of bivariate normal distributions. However, as we'll note below, the normality of x and y is not required for inferential statistics when x is a random variable.

The Sampling Distribution of *b*

We now come to a slightly untidy feature of regression as practiced in psychology and related disciplines. Regression analyses when x is a random variable are conducted using the assumptions of the regression model. This means treating the values of x as though they were fixed by the researcher. Therefore, the sampling distribution of b is based on repeated sampling from a bivariate distribution whose fixed values of x are the same as those in the sample. In reality, however, the sampling distribution of b should be based on different values of x in each sample because we are repeatedly drawing samples from a bivariate distribution in which both x and y are random variables.

The good news is that when we compute confidence intervals for b, we use the same formulas described above. Doing so produces intervals that capture β at very close to the desired level. That is, if we compute a 95% confidence interval for b using the equations above, the true slope relating y to x in the population (β) will be captured almost exactly 95% of the time. Therefore, any imprecision resulting from the fact that the assumed model (x is fixed) differs from reality (x is a random variable) seems to be of little consequence.

LEARNING CHECK 3

1. State whether the following statements are true or false.

   (a) σ_b is the variance of the sampling distribution of b.

   (b) If $E(y|x) = 118 - 6(x)$, then $E(y|2) = 108$.

   (c) If $E(y|x) = 118 - 6(x)$, then $E(y|3) = 100$.

2. If x is the number of hours of television viewing and y is a cognitive development score, what can we say about our regression equation, $E(y|x) = 118 - 6(-2)$?

3. Is b a biased statistic? Explain why or why not.

Answers

1. (a) False. σ_b is the square root of the variance.

   (b) False. $E(y|x) = 118 - 6(x) = 118 - 12 = 106$.

   (c) True. $E(y|x) = 118 - 6(x) = 118 - 18 = 100$.

2. This tells us that $E(y|x) = 118 - 6(-2) = 130$. But this makes no sense because we can't have -2 hours of television viewing.

3. b is an unbiased statistic because its expected value is β, which is the parameter it estimates.

REGRESSION WHEN x IS A RANDOM VARIABLE: AN EXAMPLE

Let's return to the example from the beginning of this chapter, in which we examined the association between television exposure at age 6 months and cognitive development scores at 14 months. The results of the study showed that the mean hours of television viewing was $m_x = 2.5$ hours ($s_x = 2$) and the mean cognitive development score was $m_y = 95$ ($s_y = 12$). The correlation between the two variables was $r = -.2$. We also saw that the slope of the regression equation was

$$b = r\left(\frac{s_y}{s_x}\right) = -.2\left(\frac{12}{2}\right) = -1.2.$$

With this information, we can compute the 95% confidence interval around b in the following four steps.

Step 1. Compute s_{est}. To compute s_{est}, we need to know sample size, s_y, and r. These were given earlier: $n = 200$, $s_y = 12$, and $r = -.2$. We can put these numbers into equation 14.9 and determine s_{est}:

$$s_{est} = \sqrt{s_y^2(1-r^2)\frac{n-1}{n-2}} = \sqrt{12^2(1-.2^2)\frac{199}{198}} = 11.79.$$

Step 2. Compute s_b. To compute s_b, we need to know sample size, s_{est}, and ss_x. We just found that $s_{est} = 11.79$ and we were told that $s_x = 2$. Knowing sample size and s_x allows us to compute $ss_x = s_x^2 * (n-1) = 4*199 = 796$. Therefore, we can calculate s_b as follows:

$$s_b = \frac{s_{est}}{\sqrt{ss_x}} = \frac{11.79}{\sqrt{199*2^2}} = 0.42.$$

Step 3. Compute $t_{\alpha/2}$. In this case, there are $n-2 = 200 - 2 = 198$ degrees of freedom. If we have only our t-table, we would use the closest degrees of freedom in the table that is less than 198, which would be 100. Therefore, using 100 degrees of freedom, we find $t_{\alpha/2} = 1.984$. (We can obtain more precision using Excel.)

Step 4. Compute the 95% confidence interval around b. Having computed b, s_b, and $t_{\alpha/2}$, we can substitute these quantities into our equation to compute the confidence interval as follows:

$$CI = b \pm t_{\alpha/2}(s_b) = -1.2 \pm 1.984(0.42) = [-2.03, -0.37].$$

So, in four relatively simple steps, we have computed the 95% confidence interval around b. As always, we have 95% confidence in this interval because we know that 95% of all intervals computed in this way will capture β. Of course, having estimated β, we would like to attach some meaning to this estimate. Before doing this, however, we will quickly revisit the issue of significance testing.

Significance Tests Concerning b

If we had been interested in testing the null hypothesis

$$H_0: \beta = 0,$$

we could answer this right away by noting that 0 does not fall within our confidence interval. Or, using a *t*-test, we could set $t_{critical}$ to be ±1.984 and then compute t_{obs} as follows:

$$t_{obs} = \frac{b}{s_b} = \frac{-1.2}{0.42} = -2.87.$$

Because t_{obs} is outside the interval ±1.984, we can reject H_0: β = 0. We could also take one further step, using Excel, to calculate the two-tailed probability of obtaining t_{obs} outside the interval ±2.87. Doing so would show that $p = .0045$. So, we can say that there is a statistically significant association between the amount of *non-child-directed* television exposure at 6 months and cognitive development at 14 months.

Interpretation

How do we interpret these results? First, because the results are statistically significant, we can reject β = 0 as a plausible hypothesis about the slope of the regression equation in the bivariate normal population from which our sample was drawn. (We might have guessed this before collecting any data.) However, we can also be quite sure that β < 0, because we have 95% confidence that β is in the interval [−2.03, −0.37]. The magnitude of the correlation ($r = -.2$) between television exposure and cognitive development is between small and medium, according to Cohen's classification.

It is very difficult to say what practical significance we can attach to this finding. First, because the study was not experimental, we can't conclude that there is a causal association between television viewing and cognitive development. One can imagine dozens of factors that may have contributed to this association. For example, mothers' IQs, nutrition, or household stress could be the underlying cause or causes of the negative association between television exposure and cognitive development.

However, *even if* the negative association between television viewing and cognitive development were a causal one, we would still have to connect the television-induced cognitive delays to practical outcomes. The slope of the regression coefficient is $b = -1.2$, which means that for every additional hour of television viewing, *average* cognitive development scores drop by 1.2 units. Therefore, we can use the regression equation to predict *average* cognitive development scores from any given number of hours of television viewing.

We saw before that the regression equation for this study is

$$\hat{y} = a + b(x) = 98 - 1.2(x).$$

Therefore, a child who watches the average amount of television per day (2.5 hours) is predicted to have a cognitive development score of 95 [$\hat{y} = a + b(x) = 98 - 1.2(2.5) = 95$]. A child who watches an additional hour of television each day (i.e., 3.5 hours) is predicted to have a cognitive development score of 93.8 [$\hat{y} = a + b(x) = 98 - 1.2(3.5) = 93.8$]. What does this mean? How severe a developmental delay does this represent? What are the consequences of having a cognitive development score of 93.8 at 14 months versus a score of 95? This is the point at which statistics has no answers. The only way to make sense of this result is to be intimately familiar with the literature on cognitive development. Again, *even if* there is a causal association between television viewing and cognitive development, it would be irresponsible to decide to develop an intervention strategy based only on the finding that there is a statistically significant association between the two variables. Sound human judgment is required here. The process of choosing actions based on research results cannot be automated.

LEARNING CHECK 4

1. If $m_x = 12$, $s_x = 3$, $m_y = 250$, $s_y = 50$, $r = -.3$, and $n = 102$, compute the 95% confidence interval around the slope of the regression line.

2. What assumptions must be true for this interval to be valid?

Answers

1. Take the following steps to compute the answer:

   (i) $b = r*s_y/s_x = -.2*250/3 = -5$.

   (ii) $s_{est} = \sqrt{s_y^2(1-r^2)*(n-1)/(n-2)}$

   $= \sqrt{2500*.91*101/100} = 47.93$.

   (iii) $s_b = s_{est}/\sqrt{ss_x} = 47.93/30.15 = 1.59$.

   (iv) $CI = b \pm t_{\alpha/2}(s_b) = -5 \pm 1.984(1.59) = [-8.15, -1.85]$.

2. The xy pairs must be drawn from a bivariate population for which the conditional distributions of y are normal and homoscedastic, and the means of the conditional distributions change linearly with x. There must also be random sampling from the population of interest.

ESTIMATING THE EXPECTED VALUE OF y: $E(y|x)$

After we estimate the slope of the regression line, we may wish to estimate the expected value of y, $E(y|x)$, in the population for a given value of x. The predicted value of y from the regression equation of the sample $[\hat{y} = a + b(x)]$ is our best estimate of $E(y|x)$. Of course, $\hat{y}$ is a point estimate, and we generally prefer interval estimates so that we have some sense of the precision of our estimate. So, we now turn to the question of computing a confidence interval around $\hat{y}$.

To construct this confidence interval, we need to know something about the sampling distribution of $\hat{y}$. The interesting thing about $\hat{y}$ is that the characteristics of its sampling distribution change as a function of how far the associated x value is from m_x. To illustrate why this is so, Figure 14.5 shows 32 regression lines, each based on a random sample (of size $n = 27$) drawn from a bivariate normal distribution with $\mu_x = 3$, $\sigma_x = 1$, $\mu_y = 100$, $\sigma_y = 15$, and $\rho = -.4$. (This is the distribution of scores shown in Figure 14.4. A sample size of $n = 27$ has been used to make our point more clearly.) Each blue line represents a regression line computed from one of the 32 samples. The straight black line represents the regression equation from the population, $E(y|x) = \alpha + \beta(x)$.

For each value of x, there are 32 values of $\hat{y}$ in this example. (The white circles show the 32 $\hat{y}$ values for $x = 5$.) In theory, there would be an infinite number of possible regression lines and therefore an infinite number of $\hat{y}$ values at each value of x. Note that the variance of the $\hat{y}$ values is smallest around $\mu_x = 3$. For values of x further from μ_x, the variance of the $\hat{y}$ values increases. The curved black lines above and below the regression line show the values of y that capture 95% of $\hat{y}$ for each value of x. Again, the width of this interval increases as the difference between x and μ_x increases. Therefore, confidence intervals for $\hat{y}$ must increase as the difference between x and m_x increases.

The Estimated Standard Error of $\hat{y}$

To compute a confidence interval around $\hat{y}$, we use the following standard formula:

$$\hat{y} \pm t_{\alpha/2}(s_{\hat{y}}).$$

(14.14)

Of course, to compute this interval we need to know the estimated standard error of $\hat{y}$, which we call $s_{\hat{y}}$. The definition of $s_{\hat{y}}$ is as follows:

$$s_{\hat{y}} = s_{est}\sqrt{\frac{1}{n} + \frac{(x - m_x)^2}{ss_x}}. \tag{14.15}$$

Equation 14.15 may seem a little complex, so let's unpack it. If $x - m_x = 0$, then the second term in equation 14.15 is 0. This means $s_{\hat{y}} = \sqrt{s_{est}^2/n}$ or $s_{\hat{y}} = s_{est}/\sqrt{n}$. This is essentially the estimated standard error of the mean for m_y; remember, $\hat{y} = m_y$ for $x = m_x$. However, if $x \neq m_x$, then the second term under the square root sign becomes greater than zero and $s_{\hat{y}}$ will be greater than $s_{est}/\sqrt{n}$. How much greater depends on (i) how far x is from m_x and (ii) how big ss_x is. For large sample sizes, ss_x will be very large and $s_{\hat{y}}$ will be only slightly greater than $s_{est}/\sqrt{n}$ even when x is much greater than m_x.

Using equations 14.14 and 14.15, we can compute a confidence interval around any $\hat{y}$. In this example, we will put a 95% confidence interval around $\hat{y}$ for $x = 4.5$, which is one standard deviation above m_x.

Step 1. Compute $\hat{y}$ for $x = 4.5$.

$$\hat{y} = a + b(x) = 98 - 1.2(4.5) = 92.6.$$

Step 2. Compute $s_{\hat{y}}$ for $x = 4.5$. To do this, we apply equation 14.15, which requires knowing sample size n, s_{est}, m_x, and s_x. These values are $n = 200$, $s_{est} = 11.79$, $m_x = 2.5$, and $s_x = 2$. From s_x and n, we can compute $ss_x = s_x^2(n-1)$. When we put these in equation 14.15, we obtain the following:

$$s_{\hat{y}} = s_{est}\sqrt{\frac{1}{n} + \frac{(x - m_x)^2}{ss_x}}$$

$$= 11.79\sqrt{\frac{1}{200} + \frac{(4.5 - 2.5)^2}{2^2 * 199}}$$

$$= 1.18.$$

Step 3. Compute $t_{\alpha/2}$. In this case, there are $n-2 = 200 - 2 = 198$ degrees of freedom. Let's assume we have only our t-table. In this case, we would use the closest degrees of freedom in the table less than 198, which would be 100. Therefore, using 100 degrees of freedom, we find $t_{\alpha/2} = 1.984$.

Step 4. Compute the 95% confidence interval around $\hat{y}$. Having computed $\hat{y}$, $s_{\hat{y}}$, and $t_{\alpha/2}$, we can substitute these quantities into our equation to compute the confidence interval as follows:

$$CI = \hat{y} \pm t_{\alpha/2}(s_{\hat{y}}) = 92.6 \pm 1.984(1.18) = [90.26, 94.94].$$

Therefore, we have 95% confidence that for $x = 4.5$, $E(y|4.5)$ lies in the interval [90.26, 94.94].

As always, attaching meaning to this interval estimate requires knowledge of cognitive development that takes us outside the domain of statistics. However, our estimate of $E(y|4.5)$ seems rather imprecise. This imprecision is expected, given that $r = -.2$ and thus $r^2 = .04$, which means that hours of television viewing explains only 4% of the variability in cognitive development scores.

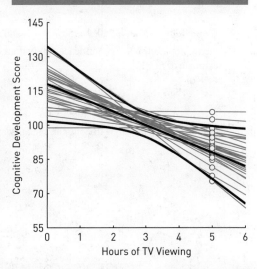

FIGURE 14.5 ■ Distribution of $\hat{y}$

Thirty-two regression lines, each based on a random sample (of size $n = 27$) from a bivariate normal distribution with $\mu_x = 3$, $\sigma_x = 1$, $\mu_y = 100$, $\sigma_y = 15$, and $\rho = -.4$. For each value of x, there are 32 values of $\hat{y}$. (The white circles show these values of $\hat{y}$ for $x = 5$.) The variability in $\hat{y}$ increases as the corresponding x value moves away from μ_x. The straight black line represents the regression line in the population, with $\alpha = 118$ and $\beta = -6$. The curved black lines above and below the regression line represent values of y at each value of x (hours of television viewing, in this example) within which 95% of $\hat{y}$ values would fall. The width of this interval increases as the difference between x and μ_x increases. We will see that the widths of the confidence intervals for $E(y|x)$ must increase as the difference between x and m_x increases.

LEARNING CHECK 5

1. If $m_x = 12$, $s_x = 3$, $m_y = 250$, $s_y = 50$, $r = -.3$, and $n = 102$, then compute the 95% confidence interval around the estimate of $E(y|10)$.

2. What assumptions must be true for this interval to be valid?

Answers

1. Take the following steps to compute the answer:

   (i) $\hat{y} = 310 - 5(10) = 260$.

   (ii) $s_{est} = 47.93$.

   (iii) $s_{\hat{y}} = s_{est}\sqrt{1/n + (x - m_x)^2/ss_x}$

   $= 47.93\sqrt{1/102 + 4/909} = 5.71$.

   (iv) $CI = \hat{y} \pm t_{\alpha/2}(s_{\hat{y}}) = 260 \pm 1.984(5.71)$

   $= [248.67, 271.33]$.

2. The xy pairs must be drawn from a bivariate population for which the conditional distributions of y are normal and homoscedastic, and the means of the conditional distributions change linearly with x. There must also be random sampling from the population of interest.

PREDICTION INTERVALS

A **prediction interval** is an interval that we can say with some confidence will contain the next value of y for a given value of x.

When we hear about associations between things such as *television viewing* and *cognitive development*, we reflexively think about their implications for individuals. For example, let's say you're a mother in the bivariate population we've been discussing and your 6-month-old child watched 4.5 hours of non-child-directed television yesterday. You would like to know what this implies about your child's cognitive development at 14 months old. In this section, we will introduce the concept of **prediction intervals**. Such predictions concern the possible values for the next y drawn from the conditional distribution corresponding to a specific value of x (4.5 hours of television viewing in our example).

Let's first think about a *bivariate population* such as that shown in Figure 14.6a. (This is the same bivariate population shown in Figure 14.4.) For each value of x, there is a conditional distribution of y scores. We will now think about the limits containing 95% of all scores of the conditional distribution for $x = 6$ in this population. I've chosen $x = 6$ because it is far (three standard deviations) above μ_x; later we will apply what we know from this example to the case of $x = 4.5$. Remember, we are reasoning about conditional distributions in the population.

The question about what limits contain 95% of all scores of the conditional distribution for $x = 6$ takes us back to the exercises we did in Chapter 4. Each conditional distribution is normal with standard deviation of σ_{est} and mean $E(y|x)$. Therefore, in the population, 95% of the scores in each conditional distribution lie in the interval $E(y|x) \pm 1.96(\sigma_{est})$. The population regression equation in this case is $E(y|x) = \alpha + \beta(x) = 118 - 6(x)$ and $\sigma_{est} = 13.75$. If $x = 6$, then

$$E(y|x) = \alpha + \beta(x) = 118 - 6(6) = 82$$

and 95% of scores lie in the interval

$$E(y|x) \pm 1.96(\sigma_{est}) = 82 \pm 1.96(13.75) = [55.05, 108.95].$$

Therefore, if we are told that an xy pair from this bivariate population has $x = 6$, then we know that 95% of y scores in the conditional distribution of y values are in the interval [55.05, 108.95]. If we predicted that the next score drawn from this conditional distribution would be in the interval [55.05, 108.95], we would be right 95% of the time. We can apply the same calculation to all values of x. The results of these calculations are shown in the parallel gray lines above and below $E(y|x)$ in Figure 14.6a.

Let's now turn to prediction intervals computed from a *bivariate sample*. To understand the logic of prediction intervals, we have to think about (i) drawing a sample from which

FIGURE 14.6 ■ Prediction Intervals

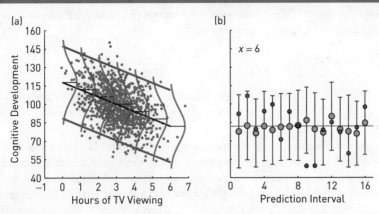

(a) A bivariate distribution is shown, with $\mu_x = 3$, $\sigma_x = 1$, $\mu_y = 100$, $\sigma_y = 15$, $\rho = -.4$, and $\sigma_{est} = 13.25$. The black line shows $E(y|x) \pm 1.96(\sigma_{est})$. (b) Prediction intervals are shown for 16 samples of 200 xy pairs drawn from the distribution in (a). The blue dots show $\hat{y}$ for $x = 6$ for each of the 16 samples. The intervals around $\hat{y}$ are 95% prediction intervals computed as $\hat{y} \pm t_{\alpha/2}(s_{y_{next}})$. The gray dots show values of y_{next}. There is much greater variability in y_{next} than in $\hat{y}$.

the regression line is computed, (ii) choosing a value of x for which we want to compute a prediction interval, and (iii) the next score (y_{next}) drawn from the conditional distribution of the population for the x value of interest. When the sample is drawn, the regression line is computed. From this, we compute $\hat{y}$ for the value of x of interest. A prediction interval is computed around $\hat{y}$ using the following equation:

$$PI = \hat{y} \pm t_{\alpha/2}(s_{y_{next}}).$$

We want the estimated standard error of y_{next} to be such that $(1-\alpha)100\%$ of the time, y_{next} falls in the prediction interval $\hat{y} \pm t_{\alpha/2}(s_{y_{next}})$. (This estimated standard error will be described later.)

The logic of prediction intervals is shown in Figure 14.6b. Sixteen samples of 200 xy pairs were drawn from the distribution in Figure 14.6a. For each sample, the regression equation was computed, from which $\hat{y}$ was computed for $x = 6$. These $\hat{y}$ values are shown as blue dots. These values vary around 82 (the horizontal gray line in Figure 14.6b), which is the expected value of y for $x = 6$; i.e., $E(y|x) = \alpha + \beta(x) = 118 - 6(x) = 82$. For each $\hat{y}$, a prediction interval has been computed using the following equation:

$$CI = \hat{y} \pm t_{\alpha/2}(s_{y_{next}}).$$

In this example, we computed 95% prediction intervals. After each prediction interval was computed, a single y value (y_{next}) was selected from the conditional distribution with $x = 6$. These values of y_{next} are shown as small gray dots in Figure 14.6b. We have 95% confidence in a prediction interval because we know that 95% of the time, the prediction interval will capture the next y score drawn from the conditional distribution of interest.

In the population, the interval capturing 95% of a conditional distribution is $E(y|x) \pm 1.96(\sigma_{est})$, so it makes sense that s_{est} would be used to estimate σ_{est}. However, as seen in the previous section, our estimate of $E(y|x)$ becomes increasingly imprecise the further x is from the sample mean, m_x. Furthermore, there is some imprecision in estimating $\hat{y}$. Therefore, s_{est} must be corrected for these imprecisions. The formula we use is as follows:

$$s_{y_{next}} = s_{est}\sqrt{1 + \frac{1}{n} + \frac{(x - m_x)^2}{ss_x}}. \tag{14.16}$$

The only difference between equation 14.16 and equation 14.15 is the addition of a 1 under the radical. The consequence of this addition is to diminish the influence of the other two terms. That is, as sample size (n) increases, the second ($1/n$) and third $[(x - m_x)^2/ss_x]$ terms go to 0, and equation 14.16 thus reduces to the following:

$$s_{y_{next}} = s_{est} \sqrt{1+0+0} = s_{est}.$$

Therefore, this *distance-from-the-mean* correction is not very great except for small sample sizes, in which case $s_{y_{next}}$ is greater than s_{est}, and increases the further x is from the sample mean, m_x.

Let's return to the question of the mother whose 6-month-old child watched 4.5 hours of non-child-directed television the previous day. Let's continue this example but instead of computing the 95% confidence interval around $\hat{y}$ as we did in the previous section, we will compute the 95% prediction interval for the next value of y when $x = 4.5$. Remember from that illustration that $\hat{y} = a + b(x) = 98 - 1.2(x)$, and $s_{est} = 11.79$. We can now proceed to compute our prediction interval.

Step 1. Compute $\hat{y}$ for $x = 4.5$.

$$\hat{y} = a + b(x) = 98 - 1.2(4.5) = 92.6.$$

Step 2. Compute $s_{y_{next}}$ for $x = 4.5$. To do this, we apply equation 14.16, which requires knowing n, s_{est}, m_x, and s_x. These values are $n = 200$, $s_{est} = 11.79$, $m_x = 2.5$, and $s_x = 2$. From s_x and n, we can compute $ss_x = s_x^2(n-1)$. When we put these in equation 14.16, we obtain the following:

$$s_{y_{next}} = s_{est} \sqrt{1 + \frac{1}{n} + \frac{(x - m_x)^2}{ss_x}} = 11.79 \sqrt{1 + \frac{1}{200} + \frac{(4.5 - 2.5)^2}{2^2 * 199}} = 11.85.$$

Step 3. Compute $t_{\alpha/2}$. In this case, there are $n - 2 = 200 - 2 = 198$ degrees of freedom. Let's assume we have only our t-table. In this case, we would use the closest degrees of freedom in the table less than 198, which would be 100. Therefore, using 100 degrees of freedom, we find $t_{\alpha/2} = 1.984$.

Step 4. Compute the 95% prediction interval. With $\hat{y}$, $s_{y_{next}}$, and $t_{\alpha/2}$ determined, our prediction interval can be computed as follows:

$$CI = \hat{y} \pm t_{\alpha/2} \left(s_{y_{next}} \right) = 92.6 \pm 1.984(11.85) = [69.10, \ 116.10].$$

Therefore, we have 95% confidence that for $x = 4.5$, the next y score will fall in the interval [69.10, 116.10].

The prediction interval computed here is extremely wide. To get some idea of just how wide it is, think about computing $m_y \pm t_{\alpha/2}(s_y)$. That is, the mean value of the sample of y scores, plus or minus roughly two times the standard deviation of the y scores. When we fill in the numbers, we have $95 \pm 1.984(12) = [71.19, 118.81]$, which is not much different from the prediction interval we just computed for a value of x one standard deviation above the mean; i.e., 95% PI [69.10, 116.10].

The width of the prediction interval is not a consequence of sample size. Rather, the prediction interval is wide because the regression equation explains relatively little variability in the outcome variable. Remember, $s_{y_{next}} \approx s_{est}$ when sample size is large, and $s_{est} \approx \sqrt{s_y^2(1 - r^2)}$ when we ignore the small correction for bias (see equation 14.10). In our example, $r = -.2$, so $r^2 = .04$. Therefore, $s_{est} \approx \sqrt{s_y^2(.96)}$, which means that $s_{y_{next}} \approx s_{est} \approx s_y$. We can see that this is so when we compare $s_y = 12$ in our scenario and

$s_{y_{next}}$ = 11.85 as calculated above. So, unless r is very large, our prediction intervals will be very wide. Cohen and Cohen (1983) stated the following: "We are forced to conclude that it is a relatively rare circumstance in the social sciences that a data-based prediction for a given individual will be a substantial improvement over simply predicting that individual at the mean" (Cohen & Cohen, 1983, p. 64).

In other words, simple regression equations very rarely make precise predictions about individuals, so a good guess about the individual's y score is simply that it is the mean of the y scores. Therefore, when you hear that there is a statistically significant correlation between two variables, keep in mind that knowing an individual's score on x will probably provide very little information about his or her score on y.

Discussion

Let's think back to the young mother who has learned that there is a statistically significant association between television viewing at 6 months and a measure of cognitive development at 14 months. She remembers that her 6-month-old child watched 4.5 hours of television the day before, and therefore worries that her child's cognitive development will be impaired. Here are some of the things she might consider before fretting that she has harmed her child.

We've already noted that the association between *hours of television viewing* and *cognitive development* might not be causal. This in itself should keep our hypothetical mother from worrying too much about her child. Furthermore, even if a causal association is assumed, the implications of the finding depend on the meaning of the cognitive development measure. For a child who watched 4.5 hours of television at 6 months, the estimated mean cognitive development score at 14 months is 92.6, 95% CI [90.26, 94.94]. For children who watched no television at 6 months, the expected cognitive development score at 14 months is 98. So, how different are the *average* life trajectories of children who had a cognitive development score of 98 at 14 months and those who had a cognitive development score of 92.6 at 14 months? Without this information, we have no way to make any sense of the data.

To these cautions we can now add the importance of distinguishing between the scores of individuals and the statistics of samples. The prediction interval we just computed shows that the next y score for a child with $x = 4.5$ will fall in the interval [69.10, 116.10]. This interval is really wide. So wide, in fact, that the association between hours of television viewing and cognitive development tells us almost nothing about what to expect at the level of the individual. If I were the parent of the child in question, I don't think I'd be terribly concerned about this result.

LEARNING CHECK 6

1. If $m_x = 12$, $s_x = 3$, $m_y = 250$, $s_y = 50$, $r = .3$, and $n = 102$, then compute the 95% prediction interval for $x = 10$.

2. What assumptions must be true for this interval to be valid?

Answers

1. The 95% prediction interval is [142.36, 337.64]. Take the following steps to compute the answer:

   (i) $\hat{y} = 190 + 5(10) = 240$.

   (ii) $s_{est} = 47.93$.

   (iii) $s_{\hat{y}} = s_{est}\sqrt{1 + 1/n + (x - m_x)^2/ss_x}$

   $= 47.93\sqrt{1 + 1/200 + 4/909} = 48.27$.

   (iv) $CI = \hat{y} \pm t_{\alpha/2}(s_{\hat{y}}) = 240 \pm 1.984(48.27)$

   $= [144.22, 335.78]$.

2. The xy pairs must be drawn from a bivariate population for which the conditional distributions of y are normal and homoscedastic, and the means of the conditional distributions change linearly with x. There must also be random sampling from the population of interest.

SUMMARY

A *bivariate distribution* is a distribution of *xy* pairs. The *regression model* assumes samples were drawn from a bivariate population for which the *x* values are fixed. The correlation coefficient relating *x* to *y* in a bivariate population is denoted ρ and computed as follows:

$$\rho = \frac{sp_{xy}}{\sqrt{ss_x ss_y}}.$$

The slope and intercept of the regression equation in a bivariate population are defined as

$$\beta = \rho \frac{\sigma_y}{\sigma_x} \text{ and}$$

$$\alpha = \mu_y - \beta(\mu_x),$$

respectively.

In the regression model, there is a distribution of *y* scores for each of the fixed *x* values. The mean of each distribution of *y* scores is called its *expected value*, which is denoted $E(y|x)$. These expected values are defined by the *regression equation*

$$E(y|x) = \alpha + \beta(x).$$

Because the distributions of *y* scores are conditional on *x*, we call them *conditional distributions*. Conditional distributions are *homoscedastic* when they all have the same standard deviation. In regression analysis we refer to the standard deviation of homoscedastic distributions as the *standard error of estimate*, which is denoted σ_{est} and is defined as

$$\sigma_{est} = \sqrt{\frac{\sum (y - E(y|x))^2}{N}}.$$

When all *y* scores from the conditional distributions are combined into a single distribution, we call this a *marginal distribution* of *y* scores. σ_{est} can also be defined as

$$\sigma_{est} = \sqrt{(1-\rho^2)\sigma_y^2},$$

where σ_y^2 is the variance of the *marginal distribution* of *y* scores.

The slope of the regression equation computed from a sample (*b*) estimates β. The sampling distribution of *b* is a probability distribution for all possible values of *b* for samples of size *n* drawn from a homoscedastic bivariate distribution. The sampling distribution of *b* has a mean of

$$\mu_b = \beta$$

and a standard error of

$$\sigma_b = \frac{\sigma_{est}}{\sqrt{ss_x}}.$$

The standard error of *b* (σ_b) is estimated by

$$s_b = \frac{s_{est}}{\sqrt{ss_x}}, \text{ where}$$

$$s_{est} = \sqrt{s_y^2(1-r^2)\frac{n-1}{n-2}}$$

estimates σ_{est}.

A confidence interval around *b* is computed as

$$CI = b \pm t_{\alpha/2}(s_b)$$

where $t_{\alpha/2}$ is based on *n*–2 *df* and *n* is the number of *xy* pairs.

When *x* is a random variable, the regression equation and its inferential statistics are computed exactly as in the case of *x* being fixed. Although the resulting standard errors are not exactly correct, they are close enough to the exact values for practical purposes.

In some cases it may be important to estimate $E(y|x)$ for a given value of *x*. To do so, we first compute $\hat{y} = a + b(x)$ and then place a confidence interval around $\hat{y}$ as follows:

$$\hat{y} \pm t_{\alpha/2}(s_{\hat{y}}),$$

where $t_{\alpha/2}$ is based on *n*–2 *df*. The estimated standard error of $\hat{y}$ is defined as follows:

$$s_{\hat{y}} = s_{est}\sqrt{\frac{1}{n} + \frac{(x-m_x)^2}{ss_x}}.$$

Finally, one might wish to use a regression equation to predict the *y* score for the next individual drawn from the conditional distribution having a specific *x* score. This *prediction interval* is computed as

$$\hat{y} \pm t_{\alpha/2}(s_{y_{next}}), \text{ where}$$

$$s_{next} = s_{est}\sqrt{1 + \frac{1}{n} + \frac{(x-m_x)^2}{ss_x}}.$$

KEY TERMS

bivariate normal distribution   363
conditional distribution   354
homoscedasticity   356

marginal distribution   355
prediction interval   370
sampling distribution of b 357

standard error of estimate
(σ_{est})   356

EXERCISES

Definitions and Concepts

Please note that asterisks (*) indicate questions relating to material covered in the appendixes or requiring insight beyond what was covered in the chapter.

1. What is a bivariate distribution?

2. What is a conditional distribution?

3. What does it mean for distributions to be homoscedastic?

4. What statistics estimate ρ, β, and α?

5. What name do we give to this quantity: $\Sigma[y - E(y|x)]^2$?

6. What name do we give to this quantity: $\Sigma[y - E(y|x)]^2/N$?

7. What is a marginal distribution?

8. What population parameter is defined as $ss_{regression}/ss_y$?

9. Explain why the sampling distribution of b is or is not a bivariate distribution.

10. What is the difference between σ_b and σ_{est}?

11. What are the five properties of a bivariate normal distribution?

12. What does it mean to have 95% confidence in the interval $b \pm t_{\alpha/2}(s_b)$?

13. What does it mean to have 95% confidence in the interval $\hat{y} \pm t_{\alpha/2}(s_{\hat{y}})$?

14. What is the difference between a confidence interval and a prediction interval?

True or False

State whether the following statements are true or false.

15. The regression model assumes scores were drawn from a bivariate normal distribution.

16. The marginal distributions of a bivariate normal distribution are normal.

17. The regression model assumes that the marginal distributions of x and y are normal.

18. The regression model assumes that the conditional distributions are heteroscedastic.

19. *When $\beta = 0$, the marginal and conditional distributions are identical.

20. *When $\rho = 0$, the marginal and conditional distributions are identical.

21. *$\sigma_{est} = \sigma_y$ when $\rho = 0$.

22. *$\sigma_{est} = \sigma_b$ when $\rho = 0$.

23. Regression analysis can never imply a causal association between x and y.

24. *$\rho = \beta \dfrac{\sigma_x}{\sigma_y}$.

Calculations

25. Compute the following for the data in the following table: (a) sp_{xy}, (b) r, (c) s_x, (d) s_y, (e) s_b, (f) s_{est}, (g) the regression equation, (h) the 95% confidence interval for b, (i) the 95% confidence interval for $\hat{y}$ when $x = 16$, and (j) the 95% prediction interval for y_{next} when $x = 16$.

x	y
16	14
12	15
11	16
11	18
11	21
11	22
10	24
10	24
10	24
10	25
9	28
m_x	m_y
11	21
ss_x	ss_y
34	212

26. Compute the following for the data in the following table: (a) sp_{xy}, (b) r, (c) s_x, (d) s_y, (e) s_b, (f) s_{est},

(g) the regression equation, (h) the 95% confidence interval for b, (i) the 95% confidence interval for $\hat{y}$ when $x = 3$, and (j) the 95% prediction interval for y_{next} when $x = 3$.

x	y
1	14
1	15
1	16
1	18
2	21
2	22
2	24
2	24
3	24
3	25
3	28
3	33
m_x	m_y
2	22
ss_x	ss_y
8	344

Scenarios

27. A researcher chose a random sample of 300 Facebook users from her department's participant pool. She administered the NPI-16 (short measure of narcissism) to each participant, and she also recorded how many Facebook friends each one had. Her results showed that $m_{FBF} = 250$ ($s_{FBF} = 100$) and $m_{NPI16} = 5$ ($s_{NPI16} = 2$). The correlation between NPI-16 scores and number of Facebook friends was $r = .1$. Compute the following from these data:

    (a) the regression equation predicting NPI-16 from Number of Facebook Friends,

    (b) the 95% confidence interval for b,

    (c) the 95% confidence interval for $\hat{y}$ when $x = 350$, and

    (d) the 95% prediction interval for y_{next} when $x = 350$.

    (e) From these results a student concluded that if you have a large number of Facebook friends, you must be narcissistic. Please comment on this conclusion.

28. A university senior was interested in examining the association between mother's warmth

and child's anxiety as part of his final-year project. To do so, he selected a random sample of 82 participants from his department's participant pool. He measured each participant's level of anxiety using the Beck Anxiety Index (BAI), which is a 21-item questionnaire that asks participants to rate on a 4-point scale (0 to 3) the degree to which they experienced symptoms of anxiety (e.g., "feeling hot," "shaky") in the last 2 weeks. He measured each participant's mother's warmth by having each participant complete a 16-item questionnaire he found on the Internet. High scores on the BAI indicate high levels of anxiety and high scores on the "warmth" questionnaire indicate high degrees of maternal warmth. His results showed that $m_{warmth} = 10$ ($s_{warmth} = 4$) and $m_{BAI} = 8$ ($s_{BAI} = 2$). The correlation between BAI scores and mother's warmth was $r = -.25$. Determine the following from these data:

    (a) the regression equation predicting BAI scores from warmth scores,

    (b) the 95% confidence interval for b,

    (c) whether the slope is statistically significantly different from 0,

    (d) the 95% confidence interval for $\hat{y}$ when $x = 6$, and

    (e) the 95% prediction interval for y_{next} when $x = 6$.

    (f) Given the negative association between warmth and BAI, the student concluded that warm mothers have children with lower anxiety. Please discuss this conclusion and look up the Beck Anxiety Index on Wikipedia to get some guidance about how to interpret its results.

29. Rosser, Lynch, Cuddihy, Gentile, Klonsky, and Merrell (2007) found that surgeons who had played video games during college made fewer errors on a test of laparoscopic skills than those who hadn't played video games. The test of laparoscopic skills was part of the Rosser Top Gun Laparoscopic Skills and Suturing Program (Top Gun). Imagine that for a random sample of 33 surgeons, researchers assessed (i) the number of hours per week (HPW) each surgeon reported playing video games in college and (ii) the number of errors each one made on the Top Gun (TG) test. Let's say that the results showed that $m_{TG} = 7.5$ ($s_{TG} = 1$) and $m_{HPW} = 3$ ($s_{HPW} = 4$). High Top Gun scores mean many errors and low scores mean few errors. The correlation between Top

Gun scores and hours per week was $r = -.35$. Compute the following from these data:

(a) the regression equation predicting Top Gun scores from hours per week of video game playing,

(b) the 95% confidence interval for b,

(c) the 95% confidence intervals for $\hat{y}$ when $x = 0$ and $x = 7$, and

(d) the 95% prediction intervals for y_{next} when $x = 0$ and $x = 7$.

(e) Discuss the implications of these results.

APPENDIX 14.1: INFERENTIAL STATISTICS FOR REGRESSION IN EXCEL

The worksheet in Figure 14.A1.1 illustrates most of the calculations described in the body of the chapter. The data are taken from the *television viewing* and *cognitive development* example. The sample statistics are provided in cells **B2** to **B4** and **B6** to **B9**. Cell **B5** shows the calculation of ss_x from s_x and n. You can calculate all

of these statistics for any xy pairs of interest using the functions described in Appendix 13.1. The worksheet is very useful for answering questions such as questions 27 to 29 in the end-of-chapter exercises. It is quite interesting that so little information (m_x, s_x, m_y, s_y, n, and r) allows you to do so much.

FIGURE 14.A1.1 ■ Confidence Intervals

	A	B	C
	Quantity	Values	Formulas
2	n	200.000	
3	m_x	2.500	
4	s_x	2.0000	
5	ss_x	796	=B4^2 * (B2-1)
6	m_y	95.000	
7	s_y	12.0000	
8	r	-.20000	
9	x	4.500	
10			
11	b	-1.20000	=B8*B7/B4
12	a	98.000	=B6-B11*B3
13			
14	s_{est}	11.7872	=SQRT(B7^2*(1-B8^2)*(B2-1)/(B2-2))
15	s_b	.4178	=B14/SQRT(B5)
16			
17	$\hat{y}$	92.6000	=B12+B11*B9
18	$s_{\hat{y}}$	1.1802	=B14*SQRT(1/B2+(B9-B3)^2/B5)
19	s_{ynext}	11.8461	=B14*SQRT(1+1/B2+(B9-B3)^2/B5)
20	α	.0500	
21	$t_{\alpha/2}$	1.9720	=T.INV.2T(B20, B2-2)
22			
23	CI for b	-2.0239	=B11-B21*B15
24		-.3761	=B11+B21*B15
25			
26	CI for $\hat{y}$	90.2726	=B17-B21*B18
27		94.9274	=B17+B21*B18
28			
29	PI for y_{next}	69.2392	=B17-B21*B19
30		115.9608	=B17+B21*B19

Calculating confidence intervals for b and $\hat{y}$ given x, and a prediction interval for y_{next} given x.

APPENDIX 14.2: INFERENTIAL STATISTICS FOR REGRESSION IN SPSS

Confidence Intervals for *b*

Figure 14.A2.1 shows the data from question 25 in the end-of-chapter exercises. These *x* and *y* scores are shown in the columns labeled *x* and *y* in the SPSS data editor. To conduct a regression analysis, we choose the Analyze→Regression→Linear . . . dialog. This dialog is overlaid on the Data Editor file. The *x* and *y* variables have been moved into the Independent(s) and Dependent regions, respectively. Although not shown in Figure 14.A2.1, I asked for SPSS to compute the 95% confidence interval around *b* through the Statistics . . . dialog. When the analysis is set up, we click OK to proceed with it.

Figure 14.A2.2 shows the SPSS output for the regression analysis. In the region labeled Unstandardized Coefficients are the intercept (Constant) and slope (*x*) of the regression equation. To the right of the coefficients are their associated estimated standard errors. The second row of Figure 14.A2.2 shows t_{obs} and the 95% confidence interval associated with *b*. You can compare these with the answer to question 25 to see that SPSS has computed the same quantities that you computed by hand. The first row of Figure 14.A2.2 shows t_{obs} and the 95% confidence interval associated with *a*. These quantities are computed for completeness but it is exceedingly rare to see these reported in research papers.

In the column labeled Standardized Coefficients is a quantity labeled Beta, associated with the slope of the regression equation. In simple linear regression, this quantity corresponds to the correlation coefficient *r*, which is computed as follows:

$$r = b\frac{s_x}{s_y}.$$

We will have much more to say about this quantity in Chapter 17 on multiple regression. In Chapter 15, we will show how to compute a confidence interval around *r*.

Confidence Intervals for $\hat{y}$ and Prediction Intervals for y_{next}

SPSS can also be used to compute confidence intervals for $\hat{y}$ and prediction intervals for y_{next}. To conduct these analyses, we click the Save . . . button in the Linear Regression dialog. When this is clicked, the dialog shown in Figure 14.A2.3 appears. The options of interest are in the regions labeled Predicted Values and Prediction Intervals. In the Predicted Values region, the two checked items are requests for SPSS to compute $\hat{y}$ (Unstandardized) and $s_{\hat{y}}$ (S.E. of mean predictions). In the Prediction Intervals region, the two checked items are requests for SPSS to compute confidence intervals for $\hat{y}$ (Mean) and y_{next} (Individual). With this done, we click

FIGURE 14.A2.1 ■ Regression in SPSS

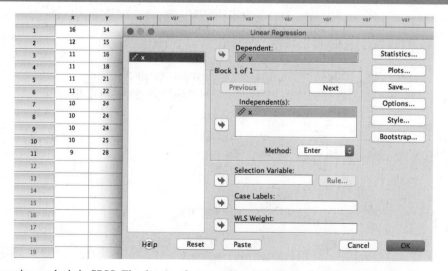

Setting up a regression analysis in SPSS. The data are from question 25 in the end-of-chapter exercises.

FIGURE 14.A2.2 ■ Regression Output

Coefficients[a]

Model		Unstandardized Coefficients		Standardized Coefficients	t	Sig.	95.0% Confidence Interval for B	
		B	Std. Error	Beta			Lower Bound	Upper Bound
1	(Constant)	43.000	5.552		7.746	.000	30.441	55.559
	x	-2.000	.498	-.801	-4.013	.003	-3.127	-.873

a. Dependent Variable: y

SPSS output from the regression analysis described in Figure 14.A2.1.

Continue to return to the Linear Regression dialog, and there we click OK to proceed with the analysis.

The quantities computed through the Save . . . dialog do not appear in the output window, as do the results of most other analyses. Instead, the results are stored in the data file in rows corresponding to individuals in the analyses. These results are shown in Figure 14.A2.4. The first two columns are the x and y values that were originally in the data file. The remaining columns are the results of the analysis specified in Figure 14.A2.3.

In question 25 we computed confidence intervals and prediction intervals for $x = 16$. These results are shown in the first row of Figure 14.A2.4. The column PRE_1 (created by SPSS) shows $\hat{y}$ for each of the 11 x scores. For $x = 16$, $\hat{y} = 11$. The next column, SEP_1, shows $s_{\hat{y}}$. You should check both of these quantities in the answers to question 25 to confirm that the same things have been computed. Columns LMCI_1 and UMCI_1 show the lower and upper limits, respectively, of the 95% confidence interval around each $\hat{y}$; the letter M in the column name denotes the estimated *mean* of the corresponding conditional distribution. Columns LICI_1 and UICI_1 show the lower and upper limits, respectively, of the 95% prediction interval for y_{next}. The first letter I in the column name denotes the prediction about the next *individual* drawn from the corresponding conditional distribution.

Once an analysis has been set up in SPSS, it can be *rerun* by simply clicking on the OK button in the Linear Regression dialog. Most output will be directed to the Output window. However, if the settings in the Save... dialog remain as they are in Figure 14.A2.3, then each time the analysis is run six more columns will appear in the data file (i.e., PRE_2, SEP_2, LMCI_2, UMCI_2, LICI_2, UICI_2). Therefore, if you're not careful, a huge number of columns may be unintentionally added to the data file.

FIGURE 14.A2.3 ■ The Save . . . Dialog

The Save . . . dialog associated with the Linear Regression dialog is used for confidence intervals for $\hat{y}$ and prediction intervals for y_{next}.

Assessing Linearity and Homoscedasticity in SPSS

The regression model assumes that a line describes the association between x and $E(y|x)$ in the population. Figure 14.A2.5a shows a sample of 200 xy pairs drawn from such a population. The least-squares regression line

FIGURE 14.A2.4 ■ Save . . . Dialog Results

	x	y	PRE_1	SEP_1	LMCI_1	UMCI_1	LICI_1	UICI_1
1	16	14	11.00	2.64	5.02	16.98	2.12	19.88
2	12	15	19.00	1.01	16.72	21.28	12.04	25.96
3	11	16	21.00	.88	19.02	22.98	14.13	27.87
4	11	18	21.00	.88	19.02	22.98	14.13	27.87
5	11	21	21.00	.88	19.02	22.98	14.13	27.87
6	11	22	21.00	.88	19.02	22.98	14.13	27.87
7	10	24	23.00	1.01	20.72	25.28	16.04	29.96
8	10	24	23.00	1.01	20.72	25.28	16.04	29.96
9	10	24	23.00	1.01	20.72	25.28	16.04	29.96
10	10	25	23.00	1.01	20.72	25.28	16.04	29.96
11	9	28	25.00	1.33	22.00	28.00	17.77	32.23

Results of the analysis to compute $\hat{y}$ (PRE_1), $s_{\hat{y}}$ (SEP_1), confidence intervals around $\hat{y}$ (LMCI_1 and UMCI_1), and prediction intervals for y_{next} (LICI_1 and UICI_1).

FIGURE 14.A2.5 ■ Homoscedasticity and Linearity

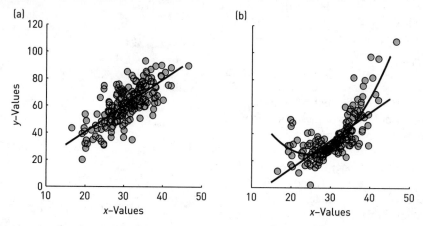

The figure plots (a) homoscedastic data that are well described by a straight line and (b) heteroscedastic data that are best described by a curved line. Despite their differences, straight lines can be fit to both data sets, and both regressions explain about the same variability in y.

through this sample of points is $\hat{y} = 1.77 + 1.94(x)$ and the proportion of variance explained by regression is $r^2 = .52$.

Figure 14.A2.5b shows a sample of 200 xy pairs drawn from a population for which the association between x and $E(y|x)$ is non-linear. The best-fitting curved line through these points is shown along with the best-fitting regression (straight) line. As we saw in Chapter 13, a straight line can be fit through any scatter of points, even if the association between x and y is not linear. In this case, the least-squares straight line

through this sample of points is $\hat{y} = -27.12 + 2.07(x)$, and the proportion of variance explained by this line is $r^2 = .49$. So, even though association between x and y is clearly curvilinear in Figure 14.A2.5b, the straight line fit explains close to the same proportion of variability as the straight line fit in Figure 14.A2.5a.

Figures 14.A2.5a and 14.A2.5b also differ in homoscedasticity. The scatterplot in Figure 14.A2.5a is homoscedastic, and the one in Figure 14.A2.5b is not. In Figure 14.A2.5b, the variability in the conditional distributions is very low, around $x = 30$, but increases for values of x further removed from 30.

The linearity and homoscedasticity assumptions can be assessed visually in SPSS. Figure 14.A2.6 shows an SPSS data file (behind two dialog boxes) that has been set up with a column of x values (which are the same in Figures 14.A2.5a and 14.A2.5b), and columns for the y values in Figure 14.A2.5. These y values have been named y1 (Figure 14.A2.5a) and y2 (Figure 14.A2.5b).

In Figure 14.A2.6 the Linear Regression dialog (that overlays the data editor window) shows that y2 has been entered as the dependent variable and x has been entered as the independent variable. One of the options in the dialog is Plots. . . . When this is clicked, the Linear Regression: Plots dialog appears. The left window of this dialog (overlaid on the Linear Regression dialog shows a number of variables, computed from the regression analysis, that are helpful when assessing linearity and homoscedasticity.

The two variables that concern us are *ZPRED and *ZRESID. As the names suggest, both variables are z-scores of some sort, and they are computed by SPSS as follows:

$$z_{\text{pred}} = \frac{\hat{y} - m_{\hat{y}}}{s_{\hat{y}}} \text{ and}$$

$$z_{\text{resid}} = \frac{\hat{y} - y}{s_{\hat{y} - \hat{y}}}.$$

That is, *ZPRED represents predicted values of y ($\hat{y}$) transformed to z-scores, and *ZRESID represents errors of prediction, or residuals ($\hat{y} - y$) transformed to z-scores. If the homoscedasticity assumption is correct, then when we plot *ZRESID on the y-axis against *ZPRED on the x-axis, we should have a scatter of points that are randomly distributed about *ZPRED = 0, and the conditional distributions of *ZRESID should all have the same standard deviation.

Figure 14.A2.7a shows a plot of *ZRESID against *ZPRED for the sample of 200 xy pairs (x, y1) shown in Figure 14.A2.5a. This is the result we expect to see when the assumptions of the regression model are satisfied; all conditional distributions are centered on *ZRESID = 0 (showing that there is no evidence of non-linearity in the data) and have the same standard deviations (showing no evidence of *hetero*scedasticity).

Figure 14.A2.7b shows a plot of *ZRESID against *ZPRED for the sample of 200 xy pairs (x, y2) shown in Figure 14.A2.5b. Figure 14.A2.7b shows massive violations of homoscedasticity and linearity. The violation of homoscedasticity is clear from the change in the variances of the conditional distributions of *ZRESID.

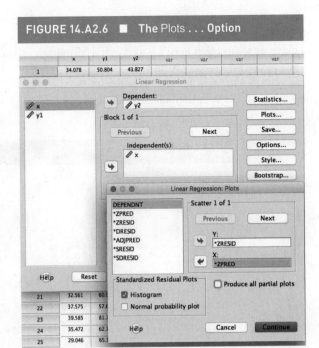

FIGURE 14.A2.6 ■ The Plots . . . Option

When we choose the Plots . . . option from the Linear Regression dialog, we have the option of plotting several variables, computed by SPSS, against each other. In this case, we plot *ZRESID (Y) against *ZPRED (X).

For values near the mean of $\hat{y}$ (i.e., *ZPRED = 0), the conditional distributions of *ZRESID are very narrow; but for values further from 0, the conditional distributions become increasingly wide.

When the association between the predictors and the outcome variable y is linear, the conditional distributions of *ZRESID are centered on 0, as they are in Figure 14.A2.7a. When the association between the predictors and the outcome variable y is non-linear, then the conditional distributions of *ZRESID are not centered on 0, as in Figure 14.A2.7b. In Figure 14.A2.7b, residuals associated with *ZPRED near 0 are below 0; they are closer to −1. Residuals associated with *ZPRED further from 0 are above 0. For example, when *ZPRED = ±3, the values of *ZRESID are closer to 3. Therefore, we can use residual plots to visually assess the assumptions of linearity and homoscedasticity. Changes in the variance of the conditional distributions of *ZRESID reveal a violation of homoscedasticity. Changes in the means of the conditional distributions of *ZRESID reveal a violation of linearity.

Assessing Normality in SPSS

The regression model assumes that the distribution of prediction errors (residuals) is normal. If you refer back to Figure 14.A2.6, you will notice a checkbox highlighted in the area titled Standardized Residual Plots. This is an instruction to SPSS to create a histogram of the residuals. Figure 14.A2.8 shows histograms of the residuals for the two linear regression analyses shown in Figure 14.A2.5. A normal distribution is overlaid on both histograms. These normal distributions have the same mean and variance as the standardized residuals. That is, the overlaid normal distributions have means of zero and standard deviations of 1. In Figure 14.A2.8a, the histogram is well fit by a normal distribution. In Figure 14.A2.8b, the distribution of residuals is clearly skewed.

FIGURE 14.A2.7 Plotting Residuals

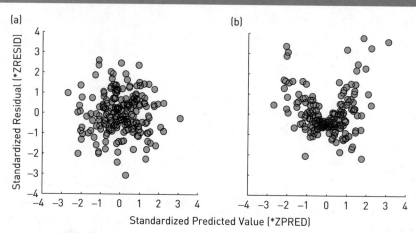

Plots of *ZRESID (*y*) against *ZPRED (*x*). These variables are plottevd for (a) *x* and *y*1 (from Figure 14.A2.5a), which show no violations of homoscedasticity or linearity, and (b) *x* and *y*2 (from Figure 14.A2.5b), which show massive violations of homoscedasticity and linearity.

FIGURE 14.A2.8 ■ Histograms of Residuals

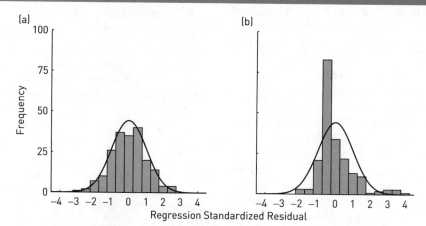

Histograms of standardized prediction errors, or residuals. (a) This histogram shows the distribution of standardized residuals for **y1** predicted from **x** (Figure 14.A2.5a) in which the assumptions of homoscedasticity and linearity have not been violated. There is no strong evidence that the normality assumption has been violated. (b) This histogram shows the distribution of standardized residuals for **y2** predicted from **x** (Figure 14.A2.5b) in which the assumptions of homoscedasticity and linearity have been violated. In this case, the distribution of residuals is highly skewed, revealing a serious violation of the assumption of normally distributed residuals.

INFERENTIAL STATISTICS FOR CORRELATION

INTRODUCTION

In Chapter 13, we saw that the correlation coefficient (r) conveys the precise degree of linear association between two scale variables. In this chapter, we will see how to use the sample correlation coefficient (r) to estimate the population correlation coefficient (ρ). For this estimation to be valid, we will assume that our sample was drawn from a *bivariate normal distribution*. We will see that the sampling distribution of r is generally skewed, which makes it difficult to use the methods from previous chapters to compute a confidence interval around r. Fortunately, a simple transformation of r produces an approximately normally distributed statistic, z_r, which allows us to use familiar techniques to compute our confidence interval. As with other confidence intervals, it is possible to use a confidence interval around r to test the null hypothesis that the population parameter (ρ in this case) is 0. Although it is widely used in research reports, the correlation coefficient is of less value than the regression coefficient in individual studies. On the other hand, the correlation coefficient is a scale-independent measure of effect size, like Cohen's δ. Therefore, r can be used to combine the results of different studies in a meta-analysis.

AN EXAMPLE

Although children inherit traits from their parents, they are not carbon copies of them. Therefore, it's reasonable to ask about the correlation between characteristics of parents and their children. Let's imagine that as part of a preliminary study on intellectual functioning, you decided to measure the association between mothers' IQs and their oldest (biological) daughters' IQs. In this case, your sample involves pairs of IQ scores: one score from each member of a mother-daughter dyad. The population in question comprises *pairs of scores* from all women who have at least one daughter. Let's say that a random sample of 259 dyads has been drawn from this population, IQ is measured for the two members of each dyad, and then the correlation coefficient is computed. The results might be reported as follows:

APA Reporting

The correlation between Mother's IQ (MIQ) and Daughter's IQ (DIQ) is estimated to be .6, approximate 95% CI [.51, .67].

(From now on, we'll use MIQ and DIQ as convenient shorthand for Mother's IQ and Daughter's IQ, respectively.)

This confidence interval seems rather unremarkable. Our best point estimate of ρ is $r = .6$, and we have *approximately* 95% confidence that the interval [.51, .67] captures ρ. We will see that this interval is approximate in the same way that confidence intervals around d (in Chapters 10 through 12) are approximate. Our confidence comes from knowing that approximately 95% of intervals constructed in this way will capture ρ. Of course, the next question is "How was the interval constructed?"

We've seen many cases in which a confidence interval is computed as *statistic ± moe*. Look at the above interval ([.51, .67]), and then stop for a moment and ask yourself if there is any reason to question whether this interval was calculated in the usual way.

If the interval *had been* computed as *statistic ± moe*, then we would expect the upper and lower limits to be symmetrical about the point estimate. In this case, they are not. The lower limit of the confidence interval is further from the point estimate $(.51 - .6 = -.09)$ than the upper limit $(.67 - .6 = .07)$. This asymmetry means that we have a mystery to solve about how this confidence interval was calculated.

As always, to understand the construction of confidence intervals, we need to know about the sampling distribution of the statistic in question. To understand the sampling distribution, we have to know what kind of distribution our samples are drawn from. Let's have a look at these two questions.

THE SAMPLING DISTRIBUTION OF r

We will restrict our focus to questions like the one discussed in the opening example. In this case, both x and y are random variables, and we will assume that our sample is a random sample from a bivariate normal distribution such as the one shown in Figure 15.1.

We now turn to the sampling distribution of r, which is a distribution of all possible values of r, computed from samples of size n, drawn from a bivariate normal distribution having a given ρ. A sample from a bivariate distribution has x and y scores for each individual, or for members of a dyad. Therefore, n refers to the number of *pairs* of scores in our sample.

Figure 15.2a illustrates sampling distributions for the statistic r for seven different values of ρ based on samples of size $n = 27$. The shape of the sampling distribution of r depends on the value of ρ and sample size (n). When $\rho = 0$, the sampling distribution of r is very close to normal. However, as ρ approaches -1 or 1, the sampling distribution of r becomes increasingly skewed. Although not shown in Figure 15.2, the sampling distribution of r becomes increasingly normal as sample size increases, but this depends on a complex relationship between sample size and ρ. Therefore, because the sampling distribution of r is typically not normal, we can't construct a confidence interval around r with the same method used to construct a confidence interval around m, $m_1 - m_2$, b, or $\hat{y}$.

The Fisher Transformation

The **Fisher transformation** converts correlation coefficients (r), which are not generally normally distributed, into z_r statistics, which are approximately normally distributed. The transformation is simply $z_r = 0.5*\ln[(1 + r)/(1 - r)]$, where ln is the natural log function.

Fortunately, there is a simple solution to this problem. Although the sampling distribution of r itself is not generally normal, Sir Ronald Fisher showed that a simple *transformation* of r produces a new statistic, called z_r, that is approximately normally distributed. The **Fisher transformation** from r to z_r is accomplished using the following formula:

$$z_r = \frac{1}{2}\ln\left(\frac{1+r}{1-r}\right),$$

(15.1)

FIGURE 15.1 ■ A Bivariate Normal Population

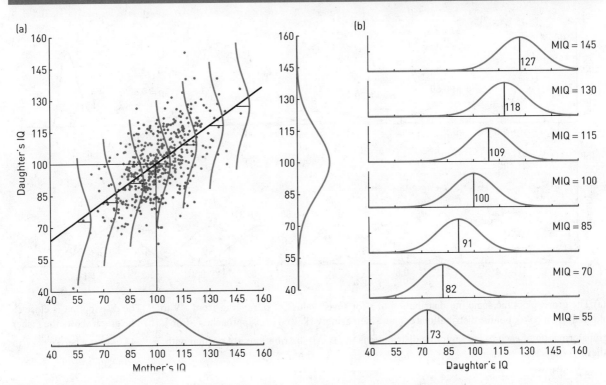

(a) The variables x and y are normally distributed, with $\mu_x = 100$ and $\sigma_x = 15$, and $\mu_y = 100$ and $\sigma_y = 15$. There is a positive linear association between x and y, and the line of best fit is plotted through the data points. Each y value on the line of best fit represents the mean of the conditional distribution for the corresponding value of x. The conditional distributions for MIQ = 55, 70, 85, 100, 115, 130, and 145 are shown as (sideways) icons of normal distributions. The correlation between x and y in this population is $\rho = .6$. (b) The seven conditional distributions from (a) are shown in their more conventional orientations. There is a conditional distribution of y (DIQ) for each value of x (MIQ). The means of the conditional distributions change linearly with x according to the formula $E(y) = 40 + 0.6(x)$. All conditional distributions have the same standard deviation of $\sigma_{est} = 12$.

where ln is the natural logarithm of the quantity in parentheses. If you are not familiar with logarithms, you can use Table 15.1, which provides z_r for values of r ranging from $-.99$ to $.995$. Figure 15.2b plots the Fisher transformation of each of the distributions of r above. All distributions in Figure 15.2b are close to normal, and the mean of each distribution is shown above its peak.

Not only are the distributions of z_r normal, but they all have the same standard error, which is defined as follows:

$$\sigma_{z_r} = \frac{1}{\sqrt{n-3}} = \sqrt{\frac{1}{n-3}}. \tag{15.2}$$

Equation 15.2 shows that (σ_{z_r}) does not depend on ρ. All distributions of z_r are normal with the same standard deviation.

FIGURE 15.2 ■ The Distributions of r and z_r

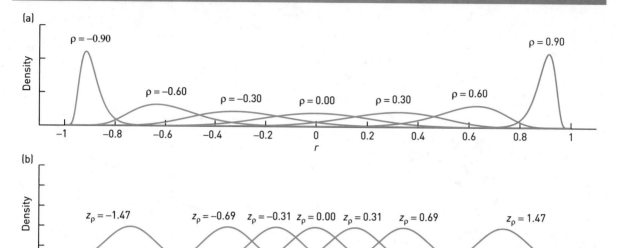

(a) The graph plots the sampling distribution of r for seven values of ρ, ranging from $-.9$ to $.9$. The distribution is close to normal when $\rho = 0$, but the distributions become increasingly skewed as ρ approaches -1 or 1. (b) The graph shows the Fisher transformation of each of the sampling distributions in (a). All distributions are normal with $\sigma_{z_r} = 1/\sqrt{n-3}$, where $n = 27$ in these examples.

The **Inverse Fisher transformation** converts z_r statistics to correlation coefficients (r). The transformation is simply $r = [\exp(2z_r) - 1]/[\exp(2z_r) + 1]$, where exp is a function that raises the base of the natural logarithm (≈ 2.7182) to some power x; i.e., $\exp(x) \approx 2.7182^x$.

The Inverse Fisher Transformation

Because the sampling distribution of z_r is normal with a known standard error (σ_{z_r}), we can convert r to z_r and place a $(1-\alpha)100\%$ confidence interval around it as follows:

$$z_r \pm z_{\alpha/2}\left(\sigma_{z_r}\right).$$

Note that for this approximate interval, we use $z_{\alpha/2}$ rather than $t_{\alpha/2}$. This does not complete the job, however, because we want the confidence interval for r, not z_r. Therefore, we transform the upper and lower confidence limits around z_r back to confidence limits around r by using the **inverse Fisher transformation** given as follows:

$$r = \frac{\exp(2z_r)-1}{\exp(2z_r)+1}. \tag{15.3}$$

Or we can simply consult Table 15.1 to find the values of r that correspond to the confidence limits on z_r.

Our Example (Continued)

Let's revisit the example in which we found the correlation (r) between MIQ and DIQ to be $r = .6$, for a sample of 259 dyads. The approximate 95% confidence interval around r was [.51, .67]. We can now work through these calculations.

    Step 1. Transform r to z_r. With the help of Table 15.1, we find that $r = .6$ corresponds to $z_r = .69$.

TABLE 15.1 ■ The Fisher r to z_r Transform							
r	z_r	r	z_r	r	z_r	r	z_r
−.99	−2.65	−.49	−0.54	.01	0.01	.51	0.56
−.98	−2.30	−.48	−0.52	.02	0.02	.52	0.58
−.97	−2.09	−.47	−0.51	.03	0.03	.53	0.59
−.96	−1.95	−.46	−0.50	.04	0.04	.54	0.60
−.95	−1.83	−.45	−0.48	.05	0.05	.55	0.62
−.94	−1.74	−.44	−0.47	.06	0.06	.56	0.63
−.93	−1.66	−.43	−0.46	.07	0.07	.57	0.65
−.92	−1.59	−.42	−0.45	.08	0.08	.58	0.66
−.91	−1.53	−.41	−0.44	.09	0.09	.59	0.68
−.90	−1.47	−.40	−0.42	.10	0.10	.60	0.69
−.89	−1.42	−.39	−0.41	.11	0.11	.61	0.71
−.88	−1.38	−.38	−0.40	.12	0.12	.62	0.73
−.87	−1.33	−.37	−0.39	.13	0.13	.63	0.74
−.86	−1.29	−.36	−0.38	.14	0.14	.64	0.76
−.85	−1.26	−.35	−0.37	.15	0.15	.65	0.78
−.84	−1.22	−.34	−0.35	.16	0.16	.66	0.79
−.83	−1.19	−.33	−0.34	.17	0.17	.67	0.81
−.82	−1.16	−.32	−0.33	.18	0.18	.68	0.83
−.81	−1.13	−.31	−0.32	.19	0.19	.69	0.85
−.80	−1.10	−.30	−0.31	.20	0.20	.70	0.87
−.79	−1.07	−.29	−0.30	.21	0.21	.71	0.89
−.78	−1.05	−.28	−0.29	.22	0.22	.72	0.91
−.77	−1.02	−.27	−0.28	.23	0.23	.73	0.93
−.76	−1.00	−.26	−0.27	.24	0.24	.74	0.95
−.75	−0.97	−.25	−0.26	.25	0.26	.75	0.97
−.74	−0.95	−.24	−0.24	.26	0.27	.76	1.00
−.73	−0.93	−.23	−0.23	.27	0.28	.77	1.02
−.72	−0.91	−.22	−0.22	.28	0.29	.78	1.05
−.71	−0.89	−.21	−0.21	.29	0.30	.79	1.07
−.70	−0.87	−.20	−0.20	.30	0.31	.80	1.10
−.69	−0.85	−.19	−0.19	.31	0.32	.81	1.13
−.68	−0.83	−.18	−0.18	.32	0.33	.82	1.16
−.67	−0.81	−.17	−0.17	.33	0.34	.83	1.19
−.66	−0.79	−.16	−0.16	.34	0.35	.84	1.22
−.65	−0.78	−.15	−0.15	.35	0.37	.85	1.26
−.64	−0.76	−.14	−0.14	.36	0.38	.86	1.29
−.63	−0.74	−.13	−0.13	.37	0.39	.87	1.33
−.62	−0.73	−.12	−0.12	.38	0.40	.88	1.38
−.61	−0.71	−.11	−0.11	.39	0.41	.89	1.42
−.60	−0.69	−.10	−0.10	.40	0.42	.90	1.47
−.59	−0.68	−.09	−0.09	.41	0.44	.91	1.53
−.58	−0.66	−.08	−0.08	.42	0.45	.92	1.59
−.57	−0.65	−.07	−0.07	.43	0.46	.93	1.66
−.56	−0.63	−.06	−0.06	.44	0.47	.94	1.74
−.55	−0.62	−.05	−0.05	.45	0.48	.95	1.83
−.54	−0.60	−.04	−0.04	.46	0.50	.96	1.95
−.53	−0.59	−.03	−0.03	.47	0.51	.97	2.09
−.52	−0.58	−.02	−0.02	.48	0.52	.98	2.30
−.51	−0.56	−.01	−0.01	.49	0.54	.99	2.65
−.50	−0.55	.00	0.00	.50	0.55	.995	2.99

Step 2. Compute the standard error of z_r. We do this using equation 15.2 as follows:

$$\sigma_{z_r} = \sqrt{\frac{1}{n-3}} = \sqrt{\frac{1}{259-3}} = .0625.$$

Step 3. Determine $z_{\alpha/2}$. Because this is the 95% confidence interval, $z_{\alpha/2} = 1.96$.

Step 4. Compute the confidence interval around z_r.

$$CI = z_r \pm z_{\alpha/2}\left(\sigma_{z_r}\right) = 0.69 \pm 1.96(.0625) = [0.5675, 0.8125].$$

Step 5. Transform the confidence limits around z_r to confidence limits around r. Using Table 15.1, we find the z_r closest to 0.5675 is 0.56, which corresponds to $r = .51$. The closest z_r to 0.8125 is 0.81, which corresponds to $r = .67$. Therefore, the approximate 95% confidence interval around $r = .6$ is [.51, .67].

From these calculations, we have approximately 95% confidence that ρ lies in the interval [.51, .67], because approximately 95% of intervals computed this way will capture ρ.

Because we used Table 15.1 to calculate our confidence interval, its precision is limited by the precision of the table. If we had used equation 15.1 to determine z_r, our answer would have been 0.6931. Therefore, the confidence interval around z_r would have been 0.6931 ± 0.1225 = [0.5706, 0.8156]. These limits would have been converted back to values of r using equation 15.3. Therefore, the confidence interval around r would be [0.5158, 0.6727].

We won't step through these calculations by hand because there is an easy way of doing them in Excel. Appendix 15.1 shows the **FISHER** and **FISHERINV** Excel functions that perform equations 15.1 and 15.3, respectively. Therefore, the best approximate intervals can be computed in Excel using these two functions.

In summary, there are a few things we should keep in mind about our approximate 95% confidence interval around r. First, this interval is approximate even when computed using equations 15.1 and 15.3, rather than using Table 15.1. The z_r scores produced by the Fisher transformation are only approximately normal. Second, using Table 15.1 to convert between r and z_r introduces further imprecision because only a limited number of r and z_r values are available. Therefore, when you compute a confidence interval for r, you should use the **FISHER** and **FISHERINV** functions in Excel. Finally, when we compute our confidence interval, we must assume that our sample is a random sample from a bivariate normal distribution of scores. This means that x and y are normally distributed, the conditional y distributions are homoscedastic and normal, and the expected values of y change linearly with x.

SIGNIFICANCE TESTS

Confidence Intervals

Confidence intervals around correlation coefficients can be used to test the null hypothesis that $\rho = 0$. In the case we've just examined, our confidence interval did not include 0, so we judge it implausible (but not impossible) that our sample was drawn from a bivariate normal distribution for which $\rho = 0$. In other words, we know that approximately 95% of the time, intervals constructed this way will capture the population parameter ρ, and approximately 5% of the time they won't. So this is either (i) one of those rare occasions in which the null hypothesis is true but the confidence interval does not capture the parameter, or (ii) the null hypothesis is false. According to the rules of significance testing, we would

LEARNING CHECK 1

1. State whether the following statements are true or false.

   (a) All distributions of r are skewed.

   (b) No distribution of r is normal.

   (c) All distributions of z_r are approximately normal.

   (d) The standard error of z_r depends only on sample size.

   (e) If $r -.07$, then $z_r = -.07$.

2. For each of the following values of r, use the Fisher transform table to determine z_r: (a) $-.78$, (b) $.36$, (c) $.92$, and (d) $-.81$.

3. For each of the following values of z_r, use the Fisher transform table to determine r: (a) -2.3, (b) 1.59, (c) $.52$, and (d) $-.66$.

4. Use the Fisher transform to compute the 95% confidence interval for each of the following: (a) $r = .8$, $n = 28$; (b) $r = -.7$, $n = 23$; (c) $r = .3$, $n = 53$; and (d) $r = .53$, $n = 67$.

Answers

1. (a) False. All distributions of r approach the normal distribution as sample size increeaes. Furthermore, when $\rho = 0$, the distribution of r is symmetrical.

   (b) False. All distributions of r approach the normal distribution as sample size increases.

   (c) True. That's why we use the Fisher transform from r to z_r.

   (d) True.

   (e) True, if we're using the table. However, if we use **FISHER** or **FISHERINV** in Excel there will be a small difference. Using Excel, we find that when $r = -.07$, $z_r = -0.070114671$.

2. (a) -1.05. (b) 0.38. (c) 1.59. (d) -1.13.

3. (a) $-.98$. (b) $.92$. (c) $.48$. (d) $-.58$.

4. (a) $[.61, .90]$. (b) $[-.86, -.41]$. (c) $[.03, .53]$. (d) $[.33, .69]$.

reject the null hypothesis and say that our point estimate ($r = .6$) is statistically significantly different from 0.

It should also be noted that 95% confidence intervals around r can test any null hypothesis about ρ. For example, the confidence interval computed above could be used to test H_0: $\rho = .4$ or H_0: $\rho = .2$. Because .4 and .2 are not included in our confidence interval, we can reject both as plausible hypotheses about ρ.

t-Tests

The traditional null hypothesis about ρ is H_0: $\rho = 0$, and this special case is typically tested with a t-statistic. We noted that when $\rho = 0$, the sampling distribution of r is essentially normal (see Figure 15.2). This sampling distribution has a mean of $\mu_r = 0$, and $\sigma_r = 1/\sqrt{n}$. This standard error is estimated by

$$s_r = \sqrt{\frac{1-r^2}{n-2}}. \tag{15.4}$$

As we've seen many times before, normal distributions of statistics can be transformed to a t-distribution. For the correlation coefficient, when $\rho = 0$, we have

$$t = \frac{r - \rho}{s_r}. \tag{15.5}$$

Because $\rho = 0$, our t_{obs} statistic is simply

$$t_{obs} = \frac{r}{s_r}. \tag{15.6}$$

We test t_{obs} for statistical significance on $n-2$ degrees of freedom, where n is the number of xy pairs in our sample.

We can continue our example about MIQ and DIQ and test the null hypothesis that $\rho = 0$. We know that $r = .6$ and $n = 259$, so we have $n-2 = 257$ degrees of freedom. If we treat this as a nondirectional test, then $t_{critical} = \pm 1.972$, according to our t-table. (The exact value of $t_{critical}$, determined using Excel, would be ± 1.969.) With this information, we can compute t_{obs} as follows:

$$t_{obs} = \frac{r}{s_r} = \frac{r}{\sqrt{\dfrac{1-r^2}{n-2}}} = \frac{.6}{\sqrt{\dfrac{1-.6^2}{259-2}}} = 12.02.$$

Because $t_{obs} = 12.02$ is clearly outside the interval ± 1.972, we can reject the null hypothesis that $\rho = 0$. The exact p-value associated with $t_{obs} = 12.02$ is about 10^{-26} (or if you like, $1/10^{26}$, which is a decimal point followed by 25 zeroes and then a 1). In other words, r outside the interval $\pm .6$ would occur *very* infrequently if the null hypothesis ($\rho = 0$) were true.

Interpreting Significance Tests

It's not exactly clear what to make of our rejection of the null hypothesis. The correlation between two psychological variables is almost never exactly zero, so we should have strongly suspected before collecting any data that the null hypothesis is false. Many authors have made this point over the years. Lykken (1968) referred to this as *ambient correlational noise*. Paul Meehl (1990) used the more colorful term *crud factor* to refer to the same idea. Meehl (1990) reported that the average correlation on 45 scales of the Minnesota Multiphasic Personality Inventory (MMPI) was about .4. Therefore, testing H_0: $\rho = 0$ is hard to justify, given that it is almost certainly false.

We encountered this problem with significance tests in previous chapters. Although it may be satisfying to reject the null hypothesis, we can be quite sure that it is false before data collection, because it is difficult to think of any two psychological variables for which ρ is precisely 0. Therefore, to reject the null hypothesis that $\rho = 0$, we only need a large enough sample. Given the near certainty that $\rho \neq 0$, it would seem to make more sense to simply estimate what it is, rather than perform a significance test.

LEARNING CHECK 2

1. If $\rho = .5$, we can estimate the standard error of r (i.e., σ_r) using $s_r = \sqrt{(1-r^2)/(n-2)}$.

2. We can test H_0: $\rho = 0$ with $t_{obs} = r/s_r$.

3. Calculate t_{obs} for each of the following: (a) $r = .8$, $n = 28$; (b) $r = -.7$, $n = 23$; (c) $r = .3$, $n = 53$; and (d) $r = .53$, $n = 67$.

Answers

1. False. This estimate applies only when $\rho = 0$.

2. True.

3. (a) 6.8. (b) −4.49. (c) 2.25. (d) 5.04.

WHAT IS A BIG CORRELATION AND WHAT IS THE PRACTICAL SIGNIFICANCE OF r?

The questions (i) "What is a big correlation?" and (ii) "What is the practical significance of a given correlation?" are easier to ask than to answer. Of course, Jacob Cohen had something to say about this. He suggested that r values of .1, .3, and .5 should be considered small, medium, and large, respectively, as shown in Table 15.2. As always, these conventions should be used as a last resort, and only when no other basis exists for judging r.

TABLE 15.2 ■ Cohen's Classifications		
Effect Size	r	r^2
Small	.1	.01
Medium	.3	.09
Large	.5	.25

Note: Cohen's classification scheme for ρ. These values are to be used as a last resort.

We've already discussed the problem of switching from p-values to effect sizes as context-independent measures of importance. Rules like these serve as crutches to give meaning to our data. A correlation of $r = .5$ might be considered large if very few correlations this large are found in a given research literature, but it might be considered small if few correlations reported in a research literature are smaller. In any case, interpretations like these are statistical, and they say nothing about the practical or substantive significance of the correlation.

Imagine that a study reported a correlation of $r = .7071$ between two random variables (x and y). This means that $r^2 = .5$, which says that 50% of the variability in y is attributable to its association with x. Should we be thrilled by this finding? Well, we can't interpret this result without knowing what x and y are, or what a theory or previous experiment suggests about what r should be. Let's say that $r = .7071$ came from a study that measured the IQs of 259 identical twins raised in the same home. I think we would be surprised to find a correlation of only .7071, 95% CI [.64, .763], between these variables. That is, we would probably expect a much larger correlation.

Let's imagine another study that measured how fast people can skate forward (x) and backward (y). We would not be impressed to find a correlation of .9, 95% CI [.89, .92], in a sample of 403 people, because both scores are measures of the same thing (i.e., both are measures of how well a person can skate).

Finally, let's say we found a correlation of $r = .1$ between coffee consumption (measured in ounces per day) and general health (assessed by family doctors) in a random sample of 1603 French adults, 95% CI [.051, .15]. This might seem like a small correlation (in terms of Cohen's classifications), but the consequences could be important. Such a correlation could mean that the average health in the population of adults in France would be improved by increasing coffee consumption. Even a small increase in overall health could lead to reductions in health-related expenditures in the country as a whole. Of course, correlation does not imply causation, so we can't conclude that increasing coffee consumption would reduce health expenditures. On the other hand, if there were even a weak causal connection between coffee consumption and general health, this could be useful when we consider the consequences for a population.

In summary, it may be possible to judge correlation coefficients as small, medium, and large within a given body of research. These classifications might differ between areas such as personality research and neuroscience. However, these classifications are separate from the practical significance of the correlation. As I've said before, statistics do not provide a substitute for thinking. To make proper use of a statistical analysis, you must deeply understand your research area. Stating only that a correlation is large, according to some universal criterion, is no better than saying that the difference between r and 0 is statistically significant.

THE CORRELATION COEFFICIENT IS A STANDARDIZED EFFECT SIZE: META-ANALYSIS

We saw in Chapter 11 that many statistics (r^2, r, t_{obs}, F) can be converted to d, which is a standardized measure of the difference between the means of two groups. In Chapter 11, we saw that conversion to *d* was useful in a meta-analysis that combines the results of studies that used different dependent variables and/or different statistics. In the case of the association between two random variables, *r* is a standardized measure of the strength of the linear association. In other words, *r*, like *d*, is an effect size.

The merit of the regression slope (*b*) is that it tells us how much change there is in *y* with a one-unit change in *x*. Therefore, *b* is generally more useful than *r*. However, if we wish to combine the results of regression analyses that addressed the same question using different variables, then a standardized effect size like *r* is useful. We saw in Chapter 13 that

$$b = r\frac{s_y}{s_x}.$$

A simple rearrangement of this equation is as follows:

$$r = b\frac{s_x}{s_y}. \tag{15.7}$$

Therefore, if a study reports the regression equation and the statistics of samples, it is then possible to recover the correlation coefficient using equation 15.7. Or, as we saw in Appendix 14.2, SPSS reports *r* as a *standardized regression coefficient* and calls it a *beta weight*.

To illustrate the value of *r* as a standardized measure of statistical association, we will imagine that five researchers investigated the association between exercise and mental agility. Each hypothetical study involved random samples of university students; within each study, the researchers determined the number of minutes that each student exercised each week from self-reports. However, the studies assessed mental agility with different tasks. These tasks involved the following measures: (i) digit span, (ii) time to complete a trail-making task, (iii) number of mental arithmetic problems solved, (iv) reading speed, and (v) number of riddles solved.

Let's first note an interesting problem that arises when one contemplates regression analyses for these five studies. For the digit span, mental arithmetic, and riddle tasks, larger scores mean higher degrees of mental agility. However, for the trial-making and reading speed measures, larger scores (slower speeds) mean lower mental agility. Therefore, a positive regression slope for the digit span, mental arithmetic, and riddle tasks means the same thing as a negative regression slope for the trail-making and reading tasks. To make these regression slopes (or correlation coefficients) compatible, we would multiply the slopes for the trail-making and reading tasks by −1. This adjustment will make negative regression slopes positive and preserves their magnitude.

The statistics of the five studies (b, s_x, s_y, and r) are shown in Table 15.3. The correlation coefficient has been recovered from s_x, s_y, and r, using equation 15.7. Notice that the magnitudes of the regression slopes (*b*) are very different across studies, whereas the magnitudes of the correlation coefficients (*r*) are quite similar. The reason for the variability in regression slopes is that the units of measurement for the dependent variables differed between studies. For example, digit span is measured in number of items recalled, and reading speed may be measured in milliseconds. Consequently, the standard deviations of the *y* scores are very different across studies. On the other hand, the standard deviations of the *x* scores are very similar across studies because the *x* variable (*minutes of exercise*) was the same in all cases. Because slope is defined as $b = r * s_y/s_x$, it becomes clear why the regression slopes are so variable. By computing $r = b * s_x/s_y$, we eliminate the effect of scale differences.

The last two rows of Table 15.3 show the results of the meta-analysis. The mean value of *r* is .16 with a standard deviation of 0.05. This analysis shows that the strength of the

TABLE 15.3 ■ Converting b to r					
Study	b	s_x	s_y	r	r^2
Digit span	0.024	11	2	.13	.02
Trail making	0.014	10	0.8	.18	.03
Mental arithmetic	0.275	12	15	.22	.05
Reading speed	24.615	13	2000	.16	.03
Riddle solving	0.017	12	2	.10	.01
Mean	4.99			.16	.03
Standard deviation	10.97			.05	.01

Note: Converting b to r is important for meta-analysis or regression results.

linear association is relating hours of exercise to mental agility is quite similar across the five studies. Of course, we have to keep in mind that correlation does not imply causation when we have two random variables.

The last column of Table 15.3 shows that only about 3% of the variability in measures of mental agility is explained by amount of weekly exercise. Considered population-wise, a causal connection between exercise and mental agility might imply gains for society as a whole, assuming that mental agility leads to increased productivity or other beneficial outcomes. However, the required increase in exercise may mean less time available for other valued activities like playing video games or cooking, even to an individual who values a potential increase in mental agility. Therefore, there may be some practical significance at the population level, but there may be little at the individual level; recall the discussion of prediction intervals from Chapter 14.

LEARNING CHECK 3

1. Two studies measured the linear association between spatial ability and verbal ability. In both studies, spatial ability was measured on a 25-point scale, where 1 indicates poor spatial ability and 25 indicates high spatial ability. Study 1 measured verbal abilities using a word recognition task: more words recognized means higher verbal ability. Study 2 measured verbal ability using a speeded reading task: faster reading times mean higher verbal ability. The results of Study 1 showed $b = 20$, $s_x = 10$, and $s_y = 250$, and the results of Study 2 showed $b = -25$, $s_x = 12$, and $s_y = 400$.

   (a) Calculate r for Study 1.

   (b) Calculate r for Study 2.

   (c) Combine these two correlations in a way that is sensible for a meta-analysis.

Answers

1. (a) $r = 20*10/250 = .8$.

   (b) $r = -25*12/400 = -.75$.

   (c) In Study 1, large scores mean high verbal ability; in Study 2, low scores mean high verbal ability. Given that both measures reflect verbal ability, they are predicted to have opposite signs on the regression slopes. To make both positive, we multiply the slope of Study 2 by -1 to produce $b = -1*-25 = 25$. This yields $r = .75$. When these two r values are averaged, we obtain $(.80 + .75)/2 = .78$.

THE GENERALITY OF CORRELATION

Correlation and the Two-Groups Design

We've emphasized throughout this part of the book that all statistics are interrelated. I've said before that the vast majority of statistical analyses used in psychology boil down to correlation, even if on the surface they don't look like correlation at all. The material covered in this section is intended to show the generality of correlation, rather than to offer an alternative way to analyze data. This material provides good preparation for Part IV. In this section, we'll revisit the independent-groups design that was covered in Chapter 11.

The Point-Biserial Correlation Coefficient

The independent-groups design might seem qualitatively different from correlational designs. The two groups in question typically represent the two values of a dichotomous qualitative variable such as sex. This seems to present an obstacle for a correlational analysis because the correlation coefficient is computed from two scale variables. How can we conduct a correlational analysis when one of the variables is qualitative? The answer is simply to code the two levels of the dichotomous variable using a pair of numbers. Any two numbers will do, but for simplicity, we can code the two levels as −1 and 1. The correlation between a coded dichotomous predictor variable and a quantitative outcome variable is called the **point-biserial correlation coefficient**.

The **point-biserial correlation coefficient** (r_{pb}) is the correlation coefficient that results from a dichotomous predictor variable (coded by any two numbers) and a quantitative outcome variable.

To illustrate the connection between the two-independent-groups design and correlation, we will consider the effects that playing two different video games have on the time it takes to solve mazes see Nelson and Strachan (2009) for a related study. The two video games will be *Portal*, which emphasizes spatial reasoning, and *Unreal Tournament*, which is a first-person shooter game. Eight participants were chosen and divided at random into two groups of four participants each. The participants played their assigned video game for 1 hour before being asked to solve four relatively complex mazes. The time (in minutes) taken to complete each maze was measured, and the dependent variable was the sum of these four completion times. These hypothetical data are shown in Table 15.4. The mean of the *Portal* group is $m_p = 15$, with a standard deviation of $s_p = 4.24$. The mean of the *Unreal Tournament* group is $m_{UT} = 23$, with a standard deviation of $s_{UT} = 3.74$. These

TABLE 15.4 ■ Example Data		
	Portal	Unreal Tournament
	15	17
	9	23
	21	25
	15	27
m	15	23
s^2	24.00	18.67
s	4.24	3.74

Note: Data for a correlation analysis of scores from two independent populations.

iStock.com Bigmouse108

results show that playing *Portal* leads to faster maze-solving times than playing *Unreal Tournament*.

The data from Table 15.4 have been rearranged in Table 15.5 so that we can approach our analysis from a correlation point of view. The column labeled "x" shows that the members of the *Portal* and *Unreal Tournament* groups are coded with a -1 and 1, respectively. The column labeled "y" (i.e., completion times in minutes) shows the dependent variable. You can confirm by checking Tables 15.4 and 15.5 that the scores from the *Portal* group are the first four entries in column y and the scores from the *Unreal Tournament* group are the next four entries in column y. Using these numbers, we can compute the *point-biserial correlation coefficient*, or r_{pb}.

To compute r_{pb} using the computational formula from Chapter 13, we need to compute $ss_x = \Sigma(x - m_x)^2$, $ss_y = \Sigma(y - m_y)^2$, and $sp_{xy} = \Sigma(x - m_x)(y - m_y)$. At the bottom of Table 15.5, we see that $ss_x = 8$, $ss_y = 256$, and $sp_{xy} = 32$. When these quantities are substituted in the computational formula for r given in Chapter 13, we obtain the following:

$$r_{pb} = \frac{sp_{xy}}{\sqrt{ss_x ss_y}} = \frac{32}{\sqrt{8 * 256}} = .71.$$

(When expressed to four decimal places, $r_{pb} = .7071$.) It is important to note again that any two numbers can be used to code for groups, because the correlation coefficient is invariant under linear transformations, and any two pairs of numbers are simply linear transformations of each other. (See Chapter 13 for a discussion of linear transformations.) For example, we could have used 20 and 24 to code for groups. These numbers are related to -1 and 1 by the linear transformation $22 + 2(x)$; i.e., $22 + 2(-1) = 20$ and $22 + 2(1) = 24$.

The Point-Biserial Correlation and d

We've just seen that we can compute r_{pb} for two groups by coding levels of the dichotomous qualitative variable with numbers. We will now see how r_{pb} is connected to the two-groups analysis covered in Chapter 11. Let's start by remembering that r^2 can be transformed into

TABLE 15.5 ■ Point-Biserial Correlation					
Group	x	y	$x - m_x$	$y - m_y$	$(x - m_x)(y - m_y)$
Portal	-1	15	-1	-4	4
Portal	-1	9	-1	-10	10
Portal	-1	21	-1	2	-2
Portal	-1	15	-1	-4	4
Unreal Tournament	1	17	1	-2	-2
Unreal Tournament	1	23	1	4	4
Unreal Tournament	1	25	1	6	6
Unreal Tournament	1	27	1	8	8
	m_x	m_y	$ss_x = \Sigma(x - m_x)^2$	$ss_y = \Sigma(y - m_y)^2$	$sp_{xy} = \Sigma(x - m_x)(y - m_y)$
	0	19	8	256	32

Note: Quantities for computing the point-biserial correlation coefficient (r_{pb}).

Cohen's d. If we plug r_{pb}^2 into the following equation (from Chapter 11) to convert from r^2 to d, we obtain

$$d = \sqrt{df_{within}\left(\frac{r^2}{1-r^2}\right)\left(\frac{1}{n_1}+\frac{1}{n_2}\right)} = \sqrt{6\left(\frac{.5}{1-.5}\right)\left(\frac{1}{4}+\frac{1}{4}\right)} = 1.73.$$

To demonstrate that we've computed the right thing, we can compute d using the means and variances of the two samples, which were given in Table 15.4:

$$d = \frac{m_{UT}-m_P}{s_{pooled}} = \frac{m_{UT}-m_P}{\sqrt{.5(s_{UT}^2)+.5(s_P^2)}} = \frac{23-15}{\sqrt{.5(18.67)+.5(24)}} = 1.73.$$

Therefore, the two ways of computing d from the same data, one using r_{pb} and the other using group means, produce exactly the same result. This illustrates how correlation underlies our analysis of the two-groups design. (In these and the following calculations, the intermediate quantities were full precisions, and not rounded to two decimals places.)

The Point-Biserial Correlation and t_{obs}

We can also show that t_{obs} statistics computed from r/s_r and from $(m_1-m_2)/s_{m_1-m_2}$ are identical. We saw in equation 15.6 that $t_{obs}=r/s_r$. Therefore, we can compute t_{obs} from r as follows:

$$t_{obs} = \frac{r}{s_r} = \frac{r}{\sqrt{\dfrac{1-r^2}{n-2}}} = \frac{.71}{\sqrt{\dfrac{1-.71^2}{8-2}}} = 2.45$$

The conventional way of computing t_{obs} divides the difference between the two sample means by the estimated standard error of m_1-m_2. That is, $t_{obs}=(m_1-m_2)/s_{m_1-m_2}$. Using this standard formula we find that

$$t_{obs} = \frac{m_{UT}-m_P}{s_{m_1-m_2}} = \frac{23-15}{\sqrt{\dfrac{s_{UT}^2}{n_{UT}}+\dfrac{s_P^2}{n_P}}} = \frac{8}{\sqrt{\dfrac{18.67}{4}+\dfrac{24}{4}}} = 2.45$$

Once again, we arrive at exactly the same statistic from a different starting point. This demonstrates again what was shown in Chapter 11 and illustrates the generality of correlation.

Regression and the Two-Groups Design

Now that we've seen how the *correlation* coefficient relates to the two-groups design, we can take a final step and connect *regression* to the two-groups design. Again, the purpose of this section is to illustrate the generality of correlation and to provide the foundations for material covered in Part IV.

To compute the regression equation, we need to know sample size, m_x, s_x, m_y, s_y, and r_{pb}. We've computed all of these quantities except s_x and s_y. We saw in Table 15.5 that $ss_x = 8$ and $ss_y = 256$. From these sums of squares, we can determine that $s_x = 1.07$ and $s_y = 6.05$; in both cases, we divide the sum of squares by $n-1 = 7$ and then take the square root. So, here are the questions: What would our regression equation be if we computed it from these quantities, and what exactly would it *predict*?

To answer these questions, we first compute the slope (b) of the regression equation

LEARNING CHECK 4

1. If $y_1 = \{2, 4, 6\}$ and $y_2 = \{4, 6, 8\}$, calculate r_{pb} and r_{pb}^2.

2. Using the data in question 1, compute d in the following ways:

   (a) $d = (m_2 - m_1)/s_{pooled}$, and

   (b) $d = \sqrt{df_{within}(r_{pb}^2/(1-r_{pb}^2))(1/n_1 + 1/n_2)}$.

3. Using the data in question 1, compute t_{obs} in the following ways:

   (a) $t_{obs} = (m_2 - m_1)/s_{m_1 - m_2}$, and

   (b) $t_{obs} = r/s_r$.

Answers

1. Take the following steps to compute these quantities:

   (i) $x = \{-1, -1, -1, 1, 1, 1\}$.

   (ii) $y = \{2, 4, 6, 4, 6, 8\}$.

   (iii) $ss_x = 6$, $ss_y = 22$.

   (iv) $sp_{xy} = 6$.

   (v) $r_{pb} = sp_{xy}/\sqrt{ss_x * ss_y} = 6/\sqrt{6*22} = .52$.

   (vi) $r_{pb}^2 = .27$.

2. (a) $d = (6 - 4)/2 = 1$.

   (b) $d = \sqrt{df_{within}\left(r_{pb}^2/\left(1-r_{pb}^2\right)\right)(1/n_1 + 1/n_2)}$

   $= \sqrt{4(.27/.73)(2/3)}$

   $= 1.00$.

3. (a) $(6 - 4)/1.633 = 1.22$.

   (b) $.52/\sqrt{.73/4} = 1.22$.

$$b = r_{pb}\frac{s_y}{s_x} = .71\frac{6.05}{1.07} = 4,$$

and then the intercept (a):

$$a = m_y - b(m_x) = 19 - 4(0) = 19.$$

Therefore, our regression equation is

$$\hat{y} = a + b(x) = 19 + 4(x).$$

We mastered these calculations in Chapters 13 and 14, but now we have to ask what $\hat{y}$ corresponds to in this situation. There's only one way to find out, so let's calculate $\hat{y}$ for our two values of x, which are $x_P = -1$, and $x_{UT} = 1$. When we do this, we find

$$\hat{y}_P = a + b(x_P) = 19 + 4(-1) = 15 \text{ and}$$

$$\hat{y}_{UT} = a + b(x_{UT}) = 19 + 4(1) = 23.$$

So, $\hat{y}_P = 15$ and $\hat{y}_{UT} = 23$. Have we seen these numbers before? Yes! If you look back to Table 15.4, you will see that $m_P = 15$ and $m_{UT} = 23$! Therefore, our regression equation "predicts" the means of the two groups.

We could perform statistical analyses on our regression equation. Specifically, if we were to compute t_{obs} we would find it to be identical to the one computed from $m_{UT} - m_P$ and the one computed from r_{pb}. We won't venture into these calculations in this section because they are special cases of a more general technique that we'll cover in Part IV. The critical point is to appreciate the interrelatedness of all statistics covered in this part of the book.

LEARNING CHECK 5

1. Use the information in the following table to perform a regression analysis that relates the codes for the two groups (A and B) to the outcome variable. Report the following: (a) r_{pb}, (b) b, (c) a, (d) s_b, (e) t_{obs}, (f) d, and (g) F.

Group	x	y
A	0	16
A	0	10
A	0	22
A	0	16
B	10	18
B	10	24
B	10	26
B	10	28
m_x		m_y
	5	20
ss_x		ss_y
	200	256

Answers

1. (a) .71. (b) 0.80. (c) 16. (d) 0.33. (e) 2.45. (f) 1.73. (g) 6.
   Remember, $F = t^2$.

SUMMARY

The sample correlation coefficient (r) conveys the precise degree of linear association between pairs of xy values, and it estimates the population correlation coefficient ρ. To make inferences about ρ from r, we assume that our bivariate sample was drawn from a *bivariate normal distribution*.

It is a challenge to compute a confidence interval around r because the sampling distribution of r is often skewed. To overcome this problem, we use the *Fisher transform* that converts values of r to z_r values using the following formula:

$$z_r = \frac{1}{2}\ln\left(\frac{1+r}{1-r}\right).$$

The sampling distribution of z_r is close to normal with a standard error of

$$\sigma_{z_r} = \frac{1}{\sqrt{n-3}} = \sqrt{\frac{1}{n-3}}.$$

We can compute a confidence interval around z_r as follows:

$$CI = z_r \pm z_{\alpha/2}\left(\sigma_{z_r}\right).$$

The limits of this confidence interval are in z_r units, and these can be converted back to r units through the *inverse Fisher transform*:

$$r = \frac{\exp(2z_r) - 1}{\exp(2z_r) + 1}.$$

We can use a confidence interval to test a null hypothesis about ρ. If ρ specified by the null hypothesis is in the interval, we retain H_0; otherwise we reject H_0. Researchers often test the null hypothesis that H_0: ρ = 0. Because the sampling distribution of r is close to normal when ρ = 0, we can compute t_{obs} as follows:

$$t_{obs} = \frac{r}{s_r}, \text{ where}$$

$$s_r = \sqrt{\frac{1 - r^2}{n - 2}}.$$

If t_{obs} exceeds $t_{critical}$ based on $n-2$ degrees of freedom, we can reject H_0; otherwise we retain it.

Correlation is an extremely general tool that underlies most of the statistical analyses covered in this part of the book. The two-independent-groups design can be cast in terms of correlation. If we code the two groups with any two numbers, we can compute the correlation coefficient between these codes and the dependent variable. We call this the *point-biserial correlation coefficient*, r_{pb}. To obtain identical results, we can compute t_{obs} either as described above or as shown in Chapter 12:

$$t_{obs} = \frac{m_1 - m_2}{\sqrt{\frac{s_{pooled}^2}{n_1} + \frac{s_{pooled}^2}{n_2}}}.$$

The square of the point-biserial correlation coefficient r_{pb}^2 is another type of effect size that expresses the proportion of variability in the dependent variable that is explained by means. r_{pb}^2 has a simple connection to d, as we saw in Chapter 12:

$$d = \sqrt{df_{within}\left(\frac{r_{pb}^2}{1 - r_{pb}^2}\right)\left(\frac{1}{n_1} + \frac{1}{n_2}\right)}.$$

Finally, we can use our codes for groups (e.g., −1 and 1) and the outcome variable to compute a regression equation. To do so, we compute m_x and s_x, which are the mean and standard deviation of the column of codes. We also compute m_y and s_y, which are the mean and standard deviation of the column containing the dependent variable. From these two columns we can compute r_{pb}. From r_{pb}, m_x, s_x, m_y and s_y, the slope and intercept of the regression equation are computed as

$$b = r_{pb}\frac{s_y}{s_x} \text{ and}$$

$$a = m_y - b(m_x).$$

This regression equation [$\hat{y} = a + b(x)$] "predicts" the means of our two groups.

KEY TERMS

Fisher transformation   384

inverse Fisher transformation   386

point-biserial correlation coefficient (r_{pb})   394

EXERCISES

Definitions and Concepts

1. What is the difference between r and ρ?
2. What is the difference between a bivariate normal distribution and the sampling distribution of r?
3. How does the shape of the sampling distribution of r depend on sample size?
4. How does the shape of the sampling distribution of r depend on ρ?
5. What is the Fisher transform? What is it used for? Why is it necessary?
6. What is the point-biserial correlation coefficient?

True or False

State whether the following statements are true or false.

7. A confidence interval around r is computed as $r \pm t_{\alpha/2}(s_r)$.
8. A confidence interval around r is computed as $z_r \pm t_{\alpha/2}(s_r)$.
9. A confidence interval around r is computed as $z_r \pm z_{\alpha/2}(s_r)$.

10. A confidence interval around r is computed as $z_r \pm z_{\alpha/2}(\sigma_r)$.

11. The Fisher transform and inverse Fisher transform allow us to compute approximate confidence intervals around r.

12. $t_{obs} = r/\sqrt{(1-r^2)/(n-2)}$.

13. $t_{obs} = r/\sqrt{(1-r^2)/df}$.

14. The regression equation for the two-independent-groups design $[\hat{y} = a + b(x)]$ will not depend on the numbers used to code for groups.

15. The two values of $\hat{y} = a + b(x)$ for the two-independent-groups design will not depend on the numbers used to code for groups.

Calculations

16. Compute the 95% confidence interval around each of the following correlation coefficients: (a) $r = .1$, $n = 12$; (b) $r = .3$, $n = 12$; (c) $r = -.3$, $n = 84$; (d) $r = -.8$, $n = 147$; (e) $r = .8$, $n = 28$; (f) $r = .44$, $n = 67$; (g) $r = .27$, $n = 228$; (h) $r = -.5$, $n = 403$; (i) $r = -.25$, $n = 1603$; and (j) $r = .64$, $n = 6403$.

17. For each of the confidence intervals you just computed, test H_0: $\rho = 0$ against the alternative hypothesis H_1: $\rho \neq 0$, assuming $\alpha = .05$.

Scenarios

18. A study of 100 college students was conducted to determine the association between physical activity and last annual GPA. Each student filled out a questionnaire (the DPAQ) about his or her daily physical activity. Scores on the questionnaire ranged from 0 to 116, with higher scores indicating greater daily physical activity. The mean DPAQ score was 59 with a standard deviation of 22, and the mean annual GPA was 3.6 with a standard deviation of .75. The correlation between the DPAQ and GPA was .23.

   (a) Calculate the linear regression equation that predicts GPA from DPAQ scores.

   (b) Compute the approximate 95% confidence interval around r.

   (c) Explain what you can conclude from this study, keeping in mind (i) the assumptions you are making when testing the confidence interval, (ii) the strength of the association between the two variables, and (iii) the question of causality.

19. A study of 57 Montreal transit workers was conducted to assess the association between cranial capacity (inferred from skull circumference) and IQ. Each transit worker completed an IQ test and a research assistant measured participants' head skull circumference. The mean IQ score was 98 with a standard deviation of 10, and the mean skull circumference was 61 cm with a standard deviation of 3.25. The correlation between the IQ and skull circumference was .41.

   (a) Calculate the linear regression equation that predicts IQ from head circumference.

   (b) Compute the approximate 95% confidence interval around r.

   (c) Explain what you can conclude from this study, keeping in mind (i) the assumptions you are making when testing the confidence interval, (ii) the strength of the association between the two variables, and (iii) the question of causality.

APPENDIX 15.1: CORRELATION ANALYSIS IN EXCEL

Figure 15.A1.1 shows 18 xy pairs. We've computed the correlation coefficient in cell **D3** (using the **CORREL** function) and the squared correlation coefficient (r^2) in cell **D4**. We have two goals here. The first is to compute a confidence interval around r, and the second is to test r for statistical significance.

To compute the approximate confidence interval around r, we transform it to z_r using the Fisher transform, put the confidence interval around z_r using $z_r \pm z_{\alpha/2}(\sigma_{z_r})$, and then transform the limits of this confidence interval back into r units using the inverse Fisher transform.

In cell **D6**, z_r has been computed using a new function (**FISHER**) that performs the Fisher transformation for us. (Therefore, we don't need to use the equations involving logarithms from the chapter.) The standard error of z_r is calculated in cell **D7**. To compute a confidence interval around z_r, we must know α, which is given in cell **D9**. Using α and **NORM.S.INV**, we computed $z_{\alpha/2}$ in cell **D10**. The confidence limits around z_r are computed in cells **D12** and **D13**. These confidence limits are converted back into r units using the inverse Fisher transform (**FISHERINV**) in cells **D15** and **D16**.

FIGURE 15.A1.1 ■ Approximate Confidence Intervals

	A	B	C	D	E
1	**x**	**y**	**Quantities**	**Values**	**Formulas**
2	89	99	n	18	=COUNT(A2:A19)
3	99	103	r	0.6309	=CORREL(A2:A19,B2:B19)
4	62	74	r^2	0.3980	=D3^2
5	86	99			
6	98	93	z_r	0.7428	=FISHER(D3)
7	101	118	σ_{zr}	0.2582	=1/SQRT(D2-3)
8	97	125			
9	84	85	α	0.05	
10	126	111	$z_{\alpha/2}$	1.96	=NORM.S.INV(1-D9/2)
11	96	99			
12	115	103	zr_{lower}	0.2368	=D6-D10*D7
13	105	116	zr_{upper}	1.2489	=D6+D10*D7
14	110	93			
15	129	114	r_{lower}	0.2325	=FISHERINV(D12)
16	113	123	r_{upper}	0.8480	=FISHERINV(D13)
17	86	86			
18	109	96	s_r	0.1940	=SQRT((1-D4)/(D2-2))
19	109	99	t_{obs}	3.2523	=D3/D18

Computing an approximate confidence interval around r, and testing r for statistical significance.

Typically, significance tests for r test the null hypothesis that $\rho = 0$. When $\rho = 0$, the standard error is estimated by $s_r = \sqrt{(1 - r^2)/(n-2)}$, which is calculated in cell **D18**. t_{obs} is computed by dividing r by s_r, as shown in cell **D19**.

APPENDIX 15.2: CORRELATION ANALYSIS IN SPSS

Figure 15.A2.1 shows the 18 xy pairs from Appendix 15.1. To conduct a correlation analysis, we choose the Analyze→Correlation→Bivariate . . . menu. The Bivariate Correlations dialog is overlaid on the Data Editor file in Figure 15.A2.1. The x and y variables have been moved into the Variables: region from the left panel. Although not shown in Figure 15.A2.1, I asked for SPSS to compute the mean and standard deviation for each variable through the Options . . . menu. With the analysis set up, we click OK to proceed.

The results of the analysis are shown in Figure 15.A2.2. The top portion shows the mean and standard deviation for each variable. The bottom portion shows a correlation table, or correlation *matrix*. In this case, we have computed the correlation between just two variables, but in general there could be many more. All variables are listed along the top row and left column. For each combination of row and column variables, three quantities are shown: (i) the correlation coefficient, (ii) the p-value associated with the statistic under the null hypothesis that $\rho = 0$, and (iii) the number of pairs from which the correlation was computed. Asterisks beside the correlation coefficients indicate statistical significance: one asterisk indicates $p < .05$, and two asterisks indicate $p < .01$.

Entries in the correlation matrix along the diagonal show the correlation of a variable with itself so r is always 1. In addition, the matrix is symmetrical so that the correlation between x and y is the same as the correlation between y and x. Because the matrix is symmetrical and redundant, and because diagonal elements are uninformative, we only look at the upper right part of a correlation table.

Confidence intervals are missing from Figure 15.A2.2. As we saw in the chapter, there is no simple way to compute an exact confidence interval around r. Therefore, we used the Fisher transform to compute approximate $(1-\alpha)100\%$ confidence intervals in Excel. Unfortunately, SPSS does not compute approximate confidence intervals for us.

FIGURE 15.A2.1 ■ Correlations in SPSS

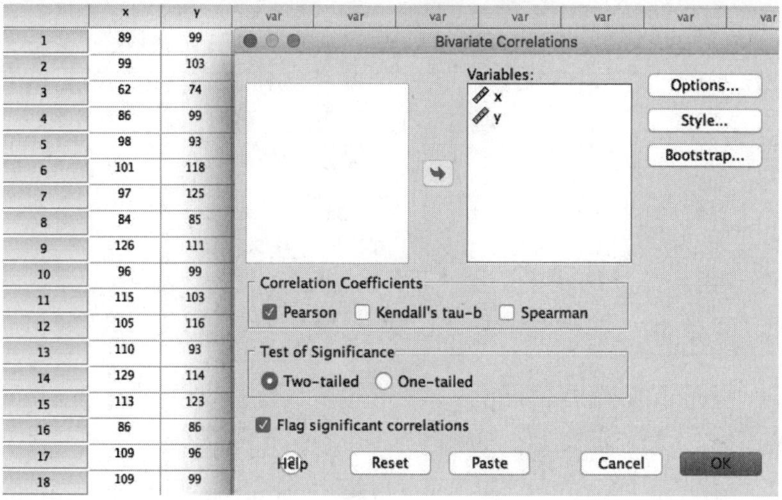

The **Bivariate Correlations** dialog in SPSS.

FIGURE 15.A2.2 ■ Correlations Table

(a)

Correlations

Descriptive Statistics

	Mean	Std. Deviation	N
x	100.78	16.145	18
y	102.00	13.746	18

(b)

Correlations

		x	y
x	Pearson Correlation	1	.631**
	Sig. (2-tailed)		.005
	N	18	18
y	Pearson Correlation	.631**	1
	Sig. (2-tailed)	.005	
	N	18	18

**. Correlation is significant at the 0.01 level (2-tailed).

(a) Descriptive statistics for variables x and y. (b) A correlation table showing the correlation between variables x and y. Correlations on the diagonal are always 1 and should be ignored.

Bootstrapping

We have seen several examples of standardized effect sizes for which exact confidence intervals are difficult to compute by hand. For example, when we estimated Cohen's δ, we had to use approximate intervals when computing by hand, or MBESS functions in **R** to compute exact intervals. There are actually many statistics whose sampling distributions are not normal and exceedingly complex. In these cases, calculating confidence intervals can be very complicated.

When the sampling distributions of statistics are complex or unknown, we can use *bootstrapping* methods to compute confidence intervals. Bootstrapping is a computationally intensive method that relies on the speed and power of modern computers. Bootstrapping methods treat a sample as a population and then repeatedly sample from this "population" with replacement. Each time a sample is drawn, the statistic of interest is computed and recorded. This process is repeated thousands of times, and the thousands of statistics are then sorted from smallest to largest. The two values of the statistic enclosing the central $(1-\alpha)100\%$ of the bootstrapped distribution define the $(1-\alpha)100\%$ bootstrapped confidence limits.

Table 15.A2.1 and Figure 15.A2.3 illustrate bootstrapping in the case of correlation coefficients, making use of the same data presented in Appendix 15.1. In this example, there are $n = 18$ xy pairs. The three leftmost columns of Table 15.A2.1 show the data from our example. The column labeled "s" contains a number for each of the 18 xy pairs. The first, second, and 10,000th samples are also shown in Table 15.A2.1, and underneath each sample is the correlation coefficient for the sample. Because we have sampled with replacement, an individual may appear more than once in a sample or not at all. For example, in Sample 1 in Table 15.A2.1, the individual denoted by 15 appears four times, and the individual denoted by 9 does not appear at all. Because of random sampling with replacement, each of our 10,000 samples is almost certain to be unique.

TABLE 15.A2.1 ■ Examples of Bootstrapped Samples

"Population"			Sample 1			Sample 2			Sample 10,000		
s	y	x	s	y	x	s	y	x	s	y	x
1	89	99	10	96	99	8	84	85	7	97	125
2	99	103	15	113	123	1	89	99	8	84	85
3	62	74	15	113	123	16	86	86	10	96	99
4	86	99	2	99	103	13	110	93	1	89	99
5	98	93	4	86	99	5	98	93	17	109	96
6	101	118	17	109	96	4	86	99	14	129	114
7	97	125	16	86	86	7	97	125	18	109	99
8	84	85	1	89	99	11	115	103	1	89	99
9	126	111	7	97	125	15	113	123	15	113	123
10	96	99	15	113	123	11	115	103	16	86	86
11	115	103	8	84	85	18	109	99	14	129	114
12	105	116	3	62	74	14	129	114	7	97	125
13	110	93	6	101	118	18	109	99	10	96	99
14	129	114	13	110	93	1	89	99	7	97	125
15	113	123	15	113	123	2	99	103	13	110	93
16	86	86	5	98	93	18	109	99	3	62	74
17	109	96	14	129	114	6	101	118	5	98	93
18	109	99	13	110	93	11	115	103	15	113	123
$r = .6309$			$r = .6750$			$r = .4284$			$r = .5780$		

Figure 15.A2.3a shows a histogram of the 10,000 values of r. This distribution is left skewed as we would expect (Figure 15.2).

Figure 15.A2.3b shows the cumulative distribution of the 10,000 values of r. That is, the values of r have been sorted from smallest to largest, and each of these values is a point on the x-axis. The y-axis shows the proportion of the 10,000 correlation coefficients below each value on the x-axis. Like most cumulative distribution functions, this one is S-shaped. The gray lines in Figure 15.A2.3b show the values of r that enclose the central 95% of the distribution. The left line has 2.5% of the distribution below it, and the right line has 2.5% of the distribution above it. (The same lines are also shown in Figure 15.A2.3a for reference.) These values are .1942 and 0.8545. Therefore, we say the bootstrapped 95% confidence interval for r is [.19, .85]. This is slightly different from the approximate 95% confidence interval computed with the Fisher transform [.23, 85]. Nevertheless, we have 95% confidence that ρ lies in this interval.

If you look back to Figure 15.A2.1, you will see that one of the options in the Bivariate Correlations dialog is Bootstrap. . . . Clicking this option allows us to compute a bootstrapped confidence interval for the correlation coefficient. Figure 15.A2.4 shows the dialog that appears when the Bootstrap . . . option is clicked. The Perform bootstrapping option is not checked when the dialog initially appears, so we checked it and specified that the bootstrapped confidence intervals should be based on 5000 samples. We chose 95% confidence intervals under Confidence Intervals, but different confidence levels can be specified in the text box. The confidence intervals will be computed using the Percentile method, exactly as described in Figure 15.A2.3. As well, the confidence intervals will be computed using the Simple Sampling method, but more complex designs can make

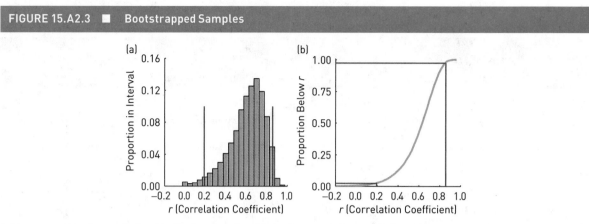

FIGURE 15.A2.3 ■ Bootstrapped Samples

A histogram (a) and cumulative distribution function (b) for the correlation coefficient (r) from 10,000 bootstrapped samples of size $n = 18$. The two gray lines in (b) enclose the central 95% of the distribution.

FIGURE 15.A2.4 ■ Bootstrapping in SPSS

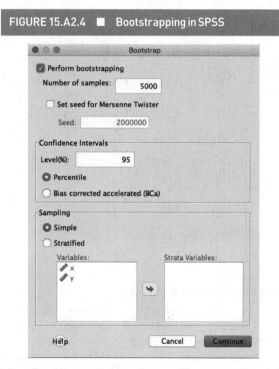

The **Bootstrap** dialog chosen from the **Bivariate Correlations** dialog.

FIGURE 15.A2.5 ■ Bootstrapped Statistics

Correlations

			x	y
x	Pearson Correlation		1	.631**
	Sig. (2–tailed)			.005
	N		18	18
	Bootstrap^c	Bias	0	-.022
		Std. Error	0	.169
		95% Confidence Interval Lower	1	.210
		Upper	1	.854
y	Pearson Correlation		.631**	1
	Sig. (2–tailed)		.005	
	N		18	18
	Bootstrap^c	Bias	-.240	0
		Std. Error	.168	0
		95% Confidence Interval Lower	.210	1
		Upper	.854	1

**. Correlation is significant at the 0.01 level (2–tailed).
c. Unless otherwise noted, bootstrap results are based on 1000 bootstrap samples

Bootstrapped confidence intervals for the correlation coefficient.

use of **Stratified** sampling. Once we've made all choices about how the bootstrapping will be done, we click **Continue** to return to the Bivariate Correlations dialog and there we click **OK** to proceed with the analysis. The output of the bootstrapping procedure is shown in Figure 15.A2.5, and the 95% bootstrapped confidence interval for r is [.210, .854], which is very close to what was shown in Figure 15.A1.1.

Bootstrapping is a useful method that can be used to compute confidence intervals for statistics that have complex or unknown sampling distributions, such as d and r, and more advanced statistics to be discussed in Part IV. Unfortunately, the student version of SPSS does not provide the option to compute bootstrapped confidence intervals for all such statistics. However, more advanced versions of SPSS allow us to write our own programs (scripts) to do these calculations. These methods are beyond the scope of this chapter. However, in later chapters, we will see that packages are available to compute bootstrapped confidence intervals for other important statistics.

PART IV

THE GENERAL LINEAR MODEL

16 INTRODUCTION TO MULTIPLE REGRESSION

INTRODUCTION

Part III introduced important statistical analyses involving two variables. In Chapters 11 and 12, we considered two groups of scores measured on interval scales. The two groups were identified with the two levels of a dichotomous variable that we called the independent variable. In Chapters 13 to 15, we considered the association between two scale variables. One variable was referred to as the predictor variable and the other as the outcome variable. In one case, the experimenter fixed the levels of the predictor; in the other case, the predictor was a random variable.

In all three cases (dichotomous, fixed, and random variable predictors), our best prediction of an individual's score on the outcome variable can be described by the following simple linear equation:

$$\hat{y} = a + b(x). \tag{16.1}$$

Statisticians refer to equation 16.1 as a *model* of the outcome variable. We say that such a model *explains* some of the variability in the outcome variable.

Of course, a model does not explain all variability in the outcome variable because an individual's score (y) rarely equals $\hat{y}$. There is always some difference between y and $\hat{y}$, and we can denote this as $\varepsilon = y - \hat{y}$. The Greek letter ε (epsilon) stands for *error*. Therefore, we can use the following equation to completely define each individual's score on the outcome variable:

$$y = a + b(x) + \varepsilon.$$

As we saw in Chapters 11 to 15, the variability in y can be decomposed into a part associated with the predictor variable [$a + b(x)$] and a part that is unique to the individual and not associated with the predictor variable (ε).

Because equation 16.1 defines a line, it is often called a *linear model*. As noted above, this linear model is very general and applies to cases in which the predictor variable (i) is a random variable measured on an interval scale, (ii) has fixed levels on an interval scale, or (iii) has numerical values chosen by the experimenter to code the levels of a dichotomous categorical variable. Therefore, we can call equation 16.1 an instance of the *general linear model*.

In the final part of this book, we will apply the general linear model to cases in which there are two or more predictor variables or three or more levels of an independent variable. In this chapter, we will extend simple linear regression, where x and y are random variables, to cases in which there are $k \geq 2$ predictors ($x_1, x_2, \ldots, x_k$) that are random variables measured on interval scales. We describe this kind of model as *multiple linear regression*, or *multiple regression* for short.

Our treatment of multiple regression will be spread over two chapters. This chapter introduces basic concepts and then shows how multiple regression is conducted in SPSS. The material is a simple extension of material covered in Appendix 14.2. Chapter 17 provides more details about multiple regression, including examples of how it is typically used in psychological research.

In Chapters 18 to 21, we will extend the general linear model to cases of independent variables having more than two levels on a categorical variable or more than two fixed levels of a scale variable. Although these analyses are a special case of multiple regression, they usually go by the name analysis of variance (ANOVA).

There are major divisions in psychology defined by the kinds of questions researchers ask, and hence the kinds of statistical analyses they use. ANOVA is most commonly used in experimental research, because experimental subjects can be randomly assigned to different treatment conditions. For example, ANOVA is commonly used in perception, cognition, social psychology, and neuroscience. When random assignment to treatment conditions is impossible, researchers may rely on multiple regression. For example, multiple regression is widely used in clinical psychology, health psychology, and personality psychology.

As we saw in Chapter 11, causal inference is much easier when we can randomly assign subjects to experimental conditions. When random assignment to conditions is impossible (for example, when comparing smokers and nonsmokers), causal inference is more difficult. This means that causal inference is more difficult in research areas that employ multiple regression than in areas that employ ANOVA with random assignment.

In Chapters 16 through 21, there will be heavy emphasis on the similarities between multiple regression and ANOVA. By now, these connections should not be difficult to make. In Chapter 14, we saw that simple linear regression can be applied to both predictors having fixed levels of a scale variable and predictors that are random variables. In Chapter 15, we saw how the two-groups design could be cast in terms of regression and correlation. These are the themes that will continue in this part of the book.

AN EXAMPLE

If you are reading this chapter, there is a good chance that you're in an advanced statistics course. If so, then you have devoted a lot of time to studying psychology and you may be thinking of going on to postgraduate studies (i.e., graduate school). When students apply to graduate programs, they are asked to include in their applications a number of documents that will help a selection committee evaluate their suitability for graduate study. Some of these documents are (i) scores on the Graduate Record Examination (GRE), (ii) an undergraduate transcript, (iii) a letter of intent, and (iv) letters of recommendation from professors the student has worked with. In this example, we'll think about how such measures relate to students' success in graduate school.

We will consider a hypothetical class of 27 graduate students in a psychology department at the University of Didd. As part of the application process, these 27 students had submitted the results of their GRE exams, which many professors consider a good predictor of success in graduate school. The GRE is a grueling, daylong exam that might be said to measure peak (or maximal) performance under pressure. GRE scores range from 130 to 170. Although GRE scores may be good predictors of academic success, as with any statistical association, they are not perfect predictors.

Some professors wonder whether personality characteristics that are not reflected in grades or GRE scores might explain some of the variability in student success in graduate school. One such personality characteristic might be an individual's curiosity, determination, and routine engagement with challenging intellectual problems. The Typical Intellectual Engagement Scale (TIE) was devised to assess these qualities (Goff & Ackerman, 1992). Chamorro-Premuzic, Furnham, and Ackerman (2006) explain that two individuals with the same GRE score (i.e., same maximal performance) may differ in the time they spend reading, thinking, solving puzzles, or engaging in self-motivated creative activities. The TIE is a self-report questionnaire on which participants use a 6-point scale to rate how well each of 59 statements like the following apply to them:

1. Almost every section of the newspaper has something in it that interests me.

2. My favorite part of school was working on research and independent projects.

The selection committee at the University of Didd thought they might better predict success in graduate school by somehow combining GRE scores with TIE scores. So one year they had each of the 27 students admitted to their program complete the TIE questionnaire during their first semester. (TIE scores typically range from 50 to 200.) The question is "Do TIE scores improve our prediction of students' success in graduate school?" Success in this case is a composite measure made up of course grades, class participation, professors' ratings, awards, presentations at scientific meetings, and publications; these measures were obtained when students finished the program.. Let's say that these composite scores range from 0 to 50. We will call this variable *success*.

The selection committee found that both *GRE* and *TIE* are positively correlated with *success* in graduate school. Figure 16.1a shows the prediction of *success* (*y*) from *GRE* scores (x_1) and Figure 16.1b shows the prediction of *success* from *TIE* scores (x_2). Each panel shows the simple regression equation that relates *x* to *y*, as well as r^2, which is the proportion of variability in *y* explained by its association with *x*. The regression lines relating x_1 and x_2 to *y* are plotted through the data points. The same individuals are shown in both panels of Figure 16.1, and each is indicated by a specific combination of symbol and color. For example, the light blue star represents the same graduate student in both panels.

It should be clear from Figure 16.1 that students with high GRE scores also tend to have high TIE high scores; conversely, students who have low GRE scores also tend to have low TIE scores. This means that there is some degree of correlation between *GRE* and *TIE*, so they might be two measures of the same thing. Therefore, we might wonder whether *TIE* scores add anything to what we already know about *success* from *GRE*. We can use multiple regression to answer this question.

Multiple linear regression extends simple linear regression from one to two or more predictor variables. Figure 16.2 shows an example of a multiple regression analysis that shares many features with the simple linear regressions shown in Figure 16.1. The same symbols are used to represent the 27 individual students. These points are plotted in a three-dimensional space to show simultaneously the association between the two predictors, *GRE* (x_1) and *TIE* (x_2), and the outcome variable *success* (*y*). GRE and TIE scores define a horizontal *plane*, and the *y* scores are shown as heights above this $x_1 - x_2$ plane.

When there are two predictor variables, the regression equation has the form

$$\hat{y} = a + b_1(x_1) + b_2(x_2).$$

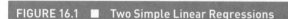

FIGURE 16.1 ■ Two Simple Linear Regressions

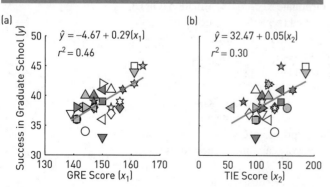

(a) $\hat{y} = -4.67 + 0.29(x_1)$
$r^2 = 0.46$

Success in Graduate School (*y*)

GRE Score (x_1)

(b) $\hat{y} = 32.47 + 0.05(x_2)$
$r^2 = 0.30$

TIE Score (x_2)

Two simple linear regressions to predict success in graduate school from scores on (a) the Graduate Record Examination (GRE) and (b) the Typical Intellectual Engagement Scale (TIE).

For our example, the regression equation is

$$\hat{y} = 0.61 + 0.23(x_1) + 0.03(x_2).$$

The regression equation defines a plane, which is shown as the semitransparent surface fit through the data in Figure 16.2. This semitransparent surface defines all possible values of $\hat{y}$. The $\hat{y}$ values on this plane are coded such that the lighter the points on the regression plane, the higher they are above the $x_1 - x_2$ plane. That is, the dark color represents low values of $\hat{y}$ and the light color represents higher values of $\hat{y}$. The vertical lines connect the symbols to the regression plane, just as they connected symbols to the regression lines in Figure 13.8.

The difference between each y and $\hat{y}$ is an error of prediction, which we now call ε. Therefore, we can describe each of these 27 scores as

FIGURE 16.2 ■ Multiple Linear Regression

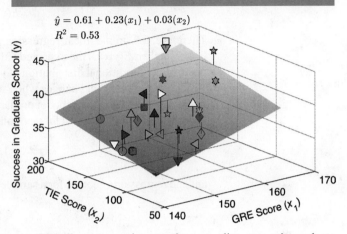

$$\hat{y} = 0.61 + 0.23(x_1) + 0.03(x_2)$$
$$R^2 = 0.53$$

A multiple linear regression equation to predict success in graduate school from scores on two tests.

$$y_i = a + b_1(x_{1,i}) + b_2(x_{2,i}) + \varepsilon_i = 0.61 + 0.23(x_{1,i}) + 0.03(x_{2,i}) + \varepsilon_i.$$

Notice that a subscript has been placed on y_i, $x_{1,i}$, $x_{2,i}$, and ε_i to make clear that we are talking about the ith individual. This was done to remind us that there is a unique value of y, x_1, x_2, and ε for each individual. As a rule, however, we will limit the use of subscripts to keep equations as clutter-free as possible. Subscripts will be used only when omitting them will cause serious ambiguity.

In Figure 16.2, R^2 represents the proportion of variability in y explained by the two predictors. R^2 plays the same role in multiple regression that r^2 plays in simple linear regression. Notice that $R^2 = .53$ in Figure 16.2, which is greater than $r^2 = .46$ in Figure 16.1a and $r^2 = .30$ in Figure 16.1b. Therefore, combining *GRE* and *TIE* provides a better prediction of *success* than either of these two predictors individually.

Having laid out the essential features of multiple regression, we will next step through the inferences we can draw from these results. There are a lot of questions to answer: What is the connection between our sample statistics and the corresponding population parameters? What is the nature of the population from which our sample was drawn, and what assumptions must be made so that valid inferences from sample to population can be made? There is a lot of material to cover before we can return to the question of interest to the selection committee ("Would adding *TIE* scores to the application process improve the selection of graduate students?"). However, we will connect each topic to this question as we move through the chapter.

PARAMETERS AND STATISTICS IN MULTIPLE REGRESSION

In the previous section, we introduced multiple regression through an example involving a sample of scores. Therefore, all of the quantities we computed were estimates of population parameters. To properly use multiple regression, we must understand the parameters estimated by our statistics.

The Regression Equation

In Chapter 13, we saw that the regression equation relating x and y for a sample is

$$\hat{y} = a + b(x).$$

In Chapter 14, we saw that the regression equation relating x and y for a population is

$$E(y|x) = \alpha + \beta(x).$$

Therefore, for simple linear regression, $\hat{y}$ estimates $E(y|x)$, a estimates α, and b estimates β. Remember, $E(y|x)$ is the expected value of y for a given value of x, which means it is the mean of the conditional distribution of y scores for a given value of x.

The only thing that changes when we move from simple linear regression to multiple linear regression is the number of predictor variables. If there are two predictors (x_1 and x_2), then the regression equation for a population is

$$E(y|x_1, x_2) = \alpha + \beta_1(x_1) + \beta_2(x_2),$$

and the regression equation for a sample (as we saw above) is

$$\hat{y} = a + b_1(x_1) + b_2(x_2).$$

Therefore, $\hat{y}$ estimates $E(y|x_1, x_2)$, a estimates α, b_1 estimates β_1, and b_2 estimates β_2.

Multiple regression is not restricted to two predictors. When there are five predictors, the population regression equation is

$$E(y|x_1, x_2, x_3, x_4, x_5) = \alpha + \beta_1(x_1) + \beta_2(x_2) + \beta_3(x_3) + \beta_4(x_4) + \beta_5(x_5).$$

In the general case, we denote the number of predictors in a regression equation with the letter k, and we can express the population regression model as

$$E(y|x_1, x_2, \ldots, x_k) = \alpha + \beta_1(x_1) + \beta_2(x_2) + \ldots + \beta_k(x_k). \tag{16.2}$$

The sample regression equation can be defined in the same way:

$$\hat{y} = a + b_1(x_1) + b_2(x_2) + \ldots + b_k(x_k). \tag{16.3}$$

The coefficients in the sample regression equation (a, b_1, b_2, ..., b_k) estimate the corresponding coefficients in the population regression equation (α, β_1, β_2, ..., β_k).

Computing the Multiple Regression Equation

In Chapter 13 we saw that it is easy to compute the simple linear regression equation from m_x, s_x, m_y, s_y, n, and r. Things are not as simple in multiple linear regression. The good news is that from now on we will use computer software to conduct regression analysis. This can be done in Excel (see Appendix 16.2, available at study.sagepub.com/gurnsey) or SPSS, as we'll see below.

Visualizing the Regression Equation

A **hyperplane** is a plane in three or more dimensions. It is the regression surface when there are three or more predictor variables.

It is easy to visualize the association between x and y in simple linear regression because the association is described by a line, as shown in Figure 16.1. It is also quite easy to visualize the association between x_1, x_2, and y, which is described by a plane, as shown in Figure 16.2. We say that a plane is a generalization of a line to two dimensions. Unfortunately, it is impossible to visualize the nature of the association between y and three or more predictor variables. However, when there are three or more predictors, the regression equation describes what mathematicians call a **hyperplane** (i.e., a plane in three or more dimensions). Although we can't visualize hyperplanes, the mechanics of multiple regression are exactly the same no matter how many predictors there are. In a later section we'll see that R^2, the proportion of variance in y explained by the regression equation, is computed in the same way for one, two, or k predictors.

Assumptions

Valid statistical inferences in regression analysis require certain assumptions to be true. University of Didd selection committee members have a sample of scores and would like to infer something about the population from which the sample was drawn. For their inferences to be valid, the following assumptions (which are modest variations on those we saw in Chapter 14) must be valid.

Linearity

In the case of simple linear regression, we assumed that the regression line correctly described the association between the predictor variable x and the outcome variable y. In multiple regression, we assume that the hyperplane $\alpha + \beta_1(x_1) + \beta_2(x_2) + \ldots + \beta_k(x_k)$ correctly describes the association between the expected value of y and the predictor variables.

Prediction Errors Are Normally Distributed and Homoscedastic

The score of each individual in a population can be expressed as

$$E(y|x_1, x_2, \ldots, x_k) + \varepsilon_i.$$

Similarly, the score of each individual in a sample can be expressed as

$$\hat{y} + \varepsilon_i.$$

Regression analysis assumes that the errors of prediction (ε_i) are normally distributed for all possible combinations of $x_1, x_2, \ldots, x_k$ in the population, and also in samples drawn from the population. Furthermore, the distributions of prediction errors are assumed to be homoscedastic (have the same variance) for all possible combinations of $x_1, x_2, \ldots, x_k$.

No Multicollinearity

The predictor variables in a multiple regression analysis are typically correlated with each other. The no-multicollinearity assumption is that the predictors are *not too correlated* with each other. When predictors are too highly correlated, we say that they are *collinear*. When one predictor can be almost perfectly predicted from the remaining predictors, we say that the predictors are *multicollinear*. When predictors are multicollinear, very small changes in the predictors can yield radically different coefficients in the regression equations. There is a much deeper discussion of collinearity in Appendix 17.2 (available online at study.sagepub.com/gurnsey)

Independence

The independence assumption is that there is no association between the prediction errors (ε). This problem is most easily understood when a predictor variable involves time. It is possible for measures taken at two consecutive time points to be correlated because they're subject to similar influences (e.g., when a negative ε at time t predicts a negative ε at time $t + 1$). In this case, we say that the independence assumption has been violated.

LEARNING CHECK 1

1. For the regression equation $E(y|x_1, x_2, x_3) = 32 + .64(x_1) + 16(x_2) + 8(x_3)$, compute $E(y|x_1, x_2, x_3)$ for the following values of x_1, x_2, and x_3, respectively:

   (a) 10, 20, 40;

   (b) −1, 18, 30; and

   (c) 20, −10, 42.

2. For the regression equation $\hat{y} = 10 + 5(x_1) + 25(x_2) - 2(x_3)$, compute the best estimate of $E(y|x_1, x_2, x_3)$ for the following values of x_1, x_2, and x_3, respectively:

   (a) 1, 1, 1;

   (b) 16, −8, −4; and

   (c) 5, 25, −2.

Answers

1. (a) $32 + .64(10) + 16(20) + 8(40) = 32 + 6.4 + 160 + 320 = 678.4$.

   (b) $32 + .64(-1) + 16(18) + 8(30) = 32 - .64 + 288 + 240 = 559.36$.

   (c) $32 + .64(20) + 16(-10) + 8(42) = 32 + 12.8 - 160 + 336 = 220.8$.

2. (a) $10 + 5(1) + 25(1) - 2(1) = 10 + 5 + 25 - 2 = 38$.

   (b) $10 + 5(16) + 25(-8) - 2(-4) = 10 + 10 - 200 - 8 = -102$.

   (c) $10 + 5(5) + 25(25) - 2(-2) = 10 + 25 + 625 + 4 = 664$.

Explained Variance

Any time a regression analysis is conducted, the researchers (like our selection committee) would like to know how well the regression equation (model) fits the data. In this section we will explain the quantity R^2, which is analogous to r^2 from simple linear regression.

Just as in the case of simple linear regression, a multiple linear regression equation with k predictors represents the best fit to the data points in the least squares sense. This means that no other regression equation would produce a smaller sum of squared deviations from the regression plane. Therefore, it is possible to compute the proportion of variability in y explained by the predictor variables, exactly as in simple linear regression.

We will continue with the example data shown in Figure 16.2. These data are shown in Table 16.1. Columns 1 to 3 show the scores for *success* (y), *GRE* (x_1), and *TIE* (x_2) that were plotted in Figure 16.2. The fourth column shows the values of $\hat{y}$. That is, $\hat{y} = 0.61 + 0.23(x_1) + 0.03(x_2)$ was computed for each individual. The fifth column shows the difference between y and $\hat{y}$ for each of the 27 individuals; i.e., $\varepsilon_i = y_i - \hat{y}_i$.

The sums of squares for y, $\hat{y}$, and $y - \hat{y}$ have the same meaning as in simple linear regression (see Chapter 13). That is,

$$ss_{total} = \sum (y - m_y)^2,$$

$$ss_{regression} = \sum (\hat{y} - m_y)^2, \text{ and}$$

$$ss_{error} = \sum (y - \hat{y})^2 = \sum \varepsilon^2.$$

At the bottom of columns 1, 4, and 5 of Table 16.1, we can see that

$$ss_{total} = ss_{y - m_y} = 232,$$

$$ss_{regression} = ss_{\hat{y} - m_y} = 122.55, \text{ and}$$

$$ss_{error} = ss_{y - \hat{y}} = ss_{\varepsilon} = 109.45.$$

TABLE 16.1 ■ Hypothetical Data[a,b]

	y	x_1	x_2	$\hat{y}$	$\varepsilon = y - \hat{y}$
●	36	141	99	36.30	−0.30
○	38	143	153	38.33	−0.33
○	34	144	131	37.92	−3.92
■	39	150	142	39.64	−0.64
▢	36	141	98	36.27	−0.27
□	45	161	185	43.45	1.55
◆	38	156	109	40.08	−2.08
◇	38	147	131	38.62	−0.62
◇	37	153	87	38.74	−1.74
▼	33	150	98	38.36	−5.36
▽	44	161	185	43.45	0.55
▽	37	139	109	36.12	0.88
△	40	147	109	37.98	2.02
▲	40	144	120	37.60	2.40
△	41	153	98	39.06	1.94
◀	41	150	142	39.64	1.36
◁	38	147	55	36.41	1.59
◁	36	150	120	39.00	−3.00
▶	38	141	109	36.59	1.41
▷	38	144	98	36.96	1.04
▷	42	150	120	39.00	3.00
★	39	147	77	37.05	1.95
☆	45	164	142	42.90	2.10
☆	38	153	131	40.02	−2.02
✶	41	157	163	41.88	−0.88
☆	42	161	120	41.56	0.44
☆	39	156	109	40.08	−1.08
m	39	150	120	39.00	0.00
ss	232	1264	23348	122.55	109.45

[a] $\hat{y} = 0.61 + 0.23(x_1) + 0.03(x_2)$.

[b] $\hat{R}^2 = 0.53$.

As we saw in Chapter 13,

$$ss_{total} = ss_{regression} + ss_{error},$$

which in this case means

$$232 = 122.55 + 109.45.$$

These numbers show, as before, that the total variability in y (ss_{total}) can be decomposed into two, nonoverlapping sources, $ss_{regression}$ and ss_{error}. Therefore, the proportion of variability in y explained by the two predictor variables (x_1 and x_1) is

$$R^2 = \frac{ss_{regression}}{ss_{total}} = \frac{ss_{total} - ss_{error}}{ss_{total}}.$$

When there are two or more predictors in a regression analysis, we denote the proportion of explained variance with R^2 rather than r^2. In this example, we find that

$$R^2 = \frac{ss_{regression}}{ss_{total}} = \frac{122.55}{232} = .53.$$

That is, 53% of the variability in the *success* scores can be explained by *GRE* and *TIE*.

R^2 Is Always Greater Than or Equal to the Largest Individual r^2

The R^2 associated with the two-predictor model is .53, and the two simple linear regressions shown in Figure 16.1 have r^2 values of .46 and .30. Therefore, in this case, R^2 is greater than both individual r^2 values. We will see in later sections that this is always the case. In general, the more predictors in our regression equation, the closer R^2 gets to 1.

Of course, R^2 is a statistic, as are the coefficients of the regression equation. Therefore, we can use R^2 to estimate the corresponding population quantity, which will be described next.

Estimating P²

P²is the proportion of variability in y explained by $x_1, x_2, \ldots, x_k$ in a multivariate population.

R^2 is a statistic computed from a sample drawn from a multivariate population of interest. R^2 estimates P^2, which is the proportion of variability in y explained by $x_1, x_2, \ldots, x_k$ in the multivariate population. P is capital ρ (rho) from the Greek alphabet.

We start by noting that R^2 is a positively biased estimate of P^2. This means that if we compute the regression equation, $a + b_1(x_1) + b_2(x_2) + \ldots + b_k(x_k)$, for all possible samples of size n drawn from our multivariate population of interest, the mean of the resulting R^2 values will be greater than P^2.

Bias Correction

A very common correction for this positive bias in R^2 is computed as follows:

$$R^2_{adj} = 1 - (1 - R^2)\frac{n-1}{n-k-1}. \tag{16.4}$$

Adjusted R^2 (R^2_{adj}) is a statistic that reduces the positive bias in R^2.

We call the resulting statistic **adjusted R^2** or R^2_{adj}. Notice in equation 16.4 that the ratio $(n-1)/(n-k-1)$ will always be greater than 1, because the numerator $(n-1)$ is always greater than the denominator $(n-k-1)$. Therefore, in equation 16.4, the unexplained variance $(1-R^2)$ is multiplied by $(n-1)/(n-k-1)$ to magnify it. We then subtract this magnified quantity from 1 to produce R^2_{adj}, which is less than R^2.

LEARNING CHECK 2

1. The following table shows 10 scores for y, x_1, x_2, and x_3. The regression equation relating y to the three predictor variables is $\hat{y} = 79.3341 - 0.1013(x_1) + 0.1274(x_2) + 0.5408(x_3)$. Use these numbers to compute R^2. You should do these calculations in Excel to compute $\hat{y}$ for each individual. These data are available at study.sagepub.com/gurnsey.

y	x_1	x_2	x_3	$\hat{y}$
54	108	−2	47	
223	187	150	289	
187	139	131	71	
210	296	302	256	
176	132	286	198	
180	61	111	136	
133	50	140	97	
260	70	133	199	
93	−60	31	67	
131	123	114	97	

Answers

1. $\hat{y}$ is computed for each individual in the last column of the following table. At the bottom of columns y and $\hat{y}$ are ss_y and $ss_{\hat{y}}$ (aka, ss_{total} and $ss_{regression}$). $R^2 = ss_{regression} / ss_{total} = 21{,}810.86 / 34{,}928.10 = 0.62$.

y	x_1	x_2	x_3	$\hat{y}$
54	108	−2	47	93.56
223	187	150	289	235.79
187	139	131	71	120.34
210	296	302	256	226.27
176	132	286	198	209.48
180	61	111	136	160.85
133	50	140	97	144.56
260	70	133	199	196.81
93	−60	31	67	125.60
131	123	114	97	133.86
34,928.10				21,810.86

For the example given in Figure 16.2 and Table 16.1, we found that $R^2 = .53$ for our two-predictor regression ($k = 2$). This is a biased estimate of P^2, so we correct this bias using equation 16.4 as follows:

$$R^2_{adj} = 1 - (1 - R^2)\frac{n-1}{n-k-1} = 1 - (1 - .53)\frac{27-1}{27-2-1} = .49.$$

Therefore, $R^2_{adj} = .49$ is a better point estimate of P^2 than R^2.

This correction for bias is one of many in the statistical literature, and none is perfect. As we'll see later, there are situations in which R^2_{adj} clearly overcorrects for positive bias.

A Confidence Interval for R²

As with all statistics, R^2 is subject to sampling error and thus has a sampling distribution. In previous chapters, we used what we know about a statistic's sampling distribution to compute confidence intervals. Unfortunately, the sampling distribution of R^2 is not normal, and it cannot be captured in a simple mathematical formula, which makes confidence intervals for R^2 difficult to compute.

The shape of the distribution of R^2 depends on the number of predictors, sample size, and P^2 itself. Figure 16.3 gives an example of the sampling distribution of R^2 when there are two predictors. For this illustration, the parameters of the population are identical to the statistics of the sample we've been discussing. That is, $E(y|x_1, x_2) = 0.61 + 0.23(x_1) + 0.03(x_2)$ and $P^2 = .53$. Of course, our sample was almost certainly drawn from a population with different parameters; these are just convenient numbers to work with. Figure 16.3 is a histogram, but because the interval width is very narrow, I have shown only the tops of the bars in order to give the impression of a continuous curve.

An **empirical sampling distribution** is a histogram of the statistic computed from a large number of samples of the same size (n), drawn at random from the population of interest.

The distribution in Figure 16.3 is an **empirical sampling distribution**, computed from 100,000 samples of size $n = 27$ drawn at random from the multivariate distribution just described. For each sample, the regression equation was estimated [$\hat{y} = a + b_1(x_1) + b_2(x_2)$] and the sample R^2 was computed. The distribution shows the proportion of times each value of R^2 fell within small intervals, whose score limits ranged from 0 to almost 1.

The calculations required to compute a confidence interval for R^2 are complex, and they are not included as a standard part of SPSS or Excel. Sadly, the standard version of SPSS does not allow us to compute bootstrapped confidence intervals for R^2. Fortunately, approximate confidence intervals for R^2 are easily computed with the MBESS function `ci.R2` in **R** as follows:

```
> ci.R2(R2 =.53, N = 27, K = 2, Random.Predictors =
TRUE, conf.level =.95)
$Lower.Conf.Limit.R2
[1] 0.1938483
$Prob.Less.Lower
[1] 0.025
$Upper.Conf.Limit.R2
[1] 0.7337642
$Prob.Greater.Upper
[1] 0.025
```

The MBESS function `ci.R2` takes five arguments: `R2`, `N`, `K`, `conf.level`, and `Random.Predictors`. `R2` is R^2, `N` is sample size, `K` is the number of predictors, `conf.level` is the confidence level (i.e., $1-\alpha$), and `Random.Predictors` is set to TRUE if the predictors are random variables and FALSE if the predictors

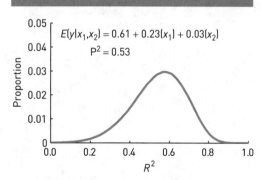

FIGURE 16.3 ■ The Distribution of R²

$E(y|x_1,x_2) = 0.61 + 0.23(x_1) + 0.03(x_2)$
$P^2 = 0.53$

Empirical sampling distribution of R^2. The equation relating $E(y)$ to x_1 and x_2 is $E(y|x_1, x_2) = 0.61 + 0.23(x_1) + 0.03(x_2)$. The population P^2 is .53. There were 1 million y, x_1, x_2, triplets in the population and 100,000 samples of size $n = 27$ were drawn at random. For each sample, the regression equation [$\hat{y} = a + b_1(x_1) + b_2(x_2)$] was computed and from this the sample R^2 was computed.

have fixed values. `ci.R2` returns four values, as shown above. The first number returned is `0.1938483`, which is the lower limit of the confidence interval. The next number is simply $\alpha/2$; because we've asked for the 95% confidence interval, $\alpha/2 = .025$. The third number returned is `0.7337642`, which is the upper limit of the confidence interval. The last number is simply $1-\alpha/2$.

This confidence interval seems very wide; i.e., $.73 - .19 = .54$. This might worry the selection committee at the University of Didd. Our best estimate of P^2 is $R^2_{adj} = 0.49$, but a P^2 of .19 is just as plausible as one of .73. This means the population regression equation might explain a lot or a little of the variability of y.

Despite its apparent importance, it is actually very rare for researchers to report confidence intervals around R^2. This is probably because tools such as the MBESS package are not as widely known as applications such as SPSS, which, as noted, provides no standard functions to compute these confidence intervals nor does it provide bootstrapping options for R^2.

Standard Error of Estimate

A final population quantity in multiple regression is the standard error of estimate (σ_{est}), which we encountered in Chapter 14 when discussing simple linear regression. The concept is essentially the same in multiple regression. In the population, the standard error of estimate is the standard deviation of the prediction errors (ε). It can be defined in terms of ss_{error} or P^2 as shown in equations 16.5a and 16.b, respectively:

$$\sigma_{est} = \sqrt{\frac{SS_{error}}{N}} \text{ or} \qquad (16.5a)$$

$$\sigma_{est} = \sqrt{(1-P^2)\sigma_y^2}. \qquad (16.5b)$$

As with simple linear regression, estimates of σ_{est} can be derived from both of these equations. For example,

$$s_{est} = \sqrt{\frac{SS_{error}}{n-k-1}} \qquad (16.6a)$$

is derived from equation 16.5a, and

$$s_{est} = \sqrt{(1-R^2)s_y^2 \frac{n-1}{n-k-1}} \qquad (16.6b)$$

is derived from equation 16.5b. Table 16.1 shows that $ss_{error} = 109.45$ for our two-predictor model. When we insert this into equation 16.6a, we find that

$$s_{est} = \sqrt{\frac{SS_{error}}{n-k-1}} = \sqrt{\frac{109.45}{27-2-1}} = 2.14.$$

Therefore, our best estimate of σ_{est} is $s_{est} = 2.14$.

Remember that in simple linear regression, s_{est} played an important role in (i) confidence intervals for the slope of the regression equation, (ii) confidence intervals for $\hat{y}$, and (iii) prediction intervals. In Chapter 17, we will see that s_{est} is involved in similar computations in multiple regression.

LEARNING CHECK 3

1. Compute R^2_{adj}, s_{est}, and the approximate 95% confidence interval for R^2 using `ci.R2`, for each of the following conditions for R^2, k, n, and s^2_y, respectively:

   (a) .75, 3, 32, 9;

   (b) .25, 3, 32, 13;

   (c) .64, 10, 100, 8;

   (d) .21, 5, 256, 16; and

   (e) .82, 4, 81, 32.

Answers

1. (a) $R^2_{\text{adj}} = .72$, $s_{\text{est}} = 1.58$, approximate 95% CI = [.50, .86].

   (b) $R^2_{\text{adj}} = .17$, $s_{\text{est}} = 3.29$, approximate 95% CI = [.00, .47].

   (c) $R^2_{\text{adj}} = .60$, $s_{\text{est}} = 1.79$, approximate 95% CI = [.46, .71].

   (d) $R^2_{\text{adj}} = .19$, $s_{\text{est}} = 3.59$, approximate 95% CI = [.11, .29].

   (e) $R^2_{\text{adj}} = .81$, $s_{\text{est}} = 2.46$, approximate 95% CI = [.72, .87].

SIGNIFICANCE TESTS

The *F*-Ratio

It is common for researchers to ask whether R^2 is statistically significant. In other words, does R^2 represent a statistically significant proportion of the total variability in the outcome variable? Although this can be answered with an approximate 95% confidence interval, researchers typically compute an *F*-statistic, similar to the ones described in Chapters 11 and 12. We will next show how an *F*-statistic can be used to test for the statistical significance of R^2. We won't see the connection between R^2 and F for a few paragraphs because we must first introduce some terminology that we will see in many SPSS outputs from now on.

We begin by defining F as follows:

$$F_{\text{obs}} = \frac{ss_{\text{regression}}/df_{\text{regression}}}{ss_{\text{error}}/df_{\text{error}}}, \tag{16.7a}$$

where $ss_{\text{regression}}$ and ss_{error} are the quantities described earlier. These two sums of squares have associated degrees of freedom. The degrees of freedom associated with regression ($df_{\text{regression}}$) is equal to the number of predictors, which we denote with k. The degrees of freedom associated with error (df_{error}) is equal to the total degrees of freedom for the outcome variable (i.e., $n-1$) minus the degrees of freedom associated with regression (i.e., k). Therefore, we can express $df_{\text{regression}}$ as

$$df_{\text{regression}} = k$$

and df_{error} as

$$df_{\text{error}} = n-1-k = df_{\text{total}} - k.$$

The total number of degrees of freedom associated with y is

$$df_{\text{total}} = df_{\text{regression}} + df_{\text{error}} = k + (n-1-k) = n-1.$$

F_{obs} defined in equation 16.7a is sometimes called the *ratio of mean squares*. This terminology comes from viewing the numerator and denominator of equation 16.7a as means. Therefore, we can define F_{obs} in terms of two **mean squares** (*ms*), as follows:

$$F_{obs} = \frac{ms_{regression}}{ms_{error}}, \tag{16.7b}$$

Mean square is the sum of squares divided by its associated degrees of freedom.

where the mean squares for regression is

$$ms_{regression} = ss_{regression}/df_{regression}$$

and the mean squares for error is

$$ms_{error} = ss_{error}/df_{error}.$$

F_{obs} can also be defined directly in terms of R^2. To see this, first note that equation 16.7a can be rearranged as follows:

$$F_{obs} = \frac{ss_{regression}}{ss_{error}} \frac{df_{error}}{df_{regression}}. \tag{16.7c}$$

This formula is identical to

$$F_{obs} = \frac{R^2}{1 - R^2} \frac{df_{error}}{df_{regression}}, \tag{16.8}$$

because $ss_{regression}/s_{error} = R^2/(1 - R^2)$. Equation 16.8 has the advantage of showing explicitly that there are two contributions to F. The first is the ratio of explained to unexplained variance $[R^2/(1 - R^2)]$ and the other is the ratio of df_{error} to $df_{regression}$ ($df_{error}/df_{regression}$).

Let's say that a two-predictor regression with $n = 13$ produces $R^2 = .2$, and a second two-predictor regression with $n = 103$ also produces $R^2 = .2$. In the first case, F_{obs} would be

$$F_{obs} = \frac{R^2}{1 - R^2} \frac{df_{error}}{df_{regression}} = \frac{.2}{.8} \frac{13 - 2 - 1}{2} = 1.25.$$

In the second case, F_{obs} would be

$$F_{obs} = \frac{R^2}{1 - R^2} \frac{df_{error}}{df_{regression}} = \frac{.2}{.8} \frac{103 - 2 - 1}{2} = 12.5.$$

Therefore, if R^2 remains constant, F_{obs} will increase as n increases. Consequently, any nonzero R^2 will produce a statistically significant F_{obs} if n is large enough.

A very similar point was made in Chapter 10 where t_{obs} was decomposed into a part based on effect size (d) and a part based on sample size ($\sqrt{n}$). That is, $t_{obs} = d\sqrt{n}$. It is really helpful to keep in mind that F and t_{obs} are both determined by effect size and sample size.

When an **F-ratio** is computed with $df_{regression} > 1$, we call it an **omnibus F-statistic**. We will see in Chapters 17 to 21 that F-statistics can be computed for individual predictors as well, and these focused tests are typically more informative than omnibus tests of R^2.

The *F*-ratio (or *F*-statistic) is the ratio of explained to unexplained variance multiplied by the ratio of error to regression degrees of freedom. The *F*-ratio increases as explained variance increases and as sample size increases.

The Central *F*-Distribution

Our selection committee would probably like to know if the obtained $R^2 = .53$ is statistically significant. Answering this question requires knowing something about the sampling

The **omnibus F-statistic** is an *F*-statistic computed when $df_{regression} > 1$.

The **multiple correlation coefficient** in a population is the correlation between y and $E(y|x_1, x_2, \ldots, x_k)$; it is denoted **P**. The multiple correlation coefficient in a sample is the correlation between y and $\hat{y}$; it is denoted **R**.

distribution of F when the null hypothesis is true. The null hypothesis in multiple regression is that the regression coefficients on the predictor variables in the population are all 0; i.e., $\beta_1 = \beta_2 = \ldots = \beta_k = 0$. This means that the correlation between y and $E(y|x_1, x_2, \ldots, x_k)$ in the population is $P = 0$. (We call P the **multiple correlation coefficient**.) Therefore, it is convenient to write this null hypothesis as

$$H_0: P = 0.$$

If indeed $P = 0$, then the sampling distribution of R would be a symmetrical distribution centered on 0. However, if all these R values are squared, then the distribution of R^2 would contain only positive values and have some nonzero mean. Therefore, we need some way to judge whether an obtained R^2 is unusually large under H_0. Typically the F-statistic is used to do this because, as shown in equation 16.8, F is directly related to R^2. The distribution of F-statistics under $H_0: P = 0$ is called the *central F-distribution*. Therefore, we can use the central F-distribution to determine the probability of F_{obs} or greater occurring by chance when $P = 0$.

To understand the sampling distribution of F, we will consider the concrete example of our two-predictor model. Each individual in the population of interest is associated with scores on y, x_1, and x_2. According to the null hypothesis, *GRE* and *TIE* do not explain any of the variability in *success*. Therefore, P is 0 in the population.

To conceptualize the sampling distribution of F under the null hypothesis, we will consider three independent ($P = 0$) normal distributions. We will call one distribution y, one x_1, and the other x_2. The sampling distribution of F under H_0 (for $k = 2$, and $n = 27$) can be thought of in the following way:

For all possible samples of $n = 27$ individuals having scores y, x_1, and x_2,

1. Compute the regression equation to predict y from x_1 and x_2,

2. Compute R^2 for each of these regression equations, and

3. Convert these R^2 statistics to F according to equation 16.8.

The result of this process is the sampling distribution of F with $df_{regression} = 2$ and $df_{error} = 24$. For our two-predictor example, the sampling distribution of F will be the right-skewed distribution in Figure 16.4a. Figure 16.4b shows the corresponding cumulative distribution function.

FIGURE 16.4 ■ A Central *F*-Distribution

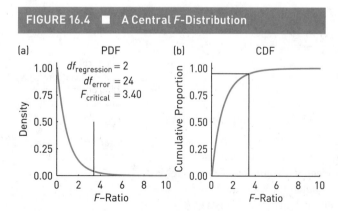

(a) PDF — $df_{regression} = 2$, $df_{error} = 24$, $F_{critical} = 3.40$. Density vs F–Ratio.

(b) CDF — Cumulative Proportion vs F–Ratio.

The sampling distribution of F with two predictor variables for samples of size 27. (a) The probability density function for F. (b) The cumulative distribution function for F.

The question of statistical significance is, as always, a matter of unusualness. An observed F-statistic (F_{obs}) is judged statistically significant when there is a low probability of that statistic, or one more extreme, occurring by chance when the null hypothesis is true. Low probability, by convention, means $p < .05$. For the case of two predictors ($df_{regression} = 2$) and sample size $n = 27$ ($df_{error} = 24$), 5% of this central F-distribution lies above 3.4, which we will call $F_{critical}$. The vertical lines in Figure 16.4 show $F_{critical}$. When we compute F_{obs} for our example, we find

$$F_{obs} = \frac{R^2}{1 - R^2} \frac{df_{error}}{df_{regression}} = \frac{.53}{.47} \frac{24}{2} = 13.53.$$

Because $F_{obs} = 13.53$ exceeds $F_{critical} = 3.4$, we can reject the null hypothesis at the $p < .05$ level and conclude that $P \neq 0$.

Critical Values of F ($F_{critical}$)

The distribution shown in Figure 16.4a is just one of infinitely many central F-distributions, each defined by $df_{regression}$ and df_{error}. For example, Figures 16.5a and 16.5b each show three central F-distributions for $k = 4$, 8, or 16 predictors. The distributions in Figure 16.5a are for sample size $n = 32$, and the three in Figure 16.5b are for sample size $n = 512$. In Figure 16.5a, $df_{regression} = 4$, 8, and 16, and $df_{error} = 27$, 23, and 15, respectively. In Figure 16.5b, $df_{regression} = 4$, 8, and 16, and $df_{error} = 507$, 503, and 495, respectively. It is clear from Figure 16.5 that the shape of a central F-distribution depends on both $df_{regression}$ and df_{error}.

When we moved from the z-distribution to the t-distribution in Chapter 10, we noted that the shape of the t-distribution depends on sample size. Therefore, our t-table listed critical values of t for a small number of degrees of freedom. Now we can no longer use a table like the t-table to test the statistical significance of a given F_{obs} because we have both $df_{regression}$ and df_{error}. Instead we use Table 16.2, which shows critical values of F for selected pairs of $df_{regression}$ and df_{error}. Numbers in the top row show $df_{regression}$, which equals the number of predictors, k. Numbers in the left column show df_{error}, which equals $n-k-1$. For each pair of $df_{regression}$ and df_{error}, the table shows $F_{critical}$ for $\alpha = .05$ (dark shaded rows) and $\alpha = .01$ (light shaded rows). Therefore, to determine whether a given F_{obs} is statistically significant, we determine whether it exceeds $F_{critical}$ for the given degrees of freedom.

Because this F-table displays only a subset of all possible values of $df_{regression}$ and df_{error}, we may do exercises for which $df_{regression}$ and df_{error} are not in the table. In such cases, we will always take the next lower $df_{regression}$ and the next lower df_{error} in the table. This rule makes tests of statistical significance more conservative. That is, following this rule makes it harder to reject the null hypothesis when $df_{error} > 2$. To see why this is so, note that when $df_{error} > 2$, the critical values decrease as both $df_{regression}$ and df_{error} increase.

For example, if $df_{regression} = 14$ and $df_{error} = 29$, which are not in the table, we would choose $df_{regression} = 10$ and $df_{error} = 25$. In this case, $F_{critical} = 2.2$, when $\alpha = .05$. If we had taken the next higher values of $df_{regression}$ and df_{error} (i.e., 15 and 30), we would find $F_{critical} = 2.0$, when $\alpha = .05$. Once again, choosing values of $df_{regression}$ and df_{error} according to this "next lower" rule will make our test more conservative. Of course, our answers will be more precise when we use statistical software such as SPSS and Excel.

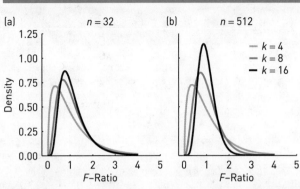

FIGURE 16.5 ■ Six Central F-Distributions

The sampling distribution of the F-ratio under the null hypothesis for regression equations with 4, 8, and 16 predictor variables and sample sizes of 32 (a) and 512 (b).

USING SPSS TO CONDUCT MULTIPLE REGRESSION

We've just introduced several important calculations that are routinely done in multiple regression analyses. These have not been much more complicated than any of the calculations done in previous chapters. However, to introduce these computations, we hid the complexity of computing the regression equation itself. Although it is possible to compute the regression equation by hand, the calculations are tedious, particularly when the number of predictors increases beyond $k = 2$. Therefore, researchers typically use specialized software, such as SPSS, to do regression-related analyses.

TABLE 16.2   ■   The F-Table: Critical Values of F Under the Null Hypothesis That $P^2 = 0$, for $\alpha = .05$ and $\alpha = .01$

df_{error}	$df_{regression}$													
	1	2	3	4	5	6	7	8	9	10	15	20	25	30
1	161	200	216	225	230	234	237	239	241	242	246	248	249	250
	4052	5000	5403	5625	5764	5859	5928	5981	6022	6056	6157	6209	6240	6261
2	18.5	19.0	19.2	19.2	19.3	19.3	19.4	19.4	19.4	19.4	19.4	19.4	19.5	19.5
	98.5	99.0	99.2	99.2	99.3	99.3	99.4	99.4	99.4	99.4	99.4	99.4	99.5	99.5
3	10.1	9.6	9.3	9.1	9.0	8.9	8.9	8.8	8.8	8.8	8.7	8.7	8.6	8.6
	34.1	30.8	29.5	28.7	28.2	27.9	27.7	27.5	27.3	27.2	26.9	26.7	26.6	26.5
4	7.7	6.9	6.6	6.4	6.3	6.2	6.1	6.0	6.0	6.0	5.9	5.8	5.8	5.7
	21.2	18.0	16.7	16.0	15.5	15.2	15.0	14.8	14.7	14.5	14.2	14.0	13.9	13.8
5	6.6	5.8	5.4	5.2	5.1	5.0	4.9	4.8	4.8	4.7	4.6	4.6	4.5	4.5
	16.3	13.3	12.1	11.4	11.0	10.7	10.5	10.3	10.2	10.1	9.7	9.6	9.4	9.4
6	6.0	5.1	4.8	4.5	4.4	4.3	4.2	4.1	4.1	4.1	3.9	3.9	3.8	3.8
	13.7	10.9	9.8	9.1	8.7	8.5	8.3	8.1	8.0	7.9	7.6	7.4	7.3	7.2
7	5.6	4.7	4.3	4.1	4.0	3.9	3.8	3.7	3.7	3.6	3.5	3.4	3.4	3.4
	12.2	9.5	8.5	7.8	7.5	7.2	7.0	6.8	6.7	6.6	6.3	6.2	6.1	6.0
8	5.3	4.5	4.1	3.8	3.7	3.6	3.5	3.4	3.4	3.3	3.2	3.2	3.1	3.1
	11.3	8.6	7.6	7.0	6.6	6.4	6.2	6.0	5.9	5.8	5.5	5.4	5.3	5.2
9	5.1	4.3	3.9	3.6	3.5	3.4	3.3	3.2	3.2	3.1	3.0	2.9	2.9	2.9
	10.6	8.0	7.0	6.4	6.1	5.8	5.6	5.5	5.4	5.3	5.0	4.8	4.7	4.6
10	5.0	4.1	3.7	3.5	3.3	3.2	3.1	3.1	3.0	3.0	2.8	2.8	2.7	2.7
	10.0	7.6	6.6	6.0	5.6	5.4	5.2	5.1	4.9	4.8	4.6	4.4	4.3	4.2
12	4.7	3.9	3.5	3.3	3.1	3.0	2.9	2.8	2.8	2.8	2.6	2.5	2.5	2.5
	9.3	6.9	6.0	5.4	5.1	4.8	4.6	4.5	4.4	4.3	4.0	3.9	3.8	3.7
14	4.6	3.7	3.3	3.1	3.0	2.8	2.8	2.7	2.6	2.6	2.5	2.4	2.3	2.3
	8.9	6.5	5.6	5.0	4.7	4.5	4.3	4.1	4.0	3.9	3.7	3.5	3.4	3.3
16	4.5	3.6	3.2	3.0	2.9	2.7	2.7	2.6	2.5	2.5	2.4	2.3	2.2	2.2
	8.5	6.2	5.3	4.8	4.4	4.2	4.0	3.9	3.8	3.7	3.4	3.3	3.2	3.1
18	4.4	3.6	3.2	2.9	2.8	2.7	2.6	2.5	2.5	2.4	2.3	2.2	2.1	2.1
	8.3	6.0	5.1	4.6	4.2	4.0	3.8	3.7	3.6	3.5	3.2	3.1	3.0	2.9
20	4.4	3.5	3.1	2.9	2.7	2.6	2.5	2.4	2.4	2.3	2.2	2.1	2.1	2.0
	8.1	5.8	4.9	4.4	4.1	3.9	3.7	3.6	3.5	3.4	3.1	2.9	2.8	2.8
25	4.2	3.4	3.0	2.8	2.6	2.5	2.4	2.3	2.3	2.2	2.1	2.0	2.0	1.9
	7.8	5.6	4.7	4.2	3.9	3.6	3.5	3.3	3.2	3.1	2.9	2.7	2.6	2.5
30	4.2	3.3	2.9	2.7	2.5	2.4	2.3	2.3	2.2	2.2	2.0	1.9	1.9	1.8
	7.6	5.4	4.5	4.0	3.7	3.5	3.3	3.2	3.1	3.0	2.7	2.5	2.5	2.4
35	4.1	3.3	2.9	2.6	2.5	2.4	2.3	2.2	2.2	2.1	2.0	1.9	1.8	1.8
	7.4	5.3	4.4	3.9	3.6	3.4	3.2	3.1	3.0	2.9	2.6	2.4	2.3	2.3
40	4.1	3.2	2.8	2.6	2.4	2.3	2.2	2.2	2.1	2.1	1.9	1.8	1.8	1.7
	7.3	5.2	4.3	3.8	3.5	3.3	3.1	3.0	2.9	2.8	2.5	2.4	2.3	2.2
45	4.1	3.2	2.8	2.6	2.4	2.3	2.2	2.2	2.1	2.0	1.9	1.8	1.8	1.7
	7.2	5.1	4.2	3.8	3.5	3.2	3.1	2.9	2.8	2.7	2.5	2.3	2.2	2.1
50	4.0	3.2	2.8	2.6	2.4	2.3	2.2	2.1	2.1	2.0	1.9	1.8	1.7	1.7
	7.2	5.1	4.2	3.7	3.4	3.2	3.0	2.9	2.8	2.7	2.4	2.3	2.2	2.1
55	4.0	3.2	2.8	2.5	2.4	2.3	2.2	2.1	2.1	2.0	1.9	1.8	1.7	1.7
	7.1	5.0	4.2	3.7	3.4	3.1	3.0	2.9	2.7	2.7	2.4	2.2	2.1	2.1
60	4.0	3.2	2.8	2.5	2.4	2.3	2.2	2.1	2.0	2.0	1.8	1.7	1.7	1.6
	7.1	5.0	4.1	3.6	3.3	3.1	3.0	2.8	2.7	2.6	2.4	2.2	2.1	2.0
65	4.0	3.1	2.7	2.5	2.4	2.2	2.2	2.1	2.0	2.0	1.8	1.7	1.7	1.6
	7.0	4.9	4.1	3.6	3.3	3.1	2.9	2.8	2.7	2.6	2.3	2.2	2.1	2.0
70	4.0	3.1	2.7	2.5	2.3	2.2	2.1	2.1	2.0	2.0	1.8	1.7	1.7	1.6
	7.0	4.9	4.1	3.6	3.3	3.1	2.9	2.8	2.7	2.6	2.3	2.2	2.1	2.0
100	3.9	3.1	2.7	2.5	2.3	2.2	2.1	2.0	2.0	1.9	1.8	1.7	1.6	1.6
	6.9	4.8	4.0	3.5	3.2	3.0	2.8	2.7	2.6	2.5	2.2	2.1	2.0	1.9

Note: All values calculated in Excel by the author.

LEARNING CHECK 4

1. What is the critical value of F, for $\alpha = .05$, given the following degrees of freedom for $df_{regression}$ and df_{error}, respectively: (a) 10, 30; (b) 4, 25; (c) 3, 44; (d) 8, 101; and (e) 6, 34?

2. For each of the following conditions of R^2, k, and n, respectively, compute F_{obs} and state whether it is statistically significant at the $\alpha = .05$ and $\alpha = .01$ levels: (a) .75, 3, 32; (b) .25, 3, 32; (c) .64, 10, 100; (d) .21, 5, 256; and (e) .10, 4, 81.

Answers

1. (a) 2.2. (b) 2.8. (c) 2.8. (d) 2.0. (e) 2.4.

2. (a) $F_{obs} = 28$, statistically significant at $\alpha = .05$ and $\alpha = .01$.

   (b) $F_{obs} = 3.11$, statistically significant at $\alpha = .05$, not statistically significant at $\alpha = .01$.

   (c) $F_{obs} = 15.82$, statistically significant at $\alpha = .05$ and $\alpha = .01$.

   (d) $F_{obs} = 13.29$, statistically significant at $\alpha = .05$ and $\alpha = .01$.

   (e) $F_{obs} = 2.11$, not statistically significant at $\alpha = .05$, not statistically significant at $\alpha = .01$.

Running a Regression Analysis

In this section, we'll see that a multiple regression analysis in SPSS is much like the simple linear regression described in Appendix 14.2. The only difference is that we enter more than one predictor. The regression.sav data file (available at study.sagepub.com/gurnsey) contains the data from Table 16.1 with columns labeled "Success," "GRE," and "TIE." With the data entered and formatted, we can conduct a regression analysis.

To conduct a multiple regression analysis, we choose the Analyze→ Regression→Linear . . . dialog, as we did in Chapter 14. Figure 16.6 shows the Linear Regression dialog. The three variables defined were Success, GRE, and TIE. Success has been moved into the Dependent: panel and GRE and TIE have been moved into the

Independent(s): panel. The results of our regression analysis are shown in the output window as soon as we press the OK button.

Default Regression Analysis

The default regression analysis produces the four tables shown in Figure 16.7. We will discuss each in turn.

Variables Entered/Removed

Figure 16.7a shows the variables that were entered into the analysis. The first column in this table is labeled Model. As noted in the introduction to this chapter, a model is a regression equation used to explain the variance in the outcome variable. So, our regression equation with predictors *GRE* and *TIE* constitutes a model of the outcome variable *success*. There are two footnotes at the bottom of the table. Note "a" tells us the dependent variable is *success*. Note "b" tells us that no other variables remain to be entered into our model. (This point will make more sense later.)

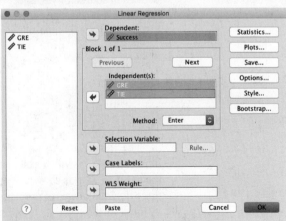

FIGURE 16.6 ■ Linear Regression Dialog

When variables have been moved into the **Dependent** and **Independent(s)** boxes, the OK button turns blue to indicate that SPSS is ready to conduct the analysis. We will see later that many other options can be selected.

FIGURE 16.7 ■ Regression Output

(a) **Variables Entered/Removed[a]**

Model	Variables Entered	Variables Removed	Method
1	TIE, GRE[b]	.	Enter

a. Dependent Variable: Success
b. All requested variables entered.

(b) **Model Summary**

Model	R	R Square	Adjusted R Square	Std. Error of the Estimate
1	.727[a]	.528	.489	2.13549

a. Predictors: (Constant), TIE, GRE

(c) **ANOVA[a]**

Model		Sum of Squares	df	Mean Square	F	Sig.
1	Regression	122.552	2	61.276	13.437	.000[b]
	Residual	109.448	24	4.560		
	Total	232.000	26			

a. Dependent Variable: Success
b. Predictors: (Constant), TIE, GRE

(d) **Coefficients[a]**

Model		Unstandardized Coefficients		Standardized Coefficients	t	Sig.
		B	Std. Error	Beta		
1	(Constant)	.606	9.465		.064	.949
	GRE	.233	.068	.543	3.424	.002
	TIE	.029	.016	.292	1.838	.078

a. Dependent Variable: Success

(a through d) The four tables produced by the default regression analysis in SPSS.

Model Summary

Figure 16.7b shows several characterizations of how well the model fits the data. We see that the multiple correlation coefficient $(R) = .727$, $R^2 = .528$, $R^2_{adj} = .489$, and $s_{est} = 2.13549$. All of these quantities were computed by hand earlier. Note "a" at the bottom of the Model Summary table shows the predictors in the model. The term (Constant) refers to the intercept of the regression equation, a.

ANOVA

Figure 16.7c shows an analysis of variance (ANOVA). The term *ANOVA* comes from the fact that we are analyzing (decomposing) the variance in our dependent variable. In this case, the ANOVA is an omnibus test of whether the two-predictor model (*GRE* and *TIE*) explains a statistically significant proportion of the variability in the outcome variable, *success*. The ANOVA in the table describes F_{obs} as a ratio of mean squares, as shown in equation 16.7b.

The first column of the ANOVA table shows the sources of variance: Regression, Residual, and Total. Statistics is plagued by a variable terminology, and it is common for error variance to be described as residual variance (i.e., the variance that can't be explained by the model). As you can see, what we have been calling *error*, SPSS calls Residual.

The second column of the ANOVA table shows the sum of squares associated with each source of variance. As we've seen before, $ss_{total} = ss_{regression} + ss_{error}$. In our example, this means $232 = 122.552 + 109.448$. The third column shows the degrees of freedom associated with each source of variance. As we saw earlier, $df_{regression} = k = 2$, $df_{error} = n-k-1 = 24$, and $df_{total} = n-1 = 26$. The fourth column shows $ms_{regression}$ and ms_{error}, which were previously defined as

$$ms_{regression} = ss_{regression}/df_{regression} \text{ and}$$

$$ms_{error} = ss_{error}/df_{error}.$$

Therefore, $ms_{regression} = 122.552/2 = 61.276$ and $ms_{error} = 109.448/24 = 4.56$.

The fifth column shows $F_{obs} = ms_{regression}/ms_{error} = 61.276/4.56 = 13.437$. The sixth column shows the p-value associated with $F_{obs} = 13.437$. Because the p-value is less than .001, the value of .000 is shown; with default settings in SPSS most numbers are output to only three decimal places of precision. Once again, these are exactly the quantities we computed by hand earlier.

Coefficients

Figure 16.7d shows the regression coefficients. We will have much more to say about these in Chapter 17, but for now we will simply step through the table to note its contents.

The first column shows the predictor variables and the second column shows the regression coefficient associated with each predictor. As was shown in Figure 16.2, the regression equation relating *GRE* and *TIE* to *success* is

$$\hat{y} = a + b_1(x_1) + b_2(x_2) = a + b_1(GRE) + b_2(TIE) = 0.606 + 0.233(GRE) + 0.029(TIE).$$

(We are using the numbers from the SPSS output, which are given to three decimal places of precision.) Every statistic has an estimated standard error, and those associated with the regression coefficients are shown in the third column of the Coefficients table. The fourth column, labeled Beta, shows the standardized regression coefficients that were described in Appendix 14.2. These are the regression coefficients that result when all variables (y and $x_1, x_2, \ldots, x_k$) are transformed to z-scores. In our example, the resulting equation has the following form:

$$z_{\hat{y}} = \text{Beta}_{GRE}(z_{GRE}) + \text{Beta}_{TIE}(z_{TIE}) = 0.543(z_{GRE}) + 0.292(z_{TIE}).$$

Each standardized regression coefficient shows the strength of the association between y and one of the predictors, $x_1, x_2, \ldots, x_k$, when differences in scale have been removed.

Each regression coefficient can be tested for statistical significance by dividing it by its estimated standard error. This yields a t-statistic, as follows:

$$t_{obs} = \frac{b_i}{s_{b_i}}.$$

The fifth column of the Coefficients table shows these t-statistics, and the sixth column (Sig.) shows the associated p-values. All of these points will be revisited in Chapter 17.

LEARNING CHECK 5

1. Enter the data from Learning Check 2 (available at study.sagepub.com/gurnsey) into the SPSS data window. Name the variables y, x_1, x_2, and x_3. Submit these numbers to a regression analysis in SPSS and report the following things:

   (a) R, R^2, R^2_{adj}, and s_{est};

   (b) $df_{regression}$, df_{error}, df_{total}, F_{obs}, and p; and

   (c) a, b_1, b_2, and b_3.

2. Does the regression equation explain a statistically significant proportion of the total variability in y?

Answers

1. (a) $R = .79$, $R^2 = .624$, R^2_{adj} .436, and $s_{est} = 46.76536$.

   (b) $df_{regression} = 3$, $df_{error} = 6$, $df_{total} = 9$, $F_{obs} = 3.324$, $p = .098$.

   (c) $a = 79.334$, $b_1 = -0.101$, $b_2 = 0.127$, $b_3 = 0.541$.

2. No. $p = .098 > .05$.

Assessing Assumptions Using SPSS

We saw in Appendix 14.2 that SPSS can be used to assess many of the assumptions underlying a simple linear regression analysis. Many of these analyses apply to multiple regression as well. We will continue with the example of predicting *success* in graduate school from *GRE* and *TIE*.

Linearity, Homoscedasticity, and Normality of Residuals

In Appendix 14.2, we saw how to plot standardized residuals (standardized errors of prediction),

$$z_{\text{resid}} = \frac{\hat{y} - y}{s_{\hat{y} - y}},$$

against standardized predicted values,

$$z_{\text{pred}} = \frac{\hat{y} - m_{\hat{y}}}{s_{\hat{y}}},$$

to check for violation of linearity and homoscedasticity. We also saw how to check the residuals for normality. We can use all of these tools to check the same assumptions for a multiple regression analysis.

Multicollinearity

The problem of multicollinear predictors is new in this chapter because this is the first time we've dealt with more than one predictor. SPSS provides several ways to assess the multicollinearity assumption. To see these options, we return to the Linear Regression dialog through the Analyze menu (see Figure 16.8). To assess the independence assumption, we can click on the Statistics . . . button to open the Linear Regression: Statistics dialog; this dialog overlays the Linear Regression dialog in Figure 16.8. When this dialog first opens, the Estimates and Model fit options are already checked (✅) by default. In addition to these checkboxes, I have checked Descriptives and Collinearity diagnostics. When Descriptives is checked, SPSS will produce a correlation table as shown in Figure 16.9. (Note that this correlation table is formatted slightly differently that the one described in Appendix 15.2.) Of particular interest is the correlation between the two predictors, *GRE* and *TIE*, which is shown to be .468. This number is not problematic. It is an example of "correlated but not too correlated."

The squared correlation between two predictors $(r^2_{x_1, x_2})$ is said to be a measure of *shared variance*. In our example, $r^2_{x_1, x_2}$ tells us how much variability in x_1 can be explained by x_2 and vice versa. There are two collinearity measures that are based on this notion of shared variance.

The first collinearity measure is tolerance (*tol*). For the two-predictor model, *tol* is simply $1 - r^2_{x_1, x_2}$, that is, 1 minus the squared correlation between the two predictors. So, if $r^2_{x_1, x_2}$ expresses how much shared variance there is between the two predictors, then *tol* is a measure of unshared variance, or the degree of independence. If the two predictors are perfectly correlated, then $r^2_{x_1, x_2} = 1$ and *tol* = 0. If the two predictors are perfectly uncorrelated, then $r^2_{x_1, x_2} = 0$ and *tol* = 1.

The second collinearity measure is the variance inflation factor (*VIF*), which is reciprocally related to *tol*; *VIF* = 1/*tol*. In multiple regression, very small values of *tol* (large values of *VIF*) may be worrisome.

Choosing Collinearity diagnostics in the Linear Regression: Statistics dialog was a request to compute

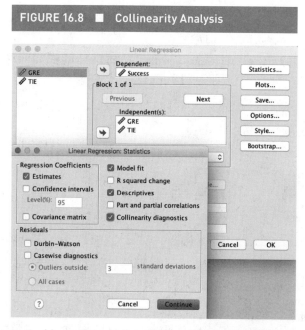

FIGURE 16.8 ■ Collinearity Analysis

In the Linear Regression dialog, click the Statistics . . . option to open the Linear Regression: Statistics dialog.

tol and *VIF*. These measures are shown in the last two columns of the augmented Coefficients table in Figure 16.10. To see how these values were computed, remember that the correlation between the two predictors was $r_{x_1, x_2} = .468$ and so $r^2_{x_1, x_2} = .219$. Therefore,

$$tol = 1 - .219 = .781 \text{ and}$$

$$VIF = 1/tol = 1/.781 = 1.28$$

as shown in Figure 16.10.

It is only slightly more complicated to compute *tol* and *VIF* when there are more than two predictors. Rather than asking how well we can predict x_1 from x_2, we ask how well we can predict x_1 from all other predictors, $x_{j \neq i}$; we return to this point in Chapter 17. Many authors suggest a rule-of-thumb criterion that tells us to worry about collinearity if *VIF* exceeds 10. Appendix 17.2 (available at study.sagepub.com/gurnsey) provides an extended discussion of whether this rule of thumb is generally useful. My view is that *VIF* is often used uncritically, and that researchers often take unnecessary steps to reduce it.

FIGURE 16.9 ■ Correlation Table

Correlations

		Success	GRE	TIE
Pearson Correlation	Success	1.000	.680	.546
	GRE	.680	1.000	.468
	TIE	.546	.468	1.000
Sig. (1-tailed)	Success	.	.000	.002
	GRE	.000	.	.007
	TIE	.002	.007	.
N	Success	27	27	27
	GRE	27	27	27
	TIE	27	27	27

Descriptive statistics for the three variables in our two-predictor multiple regression model.

FIGURE 16.10 ■ Diagnostics: *tol* and VIF

Coefficients[a]

Model		Unstandardized Coefficients		Standardized Coefficients	t	Sig.	Collinearity Statistics	
		B	Std. Error	Beta			Tolerance	VIF
1	(Constant)	.606	9.465		.064	.949		
	GRE	.233	.068	.543	3.424	.002	.781	1.280
	TIE	.029	.016	.292	1.838	.078	.781	1.280

a. Dependent Variable: Success

The coefficients table with measures of *tol* and *VIF* added.

LEARNING CHECK 6

1. Enter the data from Learning Check 2 (available at study.sagepub.com/gurnsey) into the SPSS data window. Name the variables y, x_1, x_2, and x_3. Submit these numbers to the appropriate analyses in SPSS and report the following:

   (a) skew for y, x_1, x_2, x_3;

   (b) skew/Std_Error for y, x_1, x_2, x_3;

   (c) kurtosis for y, x_1, x_2, x_3;

   (d) kurtosis/Std_Error for y, x_1, x_2, x_3;

   (e) *VIF* for x_1, x_2, x_3; and

   (f) tolerance for x_1, x_2, x_3.

Answers

1. (a) Skew for $y = -0.358$, $x_1 = 0.252$, $x_2 = 0.527$, $x_3 = 0.566$.

   (b) Skew/Std_Error for $y = -0.521$, $x_1 = 0.367$, $x_2 = 0.767$, $x_3 = 0.824$.

   (c) Kurtosis for $y = -0.319$, $x_1 = 1.641$, $x_2 = 0.194$, $x_3 = -1.119$.

   (d) Kurtosis/Std_Error for $y = -0.239$, $x_1 = 1.230$, $x_2 = 0.145$, $x_3 = -0.839$.

   (e) VIF for $x_1 = 2.007$, $x_2 = 2.440$, $x_3 = 2.188$.

   (f) Tolerance for $x_1 = 0.498$, $x_2 = 0.41$, $x_3 = 0.457$.

DEGREES OF FREEDOM

Before continuing with our analysis of the *GRE*, *TIE*, and *success* data, we will have a brief interlude to discuss the topic of degrees of freedom, which has been mentioned many times since Chapter 10. Each time this topic has come up, we have simply stated the number of degrees of freedom associated with a particular statistic without a full explanation. In our current example with 27 scores and two predictors, we have $df_{regression} = 2$ and $df_{error} = 24$. What is the logic behind this? The answer to this question is fundamental to understanding multiple regression and ANOVA. This important concept is most easily understood in the context of multiple regression, and this is our first opportunity to explain *degrees of freedom* properly.

To explain degrees of freedom, we will start with an *assertion* and then use SPSS to demonstrate that it is true:

ASSERTION

All of the variability in an outcome variable, y, can be explained by a regression model with an intercept and $n-1$ independent predictors, where n is the number of scores in y.

This assertion is very strong. It says nothing about the nature of the variables. In fact, the assertion is true even if y and all predictor variables $x_1, x_2, \ldots, x_{n-1}$ are independent random samples from the same population. Furthermore, the assertion says nothing about the shape of the distribution from which the scores in y and $x_1, x_2, \ldots, x_{n-1}$ were drawn.

Figure 16.11 shows six variables, with $n = 6$ scores in each. All $6*6 = 36$ scores are random numbers drawn from a uniform distribution of scores in the interval [0, 1]; they were generated with the **RAND** function in Excel. The first column of random numbers is designated y, and the remaining five columns are designated $x_1, x_2, \ldots, x_5$. These numbers will be submitted to a sequence of regression analyses, with each analysis in the sequence having one more predictor variable than the one before it.

Figure 16.12 shows how these models are constructed. Figure 16.12a shows the first model, in which the dependent variable is y and the predictor is x_1. Notice that the region surrounding the highlighted x_1 is titled Block 1 of 1. When the Next button is pressed, this changes to Block 2 of 2 (Figure 16.12b) and we are given an opportunity to add one or more additional predictors.

Figure 16.12b shows the second model, in which the dependent variable is y and the predictor x_2 has been added to the model. Each time we add a predictor (or predictors) in

FIGURE 16.11 ■ Thirty-Six Random Numbers

	y	x1	x2	x3	x4	x5
1	.2106	.3786	.6932	.7559	.5928	.8351
2	.5162	.4797	.9929	.2887	.7610	.3872
3	.6967	.2140	.7115	.6340	.5224	.0117
4	.6878	.6926	.8848	.8094	.2195	.0157
5	.0370	.9246	.3915	.3793	.1592	.1561
6	.2761	.2953	.0984	.4501	.7206	.1441

Six variables, each with six scores drawn from the same uniform distribution.

a block, the new model includes this new predictor (or predictors), and all predictors from previous blocks. In Block 2 of the analysis, there are thus two predictors in the model, x_1 and x_2. We continue to add predictors in this way until we finally enter x_5 in Block 5 of 5. At this point, as shown in Figure 16.12c, there are five predictors in the model, $x_1, x_2, \ldots, x_5$.

When we now click ▢ OK ▢, the regression models are computed. There will be a regression equation computed for each model (block of predictors), but we won't look at these equations for the moment. Rather, we will look at how R^2 changes as predictors are added.

Figure 16.13 summarizes the five models. The first column identifies the model in question and the second column shows the multiple correlation coefficient (R) associated with the model. Notice that the superscripted letters (a, b, c, d, e) refer to notes at the bottom of the table showing the predictors in the model. The third column shows the R^2 associated with the model.

The critical points to note are that R^2 increases as predictors are added to the model, and when all five predictors are in the model, $R^2 = 1$. That is, the five-predictor model explains 100% of the variability in our outcome variable, y. This result demonstrates the truth of the assertion we made earlier. There are $n = 6$ scores in y, and at step five there are $n-1 = 5$ predictors and an intercept. These predictors explain all of the variability in our outcome variable.

The results of Figure 16.13 may seem counterintuitive because the 36 numbers we used were all drawn from the same uniform distribution, so any correlation between y and the predictor variables is simply due to chance. Nevertheless, it is a mathematical fact that adding independent predictors will always increase the explained variance, R^2.

Degrees of Freedom: Explained and Unexplained Variance

We will now look at how the above demonstration relates to degrees of freedom. In Model 1, there is one predictor, so $df_{\text{regression}} = 1$. Because $n = 6$, $df_{\text{error}} = 6 - 1 - 1 = 4$. We see that $R^2 = .145$, which means that the proportion of variability in y explained by our one-predictor model is .145. Therefore, the proportion of variance left unexplained by our model is $1 - R^2 = 1 - .145 = .855$. *The residual degrees of freedom (df_{error}) is the number of additional predictors needed to account for this unexplained variance.* In other words, you would need four extra predictors to explain the remaining variability y. These remaining predictors can be any numbers at all, as long they are independent; i.e., they are not multicollinear. That is, the remaining predictors are free to vary.

In Model 2, there are two predictors, so $df_{\text{regression}} = 2$ and $df_{\text{error}} = 6 - 2 - 1 = 3$. Figure 16.13 shows that $R^2 = .556$, which means that the proportion of variability in y explained by the two predictors is

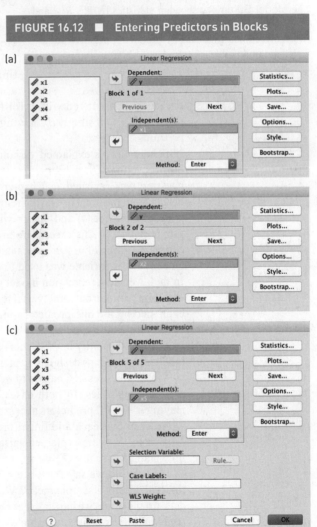

FIGURE 16.12 ■ Entering Predictors in Blocks

(a)

(b)

(c)

Predictor x_1 is entered in the first block of the analysis (a), x_2 is entered in the second block of the analysis (b), and x_5 is entered in the last block of the analysis (c).

FIGURE 16.13 ■ Five Regression Models

Model Summary

Model	R	R Square	Adjusted R Square	Std. Error of the Estimate
1	.380[a]	.145	−.069	.2802772
2	.746[b]	.556	.260	.2331225
3	.761[c]	.578	−.054	.2782599
4	.840[d]	.705	−.476	.3292756
5	1.000[e]	1.000	.	.

a. Predictors: (Constant), x1
b. Predictors: (Constant), x1, x2
c. Predictors: (Constant), x1, x2, x3
d. Predictors: (Constant), x1, x2, x3, x4
e. Predictors: (Constant), x1, x2, x3, x4, x5

The **Model Summary** showing statistics (R, R^2, R^2_{adj}, and s_{est}) associated with each model. Notice that R^2_{adj} clearly "overcorrects" in this case by producing negative values, which are impossible.

.556. The proportion of variance left unexplained by our model is $1 - R^2 = 1 - .556 = .444$. Therefore, df_{error} is the number of additional predictors needed to account for this unexplained variance. In other words, you would need three extra predictors to explain the remaining variability y.

As we move through the models, the explained variance (R^2) increases and the unexplained variance ($1 - R^2$) decreases. At step five (model 5), the last predictor is added and all variance in y is explained by the five predictors; i.e., $R^2 = 1$. We now find that $df_{error} = 6 - 5 - 1 = 0$. That is, there is no unexplained variability in y so no more predictors are needed.

Degrees of Freedom: Retrospective

We can now think back to several of the analyses covered in Part III to see where our degrees of freedom came from. From our current perspective as masters of multiple regression, we see that there are degrees of freedom associated with regression ($df_{regression} = k$) and degrees of freedom associated with error ($df_{error} = n-k-1$). Up to this point, we have referred to df_{error} as df_{within} or simply df. Having now seen two kinds of dfs, we can explain the df from previous chapters.

The notion of explained variance was introduced in Chapter 11. There we considered the proportion of variability in our sample of $n_1 + n_2$ scores explained by group means. This question was revisited in Chapter 15 when we considered the two-independent-groups design from the point of view of regression. The regression equation in this case was formed with an intercept (a) and one predictor variable, and we found that this model predicted group means. In this case, there are $n_1 + n_2$ scores in our two groups altogether, and this combined group of $n_1 + n_2$ scores has $n_1 + n_2 - 1$ total degrees of freedom associated with it. Our predictor variable was used to explain some of the variability in this group of scores. In fact, we had a regression model with $df_{regression} = k = 1$. This model explains some of the variability in our outcome variable, which we denoted with r^2. Therefore, the proportion explained by the one-predictor model is r^2, and the variability remaining unexplained is $1 - r^2$. We would need $n_1 + n_2 - k - 1$ (or $n_1 + n_2 - 2$) additional predictors to explain the remaining variability in y. So, this is why we have $n_1 + n_2 - 2$ degrees of freedom in both the independent-groups design (Chapter 11) and simple linear regression (Chapter 14).

Other analyses in Chapter 14 dealt with simple linear regression for fixed and random predictor variables. In both cases, the predictor variable used one degree of freedom, leaving $n_1 + n_2 - 2$ predictors necessary to explain the remaining variability. In this chapter, we've seen that each additional predictor variable uses a degree of freedom. The more predictors, the more explained variance. In the limit, with $n-1$ predictors, there is no more unexplained variance.

So why do we say there are $n-1$ degrees of freedom when computing a confidence interval around a sample mean? Well, we can think of the sample mean as a model of the data. In this case, we can say that our best prediction of any individual score is the sample mean. Now, if we take the sample mean as the intercept of the regression equation, then it would require $n-1$ independent predictors to explain all variability in y. Or, put the other way around, if you have $n-1$ independent predictors in your regression equation, then there is only one possible value for the intercept. With $n-1$ independent predictors specified, the intercept is not free to vary.

Degrees of Freedom and the *F*-Ratio

The ANOVA table in Figure 16.14 shows F_{obs} for each of the five regression models that we have been considering. For each model, we are given the three sources of variance (Regression, Residual, and Total), the sum of squares for each source, and the degrees of freedom for each source. You should be able to confirm that the R^2 associated with each model is computed as $ss_{regression}/ss_{total}$. For Model 1, for example,

$$R^2 = ss_{regression}/ss_{total} = 0.053/0.367 = 0.144,$$

which (except for rounding error) is shown in row 1 of Figure 16.13.

The F_{obs} values shown in Figure 16.14 can be computed as

$$F_{obs} = ms_{regression}/ms_{error} \text{ or}$$

$$F_{obs} = \frac{R^2}{1-R^2}\frac{df_{error}}{df_{regression}}.$$

FIGURE 16.14 ■ Five ANOVA Tables

ANOVA[a]

Model		Sum of Squares	df	Mean Square	F	Sig.
1	Regression	.053	1	.053	.676	.457[b]
	Residual	.314	4	.079		
	Total	.367	5			
2	Regression	.204	2	.102	1.879	.296[c]
	Residual	.163	3	.054		
	Total	.367	5			
3	Regression	.212	3	.071	.915	.560[d]
	Residual	.155	2	.077		
	Total	.367	5			
4	Regression	.259	4	.065	.597	.735[e]
	Residual	.108	1	.108		
	Total	.367	5			
5	Regression	.367	5	.073	.	.[f]
	Residual	.000	0	.		
	Total	.367	5			

a. Dependent Variable: y
b. Predictors: (Constant), x1
c. Predictors: (Constant), x1, x2
d. Predictors: (Constant), x1, x2, x3
e. Predictors: (Constant), x1, x2, x3, x4
f. Predictors: (Constant), x1, x2, x3, x4, x5

The ANOVA table for the five regression models.

For Model 1, for example,

$$F_{obs} = \frac{R^2}{1-R^2}\frac{df_{error}}{df_{regression}} = \frac{.145}{.855}\frac{4}{1} = 0.17*4 = 0.68,$$

which is shown in the first row of Figure 16.14.

F_{obs} changes from model to model for two reasons. First, $R^2/(1-R^2)$ increases as more variability is explained by the models. Second, $df_{error}/df_{regression}$ decreases as more predictors are added. The *p*-value associated with each F_{obs} tells us the proportion of the central *F*-distribution with $df_{regression}$ and df_{error} degrees of freedom lying above F_{obs}. A small *p*-value tells us that it is unusual for $k = df_{regression}$ predictors to produce the observed R^2 when H_0 is true.

In summary, the central message of this section is that degrees of freedom are intimately connected to explained variability. With *n* scores in *y*, *n*−1 predictors and an intercept are required to explain all variability in *y*. Each predictor added to a regression model increases the explained variability in *y* and uses one degree of freedom. Therefore, each predictor variable reduces the number of predictors required to explain the remaining variability in *y*. This is a fundamental fact of all regression-type analyses. And many of the analyses used in psychological research are a form of regression.

COMPARING REGRESSION MODELS

Let's return to the example of predicting success in graduate school (*success*) from GRE scores and TIE scores. Our hypothetical admissions committee wondered whether a regression model with *GRE* and *TIE* as predictors would better predict *success* than a model with just *GRE* as a predictor. Our regression analysis involved two models. The first model, with *x* = *GRE* and *y* = *success*, produced R^2 = .462, when expressed to three decimal places. The second model, with x_1 = *GRE*, x_2 = *TIE*, and *y* = *success*, produced

LEARNING CHECK 7

1. If y is a random sample of 21 scores, how many predictor variables would be required in a regression equation to explain all of the variability in y?

2. If y is a random sample of 21 scores and a regression equation with 12 predictors produces $R^2 = .4$, how many predictors would be required to explain the remaining variability in y?

3. If y is a random sample of 256 scores and a regression equation with 32 predictors produces $R^2 = .8$, how many predictors would be required to explain the remaining variability in y?

4. If y is a random sample of 256 scores, what is R^2 if a regression equation has 255 independent predictor variables?

Answers

1. 20.

2. 8.

3. 223.

4. 1.

$R^2 = .528$. Therefore, adding *TIE* as a predictor increased the explained variance from .462 to .528, which means that the two-predictor model makes better predictions than the one-predictor model.

Although R^2 increased by .066 (i.e., $.528 - .462 = .066$), we have to keep in mind that adding predictor variables to a regression equation will always increase R^2, even when the predictor is not expected to be meaningfully related to the outcome variable. Therefore, we are faced with the question of whether a 6.6-percentage-point ($.066 * 100$) increase in explained variance is meaningful. There are two ways to address this question; the first is estimation, and the second is significance testing with an F-statistic. We deal with these issues in turn.

Estimating ΔP^2

Given that our models had one and two predictors, respectively, we can refer to the R^2 values associated with these as $R^2_{smaller}$ and R^2_{larger}, respectively. Therefore, $R^2_{smaller} = .462$ and $R^2_{larger} = .528$. We will use ΔR^2 (often called "R^2 change") to denote the difference between $R^2_{smaller}$ and R^2_{larger}. Therefore,

$$\Delta R^2 = R^2_{larger} - R^2_{smaller}. \tag{16.9}$$

In our example, we've seen that

$$\Delta R^2 = R^2_{larger} - R^2_{smaller} = .528 - .462 = .066$$

$\Delta R^2 = R^2_{larger} - R^2_{smaller}$ is a statistic representing the difference in explained variance between two regression models. R^2_{larger} is the variance explained by a model with k predictors and $R^2_{smaller}$ is the variance explained by a model with a subset of these k predictors. ΔR^2 estimates $\Delta P^2 = P^2_{larger} - P^2_{smaller}$, the corresponding parameter.

ΔR^2 can be computed for any two regression models for which the larger model contains all the predictors from the smaller model, plus one or more additional predictors. For example, ΔR^2 can be computed when the smaller model has 10 predictors and the larger model has the same 10 predictors and four additional predictors.

ΔR^2 is a statistic, so we must now think about what it estimates. Earlier we saw that R^2 estimates P^2. Now that we have two models, we have two P^2 values to think about; these are P^2_{larger} and $P^2_{smaller}$. The difference between P^2_{larger} and $P^2_{smaller}$ is ΔP^2. Therefore, ΔR^2 estimates $\Delta P^2 = P^2_{larger} - P^2_{smaller}$.

Although it is easy to compute a point estimate of ΔP^2 (i.e., ΔR^2), it is not as easy to compute a confidence interval around the estimate. However, there are computationally intensive bootstrapping algorithms that provide quite good confidence intervals for ΔR^2. Appendix 16.1 describes one such bootstrapping algorithm developed by Algina, Keselman, and Penfield (2008). For the present example, we find that the 95% bootstrapped confidence interval is [0, .22]. So, our best point estimate for ΔP^2 is $\Delta R^2 = .066$, 95% CI [0, .22].

Significance Test for ΔR^2

The traditional way to assess ΔR^2 is to compute an F-statistic and test it for statistical significance using the central F-distribution. We call this statistic F_{change} and compute it as follows:

$$F_{change} = \frac{\Delta R^2}{1 - R^2_{larger}} \frac{df_{error}}{df_{change}}. \tag{16.10}$$

In equation 16.10, df_{change} is the difference in the number of predictors in the two models. It plays the same role that $df_{regression}$ played earlier. The residual or error degrees of freedom are $df_{error} = n - k_{larger} - 1$, and they reflect the residual (error) degrees of freedom associated with the larger model. For our example, we would have

$$F_{change} = \frac{\Delta R^2}{1 - R^2_{larger}} \frac{df_{error}}{df_{change}} = \frac{.066}{1 - .528} \frac{24}{1} = 3.38$$

To test F_{change} for statistical significance, we consult the F-table with df_{change} and df_{error} degrees of freedom. In this case, $df_{change} = 2 - 1 = 1$, and $df_{error} = 27 - 2 - 1 = 24$. Because $df_{change} = 1$, we look down the first column of the F-table in search of the row with $df_{error} = 24$. Because the table does not contain an entry for $df_{error} = 24$, we go to the next smaller df_{error}, which is 20. The critical value of F required to reject the null hypothesis that $\Delta P^2 = 0$ is 4.40. Because our observed value of $F_{change} = 3.38$ does not exceed 4.40, we retain the null hypothesis. To get an exact p-value associated with F_{change}, we can use the **FDIST** function in Excel as described in Appendix 16.2 (available at study.sagepub.com/gurnsey).

A Significance Test for ΔR^2 in SPSS

Figure 16.15 shows how to conduct the significance test for ΔR^2 in SPSS. As shown, variables are entered in blocks. In Block 1, *success* was specified as the dependent variable and *GRE* was entered as the predictor. *TIE* was added in Block 2. To compute and assess ΔR^2, we click the Statistics . . . button in the top right of the Linear Regression dialog. In the Linear Regression: Statistics dialog that appears, Estimates and Model fit have been checked by default. To these defaults we add R squared change by checking it (✓). We then click Continue to return to the Linear Regression dialog, and there we click OK to proceed with the analysis.

The results of this analysis are shown in the Model Summary in Figure 16.16. The relevant ΔR^2 is the one representing the difference in explained variance between Models 1 and 2. $R^2 = .462$ for Model 1, and $R^2 = .528$ for

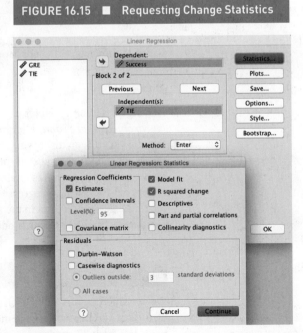

FIGURE 16.15 ■ Requesting Change Statistics

The Linear Regression: Statistics dialog that provides the option to compute ΔR^2 as predictors are added to the model.

FIGURE 16.16	■	Change Statistics

Model Summary

Model	R	R Square	Adjusted R Square	Std. Error of the Estimate	Change Statistics				
					R Square Change	F Change	df1	df2	Sig. F Change
1	.680[a]	.462	.440	2.23482	.462	21.452	1	25	.000
2	.727[b]	.528	.489	2.13549	.066	3.380	1	24	.078

a. Predictors: (Constant), GRE
b. Predictors: (Constant), GRE, TIE

The model summary for hierarchical regression models with one and two predictors. ΔR^2 is computed for both models, but it is the ΔR^2 derived from the difference between the two models ($\Delta R^2 = .066$) that is of interest.

Model 2. As shown in the Model Summary table, $\Delta R^2 = .528 - .462 = .066$. As we've seen, $df_{change}(df_1) = 2 - 1 = 1$ and $df_{error}(df_2) = 27 - 2 - 1 = 24$. We previously computed F_{change} by hand and found that it equals 3.38 as shown in the Model Summary. The p-value associated with F_{change} is $p = .078$. Assuming the usual $\alpha = .05$ criterion for statistical significance, we would conclude that adding *TIE* to our model does not lead to a statistically significant increase in explained variance.

It may seem a bit odd to have a value of ΔR^2 for Model 1, which is the model with only *GRE* as a predictor. However, we can think of this as the change in explained variance from a model with zero predictors ($R^2 = 0$) to a model with one predictor. Figure 16.16 shows that $R^2 = .462$, so $\Delta R^2 = .462 - 0 = .462$, as shown in the Change Statistics. For Model 1, there is 1 degree of freedom associated with this change ($df_{change} = 1 - 0 = 1$) and the error degrees of freedom are $df_{error} = 27 - 1 - 1 = 25$. When we calculate this F_{change}, we find

$$F_{change} = \frac{R^2}{1-R^2}\frac{df_{error}}{df_{change}} = \frac{.462}{.538}\frac{27-1-1}{1} = 21.452$$

which is shown as F_{change} for Model 1.

Hierarchical Regression

Adding predictor variables to a regression analysis in steps (or blocks) is referred to as *hierarchical regression*. Hierarchical regression is typically conducted in situations like the one we've been discussing. Given a collection of predictors we may want to know how much additional variance in the outcome variable is explained by a variable of interest, above and beyond what is explained by one or more other predictors. When reporting the results of a hierarchical regression, we typically show the correlation matrix of all variables (outcome and predictors) in the analysis. We also provide a regression table that shows the variance explained by each model.

APA Reporting

We obtained scores on the Graduate Record Examination (*GRE*), scores on the scale of typical intellectual engagement (*TIE*), and a measure of success in graduate school (*success*) from 27 graduate students in the University of Didd Department of Psychology. *GRE* was obtained at the time students applied to the program, *TIE* was obtained in their first semester, and *success* was measured in their final year in graduate school. Table 16.3 shows the correlations between the three variables.

TABLE 16.3

Correlations between the outcome and predictor variables

Variables	*success*	*GRE*	*TIE*
success	–	.68	.55
GRE		–	.47
TIE			–

Note: *success* = success in graduate school; *GRE* = scores on the Graduate Record Exam; *TIE* = scores on the Typical Intellectual Engagement Scale.

Table 16.4 shows the result of a hierarchical regression in which *GRE* was entered in step 1 and *TIE* was added on step 2. In step 1, $R^2 = .462$, 95% CI [.19, .73]; when *TIE* was added to the model, we observed $\Delta R^2 = .066$, approximate 95% CI [0, .22].

TABLE 16.4

Hierarchical regression: Relating *GRE* and *TIE* to *success*

Variables	ΔR^2	*b*	s_b	β
Prediction of success in graduate school (*success*) ($N = 27$)				
Step 1	.462			
GRE		0.291	0.063	0.680
Step 2	.066			
GRE		0.233	0.068	0.543
TIE		0.029	0.016	0.292

Note: *success* = success in graduate school; *GRE* = scores on the Graduate Record Exam; *TIE* = scores on the Typical Intellectual Engagement Scale.

(Tables 16.3 and 16.4 are formatted according to the *APA Publication Manual*, sixth edition, pages 136 and 145. You can consult the manual for further details.)

Our analysis of this small data set has so far dealt only with computing and reporting statistics. We have not yet addressed the question of practical significance. As noted many times before, statistics can only guide a discussion about practical significance, they cannot by themselves tell us what conclusion to draw. We will return to the question of practical significance at the end of the next section, after we've seen two other approaches to assessing the model.

CONFIDENCE INTERVALS FOR $\hat{y}$ AND PREDICTION INTERVALS FOR y_{NEXT}

The preceding sections focused on assessing the fit of the regression model to the data using R^2 and ΔR^2. In some areas of psychology there is a great deal of emphasis on assessing ΔR^2, as in our example of the admissions committee. If adding a variable to the regression model produces a statistically significant ΔR^2, we can be led to believe that this confers meaning on the added variable(s). However, a focus on ΔR^2 can draw attention away from the implications at the level of individuals. To judge the meaningfulness of the ΔR^2 produced by adding *TIE* to our regression equation, we might consider what the model says about two individuals who have the same *GRE* scores but different *TIE* scores. We can address such questions through confidence intervals and prediction intervals around $\hat{y}$.

LEARNING CHECK 8

1. State whether the following statements true or false.

   (a) ΔR^2 estimates $P^2_{smaller} - P^2_{larger}$.

   (b) Assuming $df_{change} = 1$, a confidence interval around $\Delta R^2 = 0.3$ will be narrower when $df_{error} = 200$ than when $df_{error} = 20$.

   (c) $df_{error} = n - 1 - 1$.

   (d) $df_{change} = n - k_{smaller} - 1$.

2. For each of the following circumstances, compute ΔR^2 and F_{change} and report the associated p-value as either $p > .05$, $p < .05$, or $p < .01$:

   (a) $n = 100, k_{smaller} = 4, k_{larger} = 5, R^2_{smaller} = .4, R^2_{larger} = .5$

   (b) $n = 25, k_{smaller} = 4, k_{larger} = 5, R^2_{smaller} = .4, R^2_{larger} = .5$

   (c) $n = 100, k_{smaller} = 3, k_{larger} = 5, R^2_{smaller} = .4, R^2_{larger} = .45$

   (d) $n = 21, k_{smaller} = 3, k_{larger} = 5, R^2_{smaller} = .4, R^2_{larger} = .45$

   (e) $n = 1000, k_{smaller} = 4, k_{larger} = 5, R^2_{smaller} = .4, R^2_{larger} = .41$

Answers

1. (a) False.

   (b) True. Confidence intervals get narrower as sample size increases.

   (c) False.

   (d) False.

2. (a) $\Delta R^2 = .10, F_{change} = 18.8, p < .01$.

   (b) $\Delta R^2 = .10, F_{change} = 3.80, p > .05$.

   (c) $\Delta R^2 = .05, F_{change} = 4.27, p < .05$.

   (d) $\Delta R^2 = .05, F_{change} = 0.68, p > .05$.

   (e) $\Delta R^2 = .01, F_{change} = 16.85, p < .01$.

Remember that in simple linear regression we computed a confidence interval around $\hat{y}$ for a given value of the predictor variable as

$$CI = \hat{y} \pm t_{\alpha/2}(s_{\hat{y}}),$$

and we computed a prediction interval around $\hat{y}$ as

$$PI = \hat{y} \pm t_{\alpha/2}(s_{\hat{y}_{next}}).$$

These two cases generalize to multiple regression. That is, we can compute a confidence interval and prediction interval around $\hat{y}$ for any combination of values of our predictors $(x_1, x_2, \ldots, x_k)$.

Computing these intervals by hand is prohibitively complex. Fortunately, we can use SPSS exactly as described in Appendix 14.2 to compute these intervals when there are two or more predictors. Figure 16.17 shows the Linear Regression: Save dialog that was previously shown in Figure 14.A2.3. In the Predicted Values region of the dialog we've requested SPSS to compute $\hat{y}$ (Unstandardized) and $s_{\hat{y}}$ (S.E. of mean predictions). In the Prediction Intervals region of the dialog, we asked for 95% confidence intervals (Mean) around all $\hat{y}$ and 95% prediction intervals (Individual) around all $\hat{y}$.

Figure 16.18 shows that the results of the analysis have been saved in the data file. The values of $\hat{y}$ and $s_{\hat{y}}$ are shown in columns PRE_1 and SEP_1, respectively. The lower and upper limits of the confidence intervals around the values of $\hat{y}$ are shown in columns LMCI_1 and UMCI_1, respectively. Finally, the lower and upper limits of the prediction intervals around the values of $\hat{y}$ are shown in columns LICI_1 and UICI_1, respectively.

Let's now turn our attention to the last three rows (28 to 30) of Figure 16.18 where we see values of *GRE* and *TIE* with no corresponding values for *success*, and yet values have been computed for $\hat{y}$ and $s_{\hat{y}}$ as well as the confidence and prediction intervals. When SPSS finds values for predictor variables with no corresponding outcome variable, it will ignore these values when computing the regression equation. Once the regression equation has been computed, however, SPSS will use it to compute the quantities specified in the Linear Regression: Save dialog for all values of the predictors, regardless of whether the predictors have a corresponding value of the outcome variable.

In row 28 the value of *GRE* is 157, which is one standard deviation above the mean *GRE* score, and *TIE* is 120, which is the mean *TIE* score. In row 29 the value of *GRE* is set to the mean *GRE* score, and *TIE* = 150 is one standard deviation above the mean *TIE* score. In row 30 both *GRE* and *TIE* scores are one standard deviation above their respective means.

Let's consider the values of *GRE* and *TIE* in row 28, which are 157 and 120, respectively. When these values are inserted into our regression equation we find that

$$\hat{y} = 0.606 + 0.233(157) + 0.029(120) = 40.67,$$

which is shown (more precisely) as 40.63 in Figure 16.18. For *GRE* = 157 and *TIE* = 120, the 95% confidence interval around $\hat{y}$ is [39.33, 41.93]. And the 95% prediction interval around $\hat{y}$ is [36.03, 45.22]. As we will see in the next section, numbers such as these can contribute to our reasoning about our research question. For example, these intervals allow us to consider what the model says about two individuals who have the same *GRE* scores but different *TIE* scores.

FIGURE 16.17   ■   The Save Dialog

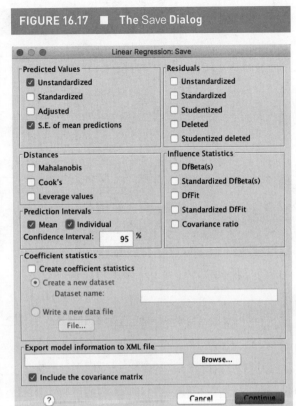

The Linear Regression: Save dialog that provides the option to compute $\hat{y}$ and its estimated standard error for all predictors in a regression model. It also allows one to compute a confidence interval and prediction interval around $\hat{y}$.

DISCUSSION OF OUR EXAMPLE: TO ADD *TIE* OR NOT TO ADD *TIE*

We've now introduced some important technical aspects of regression models using the example of predicting *success* in graduate school from *GRE* and *TIE* scores. The two-predictor model we've been discussing was shown to explain 53% of the variability in our outcome variable *success*. We saw that $R^2 = .53$ corresponds to $F_{obs} = 13.44$, $p < .001$. Our observed $R^2 = .53$ represents about a 7-percentage-point increase in explained variance over a model with only *GRE* as a predictor, $R^2 = .46$. The change in explained variance reported by SPSS was $\Delta R^2 = .066$, which corresponds to $F_{change} = 3.38$, $p = .078$. By normal conventions, this increase in explained variance is not considered statistically significant, although many psychologists would be inclined to say that there is a trend toward significance. In this section, we will try to assess the practical significance of these results.

We will imagine two sides to this discussion. One will argue that adding *TIE* scores to the application package is a good idea, and the other will argue that it isn't. You might imagine this as a debate between two individuals on the selection committee.

FIGURE 16.18 ■ Confidence Intervals and Prediction Intervals

	Success	GRE	TIE	PRE_1	SEP_1	LMCI_1	UMCI_1	LICI_1	UICI_1
1	36	141	99	36.30	.681	34.89	37.70	31.67	40.92
2	38	143	153	38.33	.949	36.37	40.29	33.51	43.15
3	34	144	131	37.92	.657	36.57	39.28	33.31	42.53
4	39	150	142	39.64	.538	38.53	40.75	35.09	44.18
5	36	141	98	36.27	.682	34.86	37.67	31.64	40.89
6	45	161	185	43.45	1.032	41.32	45.58	38.55	48.34
7	38	156	109	40.08	.657	38.72	41.43	35.47	44.69
8	38	147	131	38.62	.523	37.54	39.70	34.08	43.16
9	37	153	87	38.74	.763	37.16	40.31	34.06	43.42
10	33	150	98	38.36	.538	37.25	39.47	33.82	42.91
11	44	161	185	43.45	1.032	41.32	45.58	38.55	48.34
12	37	139	109	36.12	.798	34.47	37.77	31.42	40.83
13	40	147	109	37.98	.456	37.04	38.92	33.48	42.49
14	40	144	120	37.60	.579	36.41	38.80	33.04	42.17
15	41	153	98	39.06	.631	37.76	40.36	34.46	43.65
16	41	150	142	39.64	.538	38.53	40.75	35.09	44.18
17	38	147	55	36.41	1.035	34.28	38.55	31.51	41.31
18	36	150	120	39.00	.411	38.15	39.85	34.51	43.49
19	38	141	109	36.59	.688	35.17	38.01	31.96	41.22
20	38	144	98	36.96	.569	35.79	38.14	32.40	41.53
21	42	150	120	39.00	.411	38.15	39.85	34.51	43.49
22	39	147	77	37.05	.737	35.53	38.57	32.39	41.71
23	45	164	142	42.90	.941	40.96	44.84	38.08	47.71
24	38	153	131	40.02	.456	39.08	40.96	35.51	44.52
25	41	157	163	41.88	.745	40.34	43.42	37.21	46.55
26	42	161	120	41.56	.853	39.80	43.32	36.81	46.31
27	39	156	109	40.08	.657	38.72	41.43	35.47	44.69
28	.	157	120	40.63	.629	39.33	41.93	36.03	45.22
29	.	150	150	39.87	.628	38.58	41.17	35.28	44.47
30	.	157	150	41.50	.640	40.18	42.82	36.90	46.10

Confidence intervals and prediction intervals around $\hat{y}$. Columns PRE_1, SEP_1, LMCI_1, UMCI_1, LICI_1, and UICI_1 show, respectively, $\hat{y}$, $s_{\hat{y}}$, the lower and upper limits of $\hat{y} \pm t_{\alpha/2}(s_{\hat{y}})$, and the lower and upper limits of $\hat{y} \pm t_{\alpha/2} S_{\hat{y}_{next}}$.

Pros

The pro side to this question might be that *TIE* scores increased the explained variance in *success* by about 7 percentage points ($\Delta R^2 = .066$). Although this increase is not statistically significant at the .05 level, many researchers would say that the $F_{change} = 3.38$ associated with ΔR^2 is marginally significant, or that it shows a trend toward significance. The reasoning is that one can't say that $p = .078$ means no increase in explained variance. This is not completely unreasonable because .05 is not a magic number. Also, one might argue that many research papers find $\Delta R^2 = .01$ to be statistically significant, using much larger samples, of course. So, if there are those who argue that $\Delta R^2 = .01$ is meaningful to some researchers, then $\Delta R^2 = .066$ must be 6.6 times better. Of course, the pro side of the issue is a bit uncomfortable with relying so heavily on statistical significance because many authors have been pointing out flaws in significance tests lately (e.g., Cumming, 2012; Gigerenzer, 2004; Kline, 2013). But, with nothing else to rely on, in the end, the pro side favors requiring *TIE* scores as part of the application package because ΔR^2 shows a trend toward statistical significance.

Cons

The con side of this argument does not dispute that $\Delta R^2 = .066$ could be meaningful in some contexts. So, if the question is simply whether adding *TIE* scores increases explained variance, the answer is that it does. However, the estimate of the true increase in the population is rather imprecise, 95% CI [0, .22].

Those who have reservations about requiring *TIE* scores as part of the application package are less inclined to make a decision based on statistical significance or a trend toward statistical significance. Rather, the focus is more on how much predictions are improved in absolute terms. They note that a simple linear regression that predicts *success* from *GRE* alone yields the regression equation $\hat{y} = -4.67 + 0.291(GRE)$. For the mean *GRE* score of 150, this equation yields predicted success of $\hat{y} = -4.67 + 0.291(150) \approx 39$. The question then becomes how this prediction would be improved for an applicant that had the same *GRE* score (150) but a very good *TIE* score that is one standard deviation above the mean; i.e., *TIE* = 150. The two-predictor regression equation [$\hat{y} = 0.61 + 0.23(GRE) + 0.03(TIE)$] predicts that this student would score about 39.87 on *success* (see row 29 in Figure 16.18). The difference between these two predicted values of *success* is 39.87 − 39.00 = 0.87. Given that the standard deviation of the *success* scores is approximately 3, the best estimate is that a very good *TIE* score improves the predicted *success* by less than one-third of a standard deviation, in this case. In addition, the 95% confidence interval for $\hat{y} = 39.87$ for the two-predictor model [38.58, 41.47] shows that this modest improvement over the prediction of the simple linear regression ($\hat{y} = 39$) is quite imprecise.

At this point, the pro side might interject that the imprecision in this estimate means that the improvement could actually be quite a bit better than 39.87 − 39 = 0.87. They note that the upper limit of the confidence interval around $\hat{y} = 39.87$ from the two-predictor model is 41.47. Maybe if the study had been done with more participants, we'd have found a more precise estimate that is much closer to 41.47 (the upper limit of the interval) than 39.87 (the middle of the interval). This would represent an improvement in predicted *success* of about 41.47 − 39 = 2.47, which is about five-sixths of a standard deviation above the mean *success* score. The con side agrees that this is reasonable but points out that 38.58 (the lower limit of the interval) is just as plausible as 41.47.

The con side maintains that adding *TIE* scores provides a marginal improvement at best over the prediction of *success* from *GRE* alone. If there were no cost to requiring *TIE* scores as part of the application package, then there would be little harm in doing so. However, there is a potential danger of requiring *TIE* scores from applicants. The University of Didd is relatively small and in competition for excellent applicants with better known and more prestigious institutions such as Mount Crumpit University. Each additional component required in the application package is likely to discourage some students from applying. With this in mind, the potential marginal gains associated with better predictions about *success* are outweighed by the potential costs of requiring an additional component to the application package.

While reading this brief discussion, you may well have thought of interesting points of your own that favor one or the other conclusion. Your points may draw on data that weren't discussed above. This is how statistics are supposed to work in science. They lend support to one view or another, but in the end people must weigh these arguments and make a decision. I know what my decision would be. But my mind is open and your arguments might change my decision.

LEARNING CHECK 9

1. State whether the following statements are true or false.

   (a) A confidence interval around $\hat{y}$ is usually narrower than a prediction interval around $\hat{y}$.

   (b) Figure 16.18 shows that for $GRE = 150$ and $TIE = 142$, $s_{\hat{y}} = 0.538$.

   (c) Figure 16.18 shows that the confidence interval around $\hat{y}$ for $GRE = 147$ and $TIE = 109$ is [33.48, 42.49].

   (d) Figure 16.18 shows that the prediction interval around $\hat{y}$ for $GRE = 147$ and $TIE = 77$ is [32.39, 41.71].

   (e) Figure 16.18 shows that for $GRE = 150$ and $TIE = 120$, $s_{\hat{y}_{next}} = 0.411$.

2. Figure 16.18 shows that $s_{\hat{y}} = 0.538$ for $GRE = 150$ and $TIE = 98$ (see row 10), and $s_{\hat{y}} = 0.411$ for $GRE = 150$ and $TIE = 120$ (see row 18). Explain why these two estimated standard errors are different.

Answers

1. (a) True. (b) True. (c) False. (d) True. (e) False. SPSS doesn't report $S_{\hat{y}_{next}} = 0.411$.

2. As with simple linear regression, $s_{\hat{y}}$ increases for scores on the predictor variable that are further from the mean. In this case, both students (row 10 and row 18) scored 150 on GRE, which is the mean GRE score. However, the student in row 10 scored well below the mean on TIE, whereas the student in row 18 scored at the mean. Because of this, $s_{\hat{y}}$ for the student in row 10 is greater than $s_{\hat{y}}$ for the student in row 18.

SUMMARY

A linear regression analysis that has one outcome variable and two or more predictor variables is called *multiple linear regression*, or simply *multiple regression*. In a population, the multiple regression equation

$$E(y|x_1, x_2, \ldots, x_k) = \alpha + \beta_1(x_1) + \beta_2(x_2) + \ldots + \beta_k(x_k)$$

defines the expected value of y for specific levels of the k predictor variables, $x_1, x_2, \ldots, x_k$. In this equation, α and $\beta_1, \beta_2, \ldots, \beta_k$ are *regression coefficients*. The regression equation defines a *plane* when there are two predictors, and a *hyperplane* when there are three or more predictors. The regression equation computed from a sample of scores also defines a plane or a hyperplane:

$$\hat{y} = a + b_1(x_1) + b_2(x_2) + \ldots b_k(x_k).$$

The coefficients in the sample regression equation $(a, b_1, b_2, \ldots, b_k)$ estimate the corresponding coefficients in the population regression equation $(\alpha, \beta_1, \beta_2, \ldots, \beta_k)$. In a population, each y can be described as

$$y_i = E(y|x_1, x_2, \ldots, x_k) + \varepsilon_i.$$

In a sample, each y can be described as

$$y_i = \hat{y} + \varepsilon_i.$$

The regression equation is the best fit to scores on the outcome variable (y) in the *least squares* sense. This means that there is no other set of regression coefficients $(a, b_1, b_2, \ldots, b_k,$ or $\alpha, \beta_1, \beta_2, \ldots, \beta_k)$ that would produce a smaller sum of squared deviations from $E(y|x_1, x_2, \ldots, x_k)$ or $\hat{y}$. For a sample, we define

$$ss_{total} = \sum(y - m_y)^2,$$

$$ss_{regression} = \sum(\hat{y} - m_y)^2, \text{ and}$$

$$ss_{error} = \sum(y - \hat{y})^2 = \sum\varepsilon^2,$$

and there are corresponding equations for populations.

The proportion of variability in y explained by the predictor variables, $x_1, x_2, \ldots, x_k$, is defined as

$$R^2 = \frac{ss_{regression}}{ss_{total}} = \frac{ss_{total} - ss_{error}}{ss_{total}}$$

just as in simple linear regression. R^2 is a *positively biased* estimator of P^2, which is the proportion of variability in y explained by $x_1, x_2, \ldots, x_k$ in a population. One correction for this bias is *adjusted* R^2, which is computed as follows:

$$R_{\text{adj}}^2 = 1 - (1 - R^2)\frac{n-1}{n-k-1}.$$

Computing a confidence interval for R^2 is complicated because the sampling distribution of R^2 can't be written as a simple mathematical formula. Therefore, confidence intervals for R^2 are often very approximate. A very good approximation is given by the `ci.R2` function in the MBESS package for **R** (Kelley, 2007).

The standard error of estimate, defined as

$$\sigma_{\text{est}} = \sqrt{\frac{ss_{\text{error}}}{N}} \text{ or}$$

$$\sigma_{\text{est}} = \sqrt{(1 - P^2)\sigma_y^2},$$

is an important quantity in regression analysis. It is a measure of the variability in prediction errors and plays an important role in the construction of confidence intervals around the regression coefficients ($b_1, b_2, \ldots, b_k$) and $\hat{y}$. σ_{est} is estimated by

$$s_{\text{est}} = \sqrt{\frac{ss_{\text{error}}}{n-k-1}}$$

or equivalently

$$s_{\text{est}} = \sqrt{(1 - R^2)s_y^2\frac{n-1}{n-k-1}}.$$

Historically, researchers don't attempt to estimate P^2. Rather, they test the null hypothesis that the multiple correlation coefficient (P) is equal to 0. To do this, they make use of the *central* F-*distribution*. The observed value of F is computed as

$$F_{\text{obs}} = \frac{R^2}{1 - R^2}\frac{df_{\text{error}}}{df_{\text{regression}}}$$

where $df_{\text{regression}} = k$ is the number of predictors, and $df_{\text{error}} = n - k - 1$. The sampling distribution of F assumes $P = 0$ and depends on $df_{\text{regression}}$ and df_{error}. (The sampling distribution of F is a *central* F-*distribution* when $P^2 = 0$ and a *noncentral* F-*distribution* when $P^2 \neq 0$.) If less than 5% of a central F-distribution would lie above F_{obs} when $P^2 = 0$, then we declare F_{obs} to be statistically significant.

For inferences about R^2 to be valid, we must assume that a plane or hyperplane correctly describes its

association between the predictors and y. Furthermore, we assume that all conditional distributions about $E(y|x_1, x_2, \ldots, x_k)$ are normal and homoscedastic. As a consequence we assume that the distribution of prediction errors (ε) are normal.

A regression equation is *unstable* if small changes in the predictors cause large changes in the regression coefficients. This instability occurs when the predictors are *collinear*, or *nonindependent*, which means that they are too highly correlated with each other. *Multicollinearity* means that one predictor can be almost perfectly predicted from the other predictors.

The total degrees of freedom for the outcome variable y is the sum of $df_{\text{regression}}$ and df_{error}. $df_{\text{regression}}$ is the number of independent predictors in the model, which explains $R^2*100\%$ of the variability in y. The residual degrees of freedom, df_{error}, is the number of independent predictors required to explain the remaining $(1 - R^2)*100\%$ of the variability in y.

When predictors are added to a regression model, explained variance will increase. We denote this increase in explained variance as

$$\Delta R^2 = R_{\text{larger}}^2 - R_{\text{smaller}}^2.$$

Approximate confidence intervals for ΔR^2 can be computed with computationally intensive bootstrapping procedures as described in Appendix 16.1. To test whether ΔR^2 is statistically significant, we can compute

$$F_{\text{obs}} = \frac{\Delta R^2}{1 - R_{\text{larger}}^2}\frac{df_{\text{error}}}{df_{\text{change}}}$$

where df_{error} is associated with the unexplained variance in the larger model, and df_{change} is the difference between the number of predictors in the larger and smaller models.

Confidence and prediction intervals for $\hat{y}$ and y_{next} can be computed in SPSS. These are important quantities because they keep us connected with the implications of the regression analysis at the level of individuals.

Prospective power analysis for R^2 and ΔR^2 can be conducted in G∗Power as described in Appendix 16.3 (available at study.sagepub.com/gurnsey). To avoid pointless, underpowered hierarchical regressions, a prospective power analysis should be conducted whenever there is sufficient information in the literature to provide estimates of P_{smaller}^2 and when one knows what value of ΔP^2 would be meaningful. A "shot-in-the-dark" approach could be justified if there is insufficient guidance in the literature about what these quantities might be.

KEY TERMS

adjusted R^2 (R^2_{adj})   414
ΔR^2 432
empirical sampling
  distribution 416

F-ratio 419
F-statistic 419
hyperplane 410
mean square 419

multiple correlation
  coefficient 420
omnibus F-statistic 419
P^2 414

EXERCISES

Definitions and Concepts

1. What does ε_i represent in this regression equation $y_i = a + b_1(x_1) + b_2(x_2) + \varepsilon_i$?

2. What is the relationship between $E(y|x_1, x_2) = \alpha + \beta_1(x_1) + \beta_2(x_2)$ and $\hat{y} = a + b_1(x_1) + b_2(x_2)$?

3. If $E(y|x_1, x_2, x_3) = \alpha + \beta_1(x_1) + \beta_2(x_2) + \beta_3(x_3)$, what is k?

4. What is a hyperplane?

5. Put into words what this means: $ss_{regression} = \sum (\hat{y} - m_y)^2$.

6. Fill in the blanks with the terms ss_{error}, $ss_{regression}$, and ss_{total}: ____ = ____ + ____

7. If $R^2 = .6$ and $ss_{total} = 100$, what are $ss_{regression}$ and ss_{error}?

8. What values are on the diagonal of a correlation matrix, and why is this?

9. Why is it difficult to compute an exact confidence interval around R^2?

10. Put into words the meaning of

$$\sigma_{est} = \sqrt{(1 - P^2)\sigma_y^2}.$$

11. What is an F-ratio?

12. Put into words the meaning of $F_{critical}$.

13. What does homoscedasticity mean in multiple regression?

True or False

State whether the following statements are true or false.

14. If $\hat{y} = 2 + 3(x_1) - 4(x_2)$, then $\hat{y} = -3$ when $x_1 = 5$ and $x_2 = 5$.

15. Sometimes $R^2 < r^2$.

16. It is impossible to visualize the hyperplane corresponding to a three-predictor regression model.

17. $R^2_{adj} < R^2$ in all cases.

18. $F_{obs} = \dfrac{ss_{regression}/df_{error}}{ss_{error}/df_{regression}}$.

19. $F_{obs} = \dfrac{R^2}{1 - R^2} \dfrac{df_{error}}{df_{regression}}$.

20. $ms_{error} = ss_{error}/(n-k-1)$.

21. $R^2 = .8$, $n = 31$, and $k = 20$; therefore, $F_{obs} = 4$.

22. The central F-distribution is computed assuming $P \leq 1$.

23. If $df_{regression} = 10$, $df_{error} = 40$, and $\alpha = .01$, then $F_{critical} = 2.1$.

24. When the variance of the conditional distributions changes with values of the predictor variables, the conditional distributions are heteroscedastic.

Calculations

25. Compute R^2_{adj}, s_{est}, and the approximate 95% confidence interval for R^2 (using ci.R2 in **R**) for each of the following conditions:

    (a) $R^2 = .1$, $k = 3$, $n = 100$, $s_y^2 = 20$;

    (b) $R^2 = .2$, $k = 4$, $n = 50$, $s_y^2 = 6$;

    (c) $R^2 = .3$, $k = 6$, $n = 96$, $s_y^2 = 14$;

    (d) $R^2 = .4$, $k = 7$, $n = 114$, $s_y^2 = 25$; and

    (e) $R^2 = .5$, $k = 8$, $n = 129$, $s_y^2 = 4$.

26. Find the critical value of F for $\alpha = .05$, given the following degrees of freedom for $df_{regression}$ and df_{error}, respectively:

    (a) 1, 111;

    (b) 12, 51;

    (c) 7, 60;

    (d) 4, 120; and

    (e) 15, 2.

27. For each of the following conditions, compute F_{obs} and state whether it is statistically significant at the $\alpha = .05$ and $\alpha = .01$ levels:

(a) $R^2 = .1, k = 3, n = 32$;

(b) $R^2 = .1, k = 3, n = 304$;

(c) $R^2 = .9, k = 1, n = 5$;

(d) $R^2 = .9, k = 1, n = 17$; and

(e) $R^2 = .5, k = 4, n = 16$.

28. It is remarkable how little information is required to compute a hierarchical regression by hand. Complete the following table for each of these situations.

(a) There are three variables, y, x_1, and x_2, each having $n = 81$ scores. The sum of squares for y is $ss_y = 200$. Variable x_1 was entered on the first step of the analysis and $ss_{\hat{y}}$ was 100. On step 2, x_2 was added and $ss_{\hat{y}} = 125$.

(b) There are six variables, y and x_1, x_2, ..., x_5, each having $n = 200$ scores. The sum of squares for y is $ss_y = 1000$. Variables x_1, x_2, ..., x_4 were entered on the first step of the analysis and $ss_{\hat{y}}$ was 500. On step 2, x_5 was added and $ss_{\hat{y}} = 600$.

(c) There are nine variables, y and x_1, x_2, ..., x_8, each having $n = 122$ scores. The sum of squares for y is $ss_y = 2000$. Variables x_1, x_2, ..., x_4 were entered on the first step of the analysis and $ss_{\hat{y}}$ was 400. On step 2, x_5, x_6, ..., x_8 were added and $ss_{\hat{y}} = 500$.

Model	R	R^2	R^2_{adj}	s_{est}	Change Statistics				
					ΔR^2	ΔF	df_1	df_2	p
1									
2									

Scenarios

29. Many factors predict insomnia, including caffeine and alcohol consumption, anxiety, depression, and stress. There are also physiological factors associated with insomnia, one of which is high-frequency heart-rate variability (HF-HRV). Low HF-HRV scores are associated with poor sleep (Gouin, Wenzel, Boucetta, O'Byrne, Salimi, & Dang-Vu, 2015).

Imagine that data for 16 individuals were collected for the variables *caffeine consumption*, *anxiety*, *depression*, and *HF-HRV*. These variables will be used to predict scores on the Ford Insomnia Response to Stress Test (*First*), which is a nine-item questionnaire whose scores range from 9 to 36. High scores mean greater vulnerability to insomnia. All scores were obtained in a 2-hour session at a university research lab.

The Insomnia.sav data file (available at study .sagepub.com/gurnsey) contains scores for the five variables described above. The question is whether *HF-HRV* adds anything to the prediction of *First* scores beyond what is predicted by *caffeine consumption*, *anxiety*, and *depression*. Conduct a hierarchical regression in SPSS and use the results to answer the following questions:

(a) Do the variables *caffeine consumption*, *anxiety*, and *depression* explain a statistically significant portion of the variability in *First* scores?

(b) Determine the approximate 95% confidence interval around the R^2 for the model involving *caffeine consumption*, *anxiety*, and *depression* using the `ci.R2` function in MBESS.

(c) Does the addition of *HF-HRV* produce a statistically significant increase in explained variance in *First* scores?

(d) Assume that P^2 and ΔP^2 are the values of R^2 and ΔR^2 found in the second step of your hierarchical regression. What sample size would be required to achieve power = .8 for this ΔP^2 when $\alpha = .05$? (This question requires you to have read Appendix 1.3, available at study .sagepub.com/gurnsey.)

(e) How would you answer someone who says these results show that having low *HF-HRV* produces an increased likelihood of insomnia?

(f) Determine the mean and standard deviation for each variable in the analysis. Use the second model (with four predictors) to determine the 95% confidence interval $\hat{y}$ and the 95% prediction interval around for y_{next} for

   (i) an individual who scores at the mean of the four predictors and

   (ii) an individual who scores one standard deviation below the mean of *HF-HRV* and at the mean of *caffeine consumption*, *anxiety*, and *depression*.

30. There are many variables that affect how children perform in school. These include nutrition, genetics, parental investment, and home stability, among many others. Researchers would like to understand the associations between these variables because if the causal connections are understood, there is hope for interventions that could improve school outcomes for children. Two factors known to predict children's school performance are their parents' level of education and parents' school involvement (e.g., attending parent/teacher nights, participation on school committees, and volunteering time to go on school outings). Stevenson and Baker (1987) investigated the relative contributions of mothers' education and mothers' school involvement on their children's success in school. Mother's education was scored on a 7-point scale. Mother's involvement and child's school performance were both scored on 5-point scales. The latter two measures were based on questionnaires that were completed by each child's teacher.

The SchoolPerformance.sav data file (available at study.sagepub.com/gurnsey) includes three columns of 179 numbers representing child's school performance (*SchoolPer*), mother's level of education (*MEd*), and mother's involvement (*MInvolve*). Conduct two hierarchical regressions. For the first hierarchical regression, let *SchoolPer* be the outcome variable, enter *MEd* on step 1, and then add *MInvolve* on step 2. For the second hierarchical regression, let *SchoolPer* be the outcome variable, enter *MInvolve* on step 1, and add *MEd* on step 2. Use SPSS to conduct these analyses and answer the following questions.

First Analysis

(a) Does *MEd* explain a statistically significant portion of the variability in *SchoolPer* scores on step 1? (Report R^2, F, and p.) Use the `ci.R2` function in MBESS to determine the approximate 95% confidence interval around the R^2 for the model involving only *MEd*.

(b) Assume that P^2 and ΔP^2 are the values of R^2 and ΔR^2 found in the second step of your hierarchical regression. What sample size would be required to achieve power = .8 for this ΔP^2 when $\alpha = .05$? (This question requires you to have read Appendix 1.3, available at study.sagepub.com/gurnsey.)

(c) Determine the mean and standard deviation for each variable in the analysis. Use the second model (with two predictors) to determine the 95% confidence interval around $\hat{y}$ and the 95% prediction interval for y_{next} for

   (i) children whose mothers score at the mean of the two predictors and

   (ii) children whose mothers score at the mean of *MEd* and one standard deviation above the mean of *MInvolve*.

(d) We said little about the regression coefficients in the chapter. However, have a look at the regression coefficient on *MEd* for steps 1 and 2. How do they change? Think about what this change means.

Second Analysis

(a) Does *MInvolve* explain a statistically significant portion of the variability in *SchoolPer* scores on step 1? (Report R^2, F, and p.) Use the `ci.R2` function in MBESS to determine the approximate 95% confidence interval around the R^2 for the model involving only *MInvolve*.

(b) Assume that P^2 and ΔP^2 are the values of R^2 and ΔR^2 found in the second step of your hierarchical regression. What sample size would be required to achieve power = .8 for this ΔP^2 when $\alpha = .05$? (This question requires you to have read Appendix 1.3, available at study.sagepub.com/gurnsey.)

31. One of the pleasures of psychological research is that almost anyone can formulate research questions based on his or her experiences of everyday life. (Some might say that this is also one of the downsides of psychology.) Here's one that I've wondered about: What predicts musical aptitude? In particular, do seemingly unrelated factors such as digit span, reading fluency, and spatial memory relate to musical ability? There is some evidence for this in children. Anvari, Trainor, Woodside, and Levy (2002) found that in a sample of fifty 5-year-old children there were statistically significant correlations between pitch perception and (i) digit span ($r = .46$) and (ii) between pitch perception and reading scores on the Wechsler Intelligence Scale for Children–Revised (WISC-R) ($r = .45$).

    Let's imagine that we have a random sample of 100 participants drawn from a psychology department participant pool. We've measured *digit span*, *spatial reasoning* (e.g., fibonicci.com/spatial-reasoning/test/), and *reading fluency* for each participant. For these three measures, high numbers mean strong abilities. To measure *musical aptitude*, we present individuals with two-tone sequences that differ in one to eight of their tones, and we measure how many tones have to differ for the sequences to be distinguished. For this measure, high numbers mean weak ability. The Music.sav data file (available at study.sagepub.com/gurnsey) includes these data, with variables in columns *MusicScores*, *DigitSpan*,

    Spatial, and *Fluency*. Run a hierarchical regression with *DigitSpan* and *Fluency* entered on the first step and *Spatial* entered on the second.

    (a) Do *Fluency* and *DigitSpan* explain a statistically significant portion of the variability in *MusicScores* on step 1? (Report R^2, F, and p.) Use the `ci.R2` function in MBESS to determine the approximate 95% confidence interval around the R^2 for the model involving only *Fluency* and *DigitSpan*.

    (b) Assume that P^2 and ΔP^2 are the values of R^2 and ΔR^2 found in the second step of your hierarchical regression. What sample size would be required to achieve power = .8 for this ΔP^2 when $\alpha = .05$? (This question requires you to have read Appendix 1.3, available at study.sagepub.com/gurnsey.)

    (c) Determine the mean and standard deviation for each variable in the analysis. Use the second model (with three predictors) to determine the 95% confidence interval around $\hat{y}$ and the 95% prediction interval for y_{next} for

    (i) an individual who scores at the mean of the three predictors and

    (ii) an individual who scores one standard deviation above the mean of *Spatial* and at the mean of *Fluency* and *DigitSpan*.

    (d) Do high scores on the variable *Spatial* predict strong musical ability? Explain your answer.

APPENDIX 16.1: BOOTSTRAPPED CONFIDENCE INTERVALS FOR ΔR^2

This appendix covers bootstrapped confidence intervals for ΔR^2 and assumes that you've read the section on bootstrapping in Appendix 15.2 of Chapter 15.

Bootstrapped confidence intervals are computed when the sampling distribution of the statistic in question is either very complex or unknown. ΔR^2 provides an excellent example of this. We saw in Chapter 15 that bootstrapping procedures use the sample as a population and compute the empirical sampling distribution of the statistic through repeated sampling with replacement from this sample-*cum*-population. In the case of ΔR^2,

R^2_{smaller} and R^2_{larger} are computed for thousands of samples and we record $\Delta R^2 = R^2_{\text{larger}} - R^2_{\text{smaller}}$ for each sample.

The histogram in Figure 16.A1.1a shows the empirical sampling distribution of ΔR^2 for the data shown in Table 16.1 in the chapter. This histogram was derived from 10,000 bootstrapped samples. Figure 16.A1.1b shows the cumulative distribution function that plots the proportion of the distribution below each obtained ΔR^2. The 95% confidence interval is defined by the two values of ΔR^2 that contain the central 95% of the empirical sampling distribution.

FIGURE 16.A1.1 ■ Bootstrapping

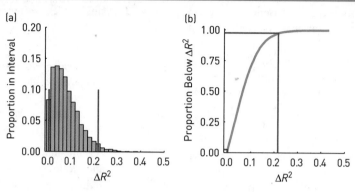

Bootstrapped confidence interval for ΔR^2 for the example given in this chapter. (a) A histogram of 10,000 ΔR^2 statistics is shown. (b) The cumulative distribution function corresponding to these 10,000 statistics is shown.

Bootstrapped confidence intervals for small values of ΔR^2 present a problem that we didn't meet in this chapter. The problem is that ΔR^2 can never be less than 0. Therefore, even if ΔP^2 in the population is 0, a computed value of ΔR^2 will almost never be 0. Therefore, almost no confidence intervals will capture $\Delta P^2 = 0$. Algina et al. (2008) showed that setting the lower limit of confidence intervals to 0 whenever ΔR^2 was not statistically different from 0 solves this problem. When ΔP^2 is very small, bootstrapped confidence intervals for ΔR^2 will capture ΔP^2 very close to 95% of the time using this simple rule. However, the accuracy of this method depends on the number of predictors in the full (larger) model. As the number of predictors increases, sample sizes must increase to achieve close to a 95% capture rate.

The capture rate of a procedure is referred to as *coverage*. So, if you are hoping to capture a parameter 95% of the time but your procedure captures it 91% of the time, we would say that the *nominal coverage* is 95% and the *empirical coverage* is 91%. Algina et al. (2008) considered bootstrapping performance to be acceptable if the empirical coverage was between 92.5% and 97.5% when the nominal coverage was 95%. If the number of predictors in the full model was four or fewer, then acceptable coverage required at least 50 scores in the sample. (Actually, 50 was the smallest sample size they examined, so acceptable coverage may be obtained with even smaller samples.) If there were five or six predictors, then a sample size of at least 150 was required for adequate coverage. If there were seven to nine predictors, then 250 scores were required to achieve acceptable coverage.

APPLYING MULTIPLE REGRESSION

INTRODUCTION

The basic concepts of multiple regression analysis were introduced in Chapter 16. We limited our attention to cases in which all variables were random variables, and we continue the same focus in this chapter. Our emphasis in Chapter 16 was on how well a regression model explains the variability in the outcome variable. This led to a discussion of R^2, how to estimate it, and how to test it for statistical significance. Furthermore, we introduced the concept of hierarchical regression in which predictor variables are entered into the regression analysis in blocks. This led to a discussion of ΔR^2, how to estimate it, and how to test it for statistical significance. Finally, we looked at the predictive power of a regression equation at the mean level [estimating $E(y|x_1, x_2, \ldots, x_k)$ with confidence intervals] and at the level of individuals using prediction intervals.

In this chapter, we move from model-level concerns in regression to a focus on the regression coefficients themselves. There are four main topics. The first concerns the nature of regression coefficients in multiple regression and how they differ from those in simple linear regression. This discussion will lead us to constructing confidence intervals around regression coefficients.

The second topic concerns *statistical control*. In Chapter 13 we pointed out that the correlation between random variables does not imply a causal association between them, because a third (confounding) variable may explain this association. We will now see how multiple regression is used to statistically control for confounding variables that may explain a spurious correlation.

The third topic concerns *mediation* analysis, which is an application of multiple regression that can help us reason about hypothetical causal associations between variables. With mediation analysis, we can assess a hypothetical chain of causal connections of the sort $x \rightarrow med \rightarrow y$. The model holds that variable x has a causal effect on variable y through an intervening variable *med*. Although mediation analysis can't prove such a causal link, it can be part of a larger set of considerations that support the plausibility of the link.

Finally, another application of multiple regression analysis, called *moderation* analysis, assesses the degree to which an association between two variables (x and y) changes as a function of individuals' scores on a third variable, denoted *mod*, for moderation. For example, the association between variables x and y might be different for males and females. In this example, *sex* would be the moderating variable. We will show how to conduct mediation and moderation analyses in SPSS using a very powerful and freely available suite of add-on routines.

THE REGRESSION COEFFICIENTS

What the Regression Coefficients Represent

At the beginning of Chapter 16 we said that SPSS would take over the hard work of computing regression coefficients for us. Although we won't have to compute them by hand, it is still important to understand what regression coefficients represent. In this section, we will explain the nature of regression coefficients in two ways.

To start this discussion, we should note that the regression coefficient associated with a given variable is not fixed; rather, it changes when other predictor variables are added to the model. Consider the following three regression equations that we encountered in Chapter 16. The first predicts *success* from *GRE* alone, the second predicts *success* from *TIE* alone, and the third predicts *success* from *GRE* and *TIE*:

$$\hat{y} = -4.67 + 0.29(GRE)$$

$$\hat{y} = 32.47 + 0.05(TIE)$$

$$\hat{y} = 0.61 + 0.23(GRE) + 0.03(TIE)$$

Notice that the regression coefficients on *GRE* and *TIE* in the simple regression equations are different from those in the multiple regression equation. This will always be the case when the predictors are correlated with each other, and raises the question of what information is conveyed by the regression coefficients in multiple regression.

Interpreting the regression slope in simple linear regression is straightforward; the regression slope (*b*) conveys the change in $\hat{y}$ with a given change in *x*. However, in the case of multiple regression, the interpretation is a little more complex. In multiple regression, a regression coefficient (b_i) conveys the change in $\hat{y}$ *with all other predictors held constant*.

Figure 17.1 illustrates what it means to hold other predictors constant. Three gray lines parallel to the *GRE*-axis are superimposed on the three-dimensional plot of the data and the regression plane. Each gray line shows how the predicted value of *success* changes as a function of *GRE* for three arbitrary values of *TIE*. The three values of *TIE* are 60, 120, and 180. Let's look at the regression equation for *TIE* = 60, which is

$$\hat{y} = 0.61 + 0.23(GRE) + 0.03(60).$$

As values of *GRE* change, *TIE* remains constant at 60. Therefore, this equation can be rewritten as

$$\hat{y} = 0.61 + 0.23(GRE) + 1.8.$$

When we add the two constants in the above equation (0.61 and 1.8), we obtain

$$\hat{y} = 2.41 + 0.23(GRE).$$

This equation describes the association between *success* and *GRE* when *TIE* is held constant at 60.

If we go through the same exercise with *TIE* held constant at 120 and 180, the regression equations relating *success* and *GRE* would be

$$\hat{y} = 4.21 + 0.23(GRE) \text{ and}$$

$$\hat{y} = 6.01 + 0.23(GRE),$$

respectively. Therefore, the slopes of these equations remain the same; only the intercepts change. This pattern is true of any value of *TIE*. Although they are not shown, similar lines would describe how the predicted value of *success* changes as a function of *TIE* for fixed values of *GRE*. These equations would all have the same slopes (0.03) and differ only in their intercepts.

The same general point holds true when there are more than two predictors. If there were three predictors, for example, then the regression coefficient b_2 would show how $\hat{y}$ changes with a change in x_2 when the values of x_1 and x_3 are held constant.

Partial Regression Coefficients (Two Predictors)

There is a slightly more complex but extremely useful explanation of regression coefficients that we will now consider. Regression coefficients are often called *partial regression coefficients* because they show the association between the outcome variable and one of the predictor variables when the influence of all other predictors is removed from this association.

Figure 17.2 illustrates this concept for the simple two-predictor model shown in Figure 17.1. Figure 17.2a shows the association between *TIE* and *success*. We saw this association previously in Figure 16.1b. For the current illustration, we will denote the predicted value of *y* (*success*) as $\hat{y}_{TIE}$, which means "*y* predicted from *TIE*."

We noted in Chapter 16 that *GRE* and *TIE* are correlated. This can be seen in Figure 17.2b, where *GRE* is plotted against *TIE*. Because *GRE* and *TIE* are correlated, we can form a regression equation to predict *GRE* from *TIE*. We will denote the predicted value of *GRE* as $\widehat{GRE}_{TIE}$, which means "*GRE* predicted from *TIE*." Figure 17.2b shows that $\widehat{GRE}_{TIE} = 136.94 + 0.11(TIE)$.

FIGURE 17.1 ■ Regression Coefficients

$$\hat{y} = 0.61 + 0.23(x_1) + 0.03(x_2)$$
$$R^2 = 0.53$$

Each gray line shows how the predicted value of *success* changes as a function of *GRE* for a fixed value of *TIE*. All lines have the same slope and differ only in their intercepts. In this illustration, the fixed values of *TIE* are 60, 120, and 180, but the same principle holds for any value of *TIE*.

FIGURE 17.2 ■ Partial Regression Coefficients

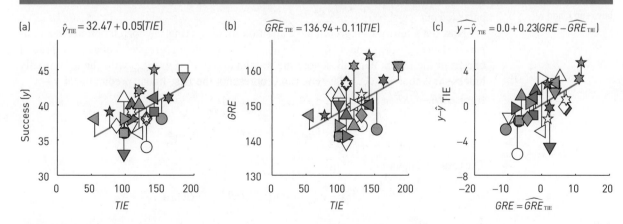

(a) Plot of *success* as a function of *TIE*, along with the regression line. (b) Plot of *GRE* as a function of *TIE*, along with the regression line. (c) Plot of prediction errors from (a) as a function of prediction errors from (b). This shows the association between *GRE* and *success* with *TIE* partialed out.

In Figures 17.2a and 17.2b, the vertical lines connecting the plot symbols to the regression lines represent errors of prediction, or residuals. In Figure 17.2a, these prediction errors represent the part of *success* that is not associated with *TIE*. In Figure 17.2b, these prediction errors represent the part of *GRE* that is not associated with *TIE*. We will denote these prediction errors as $y - \hat{y}_{TIE}$ and $GRE - \widehat{GRE}_{TIE}$.

Figure 17.2c plots these prediction errors (or residuals) against each other. The *x*-axis shows the errors of predicting *GRE* from *TIE*, and the *y*-axis shows the errors of predicting *y* from *TIE*. Any association between $y - \hat{y}_{TIE}$ and $GRE - \widehat{GRE}_{TIE}$ must be completely independent of *TIE* because these residuals are the parts of *y* and *GRE* that are not associated with *TIE*. In this case, there is clearly an association between *GRE* and *y* when the association with *TIE* has been removed from both variables.

When we remove the influence of one or more predictor variables from the association between a predictor and the outcome variable, we say the effects of these other predictors have been *partialed out* of the association. This terminology reflects the fact that we have taken something away from the two sets of scores and only a part of each score remains.

Table 17.1 explains this notion of partialing and how the data in Figure 17.2 were generated. The first column of Table 17.1 shows the plot symbols for easy reference. The next three columns show the *success* (*y*), *GRE*, and *TIE* scores that we've been considering since the beginning of Chapter 16. These are the unaltered scores that we start with. The fifth column shows $\hat{y}_{TIE}$, which is the best prediction of success that can be made for each value of *TIE*. These are points along the regression line in Figure 17.2a. The sixth column shows $\widehat{GRE}_{TIE}$, which is the best prediction of *GRE* that can be made for each value of *TIE*. These are points along the regression line in Figure 17.2b.

The last two columns in Table 17.1 show the errors of prediction (residuals) for $\hat{y}_{TIE}$ and $\widehat{GRE}_{TIE}$, which are denoted $y - \hat{y}_{TIE}$ and $GRE - \widehat{GRE}_{TIE}$, respectively. These are the residuals that are plotted in Figure 17.2c. They show what's left of *success* and *GRE* when their association with *TIE* has been removed. Because only part of *success* and *GRE* remain, we call them *partial scores*. The correlation between these partial scores is called the **partial correlation coefficient**.

A **partial correlation coefficient** is the correlation between x_i and *y* when the effects of all other predictors have been removed from the association between x_i and *y*.

Table 17.2 shows the correlation of the partial scores with each other and with *TIE*. The correlation between $y - \hat{y}_{TIE}$ and *TIE* is 0, as is the correlation between $GRE - \widehat{GRE}_{TIE}$ and *TIE*. These correlations show that there is no association between the partial scores and *TIE*. The partial correlation between $y - \hat{y}_{TIE}$ and $GRE - \widehat{GRE}_{TIE}$ is .57. This correlation shows the association between *success* and *GRE* when the influence of *TIE* has been partialed out.

Now, let's return to regression. In addition to computing the correlation between our partial scores $y - \hat{y}_{TIE}$ and $GRE - \widehat{GRE}_{TIE}$, we can create a regression equation to predict one from the other. We will denote the predicted value of $y - \hat{y}_{TIE}$ with the (admittedly hideous) symbol $\widehat{y - \hat{y}_{TIE}}$. The equation represents the best linear prediction of $y - \hat{y}_{TIE}$ from $GRE - \widehat{GRE}_{TIE}$; i.e.,

$$\widehat{y - \hat{y}_{TIE}} = a + b(GRE - \widehat{GRE}_{TIE}).$$

The resulting equation is

$$\widehat{y - \hat{y}_{TIE}} = 0 + 0.23(GRE - \widehat{GRE}_{TIE}).$$

The intercept of this equation is 0 and the regression coefficient is 0.23! Why the exclamation mark? Well, notice that 0.23 is the regression coefficient on *GRE* in the two-predictor regression equation:

$$\hat{y} = 0.61 + 0.23(GRE) + 0.03(TIE).$$

TABLE 17.1 ■ An Illustration of Partialing *TIE* Out of *Success* and *GRE*

	y	GRE[a]	TIE[a]	$\hat{y}_{TIE}$[b]	$\widehat{GRE}_{TIE}$[c]	$y-\hat{y}_{TIE}$	$GRE-\widehat{GRE}_{TIE}$
●	36	141	99	37.86	147.71	−1.86	−6.71
◐	38	143	153	40.80	153.59	−2.80	−10.59
○	34	144	131	39.60	151.19	−5.60	−7.19
■	39	150	142	40.20	152.39	−1.20	−2.39
◻	36	141	98	37.80	147.60	−1.80	−6.60
□	45	161	185	42.54	157.07	2.46	3.93
◆	38	156	109	38.40	148.80	−0.40	7.20
◇	38	147	131	39.60	151.19	−1.60	−4.19
◇	37	153	87	37.21	146.41	−0.21	6.59
▼	33	150	98	37.80	147.60	−4.80	2.40
▽	44	161	185	42.54	157.07	1.46	3.93
▽	37	139	109	38.40	148.80	−1.40	−9.80
▲	40	147	109	38.40	148.80	1.60	−1.80
△	40	144	120	39.00	150.00	1.00	−6.00
△	41	153	98	37.80	147.60	3.20	5.40
◀	41	150	142	40.20	152.39	0.80	−2.39
◁	38	147	55	35.46	142.92	2.54	4.08
◁	36	150	120	39.00	150.00	−3.00	0.00
▶	38	141	109	38.40	148.80	−0.40	−7.80
▷	38	144	98	37.80	147.60	0.20	−3.60
▷	42	150	120	39.00	150.00	3.00	0.00
★	39	147	77	36.66	145.32	2.34	1.68
☆	45	164	142	40.20	152.39	4.80	11.61
☆	38	153	131	39.60	151.19	−1.60	1.81
✶	41	157	163	41.34	154.67	−0.34	2.33
✬	42	161	120	39.00	150.00	3.00	11.00
✩	39	156	109	38.40	148.80	0.60	7.20
m	39	150	120	39.00	150.00	0.00	0.00
ss	232	1264	23348	69.10	276.38	162.92	987.46

[a] $r_{GRE.TIE}=0.4677$.

[b] $\hat{y}_{TIE}=32.47+0.05(TIE)$.

[c] $\widehat{GRE}_{TIE}=136.94+0.11(TIE)$.

TABLE 17.2 ■ A Correlation Matrix[a]			
	TIE	$y - \hat{y}_{TIE}$	$GRE - \widehat{GRE}_{TIE}$
TIE	–	.00	.00
$y - \hat{y}_{TIE}$	.00	–	.57
$GRE - \widehat{GRE}_{TIE}$	.00	.57	–

[a] *y* = success, *GRE* = graduate record exam scores, *TIE* = typical intellectual engagement scores.

Therefore, the regression coefficient on *GRE* reflects the association between *y* (*success*) and *GRE* with the influence of *TIE* on both removed (i.e., partialed out). This is why we call the regression coefficients *partial regression coefficients* in multiple regression.

Partial Regression Coefficients (More Than Two Predictors)

Partialing can be described in exactly the same way when there are more than two predictors. Consider, for example, a regression equation with four predictors (x_1, x_2, x_3, and x_4) and one outcome variable (*y*). To partial out the effect of variables x_1, x_3, and x_4 from the association between x_2 and *y*, we need two regression equations. The first regression equation predicts *y* from x_1, x_3, and x_4. We can express this prediction as $\hat{y}_{1,3,4}$. The second regression equation predicts x_2 from x_1, x_3, and x_4. We can express this prediction as $\hat{x}_{2.1,3,4}$. (The dot between 2 and 1,3,4 means "predicted from.") When we subtract $\hat{y}_{1,3,4}$ from *y* and $\hat{x}_{2.1,3,4}$ from x_2, we create partial scores that are not associated with variables x_1, x_3, and x_4. The coefficient on the regression equation that predicts $y - \hat{y}_{1,3,4}$ from $\hat{x}_{2.1,3,4}$ is a partial regression coefficient because it expresses the association between x_2 and *y* when their association with predictors x_1, x_3, and x_4 has been partialed out.

In the general case, we can talk about predictor *i* (e.g., 2) and predictors $j \neq i$ (e.g., 1, 3, 4). Therefore, a **partial regression coefficient** is the regression coefficient relating x_i to *y* when the effects of all other predictors ($x_{j \neq i}$) have been removed or partialed out of the association between x_i and *y*.

A **partial regression coefficient** is the regression coefficient relating x_i to *y* when the effects of all other predictors ($x_{j \neq i}$) have been removed from the association between x_i and *y*.

Confidence Intervals for Regression Coefficients

The sample regression coefficient b_i is our best estimate of β_i, the partial regression coefficient for variable x_i in the population. A confidence interval around b_i is defined in the usual way:

$$CI = b_i \pm t_{\alpha/2}(s_{b_i}).$$

This confidence interval has a familiar form because we've seen most of its components before. The only thing to be explained is the estimated standard error of b_i.

The Estimated Standard Error of the Partial Regression Coefficient

The general formula for s_{b_i} involves a small modification to the way s_b was computed in simple linear regression:

$$s_{b_i} = \frac{s_{est}}{\sqrt{ss_{x_i}(1 - R^2_{i.j \neq i})}}. \tag{17.1}$$

The numerator of equation 17.1 is the estimated standard error of estimate (s_{est}), which was defined in Chapter 16 (equation 16.6b) as

$$s_{est} = \sqrt{s_y^2(1 - R^2)\frac{n-1}{n-k-1}} \tag{17.2}$$

where s_y^2 is the variance of the outcome variable, R^2 is the proportion of variance in the outcome variable explained by the full regression model, *n* is the number of individuals in our study, and *k* is the number of predictor variables. In equation 17.1, ss_{x_i} is the sum of squares for the *i*th predictor variable, just as in the case of simple linear regression. The new term introduced in equation 17.1 is $R^2_{i,j \neq i}$, which we will now unpack. (As we go on, keep in mind that R^2 and $R^2_{i,j \neq i}$ are *not* the same thing.)

To understand $R^2_{i,j \neq i}$, we will use the terminology developed in the previous section. We can divide our predictor variables into the predictor of interest (x_i) and all the rest ($x_{j \neq i}$). We can compute a regression equation in which x_i is the outcome variable and $x_{j \neq i}$ are the predictors. $R^2_{i,j \neq i}$ is the proportion of variability in x_i explained by $x_{j \neq i}$. There is a different $R^2_{i,j \neq i}$ for each of our k predictors.

To understand the role of $R^2_{i,j \neq i}$ we will consider two extreme cases.

1. If $R^2_{i,j \neq i} = 0$, then *none* of the variability in x_i can be explained by the remaining predictors $x_{j \neq i}$. Equation 17.1 shows that in this case, $1 - R^2_{i,j \neq i} = 1$, and so s_{b_i} is just $s_{\text{est}} / \sqrt{ss_{x_i}}$.

2. If $R^2_{i,j \neq i} = 1$, then *all* of the variability in x_i can be explained by the remaining predictors $x_{j \neq i}$. (This is multicollinearity.) Equation 17.1 shows that in this case, $1 - R^2_{i,j \neq i} = 0$, and so s_{b_i} is $s_{\text{est}} / \sqrt{ss_{x_i}(0)} = s_{\text{est}} / 0 = \infty$.

So, the larger $R^2_{i,j \neq i}$ gets, the larger s_{b_i} gets. This means that the more highly correlated x_i is with the remaining predictors ($x_{j \neq i}$), the greater its estimated standard error and the less precise our estimate of β_i.

Tolerance and the Variance Inflation Factor

The quantity $1 - R^2_{i,j \neq i}$ is referred to as tolerance (*tol*) in the regression literature. (We introduced tolerance briefly in Chapter 16.) The variance inflation factor (*VIF*) is the reciprocal of *tol*; i.e., $VIF = 1/tol$. The variance in question is the square of s_{b_i}; i.e., $s^2_{b_i}$. (We obtain $s^2_{b_i}$ by squaring the numerator and denominator of equation 17.1.) The smaller *tol* is, the more variance x_i shares with the remaining predictors, $x_{j \neq i}$. With this in mind, the following sequence of transformations shows that we could have defined $s^2_{b_i}$ in terms of *tol* or *VIF*:

$$s^2_{b_i} = \frac{s^2_{\text{est}}}{ss_{x_i}(1 - R^2_{i,j \neq i})} = \frac{s^2_{\text{est}}}{ss_{x_i}} \frac{1}{(1 - R^2_{i,j \neq i})} = \frac{s^2_{\text{est}}}{ss_{x_i}} \frac{1}{tol} = \frac{s^2_{\text{est}}}{ss_{x_i}} VIF.$$

The term $s^2_{\text{est}} / ss_{x_i}$ specifies what the variance of the partial regression coefficient (b_i) would be if $R^2_{i,j \neq i} = 0$; i.e., if none of the variance in x_i can be explained by the remaining predictors $x_{j \neq i}$. This variance is increased (inflated) as *VIF* increases. Therefore, *VIF* is the factor by which the variance associated with b_i increases in the presence of the remaining predictors.

We noted in Chapter 16 that *tol* and *VIF* are somewhat contentious measures. Many authors suggest that we should worry about multicollinearity if *VIF* exceeds 10. Appendix 17.2 (available at study.sagepub.com/gurnsey) provides an extended discussion of whether this rule of thumb is generally useful.

Computing a Confidence Interval for b When There Are Two Predictors

We will now compute a confidence interval around the regression coefficient relating *GRE* to success in our two-predictor model (b_1). To do this we need the following quantities: s^2_y, ss_{x_i}, n, k, R^2, and $R^2_{i,j \neq i}$. From the analyses conducted in Chapter 16, we know that $n = 27$, $k = 2$, $R^2 = .528$, and $ss_{x_1} = 1264$ (as shown in Table 17.1). Table 17.1 also shows that $ss_y = 232$, and from this we can compute

$$s^2_y = \frac{ss_y}{n-1} = \frac{232}{26} = 8.92.$$

Finally, when there are only two predictors in a multiple regression, $R^2_{i.j \neq i}$ is simply the squared correlation between the predictors; i.e., $R^2_{1.2} = R^2_{1.2}$. Table 17.1 (note a) shows that the correlation between *GRE* and *TIE* is 0.4677. Therefore, the squared correlation between the two predictors is $R^2_{1.2} = 0.4677^2 = .218782$. With this information, we can compute the 95% confidence interval around b_1.

Step 1. Compute s_{est}. To do this, we need s^2_y, R^2, n, and k. When we insert these quantities into equation 17.2, we have

$$s_{est} = \sqrt{s^2_y(1-R^2)\frac{n-1}{n-k-1}} = \sqrt{8.92(1-.528)\frac{27-1}{27-2-1}} = 2.1355.$$

Step 2. Compute s_{b_i}. To do this, we need s_{est}, ss_{x_i}, and $R^2_{1.2}$. Remember, $R^2_{1.2} = .218782$. When we insert these quantities into equation 17.1 for $i = 1$, we have

$$s_{b_i} = \frac{s_{est}}{\sqrt{ss_{x_i}(1-R^2_{i.j \neq i})}} = \frac{2.1355}{\sqrt{1264(1-.218782)}} = 0.068.$$

Step 3. Determine $t_{\alpha/2}$. This is the 95% confidence interval, so $\alpha = .05$. The degrees of freedom are the same as for the full model; i.e., $n-k-1 = 24$. When we consult the t-table, we find that $t_{\alpha/2} = 2.064$.

Step 4. Compute the 95% confidence interval. The regression coefficient is $b_1 = 0.2327$. With this and the quantities previously computed, we can complete the calculation as follows:

$$CI = b_1 \pm t_{\alpha/2}(s_{b_i}) = 0.23 \pm 2.064(0.068) = [0.09, \ 0.37].$$

FIGURE 17.3 ■ Confidence Intervals

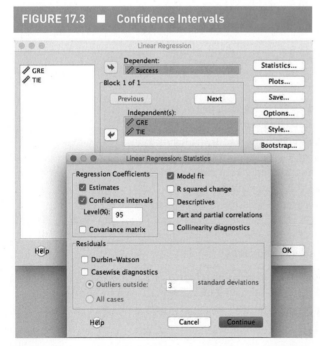

Requesting confidence intervals for regression coefficients in the **Linear Regression: Statistics** dialog.

We have 95% confidence that β_{GRE} falls in this interval because we know that 95% of all intervals computed this way will capture β_{GRE}.

Assumptions

For our confidence interval to be valid, we assume that our scores are a random sample from a multivariate distribution with the following properties: (i) the means of the conditional distributions are described by the regression equation, $E(y|x_1,x_2) = \alpha + \beta_1(x_1) + \beta_2(x_2)$; the prediction errors $[\varepsilon = y - E(y|x_1,x_2)]$ are (ii) normally distributed, (iii) homoscedastic, and (iv) independent. If these assumptions are valid, then the sampling distribution of b_1 will be a normal distribution with a mean β_1 and a standard error σ_{b_1}, which is estimated by s_{b_1}.

Computing the Confidence Intervals in SPSS

To compute confidence intervals around the regression coefficients in SPSS, we choose the **Statistics . . .** option from the **Linear Regression** dialog as shown in Figure 17.3. In the **Regression Coefficients** panel

FIGURE 17.4 ■ Confidence Intervals for Regression Coefficients

Coefficients[a]

Model		Unstandardized Coefficients B	Std. Error	Standardized Coefficients Beta	t	Sig.	95.0% Confidence Interval for B Lower Bound	Upper Bound
1	(Constant)	.606	9.465		.064	.949	-18.928	20.141
	GRE	.233	.068	.543	3.424	.002	.092	.373
	TIE	.029	.016	.292	1.838	.078	-.004	.062

a. Dependent Variable: Success

(top left), we check Confidence intervals and specify the level of confidence we wish to have. When this is done, we click Continue to return to the Linear Regression dialog, and there we click OK to proceed with the analysis.

The confidence intervals for the regression coefficients are shown in Figure 17.4. We saw most of this table in Chapter 16 (Figure 16.13), but now the confidence intervals for a (Constant), *GRE*, and *TIE* are shown in the right two columns. We've worked through the calculations for *GRE* (x_1) by hand and found that the estimated standard error and confidence interval $(0.233 \pm 2.064(0.068) = [0.092, 0.373])$ are exactly as shown in the table. The Coefficients table also shows the 95% confidence interval for *TIE* to be $[-0.004, 0.062]$.

Standardized Regression Coefficients

We noted in Chapter 16 that the standardized regression coefficients (sometimes known as beta weights) are the regression coefficients one would obtain if the outcome and predictor variables were converted to z-scores. The standardized regression coefficients can also be explained in terms of the regression coefficient relating x_i to y, and deviations of x_i and y. We compute standardized regression coefficient using the formula

$$\beta_i = b_i \frac{s_{x_i}}{s_y},$$

(17.3)

which is structurally identical to the one that related r to b in a simple linear regression

$$r = b \frac{s_x}{s_y}$$

(see Appendix 14.2). A standardized regression coefficient is a units-independent measure of the strength of association between x_i and y when the effects of other predictors are partialed out. Standardized regression coefficients normally range from –1 to 1. However, when there are problems with multicollinearity (predictors are too highly correlated), standardized regression coefficients outside this range can be obtained. Such results are impossible to interpret.

A confidence interval around a standardized regression coefficient can be computed using the `ci.src` function in the **R** MBESS package. To compute the 95% confidence interval for the standardized regression coefficient for *GRE*, the following command is entered at the **R** prompt (>):

```
ci.src(beta.k = .543, conf.level = 0.95, K = 2, N = 27, t.value = 3.424)
```

In this command, `beta.k` is the standardized regression coefficient, `conf.level` is the level of confidence we wish to have in the interval, `K` is the number of predictors, `N` is the number of scores in the sample, and `t.value` is the *t*-score used to test the null hypothesis that $\beta_i = 0$;

this quantity is taken from Figure 17.4. The 95% confidence interval returned is [0.192, 0.885], which is obviously quite wide because of the small sample size. The 95% confidence interval for the standardized regression coefficient for *TIE* is [−0.033, 0.611], which is also quite wide.

LEARNING CHECK 1

1. What does a partial regression (b_i) coefficient tell us about the association between x_i and y?

2. The multiple regression equation computed from a sample is $\hat{y} = 0.61 + 0.23(GRE) + 0.03(TIE)$. Show the simple linear regression equation that relates $\hat{y}$ to *TIE* when *GRE* is fixed at (a) 140, (b) 150, and (c) 160.

3. A multiple regression equation computed from a sample of $n = 28$ scores is $\hat{y} = 2 + 1.2(x_1) - 0.5(x_2)$. This equation produces $R^2 = .7$. The standard deviations of y, x_1, and x_2 are 4, 2, and 3, respectively. The squared correlation between x_1 and x_2 is .5. Compute the 95% confidence interval around (a) b_1 and (b) b_2.

4. A multiple regression equation computed from a sample of $n = 100$ scores is $\hat{y} = 3 + 0.3(x_1) - 0.9(x_2) + 0.5(x_3)$. This equation produces $R^2 = .60$. The standard deviations of y, x_1, x_2, and x_3 are 6, 3, 1, and 2, respectively. The values of $R^2_{i,j \neq i}$ are $R^2_{1.2,3} = .5$, $R^2_{2.1,3} = .7$, and $R^2_{3.1,2} = .6$. Compute the 95% confidence interval around (a) b_1, (b) b_2, and (c) b_3.

Answers

1. A partial regression coefficient is the slope of the regression equation relating x_i to y when the values of all other predictor variables are held constant.

2. (a) $\hat{y} = 32.81 + 0.03(TIE)$. (b) $\hat{y} = 35.11 + 0.03(TIE)$. (c) $\hat{y} = 37.41 + 0.03(TIE)$.

3. In this question,
$s_{est} = \sqrt{s_y^2(1-R^2)(n-1)/(n-k-1)} = 2.28$, $ss_{x_1} = 108$, and $ss_{x_2} = 243$.

   (a) $s_{b_1} = s_{est}/\sqrt{ss_{x_1}(1-r_{1.2}^2)} = 0.31$;
   95% CI $= 1.2 \pm 2.060(0.31) = [0.56, 1.84]$.

   (b) $s_{b_2} = s_{est}/\sqrt{ss_{x_2}(1-r_{1.2}^2)} = 0.21$;
   95% CI $= -0.5 \pm 2.060(0.21) = [-0.93, -0.07]$.

4. In this question,
$s_{est} = \sqrt{s_y^2(1-R^2)(n-1)/(n-k-1)} = 3.85$, $ss_{x_1} = 891$, $ss_{x_2} = 99$, and $ss_{x_3} = 396$.

   (a) $s_{b_1} = s_{est}/\sqrt{ss_{x_1}(1-R_{1.2,3}^2)} = 0.18$;
   95% CI $= 0.30 \pm 1.987(0.18) = [-0.06, 0.66]$.

   (b) $s_{b_2} = s_{est}/\sqrt{ss_{x_2}(1-R_{2.1,3}^2)} = 0.71$;
   95% CI $= -0.90 \pm 1.987(0.71) = [-2.31, 0.51]$.

   (c) $s_{b_3} = s_{est}/\sqrt{ss_{x_3}(1-R_{3.1,2}^2)} = 0.31$;
   95% CI $= 0.5 \pm 1.987(0.31) = [-0.11, 1.11]$.

STATISTICAL CONTROL

In Chapter 13 we noted that a correlation between two random variables does not imply a causal association. We used the example of an association between the number of instances of food poisoning and the number of street vendors. Although one could easily make up a story that blames street vendors for the cases of food poisoning, it seems far more plausible to think that an increase in both the number of cases of food poisoning and the number of street vendors is explained by population size.

In Table 17.3 there are hypothetical data for 16 cities for which we have recorded the number of cases of food poisoning (*poisonings*), the number of street vendors (*street vendors*), and the *population* (in thousands). A hierarchical regression has been performed on these data, with *poisonings* as the outcome variable. On step 1, *street vendors* was the predictor. On step 2, *population* was added as a predictor. Figure 17.5 shows that on step 1, the regression coefficient relating *street vendors* to *poisonings* was 0.349, 95% CI [0.121, 0.578]. When *population* was added on step 2, the regression coefficient relating *street vendors* to *poisonings* was reduced to

0.134, 95% CI [−0.052, 0.319]. The corresponding standardized regression coefficients (beta weights) were 0.659 and 0.252, respectively. Therefore, the association between *street vendors* and *poisonings* was substantially weakened when *population* was added as a predictor.

This analysis illustrates the notion of **statistical control**. When the strength of association between two variables is thought to be explained by a third variable, this can be assessed by adding the third variable into the regression, as we've just seen. In this case, we can say that the strength of association between *poisonings* and *street vendors* is weakened when *population* is included in the model.

We can report these results as follows:

APA Reporting

The simple regression coefficient predicting the number of cases of food poisoning (*poisonings*) in 16 cities from the number of street vendors (*street vendors*) in the cities was 0.35, 95% CI [0.12, 0.59]. However, when city population (*population*) was added as a predictor, the regression coefficient for *street vendors* dropped to 0.13, 95% CI [−0.05, 0.32]. Therefore, the association between *street vendors* and *poisonings* is largely explained by the fact that both are associated with *population*. It is worth noting, however, that even when controlling for *population*, there remains a modest association between *street vendors* and *poisonings* (standardized regression coefficient = 0.25, 95% CI [−0.08, 0.58]). Perhaps a more direct approach is in order in future studies. Along with each recorded case of food poisoning, one should record whether the victim had eaten food purchased from a street vendor within the previous 24 hours.

TABLE 17.3 ■ Hypothetical Data			
City	Poisonings	Street Vendors	Population
City 1	12	56	521
City 2	20	47	580
City 3	15	31	497
City 4	15	58	473
City 5	7	23	339
City 6	20	42	576
City 7	9	43	381
City 8	6	26	337
City 9	12	45	417
City 10	2	17	313
City 11	12	37	356
City 12	2	28	214
City 13	0	25	392
City 14	6	34	408
City 15	18	36	631
City 16	13	48	480

Note: Hypothetical data from 16 cities showing the number of cases of food poisoning, the number of street vendors, and population (in thousands).

FIGURE 17.5 ■ An Illustration of Statistical Control

Coefficients[a]

Model		Unstandardized Coefficients		Standardized Coefficients	t	Sig.	95.0% Confidence Interval for B	
		B	Std. Error	Beta			Lower Bound	Upper Bound
1	(Constant)	−2.449	4.156		−.589	.565	−11.363	6.465
	Street Vendors	.349	.107	.659	3.278	.006	.121	.578
2	(Constant)	−11.713	3.457		−3.388	.005	−19.181	−4.244
	Street Vendors	.134	.086	.252	1.559	.143	−.052	.319
	Population	.040	.009	.709	4.389	.001	.020	.060

a. Dependent Variable: Poisonings

Statistical control refers to partialing out the effect of one variable (e.g., *z*) from the association between two others (e.g., *x* and *y*). When this is done, we can talk about the association between *x* and *y* while controlling for *z*.

MEDIATION

Mediation analysis is similar in many respects to statistical control. Mediation analysis aims to explain why an association exists between two variables. Let's introduce a new example. There is evidence that abused children tend to be more aggressive than children who've not been abused (Lansford, Miller-Johnson, Berlin, Dodge, Bates, & Pettit, 2007). Let's assume that there is a continuum of abuse such that the greater the abuse, the greater the probability of exhibiting aggressive behaviors. Abused children are doubly

When variable *x* affects variable *y* indirectly through its effect on variable *med*, we say that *med* **mediates** the association between *x* and *y*.

unfortunate because they have to bear the psychological and physical scars of the abuse itself, but then their aggressive behaviors may isolate them from others and diminish their chances of healthy relationships. So, if an association exists between abuse and aggression, it would be worth understanding it so that interventions can be devised to reduce the risk of maladaptive aggressive behaviors.

An Example

The **total effect** of x on y is related to the strength of association between x and y without controlling for other variables. The total effect is denoted with c, which is defined as the slope of the simple regression equation that predicts y from x.

Let's imagine that a theory suggests that the greater the abuse children experience, the more likely they are to perceive threats from others. Assuming that some people respond aggressively to perceived threats, those having a lower threshold for perceiving a threat would be more likely to respond aggressively in situations that others would find neutral. Therefore, it is possible that the link between abuse and aggression is through increased sensitivity to threat. In the language of mediation analysis, we would hypothesize that sensitivity to threats mediates the association between abuse and aggression.

The Logic of Mediation Analysis

The connections we've just described are characterized in Figure 17.6. When we discuss this figure, we will use the notation typically found in the mediation literature. Figure 17.6a denotes the predictor variable as x and the outcome variable as y. The letter c represents the association between x and y. In our example, c is the regression coefficient in the simple regression equation that predicts y (*aggression*) from x (*abuse*). For reasons that will become clearer as we go along, c is referred to as the **total effect** of x on y.

Figure 17.6b introduces the mediating variable, denoted *med*, which in this case is sensitivity to threat (*threat*). The arrow labeled a represents the effect of x on *med*. In the mediation model, a is the slope of the simple regression equation that predicts *med* from x. The arrow labeled b represents the effect of *med* on y while controlling for x, and c' represents the effect of x on y while controlling for *med*. That is, b and c' are the partial regression coefficients for *med* and x in a regression model that predicts y from *med* and x.

We said earlier that c is the total effect of x on y. When the mediator variable is added to the mix, we can decompose the total effect (c) into two parts. One is called the **direct effect** of x on y, which is defined by c'. (Unfortunately, the terms "total effect" and "direct effect" can be a little confusing, so it takes some effort to keep these terms straight.) The **indirect effect** of x on y is the effect that x has on y via the mediator *med*. The strength of the indirect effect is given by the product of a and b (i.e., ab). The total effect is the sum of the direct and indirect effects of x on y, as follows:

$$\text{Total} = \text{Direct} + \text{Indirect}$$

$$c = c' + ab.$$

These concepts can be made concrete with an example. The data file Mediation.sav (available at study.sagepub.com/gurnsey) contains hypothetical data for 40 children. For each child, we have a measure of the degree of abuse experienced (*abuse*). There is also a measure of aggressive tendencies (*aggress*) corresponding to the mean rating given by three (hypothetical) independent raters. There is also a measure of sensitivity to threats (*threat*). *Threat* scores represent the number of threat-related words

---

FIGURE 17.6 ■ Mediation

(a)

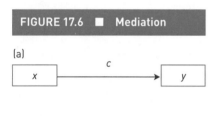

(b)

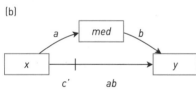

(a) The association between predictor variable x and outcome variable y. The line labeled c is referred to as the *total effect* of x on y. In mediation analysis, c is the slope of the simple regression equation that predicts y from x. (b) The role of the mediator variable (*med*). The association between x and *med* is shown by the line connecting the two. In mediation analysis, a is the slope of the simple regression equation that predicts *med* from x. The quantities b and c' are partial regression coefficients derived from a model in which x and *med* are predictors, and y is the outcome variable. In mediation analysis, c' is the partial regression coefficient for x and b is the partial regression coefficient for *med*. The coefficient c' is called the *direct effect* of x on y, and the product of a and b (ab) is called the *indirect effect* of x on y. The total effect (c) is the sum of the direct effect (c') and indirect effect (ab).

that children produced while describing a selection of neutral images of children interacting. We will submit these data to three regression analyses to illustrate the concepts underlying mediation.

The first analysis assessed the total effect of x on y. The predictor variable is *abuse* and the outcome variable is *aggress*. Figure 17.7 shows the result of this analysis. The regression equation is $\hat{y} = 7.503 + 0.199(abuse)$. Therefore, $c = 0.199$ is the total effect of x on y.

The second analysis assesses the effect of x on *med*. Our theory predicts that larger values of x (*abuse*) lead to larger values of *med* (*threat*). Figure 17.8 shows the result of this analysis. The regression equation is $\widehat{med} = 13.069 + 0.376(abuse)$. Therefore, $a = 0.376$.

The remaining quantities required are c' and b. These are obtained in a multiple regression analysis in which both *abuse* and *threat* predict *aggress*. Figure 17.9 shows the result of this analysis. The regression equation is $\hat{y} = 2.633 + 0.373(threat) + 0.059(abuse)$. Therefore, $b = 0.373$ and $c' = 0.059$.

Remember that the total effect can be expressed as the sum of the direct and indirect effects. Using the numbers we've just obtained, we find that

$$\text{Total} = \text{Direct} + \text{Indirect}$$

$$c = c' + ab$$

$$0.199 = 0.059 + 0.376*0.373$$

as promised.

Having outlined the logic of mediation analysis, we need to ask whether the results we've just obtained support the theory that sensitivity to threat mediates the association between abuse and aggression. The classical approach to mediation analysis (Barron & Kenny, 1986) suggests that they do. Figure 17.7 shows that there is a statistically significant association between *abuse* and *aggression* and Figure 17.9 shows that when *threat* is added as a predictor, the association between *abuse* and *aggression* is no longer statistically significant. In this way, the effect of *abuse* on *aggression* is weakened in the presence of *threat*, which is consistent with the idea that *abuse* exerts its influence on *aggression* through the mediating variable *threat*.

Therefore, going from statistically significant in the absence of the mediator to not statistically significant in its presence is the signature of mediation in the Barron and Kenny (1986, p. 1176) model. Of course, relying on statistical significance is fraught with difficulties. The most serious problem is that going from $p = .04$ (statistically significant) to $p = .06$ (not statistically significant) would technically indicate some degree of mediation. Of course, such a conclusion is a bit of a stretch. Therefore, an alternative approach focuses on the magnitude of the indirect effect, *ab*.

A very approximate method computing a confidence interval around *ab* uses the following formula:

$$\text{CI} = ab \pm z_{\alpha/2}(s_{ab}), \text{ where}$$

$$s_{ab} = \sqrt{a^2(s_b^2) + b^2(s_a^2)}. \tag{17.4}$$

The **direct effect** (c') is the effect of x on y while controlling for *med*; c' is the partial regression coefficient on x from a multiple regression analysis that predicts y from both x and *med*.

The **indirect effect** (*ab*) is the product of the effect of x on *med* (*a*) and the effect of *med* on y, controlling for x (*b*). That is, *a* is the simple regression coefficient on x that predicts *med* from x, and *b* is the partial regression coefficient on *med* in a multiple regression analysis that predicts y from both x and *med*.

FIGURE 17.7 ■ Total Effect of x on y

Coefficients[a]

		Unstandardized Coefficients		Standardized Coefficients		
Model		B	Std. Error	Beta	t	Sig.
1	(Constant)	7.503	9.368		.801	.428
	Abuse	.199	.091	.335	2.189	.035

a. Dependent Variable: Aggress

The regression equation showing the total effect of x on y.

FIGURE 17.8 ■ The Effect of x on *med*

Coefficients[a]

		Unstandardized Coefficients		Standardized Coefficients		
Model		B	Std. Error	Beta	t	Sig.
1	(Constant)	13.069	9.165		1.426	.162
	Abuse	.376	.089	.566	4.230	.000

a. Dependent Variable: Threat

The regression equation showing the effect of x on *med*.

FIGURE 17.9 ■ Partial Regression Coefficients

Coefficients[a]

		Unstandardized Coefficients		Standardized Coefficients		
Model		B	Std. Error	Beta	t	Sig.
1	(Constant)	2.633	9.074		.290	.773
	Abuse	.059	.104	.099	.566	.575
	Threat	.373	.156	.417	2.381	.023

a. Dependent Variable: Aggress

The regression equation predicting *aggress* from *threat* and *abuse*.

In equation 17.4, s_a^2 and s_b^2 are the squared estimated standard errors of a and b, respectively; s_a is the estimated standard error associated with the effect of *abuse* on *threat* (Figure 17.8) and s_b is the estimated standard error of the partial regression coefficient associated with the effect of *threat* on *aggress* in the presence of *abuse* (Figure 17.9). This definition of s_{ab} was suggested by Sobel (1982).

In our example, we would find

$$s_{ab} = \sqrt{a^2(s_b^2) + b^2(s_a^2)} = \sqrt{0.376^2(0.156^2) + 0.373^2(0.089^2)} = 0.067,$$

and the approximate 95% confidence interval around the indirect effect would be

$$CI = ab \pm 1.96(s_{ab}) = 0.140 \pm 1.96(0.067) = [0.01, 0.27] .$$

This approximation is widely viewed as unreliable unless samples comprise at least 100 scores (MacKinnon, Warsi, & Dwyer, 1995).

Using the PROCESS Macro to Perform Mediation Analysis

Although the approximate method just described is widely used to assess indirect effects, a much better alternative is bootstrapping. This approach is included in a suite of routines written by Professor Andrew Hayes (2013). Professor Hayes's analysis package is called PROCESS, and instructions for downloading and incorporating PROCESS into SPSS are given in Appendix 17.1. For the following discussion, I will assume that you've added PROCESS to your version of SPSS.

Once PROCESS is installed, you can access it through the Analyze→ Regression . . . dialog, as illustrated in Figure 17.10. When PROCESS is selected, the dialog shown in Figure 17.11 will appear. The PROCESS dialog looks much like other SPSS dialogs we've seen. Initially, the variables in your data file are shown in the Data File Variables panel at the top left. Our three variables have been moved into the appropriate boxes on the right. *Aggress* is the Outcome Variable (Y), *abuse* is the Independent Variable (X), and *threat* is the M Variable. (In this context M stands for mediator, but we will see later that M can also stand for moderator.)

Under the Model Number heading, the number 4 has been chosen from the drop-down list. This is the name PROCESS gives to the simple mediation model that we're considering. There is an impressive number of other models—including those with multiple mediator variables—that are beyond the scope of this book. The user can choose the number of Bootstrap Samples from the drop-down list. For this example, I've chosen 1000. There are two methods that can be used to compute the bootstrap confidence intervals (Bootstrap CI Method). The first is the Percentile method and the second is a Bias Corrected bootstrapping method, which has been chosen for this example. Further details about these options can be found in Hayes (2013). The bias-corrected bootstrapping procedure is used to eliminate skew from bootstrapped distributions. The Confidence Level has been set to 95%, but other levels can be chosen from the drop-down list.

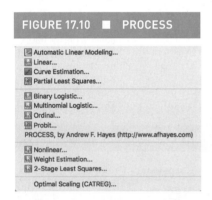

FIGURE 17.10 ■ PROCESS

Access the PROCESS routines through the **Regression** . . . dialog under the **Analyze** menu in SPSS.

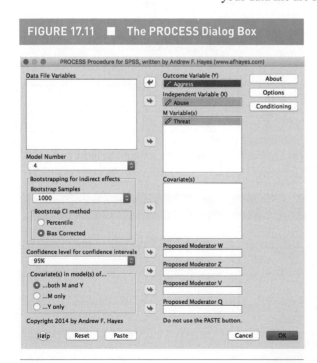

FIGURE 17.11 ■ The PROCESS Dialog Box

There are several optional settings that we can choose from when running mediation analysis in PROCESS. These are available through the Options button in the top right of the PROCESS dialog. When the Options button is clicked, the PROCESS Options dialog appears, as shown in Figure 17.12. We will mention only two of these options for now. The first is OLS/ML Confidence Intervals option; OLS/ML stands for ordinary least squares/maximum likelihood. When this option is checked, PROCESS will report the standard (non-bootstrapped) confidence intervals reported by SPSS itself. The second option, Compare indirect effects, will assess the strength of the indirect effect relative to the total and direct effects, as we'll see.

The result of the mediation analysis is shown in Figure 17.13. Note that the PROCESS output is formatted somewhat differently than the SPSS tables that we've seen previously. To make it easier to locate material in Figure 17.13, letters have been added beside the relevant sections. The areas marked (a) and (b) are the results of the two regression analyses shown in Figures 17.8 and 17.9. You can refer back to those figures to confirm that the same information is shown.

The section marked (c) shows the direct effect of *abuse* on *aggress*. This line is almost identical to the last line of section (b), except the regression slope is now under the "Effect" column. The section marked (d) is the most interesting to us. It shows the indirect effect computed as *ab* earlier. As before, the indirect effect is *ab* = 0.1401. Also shown is the 95% bias-corrected bootstrapped confidence interval, which is [0.0396, 0.2655]. From this analysis, we can see that the indirect effect (*ab* = 0.1401) is much larger than the direct effect (*c'* = 0.0588). Therefore, this analysis supports the hypothesis that sensitivity to threats mediates the association between abuse and aggression.

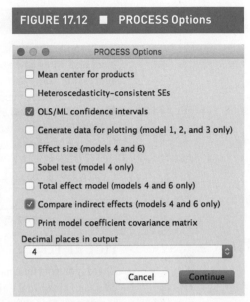

FIGURE 17.12 ■ PROCESS Options

The PROCESS Options dialog box

FIGURE 17.13 ■ PROCESS Output

```
********************************************************************************
(a) Outcome: Threat

Model Summary
         R        R-sq       MSE         F        df1        df2          p
     .5658      .3202    94.4328    17.8949     1.0000    38.0000      .0001

Model
              coeff        se          t          p       LLCI       ULCI
constant    13.0685     9.1649     1.4259      .1621    -5.4851    31.6222
Abuse         .3761      .0889     4.2302      .0001      .1961      .5560

********************************************************************************
(b) Outcome: Aggress

Model Summary
         R        R-sq       MSE         F        df1        df2          p
     .4796      .2300    87.8632     5.5271     2.0000    37.0000      .0079

Model
              coeff        se          t          p       LLCI       ULCI
constant     2.6329     9.0738      .2902      .7733   -15.7527    21.0184
Threat        .3726      .1565     2.3814      .0225      .0556      .6897
Abuse         .0588      .1040      .5655      .5751     -.1519      .2695

******************** DIRECT AND INDIRECT EFFECTS ************************

(c) Direct effect of X on Y
     Effect         SE          t          p       LLCI       ULCI
      .0588       .1040      .5655      .5751     -.1519      .2695

(d) Indirect effect of X on Y
              Effect    Boot SE    BootLLCI    BootULCI
Threat         .1401      .0580       .0396       .2655
```

The output of a mediation analysis performed by PROCESS.

We hypothesized that the association between childhood abuse and aggression was mediated by greater sensitivity to threat. We obtained measurements of abuse (*abuse*), aggression (*aggress*), and sensitivity to threat (*threat*) from 40 children. Figure 17.14 summarizes the mediation analysis. The regression equation to predict *aggress* from *abuse* showed a total effect characterized by a regression slope of $c = 0.20$, 95% CI [0.042, 1.084]. This total effect was shown to consist of a small direct effect, $c' = 0.06$, 95% CI [−0.15, 0.27], and a larger indirect effect $ab = 0.14$, 95% bootstrapped CI [0.04, 0.27].

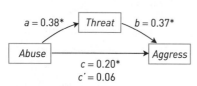

FIGURE 17.14

Regression coefficients expressing effects of abuse and threat on aggression. *$p < .05$.

These results show that the indirect path through *threat* partially mediates the association between abuse and aggression.

These results are consistent with the hypothesis that the association between childhood abuse and childhood aggression is mediated by greater sensitivity to threat. Further experimental work will be required to determine whether interventions to reduce sensitivity to threat will have the predicted effect of reducing aggression.

Interpreting Mediation

Mediation Must Be Plausible

Mediation analysis is popular because it suggests a way to assess proposals about causal links between variables. Keep in mind, however, that mediation analysis is a statistical method applied mechanically to variables of interest. One might find support for a causal link such as $x \rightarrow med \rightarrow y$, but the plausibility of this causal chain depends on what x, *med*, and y are. For example, consider the case in which x is child's IQ, y is parent's IQ, and *med* is socioeconomic status. If we found a weak direct effect and a strong indirect effect, would it make sense to say this is strong evidence for the causal model $x \rightarrow med \rightarrow y$? No, this would be implausible because it implies running time backward so that children's IQs determine parents' IQs through the mediating variable socioeconomic status. The value of the mediation analysis is no greater than the plausibility of the hypothesized causal chain.

Causality and Random Variables

The purpose of mediation analysis is to explore the chain of causal connections from the predictor variable (x) to the outcome variable (y). Our example dealt with a single mediator variable (*med*) that is hypothesized to be part of this causal chain $x \rightarrow med \rightarrow y$. In more complex mediation analyses, there may be several mediating variables acting in sequence,

$$x \rightarrow med_1 \rightarrow med_2 \rightarrow \ldots \rightarrow med_k \rightarrow y,$$

or in parallel,

$$
\begin{array}{c}
med_1 \\
x \rightarrow med_2 \rightarrow y, \\
\cdots \\
med_k
\end{array}
$$

or some combination of sequential and parallel effects.

A mediation analysis, no matter how complex, may be *consistent* with a hypothesized causal chain, but evidence of mediation *does not imply the truth* of the hypothesized causal chain. As with any study involving random variables, there may be some other variable that explains the associations between the variables under study. In our example, the indirect effect was much larger than the direct effect, which is consistent with the causal sequence $x \rightarrow med \rightarrow y$. However, another mediator variable (med_2) could provide an indirect effect on y that is far greater than the direct effect (c') or the indirect effect (ab) involving only *med*. In this case, we might have to say that the original indirect effect (ab) was spurious.

Experimental Assessment of a Causal Link

If a mediation analysis provides evidence consistent with an indirect effect, the only way to demonstrate that the indirect effect is part of a causal link from x to y is to test this link experimentally. In our example, we could do this by drawing a sample of abused children and dividing them into two equal groups. One group (treatment group) would receive an intervention that reduces sensitivity to threat, whereas the other group (control group) would receive a placebo intervention unrelated to sensitivity to threat. Ideally, our intervention would reduce the sensitivity to threat among abused children to the levels seen in nonabused children. Following the intervention, we would monitor the children in both groups for aggressive behaviors and we would use the two-groups design to analyze the results; i.e., $(m_t - m_c) \pm t_{\alpha/2} (s_{m_1 - m_2})$. If the treatment group showed a meaningful reduction in aggressive behaviors, this reduction could be reasonably attributed to the intervention and could provide experimental support for the causal model $x \rightarrow med \rightarrow y$.

Measuring the Magnitude of the Mediator

An experimental approach has the additional advantage of allowing us to determine the degree to which reducing sensitivity to threats reduces aggressive behaviors. That is, the difference between m_t and m_c tells us the mean reduction in aggressive behaviors between the two groups. Knowing how much aggressive behaviors are reduced, on average, following the intervention allows us to think about its practical significance. For example, reducing the mean number of aggressive behaviors by half a standard deviation might or might not be meaningful in different contexts.

The regression-based mediation analysis can't tell us how the marginal distribution of aggression scores would change if we were to eliminate the influence of the mediator. Rather, our regression-based model tells us about the direct (c') and indirect (ab) effects of x on y. When c' is reduced to 0, we say there is *complete mediation*; when it is reduced but not to 0, we say there is *partial mediation*. In either case, ab tells us how much change there is in y with a unit change in x through the indirect path. We noted that changing *med* should change the marginal distribution of y scores, and we saw that this could be assessed experimentally by comparing the means of the treatment and control conditions. Unfortunately, regression analysis cannot tell us what the distribution of y scores would be if the effect of the mediating variables was eliminated.

Further Reading

Mediation analysis provides a useful framework for assessing theories about causal chains of associations. However, we should use it carefully. A number of cautions about use and interpretation have been described above. Kline (2015) provides discussion of these cautions that goes beyond what can be covered here. The article provides a sobering view of how mediation analysis can be misused.

LEARNING CHECK 2

1. State whether the following statements are true or false.

   (a) In a mediation analysis, the partial regression coefficient relating x to y is the total effect.

   (b) In a mediation analysis, $c = c' + ab$.

   (c) In a mediation analysis, $c' = 0$ denotes partial mediation.

   (d) In a mediation analysis, $c' = 0$ proves the causal link $x \rightarrow med \rightarrow y$.

   (e) When there is complete mediation in a mediation analysis, we can deduce what the marginal

distribution of y scores would be if the indirect path were eliminated.

2. If $c = 10$ and $ab = 8$, what is c'?

3. If $a = 5$, $b = 6$, $s_a = 1.5$, and $s_b = 3$, compute an approximate 95% confidence interval around ab.

4. A researcher theorizes that sense of humor mediates the association between daily hassles and stress. The 95% bootstrapped confidence interval around ab is [0.15, .089]. Is the researcher's theory supported? Does this prove that humor mediates the association between daily hassles and stress?

Answers

1. (a) False. (b) True. (c) False (it denotes complete mediation). (d) False. (e) False.

2. $c' = 10 - 8 = 2$.

3. $s_{ab} = \sqrt{a^2(s_b^2) + b^2(s_a^2)} = 17.49$; 95% CI [−4.29, 64.29].

4. I would say there is at least partial mediation here so the researcher's theory is supported because there is

a nonzero indirect effect. It would be necessary to know c' because this would tell us what part of the total effect of x on y is explained by the direct effect. That is, we'd have some basis for deciding if we have complete or partial mediation. Of course, this is a study of random variables so we can't draw causal conclusions about mediation.

MODERATION

In this section we introduce the concept of **moderation**, which refers to situations in which the association between a predictor and an outcome variable depends on an individual's score on some third variable. For example, if the association between x and y is different for men and women, we would say that *sex* moderates the association between x and y. We could also say that there is a *conditional distribution* of xy pairs for each value of the moderator.

The Logic of Moderation Analysis

When the association between x and y changes as a function of a third variable *mod*, we say that *mod* **moderates** the association between x and y, or that there is an **interaction** between x and *mod*. Therefore, we call *mod* a moderator variable.

Figure 17.15 illustrates the notion of moderation. Figure 17.15a shows a plot of 96 xy pairs, where x is the predictor variable and y is the outcome variable. There is a relatively weak association between x and y; $R^2 = .032$.

Figure 17.15b shows that the 96 individuals had scores of 8, 10, or 12 on a third variable that we refer to as *mod*. Two things are apparent in Figure 17.15b. The first is that the average score on y increases as *mod* increases from 8 to 12. We say that there is a *main effect* of *mod*. In addition to this main effect, we notice that the association between x and y is different for the three values of *mod*. For *mod* = 8, y decreases as x increases; for *mod* = 10, y increases slightly as x increases; and for *mod* = 12, y increases quite sharply as x increases. Because the association between x and y is different for the three values of *mod*, we say that it moderates the association between x and y, or that there is an **interaction** between x and *mod*.

An interaction between two predictor variables means that the two predictors acting independently do not completely explain their association with *y*. In this example, the two-predictor regression equation relating *x* and *mod* to *y* would not completely capture how the association between *x* and *y* changes as a function of *mod*. However, the interaction between *x* and *mod* can be captured in a multiple regression analysis by creating a third predictor variable to represent it.

One way to create this new variable is to compute the product of *x* and *mod*. We will call this variable *xm*. *xm* is nonlinearly related to both *x* and *mod*. To see what this means, think of the simple case of *x* = [1, 2, 3, 4, 5] and *mod* = [2, 4, 6, 8, 10]. If we were to multiply *x* and *mod*, we would have *xm* = [2, 8, 18, 32, 50]. In this case, *xm* is related to *x* by the nonlinear equation $xm = 2(x^2)$ and to *mod* by the nonlinear equation $xm = 0.5(mod^2)$. Therefore, *xm* will capture something about the association between *y* and the predictors that is not captured by the predictors themselves.

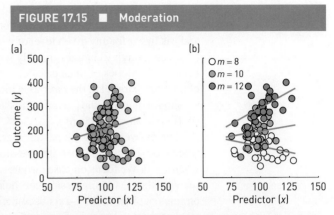

FIGURE 17.15 ■ Moderation

Moderation analysis in multiple regression. (a) The association between *x* and *y*, without considering each individual's score on a third variable, *mod*. (b) The associations between *x* and *y* for levels of *mod* = 8, 10, and 12. Because that association between *x* and *y* changes with *mod*, we say that it moderates the association between *x* and *y*.

Figures 17.16a and 17.16b show (in a three-dimensional space) the raw data from Figure 17.15. The shaded surfaces show the best fit to the data when *xm* is a predictor in the model

$$\hat{y} = 657.69 - 8.58(x) - 55.65(mod) + 0.96(xm).$$

This surface captures the change in the association between *x* and *y* for different levels of *mod*, as well as the change in the association between *mod* and *y* for different levels of *x*. The gray lines in Figure 17.16a plot the values of the regression equation ($\hat{y}$) for *mod* = 8, 10,

FIGURE 17.16 ■ Moderation

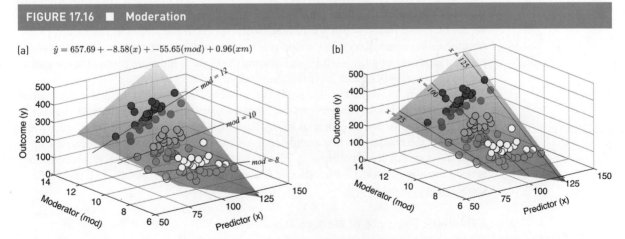

(a and b) Both panels plot (in a three-dimensional space) the same information plotted in Figure 17.15b. Once again, it is clear that the association between *x* and *y* depends on the level of *mod*. The shaded surfaces show the best-fitting surface plotted through the data. The quantity *xm* (in the equation above the figure) is the product of *x* and *mod* (described in the text), which represents the interaction between *x* and *mod*. The gray lines show how values of *y* change as a function of *x* for *mod* = 8, 10, and 12 (a) and as a function of *mod* for *x* = 75, 100, and 125 (b).

and 12. The association between x and y is linear for any given level of mod. The gray lines in Figure 17.16b plot the values of $\hat{y}$ for $x = 75$, 100, and 125. The association between mod and y is linear for any given level of x.

Even though $\hat{y}$ changes linearly with both x and mod, the surface as a whole is curved because the slopes of lines parallel to the x-axis increase as mod increases, and the slopes of the lines parallel to the mod-axis increase as x increases. This raises the question of how we should interpret the partial regression coefficient on x (-8.58) and the partial regression coefficient on mod (-55.65).

The answer is that the partial regression coefficient on x shows how $\hat{y}$ changes with x when $mod = 0$, and the partial regression coefficient on mod shows how $\hat{y}$ changes with mod when $x = 0$. We focus on cases where x and mod equal 0 because in both cases, xm is 0. To see why the partial regression coefficient on $x = -8.58$ when $mod = 0$, note that the slopes on lines parallel to the x-axis become increasingly negative as mod decreases. When mod reaches 0 (not shown in Figure 17.16), the slope of the line parallel to the x-axis will be -8.58. Similarly, the slopes on lines parallel to the mod-axis move from positive to negative as x decreases. When x reaches 0 (not shown in Figure 17.16), the slope of the line parallel to the mod-axis will be -55.65.

When neither x nor mod is 0, the change in $\hat{y}$ with a change of x depends on mod, and the change in $\hat{y}$ with a change of mod depends on x. Again, this is what it means for there to be an interaction between x and mod. Interactions can make it difficult to interpret exactly what the regression equation is telling us.

A partial solution to this problem is to subtract the mean score of each predictor variable from all scores on that variable. This is called **mean-centering**. For example, we mean-center x to produce $x_c = x - m_x$, and we mean-center mod to produce $mod_c = mod - m_{mod}$. The mean of x_c is now 0 and the mean of mod_c is now 0. That is, the 0 value on x_c corresponds to the position of m_x within the original x scores, and the 0 value on mod_c corresponds to the position of m_{mod} within the original mod scores. When the predictors x_c, mod_c, and $x_c m_c$ from our example are used to predict y, the regression equation becomes

$$\hat{y} = 203.02 + 1.02(x_c) + 39.88(mod_c) + 0.96(x_c m_c).$$

The partial regression coefficients on x_c and mod_c have now changed. They now show the slopes of the regression plane parallel to the x_c and mod_c axes, respectively, at the mean values of x_c and mod_c, which we've noted correspond to the slopes at the means of the uncentered variables. Knowing the slope of the regression plane parallel to the x-axis at the mean of mod may be more useful than knowing the slope of the regression plane parallel to the x-axis at $mod = 0$, when 0 is outside the range of mod scores.

Researchers mean-center the predictor variables in an effort to make the partial regression coefficients more interpretable. However, they also mean-center variables in the belief that doing so avoids problems of multicollinearity. Because there are simpler ways to interpret the results of a moderation analysis than mean-centering and because mean-centering does not avoid collinearity (Echambadi & Hess, 2007), I think mean-centering is not generally helpful. This question is discussed further in Appendix 17.2 (available at study.sagepub.com/gurnsey).

A Realistic Example of Moderation

In the example we've been considering, there were only three levels of mod. This made it easy to see how the association between x and y is conditional on the value of mod. And sometimes a small number of mod values is plausible if only these levels are of interest to the researcher. In this situation, one approach to making sense of the data would be to analyze the association between x and y for each level of mod. This is essentially what was done in Figure 17.16b. However, it is far more common in psychological research for both x

Subtracting the mean from a set of scores produces a new set of scores that have a mean of 0. This new set of scores is said to be a **mean-centered** version of the original set of scores.

and *mod* to be continuous random variables. Therefore, we can't simply separate the scores on *x* into a few levels of *mod* because when *mod* is continuous, each score on *mod* is probably unique in the data set. Therefore, to illustrate a moderation analysis, we will use a more realistic approach derived from a recent study of anxiety.

We all experience anxiety at one time or another, but anxiety becomes pathological for some people, which is a condition that clinicians refer to as generalized anxiety disorder (GAD). Individuals with GAD are anxious and worried all the time. A prominent feature of GAD is intolerance of uncertainty (IU), which Dugas, Gosselin, and Ladouceur (2001) define as "the excessive tendency of an individual to consider it unacceptable that a negative event may occur, however small the probability of its occurrence" (p. 552).

Chen and Hong (2010) studied the role of IU in a nonclinical sample of university students. They thought that IU might moderate the association between anxiety and the stressors that students experience daily. Therefore, they predicted that individuals who experience the same level of stress will differ in their experience of anxiety in a way that depends on their level of IU. Individuals with higher levels of IU would experience greater anxiety in response to the same stressors than individuals with lower levels of IU.

Chen and Hong (2010) obtained responses from student volunteers on the following three measures:

- The Beck Anxiety Inventory (BAI) (Beck, Epstein, Brown, & Steer, 1988) is a 21-item questionnaire that asks participants to rate on a 4-point scale (0 to 3) the degree to which they experienced symptoms of anxiety (e.g., "feeling hot," "shaky") in the last 2 weeks.

- The Inventory of College Students' Recent Life Experiences (ICSRLE) (Kohn, Lafreniere, & Gurevich, 1990) is a 55-item questionnaire that asks students to rate on a 4-point scale (0 to 3) the degree to which they've had negative life events in the last month such as "conflicts with professor(s)" and "lower grades than hoped for."

- The Intolerance of Uncertainty Scale (IUS) (Buhr & Dugas, 2002) is a 27-item questionnaire that includes statements such as "Uncertainty stops me from having a firm opinion" and "I can't stand being undecided about my future." Participants are asked to rate on a 5-point scale how well these statements describe them, with 1 being "not at all like me" and 5 being "very much like me."

In this analysis, *BAI* was the outcome variable and *ICSRLE* was the predictor variable. Chen and Hong (2010) predicted that higher *ICSRLE* scores would be associated with higher reported anxiety and that *IUS* would moderate this association. That is, the increase in *BAI* scores with increases in *ICSRLE* scores would be greater for those who also had high *IUS* scores than for those who had low *IUS* scores.

The moderationIUS.sav data file (available at study.sagepub.com/gurnsey) contains simulated data similar to those reported by Chen and Hong (2010). For each of 110 simulated participants, there were scores on *BAI*, *ICSRLE*, and *IUS*. A simple multiple regression with *ICSRLE* and *IUS* as predictors (and no interaction term) yields $R^2 = .189$. Both *ICSRLE* ($b_{ICSRLE} = 0.318$, 95% CI [0.075, 0.562]) and *IUS* ($b_{IUS} = 0.185$, 95% CI [0.098, 0.272]) make a substantial contribution to explaining variability in *BAI*, as shown in Figure 17.17. Our question is whether adding the interaction term (i) improves the prediction of *BAI* and (ii) whether the interaction between *ICSRLE* and *IUS* has the form suggested by Chen and Hong.

Moderator Analysis in PROCESS

We can approach a moderation analysis using the standard features of SPSS. However, this approach is somewhat awkward and laborious. Fortunately, the PROCESS macro is

FIGURE 17.17 ■ Moderation Analysis

(a)

Model Summary

Model	R	R Square	Adjusted R Square	Std. Error of the Estimate
1	.434[a]	.189	.174	8.321442

a. Predictors: (Constant), IUS, ICSRLE

(b)

Coefficients[a]

Model		Unstandardized Coefficients		Standardized Coefficients			95.0% Confidence Interval for B	
		B	Std. Error	Beta	t	Sig.	Lower Bound	Upper Bound
1	(Constant)	2.696	3.205		.841	.402	−3.657	9.049
	ICSRLE	.318	.123	.226	2.590	.011	.075	.562
	IUS	.185	.044	.368	4.228	.000	.098	.272

a. Dependent Variable: BAI

Model summary (a) and regression coefficients (b) for a multiple regression analysis in which the Inventory of College Students' Recent Life Experiences (*ICSRLE*) and Intolerance of Uncertainty Scale (*IUS*) are used to predict Beck Anxiety Inventory (*BAI*) scores.

tailored for moderation analysis and thus produces output that directly addresses our questions. Therefore, we will use the PROCESS macro to conduct a moderation analysis with *BAI* entered as the outcome variable (*y*), *ICSRLE* entered as the independent variable (*x*), and *IUS* entered as the moderator variable (*mod*).

Setting Up the Analysis

To conduct a moderation analysis on the data in the moderationIUS.sav data file (available at study.sagepub.com/gurnsey), we first select the PROCESS option from the Analyze→Regression menu. When the dialog shown in Figure 17.18 initially appears, the variables *BAI*, *ICSRLE*, and *IUS* were in the Data File Variables panel. Here, I've moved them into the Outcome Variable (Y), Independent (X), and M Variable (s) windows, respectively. The Model Number chosen is 1, which is how PROCESS refers to the simple moderation analysis we're conducting. From the Options dialog (Figure 17.19), I've chosen to include OLS/ML confidence intervals in the output and to Generate data for plotting, which will be discussed below.

Understanding the Output

The results of the moderation analysis are shown in Figure 17.20. The relevant parts of the output are labeled (a) to (f). The Model Summary (Figure 17.20a) shows an analysis of the overall fit of the three-predictor model. The predictors are *ICSRLE*, *IUS*, and *ICSRLE*IUS*. Together they explain 21.43% of the variability in *BAI*. This corresponds to $F_{obs} = 9.64$, which is statistically significant when tested on 3 and 106 degrees of freedom, $p < .0001$.

Next, the Model section (Figure 17.20b) shows the regression coefficients. The regression equation is

$$\hat{y} = 12.52 - 0.4225(ICSRLE) + 0.0053(IUS) + 0.0134(ICSRLE*IUS).$$

FIGURE 17.18 ■ Moderation in PROCESS

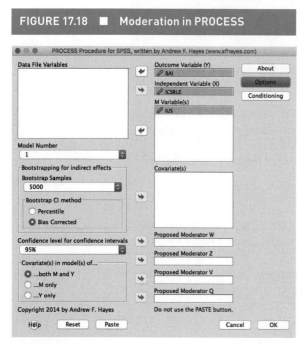

The PROCESS dialog configured to conduct a simple moderation analysis.

The 95% confidence intervals around the partial regression coefficients were computed as

$$b_i \pm t_{\alpha/2}(s_{b_i}),$$

as described earlier. That is, these are the OLS/ML confidence intervals chosen in PROCESS Options. For example, the 95% confidence interval around the partial regression coefficient for *IUS* is

$$CI = b_i \pm t_{\alpha/2}\left(s_{b_i}\right) = 0.0053 \pm 1.9826(0.1060) = [-0.2049,\ 0.2155].$$

Figure 17.20c shows that the interaction term is the product of *ICSRLE* and *IUS*. In the output, this product is referred to as int_1, or interaction term 1. The ΔR^2 analysis is shown in Figure 17.20d. $\Delta R^2 = .0256$; this means that when int_1 (*ICSRLE*∗*IUS*) was added as a predictor, it produced an increase of 2.56% explained variance. This modest increase in explained variance approaches statistical significance, $F_{obs} = 3.45, p = .066$.

If the interaction variable *ICSRLE*∗*IUS* were not in the model, then the partial regression coefficient for *ICSRLE* (i.e., b_{ICSRLE}) would tell us how much change there is in $\hat{y}$ with a one-unit change in *ICSRLE*, with *IUS* held constant (see Figure 17.17). Because we have included the interaction term *ICSRLE*∗*IUS*, the partial regression coefficients on *ICSRLE* and *IUS* cannot be interpreted this way. When *ICSRLE*∗*IUS* is in the model, the change in $\hat{y}$ with changes in *ICSRLE* is different for each value of *IUS*.

In Figure 17.20e, three rows of numbers are arranged in seven columns that help explain the nature of the interaction between *ICSRLE* and *IUS*. The first column, labeled IUS, shows three values of the moderator variable, *IUS*. These are the mean of *IUS* (58.1765) and those one standard deviation (18.2044) above and below this mean. Therefore, the three values of *IUS* shown are 39.98, 58.18, and 76.38. The column labeled Effect shows the slope of the regression line relating *ICSRLE* to $\hat{y}$ for each of these three values of *IUS*. These slopes are 0.1152, 0.3600, and 0.6049 for *IUS* = 39.98, 58.18, and 76.38, respectively. These slopes are derived from the regression equation as follows:

$$b = b_x + b_{xm}(mod),$$

where b_x is the partial regression coefficient on the independent variable, b_{xm} is the partial regression coefficient on the interaction term, and *mod* is the value of the moderator. So, in the case of *IUS* = 39.9722, for example, we would have

$$b = b_x + b_{xm}(mod) = -0.4225 + 0.0134(39.9722) = 0.1131,$$

which differs slightly from what is reported in Figure 17.20e only because the numbers used in our calculation were rounded to four decimal places.

The 95% confidence intervals for these three slopes are shown to the right in columns LLCI and LICI. These confidence intervals show that the association between *BAI* and *ICSRLE* changes as a function of *IUS*. As predicted by Chen and Hong (2010), slopes increase as *IUS* increases.

To help visualize these slopes, Figure 17.20f provides data that can be plotted in any graphing application, including SPSS or Excel. The numbers show $\hat{y}$ for all combinations of the mean of *IUS* and mean of *IUS* ± *s*, as well as the mean of *ICSRLE* and *ICSRLE* ± *s*. These values are plotted in Figure 17.21 and make clear that the predicted interaction was found.

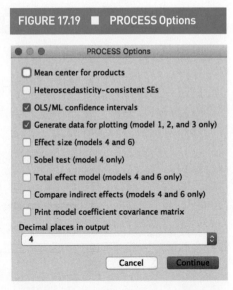

FIGURE 17.19 ■ PROCESS Options

The PROCESS Options dialog configured to conduct a simple moderation analysis.

FIGURE 17.20 ■ Moderation Output

```
*******************************************************************************
Outcome: BAI
```

(a) Model Summary

R	R-sq	MSE	F	df1	df2	p
.4629	.2143	67.6957	9.6372	3.0000	106.0000	.0000

(b) Model

	coeff	se	t	p	LLCI	ULCI
constant	12.5188	6.1646	2.0308	.0448	.2969	24.7407
IUS	.0053	.1060	.0501	.9601	-.2049	.2155
ICSRLE	-.4225	.4168	-1.0135	.3131	-1.2488	.4039
int_1	.0134	.0072	1.8577	.0660	-.0009	.0278

(c) Interactions:

```
 int_1    ICSRLE      X      IUS
```

(d) R-square increase due to interaction(s):

	R2-chng	F	df1	df2	p
int_1	.0256	3.4510	1.0000	106.0000	.0660

```
*******************************************************************************
```

(e) Conditional effect of X on Y at values of the moderator(s):

IUS	Effect	se	t	p	LLCI	ULCI
39.9722	.1152	.1634	.7047	.4825	-.2088	.4392
58.1765	.3600	.1235	2.9139	.0044	.1151	.6050
76.3809	.6049	.1964	3.0800	.0026	.2155	.9942

```
Values for quantitative moderators are the mean and plus/minus one SD from mean.
Values for dichotomous moderators are the two values of the moderator.

*******************************************************************************
```

(f) Data for visualizing conditional effect of X on Y
Paste text below into a SPSS syntax window and execute to produce plot.

```
DATA LIST FREE/ICSRLE IUS BAI.
BEGIN DATA.

     8.2514     39.9722     13.6814
    14.7392     39.9722     14.4285
    21.2270     39.9722     15.1757
     8.2514     58.1765     15.7984
    14.7392     58.1765     18.1340
    21.2270     58.1765     20.4697
     8.2514     76.3809     17.9154
    14.7392     76.3809     21.8395
    21.2270     76.3809     25.7637
```

The output from the PROCESS moderation analysis.

For low levels of *IUS* (i.e., 39.97), there is only a slight increase of *BAI* as *ICSRLE* increases. However, the increases (slopes) become greater as *IUS* increases from 58.18 to 76.38.

PROCESS does not compute the 95% confidence interval around each of the $\hat{y}$ values given in the third column of the table. These are easily computed using the standard analysis routines in SPSS by following these steps:

Step 1. Paste the nine values of the independent and moderator variables (given in the PROCESS output) at the bottom of the corresponding columns in the SPSS data file.

Step 2. Use the Compute Variable . . . dialog under the Transform menu item to compute a new variable that is the product of the independent variable and the moderator. For this example, this new variable could be called ICSRLExIUS.

Step 3. In a standard linear regression analysis (Analyze→ Regression→Linear), set the dependent variable (y) and enter the predictor (x), moderator (*mod*), and product term ($x*mod$) all in one block.

Step 4. From the Save . . . dialog, ask SPSS to generate the (i) unstandardized predicted value and (ii) 95% confidence interval around $\hat{y}$ for each combination of x, *mod*, and $x*mod$.

FIGURE 17.21 ■ Depicting Moderation

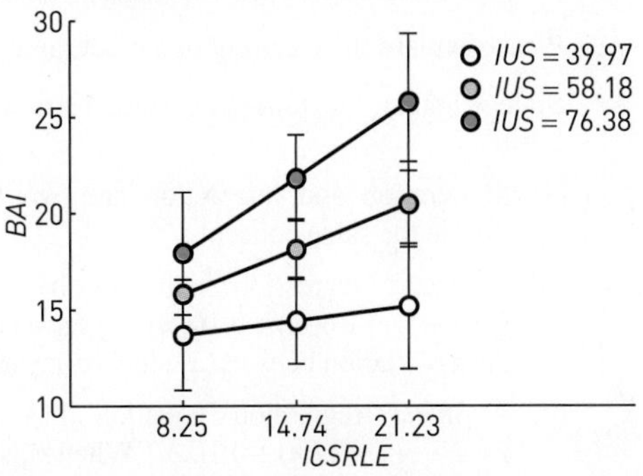

Plots of the regression equation that includes the interaction term (*ICSRLE*IUS*) for values of *ICSRLE* at the mean (14.74) and 14.74 ± 6.4878 when *IUS* is fixed at 39.97, 58.18, and 76.38.

When these steps have been completed, click Continue to return to the Linear Regression dialog, and there click OK to proceed with the analysis. For each of the nine combinations of x and *mod* (shown in Figure 17.20f), the data file will contain $\hat{y}$ and its 95% confidence interval. These five columns of numbers (x, *mod*, $\hat{y}$, lower, upper) can be used to generate a graph like the one shown in Figure 17.21.

The results of the study we've just worked through could be reported as follows:

APA Reporting

A study was conducted to determine whether intolerance of uncertainty, measured on the Intolerance of Uncertainty Scale (*IUS*), moderates the association between daily stressors, measured on the Inventory of College Students' Recent Life Events (*ICSRLE*), and anxiety, measured on the Beck Anxiety Inventory (*BAI*). *BAI* was the outcome variable, *ICSRLE* the predictor variable, and *IUS* the hypothesized moderator.

A multiple regression analysis with *ICSRLE* and *IUS* as predictors explained 19% of the variability in *BAI* ($R^2 =$. 19, approximate 95% CI [0.06, 0.32] computed using the MBESS package in **R** [Kelley, 2007]). When the interaction term (*ICSRLE*IUS*) was added, R^2 increased to .21, $\Delta R^2 = .03$, and the approximate 95% CI [0, 0.08] was computed using the bootstrapping procedure described by Algina et al. (2008).

The nature of the *ICSRLE*IUS* interaction was characterized by the slope of the regression surface relating *ICSRLE* to *BAI* for values of *IUS* at $M - SD$, M, and $M + SD$, which were 39.97, 58.18, and 76.38, respectively. These slopes were 0.12 (95% CI [−0.21, 0.44]), 0.36 (95% CI [0.12, 0.61]), and 0.60 (95% CI [0.22, 0.99]), respectively.

The interaction between *ICSRLE* and *IUS* is consistent with Chen and Hong's (2010) suggestion that *IU* moderates the association between stressors and anxiety. For those with low *IUS* scores, there was a weak association between *ICSRLE* and *BAI*. For those with higher *IUS* scores, there was a stronger association between *ICSRLE* and *BAI*. That is, the slopes relating *BAI* to *ICSRLE* increase as *IUS* increases. However, it must be stressed that the interaction term accounted for very little variability in the data. There was only a 3% increase in explained variance when *ICSRLE*IUS* was added to the regression equation.

LEARNING CHECK 3

1. Please explain the meaning of a moderator variable.

2. State whether the following statements are true or false.

   (a) Moderation and interaction are two terms that denote the same concept.

   (b) A linear regression equation has the form $\hat{y} = 2 + 5(x) + 6(m) + 0.1(xm)$. When m is fixed at 5, the association between x and $\hat{y}$ is linear.

   (c) A linear regression equation has the form $\hat{y} = 2 + 5(x) + 6(m) + 0.1(xm)$. When m is fixed at 5, the slope of the line relating x to $\hat{y}$ is 5.5.

   (d) A linear regression equation has the form $\hat{y} = 2 + 5(x) + 6(m) + 0.1(xm)$. When m is fixed at 5, the slope of the line relating x to $\hat{y}$ is 5.6.

3. A linear regression equation has the form $\hat{y} = 2 + 5(x) + 6(m) + 0.1(xm)$. Calculate the slope of the line relating x to $\hat{y}$ when m is fixed at 12.

4. A statistics instructor assumed that the number of hours students studied for a test was associated with the grades received. However, he suspected that studying distributed over several days would also lead to better grades. Furthermore, he suspected that the number of days over which studying was distributed would moderate the association between hours of study and test grade. Specifically, he thought grades would improve more rapidly with hours of study for students who distributed their studying over several days than for those who crammed a day or two before the test. The ModerationLC.sav data file (available at study.sagepub.com/gurnsey) contains the data for this question. The file contains three columns of numbers for the variables *grades*, *hours*, and *days*. Conduct analyses in SPSS to answer the following questions.

   (a) What are the means and standard deviations of the three variables?

   (b) What is the regression equation that relates *hours* and *days* to *grades*?

   (c) Does the two-predictor model explain a statistically significant proportion of the variability in *grade*? Support your answer.

   (d) When the interaction term is added to the model, what is ΔR^2 and is it statistically significant?

   (e) When *days* is fixed at 5, what is the slope of the line relating *hours* to $\hat{y}$?

   (f) Use the moderation analysis in PROCESS to determine whether the data support the instructor's prediction about how *days* moderates the association between *hours* and *grades*.

   (g) Conduct a hierarchical regression analysis with *hours* entered on step 1 and *days* entered on step 2. Do the results of this hierarchical regression help us assess the practical significance of the moderation analysis?

Answers

1. When the association between x and y changes for different values of a third variable *mod*, we say that *mod* moderates the association between x and y. We could also say that *mod* is a moderator or that there is an interaction between x and *mod*.

2. (a) True. (b) True. (c) True. (d) False.

3. $b = b_x + b_{xm}(m) = 5 + .1(12) = 6.2$.

4. (a) $m_{grades} = 70.66$, $m_{hours} = 9.77$, $m_{days} = 4.64$, $s_{grades} = 10.89$, $s_{hours} = 3.07$, $s_{days} = 2.05$.

   (b) $\hat{y} = 45.13 + 2.05(hours) + 1.8(days)$.

   (c) Yes. $R^2 = .54$, $F_{2,61} = 36.18$, $p < .001$.

   (d) Yes. $\Delta R^2 = .03$, $F_{1,60} = 4.68$, $p = .03$.

   (e) The regression equation is $\hat{y} = 58.20 + 0.65(hours) - 1.59(days) + 0.27(hours*days)$; therefore, $b = b_x + b_{xm}(m) = b = 0.65 + 0.27(5) = 2.02$.

   (f) Yes. The data support the instructor's prediction. Adding the interaction term (*hours*days*) produced a statistically significant increase in explained variance; $\Delta R^2 = .03$, $p = .03$, 95% bootstrapped CI [0, 0.11] (Algina et al., 2008). Furthermore, when *days* was fixed, at $m_{days} - s_{days}$, m_{days}, and $m_{days} + s_{days}$ = 2.59, 4.64, and 6.69, respectively, the slopes of the regression lines relating *hours* to *grades* were 1.36, 1.93, and 2.49. The 95% confidence intervals around these estimates were [0.37, 2.35], [1.16, 2.69], and [1.64, 3.34]. This means that the increase in grades with hours studied is greater when studying is distributed over more days. Therefore, there is evidence that there is an advantage to distributing study hours over several days.

   (g) Although the instructor's prediction was supported, the results of a hierarchical analysis show that the number of *hours* studied is the strongest predictor of *grades* ($R^2 = .51$) with *days* adding small, but statistically significant, increases in explained variance (both $\Delta R^2 = .03$). Of course, *hours* and *days* are random variables, so we can't conclude that either is causally related to *grades*. However, even if we were to make this assumption, it seems that *hours* is the most important factor relating to grades, with small increases in predictive power coming from *days* and *hours*days*.

SUMMARY

The regression coefficients computed from a sample,

$$b_1, b_2, \ldots, b_k$$

estimate the corresponding coefficients in a population,

$$\beta_1, \beta_2, \ldots, \beta_3.$$

In samples, each regression coefficient (b_i) expresses the change in $\hat{y}$ associated with a change in x_i, with all other predictors ($x_{j \neq i}$) held constant. Regression coefficients are also called *partial regression coefficients* because they express the association between $\hat{y}$ and x_i with the effects of all other predictors ($x_{j \neq i}$) partialed out.

A confidence interval around a sample regression coefficient, b_i, is computed as

$$CI = b_i \pm t_{\alpha/2} \left(s_{b_i} \right), \text{ where}$$

$$s_{b_i} = \frac{s_{est}}{\sqrt{ss_{x_i}(1 - R_{i.j \neq i}^2)}} \text{ and}$$

$$s_{est} = \sqrt{s_y^2(1 - R^2)\frac{n-1}{n-k-1}}.$$

The main difference between s_b in simple linear regression and s_{b_i} in multiple regression is that ss_{x_i} is multiplied by $1 - R_{i,j \neq i}^2$ in multiple regression but not simple regression. $R_{i,j \neq i}^2$ is the proportion of variance in x_i explained by the other predictors $x_{j \neq i}$, and $1 - R_{i,j \neq i}^2$ is the proportion of variance in x_i not explained by the other predictors $x_{j \neq i}$. $1 - R_{i,j \neq i}^2$ is called *tol* (tolerance) in the statistics literature. As *tol* decreases, s_{b_i} increases. The reciprocal of *tol* is the variance inflation factor ($VIF = 1/tol$). *VIF* conveys how much s_{b_i} is magnified in the presence of predictors $x_{j \neq i}$ compared with what s_{b_i} would be if $R_{i,j \neq i}^2$ were 0. Therefore, we could also write s_{b_i} as follows:

$$s_{b_i} = \sqrt{\frac{s_{est}^2}{ss_{x_i}} VIF}.$$

We reviewed three applications of multiple regression that focus on the regression coefficients rather than the predictive power of the model as a whole. These applications were *statistical control*, *mediation*, and *moderation*.

A simple correlation between two variables (e.g., x_1 and y) may be explained by the influence of a third variable (x_2). To assess the influence of x_2 on the association between x_1 and y, we can conduct a hierarchical regression with x_1 entered on step 1 and x_2 added on step 2. This allows us to examine the association between x_1 and y while *controlling for x_2*. If the contribution that x_1 makes to explained variance in y decreases substantially or disappears in the presence of x_2, this supports the idea that it is x_2 that explains the original association between x_1 and y.

Mediation analysis allows us to assess proposed causal sequences of associations among variables. In a simple mediation analysis, we ask whether the predictor variable (x) exerts its influence on the outcome variable (y) through a mediating variable (*med*). That is, we examine the proposal that x leads to *med* leads to y (i.e., $x \to med \to y$). The analysis decomposes that association between x and y into a direct path and an indirect path. The direct path from x to y is conveyed by the partial regression coefficient c' on x from a regression analysis predicting y from x and *mod*. The indirect path is the product of two regression coefficients. The first (a) is the regression coefficient on x in the equation that predicts *med* from x. The second (b) is the partial regression coefficient on *med* from a regression analysis predicting y from x and *med*. The total effect of x on y is the sum of the direct effect (c') and indirect effect (ab). A strong indirect effect (ab) and weak direct effect (c') supports a plausible hypothetical causal chain $x \to med \to y$. The sampling distribution of ab is not simple, so a bootstrapping procedure is used to compute a confidence interval around ab.

Moderation refers to situations in which the association between two variables (e.g., x and y) depends on the level of a third variable called the moderator (*mod*). We may also say that there is an interaction between x and *mod*. When this is true, a multiple regression, with only x and *mod* predicting y, will not fully capture the nature of the association. To test for moderation, a third variable (xm) is created, which is the product of x and *mod*. The product, xm, is often referred to as the interaction term. A hierarchical regression can determine whether adding xm yields a statistically significant increase in explained variance.

The following equation relates x, *mod*, and xm to $\hat{y}$:

$$\hat{y} = a + b_x(x) + b_{mod}(mod) + b_{xm}(xm).$$

The slope of the regression surface in the x direction for a given value of *mod* is given by

$$b = b_x + b_{xm}(mod).$$

To depict the interaction graphically, it is common to plot the regression line relating x to y for values of *mod* set to $m_{mod} - s$, m_{mod}, and $m_{mod} + s$. This allows us to determine whether the interaction is in agreement with the prediction we've made.

KEY TERMS

direct effect (c')   458
indirect effect (ab)   458
interaction   464
mean-centering   466

mediation   457
moderation   464
partial correlation coefficient   450
partial regression coefficient   452

statistical control   457
total effect   458

EXERCISES

Definitions and Concepts

1. Put into words what is meant by the sampling distribution of b_i.

2. What does it mean for a variable to be a mediator?

3. What does it mean for a variable to be a moderator?

4. What is a partial score?

5. What is a partial correlation?

6. What is a partial regression coefficient?

7. What is the variance inflation factor?

8. Explain tolerance.

True or False

State whether the following statements are true or false.

9. Predictor variables in multiple regression are typically uncorrelated with each other.

10. $\hat{y} = 2 + 3(x_1) - 4(x_2)$; therefore, $\hat{y} = 10 + 3(x_1)$ when $x_2 = -2$.

11. If z has been partialed out of the association between x and y, then the correlation between the partial scores for x and y must be 0.

12. $s_{b_i} = \sqrt{s_{est}^2 / ss_{x_i} VIF}$.

13. $tol = 1/VIF$.

14. In a two-predictor model, when $tol = 0$, $r_{x_1 . x_2}^2 = 1$.

15. Bootstrapping can be used only when the sampling distribution of the statistic is complex or unknown.

16. If the 95% bootstrapped confidence interval for b_i = [9, 15], then 95% of the bootstrapped values of b_i fall below 9.

17. A sampling distribution is a probability distribution of all possible values of a sample statistic based on samples of the same size.

18. $x_i - \hat{x}_{i,j \neq i}$ defines a partial score.

19. A partial correlation coefficient is the correlation between two sets of partial scores.

20. A partial regression coefficient for variable x_i shows the association between y and x_i while controlling for $x_{j \neq i}$.

21. If $k = 10$, then there are eight variables defined by $x_{j \neq i}$.

22. We use statistical control to determine whether the association between two variables is weakened in the presence of other predictors.

23. In a mediation analysis, $c' = c + ab$.

24. In a mediation analysis, c describes the direct effect of x on y.

25. In a mediation analysis, b is a partial regression coefficient.

26. In a mediation analysis, a is a partial regression coefficient.

27. In a moderation analysis, *med* is a moderator variable.

28. Mean-centering variables x and y will decrease the correlation between them.

Calculations

29. The multiple regression equation computed from a sample is $\hat{y} = 10 + 0.25(x_1) + 0.64(x_2) - 0.56(x_3)$. Show the simple linear regression equation that relates $\hat{y}$ to x_1 when

    (a) $x_2 = 1$ and $x_3 = 4$,

    (b) $x_2 = -2$ and $x_3 = -3$, and

    (c) $x_2 = 5$ and $x_3 = -2$.

30. A multiple regression equation computed from a sample of 28 scores is $\hat{y} = 2 + 1.2(x_1) - 0.5(x_2)$. This equation produces $R^2 = .7$. The standard deviations of y, x_1, and x_2 are 4, 2, and 3, respectively. The squared correlation between x_1 and x_2 is .5. Compute the 95% confidence interval around

    (a) b_1 and

    (b) b_2.

31. A multiple regression equation computed from a sample of $n = 100$ scores is $\hat{y} = 10 + 0.25(x_1) + 0.64(x_2) - 0.56(x_3)$. This equation produces $R^2 = .60$. The standard deviations of y, x_1, x_2, and x_3 are 6, 3, 1, and 2, respectively. The values of $R^2_{i.j \neq i}$ are $R^2_{1.2,3} = .5$, $R^2_{2.1,3} = .7$, and $R^2_{3.1,2} = .6$. Compute the 95% confidence intervals around

    (a) b_1,

    (b) b_2, and

    (c) b_3.

32. Use Sobel's method to compute the approximate 95% confidence interval around ab for each of the following situations:

    (a) $a = 4, b = 2, s_a = 1.5$, and $s_b = .4$;

    (b) $a = 8, b = 0.6, s_a = 2$, and $s_b = 0.20$; and

    (c) $a = 0.15, b = 1.34, s_a = 0.05$, and $s_b = 0.75$.

33. For each of the following situations one component of the mediation analysis is missing. Use the information given to solve for the missing information:

    (a) $a = 0.5, b = 0.6, c = 0.7$;

    (b) $a = 0.5, b = 0.6, c' = 0.7$;

    (c) $a = 0.5, c = 0.6, c' = 0.7$; and

    (d) $b = 0.5, c = 0.6, c' = 0.7$.

34. For each situation in question 33, find how much change there is in $\hat{y}$ with a unit change in x through the direct effect and through the indirect effect.

35. Each of the scenarios below provides enough information to conduct a hierarchical regression by hand. Use the information in each scenario to complete the tables on the next page. You should do your calculations in Excel.

    (a) There are three variables, y, x_1, and x_2, each with $n = 81$ scores. The sums of squares for these variables are $ss_y = 2335.57$, $ss_{x_1} = 38771.76$, and $ss_{x_2} = 8240.01$. The correlation between x_1 and x_2 is $r = .5287$. Variable x_1 was entered on the first step of the analysis. The slope of the regression equation was $b_1 = 0.1596$ and $ss_{\hat{y}}$ was 987.35. On step 2, x_2 was added. The two regression coefficients were $b_1 = 0.1346$ and $b_2 = 0.1025$, and $ss_{\hat{y}} = 1049.68$.

    (b) There are three variables, y, x_1, and x_2, each with $n = 51$ scores. The sums of squares for these variables are $ss_y = 26,726.72$, $ss_{x_1} = 14,145.62$, and $ss_{x_2} = 3889.62$. The correlation between x_1 and x_2 is $r = .589$. Variable x_1 was entered on the first step of the analysis. The slope of the regression equation was $b_1 = 0.7184$ and $ss_{\hat{y}}$ was 7299.46. On step 2, x_2 was added. The two regression coefficients were $b_1 = 0.4040$ and $b_2 = 1.0168$, and $ss_{\hat{y}} = 9921.90$.

    (c) There are three variables, y, x_1, and x_2, each with $n = 61$ scores. The sums of squares for these variables are $ss_y = 743.42$, $ss_{x_1} = 1220.41$, and $ss_{x_2} = 1325.4$. The correlation between x_1 and x_2 is $r = .2173$. Variable x_1 was entered on the first step of the analysis. The slope of the regression equation was $b_1 = 0.3738$ and $ss_{\hat{y}}$ was 170.6. On step 2, x_2 was added. The two regression coefficients were $b_1 = 0.2882$ and $b_2 = 0.3781$, and $ss_{\hat{y}} = 351.01$.

Model Summary

Model	R	R^2	R^2_{adj}	s_{est}	Change statistics				
					ΔR^2	ΔF	df_1	df_2	p
1									
2									

Coefficients

Model		b	s_{b_i}	Beta	t	p	95% lower	95% upper
1	b_1							
2	b_1							
	b_2							

Scenarios

36. The GPAs.sav data file (available at study.sagepub .com/gurnsey) contains data for the following variables: GPA at the end of first year university (*Univ_GPA*), high school GPA (*HS_GPA*), and ratings from high school teachers (*HS_Ratings*). Let *Univ_GPA* be the outcome variable and let *HS_GPA* and *HS_Ratings* be the predictors. Run a hierarchical regression with *HS_GPA* entered on the first step and *HS_Ratings* entered on the second step.

    (a) Is ΔR^2 statistically significant?

    (b) Report the regression coefficients and standardized regression coefficients for *HS_GPA* on steps 1 and 2.

    (c) You should find that the two-predictor model explains a statistically significant proportion of variability in *y*; $\Delta R^2 = .27$, $F_{1,22} = 4.13$, $p = .03$. Are the individual predictors statistically significant? Do you see any contradiction between the model level of analysis and the coefficient level of analysis?

    (d) If the admissions committee wished to reduce the number of performance indicators submitted by students, would they prefer to drop high school GPA or ratings from high school teachers? Explain your answer.

37. The Mystery.sav data file (available at study .sagepub.com/gurnsey) contains variables *y*, x_1, and x_2. Run two hierarchical regressions: the first with x_1 entered on step 1 and x_2 on step 2, and the second with x_2 entered on step 1 and x_1 on step 2. There is something unusual about the coefficients in these regressions that we have not seen before. Explain what this unusual behavior is and why it happens.

38. A developmental researcher had collected extensive data related to school performance in elementary school students. He asked one of his senior students to investigate the question of whether cognitive self-control mediates the association between language and math abilities. The School.sav data file (available at study .sagepub.com/gurnsey) contains the variables *PPV*, *math*, and *CSC*. *PPV* represents scores of 100 kindergarten children on the Peabody Picture Vocabulary Test, which is a measure of verbal ability and scholastic aptitude. *Math* represents the final grades in math for the same 100 students at the end of first grade. *CSC* is a measure of cognitive self-control. These *CSC* scores are from a teacher-completed questionnaire in which students are rated on the following types of items: "Does the child work for long-term goals?" and "Does the child have to be reminded several times to do something before he or she does it?" Use SPSS to assess the prediction that *CSC* mediates the association between *PPV* and *math*, where math is the outcome variable and PPV is the predictor variable. Mention the total effect, direct effect, and indirect effect. Also comment on the plausibility of the proposed mediation.

39. Let's say a researcher hypothesizes that the number of hours that children devote to practicing skills like music, painting, or sports instills discipline and a sense of competence and positive self-image. Data were collected for a sample of 64 university student volunteers from a psychology department participant pool. The Discipline.sav data file (available at study.sagepub.com/gurnsey) contains scores on three variables for each of these 64 students. These variables are *hours*, *Self_Im*, and *breaks*. *Hours* is the number of hours per week that each student self-reported participating in structured practice during childhood. *Self_Im* is a measure of self-image obtained from a multi-item questionnaire. To assess discipline, the researchers observed each of the 64 students while they completed a difficult series of logic problems. Three observers independently counted the number of times each participant disengaged from the task within a 1-hour period. The average of the three observers' counts is given in the column labeled *breaks*.

    (a) Does the association between *hours* and *Self_Im* support the researcher's theory?

    (b) What is the total effect of *hours* on *Self_Im*?

    (c) What is the direct effect of *hours* on *Self_Im*? How much change is there in *Self_Im* for each hour of practice via the direct effect?

    (d) What is the indirect effect of *hours* on *Self_Im*? How much change is there in *Self_Im* for each hour of practice via the indirect effect?

    (e) Does your analysis support the notion that discipline mediates the association between *hours* and *Self_Im*?

    (f) Comment on the plausibility of the researcher's hypothesis of mediation, and any limitations you see in this study.

40. How does parental pressure affect grades? Imagine that a researcher believes that parental pressure does not always have a positive effect on children's grades. She wonders if the effect depends on students' IQ. To assess the interaction between IQ and parental pressure, she chose a random sample of 68 statistics students at a large European university. For each student, she measured IQ (*IQ*), grade on the most recent statistics test (*grade*), and the student's self-reported assessment of how much pressure the student's parents exerted on him or her to achieve high grades (*pressure*). The Pressure.sav data file (available at study.sagepub.com/gurnsey) contains scores for *IQ*, *grade*, and *pressure* for each student. Conduct a regression analysis to determine whether there is evidence of an interaction between *IQ* and *pressure* relating to student grades.

    (a) Explain how you determined whether the interaction exists.

    (b) Put into words what the interaction shows.

    (c) Report the slope relating *IQ* to *grade* for values of *pressure* at $m_p - s$, m_p, and $m_p + s$. For each slope, report its 95% confidence interval.

    (d) Report the 95% confidence intervals for predicted values of *grade*, for all combinations of $m_p - s$, m_p, and $m_p + s$, and $m_{IQ} - s$, m_{IQ}, and $m_{IQ} + s$.

    (e) Make a graph of the interaction between *IQ* and *pressure* and show the 95% confidence interval for all combinations of $m_p - s$, m_p, and $m_p + s$, and $m_{IQ} - s$, m_{IQ}, and $m_{IQ} + s$.

    (f) In a few sentences, explain how parental pressure interacts with IQ.

41. A graduate student in an industrial/organizational psychology program wondered about how personality and situational factors influence performance on an attention-demanding task. Over a 2-month period, she visited a college cafeteria and asked volunteers to complete two forms. The first was an introversion/extroversion questionnaire with 25 items; high scores on the questionnaire mean a high degree of extroversion, and low scores mean a high degree of introversion. The second form was a page of logic puzzles. While her volunteers were completing the questionnaire and the logic puzzles, the graduate student counted the number of people in the cafeteria. Her results are in the Cafeteria.sav data file (available at study.sagepub.com/gurnsey). The first column shows the number of logic puzzles solved (*NSolved*), the second shows the number of people in the room (*NPeople*), and the third shows each volunteer's score on the introversion/extroversion scale (*IE*). The question is whether there is an interaction between *IE* and *NPeople*, where *IE* is treated as the moderator.

    (a) Explain how you determined the strength of the interaction.

    (b) Put into words what the interaction shows.

(c) Report the slope relating *NSolved* to *NPeople* for values of *IE* at $m_{IE} - s$, m_{IE}, and $m_{IE} + s$. For each slope, report its 95% confidence interval.

(d) Report the 95% confidence intervals for all combinations of $m_{NP} - s$, m_{NP}, and $m_{NP} + s$, and $m_{IE} - s$, m_{IE}, and $m_{IE} + s$.

(e) Make a graph of the interaction between *IE* and *NPeople* and show the 95% confidence interval for all combinations of $m_{NP} - s$, m_{NP}, and $m_{NP} + s$, and $m_{IE} - s$, m_{IE}, and $m_{IE} + s$.

(f) In a few sentences, explain how *NPeople* interacts with *IE*.

42. A psychology student in her senior year has begun a research project that investigates the association between scientific literacy and views on anthropogenic global warming. She suspects that this association may be moderated by interest in science fiction movies. She chose a random sample of 59 students from her department's participant pool, and she obtained three measures from each participant. The first was a 25-item questionnaire that assessed scientific literacy (*literacy*), the second was a 20-item questionnaire assessing the degree to which the respondent believed in anthropogenic global warming (*AGW*), and the third was a count of the number of science fiction movies each participant had seen in the last year (*SciFi*). The question is whether there is an interaction between *AGW* and *SciFi*, when *SciFi* is treated as the moderator.

(a) Explain how you determined whether the interaction exists.

(b) Put into words what the results show.

(c) Report the slope relating *literacy* to *AGW* for values of *SciFi* at $m_{SF} - s$, m_{SF}, and $m_{SF} + s$. Report the 95% confidence interval for each slope.

(d) Make a graph of the interaction between *literacy* and *SciFi* and show the 95% confidence interval for all combinations of $m_{Lit} - s$, m_{Lit}, and $m_{Lit} + s$, and $m_{SF} - s$, and m_{SF}, $m_{SF} + s$.

(e) To what population do these results apply?

APPENDIX 17.1: INSTALLING THE PROCESS MACRO IN SPSS

Professor Hayes's PROCESS macro can be downloaded online (processmacro.org/download.html). There is a very obvious red button on this webpage that says Click to download PROCESS v2.16 (Figure 17.A1.1a). If you click on this link, a compressed (.zip) file will be downloaded to your location of choice on your computer (e.g., the desktop or some other folder). Once downloaded, the icon of a zip file will appear; the file is called process.zip (Figure 17.A1.1b). Double-clicking on the process.zip file will cause it to uncompress, resulting in a new folder in your chosen location. Opening the folder will reveal a large number of files (Figure 17.A1.1c).

Now that the PROCESS files are on your computer, you can add the PROCESS macro to your version of SPSS. If you are using SPSS Statistics 23 or earlier, launch SPSS and then click on the Utilities menu item, as shown in Figure 17.A1.2a. At the bottom of this menu, you will see the item Custom Dialogs that offers you the option to Install Custom Dialogs. . . . When this is chosen, a window will appear that allows you to navigate to the process folder shown in Figure 17.A1.1c. Choose the file process.spd from within this folder to complete the installation. Note, if you are on a Windows system, you must be running Windows in administrator mode to do the installation.

FIGURE 17.A1.1 ■ Downloading PROCESS

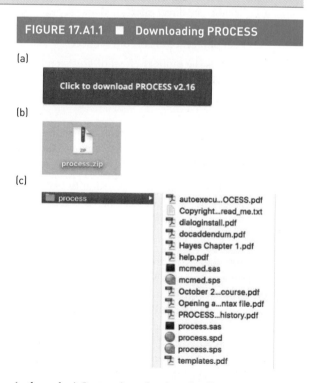

(a through c) Instructions for downloading the PROCESS macro for SPSS.

FIGURE 17.A1.2 ■ Installing PROCESS

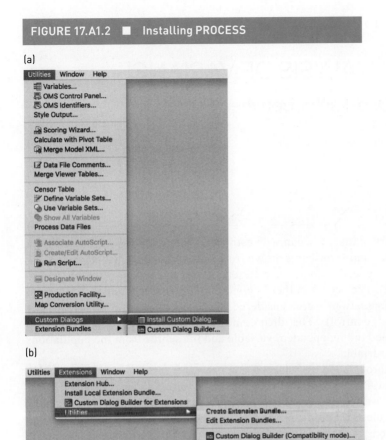

Instructions for installing the PROCESS macro for SPSS.

(Instructions for how to do this are given in the file dialoginstall.pdf in the process folder.) If you are running SPSS Statistics 24, the Install Custom Dialogs . . . item is under the extensions menu item as shown in Figure 17.A1.2b.

As noted in the chapter, once PROCESS has been installed, it will appear as one of the options available from the Analyze→Regression menu (see Figure 17.A1.3). The chapter describes how to use PROCESS to conduct mediation and moderation analyses.

FIGURE 17.A1.3 ■ Using PROCESS

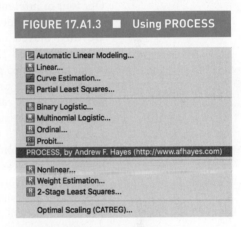

The PROCESS macro in SPSS.

18 ANALYSIS OF VARIANCE
One-Factor Between-Subjects

INTRODUCTION

Our tour of inferential statistics began with estimating the mean of a population from the mean of a random sample of scores drawn from the population (Chapters 6 and 10). We then moved on to estimating the difference between two population means (Chapters 11 and 12). The scores in these populations were measured on an interval scale, and the two populations were considered two levels of a dichotomous variable (e.g., treatment and control). The dichotomous variable that defined the two populations was called the independent variable and the scores from the populations were the dependent variable.

In Chapters 13 through 15, we widened our view and considered the association between two random variables. Initially the two variables were both measured on interval scales, in which case we referred to the dependent variable as the outcome variable and the independent variable as the predictor variable. The analysis of the association between two continuous random variables falls under the heading of correlation and regression. The universality of correlational analysis was shown when we considered the two-groups design from the perspective of correlation and regression.

In Chapters 16 and 17, the regression model was expanded to include two or more predictors, which could be interval-scaled or dichotomous. We called this multiple regression. We saw that a multiple regression analysis can be viewed from two levels. The model level addresses the variance in the outcome variable explained by the combined effects of the predictors. The predictor level examines the contributions that individual predictors make to explained variance in the outcome variable, as well as how they are affected by the presence of other predictors.

In this chapter, we will expand the two-groups design to three or more groups. As with multiple regression, we will see that the analysis of group means can be viewed at the model level and at the level of predictors. It may seem strange at this point to talk about predictors in the context of the analysis of group means, but we did this before when we viewed the two-groups design from the regression perspective.

THE ONE-FACTOR, BETWEEN-SUBJECTS ANOVA

The analysis covered in this chapter is called analysis of variance, or ANOVA. There is quite a bit of new terminology here, and some of it overlaps with what was covered in multiple regression. Therefore, we will start by establishing a basic vocabulary before moving on to details of the analysis.

Independent Variable = Factor = Way

In ANOVA terminology we often refer to an independent variable as a **factor**. For example, if we were interested in how fitness levels differ between students in different university departments, then we might obtain fitness scores from students in psychology, math, languages, political science, and biology. In this case the independent variable *department* can be called a factor, and it has values of *psychology*, *math*, *languages*, *political science*, and *biology*. The dependent variable is fitness level (*fitness*). The values of the independent variable are also called **levels** of the independent variable. When there is one independent variable, we say it is a *one-factor ANOVA*. We also say that such an ANOVA is a *one-way ANOVA*.

The one-factor ANOVA examining fitness levels across university departments could be expanded to include a second independent variable (factor, way) such as sex. That is, we might ask about how *fitness* changes as a function of both *department* and *sex*. Such a study would be a two-factor ANOVA.

For simplicity of notation, we often identify the factors in an ANOVA with the letters *A*, *B*, *C*, and so on. So, in the two-factor ANOVA just described, we might denote university department with the letter *A* and sex with the letter *B*. Because there are five levels of *A* and two levels of *B*, we can refer to this as a 5 (*departments*) by 2 (*sexes*) ANOVA. Because there are 5 × 2 = 10 combinations of the levels of the two independent variables, we say there are 10 **conditions** in the experiment, each representing a different combination of the levels of the two factors.

Between-Subjects, Within-Subjects, and Mixed Designs

In Chapter 11, we looked at the independent-groups design, in which the scores in the two groups were completely independent of each other. We noted that an independent-groups design is often called a **between-subjects** design. The two-way design investigating *fitness* as a function of *department* and *sex* is an example of a between-subjects design because each individual is in one and only one condition of the study.

In Chapter 12, we examined dependent-groups designs, which involved two samples of scores that were related to each other in some way. One type of dependent-samples design involves repeated measures (e.g., before- and after-lunch weights) and another involves matched samples (e.g., individuals matched for IQ, age, sex, or fitness). In ANOVA, when individual participants are tested at each level of an independent variable (factor), we say that the design is a **within-subjects** design. A memory test given to individuals after they were sleep deprived for 24, 36, or 48 hours is an example of a within-subjects design. A within-subjects ANOVA is a simple extension of the repeated-measures design when there are two groups.

It is not uncommon to have a two-factor design with one between-subjects factor and one within-subjects factor. For example, sleep deprivation could be the within-subjects factor and sex could be the between-subjects factor. This is an example of a **mixed design**. In this example, we might ask how memory scores change as a function of sleep deprivation for males and females.

In this chapter, we will consider the one-way between-subjects ANOVA. This means that the number of conditions in the study is equal to the number of levels of the independent variable.

An Example

We will introduce ANOVA with a hypothetical experiment that we will return to throughout the chapter. Let's say that two new treatments for Alzheimer's disease have been

---

A **factor** in an ANOVA is an independent variable. A one-factor ANOVA (or one-way ANOVA) has one independent variable, a two-factor ANOVA has two independent variables, a three-factor ANOVA has three independent variables, and so on.

Each independent variable (factor, way) in an ANOVA has two or more discrete values. These values are said to be **levels** of the independent variable (factor, way).

Each **condition** of an ANOVA represents a specific combination of the levels of the independent variables. The number of conditions is the product of the number of levels in the factors.

A **between-subjects factor** in an ANOVA involves scores on the dependent variable obtained from individuals that are unrelated to each other in any way. That is, scores are randomly selected from independent populations.

A **within-subjects factor** in an ANOVA involves scores obtained from the same individuals at different levels of the independent variable. Therefore, scores are randomly selected from dependent populations.

A **mixed** ANOVA involves one or more between-subjects factors and one or more within-subjects factors.

developed and a research team would like to assess their effectiveness. The researchers chose a random sample of 15 men ages 70 to 75 who were diagnosed with Alzheimer's disease within the last 6 months. The 15 participants were then assigned at random to one of three conditions. In one condition, participants took a placebo pill for 6 months; in the other two conditions, participants took either Drug 1 or Drug 2 for 6 months. At the end of this 6-month period, all participants were given the Montreal Cognitive Assessment (MoCA), which is a physician- or psychologist-administered test of memory, spatial reasoning, and executive function (Nasreddine et al., 2005). The MoCA yields scores that range from 0 to 30. Average scores are 27.4 for normal control participants, 22.1 for individuals with mild cognitive impairment (MCI), and 16.2 for individuals with Alzheimer's disease.

Figure 18.1 shows the mean results (and 95% CIs) for the three groups at the end of the study. The results seem very clear. The mean scores for participants in the two drug conditions were higher than the mean scores for those in the placebo condition. We will explain the logic of ANOVA using these data.

ANOVA is all about partitioning variance, a topic that was introduced in Chapters 11 and 12. In those chapters, we asked about the proportion of total variability in the dependent variable attributable to the difference between the two group means. Therefore, the first question we address is the same: What proportion of the total variability in our set of 15 scores is explained by differences in the three group means? To answer this question, we will partition the variability in these 15 scores into two sources: variability attributable to means, and variability not attributable to means.

We will refer to the three *conditions* of the experiment as a_1, a_2, and a_3, which correspond to Placebo, Drug 1, and Drug 2, respectively. Columns a_1, a_2, and a_3 of Table 18.1 show the raw data from which the means and confidence intervals in Figure 18.1 were computed. To partition the variability in this set of 15 scores, we must first compute the overall mean (sometimes called the *grand mean*) and then the sum of squared deviations about this mean. To compute the grand mean, we use the following simple formula:

$$m_y = \frac{\sum_{i=1}^{k}\sum_{j=1}^{n_i} y_{i,j}}{\sum_{i=1}^{k} n_i}, \tag{18.1}$$

where k is the number of means (conditions) and n_i is the number of participants in the ith condition. In this example, $\sum_{i=1}^{k} n_i = kn = 3*5 = 15$.

This is the first time that we've seen a double summation sign. To make clear what is being summed, indexes have been used and associated with different quantities. The subscript i ranges from 1 to $k = 3$ and represents the three groups in the experiment. The subscript j ranges from 1 to $n = 5$ and represents the participants in each condition. We can think of i and j as column and row indexes, respectively, for the 15 scores in the first three columns of Table 18.1. For example, the score in column a_3, of row 4 is 22. This corresponds to the fourth participant in the third treatment condition; i.e., $y_{3,4} = 22$. (Similarly, $y_{1,5} = 16$, and $y_{2,1} = 19$.) If you do the calculations, you will find that the mean of the 15 scores is $m_y = 18$.

The total sum of squared deviations from m_y is defined as follows:

$$ss_{total} = \sum_{i=1}^{k}\sum_{j=1}^{n}(y_{i,j} - m_y)^2. \tag{18.2}$$

Equation 18.2 tells us to subtract 18 from each of the 15 scores, square these deviations, and then sum the squared deviations. This

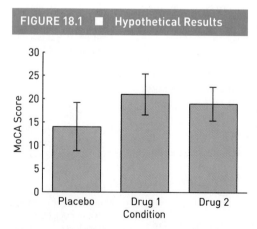

FIGURE 18.1 ■ Hypothetical Results

Mean Montreal Cognitive Assessment (MoCA) score for each of the three conditions. Error bars = 95% CI.

TABLE 18.1 ■ Data for Three Groups[a]

j	a_1	a_2	a_3	m_1	m_2	m_3	$y_{1,j} - m_1$	$y_{2,j} - m_2$	$y_{3,j} - m_3$
1	17	19	20	14	21	19	3	−2	1
2	18	21	15	14	21	19	4	0	−4
3	9	18	21	14	21	19	−5	−3	2
4	10	27	22	14	21	19	−4	6	3
5	16	20	17	14	21	19	2	−1	−2
Mean	14	21	19	14	21	19	0	0	0
	$ss_{total} = \sum_{i=1}^{k}\sum_{j=1}^{n}(y_{i,j} - m_y)^2$			$ss_a = n\sum_{i=1}^{k}(m_i - m_y)^2$			ss_1	ss_2	ss_2
	284			130			70	50	34
								$ss_{within} = ss_1 + ss_2 + ss_3$	
								154	

[a] a_1 = Placebo, a_2 = Drug 1, a_3 = Drug 2.

produces ss_{total}, which is the quantity to be partitioned. Using the **DEVSQ** function in Excel, you should be able to confirm that $ss_{total} = 284$.

We've seen before that partitioning variance is based on decomposing each score into two parts. In the case of the between-subjects ANOVA, we decompose each score into the mean of its group (m_i) and its deviation from the mean of its group ($\varepsilon_{i,j} = y_{i,j} - m_i$). At the bottom of columns 1 to 3, we see the three group means: $m_1 = 14$, $m_2 = 21$, and $m_3 = 19$. These means replace the raw scores in the center three columns of Table 18.1. When sample sizes are equal, the variability attributable to group means is computed as

$$ss_a = n\sum_{i=1}^{k}\left(m_i - m_y\right)^2. \tag{18.3}$$

This means we subtract m_y from each group mean (m_i), sum the squared differences, and then multiply by the number of scores contributing to each mean. When we substitute the three group means (m_i) and the overall mean (m_y) into equation 18.3, we find

$$ss_a = n\sum_{i=1}^{k}(m_i - m_y)^2 = 5\sum_{i=1}^{k}(\{14,21,19\} - 18)^2 = 5(-4^2 + 3^2 + 1^2) = 130.$$

The variability not attributable to group means is the variability in the residuals, $y_{i,j} - m_i$. This can be computed simply as

$$ss_{error} = ss_{total} - ss_a. \tag{18.4a}$$

Alternatively, we can sum the sums of squares within each condition as follows:

$$ss_{error} = \sum_{i=1}^{k} ss_i. \tag{18.4b}$$

These within-groups (or within-conditions) sums of squares are computed from the deviations of scores from their group means; these deviations are shown in the last three columns of Table 18.1. The within-groups sums of squares are $ss_1 = 70$, $ss_2 = 50$, and

$ss_3 = 34$. Therefore, $ss_{error} = 154$. This can be confirmed from what we've seen previously: $ss_{total} = 284$ and $ss_a = 130$; therefore, $ss_{error} = ss_{total} - ss_a = 284 - 130 = 154$.

We partition variance to determine the proportion of variability in our kn scores that is attributable to group means. This is easily determined to be

$$R^2 = \frac{ss_a}{ss_{total}} = \frac{130}{284} = .46. \tag{18.5}$$

From this analysis we see that about 46% of the variability in our 15 scores is attributable to the differences between group means. In a moment, we will see two types of inferential statistics that can be applied to R^2.

The Connection Between ANOVA and Multiple Regression

Before moving on to discuss inferential statistics for R^2, we will take a moment to connect ANOVA and multiple regression. Table 18.2 rearranges the data from Table 18.1 into the same format we saw in Chapter 15 where we connected the two-groups design, correlation, and regression.

The first column of Table 18.2 labels the conditions of the experiment (a_i, for $i = 1,2,3$). The second column shows the indexes (j) denoting the individuals within each group. The third column shows the data from Table 18.1 ($y_{i,j}$) rearranged into column form. In Chapter 15, we saw that we could arrange the scores from two groups into a column and treat them as the outcome variable in a simple linear regression. This is what's been done in column $y_{i,j}$.

The columns labeled x_1 and x_2 provide *pairs* of numbers that code for groups. These codes are completely arbitrary. The code for placebo condition is [1,2], the code for the Drug 1 condition is [1,3], and the code for the Drug 2 condition is [2,1]. The columns $y_{i,j}$, x_1, and x_2 look exactly like the two-predictor multiple regression problems we examined in Chapters 16 and 17. When these three columns of numbers are pasted into an SPSS data file and submitted to a regression analysis, we find that the regression equation relating $y_{i,j}$ to x_1 and x_2 is

$$\hat{y}_{i,j} = -12 + 12(x_1) + 7(x_2).$$

(You should do this as an exercise right now to confirm this result. If you don't have SPSS available, you can use the **LINEST** function in Excel as described in Appendix 16.2, available at study.sagepub.com/gurnsey.) So, what does this regression equation predict? The predictions are shown in the last column of Table 18.2 under the heading $\hat{y}_{i,j}$. These numbers should look familiar because they are the means of the three groups (a_i, for $i = 1,2,3$).

In Chapter 15 we saw that when there are two groups of scores, a single column of any two numbers coding for groups is sufficient to "predict" the two group means. We've just seen that when there are three groups, two columns of numbers coding for groups are sufficient to "predict" the three group means. In general, if there are k means, then $k - 1$ columns of numbers coding for groups are sufficient to "predict" the k group means, as long as these columns are not multicollinear. The variability in these predicted scores ($\hat{y}$) represents the variability in y associated with the group means.

The last row of Table 18.2 shows that the sum of squares for $y_{i,j}$ is 284 and the sum of squares for $\hat{y}_{i,j}$ is 130. These are ss_{total} and ss_a, respectively. So, as before, $R^2 = ss_a/ss_{total} = .46$. Therefore, the regression equation allows us to determine the proportion of ss_{total} attributable to differences in group means.

As just noted, when there are k means, then $k - 1$ columns of numbers coding for groups are sufficient to capture the variability in y attributable to group means. The explanation for why this is so was given in Chapter 16, where we saw that all of the variability in

n scores could be captured by a regression model with $n-1$ predictors. In ANOVA, a regression model with $k-1$ predictors explains all of the variability in the k means.

Parameters and Statistics in ANOVA

No matter how we compute R^2, it is clearly a statistic, so we must now ask what it estimates. To deal with this estimation question, we will consider the population assumptions underlying the *one-factor between-subjects* ANOVA. The first assumption is that each of the k samples is drawn from a different population. These populations may have different means (μ_i), but they are assumed to have the same standard deviation, σ. We refer to this as the **homogeneity of variance** assumption, which is simply another name for the homoscedasticity assumption in regression analysis. Inferential statistics for R^2 also rest on the assumption that these homoscedastic populations are normal.

In the one-factor between-subjects ANOVA, we consider a population that comprises several subpopulations. For simplicity, we will assume these subpopulations have the same number of scores. The mean of this larger population is the mean of the subpopulation means:

$$\mu = \frac{\sum_{i=1}^{k}\mu_i}{k}. \tag{18.6}$$

Just as we did with samples, we can compute the total variability of all scores about μ as follows:

$$ss_{\text{total}} = \sum_{i=1}^{k}\sum_{j=1}^{N}(y_{i,j}-\mu)^2 \tag{18.7}$$

where $y_{i,j}$ is the jth score in the ith population and N is the number of scores in each population. We can partition ss_{total} into ss_{between} and ss_{error} as follows:

$$ss_{\text{between}} = \sum_{i=1}^{k}N(\mu_i-\mu)^2 \text{ and} \tag{18.8}$$

$$ss_{\text{error}} = ss_{\text{total}} - ss_{\text{between}}. \tag{18.9}$$

Note that we could also refer to ss_{error} as ss_a. The ratio of ss_{between} to ss_{total} gives

$$P^2 = \frac{ss_{\text{between}}}{ss_{\text{total}}}. \tag{18.10}$$

Therefore, R^2 estimates P^2, the proportion of total variance explained by population means.

Estimating P²

The connection between R^2 and P^2 in ANOVA is the same as the connection between R^2 and P^2 in multiple regression. The only difference is that in multiple regression, the predictor variables were assumed to be random variables; in ANOVA (considered as a special case of multiple regression), our predictors are fixed. That is, the predictors in columns

TABLE 18.2 ■ Data for Three Groups[a]					
Condition	j	$y_{i,j}$	x_1	x_2	$\hat{y}_{i,j}$ [b]
a_1	1	17	1	2	14
a_1	2	18	1	2	14
a_1	3	9	1	2	14
a_1	4	10	1	2	14
a_1	5	16	1	2	14
a_2	1	19	1	3	21
a_2	2	21	1	3	21
a_2	3	18	1	3	21
a_2	4	27	1	3	21
a_2	5	20	1	3	21
a_3	1	20	2	1	19
a_3	2	15	2	1	19
a_3	3	21	2	1	19
a_3	4	22	2	1	19
a_3	5	17	2	1	19
ss		284			130

[a] a_1 = Placebo, a_2 = Drug 1, a_3 = Drug 2.
[b] $\hat{y}_{ij} = -12 + 12(x_1) + 7(x_2)$.

Homogeneity of variance means that populations of interest have the same variance. This is exactly the same as saying the distributions are homoscedastic.

x_1 and x_2 of Table 18.2 are not random variables but codes for fixed levels of the independent variable set by the experimenter.

We saw that in the case of multiple regression, there was no easy way to place a confidence interval around R^2 by hand. However, in Chapter 16, we saw how to use the MBESS routines (Kelley, 2007) in **R** to compute a confidence interval around R^2 when our predictors were random variables. We can use the same routine (`ci.R2`) to compute a confidence interval around R^2 when we have fixed rather than random predictors.

At the R prompt (>), enter the code shown in the first line below and then press return:

```
>    ci.R2(R2 = .46, N = 15, K = 2, conf.level = .95, Random.Predictors =
FALSE)
       $Lower.Conf.Limit.R2
       [1] 0.0003990625
       $Prob.Less.Lower
       [1] 0.025
       $Upper.Conf.Limit.R2
       [1] 0.6580306
       $Prob.Greater.Upper
       [1] 0.025
```

The numbers entered are from our example: `R2` = .46, `N` = 15, `K` = 2, `conf.level` = .95, and `Random.Predictors` = FALSE. `K` = 2 represents the number of fixed predictors in the ANOVA/regression model. That is, `K` = $k - 1$. The results show that the approximate 95% CI is [0.00, 0.66]. This confidence interval is very wide and thus very imprecise, because the total number of scores is very small. As always, confidence intervals become narrower as sample size increases.

Testing R^2 for Statistical Significance

Most researchers conduct ANOVAs to test the null hypothesis that all population means are the same,

$$H_0: \mu_1 = \mu_2 = \ldots = \mu_k,$$

against the alternative hypothesis that this is not true. This null hypothesis—that there is no difference between the k population means—is tested by computing an **omnibus F-statistic** as follows:

An *F*-statistic in ANOVA computed with $df_a > 1$, is called an omnibus *F*-statistic.

$$F_{obs} = \frac{R^2}{1 - R^2} \frac{df_{error}}{df_a}.$$

$$(18.11)$$

F_{obs} is declared statistically significant when statistics greater than it would occur less than 5% of the time under H_0.

In our example, there were three groups of five scores (15 total). In a one-factor between-subjects ANOVA,

$$df_a = k - 1,$$

$$(18.12)$$

$$df_{error} = df_{total} - df_a,$$

$$(18.13)$$

and

$$df_{total} = \left(\sum_{i=1}^{k} n_i\right) - 1,$$

$$(18.14)$$

where k is the number of groups (means) and n_i in the number of scores in the ith group. When sample sizes are equal, $df_{total} = kn-1$. As in multiple regression, degrees of freedom relate to the number of predictors used to explain variability in a set of scores. In a one-factor between-subjects ANOVA, $df_a = k - 1$ predictors are required to explain variability attributable to the k means, and $df_{error} = df_{total} - df_a$ predictors are required to explain all remaining variability in the kn scores.

In our example, $df_a = 2$ and $df_{error} = 12$. With this information, we can compute the following:

$$F_{obs} = \frac{R^2}{1-R^2} \frac{df_{error}}{df_a} = \frac{.46}{.54} \frac{12}{2} = 5.07.$$

To determine whether F_{obs} is statistically significant, we consult the same F-table used in Chapter 16. There we find that $F_{critical} = 3.9$ when $\alpha = .05$, $df_a = 2$, and $df_{error} = 12$. Because $F_{obs} = 5.07$ exceeds $F_{critical} = 3.9$, we reject the null hypothesis and conclude that there is a statistically significant difference between the means of our three groups.

Conducting a One-Factor Between-Subjects ANOVA in SPSS

Conducting a one-factor, between-subjects ANOVA in SPSS is very much like conducting a t-test. When we select the Analyze→Compare Means→One-Way ANOVA . . . menu item, the One-Way ANOVA dialog appears, as shown in Figure 18.2. The dialog asks for one variable to be treated as the dependent variable (Dependent List:) and one to be treated as the independent variable (Factor:). These two variables are entered as two columns in the data file. For this example, the numbers in the column MoCA are entered as a single column exactly as shown in Table 18.2. The variable coding for Groups is a single column of numbers used to distinguish the groups. All scores from the same group are associated with the same number, or code. In this example, the codes for groups were simply 1, 2, and 3.

The results of the ANOVA are shown in Figure 18.3. The quantities are essentially the same as those we computed by hand. They differ slightly because the hand calculations rounded R^2 (and $1 - R^2$) to two decimal places.

This ANOVA could have been performed as a regression, just like the regression analyses discussed in Chapters 16 and 17. The variables x_1 and x_2 shown in the top left of Figure 18.2 are the codes shown in Table 18.2. The results of the regression analysis are shown in Figure 18.4. Notice that the ANOVA table is identical to the ANOVA table in Figure 18.3. The coefficients table shows that the regression equation is $\hat{y}_{i,j} = -12 + 12(x_1) + 7(x_2)$, just as we saw earlier.

FIGURE 18.2 ■ ANOVA in SPSS

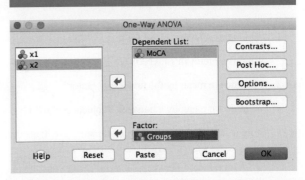

Conducting a one-factor between-subjects ANOVA in SPSS through the **One-Way ANOVA** dialog from the **Analyze** menu. The dependent variable (**MoCA**) has been moved into the **Dependent List** and the independent variable (**Groups**) has been specified as the **Factor**. Groups is simply a column of numbers such that all scores in the dependent variable from the same group are associated with the same number.

FIGURE 18.3 ■ ANOVA Table

ANOVA

MoCA

	Sum of Squares	df	Mean Square	F	Sig.
Between Groups	130.000	2	65.000	5.065	.025
Within Groups	154.000	12	12.833		
Total	284.000	14			

An ANOVA table showing the between, within, and total sum of squares, along with the degrees of freedom for the between and within sums of squares and the mean squares. The F-ratio is computed as the ratio of mean squares, and its p-value under the null hypothesis is shown to the right (**Sig.**).

FIGURE 18.4 ■ ANOVA as Regression

(a)

ANOVA[a]

Model		Sum of Squares	df	Mean Square	F	Sig.
1	Regression	130.000	2	65.000	5.065	.025[b]
	Residual	154.000	12	12.833		
	Total	284.000	14			

a. Dependent Variable: MoCA
b. Predictors: (Constant), x2, x1

(b)

Coefficients[a]

Model		Unstandardized Coefficients		Standardized Coefficients	t	Sig.
		B	Std. Error	Beta		
1	(Constant)	-12.000	9.478		-1.266	.230
	x1	12.000	3.924	1.300	3.058	.010
	x2	7.000	2.266	1.314	3.090	.009

a. Dependent Variable: MoCA

An ANOVA table when the ANOVA is conducted as a regression analysis with x_1 and x_2 from Table 18.2 as predictors. The regression equation predicts the mean of each group.

LEARNING CHECK 1

1. What does ANOVA stand for?

2. What is a factor in an ANOVA?

3. What do we mean by the levels of a factor?

4. What do we mean by a between-subjects design?

5. State whether the following statements are true or false.

   (a) A one-factor between-subjects design has three groups with 10 scores in each group; therefore, $df_{error} = 27$.

   (b) A one-factor between-subjects design has five groups with 12 scores in each group; therefore, $df_a = 4$.

   (c) A one-factor between-subjects design has six groups with three scores in each group; therefore, $df_{total} = 16$.

6. Use the data set below to compute the following quantities:

(a) ss_{total},

(b) ss_a,

(c) ss_{error},

(d) R^2 and the approximate 95% CI around R^2, and

(e) F_{obs} (also use Excel to determine the exact p-value associated with F_{obs}).

(f) Use SPSS or Excel to find the regression equation that predicts the dependent variable from codes for the groups when the codes for groups $a_1, a_2, a_3,$ and a_4 are $[1, 2, 1], [2, 1, 2], [3, 1, 1],$ and $[4, 2, 2]$, respectively.

a_1	a_2	a_3	a_4
1	7	9	4
3	9	11	4
3	9	11	5
5	11	13	7

Answers

1. Analysis of variance.

2. A factor is an independent variable with several levels.

3. A level is one of the values of the independent variable.

4. A between-subjects design is one in which participants in the groups are independent of participants in all other groups.

5. (a) True. $df_{error} = 30 - 2 - 1 = 27$.

   (b) True. $df_a = k - 1 = 5 - 1 = 4$.

   (c) False. $df_{total} = k*n - 1 = 6*3 - 1 = 17$.

6. (a) $ss_{total} = 190$.

   (b) $ss_a = 160$.

   (c) $ss_{error} = 30$.

   (d) $R^2 = .84$, approximate 95% CI [0.50, 0.89], computed using `ci.R2`.

   (e) $F_{obs} = 21.33, p = .000042$.

   (f) $\hat{y}_{i,j} = 15 + 1(x_1) - 6(x_2) - 1(x_3)$.

PLANNED CONTRASTS

If you think about the example we've been working with, you might notice that R^2 and F_{obs} are not particularly useful quantities. They tell us something about how much variability in the *dependent variable* is explained by the *independent variable*, but R^2 and F_{obs} would have been exactly the same if the means were shuffled. For example, we would have obtained exactly the same R^2 and F_{obs} if the mean of the Placebo group (m_1) was 21, the mean of the Drug 1 group (m_2) was 19, and the mean of the Drug 2 group (m_3) was 14. However, this pattern would mean something very different from the pattern shown in Figure 18.1.

Although the researcher may want to know *that* the means differ, he or she is most interested in *how* and by *how much* they differ. For example, the researcher may want to know by how much the mean of the placebo group differs from the means of the two drug groups, or by how much the means of the two drug groups differ from each other. Throughout this chapter and Chapter 19, we will refer to comparisons between group means as *contrasts*. As a general rule, it is very important to plan the contrasts you wish to make in your experiment. In this section, we will see how to analyze planned contrasts.

Simple and Complex Contrasts

In ANOVA, there are two kinds of contrasts. **Simple contrasts** involve just two means. From our example, comparing the means of the two drug groups is a simple contrast. Therefore, we can define a contrast (call it c_1) as

$$c_1 = m_2 - m_3.$$

Complex contrasts involve more than two means. From our example, comparing the mean of the placebo group with the average of the two drug group means is a complex contrast. We can define this contrast (call it c_2) as

$$c_2 = m_1 - \frac{m_2 + m_3}{2} = m_1 - 0.5(m_2) - 0.5(m_3).$$

It is convenient to define contrasts with a set of k weights (w_i, for $i = 1$ to k). For example, the following three numbers contrast the second and third of $k = 3$ means and ignore the first:

$$w = [0, 1, -1].$$

We can apply these weights to our three group means in the following way:

$$c_1 = \sum_{i=1}^{k} w_i m_i. \tag{18.15}$$

When we do the calculations, we find

$$c_1 = \sum_{i=1}^{k} w_i m_i = 0(14) + 1(21) - 1(19) = 0 + 21 - 19 = 2.$$

Complex contrasts can be defined in a similar way. The following three numbers contrast the first mean with the second and third means:

$$w = [1, -0.5, -0, 5].$$

When we do the calculations, we find

$$c_2 = \sum_{i=1}^{k} w_i m_i = 1(14) - 0.5(21) - 0.5(19) = 14 - 10.5 - 9.5 = -6.$$

In the general case, we form contrasts by assigning a 1 to each mean in one group (positives), −1 to each mean in the contrasting group (negatives), and 0 to means that we wish to ignore. We then divide these 1s and −1s by the number of means in each group. That is, the weights for the positive group are $1/n_{positive}$ and the weights for the negative group are $-1/n_{negative}$. If we had six means and wanted to compare means 1 and 3 against means 2, 4, 5, and 6, then means 1 and 3 would get positive weights ($n_{positive} = 2$) and means 2, 4, 5, and 6 would get negative weights ($n_{negative} = 4$). In this case, w would be

$$w = \left[\frac{1}{2}, \frac{-1}{4}, \frac{1}{2}, \frac{-1}{4}, \frac{-1}{4}, \frac{-1}{4} \right] = [0.5, -0.25, 0.5, -0.25, -0.25, -0.25].$$

A **simple contrast** compares two means. A **complex contrast** compares one or more means with another group of means. Properly constructed contrasts have negative weights that sum to −1 and positive weights that sum to +1.

Summing the product of w and m (using equation 18.15) is equivalent to

$$\frac{m_1 + m_3}{2} - \frac{m_2 + m_4 + m_5 + m_6}{4}.$$

When we follow this procedure, w will sum to 0, and the sum of the absolute values will be 2. The result is the difference between the average of one set of means and the average of the other set of means.

If the absolute values of the weights do not sum to 2, then the contrast will not show the difference between the means of two groups of scores. Consider contrast c_2. If we had defined this contrast using $w = [2, -1, -1]$, then when applied to our three means it would have produced a value of $28 - 21 - 19 = -12$, which is clearly not equal to $[14 - (21 + 19)/2]$. Therefore, it is *absolutely critical* that the negative weights sum to -1 and the positive weights sum to $+1$, meaning that their absolute values sum to 2. This point must be kept in mind when designing contrasts.

LANGUAGE ALERT

In this chapter and Chapters 19 to 21, it will be convenient to refer to weights as weight **vectors**. In mathematics, a vector is simply a single column or a single row of numbers. Nothing more. I introduce this term only because it can make many sentences syntactically simpler.

Contrasts Estimate Differences Between Population Means

The contrasts computed from sample means estimate the weighted difference between two groups of population means. Therefore, in a population, we can define a contrast as

$$\psi = \sum_{i=1}^{k} w_i \mu_i. \tag{18.16}$$

The Greek letter Ψ (pronounced psi) does not correspond to the roman letter c. The Greek letter for c is χ, which has an almost universal meaning in statistics so I don't want to introduce confusion by using it in this context. (See Appendix 5.4 for comments on the χ^2 distribution.) I will use Ψ because this is the symbol that Kevin Bird (2004) uses in his very useful book on ANOVA.

Confidence Intervals for Contrasts

Both simple and complex contrasts boil down to a single difference: either a difference between two means or the difference between one or more means with a second group of means. Therefore, we can compute a confidence interval around a contrast using the following familiar formula:

$$\mathrm{CI} = c \pm t_{\alpha/2}(s_c).$$

As with the confidence intervals around regression coefficients, the degrees of freedom for $t_{\alpha/2}$ is df_{error}, which is $df_{total} - df_a = 14 - 2 = 12$ for this example.

As always, we're left with the question of how to compute the estimated standard error of the statistic, which is s_c. The estimated standard error for a contrast is given by

$$s_c = \sqrt{ms_{error} \sum_{i=1}^{k} \frac{w_i^2}{n_i}}. \tag{18.17}$$

We discussed ms_{error} in Chapter 16 but haven't said much about it since, even though it has appeared in all of the SPSS ANOVA tables produced by regression analyses in Chapters 16 and 17. So, just to remind us,

$$ms_{error} = \frac{ss_{error}}{df_{error}}.$$

In ANOVA, ms_{error} is an estimate of σ^2, which is the variance that is assumed to be common to all k distributions from which our samples were drawn.

To understand why ms_{error} estimates σ^2 in ANOVA, we will think back to Chapter 11 where we estimated the difference between two independent means. In that case, we computed the quantity s^2_{pooled} as

$$s^2_{pooled} = \frac{ss_1 + ss_2}{n_1 + n_2 - 2}.$$

In ANOVA, s^2_{pooled} is computed in the same way, but with more sums of squares combined:

$$s^2_{pooled} = \frac{ss_1 + ss_2 + ,..., + ss_k}{n_1 + n_2 + ,..., + n_k - k}.$$

Because $ss_1 + ss_2 + ... + ss_k = ss_{error}$ and $n_1 + n_2 + ... + n_k - k = df_{error}$, we can see that s^2_{pooled} and ms_{error} are two names for the same thing; both are estimates of σ^2. However, depending on the context, one of these terms may be more natural.

We now have everything necessary to compute confidence intervals around c_1 and c_2, which are the simple and complex contrasts computed earlier. Let's compute the 95% confidence interval around c_2 in a step-by-step fashion, putting together things we calculated earlier.

Step 1. Compute the contrast. This requires the sample means and the weights for the contrast, which we saw earlier were $m = [14, 21, 19]$, and $w = [1, -0.5, -0.5]$. Using these, we compute the contrast:

$$c_2 = \sum_{i=1}^{k} w_i m_i = 1(14) - 0.5(21) - 0.5(19) = -6.$$

Step 2. Compute ms_{error}. First we compute ss_{error} from the raw data and then divide this by df_{error}.

$$ms_{error} = \frac{ss_{error}}{df_{error}} = \frac{154}{12} = 12.83.$$

Step 3. Compute s_c using equation 18.17. Because we are computing s_c for c_2, we need the sample sizes and the corresponding weights, $w = [1, -0.5, -0.5]$. With these and ms_{error}, we compute

$$s_c = \sqrt{ms_{error} \sum_{i=1}^{k} \frac{w_i^2}{n_i}} = \sqrt{12.83\left(\frac{1}{5} + \frac{.25}{5} + \frac{.25}{5}\right)} = 1.96.$$

Step 4. Determine $t_{\alpha/2}$. Because $df_{error} = 12$ and $\alpha = .05$, we determine from the t-table that $t_{\alpha/2} = 2.179$.

Step 5. Compute the confidence interval.

$$CI = c_2 \pm t_{\alpha/2}(s_c) = -6 \pm 2.179(1.96) = [-10.28, -1.72].$$

Therefore, we have 95% confidence that the true difference is in the interval $[-10.28, -1.72]$, because we know that 95% of all such intervals will capture $\Psi_2 = \mu_1 - (\mu_2 + \mu_3)/2$. I will leave the calculation of a confidence interval around c_1 as an exercise. The answer is $CI = 2 \pm 2.179(2.27) = [-2.95\ 6.95]$, where $s_c = 2.27$.

Orthogonal Contrasts

The contrast weights that we've been discussing possess the important property of **orthogonality**. Two weight vectors are orthogonal if they are completely uncorrelated; i.e., $r = 0$. When two weight vectors each sum to 0, as do our weight vectors, they are orthogonal if the sum of their products is also 0:

$$\sum_{i=1}^{k} w_{1_i} w_{2_i} = 0. \tag{18.18}$$

Note that equation 18.18 is the numerator of the correlation coefficient (sp_{xy}; see Chapter 13) when both sets of scores have a mean of 0.

If there are k sample means, then there are only $k - 1$ possible orthogonal contrasts. In the example we've been considering, we have $k = 3$ means so there are only $k - 1 = 2$ possible orthogonal contrasts in the orthogonal set. When we put the contrast weights we've been using into equation 18.18, we get

$$\sum_{i=1}^{k} w_{1_i} w_{2_i} = \sum[0, 1, -1] * [1, -0.5, -0.5] = \sum[0, -0.5, 0.5] = 0,$$

meaning that they are orthogonal. When all possible pairs of the $k - 1$ weight vectors are orthogonal, we say they form an **orthogonal set**.

Orthogonal contrasts act like a scalpel to partition the variability in ss_a (the sum of squares associated with the variability among the means) into nonoverlapping components. Therefore, we can say that

$$ss_a = \sum_{i=1}^{k-1} ss_{c_i}, \tag{18.19}$$

where ss_{c_i} is the sum of squares associated with the ith contrast. The sum of squares associated with any given contrast is computed as follows:

$$ss_c = \frac{c^2}{\sum_{i=1}^{k} \frac{w_i^2}{n_i}}. \tag{18.20}$$

In the example we've been working with, $w_1 = [0, 1, -1]$, $c_1 = 2$, $w_2 = [1, -0.5, -0.5]$, and $c_2 = -6$. With this information, we can calculate ss_{c_1} and ss_{c_2} using equation 18.20:

$$ss_{c_1} = \frac{c_1^2}{\sum_{i=1}^{k} \frac{w_i^2}{n_i}} = \frac{4}{\dfrac{0}{5} + \dfrac{1}{5} + \dfrac{1}{5}} = \frac{4}{.4} = 10 \text{ and}$$

$$ss_{c_2} = \frac{c_2^2}{\sum_{i=1}^{k} \frac{w_i^2}{n_i}} = \frac{36}{\dfrac{1}{5} + \dfrac{.25}{5} + \dfrac{.25}{5}} = \frac{36}{.3} = 120.$$

We saw in Tables 18.1 and 18.2 and Figure 18.3 that $ss_a = 130$. Therefore, as stated in equation 18.19, $ss_a = ss_{c_1} + ss_{c_2} = 10 + 120 = 130$, which means that our orthogonal contrasts have indeed sliced the variability in our three treatment means into nonoverlapping components.

Designing a Set of Orthogonal Contrasts

TABLE 18.3 ■ Orthogonal Contrasts: $k = 3$

	Set 1			Set 2			Set 3		
	a_1	a_2	a_3	a_1	a_2	a_3	a_1	a_2	a_3
w_1	−1	½	½	½	½	−1	½	−1	½
w_2	0	1	−1	−1	1	0	−1	0	1

In our hypothetical experiment involving patients with Alzheimer's disease, we designed a set of orthogonal contrasts that seemed quite reasonable given the experimental manipulation. It is important to note that many other orthogonal sets could have been constructed, and some of these are shown in Table 18.3.

When designing contrasts, it is critical to make sure that they are relevant to the research question. This is not a statistical issue because for k conditions of an experiment, there are many different orthogonal sets with $k − 1$ contrasts. The set you choose or design depends on the research question.

Table 18.4 provides an illustration of an orthogonal set for $k = 10$ group means, from which we will be able to derive some rules to follow when building a set of contrasts relevant to your research. We build the weights from largest to smallest. For example, the weights for contrast 1 (w_1) divide the 10 means into two groups with weights ⅙ and −¼.

The next contrast (w_2) will contain positive and negative numbers *and possibly zeros*. All nonzero numbers in w_2 must be multiplied by weights of the same sign in w_1. Therefore, the weights ⅓ and −⅓ in w_2 are multiplied by ⅙ in w_1. All other entries are set to 0. Because the positive and negative numbers in w_2 sum to zero, the product of these and any constant (e.g., ⅙) will also sum to 0. Of course, the remaining products (−¼*0) will also sum to 0. Therefore, $\sum_{i=1}^{k} w_{1_i} w_{2_i} = 0$, i.e., w_1 and w_2 are orthogonal.

The weights for the next contrast, w_3, are built the same way with the positive values in w_2 as a reference. The weights I've chosen ensure that $\sum_{i=1}^{k} w_{2_i} w_{3_i} = 0$ and $\sum_{i=1}^{k} w_{1_i} w_{3_i} = 0$. Therefore, w_1, w_2, and w_3 are orthogonal.

TABLE 18.4 ■ Orthogonal Contrasts: $k = 10$

	a_1	a_2	a_3	a_4	a_5	a_6	a_7	a_8	a_9	a_{10}
w_1	⅙	⅙	⅙	⅙	⅙	⅙	−¼	−¼	−¼	−¼
w_2	⅓	⅓	⅓	−⅓	−⅓	−⅓	0	0	0	0
w_3	½	½	−1	0	0	0	0	0	0	0
w_4	0	0	0	1	−½	−½	0	0	0	0
w_5	1	−1	0	0	0	0	0	0	0	0
w_6	0	0	0	0	1	−1	0	0	0	0
w_7	0	0	0	0	0	0	1	−⅓	−⅓	−⅓
w_8	0	0	0	0	0	0	0	½	½	−1
w_9	0	0	0	0	0	0	0	1	−1	0

The weights for contrast 4, w_4, are built with the negative values in w_2 as a reference. There are three possible arrangements of 1, $-\frac{1}{2}$, and $-\frac{1}{2}$ in this contrast, and I've chosen $[1, -\frac{1}{2}, -\frac{1}{2}]$. All remaining weights are set to zero. Without doing the calculation, you should be able to confirm that w_4 is orthogonal to w_1, w_2, and w_3. The weights in w_5 and w_6 are the only possible contrasts, related to the positive values in w_1, that are orthogonal to the first four.

The rule we follow is very simple: when we build contrasts from top to bottom, all nonzero weights in a given contrast must be multiplied by weights of one sign (positive or negative) in the contrasts above it.

The contrasts associated with the negative values of w_1 have been constructed following this rule. This entire set is orthogonal. It is one of many possible orthogonal sets that could have been constructed for $k = 10$ sample means.

Non-orthogonal Contrasts

We've seen that orthogonal contrasts slice up ss_a into nonoverlapping components. Non-orthogonal contrasts divide up the variability into overlapping components. This means that one contrast many not be independent of one or more other contrasts. Thus, from a purely statistical point of view, there is a good argument that we should prefer orthogonal contrasts. However, there are situations in which non-orthogonal contrasts make sense.

In our opening example, we considered a control group and two groups treated with new drugs for Alzheimer's disease. We could have asked how each of the two drug groups differ from the placebo group. The contrasts required to answer this question are

$$w_1 = [-1, 1, 0] \text{ and}$$

$$w_2 = [-1, 0, 1].$$

(That is, $m_2 - m_1$ and $m_3 - m_1$.) You should be able to confirm that $\sum w_1 w_2 = 1$, meaning that these contrasts are not orthogonal.

When these contrasts are applied to our sample means, we find that

$$c_1 = -14 + 21 = 7 \text{ and}$$

$$c_2 = -14 + 19 = 5.$$

We saw in equation 18.20 that to compute ss_c, we divide c^2 by $\sum_{i=1}^{k} w_i^2/n_i$, which is $\frac{1}{5} + \frac{1}{5} = 0.4$ for both contrasts. Therefore,

$$ss_{c_1} = \frac{7^2}{.4} = 122.5 \text{ and}$$

$$ss_{c_2} = \frac{5^2}{.4} = 62.5.$$

Remember from Table 18.1 that $ss_a = 130$; when we used orthogonal contrasts, we found that $ss_{c_1} + ss_{c_2} = 10 + 120 = 130$. With our two non-orthogonal contrasts, we find that $ss_{c_1} + ss_{c_2} = 122.5 + 62.5 = 185$, which clearly exceeds $ss_a = 130$.

If we can accept the unseemliness of our contrast sums of squares exceeding ss_a, the estimated standard error of each of these contrasts is $ss_c = 2.27$, just as it was for the previous contrast that compared m_2 and m_3; if you didn't work this out before, it would be a good idea to do it now. Therefore, the 95% confidence intervals around these two contrasts would be

$$CI = 7 \pm 2.179(2.27) = [2.06, 11.4] \text{ and}$$

$$CI = 5 \pm 2.179(2.27) = [0.06, 9.94].$$

Many researchers with an interest in statistics find themselves in a deep quandary when it comes to the question of orthogonal versus non-orthogonal contrasts. Our statistical superegos tell us to use orthogonal contrasts because it makes no sense to account for more variability than exists in our treatment means. That is, it seems just wrong to have our contrast sums of squares exceed ss_a. However, our research ids want answers to specific research questions, and who cares if the contrasts are not orthogonal. Our egos have to moderate these conflicting impulses. As with any negotiation, there is rarely a solution that perfectly satisfies all parties. One must always specify the contrasts of interest before data collection; in the best of all worlds, these will be orthogonal. If the questions of interest cannot be expressed as orthogonal contrasts, then non-orthogonal contrasts will have to do.

Even though we may choose to use non-orthogonal contrasts, we must obey the following rules:

NEVER try a large number of contrasts and then choose which ones to report as part of a research paper. Whether you are computing confidence intervals or significance tests, this is *p*-hacking and it is disastrous for psychology.

AVOID using a large number of planned contrasts. Keeping the number of contrasts below $k - 1$ imposes discipline on our contrasts. We will see below how proliferating the number of contrasts reduces the reliability of each one. So choose carefully.

Standardized Contrasts

We've seen in many previous chapters that statistics can be standardized. For example, d and $Beta_i$ are standardized versions of $m_1 - m_2$ and b_i, respectively. Standardizing a contrast (c_i) is very similar to standardizing the difference between two sample means. In this case, we simply divide c_i by our estimate of σ, which is the standard deviation common to the k homoscedastic populations/distributions under study.

To standardize a contrast, we simply divide it by s_{pooled}:

$$d_i = \frac{c_i}{s_{pooled}}, \tag{18.21}$$

where $s_{pooled} = \sqrt{ms_{error}}$. Let's consider contrast $c_2 = -6$ from our example of patients with Alzheimer's disease. This contrast compared the placebo group against the mean of the two drug groups. We saw earlier that $ms_{error} = 12.83$. Therefore,

$$d_2 = \frac{c_2}{s_{pooled}} = \frac{-6}{\sqrt{12.83}} = -1.68.$$

This means we estimate that the placebo group scores, on average, are 1.68 standard deviations below the mean of the two drug groups. Equivalently, the mean of the two drug groups is estimated to be 1.68 standard deviations above the mean of the placebo group. Using Cohen's U_3, our best estimate is that about 95% of the population comprising those taking one of the two drugs scores above the mean of the placebo group. This sounds very exciting.

We must always keep in mind that point estimates are subject to sampling error, so we should consider a confidence interval around our estimated effect size. Placing a confidence interval around d_i raises the same complications we saw in Chapter 11. There is no simple formula that allows us to compute the confidence interval by hand. Instead, we can use specialized software, such as the MBESS routines in **R**.

The MBESS function we use is `ci.sc` (confidence interval for standardized contrast). To compute an exact confidence interval around d_i, we provide `ci.sc` with six arguments:

- `means = c(14, 21, 19)` conveys the means of the three groups,

- `s.anova = sqrt(12.83)` conveys $s_{pooled} = \sqrt{ms_{error}}$,

- `weights = c(1, −0.5, −0.5)` conveys the weights associated with our contrast,

- `n = c(5, 5, 5)` conveys the number of scores in each group,

- `N = 15` is the total number of scores, and

- `conf.level = .95` conveys the confidence we wish to have in the interval.

When the following line of code is submitted to **R**, we find that the standardized effect size is $d_2 = -1.68$, 95% CI [−2.91, −0.39], as shown below:

```
> ci.sc(means = c(14,21,19), s.anova = sqrt(12.83), c.weights = c(1,
    −0.5, −0.5), n = c(5,5,5), N = 15,conf.level = .95)
    $Lower.Conf.Limit.Standardized.Contrast
    [1] −2.911485
    $Standardized.contrast
    [1] −1.674893
    $Upper.Conf.Limit.Standardized.Contrast
    [1] −0.3883553
```

We have 95% confidence in this interval because we know that 95% of all intervals computed this way will capture δ.

Significance Tests

Although our focus has been on estimation, contrasts can be used to test the null hypothesis

$$H_0: \psi = 0.$$

Of course, the simplest way to test this hypothesis is to ask whether 0 falls in the interval $c \pm t_{\alpha/2}(s_c)$. If it does, we retain H_0; if it doesn't, we reject H_0.

Significance tests can also be conducted with t- or F-statistics, which are directly related, as we've seen before. To compute a t-statistic, we simply divide c_i by s_{c_i}. We saw earlier that $c_1 = 2$ and $s_{c_1} = 2.27$. Therefore, the corresponding t-statistic is

$$t_{obs} = \frac{c_1}{s_{c_1}} = \frac{2}{2.27} = 0.88.$$

The exact two-tailed p-value associated with $t_{obs} = 0.88$ is .40, for 12 degrees of freedom.

An F-statistic is the square of a t-statistic, so we can easily determine $F_{contrast} = 0.88^2 = 0.78$. The F-statistic can also be computed as

$$F_{contrast} = \frac{r_c^2}{1 - R^2} df_{error},$$ (18.22a)

where $r_c^2 = ss_c / ss_{total}$ is the proportion of total variability explained by the contrast, and $1 - R^2 = ss_{error} / ss_{total}$ is the proportion of total variability not explained by differences in the treatment means. Therefore, the F-statistic can be expressed more simply as

$$F_{contrast} = \frac{ss_c}{ss_{error}} df_{error}.$$ (18.22b)

Earlier we found that $ss_{c_1} = 10$ and $ss_{error} = 154$, and we've noted several times that $df_{error} = 12$. When these quantities are inserted into equation 18.22b, we find

$$F_{contrast} = \frac{ss_c}{ss_{error}} df_{error} = \frac{10}{154} 12 = 0.78.$$

The exact p-value associated with $F_{1,12} = 0.78$ is .40, exactly as in the case of $t_{obs} = 0.88$.

Using SPSS to Conduct Planned Contrasts

If you look back to Figure 18.2, you will see the Contrasts . . . button in the top right corner of the One-Way ANOVA dialog. Clicking on Contrasts . . . brings up the One-Way ANOVA: Contrasts dialog shown in Figure 18.5. The text box labeled Coefficients: allows you to enter the contrast weights you wish to use for your analysis. After each weight is entered, press the Add button. In Figure 18.5, the weights 0 and −1 have been added, and the final weight (1, shown to the right of Coefficients) is waiting to be added. When the Add button is pressed, the final weight will be added; i.e., contrast [0, −1, 1] has been defined. To enter a second contrast, press the Next button and enter the weights. A second contrast with weights [1, −0.5, −0.5] was added, although this is not shown. When all contrasts have been entered, press ▣ Continue ▣ to return to the One-Way ANOVA dialog, and then ▣ OK ▣ to start the analysis.

When contrasts have been entered as in Figure 18.5, SPSS produces the two additional tables shown in Figure 18.6. Figure 18.6a shows the contrasts associated with the groups (which were coded as 1, 2, and 3). Figure 18.6b shows the contrasts, their standard errors, t-statistics, degrees of freedom, and p-values. Unfortunately, there is no option to provide confidence intervals, but they are easily computed from the contrasts and standard errors given in the table. Notice that SPSS provides information for cases in which equal variances are and are not assumed. These calculations are described in detail in Chapter 21.

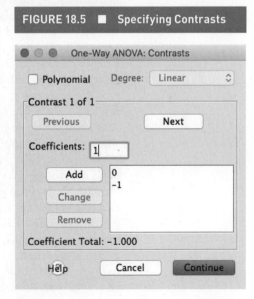

FIGURE 18.5 ■ Specifying Contrasts

The Contrasts dialog from the One-Way ANOVA dialog.

Interpretation of Our Example

We've spent many pages working with our Alzheimer's example because it has helped to introduce the basic concepts of ANOVA, planned contrasts, and standardized contrasts. These tools help

FIGURE 18.6 ■ SPSS Contrast Analysis

(a) **Contrast Coefficients**

Contrast	Groups		
	1.00	2.00	3.00
1	0	−1	1
2	1	−.5	−.5

Contrast Tests

(b)

		Contrast	Value of Contrast	Std. Error	t	df	Sig. (2-tailed)
MoCA	Assume equal variances	1	−2.0000	2.26569	−.883	12	.395
		2	−6.0000	1.96214	−3.058	12	.010
	Does not assume equal variances	1	−2.0000	2.04939	−.976	7.720	.359
		2	−6.0000	2.13307	−2.813	6.459	.028

(a) The Contrast Coefficients table shows the weights associated with the two contrasts that have been defined. (b) The Contrast Tests table shows the contrasts, their standard errors, t-statistics, degrees of freedom, and p-values.

us address the question of how much our two new drugs affect cognitive functioning in patients diagnosed with Alzheimer's disease. So we can now ask how our statistical analyses might guide our thinking about the practical significance of the results.

A positive feature of the MoCA is that normative data have been reported (Nasreddine et al., 2005). As mentioned earlier, normal control subjects score 27.4 on average, with a standard deviation of 2.2; individuals with MCI score 22.1 on average, with a standard deviation of 3.1; and individuals with Alzheimer's disease score 16.2 on average, with a standard deviation of 4.8. In our hypothetical study, the scores of the placebo group had an average of 14, which is clearly in the Alzheimer's range, as one would expect. The means of the two treatment groups (21 and 19) were closer to the mean of MCI range after 6 months of treatment, which seems like a substantial improvement.

The orthogonal planned contrasts showed that the difference between the means of the placebo group and the drug groups was −6, 95% CI [−10.28, −1.72], and the difference between the means of the two drug groups was 2, 95% CI [−2.95, 6.95]. Therefore, there is strong evidence that the drugs produce substantial increases in average MoCA scores, but there seems to be little difference between them on average.

Of course, our analysis has focused on averages and not individual cases, and these results do not guarantee that all patients with Alzheimer's disease will move from an Alzheimer's range to the MCI range. For instance, Table 18.1 shows that two participants who took Drug 2 had MoCA scores in the Alzheimer's range (15 and 17).

Cohen's U_3 (see Chapter 8 for a review) can provide some insight into the implications of the drugs at the level of populations. The effect size associated with the contrast between the placebo and drug groups was 1.68 (or −1.68, if the difference is taken the other way). Our best estimate is that about 95% of the Placebo distribution falls below the mean of the two drug conditions; i.e., below $(\mu_{drug1} + \mu_{drug2})/2$. This means that if half of all patients with recently diagnosed Alzheimer's disease received Drug 1 and the other half received Drug 2, then 95% of these patients would score above the current mean of the Placebo population $(\mu_{placebo})$. An effect of this size could be enormously meaningful, given the monetary and emotional costs of Alzheimer's disease.

With such small sample sizes, our estimates are very imprecise, and the confidence interval around the estimated effect size that we're discussing is quite wide, 95% CI [−2.91, −0.39] or [0.39, 2.91]. If we compute U_3 for these confidence limits, we find they run from .65 to 1.0, which represent increases of 15% to almost 50% of individuals scoring above the current mean of the placebo group. Even an increase of 15% above the current mean seems like a very big improvement, given the number of people currently afflicted by Alzheimer's disease. All things considered, these results would be very promising if they were real.

SOURCES OF VARIANCE

We've established that ANOVA is all about partitioning the variability in a collection of scores. We've seen that the ss_{total} can be decomposed into ss_a and ss_{error}, and that ss_a can be decomposed into parts associated with different orthogonal contrasts. When reporting an ANOVA, it is common to express a source of variance as a proportion of ss_{total}. The symbol η^2, pronounced eta square, is often used to denote this proportion for a given source. Unfortunately, this symbol is often used to denote a statistic, rather than a parameter. Therefore, we will adopt a common convention in statistics and put a hat over η^2 to produce $\hat{\eta}^2$. As mentioned in Chapter 3, a hat on a Greek letter denotes a statistic that estimates the corresponding parameter. In what follows, subscripts on $\hat{\eta}^2$ indicate the source of variance in question. For example, $\hat{\eta}_a^2$ denotes the proportion of ss_{total} explained by the *overall effect of a* and $\hat{\eta}_{c_1}^2$ denotes the proportion of ss_{total} explained by contrast c_1.

A **source table** shows the sources of variance, degrees of freedom, and some indication of the relative importance of each source.

LEARNING CHECK 2

1 A set of means is $m = [1, 2, 9, 2, 5, 3, 1, 8]$. Contrast 1 ($c_1$) compares m_1 against the average of means m_2, m_3, m_4, m_5, and m_6. Contrast 2 (c_2) compares m_7 to m_8.

(a) Show the weights for c_1.

(b) Show the weights for c_2.

(c) Compute c_1.

(d) Compute c_2.

(e) Are w_1 and w_2 orthogonal?

2. An experiment is conducted with four groups with 10 participants in each group. The group means were 20, 10, 18, and 16, $ss_a = 200$, and $ss_{error} = 1800$. The following three contrasts were used to analyze the data: $w_1 = [-0.5, -0.5, 0.5, 0.5]$, $w_2 = [-1, 1, 0, 0]$, and $w_3 = [0, 0, 1, -1]$.

(a) Compute the 95% confidence intervals around c_1, c_2, and c_3.

(b) Compute the standardized contrasts for c_1, c_2, and c_3.

(c) Are the unstandardized contrasts statistically significantly different from 0?

(d) Do these contrasts form an orthogonal set?

(e) Compute R^2 and F_{obs}.

Answers

1. (a) $w_1 = [-1, 0.2, 0.2, 0.2, 0.2, 0.2, 0, 0]$ or $w_1 = [1, -0.2, -0.2, -0.2, -0.2, -0.2, 0, 0]$.

(b) $w_2 = [0, 0, 0, 0, 0, 0, -1, 1]$ or $w_2 = [0, 0, 0, 0, 0, 0, 1, -1]$.

(c) $c_1 = \sum_{i-1}^k w_{1_i} m_i = 3.2$ (or -3.2).

(d) $c_2 = \sum_{i-1}^k w_{2_i} m_i = 7$ (or -7).

(e) $\sum_{i-1}^k w_{2_i} w_{2_i} = 0$; therefore, the contrasts are orthogonal.

2. (a) $c_1 = 2$, $ss_{c_1} = 2.24$, 95% CI = $[-2.53, 6.53]$; $c_2 = -10$, $ss_{c_2} = 3.16$, 95% CI = $[-16.41, -3.59]$; and $c_3 = 2$, $ss_{c_3} = 3.16$, 95% CI = $[-4.41, 8.41]$.

(b) $s_{pooled} = 7.0711$, $d_1 = 2/7.0711 = 0.28$, $d_2 = -10/7.0711 = -1.41$, $d_3 = 2/7.0711 = 0.28$.

(c) Only c_2 is statistically different from 0 because its CI does not capture 0. We could have also computed t_{obs} for the three contrasts. In this case, we would have found $t_{obs} = 0.89, -3.16$, and 0.63 for c_1, c_2, and c_3, respectively.

(d) Yes, these contrasts form an orthogonal set.

(e) $R^2 = ss_a / ss_{total} = 200/2000 = .1$; $F_{obs} = R^2/(1 - R^2) df_{error}/df_a = 1.3$.

TABLE 18.5 ■ Source Table

Source[a]		SS	df	MS	F	$\hat{\eta}^2$
a		130	2	65	5.07[b]	0.46
	c_1	10	1	10	0.78[c]	0.04
	c_2	120	1	120	9.35[d]	0.42
Error		154	12	12.83		0.54
Total		284	14			

Note: Table for a One-Factor, Between-Subjects ANOVA

[a] a = Conditions, c_1 = [0, –1, 1], c_2 = [1, –0.5, –0.5].

[b] $p = .02$.

[c] $p = .39$.

[d] $p = .003$.

A polynomial trend is a pattern of change across levels of the independent variable. The order of a polynomial trend is related to which coefficients in a polynomial equation can have nonzero weight.

FIGURE 18.7 ■ Polynomial Trends

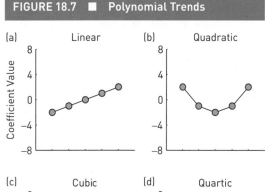

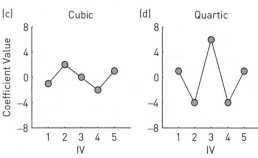

Linear (a), quadratic (b), cubic (c), and quartic (d) trends for $k = 5$. IV = independent variable.

We often make use of a **source table** when reporting the results of an ANOVA. A source table is very much like the ANOVA tables we've seen in the SPSS outputs. Source tables show the source of variance (e.g., a, c_1, or c_2), degrees of freedom, and some indication of the relative importance of each source, such as F or $\hat{\eta}^2$. Table 18.5 provides a source table for our Alzheimer's example.

TREND ANALYSIS

Up to this point, we have applied ANOVA to the means of the qualitative independent variable *drug type*. ANOVA can also be applied to treatment means associated with fixed levels of quantitative variables such as *drug dose*, study time, or hours of sleep. In such analyses, we can assess R^2 for the omnibus treatment effect; however, we've seen that omnibus effects are of limited value. Instead, we could use meaningful, planned contrasts to decompose ss_a. However, when our independent variable is quantitative, we often want to know how the means of the dependent variable change with levels of the independent variable, rather than how specific groups differ from each other. In cases such as these, we can use *trend analysis*.

A trend is a pattern of change in y with a change in x. Trend analysis is conducted exactly like the planned orthogonal contrasts we dealt with earlier, but the weights we use define $k - 1$ orthogonal patterns of change. The goal is to determine which trends best characterize the association between levels of the independent variable and the group means.

Figure 18.7 shows the four possible **polynomial trends** when $k = 5$. They are called polynomial trends because they relate to polynomial equations having the form

$$y = a + b_1(x^1) + b_2(x^2) + \ldots + b_{k-1}(x^{k-1}),$$

where k is the number of levels of x. This is like a regression equation except the coefficients ($b_1, b_2, \ldots, b_{k-1}$) are multiplied by values of an x variable raised to some power; e.g., x^2. The order of a polynomial is determined by which coefficients can have nonzero values. For a first-order polynomial only b_1 can have a nonzero value. For a second-order polynomial only b_1 and b_2 can have nonzero values. For a third-order polynomial only b_1, b_2 and b_3 can have nonzero values. And so on. The x-axis of Figure 18.7 shows arbitrary levels of the independent variable that increase in equal steps. The y-axis shows the weights of the *linear (first order)*, *quadratic (second order)*, *cubic (third order)*, and *quartic (fourth order)* trends.

The trend weights shown in Figure 18.7 are also shown in Table 18.6. These trend weights form an orthogonal set. This means that each trend "extracts" the strength of a particular pattern of change in the group means across levels of the independent variable. That is, when we compute a trend using the usual formula, $c_{\text{trend}} = \Sigma w_i m_i$, the magnitude of the result is related to the strength of that trend (pattern) in the means. Because the trends are orthogonal, each picks out a different pattern of change in the data.

It is important to note that to use the trend coefficients shown in Table 18.6, the levels of the independent variable must be *equally spaced on a linear scale*. For example, if the independent variable is drug dose, then the trend weights in Table 18.6 can be used if drug doses are 2, 4, 6, 8, and 10, but not if doses are 2, 4, 7, 11, and 16. (We will relax this requirement in Chapter 21, where the values of our variable are equally spaced on a logarithmic scale.)

The logic of trend analysis is to partition ss_a into components associated with each of these orthogonal trends and then assess the proportion of ss_{total} associated with each trend. The sum of squares associated with each trend is computed exactly as in equation 18.20:

$$ss_{trend} = \frac{c^2_{trend}}{\sum_{i=1}^{k} \frac{w_i^2}{n_i}}, \text{ where}$$

$$c_{trend} = \sum_{i=1}^{k} w_i m_i.$$

To see how this works, we will step through an example.

An Example

Let's say we're interested in how much work rats will do to receive different doses of heroin infused directly into their bloodstream (Roberts & Bennett, 1993). To assess this question, we decide to administer five, equally spaced doses of heroin (12, 34, 56, 78, or 100 µg/injection). Fifteen randomly chosen male rats are assigned at random to the five treatment levels, with three rats per condition. Prior to the experiment, all rats were trained to press a bar to receive the drug reward; a computer counted the number of bar presses and the drug was delivered when the required number was met. In the experimental conditions, rats were put on a progressive ratio schedule. They had to press the bar once to receive reward in the first trial, they had to press the bar twice to receive reward in the second trial, and so on. We will determine the maximum number of bar presses that each rat will produce to receive a reward. The (hypothetical) raw data are shown in Table 18.7 and plotted in Figure 18.8. In Table 18.7, the mean and sum of squares for each group is shown at the bottom of columns a_1 to a_5. The association between drug dose and maximum response rate is clearly non-linear.

To assess the trends in the data, we must determine ss_{error} and the sum of squares associated with each trend. To compute ss_{error}, we sum the within-group sums of squares (shown in the last row of Table 18.7). Therefore, $ss_{error} = 722 + 582 + 578 + 546 + 386 = 2814$.

As we saw earlier, ss_a is the sum of the individual trend sums of squares. This means that $ss_a = ss_{linear} + ss_{quadratic} + ss_{cubic} + ss_{quartic}$, each of which is computed using equation 18.20. The components of these calculations are shown in Table 18.8. We will use the linear trend to illustrate how to calculate trend sum of squares.

TABLE 18.6 ■ Polynomial Trends

Trend		a_1	a_2	a_3	a_4	a_5
Linear	w_1	−2	−1	0	1	2
Quadratic	w_2	2	−1	−2	−1	2
Cubic	w_3	−1	2	0	−2	1
Quartic	w_4	1	−4	6	−4	1

TABLE 18.7 ■ Data for Five Groups[a]

j	a_1	a_2	a_3	a_4	a_5
1	8	80	65	75	38
2	45	47	82	93	40
3	34	71	99	60	63
Mean	29	66	82	76	47
ss	722	582	578	546	386

[a] $a_1 = 12$, $a_2 = 34$, $a_3 = 56$, $a_4 = 70$, $a_5 = 100$.

FIGURE 18.8 ■ Hypothetical Data

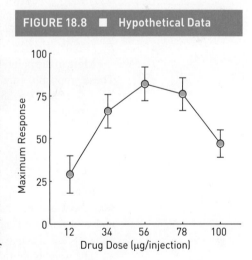

Plot of hypothetical data for five groups of three rats injected with 12, 34, 56, 78, or 100 µg/kg of heroin. The dependent variable was the maximum number of bar presses made for reward; error bars ±SEM.

Computing ss_{linear} requires first computing c_{linear} as follows:

$$c_{\text{linear}} = \sum_{i=1}^{k} w_i m_i = -2(29) - 1(66) + 0(82) + 1(76) + 2(47) = 46.$$

We then compute

$$ss_{\text{linear}} = \frac{c_{\text{linear}}^2}{\sum_{i=1}^{k} \frac{w_i^2}{n_i}} = \frac{46^2}{\frac{4}{3} + \frac{1}{3} + \frac{0}{3} + \frac{1}{3} + \frac{4}{3}} = \frac{2116}{3.33} = 634.8.$$

(It would be a worthwhile exercise to compute $ss_{\text{quadratic}}$, ss_{cubic}, and ss_{quartic} by hand.)

TABLE 18.8 ■ Computation of Trends

Source	c	c^2	$\sum_{i=1}^{k} w_i^2 / n_i$	ss_{trend}	η^2	$F_{1,10}$	p
Linear	46	2116	3.33	634.80	0.07	2.26	.16
Quadratic	−154	23,716	4.67	5082.00	0.60	18.06	.00
Cubic	−2	4	3.33	1.20	0.00	0.00	.95
Quadratic	0	0	23.33	0.00	0.00	0.00	1.00
Sum				5718.00	0.67		

The last row of Table 18.8 (in the column ss_{trend}) shows that $ss_a = 5781$. Therefore, $ss_{\text{total}} = ss_a + ss_{\text{error}} = 5718 + 2814 = 8532$. From these numbers we determine that the proportion of variance attributable to means is $R^2 = ss_a/ss_{\text{total}} = 5718/8532 = .67$, approximate 95% CI [0.04, 0.76] computed using the MBESS function `ci.R2` (Kelley, 2009). This corresponds to $F_{4,10} = 5.08, p = .02$.

More to the point, however, we can compute $\hat{\eta}^2 = ss_{\text{trend}}/ss_{\text{total}}$ for each of the four trends. These proportions are shown in the column labeled $\hat{\eta}^2$ in Table 18.8. (Notice once again that $\sum_{i=1}^{k} \hat{\eta}_i^2 = R^2 = .67$, as shown in the last row of Table 18.8.) We can compute the following for each trend:

$$F = \frac{ss_{\text{trend}}}{ss_{\text{error}}} \frac{df_{\text{error}}}{df_{\text{trend}}} = \frac{\hat{\eta}_{\text{trend}}^2}{1 - R^2} \frac{df_{\text{error}}}{df_{\text{trend}}}.$$

In all cases, $df_{\text{trend}} = 1$, because each trend uses one degree of freedom. (More will be said about this in Chapter 20.) For each of these F-statistics, we can compute a p-value under the null hypothesis from the central F-distribution, as shown in the last two columns of Table 18.8. This table makes clear that the quadratic trend is the strongest trend in the data.

The weights for the quadratic trend form a ∪-shaped function, as shown in Figure 18.7, which might seem inconsistent with the roughly ∩-shaped function in Figure 18.8. There is no contradiction here. The $\hat{\eta}^2$ values shown in Table 18.8 are really measures of how strongly the trend weights *correlate* with the pattern in the means. There is a strong negative correlation between the sample means and the weights in the quadratic trend (note that $c_{\text{quadratic}} = -154$). The $\hat{\eta}^2$ values are related to the squared correlation between the means and trend weights, so whether the correlation is positive or negative is a non-issue regarding the strength of the association.

Interpretation

In our example, we wondered about how hard rats would work for each of five doses of heroin. Figure 18.8 shows a non-linear association between drug dose and maximum response rate (work). The response rate increases up to a point (56 μg/injection) but decreases thereafter. Figure 18.8 on its own makes a strong case for the non-linear association between drug dose and response rate. Table 18.8, however, adds to the case by characterizing the contribution of each trend to the association between drug dose and bar pressing. The results of this study might be written up as follows:

APA Reporting

To determine how hard rats will work for varying doses of a heroin reward, we ran a one-factor, between-subjects experiment in which rats were administered heroin rewards of 12, 34, 56, 78, or 100 μg/injection. The dependent variable was the maximum number of bar presses each rat was willing to produce to receive the reward. Each group comprised three adult male rats that had previously learned to bar press to receive a heroin injection.

The mean scores for the five groups were 29, 66, 82, 76, and 47 for the 12, 34, 56, 78, and 100 μg/injection conditions, respectively (see Figure 18.8). Group means explained 67% of the variability in the dependent variable, $R^2 = .67$, approximate 95% CI [0.04, 0.76], $F_{4,10} = 5.08$, $p = .02$. A trend analysis showed a strong quadratic trend in the data, $\eta^2_{quadratic} = .60$, $F_{1,10} = 18.06$, $p = .002$. The remaining trends were much weaker: $\eta^2_{linear} = .07$, $F_{1,10} = 2.26$, $p = .16$; $\eta^2_{cubic} = .00$, $F_{1,10} = 00$, $p = .95$; and $\eta^2_{quartic} = .00$, $F_{1,10} = 0$, $p = 1$.

These results show that there is an optimally rewarding dose of heroin. Rats will not work as hard to receive rewards above or below this dose.

Trend Weights for $k = 3$ to 10

There were $k = 5$ levels of the independent variable in our example, meaning that there were $k - 1 = 4$ possible trends. Table 18.9 provides weights for polynomial trends for $k = 3$ to 10 levels of the independent variable. The highest-order trend in the table is the quintic trend. Researchers are generally interested in lower-order trends and hope that higher-order trends account for only a small part of ss_{total}.

Using SPSS to Conduct Trend Analysis

To conduct a trend analysis in SPSS, we go to Contrasts . . . in the top right corner of the One-Way ANOVA dialog, just as we did when using planned contrasts. At the top of the One-Way ANOVA Contrasts dialog (Figure 18.9), there is a check box beside the word Polynomial. When this is checked, SPSS will conduct a polynomial trend analysis on the means, but the user must choose which trends to assess. The drop-down list to the right titled Degree allows you to specify which trends to test. As with Table 18.9, the trends go up to quintic, which SPSS calls 5th order. When you have checked the trends you wish to assess, press Continue to return to the One-Way ANOVA dialog and then OK to start the analysis.

The data from Table 18.6 have been used to illustrate trend analysis. The 15 scores have been entered in a column and the variable is named MaxResp. Codes for groups were put in a column labeled Groups. These codes were the numbers 1 to 5. Figure 18.10 shows the result of the trend analysis. Most of the numbers in the table were calculated earlier. For example, the first row in the table shows $ss_a = 5718$ and the last two rows show $ss_{error} = 2814$ and $ss_{total} = 8532$. The sums of squares associated with each of the four trends are shown in the column Sums of Squares. We have seen these before, too.

TABLE 18.9 ■ Linear, Quadratic, Cubic, Quartic, and Quintic Trends for $k = 3$ to 10

k											
3	Linear	−1	0	1							
	Quadratic	1	−2	1							
4	Linear	−3	−1	1	3						
	Quadratic	1	−1	−1	1						
	Cubic	−1	3	−3	1						
5	Linear	−2	−1	0	1	2					
	Quadratic	2	−1	−2	−1	2					
	Cubic	−1	2	0	−2	1					
	Quartic	1	−4	6	−4	1					
6	Linear	−5	−3	−1	1	3	5				
	Quadratic	5	−1	−4	−4	−1	5				
	Cubic	−5	7	4	−4	−7	5				
	Quartic	1	−3	2	2	−3	1				
	Quintic	−1	5	−10	10	−5	1				
7	Linear	−3	−2	−1	0	1	2	3			
	Quadratic	5	0	−3	−4	−3	0	5			
	Cubic	−1	1	1	0	−1	−1	1			
	Quartic	3	−7	1	6	1	−7	3			
	Quintic	−1	4	−5	0	5	−4	1			
8	Linear	−7	−5	−3	−1	1	3	5	7		
	Quadratic	7	1	−3	−5	−5	−3	1	7		
	Cubic	−7	5	7	3	−3	−7	−5	7		
	Quartic	7	−13	−3	9	9	−3	−13	7		
	Quintic	−7	23	−17	−15	15	17	−23	7		
9	Linear	−4	−3	−2	−1	0	1	2	3	4	
	Quadratic	28	7	−8	−17	−20	−17	−8	7	28	
	Cubic	−14	7	13	9	0	−9	−13	−7	14	
	Quartic	14	−21	−11	9	18	9	−11	−21	14	
	Quintic	−4	11	−4	−9	0	9	4	−11	4	
10	Linear	−9	−7	−5	−3	−1	1	3	5	7	9
	Quadratic	6	2	−1	−3	−4	−4	−3	−1	2	6
	Cubic	−42	14	35	31	12	−12	−31	−35	−14	42
	Quartic	18	−22	−17	3	18	18	3	−17	−22	18
	Quintic	−6	14	−1	−11	−6	6	11	1	−14	6

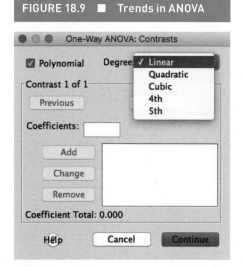

FIGURE 18.9 ■ Trends in ANOVA

Setting up a trend analysis in **One-Way ANOVA: Contrasts** dialog in SPSS.

Under each contrast is a quantity described as Deviation. This represents the difference between ss_a and the sums of squares associated with the trends entered to that point. For example, the Deviation sum of squares for linear is $ss_a - ss_{\text{linear}} = 5718 - 634.8 = 5083.2$. The Deviation sum of squares for quadratic is $ss_a - ss_{\text{linear}} - ss_{\text{quadratic}} = 5718 - 634.8 - 5082 = 1.2$. These deviation sums of squares can be tested for statistical significance. As we noted earlier, ss_{linear} and $ss_{\text{quadratic}}$ account for almost all of ss_a. To some researchers, it would be useful to know at each point whether a statistically significant proportion of variance remains unaccounted for.

CORRECTIONS FOR MULTIPLE CONTRASTS

It is common practice among those who conduct ANOVAs to test the omnibus F for statistical significance and then use **unplanned (post hoc)** simple and complex **contrasts** in an effort to understand the omnibus effect. In my view, this is a flawed practice because it results in underpowered research. Although multiple contrasts are most often

FIGURE 18.10 ■ Trend Analysis

ANOVA

MaxResp

			Sum of Squares	df	Mean Square	F	Sig.
Between Groups	(Combined)		5718.000	4	1429.500	5.080	.017
	Linear Term	Contrast	634.800	1	634.800	2.256	.164
		Deviation	5083.200	3	1694.400	6.021	.013
	Quadratic Term	Contrast	5082.000	1	5082.000	18.060	.002
		Deviation	1.200	2	.600	.002	.998
	Cubic Term	Contrast	1.200	1	1.200	.004	.949
		Deviation	.000	1	.000	.000	1.000
	4th-order Term	Contrast	.000	1	.000	.000	1.000
Within Groups			2814.000	10	281.400		
Total			8532.000	14			

Results of a trend analysis with $k = 5$ means.

LEARNING CHECK 3

1. Yerkes and Dodson (1908) showed that task performance in rats follows an inverted-U function of arousal. This pattern has been seen in many studies since then. A recent example was provided by Arent and Landers (2003). They had participants perform a simple reaction time task while riding a "bicycle ergometer," which allowed the experimenter to control participants' heart rates. There were eight groups of 13 participants. Participants in each group performed the task at 20% to 90% of their heart rate reserve (HHR). The greater the HHR, the greater the arousal. The mean data are shown in the following table. For this data set, $m_y = 439$ and $ss_{total} = 285,538$.

   (a) Compute the following: ss_a, R^2, and $F_{omnibus}$.

   (b) For the linear and quadratic trends, compute the following quantities: c_{trend}, ss_{trend}, η^2_{trend}, and F_{trend}.

   (c) What proportion of ss_a is explained by the linear and quadratic trends?

	a_1	a_2	a_3	a_4	a_5	a_6	a_7	a_8
HRR	20	30	40	50	60	70	80	90
mean	486	454	435	416	418	423	429	451

Answers

1. (a) $ss_a = n\sum_{i=1}^{k}(m_i - m_y)^2 = 13*3920 = 50,960$;

   $R^2 = ss_a/ss_{total} = 50,960/285,538 = 0.18$;

   $F_{omnibus} = ss_a/ss_{error} * df_{error}/df_a$
   $= 50,960/234,578 * 96/7 = 2.98$.

   (b) $c_{linear} = -404$, $c_{quadratic} = 698$, $ss_{linear} = 12,629.81$,
   $ss_{quadratic} = 37,700.31$, $\hat{\eta}^2_{linear} = .04$, $\hat{\eta}^2_{quadratic} = .13$,

   $F_{linear} = ss_{linear}/ss_{error} * df_{error}/df_{linear} = 5.17$, $F_{quadratic}$
   $= 15.43$.

   (c) $(ss_{linear} + ss_{quadratic})/ss_a = (12,629.81 + 37,700.31)/50,960 = .99$.

associated with *significance tests*, we will begin this discussion by considering the more intuitive case of multiple unplanned *confidence intervals*.

Confidence Intervals

Throughout Parts II and III we considered situations involving a single sample statistic to estimate a population parameter. Let's think back to estimating the difference between two population means ($\Psi = \mu_1 - \mu_2$). If we draw a random sample from each of the populations

Unplanned contrasts are also called **post hoc contrasts**. They are contrasts conducted after a statistically significant omnibus effect has been found.

and compute the 95% confidence interval around $m_1 - m_2$, we will have 95% confidence in the interval because we know that 95% of all intervals calculated this way will capture the parameter, which in this case is $\Psi = \mu_1 - \mu_2$.

Now imagine taking two more random samples from these two populations and computing a second 95% confidence interval around the new difference, $m_1 - m_2$. What confidence should we have that *both* confidence intervals will capture $\Psi = \mu_1 - \mu_2$? Well, if these are independent events, then, as we saw in Appendix 2.3 (available at study.sagepub .com/gurnsey), the probability of both occurring is given by the product of their individual probabilities, which would be $.95*.95 \approx .90$. This means that if we compute $m_1 - m_2$ for two pairs of samples, then about 90% of the time both will capture Ψ. Therefore, we should have 90% confidence that both intervals will capture Ψ.

If we had computed three independent confidence intervals, our confidence that all three have captured the parameter would be $.95*.95*.95 \approx .86$. In the general case, our confidence that n independent intervals will all capture the parameter is $(1-\alpha)^n * 100\%$.

Computing multiple confidence intervals causes problems because the number of possible contrasts increases exponentially as the number of means in the study increases. For example, when $k = 5$, there are $(k^2 - k)/2 = (25 - 5)/2 = 10$ possible simple contrasts and 80 possible complex contrasts. If we were to compute the 95% confidence interval for all 10 simple contrasts, then our confidence that all 10 would capture the parameter of interest would be $(1-\alpha)^n * 100\% = .95^{10} * 100\% = 60\%$. This is a rather low degree of confidence. Because we don't know which confidence intervals fail to capture the parameter, they all become suspect. Our confidence in any one interval is no greater than our confidence in the entire set.

To maintain a desired level of confidence in the entire set of intervals, we can increase our confidence in the individual intervals. To make this clear, we will distinguish between per-contrast confidence (pc), which is the confidence we have in a single interval, and family-wise confidence (fw), which is the confidence we have in multiple intervals. The term $(1-\alpha_{pc})100\%$ defines confidence for a single interval (when no others have been computed), and $(1-\alpha_{fw})100\%$ defines the confidence we have in our set of n confidence intervals. Changing α_{pc} will affect α_{fw}.

If we want α_{fw} to be .05 (i.e., to produce 95% confidence in our collection of intervals), then we set α_{pc} to α_{fw}/n. The result is that α_{fw} is approximately equal to $(1-\alpha_{pc})^n$. For example, if $\alpha_{fw} = .05$, then

$$(1-.05) \approx \left(1 - \frac{.05}{n}\right)^n.$$

This means that if we were to compute 10 confidence intervals and want to achieve $\alpha_{fw} = 05$, we would have to set α_{pc} to $\alpha_{fw}/10 = .05/10 = .005$. When we work through this, we see that

$$(1-\alpha_{fw}) = .95 \approx (1 - \alpha_{fw}/n)^n \approx (1 - .05/10)^{10} \approx (1 - .005)^{10} \approx (.995)^{10} \approx .9511.$$

Therefore, if we compute the $(1-\alpha_{fw}/n)100\%$ confidence interval for all n contrasts, then our confidence that all n intervals will contain the parameter they are estimating is very close to $(1-\alpha_{fw})100\%$.

The correction method just described is called the **Bonferroni correction**. There are many other corrections that have the same objective but are based on different assumptions. We will not examine these alternatives in detail.

Let's return to the three means in our Alzheimer's example and think about the three simple (non-orthogonal) contrasts that can be conducted; i.e., $m_1 - m_2 = -7$, $m_1 - m_3 = -5$, and $m_2 - m_3 = 2$. To ensure a family-wise confidence level of .95, we would have to compute the $(1 - .05/3)100\% = 98.33\%$ confidence interval for each contrast. Each contrast would

The **Bonferroni correction** for multiple contrasts achieves $(1-\alpha_{fw})100\%$ confidence in a family of n contrasts by setting the per-contrast α level (α_{pc}) to α_{fw}/n.

have the same standard error ($s_c = \sqrt{0.4(12.83)} = 2.27$). Using Excel, we determine that $t_{\alpha/2}$ is 2.779, based on $\alpha_{pc} = .05/3$ and 12 degrees of freedom. The three confidence intervals would be

$$CI_1 = -7 \pm 2.779(2.27) = [-13.30, -0.70],$$

$$CI_2 = -5 \pm 2.779(2.27) = [-11.30, 1.30], \text{ and}$$

$$CI_3 = 2 \pm 2.779(2.27) = [-4.30, 8.30].$$

If we had not corrected for multiple contrasts, $t_{\alpha/2}$ would have been 2.179 rather than 2.779. Therefore, the corrected intervals are $2.779/2.179 = 1.275$ times wider than the uncorrected intervals.

Significance Tests

Historically, studies using contrasts compute p-values rather than confidence intervals. However, exactly the same problem with multiple contrasts arises. When we compute a single contrast, the probability of a Type I error is $\alpha = .05$. However, if we compute multiple contrasts, then the probability of at least one producing a Type I error is greater than .05. If $\alpha_{pc} = .05$ is the significance level used for each contrast, then the probability that at least one contrast will produce a Type I error is

$$\alpha_{fw} = 1 - (1 - \alpha_{pc})^n.$$

So, if there are three means and three simple contrasts, the probability that at least one of the contrasts will produce a Type I error is

$$\alpha_{fw} = 1 - (1 - \alpha_{pc})^n = 1 - (1 - .05)^3 = 1 - (.95)^3 = .14.$$

There are two equivalent ways of using the Bonferroni correction to control the family-wise Type I error rate. The first is to set $\alpha_{pc} = \alpha_{fw}/n$ so that α_{fw} remains at .05. In this case, $t_{critical}$ is based on $\alpha_{pc} = .05/3 = .017$ rather than .05. For a two-tailed test of the null hypothesis in our example, $t_{critical}$ would be ± 2.779. Let's reconsider the three contrasts ($m_1 - m_2 = -7$, $m_1 - m_3 = -5$, $m_2 - m_3 = 2$) described in the previous section. The observed t-statistics would be

$$t_{obs_1} = -7/2.27 = -3.08,$$

$$t_{obs_2} = -5/2.27 = -2.20, \text{ and}$$

$$t_{obs_3} = 2/2.27 = 0.88.$$

Only $t_{obs_1} = -3.08$ exceeds $t_{critical} = \pm 2.779$. If $t_{critical}$ had not been corrected for multiple contrasts, it would have been 2.179 and $t_{obs_2} = -2.20$ would have also been statistically significant.

The second application of the Bonferroni method is to compute the associated p-value and then multiply it by the number of contrasts conducted (i.e., $p_{corrected} = p_{uncorrected} * n_{contrasts}$). The statistic is considered statistically significant only if $p_{corrected}$ is less than .05. If we use Excel to compute exact, two-tailed $p_{uncorrected}$-values for the three t_{obs} computed above, we would find that they are .0094, .0475, and .3947. To correct these p-values, we multiply by $n = 3$ and find that they are .028, .143, and 1.184, respectively. Of course, $p = 1.184$ makes no sense because it exceeds 1. Whenever this occurs, we simply set p to 1.

Post Hoc Contrasts in SPSS

Many methods of computing post hoc contrasts are available through the Post Hoc . . . button on the right side of the One-Way ANOVA dialog. When this is clicked, the dialog box in Figure 18.11 appears. The checkbox next to the name Bonferroni will produce the adjusted confidence intervals and significance tests that we have described in this section. To illustrate the Bonferroni contrasts, we will use the data from the Alzheimer's experiment that we've been discussing.

Figure 18.12 shows the three unplanned contrasts. If you look at the contrasts, p-values, and confidence intervals, you will see that they are exactly the same as those we computed by hand above. Note that SPSS does not prune redundant contrasts and thus produces twice as many as needed. For example, the first set of contrasts shows $m_1 - m_2$ and $m_1 - m_3$, and the second set shows $m_2 - m_1$ and $m_2 - m_3$. The only difference between $m_1 - m_2$ and $m_2 - m_1$, for example, is the sign of the contrast.

Figure 18.11 shows that many different post hoc tests can be applied to data. Typically these methods are applied only if the omnibus F is statistically significant. We won't explore the correction methods because the Bonferroni method provides a sensible correction that keeps family-wise confidence at the desired level. We will say more about problems with post hoc tests below.

Before we turn to problems with post hoc tests, we will address a practice in psychological research that is reflected in the design of SPSS. Logically, corrections for multiple contrasts should be an issue whether the contrasts have been planned or not. However, SPSS does not correct for multiple planned contrasts the way it does for post hoc contrasts. To illustrate, Figure 18.13 shows the same three contrasts shown in Figure 18.12 but entered through the Contrasts . . . dialog rather than the Post Hoc . . . dialog. The contrast weights are shown in Figure 18.13a, and the contrast analysis is shown in Figure 18.13b. Notice that the p-values in Figure 18.12 are three times those in the equal variance part of Figure 18.13. This is because the p-values in Figure 18.12 have been Bonferroni-corrected and those in Figure 18.13 have not.

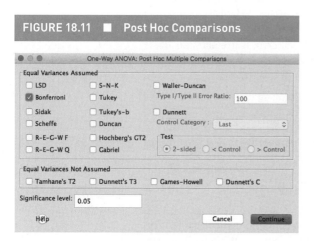

FIGURE 18.11 ■ Post Hoc Comparisons

The Post Hoc Multiple Comparisons dialog accessed through the One-Way ANOVA dialog.

FIGURE 18.12 ■ Bonferroni-Corrected Contrasts

Multiple Comparisons

Dependent Variable: MoCA
Bonferroni

(I) Groups	(J) Groups	Mean Difference (I–J)	Std. Error	Sig.	95% Confidence Interval Lower Bound	95% Confidence Interval Upper Bound
1.00	2.00	−7.00000*	2.26569	.028	−13.2974	−.7026
	3.00	−5.00000	2.26569	.143	−11.2974	1.2974
2.00	1.00	7.00000*	2.26569	.028	.7026	13.2974
	3.00	2.00000	2.26569	1.000	−4.2974	8.2974
3.00	1.00	5.00000	2.26569	.143	−1.2974	11.2974
	2.00	−2.00000	2.26569	1.000	−8.2974	4.2974

*. The mean difference is significant at the 0.05 level.

FIGURE 18.13 ■ Uncorrected Planned Contrasts

(a)

Contrast Coefficients

	Groups		
Contrast	1.00	2.00	3.00
1	1	–1	0
2	1	0	–1
3	0	–1	1

(b)

Contrast Tests

		Contrast	Value of Contrast	Std. Error	t	df	Sig. (2–tailed)
MoCA	Assume equal variances	1	–7.0000	2.26569	–3.090	12	.009
		2	–5.0000	2.26569	–2.207	12	.048
		3	–2.0000	2.26569	–.883	12	.395
	Does not assume equal variances	1	–7.0000	2.44949	–2.858	7.784	.022
		2	–5.0000	2.28035	–2.193	7.144	.064
		3	–2.0000	2.04939	–.976	7.720	.359

Planned contrasts are uncorrected in SPSS.

When we report contrasts, we should always mention whether (and what) correction for multiple contrasts has been applied. If you have followed convention and not applied a correction to planned contrasts, readers will be able to make the adjustment themselves.

What's Wrong With Multiple Post Hoc Contrasts?

Reduced Power

At the beginning of this section I said that using multiple post hoc contrasts is a flawed practice because it results in underpowered research. One part of the problem is that programs like SPSS apply post hoc corrections too broadly. For example, we noted before that with five levels of an independent variable, there are 10 possible simple contrasts. If you request Bonferroni through the post hoc dialog, it will compute all 10 simple contrasts and correct them accordingly. This would be overkill if only two of those contrasts are meaningful to the researcher. Therefore, this is one of the ways that post hoc contrasts reduce power.

Computing contrasts only if the omnibus F is statistically significant is another way to reduce power. Imagine a simple experiment with three levels of the independent variable and five participants in each group. This experiment could be analyzed with two planned orthogonal contrasts. What would happen if finding a significant omnibus F were a precondition for examining the contrasts? Let's say that $\hat{\eta}^2_{c_1} = 30$ and $\hat{\eta}^2_{c_2} = .05$ and $df_{error} = 12$. This means that $R^2 = .35$. The omnibus F-statistic would be

$$F_{2,12} = \frac{R^2}{1 - R^2} \frac{df_{error}}{df_a} = \frac{.35}{.65} \frac{12}{2} = 3.23,$$

which is not statistically significant, $p = .08$, and means we can't look at the contrasts. However, without this precondition, contrast 1 would produce

$$F_{1,12} = \frac{\hat{\eta}^2_{c_1}}{1 - R^2} \frac{df_{error}}{df_{c_1}} = \frac{.3}{.65} \frac{12}{1} = 5.54,$$

which is statistically significant, $p = .04$. It's not that I'm recommending significance tests; however, if they are to be used, we shouldn't work against ourselves by doing low-powered tests.

Instead of wasting degrees of freedom on an uninteresting omnibus test, we can use planned orthogonal contrasts to partition ss_a into meaningful components. Doing so will reduce the number of contrasts we make and thus increase the precision of our confidence intervals and the power of our significance tests. A little thinking ahead of time makes all the difference.

p-Hacking (Again)

Of course with so many possible contrasts available in an ANOVA we always face the temptation of p-hacking, and post hoc contrasts feature prominently in this practice. If a researcher is driven by the quest for statistical significance, she may perform all possible contrasts, select those out of which an interesting story can be made, and then claim that they were planned contrasts all along, allowing smaller effects to be treated as statistically significant. Or a researcher could try all of the post hoc options in Figure 18.11 and choose to report the analysis that produces the pattern of significant and nonsignificant results he was hoping to find.

Researcher degrees of freedom refers to the number of undisclosed decisions that researchers make about what statistics are reported.

Simmons, Nelson, and Simonsohn (2011) described all of these behind-the-scenes tests using the figurative term **researcher degrees of freedom**. Researcher degrees of freedom refers to the number of undisclosed decisions that researchers make about what statistics are reported. Without full knowledge of how many tests were actually computed, a reader has no idea what the true probability of a Type I error is or what the true confidence should be. This kind of p-hacking undermines the validity of the conclusions drawn and results in the publication of meaningless research, which, as we saw in Chapter 9, may be difficult to dislodge from the literature.

The following is a sobering passage from a recent report by David Peterson (2016), who visited a number of developmental research labs as part of a larger effort to get a sense of the research practices in different research areas.

> When a clear and interesting story could be told about significant findings, the original motivation was often abandoned. I attended a meeting between a graduate student and her mentor at which they were trying to decipher some results the student had just received. Their meaning was not at all clear, and the graduate student complained that she was having trouble remembering the motivation for the study in the first place. Her mentor responded, "You don't have to reconstruct your logic. You have the results now. If you can come up with an interpretation that works, that will motivate the hypothesis."
>
> A blunt explanation of this strategy was given to me by an advanced graduate student: "You want to know how it works? We have a bunch of half-baked ideas. We run a bunch of experiments. Whatever data we get, we pretend that's what we were looking for." Rather than stay with the original, motivating hypothesis, researchers in developmental science learn to adjust to statistical significance. They then "fill out" the rest of the paper around this necessary core of psychological research. (p. 6)

We should not use poor advice from one "mentor" and an ill-advised comment from a graduate student to indict whole fields of research. However, my experience suggests the strategies described in this passage are not unusual. The more widespread these practices are, the more seriously compromised psychological research is as a whole. At the heart of such questionable practices is the quest for statistical significance. This is clearly a case of the tail wagging the dog.

If we are to use statistics, we must use them properly. Above all, we must have clear hypotheses about the expected results of our experiment, we must have a clear plan for the analyses before data are collected, and we must then stick to our analysis plan once data have been collected. We must then report how the results compare with our predictions. If interesting and unexpected results occur, then it is completely legitimate to describe these. However, the full extent of these unplanned analyses must be presented so that the reader can assess how many researcher degrees of freedom were used to produce them. These unexpected results could be really important to other researchers, but it is critically important for the theoretical reasoning that leads to an experiment (the original predictions) to be assessed in light of the data collected.

LEARNING CHECK 4

1. Below are the data from Arent and Landers (2003) introduced in Learning Check 3. For this data set, $m_y = 439$, $ss_{total} = 285{,}538$, and $n = 13$. Compute unplanned 95% confidence intervals for (a) $m_5 - m_1$, (b) $m_5 - m_8$, (c) $m_2 - m_7$, and (d) $m_1 - m_8$.

	a_1	a_2	a_3	a_4	a_5	a_6	a_7	a_8
HRR	20	30	40	50	60	70	80	90
Mean	486	454	435	416	418	423	429	451

2. For each of the contrasts just computed, compute t_{obs} and the two-tailed p-value corrected for multiple contrasts. (Computing these p-values requires Excel.)

3. Count the number of simple contrasts in this set of eight means.

Answers

1. $ms_{error} = 234578/96 = 2443.52$. For all confidence intervals, the estimated standard error is $\sqrt{ms_{error} * 2/13} = 19.39$ and the corrected $t_{\alpha/2} = 2.546$.

   (a) $c = -68$, 95% CI $= -68 \pm 49.36 = [-177.36, -18.64]$.

   (b) $c = -33$, 95% CI $= -33 \pm 49.36 = [-82.36, 16.36]$.

   (c) $c = 25$, 95% CI $= 25 \pm 49.36 = [-24.36, 74.36]$.

   (d) $c = 35$, 95% CI $= 35 \pm 49.36 = [-14.36, 84.36]$.

2. For all t statistics, the estimated standard error is $\sqrt{2/13 * ms_{error}} = 19.39$.

   (a) $t_{obs} = -68/19.39 = -3.51$, $p = 0.003$

   (b) $t_{obs} = -33/19.39 = -1.70$, $p = 0.368$

   (c) $t_{obs} = 25/19.39 = 1.29$, $p = 0.801$

   (d) $t_{obs} = 35/19.39 = 1.81$, $p = 0.297$

3. There are $(k^2 - k)/2 = (64 - 8)/2 = 28$ simple contrasts.

REGRESSION AND ANOVA ARE THE SAME THING

In Chapters 16 and 17 we saw that regression analysis can be performed at the omnibus level or at the level of individual predictors. For example, we can test the statistical significance of a regression model as a whole (based on R^2) and we can test the statistical significance of individual regression coefficients. In this chapter we've seen a parallel distinction in ANOVA. That is, we can test the statistical significance of the omnibus R^2 based on variability attributable to means, and we can test the statistical significance of individual contrasts. We will now see that an ANOVA can be conducted as a regression that yields exactly the same results at the omnibus level and at the level of the predictors. To some extent this will be a review of things we've seen already, but these points will be brought together in one place, with additional points made to consolidate the connection.

Table 18.10 shows the data from our Alzheimer's experiment one last time. The data are laid out in "regression style" exactly as in Table 18.2. Column y shows scores on the MoCA for the 15 subjects in the experiment. Columns w_1 and w_2 replace x_1 and x_2 from Table 18.2. In fact, columns w_1 and w_2 are expanded versions of the orthogonal contrast weights, $w_1 = [0, 1, -1]$ and $w_2 = [1, -0.5, -0.5]$, that were introduced earlier. However, we now associate

TABLE 18.10 ■ Data for Three Groups[a]				
Condition	y (MoCA)	w_1	w_2	$\hat{y}$[b]
a_1	17	0	1	14
a_1	18	0	1	14
a_1	9	0	1	14
a_1	10	0	1	14
a_1	16	0	1	14
a_2	19	1	−0.5	19
a_2	21	1	−0.5	19
a_2	18	1	−0.5	19
a_2	27	1	−0.5	19
a_2	20	1	−0.5	19
a_3	20	−1	−0.5	21
a_3	15	−1	−0.5	21
a_3	21	−1	−0.5	21
a_3	22	−1	−0.5	21
a_3	17	−1	−0.5	21
	ss_{total}	$r^2_{c_1}$	$r^2_{c_2}$	ss_a
	284	0.035	0.423	130
		$\Sigma w_1 y$	$\Sigma w_2 y$	
		10.00	−30.00	
		$\Sigma |w_1|/2$	$\Sigma |w_2|/2$	
		5	5	

[a] a_1 = Placebo, a_2 = Drug 1, a_3 = Drug 2.
[b] $\hat{y}_{i,j} = 18 − 1(x_1) − 4(x_2)$.

the contrast weights with *individuals* within the groups (a_1, a_2, and a_3) rather than with the *group means*. When we submit columns y, w_1, and w_2 to a regression analysis we obtain the ANOVA and coefficients tables shown in Figure 18.14. We will now show that the information in this figure mirrors quantities we computed earlier using contrasts.

Figure 18.14a shows that the regression equation produces $R^2 = .46$, exactly as we saw earlier (see Table 18.4). When R^2 is tested for statistical significance in Figure 18.14b, we find that $F_{2,12} = 5.07$, exactly as shown in Figures 18.3 and 18.4. This means that ANOVA and regression yield exactly the same results at the level of the model. As noted earlier, any predictors used to code groups will produce exactly the same R^2, as long as they are not multicollinear.

We now consider the analysis at the level of the predictors. Because the predictors are w_1 and w_2 rather than x_1 and x_2 used in Table 18.2, the regression equation has changed, from $\hat{y}_{i,j} = −12 + 12(x_1) + 7(x_2)$ in Figure 18.4b to $\hat{y}_{i,j} = 18 + 1(w_1) + −4(w_2)$ in Figure 18.14c. (Both regression equations "predict" group means.) Previously we tested the statistical significance of our two orthogonal contrasts and found that for c_1, $t_{obs} = 0.88$, and for c_2, $t_{obs} = −3.06$ (see Figure 18.6). Notice that these are exactly the same t_{obs} values for w_1 and w_2 shown in Table 18.14c. Therefore, whether the contrast weights are applied to group means or used as a predictor in a regression analysis, they produce exactly the same t_{obs}. This shows that ANOVA and regression are doing the same things at the predictor/contrast level, even though on the surface they look quite different. Of course, if different predictors were used, the regression coefficients, as well as the associated t_{obs}, would change because different predictors would represent different contrasts between means.

Finally, we saw earlier that orthogonal contrasts carve up ss_a into nonoverlapping slices, which we called ss_{c_1} and ss_{c_2}. When these are divided by ss_{total}, we obtain $\hat{\eta}^2_{c_1} = .04$ and $\hat{\eta}^2_{c_2} = .42$, which represent the proportion of total variance in y explained by contrasts c_1 and c_2 respectively (see Table 18.5). At the bottom of columns w_1 and w_2 in Table 18.10, we see the squared correlation between y and w_1 and between y and w_2. These are also $r^2_{c_1} = .04$ and $r^2_{c_2} = .42$, which represent the proportion of total variance in y explained by contrasts w_1 and w_2, respectively. Therefore, with orthogonal contrast weights, we can decompose the variability in our outcome variable by computing sums of squares derived from contrasts applied to means or from correlating the contrast weights (expressed as predictors) with the outcome variable. Again, ANOVA and regression are two ways of accomplishing the same thing.

The last point made in Table 18.10 will bring us back to our real interest, which is the size of the contrast. So far we've computed contrasts as

$$c = \sum_{i=1}^{k} w_i m_i.$$

However, the numbers in columns w_1 and w_2 in Table 18.10 are the same as those in the contrasts applied to means. Instead of multiplying contrast weights and group means, we

FIGURE 18.14 ■ ANOVA Conducted as Regression

(a)

Model Summary

Model	R	R Square	Adjusted R Square	Std. Error of the Estimate
1	.677[a]	.458	.367	3.58236

a. Predictors: (Constant), w2, w1

(b)

ANOVA[a]

Model		Sum of Squares	df	Mean Square	F	Sig.
1	Regression	130.000	2	65.000	5.065	.025[b]
	Residual	154.000	12	12.833		
	Total	284.000	14			

a. Dependent Variable: MoCA

b. Predictors: (Constant), w2, w1

(c)

Coefficients[a]

Model		Unstandardized Coefficients		Standardized Coefficients	t	Sig.
		B	Std. Error	Beta		
1	(Constant)	18.000	.925		19.460	.000
	w1	1.000	1.133	.188	.883	.395
	w2	4.000	1.308	-.650	-3.058	.010

a. Dependent Variable: MoCA

can multiply the contrast weights and the individual scores within the groups. With these contrast weights expanded as in Table 18.10 and applied to scores, they can be used to compute a contrast using the following formula:

$$c = \frac{\sum w^*y}{\sum |w|/2}.$$

(18.23)

(I have reduced some clutter by not using subscripts in equation 18.23.) The numerator of equation 18.23 says to multiply all scores in y by a corresponding weight (e.g., the numbers in column w_1) and sum the products; i.e., $\sum w^*y$. The denominator of equation 18.23 says to sum the absolute values of the weights and divide this sum by 2; i.e., $\sum |w|/2$.

The quantities in the numerator of equation 18.23 are shown under $\sum w_1^*y$ and $\sum w_2^*y$ in Table 18.10, and those in the denominator are shown under $\sum |w_1|/2$ and $\sum |w_2|/2$. When equation 18.23 is evaluated for our two contrasts, we find

$$c_1 = \frac{\sum w^*y}{\sum |w|/2} = \frac{10}{5} = 2 \text{ and}$$

$$c_2 = \frac{\sum w^*y}{\sum |w|/2} = \frac{-30}{5} = -6.$$

That is, equation 18.23 yields exactly the same contrasts previously obtained from means. The denominator of equation 18.23 simply ensures that the weights sum to 2.

Equation 18.23 makes the general point that we don't need to worry about the scale of the contrast weights when we construct contrasts, but the weights must sum to 0. If we define a contrast as

$$c = \sum_{i=1}^{k} w_i m_i,$$

then we only need to divide c by $\Sigma|w|/2$ to ensure that the contrast represents the difference between two groups of means. This point will be important in Chapter 20.

In summary, Table 18.10 connects two ways of thinking about orthogonal contrasts. One involves the application of weights to means and the other is the correlation of weights with individual scores. Both partition variance to produce the same results. This is why I said earlier that most statistical analyses used in psychological research boil down to correlation, in one form or another.

In Chapters 19 and 20, we will look at versions of ANOVA that increase in complexity. To explain these analyses we will make use of tables like Table 18.10 to partition the variability in y into nonoverlapping components using orthogonal vectors. These will allow us to understand the sources of variance involved in an ANOVA at a conceptual level without having to memorize elaborate equations. A critical point to keep in mind is that each vector used to explain the variability among means is really a contrast that can tell us about how the treatment means differ.

POWER

We've noted at many points that significance tests are frequently underpowered, thus wasting time and resources. Obviously we should aim for high power when conducting ANOVAs, but there are two perspectives on this question that have to do with the two levels of analysis that we have considered. That is, we can think of power in terms of the omnibus test of

$$H_0: \mu_1 = \mu_2 = \ldots = \mu_k$$

or in terms of specific contrasts that test

$$H_0: \psi_i = 0.$$

The G∗Power program makes it very easy to do a power analysis at both levels. Appendix 18.1 (available at study.sagepub.com/gurnsey) addresses these two perspectives in turn. However, ensuring adequate power for specific contrasts is far more important than ensuring adequate power for the omnibus test.

SUMMARY

ANOVA is a special case of multiple regression in which there are $k = 3$ or more fixed levels of an independent variable. The independent variables may be categorical or quantitative variables with levels fixed by the experimenter. Each of the k levels of the independent variable corresponds to a normally distributed population. These populations are assumed to have the same standard deviation (σ) but possibly different means (μ_j). When these k populations are considered as one, the population will have a mean of

$$\mu = \frac{\sum_{i=1}^{k} \mu_i}{k},$$

assuming that all populations are the same size. The total variability in this superpopulation can be defined as

$$ss_{\text{total}} = \sum_{i=1}^{k}\sum_{j=1}^{N}\left(y_{i,j} - \mu\right)^2,$$

where $y_{i,j}$ is the jth score in the ith population, and N is the number of scores in each population. Because each score ($y_{i,j}$) can be decomposed into a component associated with a population mean (μ_i) and a component *not* associated with a population mean (i.e., $y_{i,j} - \mu_i$), we can decompose ss_{total} into *between-group* and *within-group* variability as follows:

$$ss_{\text{between}} = \sum_{i=1}^{k} N(\mu_i - \mu)^2 \text{ and}$$

$$ss_{\text{within}} = \sum_{i=1}^{k}\sum_{j=1}^{N}\left(y_{i,j} - \mu_i\right)^2 = ss_{\text{total}} - ss_{\text{between}}.$$

Therefore, the proportion of variance explained by the difference between the k population means is

$$P^2 = \frac{ss_{\text{between}}}{ss_{\text{total}}}.$$

When random samples of size n are drawn from each of the k populations, we can derive equations similar to those shown above for the kn sample scores. That is,

$$m_y = \frac{\sum_{i=1}^{k} m_i}{k} \text{ and}$$

$$ss_{\text{total}} = \sum_{i=1}^{k}\sum_{j=1}^{n_i}\left(y_{i,j} - m_y\right)^2.$$

The between-group sum of squares for the samples is

$$ss_{\text{between}} = \sum_{i=1}^{k} n_i\left(m_i - m_y\right)^2.$$

The within-group sum of squares for the samples is

$$ss_{\text{within}} = \sum_{i=1}^{k}\sum_{j=1}^{n_i}\left(y_{i,j} - m_i\right)^2 = \sum_{i=1}^{k} ss_i = ss_{\text{total}} - ss_{\text{between}}.$$

From ss_{between} and ss_{total}, we can compute the following to estimate P^2:

$$R^2 = \frac{ss_{\text{between}}}{ss_{\text{total}}}.$$

An approximate $(1-\alpha)100\%$ confidence interval around R^2 can be computed using the MBESS routines in **R**.

Typical applications of ANOVA test the null hypothesis that all population means are the same,

$$H_0: \mu_1 = \mu_2 = \ldots = \mu_k,$$

against the alternative hypothesis that at least one of the population means is different from the rest. This null hypothesis is tested by computing an *omnibus* F-statistic as follows:

$$F_{\text{obs}} = \frac{R^2}{1 - R^2}\frac{df_{\text{error}}}{df_a},$$

where $df_a = k - 1$ and $df_{\text{error}} = df_{\text{total}} - df_a$. An F_{obs} is declared statistically significant when statistics greater to it would occur less than 5% of the time under H_0.

Testing whether R^2 is statistically significant is rarely the focus of ANOVA. Instead, researchers want to know how the means differ. The best way to compare means is with *planned orthogonal contrasts*. In the population, a contrast (Ψ) is the sum of the products of k weights (w) and k means:

$$\Psi = \sum_{i=1}^{k} w_i \mu_i.$$

We assume that the weights associated with a contrast sum to 0,

$$\sum_{i=1}^{k} w_i = 0.$$

Therefore, the weights are positive and negative and thus compare a mean (or several means) against another mean (or several other means). So that a contrast represents a comparison between two groups of means in the original units, we design our weights such that

$$\sum_{i=1}^{k}\left|w_i\right| = 2.$$

An *orthogonal set* consists of $k-1$ orthogonal contrasts.

A population contrast (Ψ) is estimated from sample means by

$$c = \sum_{i=1}^{k} w_i m_i.$$

The estimated standard error of c is

$$s_c = \sqrt{ms_{\text{error}} \sum_{i=1}^{k}\frac{w_i^2}{n_i}}.$$

Therefore, a $(1-\alpha)100\%$ confidence interval around c is given by

$$c \pm t_{\alpha/2}(s_c),$$

where $t_{\alpha/2}$ is based on df_{error} degrees of freedom. Of course, the contrast can also be tested for statistical significance by computing a t-statistic:

$$t_{obs} = \frac{c}{s_c}.$$

A contrast can be standardized by dividing it by $\sqrt{ms_{error}}$:

$$d = \frac{c}{\sqrt{ms_{error}}}.$$

Confidence intervals around standardized contrasts can be performed using the **R** MBESS package.

Polynomial trends are like orthogonal contrasts. The weight vectors (w) for polynomial trends are related to polynomial equations. The coefficients (weights) in polynomial trends describe orthogonal patterns. Computing $c_{trend} = \Sigma w_i m_i$ yields a measure of the extent to which a given trend (pattern) exists in the sample means. Trend analysis can be used only when the independent variable is quantitative and intervals between levels are equally spaced.

Although orthogonal contrasts are recommended, many researchers conduct significance tests for unplanned, non-orthogonal contrasts, but only if the omnibus F is declared statistically significant. As the number of unplanned comparisons increases, the probability that one of them will prove statistically significant by chance (a Type I error) also increases. If we conduct n unplanned significance tests, then the probability is

$$\alpha_{fw} = 1 - (1 - \alpha_{pc})^n$$

that at least one of them will be statistically significant when the null hypothesis is true. In this case, α_{fw} is the family-wise Type I error rate, and α_{pc} is the per-contrast Type I error rate. To maintain α_{fw} at .05, a contrast is declared statistically significant only if its p-value is less than α_{fw}/n. Similar comments hold for the calculation of multiple confidence intervals. In this case, we choose $t_{\alpha/2}$ based on α_{fw}/n to ensure that $(1-\alpha_{fw})100\%$ of all n intervals will capture the population contrast (Ψ) being estimated.

Finally, regression analysis and ANOVA are the same thing. In both cases, we can use orthogonal contrasts to determine the omnibus R^2 associated with treatment means, and, more importantly, to determine how treatment means differ.

KEY TERMS

between-subjects factor   481
Bonferroni correction   506
complex contrasts   489
condition   481
factor   481
homogeneity of variance   485

levels   481
mixed design   481
omnibus F-statistic   486
orthogonal set   492
orthogonality   492
polynomial trends   500

post hoc contrasts   504
researcher degrees of freedom   510
simple contrasts   489
source table   500
unplanned contrasts   504
within-subjects factor   481

EXERCISES

Definitions and Concepts

1. What does ANOVA stand for?

2. What is a factor in ANOVA?

3. What is a condition in an ANOVA?

4. What does it mean to do a four-way ANOVA?

5. How many conditions are there in a $5 \times 6 \times 3 \times 2$ between-subjects ANOVA?

6. How many conditions are there in a $5 \times 6 \times 3 \times 2$ within-subjects ANOVA?

7. What is a mixed design?

8. What is a sampling distribution?

9. What does it mean to have 95% confidence in an interval?

10. What is the correct definition of a p-value?

11. What is the difference between a simple and a complex contrast?

12. What does it mean to say that two sets of numbers are orthogonal?

13. Why is power reduced when we use many unplanned contrasts?

True or False

State whether the following statements are true or false.

14. A mixed ANOVA can involve between-subjects and within-subjects factors, or just between-subjects or just within-subjects factors.

15. In a mixed ANOVA with five levels of neuroticism and 10 different test days, neuroticism is most likely to be the within-subjects factor.

16. A mixed ANOVA with five levels of neuroticism and 10 test days has 36 conditions.

17. Homogeneity of variance applies only to ANOVA and homoscedasticity applies only to multiple regression.

18. The grand mean is the average score within a condition.

19. The grand mean is the average of all kn scores.

20. $m_1 - m_8$ is a simple contrast.

21. $s_c = \sqrt{ms_{error} \sum_{i=1}^{k} w_i / n_i}$.

22. $ms_{error} = ss_{error} / df_{error}$.

23. $ms_{error} = s_{pooled}^2$.

24. $[3, -1, -1, -1]$ and $[0, 2, -1, -1]$ are orthogonal.

Calculations

25. If a one-factor between-subjects ANOVA has 10 levels of the independent variable and 120 participants, how many participants are in each condition?

26. The table below shows the scores for a one-way ANOVA with four levels of the independent variable.

    (a) Compute m_y, $ss_{between}$, ss_{error}, ss_{total}, R^2, df_a, df_{error}, and F.

    (b) Using SPSS or Excel, use the following codes for groups and compute the regression equation to predict $y_{i,j}$: $[3, 0, 0]$, $[-1, 2, 0]$, $[-1, -1, 1]$, and $[-1, -1, -1]$.

    (c) Use `ci.R2` to compute an approximate 95% confidence interval around R^2.

a_1	a_2	a_3	a_4
8	5	20	5
10	11	11	10
8	16	12	18
5	10	16	14
8	14	8	16
6	6	17	7
4	8	21	14

27. A one-way ANOVA with five levels of the independent variable has the following scores:

a_1	a_2	a_3	a_4	a_5
8	6	19	5	3
10	11	9	10	4
8	18	12	18	9
6	9	16	15	4

    (a) Compute m_y, $ss_{between}$, ss_{error}, ss_{total}, R^2, df_a, df_{error}, and F.

    (b) Using SPSS or Excel, use the following codes for groups and compute the regression equation to predict $y_{i,j}$: $[4, 0, 0, 0]$, $[-1, 3, 0, 0]$, $[-1, -1, 2, 0]$, $[-1, -1, -1, 1]$, and $[-1, -1, -1, -1]$.

    (c) Use `ci.R2` to compute an approximate 95% confidence interval around R^2.

28. A one-factor between-subjects ANOVA with four levels of the independent variable produced the following means: $m = [7, 10, 15, 12]$. There were seven subjects in each condition and $ss_{total} = 640$. Compute the 95% confidence interval, t_{obs}, $\hat{\eta}^2$, and F for each of the following contrasts:

    (a) $(m_1 + m_2)/2 - (m_3 + m_4)/2$

    (b) $m_1 - m_2$

    (c) $m_3 - m_4$

29. A one-factor between-subjects ANOVA with five levels of the independent variable produced

the following means: $m = [8, 11, 14, 12, 5]$. There were four subjects in each condition and $ss_{total} = 464$. Compute the 95% confidence interval, t_{obs}, $\hat{\eta}^2$, and F for each of the following contrasts:

(a) $m_1 - (m_2 + m_3\ m_4 + m_5)/4$

(b) $(m_2 + m_3)/2 - (m_4 + m_5)/2$

(c) $m_2 - m_3$

(d) $m_4 - m_5$

30. Do the following columns of numbers form an orthogonal set? Explain why or why not.

w_1	w_2	w_3	w_4
4	0	0	0
−1	3	0	0
−1	−1	2	0
−1	−1	−1	−1
−1	−1	−1	1

31. Do the following columns of numbers form an orthogonal set? Explain why or why not.

w_1	w_2	w_3	w_4
−2	2	−1	1
−1	−1	2	−4
0	−2	0	6
1	−1	−2	−4
2	2	1	1

32. Consider a set of 10 means analyzed with nine contrasts. If there were eight scores contributing to each of the 10 means, how much wider would the 95% confidence intervals around the contrasts be when corrected versus not corrected for multiple contrasts?

Scenarios

33. Imagine that a researcher would like to compare the effectiveness of cognitive behavioral therapy (CBT) and schema therapy (ST) in the treatment of narcissistic personality disorder (NPD). Without doing a power analysis, the researchers selected 60 individuals meeting the clinical criteria for NPD and randomly assigned 20 to each of three conditions; CBT, ST, or a wait-list control condition in which participants received no therapy but were promised therapy at a later time. After the two treatment groups had undergone 6 months of treatment, those in all three conditions were administered the narcissistic personality inventory. The results are shown in the following table, along with the means and sums of squares for each of the three conditions. (I recommend doing the questions below by hand and checking your answer by running the analyses in SPSS.)

	Control	CBT	ST
	34	28	33
	18	15	17
	23	12	4
	38	39	19
	25	20	21
	28	25	6
	31	20	11
	29	14	18
	29	27	19
	29	31	19
	38	21	25
	27	23	19
	30	31	24
	9	19	24
	28	23	22
	30	38	17
	32	25	16
	33	32	21
	24	30	27
	25	27	18
m	28	25	19
ss	818	1028	840

Use the data in the table to answer the following questions.

(a) Compute the following for two planned contrasts that compare the mean of the control group with the mean of the two treatment means as well as the CBT mean with the ST mean:

   (i) uncorrected 95% confidence intervals around the contrasts,

   (ii) 95% confidence intervals for the standardized contrasts (using `ci.sc` in MBESS), and

   (iii) t_{obs} and p.

(b) Compute

   (i) R^2 and the approximate 95% confidence interval (using `ci.R2`) and

   (ii) F_{obs} and its exact p-value.

(c) Compute the following for all three possible contrasts between means:

   (i) Bonferroni-corrected 95% confidence intervals and

   (ii) t_{obs} and Bonferroni-corrected p-values.

(d) The "dodo bird verdict" is a term from *Alice in Wonderland* (Carroll, 1865) in which the dodo declared at the end of the caucus race that "Everybody has won and all must have prizes." This term was introduced by Saul Rosenzweig in 1936 when he observed that any psychotherapy seemed to work about as any other. Luborsky et al. (2002) did a meta-analysis of 17 meta-analyses (i.e., a meta-meta-analysis) of studies that compared one treatment against another. This meta-meta-analysis involved 130 original studies. The average effect size was $d = .2$ and was not statistically significant. This effect grew even smaller when the authors corrected the effect sizes for author bias (e.g., the authors disposed to CBT who found CBT superior to some other treatment). How do the current results relate to the dodo bird verdict?

34. When drug companies want to bring a drug to market, they must first have some evidence that it works as it is supposed to in humans. Let's imagine that a drug company is working on a new type of stimulant that they believe will improve concentration and lead to improved scores on a standardized test of cognitive ability. The company's senior scientist thinks that the improvements will be dose dependent. The company solicited 20 volunteers from the community and randomly assigned them to the five different doses, which were 50, 60, 70, 80, and 90 µg/kg. Each participant was given the test of cognitive ability and the results are shown in the following table. To assess the predictions, the researchers conducted significance tests using trend analysis.

Dose	0	25	50	75	100
	252	251	185	280	120
	162	197	237	185	138
	103	215	302	89	65
	159	141	256	162	61
m	169	201	245	179	96
ss	11,394	6312	7034	18,626	4526

Use the data in the table to answer the following questions.

(a) Plot the means and 95% confidence interval in a graph formatted exactly like Figure 18.8.

(b) Conduct a trend analysis on the data treating the four trends and do not correct to multiple contrasts. Present the results as a source table with columns that show

   (i) the trend contrast,

   (ii) the estimated standard error of the contrast,

   (iii) t_{obs},

   (iv) p, and

   (v) $\hat{\eta}^2$.

(c) Was the researcher's prediction about the linear trend supported? Explain your reasoning.

(d) Explain why you do or do not think these data should be published.

(e) If you were to publish these data, what story do you think they tell and how would you support your argument?

35. Imagine yourself sitting in a room facing 16 speakers arrayed in a semicircle around you. Your left ear is plugged so that you hear sounds through only your right ear. In this case, we say that you are hearing sounds *monaurally*. When one of the speakers issues a sound burst, you have to point to it. This is an auditory localization task in which researchers measure your pointing errors, which are the absolute difference between the direction you point and the true direction of the speaker. These pointing or monaural localization errors are measured in degrees. Lessard, Paré, Lepore, and Lassonde (1998) wanted to know how different monaural localization accuracy was in three groups of participants:

    (i)   normally sighted individuals,

    (ii) those that were totally blind from birth, and

    (iii) those who developed severely reduced vision, with some preserved peripheral vision, later in life. The following table shows the mean absolute localization errors (measured in degrees) and sums of squares for three groups with 10 participants in each. These are not the original data from Lessard et al., but they are qualitatively similar.

	Normal	Totally Blind	Partially Blind
m	52	35	84
ss	1448	20,312	660

Conduct an ANOVA on these data and compute the following:

(a) The omnibus F.

(b) Bonferroni corrected 95% confidence intervals to compare all three pairs of means.

(c) Do these data seem to violate any of the assumptions of ANOVA?

(d) What do these results suggest about the role of vision loss in monaural auditory localization? (This is not a statistical question.)

19 ANALYSIS OF VARIANCE
One-Factor Within-Subjects

INTRODUCTION

In Part III we distinguished between independent-samples designs (Chapter 11) and dependent-samples designs (Chapter 12). The between-subjects ANOVA covered in Chapter 18 is an extension of the independent-samples design, and the repeated-measures design covered in this chapter is an extension of the dependent-samples design. Most of the terminology and conceptual content developed in Chapter 18 is continued in this chapter. The main change is that scores are obtained from the same participants in all conditions of the analysis. Therefore, the analyses covered in this chapter are repeated-measures analyses.

As we saw in Chapter 12, the advantage of a repeated-measures design is that we achieve greater precision in our estimates by eliminating subject variability that is not relevant to the research question. For example, in Chapter 18, we started with the example of two new drugs for patients with Alzheimer's disease and we studied the effects of the two drugs in a between-groups design. One would expect a lot of between-subject variability in this design, which is confirmed by the errors in the bar graph shown in Figure 18.1. An alternative approach would be to use the so-called ABA design. The ABA method involves testing each individual before treatment (A_1), then again at some point during treatment with the drug (B), and then once again at some point after treatment stops (A_2). This method provides three measures from each subject, from which we can make precise estimates of improvements from A_1 to B and then losses from B to A_2.

The within-subjects design is useful in many other situations. For example, one could measure the effect of several doses of some drug on rats' behavior. Or one could measure improvements in accuracy across several days of training in a learning task, for either animals or humans. Memory tests could be given every hour for subjects participating in a sleep deprivation experiment. In a perception task, we could measure the effect of stimulus duration on the accuracy of stimulus detection. In a visual search task, we could measure accuracy as a function of the number of distractors. Of course, there are always dangers of carryover effects in repeated-measures designs, and these possibilities cannot be ignored. However, in many cases (such as perceptual discrimination tasks), these risks are small relative to the efficiency of the method and the precision of the obtained results.

Our coverage of the one-factor within-subjects design will start with a discussion of omnibus effects. Although omnibus effects are not particularly meaningful, they help us orient to the within-subjects design. Our main focus will be on confidence intervals for planned contrasts. We must also cover significance tests for contrasts. As always, it makes little sense to conduct significance tests without first doing a power analysis, so we will say a few things about this as well.

AN EXAMPLE: THE POSNER CUING TASK

The Posner cuing task (Posner, 1980) is a laboratory method used to assess the effect of attentional cues on stimulus detection. Each trial begins with a fixation dot at the center of a computer screen (Time 1), followed by a cue (Time 2), and the stimulus is then presented to the right or left of center (Time 3). The participant maintains fixation at the center of the screen throughout the trial and presses a button as quickly as possible when the target appears at Time 3. The time (in milliseconds) that the participant takes to press the button after the stimulus appears is the dependent variable.

The nature of the cue is the independent variable. The cues are a plus sign (+) or arrows pointing to the left (←) or right (→) of the screen. The + sign gives no information about whether the target will appear on the left or right side of the screen, so it is called a *neutral cue* (see Figure 19.1b). The arrow cues, however, can be *valid* or *invalid*. A cue is valid when it points to the side of the screen where the target appears, and a cue is invalid when it points to the side of the screen opposite to where the target appears. Typically 80% of the arrow cues are valid and 20% are invalid (see Figures 19.1a and 19.1c).

Imagine that responses were collected from six participants, and each participant went through 600 experimental trials. For each participant, 200 trials involved a neutral cue, and the remaining 400 trials involved 320 valid cues (80%) and 80 invalid cues (20%). The 600 trials were presented in a different random order to each of the six participants. The results in Figure 19.2 show mean reaction times for valid, neutral, and invalid cues. The error bars are 95% confidence intervals. The independent variable is cue type, with three levels: valid, neutral, and invalid. The dependent variable is average reaction time (in milliseconds).

This experiment is a within-subjects experiment because all subjects provided data for each level of the independent variable. We will use this example to develop the one-factor, within-subjects ANOVA.

THE OMNIBUS ANALYSIS

Table 19.1 shows the raw data from the hypothetical experiment described above. The average response times for the valid, neutral, and invalid conditions are shown for each

FIGURE 19.1 ■ The Posner Cuing Paradigm

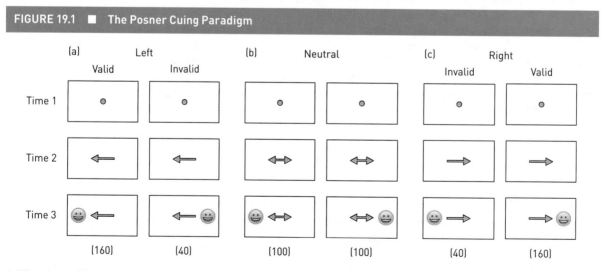

At Time 1, participants steady their gaze on the fixation dot at the center of the computer screen. About 500 ms later (Time 2), the black fixation dot is replaced by one of three cues: ←, ↔, or →. At Time 3, a target appears and participants press a button as soon as the target is detected. Reaction time is the dependent variable.

participant. Values in the "*m*" row are the means for the three conditions (267, 289, and 329). Values in the "*s*" column are the mean scores for each of the six subjects in the experiment (403, 166, 254, 327, 320, and 300).

Partitioning Variance

Partitioning variance for the omnibus ANOVA is a simple extension of the material covered in Chapter 12. There are three sources of variance of interest in the one-factor within-subjects ANOVA:

- variability attributable to condition means (ss_a),
- variability attributable to subjects (ss_s), and
- error variability remaining when the first two sources have been accounted for (ss_{error}).

We will extract these three sources of variability in much the same way we did in Chapter 12.

The first task is to compute the grand mean of the scores in all *k* (conditions) by *n* (subjects) in the experiment. This is done exactly as in Chapter 18:

$$m_y = \frac{\sum_{i=1}^{k}\sum_{j=1}^{n} y_{i,j}}{kn}. \tag{19.1}$$

The total sum of squares in our data set is the sum of squared deviations from m_y, computed as follows:

$$ss_{total} = \sum_{i=1}^{k}\sum_{j=1}^{n}(y_{i,j}-m_y)^2. \tag{19.2}$$

Table 19.1 shows that $ss_{total} = 107{,}632$. As always, ss_{total} is the quantity to be partitioned.

FIGURE 19.2 ■ Hypothetical Results

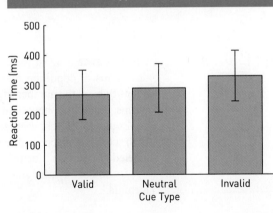

Results of a hypothetical Posner experiment. Reaction times are shown for each of the three cue types. Error bars represent 95% confidence intervals.

TABLE 19.1 ■ Data From a Hypothetical Posner Experiment

Subject	Valid (a_1)	Neutral (a_2)	Invalid (a_3)	s	Condition Means			Subject Means		
1	375	393	441	403	267	289	329	403	403	403
2	137	166	195	166	267	289	329	166	166	166
3	234	246	282	254	267	289	329	254	254	254
4	284	324	373	327	267	289	329	327	327	327
5	299	317	344	320	267	289	329	320	320	320
6	273	288	339	300	267	289	329	300	300	300
m	267	289	329		$ss_a = 11{,}856$			$ss_s = 94{,}980$		
	$m_y = 295$									
	$ss_{total} = 107{,}632$									
	$ss_{error} = ss_{total} - ss_a - ss_s = 796$									

To obtain ss_a (the sum of squares associated with the condition means), we use the following formula:

$$ss_a = n\sum_{i=1}^{k}(m_i - m_y)^2, \tag{19.3}$$

which is exactly what we did in Chapter 18. Equation 19.3 computes the sum of squared deviations of condition means (m_i) from the grand mean (m_y) and then multiplies this quantity by n, the number of subjects. When equation 19.3 is applied to the three condition means, Table 19.1 shows that $ss_a = 11,856$.

Equation 19.3 shows that ss_a is the quantity we would obtain from equation 19.2 when each score is replaced by the mean of the condition it is in. This is illustrated in the columns labeled "Condition Means" in Table 19.1.

To obtain ss_s (the sum of squares associated with subjects), we use the following formula:

$$ss_s = k\sum_{j=1}^{n}(m_{s_j} - m_y)^2. \tag{19.4}$$

Equation 19.4 computes the sum of squared deviation of subject means (m_{s_j}) from the grand mean (m_y) and then multiplies this quantity by k, the number of conditions in the experiment. When equation 19.4 is applied to the six subject means, Table 19.1 shows that $ss_s = 94,980$.

Equation 19.4 shows that ss_s is the quantity we would obtain from equation 19.2 when each score is replaced by the mean score of the subject that produced it. This is illustrated in the columns labeled "Subject Means" in Table 19.1.

Once ss_a and ss_s have been computed, ss_{error} is what is left over. Therefore, we compute ss_{error} as follows:

$$ss_{error} = ss_{total} - ss_a - ss_s. \tag{19.5}$$

Table 19.1 shows that $ss_{error} = 796$. From these calculations, we find that

$$ss_{total} = ss_a + ss_s + ss_{error} = 11,856 + 94,980 + 796 = 107,632.$$

The *F*-Statistic

In Chapter 18, ss_{total} for the between-subjects ANOVA was partitioned into ss_a and ss_{error} and we computed $F = ss_a/ss_{error} * df_{error}/df_a$. In that case, $ss_{error} = ss_{total} - ss_a$ and $df_{error} = df_{total} - df_a$. Were we to compute F this way for the data in Table 19.1, we would find that $F_{obs} = 0.62$, which is pretty darn small.

We compute F for the within-subjects ANOVA essentially the same way, but ss_{error} and df_{error} are different. For the within-subjects design, $ss_{error} = ss_{total} - ss_a - ss_s$ and $df_{error} = df_{total} - df_a - df_s$. Therefore, ss_{error} is smaller in the within-subjects design because subject variability is eliminated.

In the within-subjects design, F_{obs} is computed as follows:

$$F_{obs} = \frac{ss_a}{ss_{error}} \frac{df_{error}}{df_a}. \tag{19.6}$$

In our example,

$$df_{total} = n*k - 1 = 18 - 1 = 17,$$
$$df_a = k - 1 = 3 - 1 = 2, \text{ and}$$
$$df_s = n_s - 1 = 6 - 1 = 5.$$

Therefore,

$$df_{error} = df_{total} - df_a - df_s = 17 - 2 - 5 = 10.$$

When we compute F_{obs} using $ss_a = 11{,}856$, $ss_{error} = 796$, $df_a = 2$, and $df_{error} = 10$, we find that

$$F_{obs} = \frac{ss_a}{ss_{error}} \frac{df_{error}}{df_a} = \frac{11{,}856}{796} \frac{10}{2} = 74.47.$$

Obviously, the F_{obs} obtained when subject variability is eliminated (74.47) is much larger than the F_{obs} obtained when subject variability is not eliminated (0.62). In general, the within-subjects F-test is more sensitive to the differences among condition means than the between-subjects F-test because subject variability is eliminated. To test F_{obs} for statistical significance, we consult the F-table with df_a and df_{error} degrees of freedom. As we've seen, $df_a = 2$ and $df_{error} = 10$.

There is a really important point about degrees of freedom that I will raise now, but not explain until Chapter 20. For now I will only say that it is not a coincidence that $df_{error} = df_a * df_s = 2*5 = 10$. That is, we can express df_{total} as $df_a + df_s + df_a * df_s$.

Partial Eta Squared ($\hat{\eta}_p^2$)

In Chapter 18, we used the symbol $\hat{\eta}^2$ to denote the proportion of ss_{total} explained by a particular source in a between-subjects design. We can do the same in a within-subjects design:

$$\hat{\eta}_a^2 = \frac{ss_a}{ss_{total}}, \tag{19.7}$$

$$\hat{\eta}_s^2 = \frac{ss_s}{ss_{total}}, \text{ and} \tag{19.8}$$

$$\hat{\eta}_{error}^2 = \frac{ss_{error}}{ss_{total}}. \tag{19.9}$$

For the between-subjects ANOVA, $\hat{\eta}_a^2$ is a units-independent effect size, reflecting the proportion of variability in the dependent variable attributable to differences between group means. However, $\hat{\eta}_a^2$ is a poor measure of effect size in the within-subjects ANOVA because it is affected by subject variability. For example, if ss_a remains fixed while ss_s increases, then ss_a will account for a smaller and smaller portion of ss_{total}.

We can alter the data in Table 19.1 to illustrate this problem. Table 19.2 is identical to Table 19.1 in all respects except that 1002 ms has been added to the three scores of subject 1. As a consequence, ss_s has increased from 94,980 in Table 19.1 to 3,254,286 in Table 19.2. Notice, however, that ss_a is still 11,856 and ss_{error} is still 796. This means that F_{obs} would still be 74.47, as computed above.

When $\hat{\eta}_a^2$ is computed from the data in Table 19.1, we obtain

$$\hat{\eta}_a^2 = \frac{ss_a}{ss_{total}} = \frac{11856}{107632} = .11.$$

However, when $\hat{\eta}^2$ is computed from the data in Table 19.2, we obtain a far smaller value:

$$\hat{\eta}_a^2 = \frac{ss_a}{ss_{total}} = \frac{11{,}856}{3{,}254{,}286} = .0036.$$

TABLE 19.2 ■ The Effect of Subject Variability									
Subject	Valid (a_1)	Neutral (a_2)	Invalid (a_3)	s	Condition Means			Subject Means		
1	1377	1395	1443	1405	434	456	496	1405	1405	1405
2	137	166	195	166	434	456	496	166	166	166
3	234	246	282	254	434	456	496	254	254	254
4	284	324	373	327	434	456	496	327	327	327
5	299	317	344	320	434	456	496	320	320	320
6	273	288	339	300	434	456	496	300	300	300
m	434	456	496		$ss_a = 11{,}856$			$ss_s = 3{,}254{,}286$		
$m_y = 462$										
$ss_{total} = 3{,}266{,}938$										
$ss_{error} = ss_{total} - ss_a - ss_s = 796$										

Once again, because $\hat{\eta}^2$ is affected by subject variability, it is a poor measure of effect size in a design whose purpose is to eliminate subject variability.

A common alternative to $\hat{\eta}^2$ for within-subjects designs is **partial eta squared**, which we denote as $\hat{\eta}_p^2$ and define as follows:

$$\hat{\eta}_p^2 = \frac{ss_a}{ss_a + ss_{error}}. \tag{19.10a}$$

We would obtain the following for the data in both Tables 19.1 and 19.2:

$$\hat{\eta}_p^2 = \frac{ss_a}{ss_a + ss_{error}} = \frac{11{,}856}{11{,}856 + 796} = .94.$$

This shows that ss_a is very large relative to ss_{error}. Note that $\hat{\eta}_p^2$ can also be computed as

$$\hat{\eta}_p^2 = \frac{F}{F + \dfrac{df_{error}}{df_a}}. \tag{19.10b}$$

One could compute $\hat{\eta}_p^2$ for a one-factor, between-subjects design; but in this case, it is exactly equal to $\hat{\eta}^2$. In a one-factor between-subjects design, $\hat{\eta}_p^2 = ss_a/(ss_a + ss_{error}) = \hat{\eta}^2 = ss_a/ss_{total}$. Therefore, $\hat{\eta}_p^2$ should be reserved for within-subjects designs.

These days, it is very common for researchers to report $\hat{\eta}_p^2$, along with F_{obs} and its associated p-value. This practice is both good and bad. It is good because researchers recognize that significance tests should be augmented with reports of effect size. Reporting $\hat{\eta}_p^2$ puts an F_{obs} in perspective when it is large and statistically significant only because sample size is large. But it is bad because $\hat{\eta}_p^2$ is difficult to interpret. Therefore, tacking $\hat{\eta}_p^2$ on to a reported F_{obs} suggests to editors, and reviewers (see Figure 9.1), and readers that you are on top of the latest thinking about significance tests. But it doesn't add much to aid interpretation.

We've seen before that Cohen's d is tremendously useful in meta-analysis. And, on its own, Cohen's d can be used to estimate U_3 and thus tell us something about the effect of a treatment on individuals within a population. Unfortunately, $\hat{\eta}_p^2$ does not have these virtues. Part of the problem is that $\hat{\eta}_p^2$ associated with an omnibus F_{obs} has all the limitations of the omnibus F_{obs}. We will see that $\hat{\eta}_p^2$ can also be computed for simple and complex contrasts in within-subjects ANOVAs and thus has the potential to be more useful in this context. However, even in the context of contrasts, $\hat{\eta}_p^2$ does not have the virtues of Cohen's d.

*The Sphericity Assumption

The validity of a p-value associated with an F-statistic requires certain assumptions to be true. When the omnibus F_{obs} is computed for a within-subjects design, we have to assume that the sphericity assumption is valid. If it is not, we have to take steps to correct the violation. Sphericity replaces the homoscedasticity assumption from the between-subjects design. Discussions of sphericity can get quite technical, so we will take an intuitive approach to illustrating it.

> The sphericity assumption is that the variances of difference scores for all possible simple contrasts are the same in the population.

Sphericity requires us to think about all possible difference scores (simple contrasts) computed for each subject in the experiment. This is illustrated in Table 19.3, which shows the raw data we've been working with. To the right of the raw data are columns of difference scores computed for each subject. For our example, these differences are $a_1 - a_2, a_1 - a_3$, and $a_2 - a_3$. In general, there are $(k^2 - k)/2$ possible difference scores, just as there are $(k^2 - k)/2$ possible simple contrasts among means. The sphericity assumption is that the populations of all difference scores have the same variance; that is, $\sigma_{a_1 - a_2}^2 = \sigma_{a_1 - a_3}^2 = \sigma_{a_2 - a_3}^2$. The sample difference-score variances (i.e., $s_{a_1 - a_2}^2, s_{a_1 - a_3}^2$, and $s_{a_2 - a_3}^2$) are shown at the bottom of each column of difference scores in Table 19.3. These variances are not identical, but are they different enough to cause problems? If so, what should we do?

Violations of the sphericity assumption increase the probability of a Type I error in a significance test. There are methods that measure how much the variances violate the sphericity assumption and then correct for inflated Type I error rates. The two most common approaches to correcting violations of sphericity are the *Greenhouse-Geisser* procedure and the *Huynh-Feldt* procedure. Both procedures reduce the degrees of freedom associated with the test (df_a and df_{error}) in proportion to the measured degree of violation. (This type of correction is similar to the Welch-Satterthwaite correction shown in Appendix 11.4, available at study.sagepub.com/gurnsey.) For extreme violations of sphericity, df_a is reduced to 1 and df_{error} is reduced to $n-1$. This is sometimes called the *lower-bound* correction. We will see examples of these corrections when we consider the outputs of SPSS analyses of the one-factor within-subjects design.

TABLE 19.3 ■ The Sphericity Assumption

Subject	Valid (a_1)	Neutral (a_2)	Invalid (a_3)	$a_1 - a_2$	$a_1 - a_3$	$a_2 - a_3$
1	375	393	441	−18	−66	−48
2	137	166	195	−29	−58	−29
3	234	246	282	−12	−48	−36
4	284	324	373	−40	−89	−49
5	299	317	344	−18	−45	−27
6	273	288	339	−15	−66	−51
			s^2	110.8	252.4	114.4

LEARNING CHECK 1

1. State whether the following statements are true or false.

   (a) If there are six levels in a one-factor within-subjects ANOVA with 12 subjects, then $df_{error} = 55$.

   (b) If there are eight levels in a one-factor within-subjects ANOVA with 4 subjects, and $F_{obs} = 6$, then $\hat{\eta}_p^2 = .6$.

   (c) The sphericity assumption is that the variance of the dependent variable is the same at all levels of the independent variable.

   (d) In a one-factor within-subjects ANOVA, $ss_a = 10$, $ss_s = 2$, and $ss_{error} = 40$. Therefore, $\hat{\eta}_a^2 = .19$.

2. In a one-factor, within-subjects ANOVA with five levels of the independent variable and four subjects,

$ss_{total} = 290$. The means for the five conditions are [1, 3, 5, 7, 9] and the means for the four subjects are [2, 3, 5, 6]. Using this information, answer the following:

   (a) Compute F_{obs}.

   (b) Compute the exact p-value associated with F_{obs} using the **F.DIST.RT** function in Excel.

   (c) If sphericity were maximally violated, would F_{obs} be statistically significant?

   (d) Compute $\hat{\eta}_p^2$.

   (e) Compute $\hat{\eta}^2$.

3. In the experiment described in question 2, is it possible for ss_{total} to be 200? Explain why or why not.

Answers

1. (a) True. $(6 - 1)*(12 - 1) = 55$.

   (b) False. $6/(6 + 21/7) = .67$.

   (c) False. The sphericity assumption is that variance is the same for the difference scores for all possible simple contrasts.

   (d) True. $10/52 = .19$.

2. (a) $F_{obs} = 160/80*12/4 = 6$.

   (b) $p = .007$.

   (c) If sphericity were maximally violated, then F_{obs} would be best tested on 1 and 3 degrees of freedom (i.e., the lower-bound correction), and using the **F.DIST.RT** function in Excel shows that the exact p-value would be .092, which is not statistically significant.

   (d) $\hat{\eta}_p^2 = 160/(160 + 80) = .67$.

   (e) $\hat{\eta}^2 = 160/290 = .55$.

3. It is not possible for ss_{total} to be 200 because ss_{total} must be greater than $ss_a + ss_s$; in this case, it is not.

CONFIDENCE INTERVALS AND SIGNIFICANCE TESTS FOR CONTRASTS

In general, the design of an experiment specifies the questions of interest to the researcher. In the Posner task, there are really two main questions: (i) how great are the benefits of valid cues and (ii) how great are the costs of invalid cues? This means we want to (i) contrast scores in the valid condition to scores in the neutral condition and (ii) contrast scores in the invalid condition to scores in the neutral condition. (It would be worth the effort to stop for a moment to think about whether these two simple contrasts are orthogonal.)

Planned Contrasts

There are two ways to approach planned contrasts for the one-factor, within-subjects ANOVA. The first, which is absolutely not recommended, is essentially like the approach used in Chapter 18. That is, we express these contrasts using weights and compute a contrast as follows:

$$c = \sum_{i=1}^{k} w_i m_i.$$

(19.11)

We could then compute the estimated standard error of the contrast from ss_{error} and then a confidence interval around c. Unfortunately, this method requires very strong assumptions about sphericity, and even very small violations of the assumption can lead to very inexact confidence intervals. Therefore, this approach is of little use, so it won't be discussed further.

The preferred—and more straightforward—approach involves weights applied to the scores themselves, rather than means. The result is a **contrast score** for each subject. For subject j a contrast score (cs_j) is computed as

$$cs_j = \sum_{i=1}^k w_i y_{i,j}.$$ (19.12)

A **contrast score** is the sum of the products of a weight vector (contrast vector) and the scores of a subject at each level of the independent variable; i.e., $cs_j = \sum_{i=1}^k w_i y_{i,j}$.

Because contrast scores are computed for each subject, a new "contrast variable" is created. The distribution of this variable in the population has a mean of

$$\psi = \frac{\sum_{j=1}^N cs_j}{N}.$$ (19.13)

When we compute the mean contrast score from a sample as

$$c = \frac{\sum_{j=1}^n cs_j}{n},$$ (19.14)

we have an estimate of Ψ. The variance of a sample of contrast scores can be denoted

$$s_{cs}^2 = \frac{\sum_{j=1}^n (cs_j - c)^2}{n-1}$$ (19.15)

and the estimated standard error of c is given by

$$s_c = \sqrt{\frac{s_{cs}^2}{n}}.$$ (19.16)

From the sample contrast (c) and its estimated standard error (s_c), we can compute a $(1-\alpha)100\%$ confidence interval in the usual way:

$$CI = c \pm t_{\alpha/2}(s_c).$$ (19.17)

We have $(1-\alpha)100\%$ confidence in this interval because we know that $(1-\alpha)100\%$ of all intervals computed this way will capture Ψ.

Table 19.4 shows, once again, the raw data from the hypothetical Posner experiment described above. For each of the six participants, the table shows average response times for the valid, neutral, and invalid conditions in columns 2 to 4. The means and variances for each condition are shown below the scores.

Below the sample statistics are three weight vectors. The first, w_1, contrasts the valid (a_1) and neutral (a_2) conditions, and w_2 contrasts the neutral (a_2) and invalid (a_3) conditions. (We will discuss w_3 later.) The contrast score cs_1 for the first participant is

$$cs_{1,1} = \sum_{i=1}^k w_{1,i} y_{i,j} = \sum_{i=1}^k [1,-1,0][375,393,441] = 375-393+0 = -18.$$

The column labeled "$c_1 = a_1 - a_2$" shows this computation for all six participants. Having computed the contrasts for each participant, a 95% confidence interval is computed as $CI = c_1 \pm t_{\alpha/2}(s_{c_1})$ exactly as in Chapter 12.

TABLE 19.4 ■ Data From a Hypothetical Posner Experiment							
Participant	Valid (a_1)	Neutral (a_2)	Invalid (a_2)		$c_1 = a_1 - a_2$	$c_2 = a_3 - a_2$	$c_3 = a_1 - (a_2 + a_3)/2$
1	375	393	441		−18	48	−42.0
2	137	166	195		−29	29	−43.5
3	234	246	282		−12	36	−30.0
4	284	324	373		−40	49	−64.5
5	299	317	344		−18	27	−31.5
6	273	288	339		−15	51	−40.5
m	267	289	329	c	−22	40	−42.0
s^2	6200.4	5960.8	6994.0	s^2_{cs}	110.80	114.40	153.00
				s_c	4.30	4.37	5.05
w_1	1	−1	0	$t_{\alpha/2}$	2.571	2.571	2.571
w_2	0	−1	1	lower	−33.05	28.78	−54.98
w_3	1	−0.5	−0.5	upper	−10.95	51.22	−29.02

You should be able to follow the computation of this 95% confidence interval in Table 19.4, but we will make the steps explicit. These calculations can be carried out very easily in Excel or SPSS.

Step 1. Compute the contrast score $cs_j = \sum_{i=1}^{k} w_i y_{i,j}$ for each of the j subjects. This is shown in the first six rows of column "$c_1 = a_1 - a_2$" in Table 19.4.

Step 2. Compute the contrast $c = \sum_{j=1}^{n} cs_j / n$. This is shown in column "$c_1 = a_1 - a_2$" in the row labeled "c." For the first contrast, $c_1 = -22$.

Step 3. Compute $t_{\alpha/2}$. There are $n-1 = 5$ degrees of freedom. Because this is the 95% confidence interval, $\alpha = .05$. When we consult the t-table, we find that $t_{\alpha/2} = 2.571$.

Step 4. Compute s_c using equation 19.16. Table 19.4 shows that $s^2_{cs} = 110.8$ and $n = 6$. Therefore,

$$s_c = \sqrt{\frac{s^2_{cs}}{n}} = \sqrt{\frac{110.8}{6}} = 4.30.$$

Step 5. Compute the 95% confidence interval around c.

$$\text{CI} = c \pm t_{\alpha/2}(s_c) = -22 \pm 2.571(4.30) = [-33.05, -10.95].$$

We have 95% confidence that Ψ lies in the interval $[-33.05, -10.95]$. Our confidence comes from knowing that 95% of intervals computed this way will capture Ψ.

The same calculations applied to the second contrast ($c_2 = a_3 - a_2$) show that $c_2 = 40$, 95% CI [28.78, 51.22]. The results of this contrast analysis could be reported as follows:

The mean detection times for valid, neutral, and invalid cues were $M_V = 267$, $M_N = 289$, and $M_I = 329$, respectively, with standard deviations of $s_V = 78.74$, $s_N = 77.21$, and $s_I = 83.63$. The difference between the means of the valid and neutral trials was $267 - 289 = -22$, 95% CI [−33.05, −10.95]. The difference between the means of the invalid and neutral trials was $289 - 329 = 40$, 95% CI [28.78, 51.22]. There are clearly benefits of valid cues and costs to invalid cues. Furthermore, for these data, it appears that the magnitude of the benefits (22 ms on average) is less than the magnitude of the costs (40 ms on average).

The two contrasts of interest ($c_1 = a_1 - a_2$ and $c_2 = a_3 - a_2$) are simple contrasts, and both are computed in Table 19.4. Complex contrasts can be computed in exactly the same way. Contrast 3 in Table 19.4 contrasts the valid cue with the average of the neutral and invalid cue; i.e., $c_3 = a_1 - (a_2 + a_3)/2$. For subject 1, the contrast score is

$$cs_{3,1} = \sum_{i=1}^{k} w_{1,i} y_{i,j} = \sum_{i=1}^{k} [1, -0.5, -0.5][375, 393, 441] = 375 - 196.5 - 220.5 = -42.$$

When contrast scores are computed in this way for all subjects, we can compute a confidence interval around the mean contrast score just as described above. This is not a sensible contrast in the context of the Posner task, but it shows that the method described above is completely general, in that it can be applied to both simple and complex contrasts.

Corrections for Multiple Contrasts

As we saw in Chapter 18, there are different customs in psychology about whether planned contrasts should be corrected to maintain a family-wise error rate of $\alpha = .05$. We saw that SPSS does not make this correction, even though this is not strictly logical. Although there are many ways to control the family-wise confidence, the Bonferroni correction discussed in Chapter 18 works well in most cases, including contrasts for within-subjects ANOVAs. Therefore, to maintain family-wise confidence at $(1 - \alpha_{fw})100\% = 95\%$, we choose $t_{\alpha/2}$ based on $\alpha_{pc} = \alpha_{fw}/n_{contrasts}$ as before.

Standardized Contrasts

To standardize a contrast (c), we simply divide it by the standard deviation of the contrast scores:

$$d = \frac{c}{s_{cs}}. \tag{19.18}$$

For example, Table 19.4 shows that $c_1 = -22$ with variance $s_{cs}^2 = 110.8$. Therefore, the standardized contrast is

$$d = \frac{-22}{\sqrt{110.8}} = -2.09.$$

Because contrasts for within-subjects designs are essentially the same as contrasts for the dependent-samples design (Chapter 12), we can take the same approach to computing an approximate confidence interval around the estimated effect size. That is, we can compute

$$d \pm z_{\alpha/2}(s_d), \text{ where}$$

$$s_d = \sqrt{\frac{d^2}{2*df} + \frac{1}{n}}.$$

Therefore, to compute the approximate 95% confidence interval around $d = -2.09$, we would first compute

$$s_d = \sqrt{\frac{d^2}{2*df} + \frac{1}{n}} = \sqrt{\frac{-2.09^2}{2*5} + \frac{1}{6}} = 0.7768.$$

We would then apply our standard formula to obtain

$$\text{CI} = d \pm z_{\alpha/2}(s_d) = -2.09 \pm 1.96(0.7769) = [-3.61, -0.58].$$

To compute an exact confidence interval for d, we can use the MBESS function `ci.sm`, which stands for confidence interval for a standardized mean. At the **R** prompt, simply type in

```
ci.sm(sm = -2.09, N = 6, conf.level = .95)
```

In this case, `sm = -2.09` is the effect size d, `N = 6` is the number of participants, and `conf.level = .95` is the confidence level. The exact confidence interval returned is [−3.56, −0.58], which is not very different from the approximate confidence interval computed by hand.

We've mentioned at several points that estimated effect sizes are very useful for meta-analyses. However, the same caution from Chapter 12 applies here. It is crucial not to combine standardized contrasts from between-subjects and within-subjects designs because they estimate different things. In the between-subjects design, d estimates Ψ/σ, where σ is the standard deviation common to all k distributions. In the within-subjects design, d estimates Ψ/σ_{cs}, where σ_{cs} is the standard deviation of the population of contrast scores.

Significance Tests for Planned Contrasts

Significance tests for contrasts typically test the null hypothesis

$$H_0: \psi = 0.$$

To test this hypothesis, one could ask whether the confidence interval for the non-standardized contrast interval includes 0. If it does, we retain H_0; if not, we reject H_0.

The traditional method of significance testing uses a t-statistic, which is computed as

$$t_{obs} = \frac{c}{s_c}.$$

The critical value of t for a two-tailed test is based on α and $n-1$ degrees of freedom. Therefore, for the contrasts computed above when $\alpha = .05$ and $df = 5$, $t_{critical} = \pm 2.571$. For contrasts c_1 and c_2 in Table 19.4, the values of t_{obs} would be

$$t_{obs} = \frac{-22}{4.30} = -5.12 \text{ and}$$

$$t_{obs} = \frac{-40}{4.37} = -8.32.$$

Both of these values are statistically significant at the $\alpha = .05$ level because both exceed ± 2.571.

If these tests were corrected for multiple contrasts, then we would determine $t_{critical}$ based on $\alpha_{pc} = .05/2 = .025$ to achieve $\alpha_{fw} = .05$ because there are two contrasts. In this case, $t_{critical} = \pm3.163$. Or we could compute the p-values associated with -5.12 and -8.32 and multiply each by 2. We saw in Chapter 18 that this is what SPSS does when it applies the Bonferroni correction.

LEARNING CHECK 2

1. The data in the following table are from a within-subjects experiment with four levels of the independent variable and four subjects in each condition. You can use Excel for the calculations below. However, you can also do the calculations by hand because all means and sums of squares are integers, thus making the calculations less prone to error.

	a_1	a_2	a_3	a_4
s_1	28	16	21	27
s_2	9	21	22	16
s_3	15	17	22	22
s_4	4	18	23	7

(a) Compute the 95% confidence interval around the difference between a_1 and a_2.

(b) Compute the 95% confidence interval around the difference between a_3 and a_4.

(c) Compute the 95% confidence interval around the difference between the mean of a_1 and a_2 versus the mean of a_3 and a_4.

(d) For each of the three confidence intervals just computed, compute the following:

   ● The approximate 95% confidence interval around the standardized contrast, and

   ● The exact 95% confidence interval around the standardized contrast.

(e) Compute a t-statistic to determine whether the linear trend in the data is statistically significant for a nondirectional test.

Answers

1. (a–c) The 95% confidence intervals are shown in the following table.

(d) Approximate confidence intervals for d:
   $a_1 - a_2$: [−1.35, 0.68],
   $a_3 - a_4$: [−0.61, 1.46], and
   $(a_1+ a_2)/2 - (a_3+ a_4)/2$: [−4.64, −0.26].

   Approximate confidence intervals for d:
   $a_1 - a_2$: [−1.32, 0.70],
   $a_3 - a_4$: [−0.64, 1.43], and
   $(a_1+ a_2)/2 - (a_3+ a_4)/2$: [−4.53, −0.34].

(e) The trend scores per subject (computed with [−3, −1, 1, 3]) are shown in the "Linear" column. The contrast is $c_{linear} = 16$ with standard error $s_{linear} = 5.29$. Therefore, $t_{obs} = c_{linear}/s_{linear} = 3.02$, $p = .057$. By convention, this is not statistically significant.

	$a_1 - a_2$	$a_3 - a_4$	$(a_1 + a_2)/2 - (a_3 + a_4)/2$	Linear
s_1	−12	−6	−2	2
s_2	−12	6	−4	22
s_3	−2	0	−6	26
s_4	−14	16	−4	14
c	−4.00	4.00	−4.00	16.00
s_{sc}^2	141.33	88.00	2.67	112.00
s_c	5.94	4.69	0.82	5.29
moe	18.92	14.93	2.60	
Lower	−22.92	−10.93	−6.60	
Upper	14.92	18.93	−1.40	

CONDUCTING THE ONE-FACTOR WITHIN-SUBJECTS ANOVA IN SPSS

When we conduct a within-subjects ANOVA, all scores for a given subject are entered on a single line in the data file. As a consequence, the columns correspond to the levels of the independent variable. This is an extension of how data are arranged for the dependent-samples design covered in Chapter 12.

Setting Up the Analysis

Figure 19.3 shows the data from the Posner experiment taken from Table 19.1. The left side of the figure shows the data editor with the data arranged in three columns that have been given the names Valid, Neutral, and Invalid. The analysis is conducted by choosing the Analyze→ General Linear Model→ Repeated Measures . . . dialog, as shown in Figure 19.3.

When this is chosen, the dialog shown in Figure 19.4a then appears. The first text box (Within-Subject Factor Name) asks for the name of the factor. I've called the factor

FIGURE 19.3 ■ Preparing a Within-Subjects ANOVA in SPSS

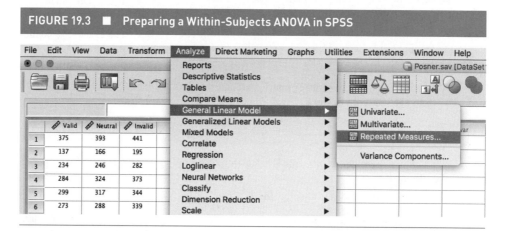

FIGURE 19.4 ■ Defining the Number of Conditions

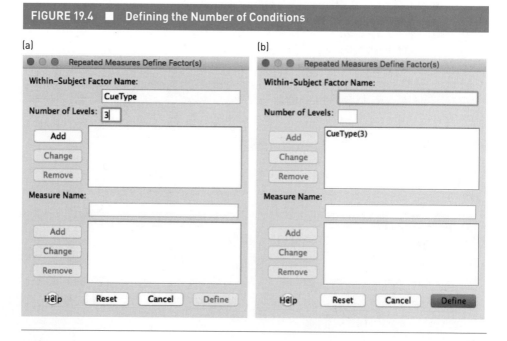

CueType. The text box below it (Number of Levels) asks for the number of levels of the factor, so I entered 3. Once the factor name and number of levels have been entered, we click on the Add button and the factor is entered into the factor list (shown in Figure 19.4b).

Clicking the Define button in Figure 19.4b brings up a new dialog (Figure 19.5) that associates variables in the data sheet with the Factor levels we just defined. Figure 19.5a shows all variables in the data sheet; we have only three but there could be many more. Clicking on a variable name selects it and clicking the ⇨ button moves the selected variable to the first available slot in the region labeled Within-Subjects Variables (Cue Type):. If all three variables are selected, then clicking ⇨ will move them into the three available slots on the right. The result of this assignment is shown in Figure 19.5b.

FIGURE 19.5 ■ Associating Variables With Conditions

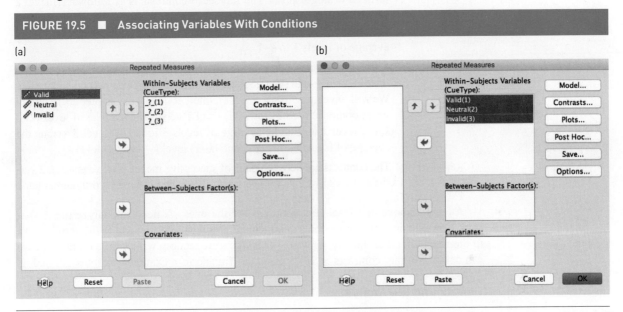

The standard version of SPSS does not allow users to enter within-subjects contrasts that they have constructed. Instead, one has to choose from six different sets of contrasts available through the Contrasts . . . button in the Repeated Measures dialog. Clicking

Contrasts . . . brings up the Repeated Measures: Contrasts dialog shown in Figure 19.6. The drop-down list named Contrast: has been clicked to show the available contrasts. The default is Polynomial, which refers to polynomial trends. To change to a different set of contrasts, we must choose a new set from the drop-down list and then click the Change button to the right. The change is not made without clicking the Change button. SPSS shows the contrast analysis to be conducted in the box above the Change Contrast region.

Figure 19.6 shows that I have selected the Repeated contrasts. Fortunately, this will calculate contrasts c_1 and c_2 described above. Before I continue with the analysis of the Posner experiment, I will describe the nature of the contrasts available in the Repeated Measures: Contrasts dialog. In all cases there will be k levels of the independent variable, where the levels are ordered from 1 to k as in Figure 19.5. In all cases, $k-1$ contrasts are computed.

FIGURE 19.6 ■ Within-Subjects Contrasts

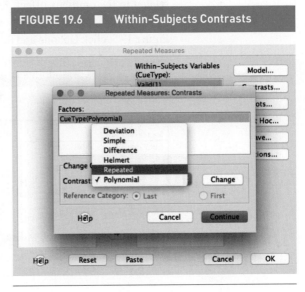

Deviation.   Each contrast represents the difference between a condition mean (conditions 1 to k–1) and the grand mean. These contrasts do not form an orthogonal set.

Simple.   Each contrast represents the difference between a condition mean and a reference mean. The reference mean is either the first or last variable in the variable list, depending on whether the Last or First radio button is chosen in the region beside Reference Category. These contrasts do not form an orthogonal set.

Difference.   There is one simple contrast and k–2 complex contrasts. These contrasts form an orthogonal set. The full set of contrasts is as follows: (i) level 2 versus level 1, (ii) level 3 versus the average of levels 1 and 2, (iii) level 4 versus the average of levels 1, 2, and 3 . . . , and (k–1) level k versus the average of levels 1 to k–1.

Helmert.   There is one simple contrast and k–2 complex contrasts. These contrasts form an orthogonal set. (Look up Friedrich Robert Helmert on Wikipedia. We will have more to say about Helmert contrasts in Chapter 20.) The full set of contrasts is as follows: (i) level 1 versus the average of levels 2 to k, (ii) level 2 versus the average of levels 3 to k, (iii) level 3 versus the average of levels 4 to k, . . . , and (k–1) level k–1 versus level k.

Repeated.   The contrasts are between pairs of successive means (i.e., 1 versus 2, 2 versus 3, 3 versus 4, . . . , k–1 versus k). These do not form an orthogonal set.

As noted above, the Repeated contrasts are the ones we need to analyze the Posner data. However, in many cases, the available contrasts won't correspond to those of interest to the researcher. Fortunately, as we'll see in the next section, with a little extra work we can define arbitrary contrasts in SPSS using the Transform→ Compute Variable . . . dialog.

Conducting the Analysis

When the analysis has been set up, clicking the ▭ OK ▭ button instructs SPSS to conduct the analysis you've described. Figure 19.7 shows the omnibus analysis of CueType. The panel is divided into two components: the top part concerns viability attributable to CueType and the bottom part concerns error variability. The first line in each part shows quantities related to the omnibus effect, resulting in the calculation of $F_{obs} = 74.47$. Also shown is $\hat{\eta}_p^2 = .937$, as computed earlier. This was an option chosen through the Options dialog in Figure 19.5; click Options and then check the box titled Estimates of effect size.

FIGURE 19.7   ■   Omnibus Within-Subjects ANOVA

Tests of Within-Subjects Effects

Measure: MEASURE_1

Source		Type III Sum of Squares	df	Mean Square	F	Sig.	Partial Eta Squared
CueType	Sphericity Assumed	11856.000	2	5928.000	74.472	.000	.937
	Greenhouse–Geisser	11856.000	1.489	7960.684	74.472	.000	.937
	Huynh-Feldt	11856.000	1.976	6001.029	74.472	.000	.937
	Lower-bound	11856.000	1.000	11856.000	74.472	.000	.937
Error(CueType)	Sphericity Assumed	796.000	10	79.600			
	Greenhouse–Geisser	796.000	7.447	106.894			
	Huynh-Feldt	796.000	9.878	80.581			
	Lower-bound	796.000	5.000	159.200			

Most of Figure 19.7 is taken up with corrections for violations of the sphericity assumption. For the Greenhouse-Geisser, Huynh-Feldt, and lower-bound corrections, the sums of squares, F_{obs}, and $\hat{\eta}_p^2$ remain the same. The main change is to the degrees of freedom associated with the treatment effect and error term. When sphericity is assumed to be valid, the degrees of freedom for CueType are 2 and 10. (These are the quantities we used in our calculations above.) These degrees of freedom are adjusted for the Greenhouse-Geisser and Huynh-Feldt corrections. Although these corrections do not affect F_{obs}, they will affect the associated p-values. For our example, F_{obs} is very large and the change in the p-value is after the third decimal place and so is not apparent in Figure 19.7. The lower-bound correction reduces the degrees of freedom for CueType to 1 and 5. As noted before, this correction is the largest reduction in degrees of freedom that could be made in response the greatest violation of sphericity. Although not shown in our example, F_{obs} may be statistically significant ($p < .05$) when sphericity is assumed, but not when a correction factor is applied.

Figure 19.8 shows the analysis of the planned contrasts that are specified in Figure 19.6. There are three points worth noting here. The first is that no corrections to the degrees of freedom have been made because the degrees of freedom for individual contrasts (1 and 5) represent the lower-bound correction shown in Figure 19.7. Therefore, there is no need to worry about violations of sphericity for contrasts.

The second point is that SPSS does not report the contrasts with their associated confidence intervals. The assumption seems to be that researchers don't want to know the magnitude of the difference represented by the contrast, just whether it is statistically significant. The $\hat{\eta}_p^2$ for the contrast is reported, but as noted before, this quantity is difficult to interpret. In this case, however, we can say that the $\hat{\eta}_p^2$ for both c_1 and c_2 (.84 and .94, respectively) are very large.

The third point is related to the second. SPSS does not provide an option for post hoc contrasts for within-subjects designs, so there are no corrections for multiple contrasts. There are advanced ways of doing this through SPSS scripts, but we won't cover these methods in this book. Instead, we will do these ourselves when necessary.

Contrasts in SPSS

Although SPSS provides few contrasts for the within-subjects design, it is very easy to specify any contrast of interest and then compute a confidence interval around it. The first step is to compute a contrast score for each subject through the Transform→ Compute Variable . . . dialog. (We first met this dialog in Appendix 1.2 when computing GPA.) The dialog is shown in Figure 19.9. Two contrasts have already been computed. Contrast c1 is Valid - Neutral and contrast c2 is Invalid - Neutral. The computation of c3 is under way. The Target Variable has been specified as c3. In the box labeled Numeric Expression:, we see the formula Valid - .5*Neutral - .5*Invalid. This is the contrast described earlier as $c_3 = a_1 - (a_2 + a_3)/2$. When the OK button is clicked, this quantity will be computed and stored in a column labeled c3.

FIGURE 19.8 ■ Planned Contrasts

Tests of Within-Subjects Contrasts

Measure: MEASURE_1

Source	CueType	Type III Sum of Squares	df	Mean Square	F	Sig.	Partial Eta Squared
CueType	Level 1 vs. Level 2	2904.000	1	2904.000	26.209	.004	.840
	Level 2 vs. Level 3	9600.000	1	9600.000	83.916	.000	.944
Error(CueType)	Level 1 vs. Level 2	554.000	5	110.800			
	Level 2 vs. Level 3	572.000	5	114.400			

FIGURE 19.9 ■ Computing Contrast Scores in Compute Variable

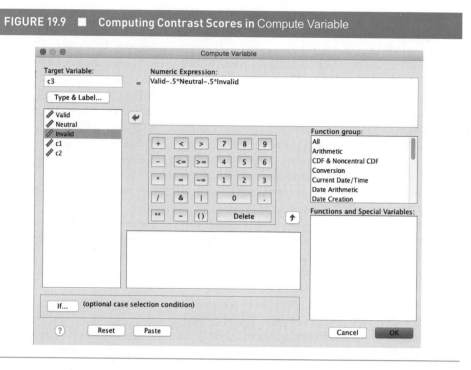

FIGURE 19.10 ■ The One-Sample T Test **dialog**

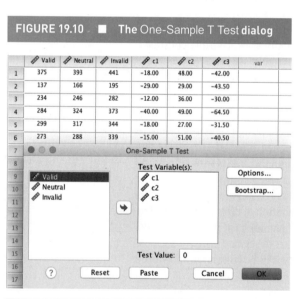

Figure 19.10a shows the current state of the data file. The original three variables are there along with the contrast score variables in columns c1, c2, and c3. These are exactly the same contrast scores shown in Table 19.4. To compute confidence intervals around the mean contrast scores, we use the Analyze→ Compare Means→ One-Sample T Test . . . dialog. This dialog is shown in the bottom part of Figure 19.10. The three contrasts have been moved into the Test Variable(s): window and the Test Value has been set to 0. This means that SPSS will test the null hypothesis that $\Psi = 0$. When the OK button is clicked, the analysis begins and the output is shown in Figure 19.11.

Figure 19.11a shows the descriptive statistics for the contrasts. Figure 19.11b shows t_{obs}, df, p, c, and the 95% confidence intervals around c for each of the contrasts. The confidence intervals are exactly as computed earlier. (Note that for contrasts 1 and 2, the t_{obs} values are -5.120 and 9.161. These correspond to the F_{obs} values of 26.209 and 83.916 associated with these contrasts in Figure 19.8. That is, squaring t_{obs} yields F_{obs}.) The confidence intervals and p-values have not been corrected for multiple comparisons. For the significance tests, this can be done by hand by multiplying the p-values by 3 (the number of contrasts). For the confidence intervals, one can use Excel to compute $t_{\alpha/2}$ for 5 degrees of freedom and $\alpha = .05/3 = .0167$. The means and standard error required to compute the confidence intervals are shown in Figure 19.11a.

Although we have used SPSS to do these calculations, they could have been done as easily (or more easily) in Excel. In fact, I would recommend Excel for this kind of thing because your data and calculations remain in place in your Excel workbook and can be easily modified without the need to restart the analysis as is necessary in SPSS.

FIGURE 19.11 ■ Contrast Analysis

(a)

One-Sample Statistics

	N	Mean	Std. Deviation	Std. Error Mean
c1	6	-22.0000	10.52616	4.29729
c2	6	40.0000	10.69579	4.36654
c3	6	-42.0000	12.36932	5.04975

(b)

One-Sample Test

	Test Value = 0					
					95% Confidence Interval of the Difference	
	t	df	Sig. (2-tailed)	Mean Difference	Lower	Upper
c1	-5.120	5	.004	-22.00000	-33.0465	-10.9535
c2	9.161	5	.000	40.00000	28.7755	51.2245
c3	-8.317	5	.000	-42.00000	-54.9808	-29.0192

Power

As always, power is an important consideration when planning significance tests. Appendix 19.1 (available at study.sagepub.com/gurnsey) shows how to conduct a prospective power analysis for contrasts.

LEARNING CHECK 3

1. The data in the table to the right are from Learning Check 2. Use these data to compute the following:

   (a) Compute the omnibus F_{obs}, $\hat{\eta}_p^2$, and the p-value that results when sphericity is assumed.

   (b) Compute the linear, quadratic, and cubic trend contrasts, and for each report F_{obs}, p, and $\hat{\eta}_p^2$.

   (c) Explain why there are no corrected p-values associated with the trends.

   (d) Provide the correct definition of a p-value.

	a_1	a_2	a_3	a_4
s_1	28	16	21	27
s_2	9	21	22	16
s_3	15	17	22	22
s_4	4	18	23	7

Answers

1. (a) $F_{obs} = 1.091$, $\hat{\eta}_p^2 = .267$, $p = .402$.

   (b) Linear: $F_{obs} = 9.143$, $p = .057$, $\hat{\eta}_p^2 = .753$; quadratic: $F_{obs} = 0.593$, $p = .497$, $\hat{\eta}_p^2 = .165$; cubic: $F_{obs} = 3.429$, $p = .161$, $\hat{\eta}_p^2 = .533$.

   (c) The lower bound is the maximum correction for violations of sphericity. In this case, the degrees of freedom are reduced to 1 and $n-1$. Because these are the degrees of freedom associated with each contrast, there is nothing to correct.

   (d) A p-value is the probability of the observed statistic, or one more extreme, occurring by chance when the null hypothesis is true.

SUMMARY

The omnibus test of significance in the one-factor within-subjects ANOVA requires partitioning variance into three sources: (i) variance attributable to the independent variable, (ii) variance attributable to subjects, and (iii) error variance. Each of these sources has associated sums of squares (ss_a, ss_s, ss_{error}) and degrees of freedom (df_a, df_s, df_{error}). The omnibus F-statistic (F_{obs}) is computed as follows:

$$F_{obs} = \frac{ss_a}{ss_{error}} \frac{df_{error}}{df_a}.$$

The p-value associated with this statistic is computed on df_a and df_{error} degrees of freedom.

The sphericity assumption in the within-subjects ANOVA replaces the homoscedasticity assumption in the between-subjects ANOVA. The assumption is that the variance of the difference scores, computed from each of the possible simple contrasts, is the same in the population. Violations of sphericity lead to increased Type I error rates. To compensate for inflated Type I error rates, three different corrections are used: the Greenhouse-Geisser, the Hyunh-Feldt, and the lower-bound correction. The Greenhouse-Geisser and Hyunh-Feldt corrections reduce df_a and df_{error} in proportion to the degree to which sphericity has been violated; the two adjustments are based on different assumptions, but we did not explore these in detail. The corrections produce fractional degrees of freedom. The maximum correction for violations of sphericity results in df_a being reduced to 1 and df_{error} being reduced to $n-1$. This is the lower-bound correction. Except in unusual circumstances, the p-values associated with these corrections are lower than the p-value obtained when the sphericity assumption is assumed to be satisfied.

In the *between*-subjects ANOVA, it is plausible to use $\hat{\eta}^2 = ss_a/ss_{error}$ as a measure of effect size, but this is not the case in the *within*-subjects ANOVA because $\hat{\eta}^2$ will decrease as ss_s increases. Therefore, partial eta squared ($\hat{\eta}_p^2$) replaces $\hat{\eta}^2$ in the within-subjects ANOVA. Partial eta squared is defined as

$$\hat{\eta}_p^2 = \frac{ss_a}{ss_a + ss_{error}}.$$

Although it often reported, partial eta squared is not easy to interpret. However, it does provide some context for a statistically significant F_{obs}, which may be large just because sample size is large. Partial eta squared can be recovered from F_{obs} as follows:

$$\hat{\eta}_p^2 = \frac{F_{obs}}{F_{obs} + \dfrac{df_{error}}{df_a}}.$$

For a fixed value of F_{obs}, $\hat{\eta}_p^2$ decreases as df_{error} increases.

Because omnibus effects say nothing about how the condition means differ, contrasts are used to answer these questions. Contrasts can be defined as weights exactly as in the between-subjects ANOVA. However, the weights are applied to scores of subjects to produce a contrast score (cs) for each subject. The average contrast score defines the contrast (c), and the cs variance (s_{cs}^2) is used to compute a confidence interval. In fact, once the contrast scores have been computed, a confidence interval is computed exactly as a confidence interval for a difference between two dependent means. That is,

$$\text{CI} = c \pm t_{\alpha/2}(s_c), \text{ where}$$

$$s_c = \sqrt{\frac{s_{cs}^2}{n}}.$$

The value of $t_{\alpha/2}$ is based on $n-1$ degrees of freedom.

A contrast can be tested for statistical significance by computing a t-statistic:

$$t_{obs} = \frac{c}{s_c},$$

which when squared produces F_{obs}. Both are tested for statistical significance on 1 and $n-1$ degrees of freedom. If significance tests are planned, then a power analysis for contrasts should be conducted. This is done exactly as for the dependent-samples t-test.

KEY TERMS

Definitions and Concepts

1. Define each of the following quantities: ss_a, ss_s, and ss_{error}.

2. What is the grand mean?

3. Show the equation that defines the omnibus F-statistic for the within-subjects design.

4. How does ss_{error} differ in the between- and within-subjects designs?

5. Use formulas to show the definitions of $\hat{\eta}^2$ and $\hat{\eta}_p^2$.

6. Why is $\hat{\eta}^2$ an inappropriate measure of effect size in a within-subjects ANOVA?

7. What is the sphericity assumption in the within-subjects ANOVA?

8. What is the main consequence of violating the sphericity assumption?

9. What is a contrast score?

10. What does it mean to have 95% confidence in the interval $c \pm t_{\alpha/2}(s_c)$?

11. What is the difference between s_c and s_{cs}?

12. Why do we use the Bonferroni correction for multiple contrasts when computing confidence intervals for contrasts?

13. What does $d = c/s_{cs}$ estimate?

14. Why can we not combine d computed in a between-subjects design with d computed in a within-subjects design in a meta-analysis?

True or False

State whether the following statements are true or false.

15. ss_{total} is almost always greater than $ss_a + ss_{error}$.

16. If $ss_{total} = ss_a + ss_{error}$, then $ss_s = 0$.

17. The grand mean is defined as $m_y = \sum_{i=1}^{k} \sum_{j=1}^{n} y_{i,j}$.

18. The Greenhouse-Geisser and Huynh-Feldt corrections for violations of sphericity produce exactly the same change to the degrees of freedom.

19. If $F_{obs} = 10$, $df_{error} = 40$, and $df_a = 4$, then $\hat{\eta}_p^2 = .5$.

20. For a between-subjects design, $\hat{\eta}^2 = \hat{\eta}_p^2$.

21. Sphericity is another name for homoscedasticity.

22. $c/s_c = c/s_{cs}$.

23. If we say we have approximately 95% confidence in the interval $d \pm z_{\alpha/2}(s_d)$, this means that approximately 95% of all such intervals will capture d.

24. No correction for multiple contrasts is required if all contrasts are planned.

25. A p-value is the probability that the null hypothesis is true.

26. For a confidence interval around c to be valid, the distributions of contrast scores must be normal in the population.

27. In a within-subjects ANOVA, $d = c/s_{cs}$ estimates $\delta = \Psi/\sigma_{cs}$.

28. If $ss_{total} = ss_s + ss_{error}$ then $ss_a = 0$.

Calculations

29. If a one-factor within-subjects ANOVA has 10 levels of the independent variable and 100 participants, how many participants are in each condition?

30. A one-way within-subjects ANOVA with four levels of the independent variable has the following scores:

a_1	a_2	a_3	a_4
8	5	20	5
10	11	11	10
8	16	12	18
5	10	16	14
8	14	8	16
6	6	17	7
4	8	21	14

(a) Use Excel or SPSS to compute m_y, ss_a, ss_s, ss_{total}, df_a, df_{error}, F_{obs}, and $\hat{\eta}_p^2$.

(b) Using Excel or SPSS, compute contrast scores with the following three weight vectors: $w_1 = [-1, \frac{1}{3}, \frac{1}{3}, \frac{1}{3}]$, $w_2 = [0, -1, \frac{1}{2}, \frac{1}{2}]$, and $w_3 = [0, 0, -1, 1]$. For each of these contrasts,

   (i) compute the 95% confidence interval around c_i,

   (ii) compute the approximate 95% confidence interval around d_i, and

   (iii) use ci.sm to compute the exact 95% confidence interval around d_i.

31. A one-way, within-subjects ANOVA with five levels of the independent variable has the following scores:

a_1	a_2	a_3	a_4	a_5
8	6	19	5	3
10	11	9	10	4
8	18	12	18	9
6	9	16	15	4

(a) Use Excel or SPSS to compute m_y, ss_a, ss_s, ss_{total}, df_a, df_{error}, F_{obs}, and $\hat{\eta}_p^2$.

(b) Use Excel or SPSS to compute the linear, quadratic, cubic, and quartic trends in the data. For each trend, compute t_{obs}, F_{obs}, p, and $\hat{\eta}_p^2$.

Scenarios

32. The ABA paradigm used in drug testing involves testing a patient before treatment (A), during treatment (B), and later once treatment stops (A). Imagine that a new drug for attention deficit hyperactivity disorder (ADHD) is being tested in five college students who were recently diagnosed with ADHD. The hope is that the new drug will improve students' ability to stay on task, so the dependent variable is the number of complex puzzles solved in a 1-hour testing session. There were three sessions: one at baseline, before the drug treatment started (A_1); 3 months into the drug treatment regime (B); and 3 months after treatment ended (A_2). The raw data are shown in the following table.

	a_1	a_2	a_3
s_1	4	13	7
s_2	10	19	13
s_3	16	20	18
s_4	17	21	19
s_5	23	32	23

(a) Use Excel or SPSS to compute m_y, ss_a, ss_s, ss_{total}, df_a, df_{error}, F_{obs}, and $\hat{\eta}_p^2$.

(b) Use $w_1 = [-1, 0, 1]$ to assess the magnitude of the change from A_1 to A_2.

   (i) Compute the 95% confidence interval around the contrast. Do not correct for multiple contrasts.

   (ii) Compute the approximate 95% confidence interval around the standardized contrast.

(c) Use $w_2 = [-0.5, 1, -0.5]$ to assess the magnitude of the difference between B and the average of A_1 and A_2.

   (i) Compute the 95% confidence interval around the contrast. Do not correct for multiple contrasts.

   (ii) Compute the approximate 95% confidence interval around the standardized contrast.

   (iii) If you were computing an exact confidence interval around the standardized contrasts for c_1 and c_2, what confidence level would be required to ensure a family-wise confidence level of 95%?

(d) The contrasts defined by weight vectors w_1 and w_2 can also be seen as linear and quadratic trends. Compute F_{obs}, p, and $\hat{\eta}_p^2$ for the two contrasts. Is trend analysis valid in this scenario? Explain why or why not.

(e) What assumptions must be made for the inferences from the sample to the population to be valid?

(f) Here are a few things to think about.

   (i) Does the drug seem to be an effective treatment for ADHD?

   (ii) Can you think of any questions you'd like answered about the drug before it is recommended as a treatment for ADHD?

   (iii) What can you say about the change from A_1 to A_2? How would you explain this?

   (iv) Do you think a carryover effect explains the change from A_1 to B?

33. There is evidence that exercise improves children's school performance (e.g., Ellemberg & St-Louis-Deschênes, 2010; Hillman, Pontifex, Raine, Castelli, Hall, & Kramer, 2009). Let's imagine an experiment that examines the effect of exercise duration on school performance in a random sample of fifth-grade students. Each student walked on a treadmill for 0, 2, 4, 8, or 16 minutes at the beginning of each day. We will call this variable time on treadmill, or ToT. Over 40 days, each student was on the treadmill eight times at each ToT (e.g., 8 times at 0 minutes, 8 times at 2 minutes, and so on). Therefore, there were eight

replications per student at each of the five ToTs. Students went through the 40 exercise trials in a different random order.

During the morning, four trained raters independently observed the children and scored their classroom behaviors, with high numbers representing positive activities like asking questions, staying on task, and generally being cooperative. The raters were blind to the amount of exercise each student had performed in the morning. Their ratings were averaged for each day and then these were averaged across each of the eight replications of each ToT condition. The following table shows the results of the experiment. Each entry represents the average of the scores from three raters, averaged over eight replications.

	a_1	a_2	a_3	a_4	a_5
s_1	2	3	4	5	6
s_2	1	2	6	8	8
s_3	1	2	5	8	9
s_4	3	6	9	8	9
s_5	0	2	2	6	5
s_6	5	6	7	8	9
s_7	0	0	3	4	3
s_8	4	3	4	9	15

(a) Submit these data to a one-factor, within-subjects ANOVA in SPSS and report ss_a, ss_s, ss_{total}, df_a, df_{error}, F_{obs}, and $\hat{\eta}_p^2$.

(b) Through the Contrasts dialog, choose the Repeated option. For each of these contrasts, report F_{obs}, p, and $\hat{\eta}_p^2$.

(c) Conduct the same contrast and analysis in Excel (or using the Compute Variable dialog in SPSS) but compute

   (i) the 95% confidence interval around c_i and

   (ii) the approximate 95% confidence interval around d_i, and

   (iii) then use `ci.sm` to compute an exact 95% confidence interval around d_i.

(d) What do you conclude from this study? That is, how would you characterize the relationship between ToT and school performance?

(e) If you were a school administrator, what questions would you like answered before making a decision about whether to add exercise to each student's day? What actions would you take, based on these results?

34. Researchers have long been interested in how orientation affects our ability to recognize objects and faces. Let's think about an experiment in which participants are asked to identify briefly flashed images of famous faces presented in the five orientations shown below; upright and 45°, 90°, 135°, and 180° counter-clockwise from upright. There are 200 famous faces, and each is presented at these five orientations for 50 ms. The participants' task is to name each of the faces immediately after it has been presented on the screen. The dependent variable is the number of correct responses, which ranges from 0 to 200. The results for seven participants are shown in the following table.

	a_1	a_2	a_3	a_4	a_5
s_1	150	140	120	90	25
s_2	160	130	110	70	30
s_3	110	90	70	50	20
s_4	190	180	150	100	70
s_5	120	100	80	40	20
s_6	180	150	120	90	60
s_7	168	134	127	113	48

(a) Use Excel (or the Compute Variable dialog in SPSS) to compute contrasts that compare upright faces (a_1) to each of the four other orientations (a_2, a_3, a_4, and a_5). For each of these contrasts, compute the 95% confidence interval, t_{obs}, and p.

(b) Use SPSS to compute the same contrasts. (Through the Contrasts dialog, choose Simple and specify the reference value as First.) For each of these contrasts, report F_{obs}, p, and $\hat{\eta}_p^2$.

(c) Plot these data using Excel or some other graphing program. Plot the 95% confidence intervals around each of the condition means.

(d) What conclusion can we draw from these data?

(e) Do the statistical analyses seem important in this data set? Explain.

20 TWO-FACTOR ANOVA
Omnibus Effects

INTRODUCTION

In Chapters 18 and 19, we examined the one-factor between- and one-factor within-subjects designs in detail. As the names remind us, these designs involve one independent variable (factor, way) having two or more levels. We will now turn our attention to two-factor designs in which there are two independent variables.

To illustrate the notion of a two-factor design, we can think about several variations of the Posner task. There are many factors that could affect the pattern of responses shown in Figure 19.2 in Chapter 19. For example, males and females might produce different patterns of reaction times across valid, neutral, and invalid trials. If we were to run the Posner task with a group of males and a group of females, then *cue type* would be one factor and *sex* would be the second factor. As another example, the cuing method used might affect the pattern of reaction times. In Figure 19.1, the cue was an arrow at fixation pointing to the left or right, but it just as easily could have been the words *right* and *left*. Or a small arrow might have appeared above the left or right locations just before the target appeared. Therefore, we can think of at least three different cuing methods. An experiment might have examined reaction times across valid, neutral, and invalid trials for each of these three cuing methods. *Cue type* (valid, neutral, invalid) would be one factor and *cuing method* (central arrows, words, peripheral arrows) would be the second factor.

There are three different versions of two-factor designs: (i) the two-factor between-subjects design, (ii) the two-factor within-subjects design, and (iii) the two-factor mixed design. The two-factor between-subjects design involves two independent variables with completely independent groups of subjects in all conditions. The two-factor within-subjects design involves two independent variables with all subjects subjected to all conditions. The two-factor mixed design involves one between-subjects factor (e.g., *sex* or *cuing method*) and one within-subjects factor (e.g., *cue type*). Although these three versions of the two-factor design are analyzed differently, we will introduce their commonalities first.

This chapter is about omnibus effects in ANOVA, which refer to F-statistics with two or more degrees of freedom associated with a source of variance of interest. For example, testing R^2 for a regression model with two or more predictors is a test of an omnibus effect. The overall effect of treatment in an ANOVA with more than two levels of the independent variable is also an omnibus effect. In this chapter, we focus on significance tests because omnibus effects are almost invariably tested this way in research. We will have little to say about estimation in ANOVA until we discuss contrasts in Chapter 21.

For seven decades or more, students (like you) have been taught the formulas required to test the statistical significance of omnibus effects in ANOVA. This is hard-won knowledge, as you may soon discover. Perhaps because of the large time investment required to understand these calculations, generations of researchers in psychology routinely put this knowledge to work testing omnibus effects for statistical significance. My view is that such tests are rarely useful, and I'll explain why at end of this chapter. However, because omnibus tests are so prevalent, you are certain to encounter them when you are reading original research papers. If you do an advanced research project, your mentor may require you to test omnibus effects. Therefore, although omnibus effects may be of limited value, you will have to understand their logic.

The approach taken to omnibus effects in this chapter is based on decomposing (partitioning) variance using orthogonal vectors. This approach does not provide a way to compute two-factor ANOVAs by hand. Rather, it shows a general way to partition variance that keeps the connection between ANOVA and multiple regression in mind. Therefore, our approach will be more conceptual than computational. Understanding how variance is partitioned with orthogonal vectors will help you understand what programs such as SPSS are doing when they perform an ANOVA.

MAIN EFFECTS AND INTERACTIONS IN A 3 × 4 DESIGN

To introduce two-factor designs, we will think about a hypothetical experiment involving rats that have been trained to bar press for a heroin reward administered directly into their bloodstreams. It has been shown that food restriction slows the extinction of this learned response, so let's say we would like to know how extinction depends on the degree of food restriction (Sedki, D'Cunha, & Shalev, 2013). Knowing how extinction depends on bodily states, such as satiety and deprivation, is relevant to understanding relapse after drug rehabilitation in humans.

In this hypothetical experiment, we will consider 30 rats that have all been trained to bar press for a heroin reward. (Thirty is used only to make this example concrete; it is not critical to anything that follows.) After all rats have learned the task, they are randomly divided into three groups of 10 rats each and each group experiences some degree of food deprivation for 2 weeks.

One group is seriously food deprived, and these rats receive 50% of their normal daily food intake. A second group is moderately food deprived, and these rats receive 75% of their normal daily food intake. A third group is not food deprived, so they consume 100% of their normal daily food intake. After 2 weeks, rats are put in test cages equipped with the same bar they previously pressed to receive a heroin reward. Now, however, pressing the bar no longer results in a reward.

The two independent variables in this experiment are *degree of food restriction* (*A*) and *days of testing* (*B*). *Degree of food restriction* is a between-subjects factor and *days of testing* is a within-subjects factor. Therefore, this is a 3 (groups) × 4 (days) mixed ANOVA.

The dependent variable in this experiment is the *number of bar presses* each rat makes during a 1-hour session in the test cage. Rats are tested on 4 consecutive days, and we would expect bar pressing to decrease over days as the rats learn that bar presses no longer deliver the drug reward. The question is, How will this change in bar pressing differ for the three groups?

Figure 20.1 shows hypothetical data for this experiment. Figure 20.1a shows all 12 conditions of the experiment. For all three groups (50%, 75%, and 100%), bar pressing decreases over days as rats learn that doing so no longer results in a reward. Furthermore, bar pressing is related to the degree of food deprivation. The most food-deprived rats make the most bar presses, and rats that are not food deprived make the fewest bar presses.

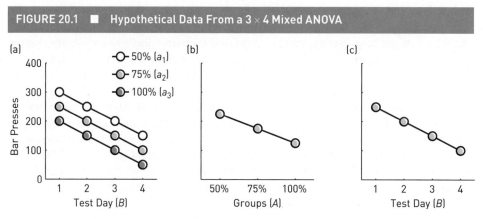

FIGURE 20.1 ■ Hypothetical Data From a 3 × 4 Mixed ANOVA

Hypothetical data from a 3 (groups) × 4 (days) mixed ANOVA. (a) Means for all 12 conditions of the experiment. (b) The main effect of groups. (c) The main effect of days.

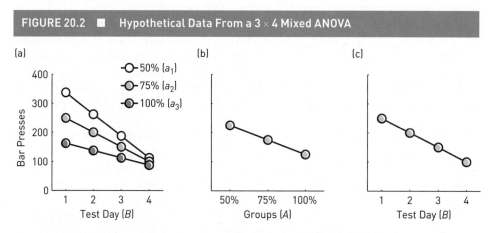

FIGURE 20.2 ■ Hypothetical Data From a 3 × 4 Mixed ANOVA

Hypothetical data from a 3 (groups) × 4 (days) mixed ANOVA. (a) Means for all 12 conditions of the experiment. (b) The main effect of groups. (c) The main effect of days.

The **main effect** of an independent variable refers to changes in the means of its levels averaged across all levels of other variables.

An **interaction** between two independent variables means that the effect of one independent variable on the dependent variable changes as a function of the level of the other independent variable.

Figure 20.1b shows bar pressing for each group averaged over days. Each symbol shows the mean number of bar presses for the 10 rats in one of the three groups averaged over days. In ANOVA terminology, we say that Figure 20.1b shows the **main effect** of groups. Figure 20.1c shows the main effect of days (*B*). Each symbol shows the mean number of bar presses on each day, averaged over groups.

Figure 20.2 shows another possible outcome for this hypothetical experiment. Figure 20.2a shows the mean number of bar presses for each of the 12 experimental conditions. Figures 20.2b and 20.2c show the main effects of groups and days, respectively. Although the two main effects are identical in Figures 20.1 and 20.2, the pattern of results in Figure 20.2a is very different from that in Figure 20.1a. Figure 20.2a shows an **interaction** between the two variables. This means that the effect of one factor changes across levels of the other factor. For example, there is a big difference in bar pressing for the three groups on day 1, but there is very little difference in bar pressing between the three groups on day 4.

It is very easy to spot two-factor interactions in graphs like those in Figures 20.1a and 20.2a. Interactions exist when the lines relating levels of the independent variables to the dependent variable are not parallel. Because the three lines in Figure 20.2a (representing different groups) *are not* parallel, we say there is an interaction between days and groups.

Because the three lines in Figure 20.1a *are* parallel, we say there is no interaction between days and groups.

Although one can examine main effects and test them for statistical significance, it should be clear that these are not very informative in the presence of an interaction. For example, if we examined only the main effects in Figures 20.1 and 20.2, we would be unaware of the fundamental differences between these two data sets. Therefore, assessing interactions is a critical new feature of data sets that we didn't have to address in one-factor designs.

If you think back to Chapter 17, you may recognize that our definition of an interaction in ANOVA is exactly the same as our definition of an interaction in multiple regression. In both cases, there is an interaction when the effect of one factor changes across levels of the other factor. In a two-predictor *multiple regression*, the two predictor variables each explain some portion of the variability in the outcome variable. When we multiply the two predictors we produce a product vector that captures variability arising from the interaction of the two predictor variables. In *ANOVA*, the two independent variables (main effects) explain some of the variability in the dependent variable. We saw at the end of Chapter 18 that variability attributable to a main effect can be captured by orthogonal vectors. In this chapter we will see that we can multiply the vectors coding for main effects, to produce new vectors that capture variability attributable to interactions. Unlike in regression, we may need more than one product vector in ANOVA to capture the interaction. We will cover this point in detail in the next section.

LEARNING CHECK 1

1. The following table of means from a 2×4 design shows four levels of Factor A, which are denoted a_1, a_2, a_3, and a_4. There are two levels of Factor B, which are denoted b_1 and b_2. Are there main effects and an interaction? Explain why or why not.

2. Are there main effects of A and B in the following table? Is there an interaction? Explain why or why not.

	a_1	a_2	a_3	a_4
b_1	1	2	3	4
b_2	2	3	4	5

	a_1	a_2	a_3	a_4
b_1	1	2	3	4
b_2	4	3	2	3

Answers

1. There is a main effect of A because the mean values of A are different when averaged over B (1.5, 2.5, 3.5, and 4.5). There is a main effect of B because the mean values of B are different when averaged over A (2.5 and 3.5). There is no interaction because the difference between b_1 and b_2 is the same for all levels of A (all differences equal −1).

2. There is a main effect of A because the mean values of A are different when averaged over B (2.5, 2.5, 2.5, and 3.5). There is a main effect of B because the mean values of B are different when averaged over A (2.5 and 3). There is an interaction because the difference between b_1 and b_2 differs across levels of A (i.e., $b_2 - b_1$ = 3, 1, −1, and −1 for $a_1, a_2, a_3,$ and a_4, respectively).

PARTITIONING VARIABILITY AMONG MEANS: ORTHOGONAL DECOMPOSITION

The analysis of main effects and interactions in a two-factor ANOVA is an exercise in partitioning variance for the purpose of significance testing. In a two-factor design, we partition variability among means into components associated with the main effects of A and B, as well

FIGURE 20.3 ■ Partitioning $ss_{treatment}$

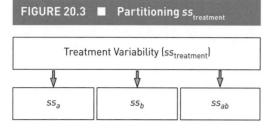

A schematic diagram illustrating how the total variability (sum of squares) among treatment means can be partitioned into sources associated with the two main effects (ss_a and ss_b) and the interaction (ss_{ab}).

as the interaction between A and B, which we denote AB. Each of these sources has an associated error term used to compute F_{obs}. The error terms for A, B, and AB depend on whether the design is a between, within, or mixed ANOVA. Therefore, we begin by partitioning variability among the treatment means, which is common to all designs, before we discuss the error terms associated with the different designs.

Figure 20.3 illustrates graphically the logic of the analysis we will pursue in this section. As just noted, we will ignore all sources of variability except for variability among treatment means. If we consider a 3 × 4 ANOVA, there will be 3*4 = 12 treatment means (as in Figures 20.1a and 20.2a). The total variability (sum of squares) among these 12 treatment means can be denoted $ss_{treatment}$. This total variability can be partitioned into components associated with A, B, and AB. We will call these components ss_a, ss_b, and ss_{ab}.

We will examine two ways of partitioning treatment variability. The first involves computing sums of squares, and the second involves orthogonal vectors. These two approaches produce identical results, but decomposition with orthogonal vectors provides a very general way to think about the nature of ANOVA. Our approach will be to first show the equivalence of these two methods so that we can use orthogonal vectors throughout the rest of the chapter.

To help us think about partitioning variance, we will use the data shown in Figure 20.4 as an illustration. These data come from the hypothetical experiment described above but show a more realistic pattern of extinction across the three groups. The main effect of groups shows an essentially linear trend associated with the degree of food deprivation. The main effect of test day shows a combination of linear and quadratic change over days. The interaction in Figure 20.4a shows that there were large differences in bar pressing between the three groups on day 1 of testing, such that bar pressing increased with the degree of food deprivation. However, over time, the three groups converged to similar levels of bar pressing on day 4.

Sums of Squares for Main Effects and Interactions

We will now look at how to decompose the total variability among the 12 means in Figure 20.4a into components associated with A, B, and AB. These means are shown as "Condition means" in Table 20.1. I should point out that Table 20.1 is structurally

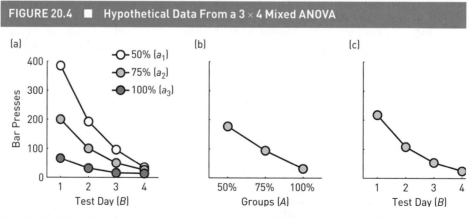

Hypothetical data from a 3 (groups) × 4 (days) mixed ANOVA. (a) Means for all 12 conditions of the experiment. (b) The main effect of days. (c) The main effect of group.

TABLE 20.1 ■ Partitioning Variance in the Means of a 3 × 4 Design[a]										
	Condition Means					**Factor A Means**			**Factor B Means**		
	a_1	a_2	a_3	**B**		a_1	a_2	a_3	a_1	a_2	a_3
b_1	384	199	65	m_{b_1}	216	176	93	31	216	216	216
b_2	191	99	31	m_{b_2}	107	176	93	31	107	107	107
b_3	95	49	15	m_{b_3}	53	176	93	31	53	53	53
b_4	34	25	13	m_{b_4}	24	176	93	31	24	24	24
	m_{a_1}	m_{a_2}	m_{a_3}								
A	176	93	31								
m_{all}	100										
$ss_{\text{treatment}}$	132,126										
ss_a	42,344										
ss_b	64,470										
$ss_{ab} = ss_{\text{treatment}} - ss_a - s_b$	25,312										

[a] $a_1 = 50\%$; $a_2 = 75\%$; $a_3 = 100\%$; b_1 to $b_4 =$ days 1 to 4, respectively.

identical to Table 19.1. The only difference is that where Table 20.1 shows Factor B, Table 19.1 showed *subjects*. Therefore, our approach to partitioning treatment variability is parallel to what we did in Chapter 19.

The values in "Condition means" columns a_1, a_2, and a_3 are the means for the three groups (A) at each of the 4 days of testing (B). Each of these columns is plotted as a line in Figure 20.4a. Below these columns are the means m_{a_i} for $i = 1$ to 3) for the three levels of A (i.e., 176, 93, and 31, respectively). These are the numbers plotted in Figure 20.4b. To the right of the condition means (in column "B") are the means (m_{b_j} for $j = 1$ to 4) for the four levels of B (i.e., 216, 107, 53, and 24, respectively). These are the numbers plotted in Figure 20.4c.

When we partition the variability in our set of 12 means, our first task is to compute the grand mean of all 12 conditions. In the two-factor design, we denote the number of levels of A as k_a and the number of levels of B as k_b. For our example, $k_a = 3$ and $k_b = 4$. In addition, we denote each sample mean as $m_{i,j}$, where i indexes levels of A and j indexes levels of B. Using these conventions, we can define the mean of all 12 means in Table 20.1 as follows:

$$m_{\text{all}} = \frac{\sum_{i=1}^{k_a}\sum_{j=1}^{k_b} m_{i,j}}{k_a * k_b}. \tag{20.1}$$

The sum of the 12 means is 1200; therefore, $m_{\text{all}} = 1200/12 = 100$.

The total sum of squares in our 12 means ($ss_{\text{treatment}}$) is the sum of squared deviations from m_{all}, which is computed as follows:

$$ss_{\text{treatment}} = \sum_{i=1}^{k_a}\sum_{j=1}^{k_b}(m_{i,j} - m_{\text{all}})^2. \tag{20.2}$$

Table 20.1 shows that $ss_{\text{treatment}} = 132{,}126$. This is the quantity to be partitioned.

Figure 20.3 shows the three components of $ss_{\text{treatment}}$. The first is ss_a, the sum of squares associated with the main effect of A. We obtain ss_a as follows:

$$ss_a = k_b \sum_{i=1}^{k_a} (m_{a_i} - m_{\text{all}})^2. \tag{20.3}$$

Equation 20.3 computes the sum of squared deviations of the means for (m_{a_i}) from the grand mean (m_{all}) and then multiplies this quantity by k_b, the number of levels of B. When equation 20.3 is applied to the three means for A, Table 20.1 shows that $ss_a = 42{,}344$.

Equation 20.3 shows that ss_a is the quantity we would obtain from equation 20.2 if we replaced each condition mean with the mean of its corresponding level of A. This is illustrated in the columns labeled "Factor A means" in Table 20.1.

The second component of $ss_{\text{treatment}}$ is ss_b, the sum of squares associated with the main effect of B. We obtain ss_b as follows:

$$ss_b = k_a \sum_{j=1}^{k_b} (m_{b_j} - m_{\text{all}})^2. \tag{20.4}$$

Equation 20.4 computes the sum of squared deviations of the means for B (m_{b_j}) from the grand mean (m_{all}) and then multiplies this quantity by k_a, the number of levels of A. When equation 20.4 is applied to the four means of B, Table 20.1 shows that $ss_b = 64{,}470$.

Equation 20.4 shows that ss_b is the quantity we would obtain from equation 20.2 if we replaced each condition mean with the mean of its corresponding level of B. This is illustrated in the columns labeled "Factor B means" in Table 20.1.

Once ss_a and ss_b have been computed, ss_{ab} is what is left over. Therefore, we compute ss_{ab} as follows:

$$ss_{ab} = ss_{\text{treatment}} - ss_a - ss_b. \tag{20.5}$$

Given that we know $ss_{\text{treatment}}$, ss_a, and ss_b, we find that

$$ss_{ab} = ss_{\text{treatment}} - ss_a - ss_b = 132{,}126 - 42{,}233 - 64{,}470 = 25{,}312.$$

That is, ss_{ab} is the sum of squares associated with the AB interaction. If ss_{ab} equals zero, then there is no interaction and all variability in the treatment means is explained by the main effects. This is the case in Figure 20.1. If ss_{ab} does not equal zero, then there is an interaction and the main effects do not explain all of the variability in the treatment means. This is the case in Figures 20.2 and 20.4.

Having computed ss_a, ss_b, and ss_{ab}, we can now determine what *proportion* of $ss_{\text{treatment}}$ each of these sources represents. When we plug in the numbers computed above, we find that

$$\hat{\eta}_a^2 = \frac{ss_a}{ss_{\text{treatment}}} = \frac{42{,}344}{132{,}126} = .32,$$

$$\hat{\eta}_b^2 = \frac{ss_b}{ss_{\text{treatment}}} = \frac{64{,}470}{132{,}126} = .49, \text{ and}$$

$$\hat{\eta}_{ab}^2 = \frac{ss_{ab}}{ss_{\text{treatment}}} = \frac{25{,}312}{132{,}126} = .19.$$

This means that 81% of the variability in the 12 means is associated with the main effects of A and B together ($.32 + .49 = .81$) and 19% is explained by the AB interaction.

In the next section, we will introduce a second way of computing $\hat{\eta}_a^2$, $\hat{\eta}_b^2$, and $\hat{\eta}_{ab}^2$. This alternative method will provide a good conceptual foundation for understanding omnibus sources of variance in multifactor ANOVAs.

Orthogonal Decomposition for Main Effects and Interactions

In Chapter 16, we saw that the total variability in n scores can be captured with $n-1$ predictors (see Figure 16.14). If these $n-1$ predictors are orthogonal, then each explains a unique part of the variability in the set of n scores. In Chapter 18, we applied this idea to condition means in a one-factor between-subjects ANOVA (e.g., Tables 18.3 and 18.10). When there are k levels of a factor, the variability in the condition means can be partitioned using $k-1$ orthogonal vectors. When there are two factors, k_a-1 vectors capture variability associated with the main effect of A, and k_b-1 vectors capture variability associated with the main effect of B. We will now see that the *products* of these orthogonal vectors capture variability attributable to the AB interaction; there will be $(k_a-1)(k_b-1)$ product vectors. To get to this point, we will first show an easy way to generate orthogonal vectors, and we will then apply this approach to partitioning variability in main effects.

Helmert Vectors

In this section we will be interested in the orthogonality of a set of vectors and not the contrasts they represent. Therefore, it will be very helpful to have a simple way to generate orthogonal sets. Helmert vectors, which were discussed briefly in Chapter 19, suit this purpose.

An orthogonal set of contrasts has k rows and $k-1$ columns, or vice versa. Table 20.2 shows Helmert vectors for $k = 2$ to 7 means. In this table, we will think about means arranged in a column, as in column B of Table 20.1. So, for each orthogonal set, there are k rows corresponding to the number of means, and $k-1$ columns corresponding to the number of vectors in the set.

When there are k means, the top left entry in a Helmert set is $k-1$. The $k-1$ numbers below this entry are -1, so entries in the *first column* sum to 0. (You can confirm this by looking at the *first column* of each of the six sets in Table 20.2.) The remaining entries in the *first row* are 0 (if $k>2$). (You can confirm this by looking at the *first row* of each of the six sets in Table 20.2.)

The second entry in the second column is $k-2$ and the $k-2$ numbers below this entry are -1. The remaining entries in the second row are 0. We proceed this way until we have $k-1$ columns. You should be able to verify (using Excel or SPSS) that each set of Helmert vectors in Table 20.2 forms an orthogonal set.

TABLE 20.2 ■ Helmert Vectors for $k = 2$ to $k = 7$

k = 2	k = 3		k = 4			k = 5				k = 6					k = 7					
1	2	0	3	0	0	4	0	0	0	5	0	0	0	0	6	0	0	0	0	0
−1	−1	1	−1	2	0	−1	3	0	0	−1	4	0	0	0	−1	5	0	0	0	0
	−1	−1	−1	−1	1	−1	−1	2	0	−1	−1	3	0	0	−1	−1	4	0	0	0
			−1	−1	−1	−1	−1	−1	1	−1	−1	−1	2	0	−1	−1	−1	3	0	0
						−1	−1	−1	−1	−1	−1	−1	−1	1	−1	−1	−1	−1	2	0
										−1	−1	−1	−1	−1	−1	−1	−1	−1	−1	1
															−1	−1	−1	−1	−1	−1

Main Effect Vectors

As a first step toward partitioning variance with orthogonal vectors, we will consider the means of A and B from Table 20.1. These means are shown in columns "A" and "B" in Table 20.3. A set of Helmert vectors is shown to the right of columns A and B. The vectors are a_{w_1} and a_{w_2} for A and b_{w_1}, b_{w_2}, and b_{w_3} for B. The subscript w in these vector names denotes *weight*. It is important to distinguish the notations for levels of A (e.g., a_1, a_2, and a_3) and B (e.g., b_1, b_2, b_3, and b_4) from the notations for the vectors that code for A (e.g., a_{w_1} and a_{w_2}) and B (e.g., b_{w_1}, b_{w_2}, and b_{w_3}).

Rather that sum the products of vectors and means (as we do when computing contrasts), we will compute the squared correlation between vectors and means. At the bottom of columns a_{w_1} and a_{w_2} is the squared correlation between these vectors and the three means in A. These squared correlations (.82 and .18) sum to 1, showing that two orthogonal vectors capture all of the variability in the three means. At the bottom of columns b_{w_1}, b_{w_2}, and b_{w_3} are the squared correlations between these vectors and the four means in B. These squared correlations (.83, .15, and .02) also sum to 1, showing that three orthogonal vectors capture all of the variability in the four means.

The analysis in Table 20.3 is extended to all 12 means in Table 20.4. These 12 means are shown as a single column (m). The coefficients in columns a_{w_1} and a_{w_2} in Table 20.3 have been expanded in Table 20.4. Consider the coefficients in a_{w_1}. All means at a_1 are associated with 2, and all means at a_2 and a_3 are associated with -1.

The coefficients in columns b_{w_1}, b_{w_2}, and b_{w_3} in Table 20.3 have also been expanded in Table 20.4. Consider the coefficients in b_{w_1}. All means at b_1 are associated with 3, and all means at b_2, b_3, and b_4 are associated with -1. It would be worth taking a moment to make sure you see the correspondence between Tables 20.3 and 20.4. It might also be helpful to review the last section of Chapter 18.

The squared correlation (r^2) between vectors and means are shown below each column. For example, the squared correlation (r^2) between columns m and a_{w_1} is shown at the bottom of column a_{w_1} to be .262, and the squared correlation between columns m and a_{w_2} is shown at the bottom of column a_{w_2} to be .058. Therefore, the proportion of variance in m (the 12 means) explained by a_{w_1} and a_{w_2} is $\hat{\eta}_a^2 = .262 + .058 = .32$. This means that the two contrasts underlying the main effect of A explain 32% of the variability in the 12 means. Remember that we saw this number before.

Exactly the same analysis applies to columns b_{w_1}, b_{w_2}, and b_{w_3}. Table 20.4 shows that the proportion of variability in m explained by b_{w_1}, b_{w_2}, and b_{w_3} together is $\hat{\eta}_b^2 = .407 + .071 + .010 = .49$ when rounded to two decimal places. This means that the three contrasts underlying the main effect of B explain 49% of the variability in the 12 means.

Interaction Vectors

We now come to a new and important point: the variability in m attributable to the interaction between A and B is captured by the products of the vectors for A and B.* In our example, two vectors code for A and three code for B, so there are $2 * 3 = 6$ product vectors in total. These are shown in in the last six columns of Table 20.4 as $a_{w_1}b_{w_1}$, $a_{w_1}b_{w_2}$, $a_{w_1}b_{w_3}$, $a_{w_2}b_{w_1}$, $a_{w_2}b_{w_2}$, and $a_{w_2}b_{w_3}$. We call these **interaction vectors**. Notice that we have now generated 11 orthogonal vectors ($2 + 3 + 2 * 3$) to capture the variability in 12 means.

The six vectors representing the interaction of A and B are guaranteed to be orthogonal to those coding for A and B. (You can verify this in Excel or SPSS.) This means that they capture variability that is not attributable to the main effects of A and B. At the bottom of

Interaction vectors are the products of the orthogonal vectors coding for the two main effects. Because df_a vectors are required to capture the main effect of A and df_b vectors are required to capture the main effect of B, we require $df_a * df_b$ vectors to capture the AB interaction.

---

*If you think back to moderation analysis in Chapter 17, we tested the interaction between two predictor variables by forming a new predictor from the product of the two. We are following exactly the same principle here, except there are more predictors to multiply together.

TABLE 20.3 ■ Using Helmert Vectors

	A	a_{w_1}	a_{w_2}		B	b_{w_1}	b_{w_2}	b_{w_3}
	176	2	0	b_1	216	3	0	0
a_1	93	−1	1	b_2	107	−1	2	0
a_2	31	−1	−1	b_3	53	−1	−1	1
a_3	r^2	.82	.18	b_4	24	−1	−1	−1
					r^2	.83	.15	.02

TABLE 20.4 ■ Partitioning Variance in the Means of a 3 × 4 design

		m	A		B			AB					
			a_{w_1}	a_{w_2}	b_{w_1}	b_{w_2}	b_{w_3}	$a_{w_1}b_{w_1}$	$a_{w_1}b_{w_2}$	$a_{w_1}b_{w_3}$	$a_{w_2}b_{w_1}$	$a_{w_2}b_{w_2}$	$a_{w_2}b_{w_3}$
a_1	b_1	384	2	0	3	0	0	6	0	0	0	0	0
a_1	b_2	191	2	0	−1	2	0	−2	4	0	0	0	0
a_1	b_3	95	2	0	−1	−1	1	−2	−2	2	0	0	0
a_1	b_4	34	2	0	−1	−1	−1	−2	−2	−2	0	0	0
a_2	b_1	199	−1	1	3	0	0	−3	0	0	3	0	0
a_2	b_2	99	−1	1	−1	2	0	1	−2	0	−1	2	0
a_2	b_3	49	−1	1	−1	−1	1	1	1	−1	−1	−1	1
a_2	b_4	25	−1	1	−1	−1	−1	1	1	1	−1	−1	−1
a_3	b_1	65	−1	−1	3	0	0	−3	0	0	−3	0	0
a_3	b_2	31	−1	−1	−1	2	0	1	−2	0	1	−2	0
a_3	b_3	15	−1	−1	−1	−1	1	1	1	−1	1	1	−1
a_3	b_4	13	−1	−1	−1	−1	−1	1	1	1	1	1	1
	r^2		.262	.058	.407	.071	.010	.128	.025	.006	.026	.005	.001
	$\hat{\eta}^2$		0.320		0.488			0.192					

$ss_{\text{treatment}}$	132,126
$ss_a = \hat{\eta}^2_a * ss_{\text{treatment}}$	42,344
$ss_b = \hat{\eta}^2_b * ss_{\text{treatment}}$	64,470
$ss_{ab} = \hat{\eta}^2_{ab} * ss_{\text{treatment}}$	25,312

these six columns are the squared correlations between the corresponding vectors and m. When these are summed, we find that the proportion of variability in m explained by the interaction of A and B is $\hat{\eta}^2_{ab}$ = .19, when rounded to two decimal places. That is, the six contrasts underlying the AB interaction explain 19% of the variability in the 12 means.

If we look back to Table 20.1, we see that the sum of squares associated with A is ss_a = 42,344 and the sum of squares associated with B is ss_a = 64,470. Knowing that $ss_{\text{treatment}}$ = 132,126, we deduced that $ss_{ab} = 25{,}132$. Table 20.4 connects the sums of squares computed in Table 20.1 to the $\hat{\eta}^2$ values associated with A, B, and AB. When we multiply $ss_{\text{treatment}}$ by the $\hat{\eta}^2$ associated with each of these three sources we find that $ss_a = 42{,}344$, $ss_b = 64{,}470$, and $ss_{ab} = 25{,}132$. Therefore, computing sums of squares from means or using orthogonal vectors produces exactly the same results.

Degrees of Freedom

We've seen before that there is an important connection between the number of vectors required to code a factor (main effect) and the degrees of freedom associated with the factor. For the main effect of A, there are $df_a = 2$ degrees of freedom. For the main effect of B, there are $df_b = 3$ degrees of freedom. We've just seen that for the AB interaction, there are $2 * 3 = 6$ degrees of freedom. That is, $df_{ab} = df_a * df_b$. For a two-factor design, there will always be $df_a + df_b + df_{ab} = k_a * k_b - 1$ degrees of freedom associated with the $k_a * k_b$ condition means.

Now that we know how to decompose the variability attributable to the main effects and interaction in a two-factor design, we will move on to see how these sources are tested for statistical significance in between, within, and mixed ANOVAs. We will continue our discussion using the conceptual framework of orthogonal decomposition.

AN EXAMPLE: THE TEXTURE DISCRIMINATION TASK

When I was a graduate student I became interested in "low-level vision." The meaning of low-level vision is not obvious unless you've taken a course on perception, but let me try to illustrate. You probably know that your eyes contain three types of photoreceptors (cones) that transduce light into neural signals. These photoreceptors don't recognize faces, but cells deeper in your brain (e.g., the inferotemporal cortex, or IT) arguably do. Neural signals go through many neural transformations between the photoreceptors and IT. When we talk about low-level vision, we mean visual processing mechanisms that are closer to photoreceptors than to the IT.

Texture segmentation is widely thought to be one such low-level visual process. Many years ago, Bela Julesz (1981) had a theory of texture segmentation (called texton theory) that I thought was wrong. Therefore, I spent long days and nights running experiments trying to prove this. One of these experiments involved the two displays shown in Figure 20.5. On each trial of the experiment, a participant was shown a display (e.g., Ls embedded in Xs) very briefly and had to report which quadrant contained the disparate texture. So, a participant would have to indicate "bottom right" when shown Figure 20.5a or "upper right" when shown Figure 20.5b. Because the disparate texture could appear in one of four quadrants, this is an example of a four-alternative forced-choice task. Of course, participants make mistakes when the displays are presented very briefly. Therefore, the dependent variable in this experiment was the number of correct responses made over a series of trials (*number correct*).*

---

*Those who propose theories give the rest of us something to think about. However, science is adversarial and scientists try to understand nature by challenging each other's theories. As you can imagine, not everyone appreciates being challenged in a public forum like a research journal. As I ground through my PhD research, I expected that Bela Julesz's response to my work would probably be quite negative so I mentally prepared for an eventual meeting. Several years after my thesis work appeared in print, Julesz visited a poster that I was giving at a conference in Florida. I expected a harsh response to some of the cocky comments in the paper. Instead Julesz said that he found the work very interesting and that he appreciated having his theories challenged. Quoting another famous Hungarian scientist, he said that at his age his adversaries were either dead or had become supporters who no longer read his papers and only praised his past contributions. He invited me to lunch with some of the distinguished researchers that he had supervised in the past. At that lunch I quickly became aware just how little I knew about vision. Bela Julesz was both a great scientist and a great person, and his impact on the field is measured by his own outstanding contributions and those of the many distinguished researchers he mentored.

LEARNING CHECK 2

1. Use Helmert vectors to code for the main effects of A and B for the means in the following table, and use the products of these vectors to code the interaction between A and B. Use Excel (or SPSS) to determine the proportion of variability associated with A, B, and AB. Confirm that your answer is correct using sums of squares. (It would also be helpful to plot the data.)

	a_1	a_2	a_3
b_1	1	2	3
b_2	2	5	8
b_3	3	2	4

Answers

1. We answer this question by creating the following table. For both A and B, we use Helmert vectors $[2, -1, -1]$ and $[0, 1, -1]$. These are shown in columns a_{w_1}, a_{w_2}, b_{w_1}, and b_{w_2}. The products are shown in the remaining four columns. Below each column is the squared correlation (r^2) between the vector and the means (m). The proportion of total variance explained by the main effect of A is .389. The proportion of total variance explained by the main effect of B is also .389. The proportion of total variance explained by the AB interaction is .222. The total sum of squares is 36; therefore, the sum of squares is $.389 * 36 = 14$ for A, $.389 * 36 = 14$ for B, and $.222 * 36 = 8$ for the AB interaction.

			A		B		AB			
A	B	m	a_{w_1}	a_{w_2}	b_{w_1}	$b_{w_{12}}$	$a_{w_1}b_{w_1}$	$a_{w_1}b_{w_2}$	$a_{w_2}b_{w_1}$	$a_{w_2}b_{w_2}$
a_1	b_1	1	2	0	2	0	4	0	0	0
a_1	b_2	2	2	0	-1	1	-2	2	0	0
a_1	b_3	3	2	0	-1	-1	-2	-2	0	0
a_2	b_1	2	-1	1	2	0	-2	0	2	0
a_2	b_2	5	-1	1	-1	1	1	-1	-1	1
a_2	b_3	2	-1	1	-1	-1	1	1	-1	-1
a_3	b_1	3	-1	-1	2	0	-2	0	-2	0
a_3	b_2	8	-1	-1	-1	1	1	-1	1	-1
a_3	b_3	4	-1	-1	-1	-1	1	1	1	1
	r^2		.222	.167	.222	.167	.007	.188	.021	.007
	$\hat{\eta}^2$		.389		.389		.222			

In this experiment, there were three questions of interest. The first was whether it is easier to detect Ls in Xs or Xs in Ls. For lack of a better term, we will call this variable *display type*. The second question was about how accuracy changes as a function of stimulus duration; we would expect accuracy to improve with longer exposure durations. The third question was about how accuracy changes with *duration* for the two display types; that is, we might ask if there is an interaction between *display type* and *duration*.

If three stimulus durations were used (50, 100, and 150 ms), then there would be $2 * 3 = 6$ conditions in the experiment. Let's say that each of the two *display types* (Ls in Xs versus Xs

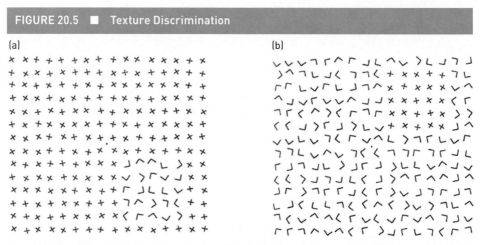

FIGURE 20.5 ■ Texture Discrimination

(a and b) Two example displays from a texture discrimination task reported by Gurnsey and Browse (1987).

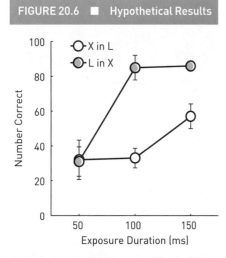

FIGURE 20.6 ■ Hypothetical Results

Hypothetical results of a texture discrimination experiment in which participants were to detect Xs in Ls or Ls in Xs at exposure durations of 50, 100, or 150 ms. Error bars represent ±s.

in Ls) was presented 100 times at each duration with the disparate region presented 25 times in each quadrant. We would expect scores in each condition to range from chance (25 out of 100) to 100. Figure 20.6 plots artificial data similar to those reported by Gurnsey and Browse (1987).

This experiment could be performed in three different ways. First, it could be run completely between subjects, meaning that different subjects were tested in each of the six conditions. Second, it could be run completely within subjects, meaning that all subjects were tested in all six conditions. Finally, the experiment could be run as a mixed design with *display type* as a between-subjects factor (one group of subjects saw Ls in Xs, and another group of subjects saw Xs in Ls) and *duration* as a within-subjects factor. We will examine each of these analyses with the same data set.

To keep our numbers manageable, we will imagine that each condition of the experiment involves just two subjects. For the 2 × 3 completely between-subjects analysis, this means we would have 2 * 6 = 12 participants in total. For the completely within-subjects analysis, we would have only two participants in total. For the mixed analysis, we would have two subjects in each *display type* condition for a total of four subjects in the experiment.

In all cases, orthogonal vectors will be used to partition the variability in our set of 12 scores. The objective is not to provide a method to be used for hand calculations. Rather, the objective is to show how sources of variance are extracted in two-factor designs and how these sources are tested for statistical significance. Although we will say a few things about contrasts as we move along (because vectors describe contrasts), we provide a full discussion of contrasts in Chapter 21. Finally, the emphasis in what follows is squarely on significance tests and not estimation.

THE TWO-FACTOR BETWEEN-SUBJECTS DESIGN

In this section, we will treat the texture discrimination experiment as a between-subjects design, which means that scores in all conditions are obtained from different individuals. In an experiment, participants are typically chosen at random from some population and then randomly assigned to experimental conditions. In quasi-experiments, participants may be selected from existing populations and then possibly assigned to different experimental conditions. For example, bar pressing for a heroin reward could be measured for three breeds of rats under three levels of food deprivation.

Sources of Variance

Figure 20.7 illustrates how the total variability in a two-factor between-subjects design is partitioned. At the most general level, we can partition the total variability in our scores into two components: one component captures variability associated with the treatment means ($ss_{treatment}$), and the other component captures variability not associated with the treatment means (ss_{error}). As we saw in Figure 20.3, $ss_{treatment}$ can be broken down into ss_a, ss_b, and ss_{ab}. Our goal is to show how to partition the variability in an actual data set into four components: ss_a, ss_b, ss_{ab}, and ss_{error}.

FIGURE 20.7 ■ Sources of Variance (Between)

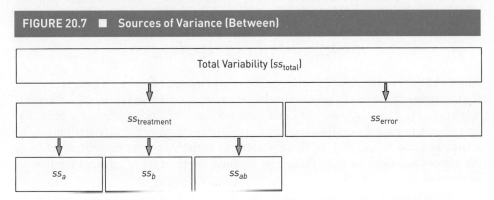

A schematic diagram illustrating how the total variability in our experiment (ss_{total}) can be partitioned into sources associated with the two main effects (ss_a and ss_b), the interaction (ss_{ab}), and the residual error (ss_{error}).

Sources of Variance Computed From Means

The raw data corresponding to Figure 20.6 are shown in Table 20.5. There are two scores within each condition of the experiment. For example, the two scores in a_2b_2 (Ls in Xs presented at 100 ms) are 90 and 80. We will first partition the variability in this set of scores into the four sources shown in Figure 20.7 using standard formulas, and we then show how the same results are achieved with orthogonal vectors. Partitioning $ss_{treatment}$ into ss_a, ss_b, and ss_{ab} will be common to all three designs covered in this chapter. Therefore, this topic will be covered extensively in this section but won't be duplicated in subsequent sections.

Sums of Squares for A. To compute the sum of squares for A, we compute the sum of squares for the means of the two levels of A (40.67 and 67.33) and then multiply by the number of scores that contributed to each mean. This is captured in the following formula:

$$ss_a = nk_b \sum_{i=1}^{k_a} (m_{a_i} - m_y)^2, \quad (20.6)$$

where n is the number of scores in each condition, k_b is number of levels of B, k_a is number of levels of A, m_{a_i} is the mean of the ith level of A, and m_y is the overall mean of the 12 scores. You

TABLE 20.5 ■ Hypothetical Data

	b_1	b_2	b_3	Factor A Means
a_1	40	37	62	40.67
	24	29	52	
a_2	37	90	88	67.33
	25	80	84	
Factor B Means				
	31.5	59.0	71.5	
AB Interaction Means				
	32.0	33.0	57.0	
	31.0	85.0	86.0	

[a] a_1 = Xs in Ls; a_2 = Ls in Xs; b_1, b_2, and b_3 = durations 50, 100, and 150, respectively.

should be able to verify that the mean of the 12 scores is $m_y = 54$ and that $\sum_{i=1}^{k_a}(m_{a_i} - m_y)^2 = 355.56$. We also know that $n = 2$ and $k_b = 3$. Therefore,

$$ss_a = nk_b\sum_{i=1}^{k_a}(m_{a_i} - m_y)^2 = 2*3*355.56 = 2133.33.$$

Sums of Squares for B. To compute the sum of squares for B, we compute the sum of squares for the means of the three levels of B (31.5, 59, and 71.5) and then multiply by the number of scores that contributed to each mean. This is captured in the following formula:

$$ss_b = nk_a\sum_{j=1}^{k_b}(m_{b_j} - m_y)^2, \tag{20.7}$$

where m_{b_j} is the mean of the jth level of B. If you do the computations in Excel, you will find that $\sum_{j=1}^{k_b}(m_{b_j} - m_y)^2 = 837.5$. From these numbers, we can compute

$$ss_b = nk_a\sum_{j=1}^{k_b}(m_{b_j} - m_y)^2 = 2*2*837.5 = 3350.$$

Sums of Squares for AB. To compute the sum of squares for the AB interaction, we first compute the sum of squares for the means of the six conditions of the experiment (32, 33, 57, 31, 85, and 86) and then multiply by the number of scores that contributed to each mean. However, this quantity contains variability attributable to A (i.e., ss_a) and B (i.e., ss_b), so these sources of variability must be subtracted from the sum of squares just described. The formula for the sum of squares for the AB interaction is

$$ss_{ab} = n\sum_{i=1}^{k_a}\sum_{j=1}^{k_b}(m_{a_ib_j} - m_y)^2 - ss_a - ss_b. \tag{20.8}$$

If you do the computations in Excel, you will find that $n\sum_{i=1}^{k_a}\sum_{j=1}^{k_b}(m_{a_ib_j} - m_y)^2 = 6896$. Therefore,

$$ss_{ab} = n\sum_{i=1}^{k_a}\sum_{j=1}^{k_b}(m_{a_ib_j} - m_y)^2 - ss_a - ss_b = 6896 - 2133.33 - 3350 = 1412.67.$$

Residual (Error) Sums of Squares. So far we have extracted sums of squares associated with A, B, and the AB interaction. Together these make up $ss_{treatment}$ (see Figure 20.7). To compute ss_{error}, we must first know ss_{total}. If you do the computations in Excel, you will find that $ss_{total} = 7236$. Therefore, we can compute ss_{error} as follows:

$$ss_{error} = ss_{total} - ss_{treatment} = ss_{total} - ss_a - ss_b - ss_{ab} = 7236 - 2133.33 - 3350 - 1412.67 = 340.$$

Sources of Variance Extracted With Orthogonal Vectors

Table 20.6 shows the raw data from Table 20.5 rearranged into a single column (y) that represents the dependent variable (*number correct*). The three columns to the left of y show subjects, levels of A (*display type*), and levels of B (*duration*). The Helmert vectors used to code A and B and their interaction (AB) are shown in their respective columns in Table 20.6. Because there are only two levels of A, we need only one vector to code it (a_w); remember, the subscript w stands for *weight*. Because there are three levels of B, we need two vectors to code it (b_{w_1} and b_{w_2}).

TABLE 20.6			■ Partitioning Variance in the Means of a 2×3 Between-Subjects Design[a]					
				A	**B**		**AB**	
			y	a_w	b_{w_1}	b_{w_2}	$a_w b_{w_1}$	$a_w b_{w_2}$
s_1	a_1	b_1	40	1	2	0	2	0
s_2	a_1	b_1	24	1	2	0	2	0
s_3	a_1	b_2	37	1	−1	1	−1	1
s_4	a_1	b_2	29	1	−1	1	−1	1
s_5	a_1	b_3	62	1	−1	−1	−1	−1
s_6	a_1	b_3	52	1	−1	−1	−1	−1
s_7	a_2	b_1	37	−1	2	0	−2	0
s_8	a_2	b_1	25	−1	2	0	−2	0
s_9	a_2	b_2	90	−1	−1	1	1	−1
s_{10}	a_2	b_2	80	−1	−1	1	1	−1
s_{11}	a_2	b_3	88	−1	−1	−1	1	1
s_{12}	a_2	b_3	84	−1	−1	−1	1	1
ss_{total}	7236		r^2	.295	.420	.043	.159	.037
			$\hat{\eta}^2$	0.295	0.463		0.195	

[a]a_1 = Xs in Ls; a_2 = Ls in Xs; b_1, b_2, and b_3 are durations 50, 100, and 150, respectively.

The vectors that code for the interaction of A and B are the products of vectors that code the main effects of A (a_w) and B (b_{w_1} and b_{w_2}). In this case, the product vectors are $a_w b_{w_1}$ and $a_w b_{w_2}$. The five vectors shown under columns A, B, and AB in Table 20.6 capture all of the variability in our dependent variable (y) attributable to treatment.

To demonstrate that our five vectors properly decompose the variability in y, we will first compute the squared correlation between each vector and y. These squared correlations (r^2) are shown at the bottom of Table 20.6; they are .295, .420, .043, .159, and .037. The sum of these five $\hat{\eta}^2$ values is $\hat{\eta}^2_{\text{treatment}} = .953$, which represents the proportion of ss_{total} explained by treatment. Therefore, the proportion of ss_{total} *not* explained by treatment is $\hat{\eta}^2_{\text{error}} = 1 - \hat{\eta}^2_{\text{treatment}} = 1 - .953 = .047$.

Previously we determined that $ss_{\text{treatment}} = 6896$ and $ss_{\text{error}} = 340$. We can now show this differently by multiplying ss_{total} by $\hat{\eta}^2_{\text{treatment}}$ and $\hat{\eta}^2_{\text{error}}$. When we do so, we find that

$$ss_{\text{treatment}} = \hat{\eta}^2_{\text{treatment}} * ss_{\text{total}} = .953 * 7236 = 6896, \text{ and}$$

$$ss_{\text{error}} = \hat{\eta}^2_{\text{error}} * ss_{\text{total}} = .047 * 7236 = 340.$$

From this, we see that our five orthogonal vectors have properly decomposed the variability in our dependent variable into sources associated with treatment and error.

The next step is to decompose $ss_{\text{treatment}}$ into components associated with A, B, and AB.

- Only one vector is required to code A. Therefore, the proportion of ss_{total} explained by the main effect of A is $\hat{\eta}^2_a = .295$.
- For the main effect of B, we must sum $\hat{\eta}^2_{b_{w_1}}$ and $\hat{\eta}^2_{b_{w_2}}$, which yields $\hat{\eta}^2_b = .420 + .043 = .463$. Therefore, the proportion of ss_{total} explained by the main effect of B is $\hat{\eta}^2_b = .463$.
- For the AB interaction, we must sum $\hat{\eta}^2_{a_w b_{w_1}}$ and $\hat{\eta}^2_{a_w b_{w_2}}$, which yields $\hat{\eta}^2_{ab} = .159 + .037 = .195$. Therefore, the proportion of ss_{total} explained by the AB interaction is $\hat{\eta}^2_{ab} = .195$.

Table 20.7 summarizes the $\hat{\eta}^2$ values associated with each of the five orthogonal vectors, along with the omnibus effects of B and AB, which involve summing $\hat{\eta}^2$ values as we've just seen. Table 20.7 also shows the degrees of freedom associated with each source.

- For the main effect of A, we require $k_a - 1 = 2 - 1 = 1$ vector, so $df_a = 1$.
- For the main effect of B, we require $k_b - 1 = 3 - 1 = 2$ vectors, so $df_b = 2$.
- For the AB interaction, we require $(k_a - 1)(k_b - 1) = 1*2 = 2$ vectors, so $df_{ab} = 2$.
- Because we require $n*k_a*k_b - 1$ vectors to explain all variability in our 12 scores, $df_{\text{total}} = 11$.
- Because five of the available vectors have been "used up" to explain treatment variability, $df_{\text{total}} - df_{\text{treatment}} = 11 - 5 = 6$ vectors are required to explain the remaining variability. Therefore, $df_{\text{error}} = 6$.

Significance Tests for Main Effects and Interactions

Significance tests can be carried out for the omnibus main effects and interactions by computing F-ratios as we did in Chapters 16 through 19. For each source of variance, we compute the F-statistic as

$$F_{\text{obs}} = \frac{ss_{\text{source}}}{ss_{\text{error}}} \frac{df_{\text{error}}}{df_{\text{source}}},$$

TABLE 20.7 ■ Source Table for a 2 × 3 Between-Subjects Design[a]

Source	$\hat{\eta}^2_{\text{source}}$	$\hat{\eta}^2_{\text{error}}$	ss_{source}	df_1	df_2	F	p	$\hat{\eta}^2_p$
A	0.295	0.047	2133.333	1	6	37.647	.001	0.863
B	0.463	0.047	3350.000	2	6	29.559	.001	0.908
b_{w_1}	0.420	0.047	3037.500	1	6	53.603	.000	0.899
b_{w_2}	0.043	0.047	312.500	1	6	5.515	.057	0.479
AB	0.195	0.047	1412.667	2	6	12.465	.007	0.806
$a_w\, b_{w_1}$	0.159	0.047	1148.167	1	6	20.262	.004	0.772
$a_w\, b_{w_2}$	0.037	0.047	264.500	1	6	4.668	.074	0.438
Error		0.047	340.000					

[a] $a_1 = $ Xs in Ls; $a_2 = $ Ls in Xs; b_1, b_2, and b_3 are durations 50, 100, and 150, respectively.

or (equivalently)

$$F_{obs} = \frac{\hat{\eta}^2_{source}}{\hat{\eta}^2_{error}} \frac{df_{error}}{df_{source}}.$$

We can use the F-table or the **F.DIST.RT** function in Excel to determine the p-value associated with each of these F-statistics.

Using the information from Table 20.7, the F-ratios for A, B, and AB are computed as follows:

$$F_a = \frac{\hat{\eta}^2_a}{\hat{\eta}^2_{error}} \frac{df_{error}}{df_a} = \frac{.295}{.047} \frac{6}{1} = 37.65,$$

$$F_b = \frac{\hat{\eta}^2_b}{\hat{\eta}^2_{error}} \frac{df_{error}}{df_b} = \frac{.463}{.047} \frac{6}{2} = 29.56, \text{ and}$$

$$F_{ab} = \frac{\hat{\eta}^2_{ab}}{\hat{\eta}^2_{error}} \frac{df_{error}}{df_{ab}} = \frac{.195}{.047} \frac{6}{2} = 12.46.$$

Note that the values of $\hat{\eta}^2$ used in these calculations were of higher precision than the three-decimal-place precision shown in the above equations. Once the F-ratios have been computed, they can be tested for statistical significance. Table 20.7 shows that the F-ratios for A, B, and AB are associated with p-values of .001, .001, and .007, respectively. That is, both of the main effects and the interaction are statistically significant, assuming the standard $\alpha = .05$ criterion.

Table 20.7 also shows $\hat{\eta}^2_p$ for the omnibus effects. Remember that $\hat{\eta}^2_p$ can be computed in two ways:

$$\hat{\eta}^2_p = \frac{\hat{\eta}^2_{source}}{\hat{\eta}^2_{source} + \hat{\eta}^2_{error}} = \frac{F_{obs}}{F_{obs} + df_{error}/df_{source}}$$

We noted in Chapter 19 that $\hat{\eta}^2_p$ is a kind of effect size that is often reported but rarely interpreted. The main value of $\hat{\eta}^2_p$ is to remind us that a small p-value can arise from a large sample size.

Significance Tests for Contrasts

In Chapter 18, considerable space was devoted to showing that each vector contributing to partitioning the variance in an omnibus effect can be tested for statistical significance. This is also true in two-factor designs. Table 20.7 also shows that each of the five vectors that account for treatment effects can be tested for statistical significance, and we can compute $\hat{\eta}^2_p$ for each. The F-ratios for these five vectors were computed as follows:

$$F_a = \frac{\hat{\eta}^2_{a_w}}{\hat{\eta}^2_{error}} \frac{df_{error}}{df_{a_w}} = \frac{.295}{.047} \frac{6}{1} = 37.65,$$

$$F_{b_{w1}} = \frac{\hat{\eta}^2_{b_{w1}}}{\hat{\eta}^2_{error}} \frac{df_{error}}{df_{b_{w1}}} = \frac{.420}{.047} \frac{6}{1} = 53.603,$$

$$F_{b_{w2}} = \frac{\hat{\eta}^2_{b_{w2}}}{\hat{\eta}^2_{error}} \frac{df_{error}}{df_{b_{w2}}} = \frac{.043}{.047} \frac{6}{1} = 5.52,$$

$$F_{a_w b_{w1}} = \frac{\hat{\eta}^2_{a_w b_{w1}}}{\hat{\eta}^2_{error}} \frac{df_{error}}{df_{a_w b_{w1}}} = \frac{.159}{.047} \frac{6}{1} = 20.26, \text{ and}$$

$$F_{a_w b_{w1}} = \frac{\hat{\eta}^2_{a_w b_{w1}}}{\hat{\eta}^2_{error}} \frac{df_{error}}{df_{a_w b_{w1}}} = \frac{.037}{.047} \frac{6}{1} = 4.67.$$

If chosen with a research question in mind, the vectors underlying omnibus effects can be treated as contrasts that answer relevant questions. In this example, we've used Helmert vectors because they are easy to generate. In real research, we would choose vectors/ contrasts to answer relevant questions. These could be trends, or contrasts constructed by the researcher. Although we won't dwell on this point here, it is extremely important. Researchers are typically less interested in omnibus effects than in specific contrasts between means. With properly constructed contrasts we can go directly to the questions of interest and bypass the analysis of omnibus effects.

The Two-Factor Between-Subjects ANOVA in SPSS

To conduct this analysis in SPSS, we enter the scores for our dependent variable in a single column, as shown in the left column (NCorrect) in Figure 20.8. For each independent variable, there is a column of numbers that associates scores on the dependent variable with levels of the independent variable. These are shown in columns D_Type and Duration in Figure 20.8. The analysis is conducted through the Analyze→General Linear Model→Univariate dialog as shown in Figure 20.8.

The Univariate . . . dialog is shown in Figure 20.9. The variable NCorrect has been specified as the Dependent Variable and D_type and Duration have been specified as the Fixed Factors. The Contrasts . . . and Post Hoc . . . dialogs allow us to choose from the range of contrasts described in Chapter 19. We will not specify contrasts in this analysis, but we will ask SPSS to report estimates of effect size, $\hat{\eta}^2_p$ through the Options . . . dialog. When the analysis has been set up, we click OK to proceed.

FIGURE 20.8 ■ Two-Factor, Between-Subjects

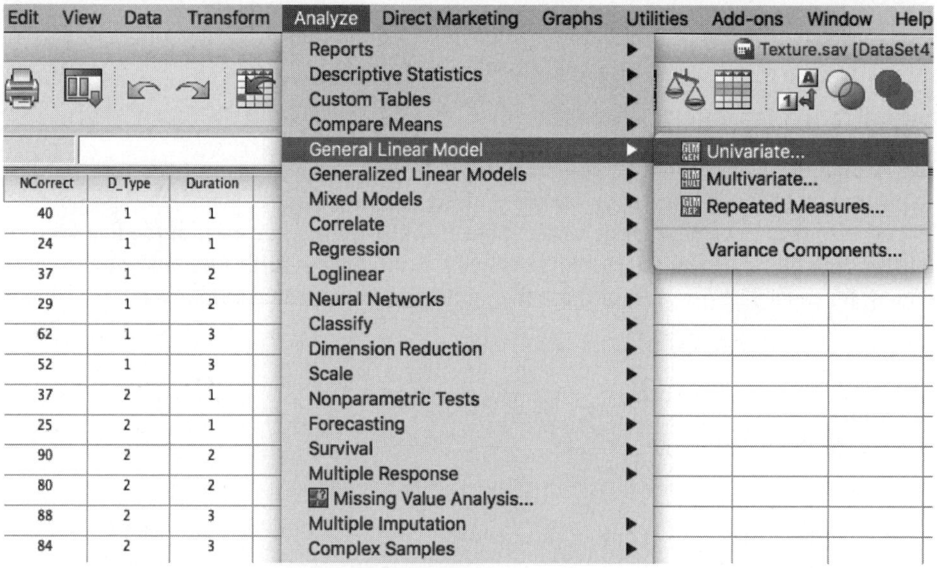

Setting up a 2 × 3 between-subjects ANOVA in SPSS.

The results of this analysis are shown in Figure 20.10. The four lines of interest are those labeled D_Type, Duration, D_Type*Duration, and Error. These are the sources of variance for the two main effects and interactions that we have been working with. If you compare the sums of squares, degrees of freedom, F-ratios, and $\hat{\eta}_p^2$ values with those computed above (and shown in Table 20.7), you will find that they are the same. Figure 20.10 shows that there is a statistically significant main effect of D_Type, $F_{1,6} = 37.65, p = .001$; a statistically significant main effect of Duration, $F_{2,6} = 29.56, p = .0001$; and a statistically significant interaction between D_Type and Duration, $F_{2,6} = 12.47, p = .007$.

There is additional information in Figure 20.10 that relates to ANOVA conducted as a regression analysis. This information is not particularly useful but deserves a brief comment because it can cause confusion. The F-ratio reported for the so-called Corrected Model is derived from the total variability explained by A, B, and AB. Table 20.7 shows that $\hat{\eta}_a^2 = .295$, $\hat{\eta}_b^2 = .463$, and $\hat{\eta}_{ab}^2 = .195$. We saw earlier that the sum of these three $\hat{\eta}^2$ values is $\hat{\eta}_{treatment}^2 = .953$. At the bottom of Figure 20.10, SPSS reports $\hat{\eta}_{treatment}^2$ as $R^2 = .953$, because that is the proportion for total variance captured by our predictors.

The sum of squares associated with the Corrected Model (first line in Figure 20.10) is $R^2 * ss_{total} = .953*7236 = 6896$, ignoring rounding error. (This is the quantity we've called $ss_{treatment}$.) Because there are 5 degrees of freedom associated with R^2 (which we will call $df_{regression}$) and 6 associated with the remaining error variability, the F-ratio computed from R^2 is

$$F = \frac{R^2}{1 - R^2} \frac{df_{error}}{df_{regression}} = \frac{.953}{.047} \frac{6}{5} = 24.34,$$

as shown in the row labeled Corrected Model in Figure 20.10.

FIGURE 20.9 ■ **The** Univariate **Dialog**

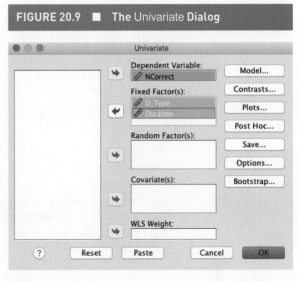

The Univariate dialog for the between-subjects ANOVA in SPSS.

FIGURE 20.10 ■ **SPSS Output, 2 × 3 Between**

Tests of Between-Subjects Effects

Dependent Variable: NCorrect

Source	Type III Sum of Squares	df	Mean Square	F	Sig.	Partial Eta Squared
Corrected Model	6896.000[a]	5	1379.200	24.339	.001	.953
Intercept	34992.000	1	34992.000	617.506	.000	.990
D_Type	2133.333	1	2133.333	37.647	.001	.863
Duration	3350.000	2	1675.000	29.559	.001	.908
D_Type * Duration	1412.667	2	706.333	12.465	.007	.806
Error	340.000	6	56.667			
Total	42228.000	12				
Corrected Total	7236.000	11				

a. R Squared = .953 (Adjusted R Squared = .914)

SPSS output for a 2 × 3 between-subjects ANOVA.

The quantity labeled *intercept* is also related to regression. It is a vestige of earlier times, and I have never seen it reported in a research article. If we were to use the five orthogonal vectors shown in Table 20.6 as predictors in a regression equation, that equation would be

$$\hat{y} = a + b_1\left(a_w\right) + b_2\left(b_{w_1}\right) + b_3\left(b_{w_2}\right) + b_4\left(a_w b_{w_1}\right) + b_5\left(a_w b_{w_2}\right)$$

$$= 54 - 13.33\left(a_w\right) - 11.25\left(b_{w_1}\right) - 6.25\left(b_{w_2}\right) + 6.92\left(a_w b_{w_1}\right) - 5.75\left(a_w b_{w_2}\right).$$

The intercept of this equation is 54, which is the mean of the 12 scores in our data set. When samples sizes are all the same, any orthogonal predictors that individually sum to 0 would produce a regression equation with an intercept equal to the mean of the scores in the data set. The sum of squares associated with *intercept* is $n_{\text{total}}*intercept^2$, where $n_{\text{total}} = \Sigma n_i$ is the number of scores in the data set. Therefore, $ss_{\text{intercept}} = 12*54^2 = 34{,}992$, as shown in Figure 20.10. The point of this analysis is to determine whether the intercept is statistically significantly different from 0. The *F*-ratio is computed as

$$F = \frac{n*intercept^2}{ss_{\text{error}}}df_{\text{error}} = \frac{12*54^2}{340}6 = 617.51.$$

For what it's worth, we can conclude that the mean of the 12 scores (54) is statistically significantly different from 0, $F_{2,6} = 617.51, p < .001$.

The SPSS notation regarding total sums of squares and total degrees of freedom can be confusing. The last row of Figure 20.10, Corrected Total, shows the quantity we call ss_{total}. Its value is 7236, with $df_{\text{total}} = 11$. The line above, labeled Total, shows $ss_{\text{total}} + ss_{\text{intercept}} = 7236 + 34{,}992 = 42{,}228$, with $df = df_{\text{total}} + df_{\text{intercept}} = 11 + 1 = 12$. You should be careful not to confuse Corrected Total with Total when reporting results in the SPSS output.

LEARNING CHECK 3

1. In a two-factor, completely between-subjects design, there are six levels of *A*, four levels of *B*, and 12 subjects in each condition of the experiment. $\hat{\eta}_a^2 = .1$, $\hat{\eta}_b^2 = .2$, and $\hat{\eta}_{ab}^2 = .1$

(a) How many degrees of freedom are associated with *A*, *B*, *AB*, and error?

(b) Compute F_{obs} and the associated *p*-value for each main effect and the interaction.

Answers

1. (a) $df_{\text{total}} = k_a*k_b*n - 1 = 6*4*12 - 1 = 287.$
$df_a = 6 - 1 = 5.$
$df_b = 4 - 1 = 3. \; df_{ab} = df_a*df_b = 5*3 = 15.$
$df_{\text{error}} = df_{\text{total}} - df_a - df_b - df_{ab} = 287 - 5 - 3 - 15$
$= 264.$

(b) To note, $\hat{\eta}_{\text{error}}^2 = 1 - \hat{\eta}_a^2 - \hat{\eta}_b^2 - \hat{\eta}_{ab}^2 = .6.$
$F_a = \hat{\eta}_a^2/\hat{\eta}_{\text{error}}^2 *df_{\text{error}}/df_a = .1/.6*264/5 = 8.80, p < .001.$
$F_b = \hat{\eta}_b^2/\hat{\eta}_{\text{error}}^2 *df_{\text{error}}/df_b = .2/.6*264/3 = 29.33, p < .001.$
$F_{ab} = \hat{\eta}_{ab}^2/\hat{\eta}_{\text{error}}^2 *df_{\text{error}}/df_{ab} = .1/.6*264/15 = 2.93, p < .001.$

THE TWO-FACTOR WITHIN-SUBJECTS DESIGN

We now turn our attention to the two-factor within-subjects design. In this design, all subjects are tested in all conditions of the experiment. We begin with an overview of the variance components of interest and then show how these are combined to conduct significance tests.

Sources of Variance

At the most general level, we can decompose ss_{total} into $ss_{treatment}$, $ss_{subjects}$, and $ss_{subjects*treatment}$, as shown in Figure 20.11. As in the one-factor within-subjects design, we measure variability attributable to subjects so that it can be ignored. (We first eliminated subject variability way back in Chapter 12.) Variability associated with treatment is of interest, of course, and as before we can extract components associated with A, B, and AB (i.e., ss_a, ss_b, and ss_{ab}).

To understand what follows, we will first look back to Chapter 19 and reconsider the quantity ss_{error}. In Chapter 19, the three sources of variance for the *one-factor within-subjects ANOVA* were (i) the main effect of A, (ii) subject variability, and (iii) whatever was left over once ss_a and ss_s had been accounted for. We didn't say very much about this residual variability; we just called it ss_{error}. However, had we coded subjects and conditions with orthogonal vectors, the products of these vectors would have provided a set of vectors that captured the interaction of subjects with treatment. The number of interaction vectors, $(k_a - 1)(n-1)$, is equal to the number of degrees of freedom associated with ss_{error}. In other words, in Chapter 19 ss_{error} represented the interaction of subjects with treatment.

In the Posner task example in Chapter 19, there were three levels of A and six subjects, resulting in 18 scores altogether. Therefore, $df_{total} = 18 - 1 = 17$. These 17 total degrees of freedom were made up of $df_a = 2$, $df_s = 5$, and $df_{error} = df_a df_s = 10$. As I noted in Chapter 19, it is not a coincidence that $df_{error} = df_a df_s$ for the one-factor, within-subjects design. Therefore, instead of calling the residual variance ss_{error}, we could have called it ss_{sa}. We used the term ss_{error} at that point because we didn't know about interactions.

We are now in a position to see that in the two-factor within-subjects ANOVA, the residual variability used to compute all F-statistics comes from the interaction of a source (main effect or interaction) with subjects. That is, each component of the overall treatment effect (i.e., A, B, and AB) can interact with subjects. There will thus be three new sums of squares to consider and we call these ss_{sa}, ss_{sb}, and ss_{sab}. Together these quantities make up $ss_{subjects \times treatment}$, as shown in Figure 20.11.

With our data set treated as a two-factor completely within-subjects design, we proceed much as we did in the two-factor between-subjects design by choosing vectors that code for main effects and interactions. Now, however, we have two main effects (A and B), subjects (S), and four interactions (AB, SA, SB, and SAB). We can compute a set of orthogonal vectors to code for each of the interactions by multiplying appropriate vectors as before.

The data and vectors for our analysis are shown in Table 20.8. As before, there are six conditions in the experiment, but unlike the between-subjects design, the same two subjects participated in all six conditions. Therefore, we still have 12 scores altogether, as

FIGURE 20.11 ■ Sources of Variance (Within)

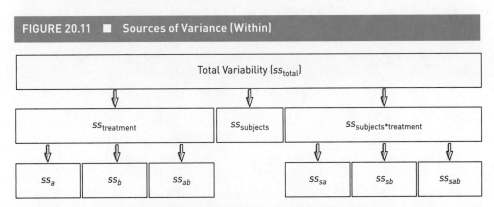

A schematic diagram illustrating how the total variability in our experiment (ss_{total}) can be partitioned into sources associated with the two main effects (ss_a and ss_b), the interaction (ss_{ab}), subjects (ss_s), and subjects by treatment interactions (ss_{sa}, ss_{sb}, and ss_{sab}).

shown in column "y." The three columns to the left of y show subjects and levels of A and B, just as in Table 20.6. The five columns to the right of y show the five vectors from the previous section that were used to code A, B and AB (i.e., a_w, b_{w_1}, b_{w_2}, $a_w b_{w_1}$, and $a_w b_{w_2}$). Again, these are the same vectors shown in Table 20.6.

The first new feature of this design is that we have a vector to code for subjects. In this case, we need only one vector, which is shown in column "S." This vector is orthogonal to the five vectors to its left that code for A, B, and AB. Our example is the simplest possible within-subjects design because there are only two subjects. In general, when we have n_s subjects, we require $n_s - 1$ Helmert vectors to capture variability attributable to subjects.

The remaining five columns in Table 20.8 show vectors coding for the interactions SA, SB, and SAB. Variability associated with interactions between subjects and factors can be extracted in the same way as interactions between factors. That is, we simply generate interaction vectors by multiplying those that code for subjects with those that code for a factor or an interaction between factors.

The vectors coding for the interactions of subjects with A, B, and AB are shown in columns "SA," "SB," and "SAB" in Table 20.8. Column SA codes the interaction between subjects and A (*display type*). Because both subjects and A have just two levels, only one vector is required to capture both and thus one product vector ($s_w a_w$) is required to capture their interaction.

The interaction between subjects and B (*duration*) requires two product vectors. The vector that codes for subjects is multiplied by the two that code for B. These two product vectors are shown in column SB (i.e., $s_w b_{w_1}$ and $s_w b_{w_2}$). Finally, subjects interact with the

TABLE 20.8 ■ Partitioning Variance in the Means of a 2 × 3 Within-Subjects Design[a]

			y	A a_w	B b_{w_1}	b_{w_2}	AB $a_w b_{w_1}$	$a_w b_{w_2}$	S s_w	SA $s_w a_w$	SB $s_w b_{w_1}$	$s_w b_{w_2}$	SAB $s_w a_w b_{w_1}$	$s_w a_w b_{w_2}$
s_1	a_1	b_1	40	1	2	0	2	0	1	1	2	0	2	0
s_2	a_1	b_1	24	1	2	0	2	0	−1	−1	−2	0	−2	0
s_1	a_1	b_2	37	1	−1	1	−1	1	1	1	−1	1	−1	1
s_2	a_1	b_2	29	1	−1	1	−1	1	−1	−1	1	−1	1	−1
s_1	a_1	b_3	62	1	−1	−1	−1	−1	1	1	−1	−1	−1	−1
s_2	a_1	b_3	52	1	−1	−1	−1	−1	−1	−1	1	1	1	1
s_1	a_2	b_1	37	−1	2	0	−2	0	1	−1	2	0	−2	0
s_2	a_2	b_1	25	−1	2	0	−2	0	−1	1	−2	0	2	0
s_1	a_2	b_2	90	−1	−1	1	1	−1	1	−1	−1	1	1	−1
s_2	a_2	b_2	80	−1	−1	1	1	−1	−1	1	1	−1	−1	1
s_1	a_2	b_3	88	−1	−1	−1	1	1	1	−1	−1	−1	1	1
s_2	a_2	b_3	84	−1	−1	−1	1	1	−1	1	1	1	−1	−1
			r^2	.295	.420	.043	.159	.037	.041	.001	.003	.000	.000	.001
s_{total}	7236		$\hat{\eta}^2$	0.295	0.463		0.195		0.041	0.001	0.004		0.001	

[a] a_1 = Xs in Ls; a_2 = Ls in Xs; b_1, b_2, and b_3 are durations 50, 100, and 150, respectively.

interaction between A and B. The two AB interaction vectors shown in column AB are multiplied by the subjects vector to produce the SAB vectors (i.e., $s_w a_w b_{w_1}$ and $s_w a_w b_{w_2}$).

The 11 orthogonal vectors in Table 20.8 explain all variability in the dependent variable in column y. The next to the last row of Table 20.8 shows the squared correlation of each vector with y. These 11 vectors are grouped to explain variability attributable to A, B, AB, S, SA, SB, and SAB. The last row in Table 20.8 shows the $\hat{\eta}^2$ value associated with each source.

Significance Tests for Main Effects and Interactions

To test for statistical significance in a within-subjects design, we compute an F-ratio for a source by dividing its associated $\hat{\eta}^2$ by the $\hat{\eta}^2$ associated with the interaction of the source with subjects. The following equation makes this explicit:

$$F_{obs} = \frac{\hat{\eta}^2_{source}}{\hat{\eta}^2_{subjects*source}} \frac{df_{subjects*source}}{df_{source}}.$$

(20.9)

Using the information from Table 20.8, the F-ratios for A, B, and AB can be computed as follows:

$$F_a = \frac{\hat{\eta}^2_a}{\hat{\eta}^2_{sa}} \frac{df_{sa}}{df_a} = \frac{.295}{.00074} \frac{1}{1} = 400$$

$$F_b = \frac{\hat{\eta}^2_b}{\hat{\eta}^2_{sb}} \frac{df_{sb}}{df_b} = \frac{.463}{.0036} \frac{2}{2} = 128.85, \text{ and}$$

$$F_{ab} = \frac{\hat{\eta}^2_{ab}}{\hat{\eta}^2_{sab}} \frac{df_{sab}}{df_{ab}} = \frac{.195}{.0012} \frac{2}{2} = 163.$$

Note that the values of $\hat{\eta}^2$ in these calculations used higher precision than shown. Once the F-ratios have been computed, they can be tested for statistical significance.

This analysis is shown in more detail in Table 20.9. To make the table easier to read, the next to the last row in Table 20.8 is duplicated at the top of Table 20.9 in the row labeled r^2. Above each r^2 is the vector to which it corresponds, and above the vectors are the identities of the omnibus sources to which they correspond.

Rows A, B, and AB in Table 20.9 show the same calculations we saw before. In addition, the sum of squares for each source is shown, along with its degrees of freedom, p-value, and $\hat{\eta}^2_p$ value. These values will be compared with those obtained when the analysis is conducted in SPSS.

Significance Tests for Contrasts

We've seen that the vectors that decompose the variability in an omnibus effect are themselves contrasts. Each of these contrasts can be tested for statistical significance in exactly the same way that we test the statistical significance of an omnibus effect using equation 20.9. Now, however, $\hat{\eta}^2_{source}$ corresponds to a single vector (e.g., b_{w_1}) and $\hat{\eta}^2_{subjects*source}$ is derived from the product of the vector and those coding for subjects.

In the general case, if vector b_{w_i} is one of the vectors coding for B, then $\hat{\eta}^2_{subjects*source}$ is the variance accounted for by all vectors $s_j b_{w_i}$, where j takes values from 1 to $n-1$ and n is the number of subjects. For example, if there were six subjects, then there would be five product vectors ($s_{w_1} b_{w_i}$, $s_{w_2} b_{w_i}$, $s_{w_3} b_{w_i}$, $s_{w_4} b_{w_i}$, and $s_{w_5} b_{w_i}$).

TABLE 20.9 ■ **Source Table for 2 × 3 Within-Subjects Design**[a][b]

	A	B		AB		S	SA	SB		SAB	
	a_w	b_{w_1}	b_{w_2}	$a_w b_{w_1}$	$a_w b_{w_2}$	s_w	$s_w a_w$	$s_w b_{w_1}$	$s_w b_{w_2}$	$s_w a_w b_{w_1}$	$s_w a_w b_{w_2}$
r^2	.2948	.4198	.0432	.1587	.0366	.0415	.0007	.0033	.0003	.0001	.0011

Source	$\hat{\eta}^2_{source}$	ss_{source}	Error	$\hat{\eta}^2_{error}$	ss_{error}	df_1	df_2	F	p	$\hat{\eta}^2_p$
A	0.295	2133.333	SA	0.0007	5.333	1	1	400.000	.032	0.998
B	0.463	3350.000	SB	0.0036	26.000	2	2	128.846	.008	0.992
b_{w_1}	0.420	3037.500	$s_w b_{w_1}$	0.0033	24.000	1	1	126.562	.056	0.992
b_{w_2}	0.043	312.500	$s_w b_{w_2}$	0.0003	2.000	1	1	156.250	.051	0.994
AB	0.195	1412.667	SAB	0.0012	8.667	2	2	163.000	.006	0.994
$a_w b_{w_1}$	0.159	1148.167	$s_w a_w b_{w_1}$	0.0001	0.667	1	1	1722.251	.015	0.999
$a_w b_{w_2}$	0.037	264.500	$s_w a_w b_{w_2}$	0.0011	8.000	1	1	33.063	.110	0.971

[a] a_1 = Xs in Ls; a_2 = Ls in Xs; b_1, b_2, and b_3 are durations 50, 100, and 150, respectively.
[b] ss_{total} = 7236.

In the present example, the vector coding the error term for b_{w_1} is the product of this vector and the subjects vector. (Because there are only two subjects in this experiment, there is only one vector coding for subjects, so the error term for b_{w_1} is $s_w b_{w_1}$.) Similarly, the vector coding the error term for b_{w_2} is $s_w b_{w_2}$.

Table 20.9 shows F-ratios, p, and $\hat{\eta}^2_p$ for the five contrasts coding for A, B, and AB. The F-ratios were computed as follows:

$$F_{a_w} = \frac{\hat{\eta}^2_a}{\hat{\eta}^2_{s_w a_w}} \frac{df_{s_w a_w}}{df_a} = \frac{.295}{.00074} \frac{1}{1} = 400,$$

$$F_{b_{w_1}} = \frac{\hat{\eta}^2_{b_{w_1}}}{\hat{\eta}^2_{s_w b_{w_1}}} \frac{df_{s_w b_{w_1}}}{df_{b_{w_1}}} = \frac{.420}{.0033} \frac{1}{1} = 126.562,$$

$$F_{b_{w_2}} = \frac{\hat{\eta}^2_{b_{w_2}}}{\hat{\eta}^2_{s_w b_{w_2}}} \frac{df_{s_w b_{w_2}}}{df_{b_{w_2}}} = \frac{.043}{.0003} \frac{1}{1} = 156.25,$$

$$F_{a_w b_{w_1}} = \frac{\hat{\eta}^2_{a_w b_{w_1}}}{\hat{\eta}^2_{s_w a_w b_{w_1}}} \frac{df_{s_w a_w b_{w_1}}}{df_{a_w b_{w_1}}} = \frac{.159}{.0001} \frac{1}{1} = 1722.251, \text{ and}$$

$$F_{a_w b_{w_2}} = \frac{\hat{\eta}^2_{a_w b_{w_2}}}{\hat{\eta}^2_{s_w a_w b_{w_2}}} \frac{df_{s_w a_w b_{w_2}}}{df_{a_w b_{w_2}}} = \frac{.037}{.0011} \frac{1}{1} = 33.063.$$

As discussed previously, these Helmert vectors/contrasts were not chosen with a research question in mind, but they could have been. When orthogonal vectors/contrasts are chosen with a research question in mind, these can be tested for statistical significance, so there is no need to compute omnibus effects.

The Two-Factor, Within-Subjects ANOVA in SPSS

To conduct a two-factor within-subjects ANOVA in SPSS, we use the Analyze→General Linear Model→Repeated Measures dialog shown in Figure 20.12. The organization of the data necessary for this analysis is shown in the portion of the Data View window to the left of the dialog. All data for each subject are entered on a single line. Because there are only two subjects in this example, there are only two lines of data. As with the one-factor within-subjects design, each column corresponds to a condition in the experiment. I have entered column names that correspond to the conditions of our example; a denotes *display type* (D_Type) and b denotes *duration* (Duration).

When Analyze→General Linear Model→Repeated Measures is chosen, the Repeated Measures Define Factor(s) dialog appears and asks the user to specify the Within-Subject Factor Names and the Number of Levels; see Figure 20.12. When a factor name and number of levels have been entered, we click Add and then either (i) add another factor name with its number of levels or (ii) click Define to associate columns in the data sheet with the conditions that have been defined. This is shown in the Repeated Measures dialog in Figure 20.13. The six columns in the data sheet have been associated with the appropriate factor levels in the dialog.

For a within-subjects analysis, SPSS will output a contrast analysis by default. The only choice we have here is which contrast analysis is applied. Therefore, I used the Contrasts . . . dialog to request Helmert contrasts so that we can compare the SPSS output with the results in Table 20.9. I also asked SPSS to report estimates of effect size, $\hat{\eta}_p^2$ through the Options . . . dialog.

Clicking   OK   initiates the analysis. The results of the omnibus tests of the main effects and interaction are shown in Figure 20.14a. Each source of variance is shown along with its error term. In all cases, the *F*-ratios and degrees of freedom in the Sphericity Assumed rows correspond to the values in the source table in Table 20.9. It is worth noting that there are some serious violations of the sphericity assumption in this analysis. The Greenhouse-Geisser correction is the same as the lower-bound correction. This means that the Greenhouse-Geisser correction is the maximum possible correction. Furthermore, because of the small sample size, the Huynh-Feldt correction could not be computed, so entries in these rows are filled with periods (".") to indicate this failure.

The results of the Helmert contrasts are shown in Figure 20.14b. All results are the same as those shown in Table 20.9. The D_Type contrast (first row) shows that there is a statistically significant difference in average accuracy for the two display types, $F_{1,1} = 400$, $p = .032$.

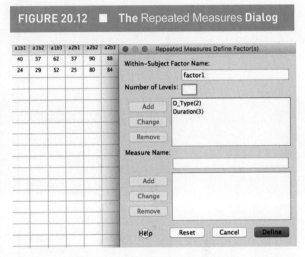

FIGURE 20.12 ■ **The** Repeated Measures **Dialog**

Defining factors and their levels in a within-subjects ANOVA in SPSS.

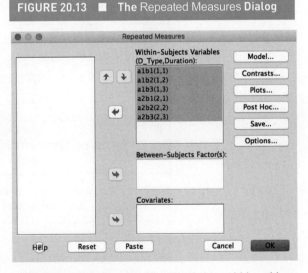

FIGURE 20.13 ■ **The** Repeated Measures **Dialog**

Associating variables with factor levels in a within-subjects ANOVA in SPSS.

FIGURE 20.14 ■ SPSS Output, 2 × 3 Within

(a)

Tests of Within-Subjects Effects

Measure: MEASURE_1

Source		Type III Sum of Squares	df	Mean Square	F	Sig.	Partial Eta Squared
D_Type	Sphericity Assumed	2133.333	1	2133.333	400.000	.032	.998
	Greenhouse-Geisser	2133.333	1.000	2133.333	400.000	.032	.998
	Huynh–Feldt	2133.333	.	.	.	.	.998
	Lower-bound	2133.333	1.000	2133.333	400.000	.032	.998
Error(D_Type)	Sphericity Assumed	5.333	1	5.333			
	Greenhouse-Geisser	5.333	1.000	5.333			
	Huynh–Feldt	5.333	.	.			
	Lower-bound	5.333	1.000	5.333			
Duration	Sphericity Assumed	3350.000	2	1675.000	128.846	.008	.992
	Greenhouse-Geisser	3350.000	1.000	3350.000	128.846	.056	.992
	Huynh–Feldt	3350.000	.	.	.	.	.992
	Lower-bound	3350.000	1.000	3350.000	128.846	.056	.992
Error(Duration)	Sphericity Assumed	26.000	2	13.000			
	Greenhouse-Geisser	26.000	1.000	26.000			
	Huynh–Feldt	26.000	.	.			
	Lower-bound	26.000	1.000	26.000			
D_Type * Duration	Sphericity Assumed	1412.667	2	706.333	163.000	.006	.994
	Greenhouse-Geisser	1412.667	1.000	1412.667	163.000	.050	.994
	Huynh–Feldt	1412.667	.	.	.	.	.994
	Lower-bound	1412.667	1.000	1412.667	163.000	.050	.994
Error (D_Type*Duration)	Sphericity Assumed	8.667	2	4.333			
	Greenhouse-Geisser	8.667	1.000	8.667			
	Huynh–Feldt	8.667	.	.			
	Lower-bound	8.667	1.000	8.667			

(b)

Tests of Within-Subjects Contrasts

Measure: MEASURE_1

Source	D_Type	Duration	Type III Sum of Squares	df	Mean Square	F	Sig.	Partial Eta Squared
D_Type	Level 1 vs. Level 2		1422.222	1	1422.222	400.000	.032	.998
Error(D_Type)	Level 1 vs. Level 2		3.556	1	3.556			
Duration		Level 1 vs. Later	2278.125	1	2278.125	126.563	.056	.992
		Level 2 vs. Level 3	312.500	1	312.500	156.250	.051	.994
Error(Duration)		Level 1 vs. Later	18.000	1	18.000			
		Level 2 vs. Level 3	2.000	1	2.000			
D_Type * Duration	Level 1 vs. Level 2	Level 1 vs. Later	3444.500	1	3444.500	1722.250	.015	.999
		Level 2 vs. Level 3	1058.000	1	1058.000	33.063	.110	.971
Error (D_Type*Duration)	Level 1 vs. Level 2	Level 1 vs. Later	2.000	1	2.000			
		Level 2 vs. Level 3	32.000	1	32.000			

(a) Main effects and interactions for a 2 × 3 within-subjects ANOVA. (b) Orthogonal Helmert contrasts for a 2 × 3 within-subjects ANOVA.

Because there are only two levels of D_Type, this contrast is exactly the same as the main effect of D_Type in Table 20.9.

The two Helmert contrasts for Duration were [2, −1, −1], which compares b_1 with b_2 and b_3, and [0, 1, −1], which compares b_2 with b_3. Both contrasts average over levels of A

(D_Type). The analyses of these two contrasts are shown in the row labeled Duration and are expressed as Level 1 vs. Later and Level 2 vs. Level 3, respectively. The two F-ratios are $F_{1,1}$ = 126.56, p = .056, and $F_{1,1}$ = 156.25, p = .051, respectively. Both of these contrasts indicate that, averaged over the two levels of D_Type, there is a change in accuracy with duration.

*Interaction Vectors

The last two contrasts in Figure 20.14b are interaction contrasts. We will have more to say about such contrasts in Chapter 21, but I will give a brief explanation here. For this explanation, we will think about our vectors as unscaled contrasts and sum the products of weights and means, rather than compute squared correlations.

Table 20.10 shows the six means from our sample data in the row labeled "Means (m)." The three rows above these means show the conditions of the experiment to which the means correspond. The row labeled "Weights (w)" shows the interaction weights that result from multiplying the vector coding for the main effect of A (a_w) by the first vector coding the main effect of B (b_{w_1}). This interaction vector asks whether $2(a_1b_1) - a_1b_2 - a_1b_3$ is different from $2(a_2b_1) - a_2b_2 - a_2b_3$. According to the null hypothesis, the products of w and μ in the population sum to 0.

Note that the first three coefficients in $a_wb_{w_1}$ are [2, −1, −1], and the second three are just sign-reversed [−2, 1, 1]. When we sum the products of w and m, we obtain

$$2(32) - 33 - 57 - 2(31) + 85 + 86 = 83.$$

This contrast is definitely not 0, showing that $2(a_1b_1) - a_1b_2 - a_1b_3$ is different from $2(a_2b_1) - a_2b_2 - a_2b_3$. In our example, this difference has a p = .015 under the null hypothesis; i.e., the contrast is statistically significant.

The second set of means in Table 20.10 [Means (h)] provides an example that is consistent with the null hypothesis. The means for Ls in Xs at the three durations are just 10 units greater than the corresponding means for Xs in Ls at the three durations. When we sum the products of w and h, we obtain

$$2(32) - 33 - 57 - 2(42) + 43 + 67 = 0.$$

A simpler way of looking at this question is to scale the weights so that they form explicit comparisons between means. When this is done, we note that for the means from our example (m), $a_1b_1 - (a_1b_2 + a_1b_3)/2 = -13$ and $a_2b_1 - (a_2b_2 + a_2b_3)/2 = -54.5$. There is a big

TABLE 20.10 ■ Interaction Weights

Display type	Xs in Ls			Ls in Xs		
Duration	50	100	150	50	100	150
Conditions	a_1b_1	a_1b_2	a_1b_3	a_2b_1	a_2b_2	a_2b_3
Means (m)	32	33	57	31	85	86
Means (h)	32	33	57	42	43	67
Weights (w) ($a_wb_{w_1}$)	2	−1	−1	−2	1	1
a_w	1	1	1	−1	−1	−1
b_{w_1}	2	−1	−1	2	−1	−1

difference between these two differences (–13 and –54.5). For the second set of means (h), $a_1b_1 - (a_1b_2 + a_1b_3)/2 = -13$ and $a_2b_1 - (a_2b_2 + a_2b_3)/2 = -13$. There is no difference between these two differences (–13 and –13).

Interaction vectors can get quite complex. However, we do not have to be chained to a main effects and interactions style of ANOVA and hence interaction vectors. In Chapter 21, we will see that we have essentially complete freedom to compute whatever contrasts make sense in a given research context.

LEARNING CHECK 4

1. In a two-factor completely within-subjects design, there are five levels of A, eight levels of B, and 15 subjects in each condition of the experiment. The results show that $\hat{\eta}_a^2 = .1$, $\hat{\eta}_b^2 = .2$, $\hat{\eta}_{ab}^2 = .1$, $\hat{\eta}_s^2 = .2$, $\hat{\eta}_{sa}^2 = .05$, and $\hat{\eta}_{sb}^2 = .05$.

(a) How many degrees of freedom are associated with A, B, AB, S, SA, SB, and SAB?

(b) Compute F_{obs} and the associated p-value for each main effect and the interaction.

Answers

1. (a) $df_{total} = k_a * k_b * n_s - 1 = 5*8*15 - 1 = 599$.
   $df_a = 5 - 1 = 4$.
   $df_b = 8 - 1 = 7$.
   $df_{ab} = df_a * df_b = 4*7 = 28$.
   $df_s = 15 - 1 = 14$.
   $df_{sa} = df_s * df_a = 14*4 = 56$. $df_{sb} = df_s * df_b = 14*7 = 98$.

   $df_{sb} = df_s * df_{ab} = 14*28 = 392$.

   (b) To note, $\hat{\eta}_{sab}^2 = 1 - \hat{\eta}_a^2 - \hat{\eta}_b^2 - \hat{\eta}_{ab}^2 - \hat{\eta}_s^2 - \hat{\eta}_{sa}^2 - \hat{\eta}_{sb}^2 = .3$.
   $F_a = \hat{\eta}_a^2 / \hat{\eta}_{sa}^2 * df_{sa}/df_a = .1/.05*56/4 = 28, p < .001$.
   $F_b = \hat{\eta}_b^2 / \hat{\eta}_{sb}^2 * df_{sb}/df_b = .2/.05*98/7 = 56, p < .001$.
   $F_{ab} = \hat{\eta}_{ab}^2 / \hat{\eta}_{sab}^2 * df_{sab}/df_{ab} = .1/.3*392/28 = 4.67, p < .001$.

THE TWO-FACTOR MIXED DESIGN

We will now approach the analysis of our small data set by treating it as a mixed design. The variable *display type* (A) will be the between-subjects factor and *duration* (B) will be the within-subjects factor. We begin with an overview of the variance components of interest and then show how these are combined to conduct significance tests.

Sources of Variance

In the mixed design, we decompose the total variability in the dependent variable into three general components, as shown in Figure 20.15. As always, the first component is treatment variability, which is decomposed into ss_a, ss_b, and ss_{ab}. The second component is the sum of squares associated with the variability among subjects within each level of the between-subjects factor. We call this "subjects within A" and denote it $ss_{subjects/a}$, or $ss_{s/a}$ for short. The third component is the interaction of the within-subjects factor (B) with "subjects within A"; we denote this quantity $ss_{b*s/a}$. These two new sources of variance require some explanation.

Let's first consider the concept of variability attributable to subjects within A. For this example, it will help to have a concrete example to think about. Table 20.11 is very similar to Table 20.5, but information specific to the mixed design has been added. The second column in the table identifies the subject that produced the data in that row. The sixth column ("Subject means") shows the mean score for each subject averaged over levels of B. When we test the statistical significance of the between-subjects factor (A), we compute the sum of squares associated with the subject means, within each level of A.

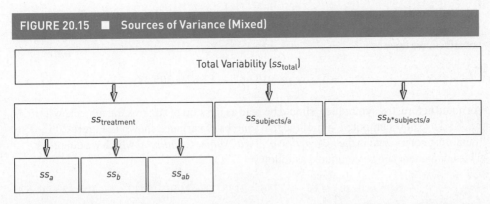

FIGURE 20.15   ■   Sources of Variance (Mixed)

A schematic diagram illustrating how the variability in a mixed ANOVA is partitioned. The between-subjects factor is denoted with a and the within-subjects factor is denoted with b. The total variability (ss_{total}) is partitioned into sources associated with the two main effects (ss_a and ss_b), the interaction (ss_{ab}), subjects within a ($ss_{s/a}$), and the interaction of b with subjects within a ($ss_{b*s/a}$).

TABLE 20.11   ■   Raw Data From a 2 × 3 Texture Discrimination Experiment[a]

		b_1	b_2	b_3	Subject Means	Factor A Means
a_1	s_1	40	37	62	46.33	40.67
a_1	s_2	24	29	52	35.00	
a_2	s_3	37	90	88	71.67	67.34
a_2	s_4	25	80	84	63.00	
		Factor B Means				
		31.5	59.0	71.5		
		AB Interaction Means				
		32.0	33.0	57.0		
		31.0	85.0	86.0		

[a] a_1 = Xs in Ls; a_2 = Ls in Xs; b_1, b_2 and b_3 are durations 50, 100, and 150, respectively.

If you think back to the one-factor within-subjects design, we computed sums of squares for subjects as

$$ss_s = k_b \sum_{j=1}^{n} (m_{s_j} - m_y)^2.$$

(See equation 19.4 in Chapter 19.) For the first level of A (a_1), this equation produces

$$ss_{s(a_1)} = k_b \sum_{j=1}^{n} (m_{s_j} - m_{a_1})^2 = 3[(46.33 - 40.67)^2 + (35 - 40.67)^2] = 192.67.$$

In this equation, m_{s_j} is the mean score for the jth subject in a given level of A, which is a_1 in this case. Therefore, ss_{s/a_1} is the sum of squared deviations for subject means in the first level of A. For the second level of A (a_2), the equation produces $ss_{s/a_2} = 112.67$. Therefore, we have computed sums of squares for subjects within the two levels of A. As noted before, we refer to these variance components as *subjects within A*. When these sums of squares are added, we have

$$ss_{s/a} = \sum_{i=1}^{k_a} ss_{s/a_i} = 192.67 + 112.67 = 305.33.$$

As noted before, the symbol $ss_{s/a}$ should be read as the sum of squares for subjects within A.

We have now computed four of the five sources of variance shown in Figure 20.15. The remaining component is *the interaction of B with subjects within A*, which we denote $ss_{b*s/a}$. This component can be computed as follows:

$$ss_{b*s/a} = ss_{\text{total}} - ss_a - ss_b - ss_{ab} - ss_{s/a} = 7236 - 2133.33 - 3350 - 1412.67 - 305.33 = 34.67.$$

The notion of the "interaction of B with subjects within A" will be made more explicit in the following paragraphs.

We will now use orthogonal contrasts to analyze our small data set as a mixed design, with *display type* (Xs in Ls versus Ls in Xs) as the between-subjects factor and *duration* as the within-subjects factor. There are two subjects in each of the between-subjects conditions for a total of four subjects altogether, as shown in the leftmost column of Table 20.12. The next eight columns are identical to those of Tables 20.6 and 20.8.

A new feature in this design is that we require two vectors to account for subject variability (one vector for each group). One might think that we should require three

TABLE 20.12 ■ Partitioning Variance in the Means of a 2 × 3 Mixed Design[a]

		y	A a_w	B b_{w_1}	B b_{w_2}	AB $a_w b_{w_1}$	AB $a_w b_{w_2}$	S/A s/a_{w_1}	S/A s/a_{w_2}	B*S/A $b_{w_1}s/a_{w_1}$	B*S/A $b_{w_2}s/a_{w_1}$	B*S/A $b_{w_1}s/a_{w_2}$	B*S/A $b_{w_2}s/a_{w_2}$
s_1 a_1 b_1		40	1	2	0	2	0	1	0	2	0	0	0
s_2 a_1 b_1		24	1	2	0	2	0	−1	0	−2	0	0	0
s_1 a_1 b_2		37	1	−1	1	−1	1	1	0	−1	1	0	0
s_2 a_1 b_2		29	1	−1	1	−1	1	−1	0	1	−1	0	0
s_1 a_1 b_3		62	1	−1	−1	−1	−1	1	0	−1	−1	0	0
s_2 a_1 b_3		52	1	−1	−1	−1	−1	−1	0	1	1	0	0
s_3 a_2 b_1		37	−1	2	0	−2	0	0	1	0	0	2	0
s_4 a_2 b_1		25	−1	2	0	−2	0	0	−1	0	0	−2	0
s_4 a_2 b_2		90	−1	−1	1	1	−1	0	1	0	0	−1	1
s_4 a_2 b_2		80	−1	−1	1	1	−1	0	−1	0	0	1	−1
s_4 a_2 b_3		88	−1	−1	−1	1	1	0	1	0	0	−1	−1
s_4 a_2 b_3		84	−1	−1	−1	1	1	0	−1	0	0	1	1
	r^2		.295	.420	.043	.159	.037	.027	.016	.00226	.00014	.00115	.00124
	$\hat{\eta}^2$		0.295	0.463		0.195		0.042		0.0048			

[a] a_1 = Xs in Ls; a_2 = Ls in Xs; b_1, b_2, and b_3 are durations 50, 100, and 150, respectively.

vectors to account for the variability of four subjects. However, the between-factor A in this example is like an independent-groups t-test. As with an independent-groups t-test, we compute the sum of squares within each group. With $n = 2$ scores in each group, there would be $n + n - 2$ degrees of freedom within groups (i.e., df_{within} would be 2). Equivalently, $df_1 = 1$ and $df_2 = 1$, so $df_{within} = df_1 + df_2 = 2$.

More generally, when there are k_a groups with n participants in each group, we would need $k_a * (n - 1)$ vectors to capture subject variability within groups (i.e., $n - 1$ vectors within each of the k_a groups). In our example, there are $k_a = 2$ groups and $n = 2$ subjects per group, requiring $k_a * (n - 1) = 2 * 1 = 2$ vectors to capture subject variability within groups.

To restrict the computation of within-group variability to a single group, we place zeroes in column entries that correspond to subjects from other groups. In our example, the numbers in s/a_1 (subjects within a_1) encode subject variability within group 1 (all entries below the horizontal line are zeroes) and the numbers in s/a_2 (subjects within a_1) encode subject variability within group 2 (all entries above the horizontal line are zeroes).

The remaining source of variability is the interaction of B with subjects within A. To capture this variability, we multiply all pairs of vectors in these two sources. For our example, the result is the four vectors shown in columns $b_{w_1}s/a_{w_1}$, $b_{w_2}s/a_{w_1}$, $b_{w_1}s/a_{w_2}$, and $b_{w_2}s/a_{w_2}$ in Table 20.12. Together these four vectors (B^*S/A) account for the interaction of B with subjects within A.

By computing the squared correlation between each of the 11 vectors and y, we account for the total variability in y. As in our two previous analyses, we find that $\hat{\eta}_a^2 = .295$, $\hat{\eta}_b^2 = .463$, and $\hat{\eta}_{ab}^2 = .195$. Table 20.12 shows that $\hat{\eta}_{s/a}^2 = .027 + .016 = .042$. Finally, $\hat{\eta}_{b*s/a}^2 = 0.00226 + 0.00014 + 0.00115 + 0.00124 = .0048$.

Significance Tests for Main Effects and Interactions

Having determined the proportion of total variability associated with each source of variance, we can now compute F-statistics for the purpose of conducting significance tests. These calculations will be shown in Table 20.13. As with Table 20.9, the top part of Table 20.13 shows the squared correlation (r^2) of each vector with the dependent variable y for reference.

The F-statistic for the between-subjects factor is computed as

$$F_{obs} = \frac{\hat{\eta}_a^2}{\hat{\eta}_{s/a}^2} \frac{df_{s/a}}{df_a},$$

where $\hat{\eta}_{s/a}^2$ corresponds to the proportion of variance accounted for by subjects within A (i.e., $ss_{s/a}$). For our example, the F for the main effect of A is

$$F_a = \frac{\hat{\eta}_a^2}{\hat{\eta}_{s/a}^2} \frac{df_{s/a}}{df_a} = \frac{.295}{.042} \frac{2}{1} = 13.97,$$

as shown in row A in Table 20.13.

The error term for both the main effect of B and the AB interaction is computed from the products of the vectors that code for B and the vectors that code for S/A. These four product vectors are shown under "$B*S/A$" in Table 20.12. We've seen that the proportion of variability in y explained by these four vectors is $\hat{\eta}_{b*s/a}^2 = .0048$. Therefore, the F-ratio for the main effect of B is computed as

$$F_b = \frac{\hat{\eta}_b^2}{\hat{\eta}_{b*s/a}^2} \frac{df_{b*s/a}}{df_b} = \frac{.463}{.0048} \frac{4}{2} = 193.27,$$

and the F-ratio for the AB interaction is computed as

$$F_{ab} = \frac{\hat{\eta}^2_{ab}}{\hat{\eta}^2_{b*s/a}} \frac{df_{b*s/a}}{df_b} = \frac{.195}{.0048} \frac{4}{2} = 81.5.$$

These two F-ratios are shown in rows B and AB in Table 20.13.

The source table in Table 20.13 summarizes the calculations for the omnibus effects that we just stepped through. In addition to the degrees freedom and $\hat{\eta}^2$ values for each source, Table 20.13 also shows the sums of squares, p, and $\hat{\eta}^2_p$ values associated with each omnibus test.

Significance Tests for Contrasts

Significance tests for contrasts are computed in a manner parallel to the main effects and interaction of which they are a part. For the between-subjects contrasts, averaged over levels of the within-subjects variable, $\hat{\eta}^2$ for the contrast is divided by $\hat{\eta}^2$ for subjects within A. For the within-subjects contrasts, averaged over levels of the between-subjects variable, $\hat{\eta}^2$ for the contrast is divided by the $\hat{\eta}^2$ representing the contrast's interaction with subjects. The interaction contrast $\hat{\eta}^2$ values are divided by the component of $B*S/A$ that includes the interaction contrast vector in question.

Table 20.13 shows F-ratios, p, and $\hat{\eta}^2_p$ for the five contrasts coding for A, B, and AB. The F-ratios were computed as

$$F_{a_w} = \frac{\hat{\eta}^2_a}{\hat{\eta}^2_{s/a}} \frac{df_{s/a}}{df_a} = \frac{.295}{.042} \frac{2}{1} = 13.974$$

$$F_{b_{w_1}} = \frac{\hat{\eta}^2_{b_{w_1}}}{\hat{\eta}^2_{b_{w_1}s/a}} \frac{df_{b_{w_1}s/a}}{df_{b_{w_1}}} = \frac{.420}{.0034} \frac{2}{1} = 246.284$$

TABLE 20.13 ■ Source Table for 2 × 3 Mixed Design[a]

	A	B		AB		S/A		B*S/A			
	a_w	b_{w_1}	b_{w_2}	$a_w b_{w_1}$	$a_w b_{w_2}$	s/a_{w_1}	s/a_{w_2}	$b_{w_1}s/a_{w_1}$	$b_{w_2}s/a_{w_1}$	$b_{w_1}s/a_{w_2}$	$b_{w_2}s/a_{w_2}$
r^2	.295	.420	.043	.159	.037	.027	.016	.00226	.00014	.00115	.00124
Source	$\hat{\eta}^2_{source}$	ss_{source}	Error	$\hat{\eta}^2_{error}$	ss_{error}	df_1	df_2	F	p	$\hat{\eta}^2_p$	
A	0.295	2133.333	S/A	0.0422	305.333	1	2	13.974	.065	0.875	
B	0.463	3350.000	B*S/A	0.0048	34.667	2	4	193.269	.000	0.990	
b_{w_1}	0.420	3037.500	$b_{w_1}*s/a$	0.0034	24.667	1	2	246.284	.004	0.992	
b_{w_2}	0.043	312.500	$b_{w_2}*s/a$	0.0014	10.000	1	2	62.500	.016	0.969	
AB	0.195	1412.667	B*S/A	0.0048	34.667	2	4	81.500	.001	0.976	
$a_w b_{w_1}$	0.159	1148.167	$b_{w_1}*s/a$	0.0034	24.667	1	2	93.095	.011	0.979	
$a_w b_{w_2}$	0.037	264.500	$b_{w_2}*s/a$	0.0014	10.000	1	2	52.900	.018	0.964	

[a] a_1 = Xs in Ls; a_2 = Ls in Xs; b_1, b_2, and b_3 are durations 50, 100, and 150, respectively.

$$F_{b_{w2}} = \frac{\hat{\eta}^2_{b_{w2}}}{\hat{\eta}^2_{b_{w2}s/a}} \frac{df_{b_{w2}s/a}}{df_{b_{w2}}} = \frac{.043}{.0014} \frac{2}{1} = 62.5$$

$$F_{a_wb_{w1}} = \frac{\hat{\eta}^2_{a_wb_{w1}}}{\hat{\eta}^2_{b_{w1}s/a}} \frac{df_{b_{w1}s/a}}{df_{a_wb_{w1}}} = \frac{.159}{.0034} \frac{2}{1} = 93.095, \text{ and}$$

$$F_{a_wb_{w2}} = \frac{\hat{\eta}^2_{a_wb_{w2}}}{\hat{\eta}^2_{b_{w2}s/a}} \frac{df_{b_{w2}s/a}}{df_{a_wb_{w2}}} = \frac{.037}{.0014} \frac{2}{1} = 52.9.$$

The Two-Factor Mixed ANOVA in SPSS

As one might expect, the analysis of the two-factor mixed ANOVA in SPSS employs elements of the within- and between-subjects analyses we saw earlier. The upper-left portion of Figure 20.16 shows how data are entered. As with the within-subjects design, each line corresponds to a subject, and the first three columns correspond to levels of the within-subjects factor. The fourth column (D_Type) shows codes for the two groups. The analysis is conducted through the Analyze→General Linear Model→Repeated Measures dialog, as shown in Figure 20.16. The within-subjects factor has been named Duration and is defined to have three levels. The appropriate columns have been moved into the three entries in the Within-Subjects Variables (Duration): region. Below this, variable D_Type has been specified as the Between-Subjects Factor. Although not shown, I used the Contrasts . . . dialog to request Helmert contrasts for the within-subjects effects, and I requested estimates of effect size ($\hat{\eta}^2_p$) through the Options . . . dialog. When the setup is complete, click **OK** to initiate the analysis. The results are shown in Figure 20.17.

Figure 20.17a (Tests of Within-Subjects Effects) shows the omnibus analysis of the within-subjects parts of the mixed design. These include an analysis of the main effect of B and the AB interaction. The analysis shows the F-ratios for B and AB. Figure 20.17a also shows the degrees of freedom (i) unaltered (assuming sphericity has not been violated), (ii) corrected for violations of sphericity with the Greenhouse-Geisser and

FIGURE 20.16 ■ **The** Repeated Measures **Dialog**

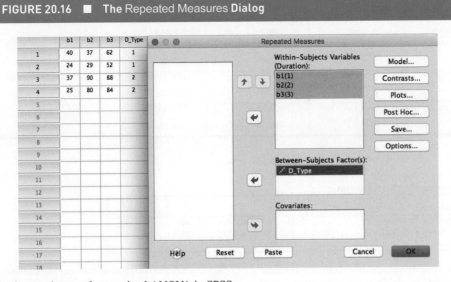

Setting up the two-factor mixed ANOVA in SPSS.

(iii) Huynh-Feldt methods, and (iv) the lower-bound correction. As always, the three correction factors vary in the degree to which they reduce the degrees of freedom associated with the within-subjects sources of variance. Because the F-ratios remain constant, the p-values increase with the degree to which the degrees of freedom have been reduced. The p-values for the cases in which sphericity is assumed are the same as those we computed in Table 20.13.

Figure 20.17c (Tests of Between-Subjects Effects) shows the between-subjects contrast. This is the single contrast that compares the difference between the means of the two display types averaged over durations. SPSS conducts the analysis of the between-subjects means as though the experiment were a one-factor between-subjects design. For this reason the sum of squares for the main effect of A is different from what is shown in Table 20.13, even though the F-ratio, p-values, and $\hat{\eta}_p^2$-values are exactly as we computed earlier.

FIGURE 20.17 ■ SPSS Output, 2 × 3 Mixed

(a)

Tests of Within-Subjects Effects

Measure: MEASURE_1

Source		Type III Sum of Squares	df	Mean Square	F	Sig.	Partial Eta Squared
Duration	Sphericity Assumed	3350.000	2	1675.000	193.269	.000	.990
	Greenhouse–Geisser	3350.000	1.600	2093.750	193.269	.000	.990
	Huynh–Feldt	3350.000	2.000	1675.000	193.269	.000	.990
	Lower-bound	3350.000	1.000	3350.000	193.269	.005	.990
Duration * D_Type	Sphericity Assumed	1412.667	2	706.333	81.500	.001	.976
	Greenhouse–Geisser	1412.667	1.600	882.917	81.500	.002	.976
	Huynh–Feldt	1412.667	2.000	706.333	81.500	.001	.976
	Lower-bound	1412.667	1.000	1412.667	81.500	.012	.976
Error(Duration)	Sphericity Assumed	34.667	4	8.667			
	Greenhouse–Geisser	34.667	3.200	10.833			
	Huynh–Feldt	34.667	4.000	8.667			
	Lower-bound	34.667	2.000	17.333			

(b)

Tests of Within-Subjects Contrasts

Measure: MEASURE_1

Source	Duration	Type III Sum of Squares	df	Mean Square	F	Sig.	Partial Eta Squared
Duration	Level 1 vs. Later	4556.250	1	4556.250	246.284	.004	.992
	Level 2 vs. Level 3	625.000	1	625.000	62.500	.016	.969
Duration * D_Type	Level 1 vs. Later	1722.250	1	1722.250	93.095	.011	.979
	Level 2 vs. Level 3	529.000	1	529.000	52.900	.018	.964
Error(Duration)	Level 1 vs. Later	37.000	2	18.500			
	Level 2 vs. Level 3	20.000	2	10.000			

(c)

Tests of Between-Subjects Effects

Measure: MEASURE_1
Transformed Variable: Average

Source	Type III Sum of Squares	df	Mean Square	F	Sig.	Partial Eta Squared
Intercept	11664.000	1	11664.000	229.205	.004	.991
D_Type	711.111	1	711.111	13.974	.065	.875
Error	101.778	2	50.889			

The one-within and one-between-subjects ANOVA in SPSS. (a) Tests of within-subjects effects. (b) Tests of within-subjects contrasts. (c) Tests of between-subjects effects.

Figure 20.17b (Tests of Within-Subjects Contrasts) shows the Helmert contrasts underlying the main effect of duration (Duration: Level 1 vs. Later, Level 2 vs. Level 3) and the interaction contrasts for the *display type* by *duration* interaction (Duration * D_Type: Level 1 vs. Later, Level 2 vs. Level 3). The interpretation of these contrasts was discussed in some detail in the previous section. For now we simply note that the degrees of freedom, *F*-ratios, *p*-values, and $\hat{\eta}_p^2$ values are exactly as calculated in Table 20.13. (The sums of squares are different because SPSS has computed intermediate quantities differently from what was shown above.)

Please note that in general, SPSS does not compute significance tests for each of the vectors (contrasts) making up the *AB* interaction. Instead, it will test the interaction of each specified within-subjects contrast with the overall effect of the between-subjects factors (*A*). This makes the mixed contrasts omnibus effects. Therefore, we won't pursue the SPSS approach to mixed contrasts because we will describe a more general approach in the next chapter.

LEARNING CHECK 5

1. In a two-factor mixed design where *A* is the between-subjects factor, there are four levels of *A*, six levels of *B*, and 10 subjects in each condition of the experiment. $\hat{\eta}_a^2 = .1$, $\hat{\eta}_b^2 = .2$, $\hat{\eta}_{ab}^2 = .1$, $\hat{\eta}_{s/a}^2 = .2$, and $\hat{\eta}_{b*s/a}^2 = .4$

(a) How many degrees of freedom are associated with *A, B, AB, S, S/A,* and *B*S/A*?

(b) Compute F_{obs} and the associated *p*-value for each main effect and the interaction.

Answers

1. (a) $df_{total} = k_a * k_b * n - 1 = 4 * 6 * 10 - 1 = 239$.

$df_a = 4 - 1 = 3$.

$df_b = 6 - 1 = 5$.

$df_{ab} = df_a * df_b = 3 * 5 = 15$.

$df_{s/a} = k_a(n-1) = 4 * 9 = 36$.

$df_{b*s/a} = df_b * df_{s/a} = 5 * 36 = 180$.

(b) $F_a = \hat{\eta}_a^2/\hat{\eta}_{s/a}^2 * df_{s/a}/df_a = .1/2 * 36/3 = 6, p = .002$.

$F_{ab} = \hat{\eta}_b^2/\hat{\eta}_{b*s/a}^2 * df_{b*s/a}/df_b = .2/.4 * 180/5 = 18$, $p < .001$.

$F_{ab} = \hat{\eta}_{ab}^2/\hat{\eta}_{b*s/a}^2 * df_{b*s/a}/df_b = .1/.4 * 180/15 = 3$, $p < .001$.

UNEQUAL SAMPLE SIZES AND MISSING DATA

The development of omnibus tests in this chapter as well as Chapters 18 and 19 assumed equal numbers of scores in each condition of the analysis. This situation is sometimes referred to as a *balanced* ANOVA, or an ANOVA with *equal n*. Unfortunately, this conceptual approach does not work well when there are missing data and thus different numbers of scores in each cell of the design. These situations are sometimes referred to as *unbalanced* ANOVAs, or ANOVAs with *unequal n*. In this section, we will discuss different reasons for missing data and then show how analyses are conducted when data are missing.

Reasons for Missing Data

Although researchers typically design experiments to have equal numbers of scores in each condition, the messiness of the real world may derail this plan when data are lost. Some cases of data loss are more serious than others. The least serious case occurs when data are lost more or less randomly. For example, a participant may fail to show up to an appointment because he was ill, was stuck in traffic, or simply forgot. In animal research, subjects may become ill (for various reasons) and are thus unable to complete the experiment. In cases like these, the analysis can be completed with the existing data, and we will describe how in subsequent subsections.

It is far more serious when the treatment itself is the cause of missing data. For example, subjects may drop out of one condition of an experiment because this condition required them to disclose information they wished to keep to themselves. Similarly, subjects may drop out of conditions that are too physically or mentally demanding. Or subjects in one condition of an experiment may withdraw before completion because they find it too boring, too anxiety provoking, or otherwise unpleasant. When we examine the effect of different drug doses on behavior in animal research, we may find that one drug dose is so high that it affects subjects' health and some may die or become unable to respond in the experiment.

In all of the cases just described, subjects were initially assigned randomly to conditions to avoid confounds with other variables. However, when something about the treatment in one condition causes subjects to drop out, those who remain in the problematic condition(s) may have different characteristics from those in the nonproblematic conditions. For example, those animals who survive a drug dose that kills other subjects may be constitutionally different. Similarly, if being asked to reveal sensitive personal information in an experiment causes some participants to withdraw, then those who remain are probably quite different from those who withdrew. In such cases, the initial random assignment of individuals to conditions is compromised, leading to confounds and the impossibility of meaningful interpretation. Consequently, no statistical solution will salvage the data.

Missing Data in Within-Subjects Designs

In this section and the next, we will assume that data are lost at random rather than as a systematic consequence of treatment.

Missing data can be problematic for within-subjects designs. Imagine a two-factor completely within-subjects experiment with $k_a = 3$, $k_b = 8$, and $n_s = 6$. There are 24 conditions in the experiment, with six scores in each cell, for a total of 144 scores. If one of these 144 scores is missing, then SPSS will conduct the ANOVA without the data for the subject with the missing score. In this case, this means that when one-144th of the data are missing, SPSS ignores one-sixth of the data set.

To illustrate how SPSS handles missing data, we will reconsider the one-factor ANOVA from Chapter 19. Figure 20.18a shows a portion of the Data View window for the

FIGURE 20.18 ■ Dealing With Missing Data

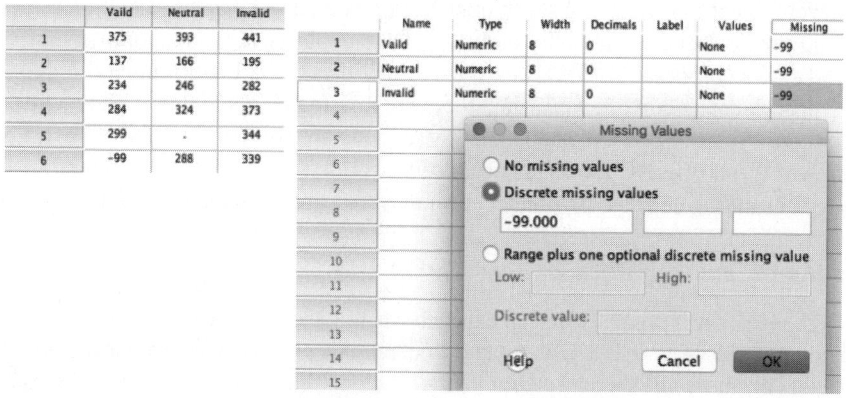

The Data View (a) and Variable View (b) of the Posner data set with one missing data point coded as a −99 by the user and a second cell left empty. The Missing Values panel shows how to convey codes for missing data to SPSS.

Posner data set, which now has two missing values. The cell in the fifth row of the Neutral condition is blank, and SPSS will recognize this as a missing data point when conducting an ANOVA. The sixth row of the Valid condition contains the value −99, which I typed in as a signal to SPSS that the score for this cell is missing. The Variable View window in Figure 20.18b shows how I established this signal. The dialog shown appears when the cell under Missing is clicked for a given variable. Clicking on the Discrete missing values radio button in the dialog allows the user to enter up to three numbers that will be recognized as codes for missing data. I've used −99 as one such code.

The Posner data with missing scores were submitted to a one-factor within-subjects ANOVA just as in Chapter 19. Through the Options . . . menu, I asked SPSS to show the descriptive statistics for the data submitted to the analysis. Figure 20.19a shows these descriptive statistics. Note that the means and standard deviations for all three variables have been computed from just four scores. This means that all data for subjects 5 and 6 were eliminated from the analysis because each has a missing score. Figure 20.19b shows the source table for the analysis. Note that the degrees of freedom for the error term is 6. This represents $(k_a - 1)(n_s - 1)$, where k_a is the number of levels of the independent variable and n_s is the number of subjects. Because of missing data, $n_s = 4$ so $(k_a - 1)(n_s - 1) = 2 * 3 = 6$.

To avoid eliminating all data for subjects with missing scores, it is possible to have SPSS estimate the missing scores. One way to do this is to set the missing score equal to the mean of the variable corresponding to a given column. The dialog in Figure 20.20a shows how this is done.

The dialog was reached through Transform→Replace Missing Values. . . . The drop-down list in the Method region of the Replace Missing Values . . . dialog shows five methods of filling in the missing score. The highlighted item (Series mean) will replace the missing score with the mean of the remaining scores in the variable. (The other options are useful when there is some inherent order in the scores, which is not the case here.) When a method has been selected, the variables on the left can be moved into the New Variable(s):

FIGURE 20.19 ■ Consequences of Missing Data

(a) **Descriptive Statistics**

	Mean	Std. Deviation	N
Vaild	257.50	99.299	4
Neutral	282.25	98.042	4
Invalid	322.75	107.221	4

(b) **Tests of Within-Subjects Effects**

Measure: MEASURE_1

Source		Type III Sum of Squares	df	Mean Square	F	Sig.
Cue_Type	Sphericity Assumed	8680.500	2	4340.250	47.219	.000
	Greenhouse–Geisser	8680.500	1.362	6373.470	47.219	.002
	Huynh–Feldt	8680.500	2.000	4340.250	47.219	.000
	Lower–bound	8680.500	1.000	8680.500	47.219	.006
Error(Cue_Type)	Sphericity Assumed	551.500	6	91.917		
	Greenhouse–Geisser	551.500	4.086	134.976		
	Huynh–Feldt	551.500	6.000	91.917		
	Lower–bound	551.500	3.000	183.833		

(a) Descriptive statistics for the four individuals with complete data sets. (b) The one-factor within-subjects ANOVA conducted on the data for the four subjects having complete data sets.

FIGURE 20.20 ■ Replacing Missing Data

(a)

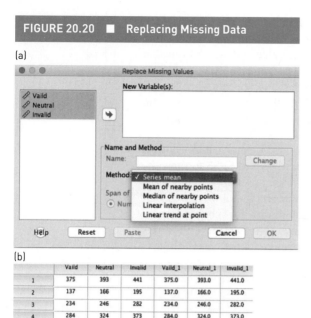

(b)

	Vaild	Neutral	Invalid	Vaild_1	Neutral_1	Invalid_1
1	375	393	441	375.0	393.0	441.0
2	137	166	195	137.0	166.0	195.0
3	234	246	282	234.0	246.0	282.0
4	284	324	373	284.0	324.0	373.0
5	299	.	344	299.0	283.4	344.0
6	-99	288	339	265.8	288.0	339.0

(a) The **Replace Missing Values** . . . dialog. Any missing data (coded by either –99 or an empty cell) will be replaced with the mean of the remaining scores for the relevant variable. (b) In this case, three new variables were created with the original data and the missing data filled in.

region, and SPSS will generate default names for new variables to store the altered variables. Once this is done, the [OK] button will become active and clicking it executes the recoding.

Figure 20.20b shows the result. The three new variables are called Valid_1, Neutral_1, and Invalid_1; these are default names generated by SPSS. The two scores that were missing have been replaced with the means of the remaining scores in the variable. The missing score in Valid was replaced with 265.8 in Valid_1, and the missing score in Neutral was replaced with 283.4 in Neutral_1. Now that the missing data have been estimated and stored in the new variables, the ANOVA can be run as before.

Of course, there is really no way of knowing whether these estimates of the missing numbers are valid. Therefore, when you write a report of the analysis, you must make clear to the reader that this is how the missing data were addressed.

Missing Data in Between-Subjects Designs

Missing data for between-subjects designs can lead to problems of interpretation. To illustrate the general problem of unequal cell sizes (unequal n), we will consider a situation in which one condition of an experiment has more scores than other conditions. You can imagine that all conditions started off with six scores in each cell, but four scores were lost for all conditions except a_2b_1. We won't fret about the plausibility of this situation. Instead we will focus on the consequences of unequal cell sizes.

Table 20.14 shows the data from our texture discrimination experiment with four extra scores added to condition a_2b_1 (Ls in Xs presented at 50 ms). Column m shows the means for the six conditions. Our interest is in the main effect of *display type* (A). Table 20.14

									Weighted	Unweighted
TABLE 20.14 ■ Computing Weighted and Unweighted Means for the Main Effect of A[a]										
A	B	s_1	s_2	s_3	s_4	s_5	s_6	m	Weighted Mean	Unweighted Mean
a_1	b_1	25	39					32	40.67	40.67
a_1	b_2	26	40					33		
a_1	b_3	48	66					57		
a_2	b_1	25	37	20	42	27	35	31	52.80	67.33
a_2	b_2	80	90					85		
a_2	b_3	90	82					86		
						$m_{a_2} - m_{a_1}$			12.13	26.67

[a] a_1 = Xs in Ls; a_2 = Ls in Xs; b_1, b_2, and b_3 are durations 50, 100, and 150, respectively.

shows two different ways to compute the mean values at a_1 (Xs in Ls) and a_2 (Ls in Xs) averaged over durations (B).

The so-called weighted mean is computed as *the arithmetic average of all scores in a given level of A*. The weighted average for Xs in Ls (averaged over durations) is $m_{a_1} = 40.67$ (based on the six scores within a_1) and the weighted average for Ls in Xs (averaged over durations) is $m_{a_2} = 52.8$ (based on the *10* scores within a_2). The difference between these two means is $m_{a_2} - m_{a_1} = 52.8 - 40.67 = 12.13$.

The so-called unweighted mean is computed as *the arithmetic average of condition means for a given level of A*. In this example, the unweighted mean for Xs in Ls (averaged over durations) is $(32 + 33 + 57)/3 = 40.67$. [Because there are the same number of scores at the three durations for a_1 (Xs in Ls), the weighted mean and the unweighted mean are exactly the same.] The interesting case is the unweighted mean for a_2 (Ls in Xs), which is $(31 + 85 + 86)/3 = 67.33$. The unweighted mean (67.33) is greater than the weighted mean (52.8). As a consequence, the difference between the two unweighted means, $m_{a_2} - m_{a_1} = 67.33 - 40.67 = 26.67$, is much greater than the difference between the weighted means. Why does this difference occur, and does it matter?

To understand why the weighted and unweighted means differ, we note that the scores at a_2b_1 (Ls in Xs at 50 ms) contribute more heavily to the weighted mean than the unweighted mean because there are simply more of them. In fact, we can see where the term *weighted mean* comes from when we calculate it as follows:

$$m_{a_2} = \frac{6(31) + 2(85) + 2(6)}{6 + 2 + 2} = 52.8.$$

In this case, we multiplied the unweighted mean at each of the three durations for a_2 (Ls in Xs) by the number of scores contributing to each mean, and we then divided this by the total number of scores in the three conditions (10). Clearly, the scores at 50 ms contributed more heavily to the weighted mean than the unweighted mean. (The unweighted mean might be better called the equally weighted mean; i.e., $[1(31) + 1(85) + 1(6)]/3 = 67.33$.)

So which of these two means is correct? In this case, the unweighted mean is obviously correct. The weighted mean *confounds* the effect of *duration* with the effect of *display type*.

One way to understand how unequal sample sizes can confound significance tests is to look at the data the way we have throughout the chapter. Column DV in Figure 20.21a shows the 16 scores from Table 20.14. Columns A and B show codes for conditions when submitting the data to a between-subjects ANOVA through Analyze→General Linear Model→Univariate. Finally, columns AW, BW1, BW2, AWBW1, and AWBW2 show the Helmert vectors used throughout the chapter to code for the main effects and interactions.

The consequence of unequal n is that the Helmert vectors are no longer orthogonal. For example, the correlation between a_w and b_{w_1} is .258. This means that the Helmert vectors are not partitioning the variability in the dependent variable into nonoverlapping pieces. Rather, some of the variability explained by a_w, for example, is also explained by b_{w_1}. We could say that some of the effect of B spills into the effect of A. Therefore, to "decontaminate" our effects, we need to partial out the effects of B from A and vice versa. We will see that this can be done using a hierarchical regression approach to ANOVA.

The source table in Figure 20.21 (Tests of Between-Subjects Effects) shows the results of the ANOVA conducted exactly as in the between-subjects design described earlier in Figure 20.9, and the output is in exactly the same *form* as in Figure 20.10. The F-ratios for A, B, and AB are 32.52, 28.58, and 11.79, respectively. We will see that these F-ratios were obtained from a sequence of regression analyses. The regression analyses compute R^2 associated with all vectors coding for A, B, and AB and then determine the change in explained variance (R^2_{change}) when the vectors coding one of these omnibus effects are

FIGURE 20.21 ■ Analysis of Unbalanced ANOVA

(a)

	DV	A	B	AW	BW1	BW2	AWBW1	AWBW2
1	27	1	1	-1	2	0	-2	0
2	37	1	1	-1	2	0	-2	0
3	26	1	2	-1	-1	1	1	-1
4	40	1	2	-1	-1	1	1	-1
5	48	1	3	-1	-1	-1	1	1
6	66	1	3	-1	-1	-1	1	1
7	25	2	1	1	2	0	2	0
8	37	2	1	1	2	0	2	0
9	20	2	1	1	2	0	2	0
10	42	2	1	1	2	0	2	0
11	27	2	1	1	2	0	2	0
12	35	2	1	1	2	0	2	0
13	80	2	2	1	-1	1	-1	1
14	90	2	2	1	-1	1	-1	1
15	90	2	3	1	-1	-1	-1	-1
16	82	2	3	1	-1	-1	-1	-1

(b)

Tests of Between-Subjects Effects

Dependent Variable: DV

Source	Type III Sum of Squares	df	Mean Square	F	Sig.
Corrected Model	8483.000[a]	5	1696.600	22.989	.000
Intercept	39366.000	1	39366.000	533.415	.000
A	2400.000	1	2400.000	32.520	.000
B	4217.857	2	2108.929	28.576	.000
A * B	1740.714	2	870.357	11.793	.002
Error	738.000	10	73.800		
Total	46470.000	16			
Corrected Total	9221.000	15			

a. R Squared = .920 (Adjusted R Squared = .880)

(c)

Model Summary

Model	R	R Square	Adjusted R Square	Std. Error of the Estimate	R Square Change	F Change	df1	df2	Sig. F Change
1	.959[a]	.920	.880	8.591	.920	22.989	5	10	.000
2	.812[b]	.660	.536	16.890	-.260	32.520	1	10	.000
3	.959[c]	.920	.880	8.591	.260	32.520	1	10	.000
4	.680[d]	.463	.328	20.322	-.457	28.576	2	10	.000
5	.959[e]	.920	.880	8.591	.457	28.576	2	10	.000
6	.855[f]	.731	.664	14.372	-.189	11.793	2	10	.002

a. Predictors: (Constant), AWBW2, AWBW1, BW2, AW, BW1
b. Predictors: (Constant), AWBW2, AWBW1, BW2, BW1
c. Predictors: (Constant), AWBW2, AWBW1, BW2, BW1, AW
d. Predictors: (Constant), AWBW2, AWBW1, AW
e. Predictors: (Constant), AWBW2, AWBW1, AW, BW2, BW1
f. Predictors: (Constant), AW, BW2, BW1

(a) Data from Table 20.14 and codes to be used for ANOVA (A and B) and codes to be used for regression (AW, BW1, BW2, AWBW1, AWBW2). (b) The ANOVA table shows the data submitted to a two-factor between-subjects ANOVA. (c) The table shows the same analysis performed as a regression. The F_{change} shown for models 2, 4, and 6 corresponds to the A, B, and AB effects in the ANOVA table.

---

removed. This R^2_{change} shows the proportion of total variability uniquely explained by the predictors that were removed.

The Model Summary in Figure 20.21c shows the result of a regression analysis with six steps. Steps 1, 3, and 5 are identical and show that $R^2_{treatment} = .92$ when all five predictors are included. Steps 2, 4, and 6 show the change in R^2 when we remove the vectors coding for A, B, and AB, respectively. Using the change in R^2, we can compute F_{change} essentially as in Chapter 16. The predictors in the model on each step are given in notes "a" through "f" at the bottom of the table. To make this explanation easier to follow, I will now describe each step.

Model 1.   On step 1, all five predictors (Helmert vectors) are part of the model. For the five-predictor model (Tests of Between-Subjects Effects), $R^2_{treatment} = .92$. This shows the total variability explained by the vectors coding for treatment.

Model 2.   On step 2, vector a_w has been removed from the model and this produces a change in explained variance of $R^2_{change} = -.26$. In Chapter 16, we computed R^2_{change} by adding predictors to a smaller model. Here we computed R^2_{change}

by subtracting predictors from a larger model. The two methods produce the same proportion of variance explained by the full model ($R^2_{treatment}$) and the same magnitude of R^2_{change}, so we can simply ignore the sign on R^2_{change}, and compute F_{change} as follows:

$$F_{change} = \frac{\left|R^2_{change}\right|}{1 - R^2_{treatment}} \frac{df_{error}}{df_{change}} = \frac{.26}{.08} \frac{10}{1} = 32.52.$$

Remember, $\left|R^2_{change}\right|$ is the proportion of variance explained by a_w with the effects of all other predictors (b_{w_1}, b_{w_2}, $a_w b_{w_1}$, and $a_w b_{w_2}$) partialed out. The computed $F_{change} = 32.52$ is the F-value associated with A in the source table.

Model 3.   On step 3, vector a_w was added back into the model, restoring $R^2_{treatment}$ to .92. (Note that adding a_w to the four-predictor model produces the same F_{change} as subtracting a_w from the five-predictor model.)

Model 4.   On step 4, vectors b_{w_1} and b_{w_2} were subtracted from the five-predictor model, producing a change in explained variance of $-.457$. From this, we can compute

$$F_{change} = \frac{\left|R^2_{change}\right|}{1 - R^2_{treatment}} \frac{df_{error}}{df_{change}} = \frac{.457}{.08} \frac{10}{2} = 28.58.$$

In this case, $\left|R^2_{change}\right|$ is the proportion of variance explained by b_{w_1} and b_{w_2} with the effects of all other predictors (a_w, $a_w b_{w_1}$, and $a_w b_{w_2}$) partialed out. $F_{change} = 28.58$ is the F-value associated with B in the source table.

Model 5.   On step 5, vectors b_{w_1} and b_{w_2} were added back into the model, restoring $R^2_{treatment}$ to .92 and producing $F_{change} = 28.58$.

Model 6.   Finally, on step 6, vectors $a_w b_{w_1}$ and $a_w b_{w_2}$ were subtracted from the five-predictor model, producing a change in explained variance of $R^2_{change} = -.189$. From this, we can compute

$$F_{change} = \frac{\left|R^2_{change}\right|}{1 - R^2_{treatment}} \frac{df_{error}}{df_{change}} = \frac{.189}{.08} \frac{10}{2} = 11.79.$$

In this case, $\left|R^2_{change}\right|$ is the proportion of variance explained by $a_w b_{w_1}$ and $a_w b_{w_2}$ with the effects of all other predictors (a_w, b_{w_1}, and b_{w_2}) partialed out. $F_{change} = 11.79$ is the F-value associated with AB in the source table.

Figure 20.21 shows in the row Corrected Total that the total sum of squares in the dependent variable is $ss_{total} = 9221$. We've just seen that $R^2_a = .26$, $R^2_b = .457$, and $R^2_{ab} = .198$. From these, we can compute

$$ss_a = R^2_a * ss_{total} = .26 * 9221 = 2400,$$

$$ss_b = R^2_b * ss_{total} = .457 * 9221 = 4217.857,$$

$$ss_{ab} = R_{ab}^2 * ss_{total} = .189 * 9221 = 1740.714, \text{ and}$$

$$ss_{error} = (1 - R_{treatment}^2) * ss_{total} = .08 * 9221 = 738.$$

(These computations used higher precision values of R_a^2, R_b^2, R_{ab}^2, and $R_{treatment}^2$ to obtain the sums of squares.) The sums of squares just computed are exactly those shown in the source table in Figure 20.21 in the column Type III Sum of Squares.

There are other methods of computing sums of squares for between-subjects designs with unequal sample sizes. These produce Type I and Type II sums of squares. However, if you are interested in omnibus effects when sample sizes are unequal, Type III sums of squares seem to me to be quite reasonable.

WHY BOTHER WITH MAIN EFFECTS AND INTERACTIONS?

ANOVA is arguably the most widely used method of significance testing in psychology. Reports of ANOVAs typically start with an account of which main effects and interactions are statistically significant. However, no analysis stops with reports of these omnibus effects. We've previously discussed why this is the case, but it bears repeating here.

Figure 20.22a shows the results of the texture discrimination experiment that we've been working with throughout the chapter. Figures 20.22b and 20.22c simply reorder the means at levels of *B* (*duration*). Compared with Figure 20.22a, the means at 50 and 100 ms have been switched in Figure 20.22b, and the means at 50 and 150 ms have been switched in Figure 20.22c. In the latter case, the order of means in Figures 20.22a and 20.22c are reversed. Each of these graphs conveys something completely different about the effects of *display type* and *duration* on discrimination accuracy. Despite these fundamental differences in meaning, the data in all three panels of Figure 20.22 would produce exactly the same main effects and interactions. No matter what design may have produced these data (two-factor between, within, or mixed), you would know nothing about the pattern of variation in the means simply by looking at the omnibus effects for the main effects and interactions in the source table.

Given the ambiguities of main effects and interactions in ANOVAs, why do researchers report them? One reason is that many researchers believe that statistically significant main

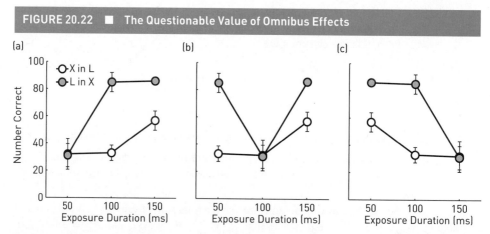

FIGURE 20.22 ■ The Questionable Value of Omnibus Effects

Data from Figure 20.6 plotted three ways. (a) This graph is identical to the plot in Figure 20.6 and shows results similar to published work. (b) The values at exposure durations 50 and 100 ms have been swapped. (c) The values at exposure durations 50 and 150 ms have been swapped.

effects or interactions are the permission slips required to assess specific contrasts. This belief may arise from researchers' tendency to rely on *post hoc* tests, for which there are legitimate reasons to require statistically significant main effects and interactions as gatekeepers. For example, imagine setting up an experiment to examine the combined effects of *A* and *B* with no clear idea of what these effects might be. Because of randomness, there may be large differences between the means of some conditions, even when the null hypothesis is true. If researchers tested only the largest differences found, then they would massively increase the number of Type I errors they report. Therefore, the requirement for a statistically significant main effect (or interaction) goes some way to controlling Type I errors.

Previous chapters have repeatedly made the case that it is better to approach an analysis with planned contrasts. This avoids the need to pass through gatekeeping omnibus tests in order to answer questions of interest. For example, Tables 20.7, 20.9, and 20.13 show how the contrasts underlying omnibus effects can be tested for statistical significance. We chose Helmert vectors to partition variance because they required no serious thinking. Although these contrasts did not address questions of interest to the research, other contrasts that do answer questions of interest could have been chosen.

Even in the case of Helmert vectors, we can imagine the following odd situation that results from giving omnibus effects more credit than they deserve. Let's say our texture discrimination experiment had been run using one of the three designs described in the previous sections. We might have wanted to know the degree to which accuracy differs between the 50-ms condition and the average of the 100- and 150-ms conditions. This question can be addressed using a contrast with weights [1, −0.5, −0.5]. Using the main effect of *duration* as a gatekeeper, we're only allowed to test this contrast after finding that *duration* is statistically significant. This is counterproductive because in Tables 20.7, 20.9, and 20.13, this contrast (b_{w_1}) is one of the orthogonal vectors used to code for the main effect of *duration*. Therefore, we could have skipped the entire gatekeeping role of the main effects and interactions analysis and tested this contrast (and others of interest) for statistical significance in the first place (i.e., *if* the contrast is meaningful and *if* significance testing is our thing).

A second point that must be emphasized is that contrasts reveal *how* means differ, whereas omnibus tests reveal (at best) *that* the means differ. Once again, researchers want to know how means differ, not just that they differ. Let's consider the main effect of *duration* in Figure 20.22. The means (averaged over *display type*) in Figures 20.22a through 20.22c are [31.5, 51, 79.5], [51, 31.5, 79.5], and [79.5, 51, 31.5], respectively. When we apply vector b_{w_1} as a contrast [1, −0.5, −0.5] to these three patterns, we obtain

$$c_1 = \sum[1, -0.5, -0.5]*[31.5, 51, 79.5] = -33.75,$$

$$c_2 = \sum[1, -0.5, -0.5]*[51, 31.5, 79.5] = -4.50, \text{ and}$$

$$c_3 = \sum[1, -0.5, -0.5]*[79.5, 51, 31.5] = 38.25.$$

These three patterns of means produce very different contrasts, each of which tells us how the means differ, whereas the omnibus effect is blind to these differences.

We will see in Chapter 21 that the plucky little *t*-test can do all the useful work of assessing the statistical significance of contrasts for us. Or, taking a more modern approach, we can place confidence intervals around computed contrasts, either in the original units of the dependent variable or in standardized units. As we've seen before, confidence intervals keep our focus on the data and don't lead us to think of *p*-values as measures of importance. So, in answer to the question that opened this section, I can see no good reason to bother with omnibus tests on main effects and interactions when contrasts are available.

SUMMARY

Two-factor designs involve two independent variables or factors. We examined designs with *two between-subjects factors*, *two within-subjects factors*, and *one between- and one-within-subjects factor*.

The term *main effect* refers to variability associated with one independent variable averaged over levels of the other variable. The term *interaction* refers to variability among means that is not associated with the main effects. The variability in the means of conditions in a two-factor design can be decomposed using orthogonal vectors. The variability attributable to a main effect can be captured with $k-1$ orthogonal vectors, where k is the number of levels of the factor. The number of orthogonal vectors required to encode a main effect defines the degrees of freedom associated with the effect. If there are k_a levels of A and k_b levels of B, then $(k_a - 1)(k_b - 1)$ vectors are required to capture variability attributable to the AB interaction. The vectors used to capture these three sources of variance (A, B, and AB) are exactly the same for between, within, and mixed designs.

The omnibus effects of A, B, and AB are tested with F-statistics. The error term for each test depends on the nature of the design. For a completely between-subjects design, all F-statistics are computed as

$$F_{obs} = \frac{\hat{\eta}^2_{source}}{\hat{\eta}^2_{error}} \frac{df_{error}}{df_{source}},$$

where source can be A, B, or AB, and $\hat{\eta}^2_{error}$ is the proportion of total variability not attributable to A, B, or AB.

For a completely within-subjects design, all F-statistics are computed as

$$F_{obs} = \frac{\hat{\eta}^2_{source}}{\hat{\eta}^2_{subjects*source}} \frac{df_{subjects*source}}{df_{source}},$$

where source can be A, B, or AB, and $\hat{\eta}^2_{subjects*source}$ is the proportion of total variability explained by the products of vectors coding for a source (A, B, or AB) and subjects; $k_s - 1$ vectors are required to capture variability attributable to subjects. These product vectors capture variability associated with *the interaction between subjects and a source*.

For a mixed design, the F-statistic for the omnibus effect of the between-groups factor (A) is computed as

$$F_{obs} = \frac{\hat{\eta}^2_{a}}{\hat{\eta}^2_{s/a}} \frac{df_{s/a}}{df_{a}},$$

where $\hat{\eta}^2_{s/a}$ is the proportion of total variability explained by vectors coding for subject variability within each level of A. The F-statistic for the omnibus effects B and AB is computed as

$$F_{obs} = \frac{\hat{\eta}^2_{source}}{\hat{\eta}^2_{b*s/a}} \frac{df_{b*s/a}}{df_{source}},$$

where source can be B or AB, and $\hat{\eta}^2_{b*s/a}$ is the proportion for variability associated with the interaction between B and subjects within A.

Missing data are a serious problem if the treatment caused data to be lost. In this case, there is no way to continue with the analysis because the effects of randomization have been defeated.

Although main effects and interactions are often reported in research papers, they typically do not answer the questions that motivated the research, because many different arrangements of the conditions means would produce the same F-statistics for these omnibus effects. Therefore, a better approach to data analysis is to compute simple and complex contrasts, or trends.

KEY TERMS

interaction   546   interaction vector   552   main effect   546

EXERCISES

Definitions and Concepts

1. What does the term *main effect* refer to?

2. What does the term *interaction* refer to?

3. Explain what each of the following quantities refers to: $\hat{\eta}^2_a$, $\hat{\eta}^2_b$, and $\hat{\eta}^2_{ab}$.

4. In a 5×4 ANOVA, how many degrees of freedom are associated with the AB interaction?

5. In a 2×2 ANOVA, how many degrees of freedom are associated with the AB interaction?

6. In a 6×3 within-subjects ANOVA, how many degrees of freedom are associated with the SAB interaction when there are 10 subjects?

7. What is an omnibus effect?

8. What are the four sources of variance in a completely between-subjects ANOVA?

9. What are the seven sources of variance in a completely within-subjects ANOVA?

10. What are the five sources of variance in a mixed ANOVA?

11. Why is an omnibus main effect of little value in an ANOVA?

12. Why is an omnibus interaction effect of little value in an ANOVA?

True or False

State whether the following statements are true or false.

13. The vectors coding for A, B, and AB in a mixed ANOVA are the same as those required in a between-subjects ANOVA.

14. A mixed ANOVA has three levels of the between-subjects factor and 10 levels of the within-subjects factor. If there are 12 subjects in each level of the between-subjects factor, then $df_{s/a} = 33$.

Calculations

15. If a two-factor within-subjects ANOVA has 10 levels of both independent variables, how many participants are in each condition if there are 100 subjects altogether?

16. In a two-factor, completely between-subjects design, there are five levels of A, three levels of B, and 10 subjects in each condition of the experiment. $\hat{\eta}_a^2 = .2$, $\hat{\eta}_b^2 = .05$, and $\hat{\eta}_{ab}^2 = .24$.

    (a) How many degrees of freedom are associated with A, B, AB, and error?

    (b) Compute F_{obs} and the associated p-value for each main effect and the interaction.

17. In a two-factor, completely within-subjects design, there are three levels of A, six levels of B, and 11 subjects in each condition of the experiment. $\hat{\eta}_a^2 = .2$, $\hat{\eta}_b^2 = .05$, $\hat{\eta}_{ab}^2 = .15$, $\hat{\eta}_s^2 = .25$, $\hat{\eta}_{sa}^2 = .05$, and $\hat{\eta}_{sb}^2 = .1$.

    (a) How many degrees of freedom are associated with A, B, AB, S, SA, SB, and SAB?

    (b) Compute F_{obs} and the associated p-value for each main effect and the interaction.

18. In a two-factor mixed design where A is the between-subjects factor, there are four levels of A,

six levels of B, and 10 subjects in each condition of the experiment. $\hat{\eta}_a^2 = .15$, $\hat{\eta}_b^2 = .25$, $\hat{\eta}_{ab}^2 = .1$, $\hat{\eta}_{s/a}^2 = .3$, and $\hat{\eta}_{b*s/a}^2 = .2$.

    (a) How many degrees of freedom are associated with A, B, AB, S/A, and $B*S/A$?

    (b) Compute F_{obs} and the associated p-value for each main effect and the interaction.

19. In a two-factor, completely between-subjects design, there are two levels of A, 12 levels of B, and 10 subjects in each condition of the experiment; $\hat{\eta}_a^2 = .05$, $\hat{\eta}_b^2 = .05$, and $\hat{\eta}_{ab}^2 = .025$.

    (a) How many degrees of freedom are associated with A, B, AB, and error?

    (b) Compute F_{obs} and the associated p-value for each main effect and the interaction.

20. In a two-factor, completely within-subjects design, there are 10 levels of A, three levels of B, and 10 subjects in each condition of the experiment. $\hat{\eta}_a^2 = .16$, $\hat{\eta}_b^2 = .04$, $\hat{\eta}_{ab}^2 = .18$, $\hat{\eta}_s^2 = .12$, $\hat{\eta}_{sa}^2 = .06$, and $\hat{\eta}_{sb}^2 = .14$.

    (a) How many degrees of freedom are associated with A, B, AB, S, SA, SB, and SAB?

    (b) Compute F_{obs} and the associated p-value for each main effect and the interaction.

21. In a two-factor, mixed design where A is the between-subjects factor, there are three levels of A, 12 levels of B, and 10 subjects in each condition of the experiment. $\hat{\eta}_a^2 = .08$, $\hat{\eta}_b^2 = .14$, $\hat{\eta}_{ab}^2 = .03$, $\hat{\eta}_{s/a}^2 = .2$, and $\hat{\eta}_{b*s/a}^2 = .55$.

    (a) How many degrees of freedom are associated with A, B, AB, S/A, and $B*S/A$?

    (b) Compute F_{obs} and the associated p-value for each main effect and the interaction.

22. The following data are from a two-factor ANOVA. Use SPSS to compute $\hat{\eta}_p^2$ for A, B, and AB, treating the design in the following ways:

    (a) as a completely between-subjects design;

    (b) as a completely within-subjects design, where the order of subjects is the same at the two levels of A; and

    (c) as a mixed design with A as the between-subjects factor. These data are available at study.sagepub.com/gurnsey.

	b_1	b_2	b_3	b_4
a_1	8	5	20	5
a_1	10	11	11	10
a_1	8	16	12	18
a_1	5	10	16	14
a_2	8	14	8	16
a_2	4	8	7	12
a_2	6	6	17	7
a_2	4	8	21	14

Scenarios

23. In the visual search task, participants are asked to report, as quickly as possible, the presence or absence of a target quickly in a variable number of nontarget items. For example, in the display to the right, the target is Я and the nontarget items are Rs. To make the task more difficult, target and nontarget items are randomly rotated. In this task, the dependent variable is the *time* (measured in milliseconds) required to press a button, after display onset, to indicate that the target is present or absent. The two independent variables are (i) *display type* (i.e., target present or target absent) and

(ii) *display size* (i.e., on each trial, there were 5, 10, 15, or 20 items in the display).

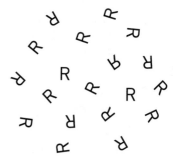

The data are presented in the following table, and are available in the VisualSearch.sav data file (available at study.sagepub.com/gurnsey). The researcher expects that (i) subjects will be able to report the presence of the target faster than its absence, (ii) response times will increase with display size, and (iii) the difference in reaction times for target-present and target-absent conditions will increase with display size. Normally the visual search task is run completely within subjects, so treat the data this way for the present exercise. Use Excel or SPSS to compute $\hat{\eta}_p^2$ for A = display type, B = display size, and AB = the display type by display size interaction.

	a_1[a]				a_2			
	b_1	b_2	b_3	b_4	b_1	b_2	b_3	b_4
	313	629	692	780	440	717	1041	1397
	675	598	871	903	778	1010	1419	1120
	560	702	722	548	539	1012	806	1731
	486	515	607	596	337	1049	1065	1290
	523	382	569	692	847	1202	1501	1388
	475	386	576	662	481	829	751	1295
	314	386	534	852	463	838	1453	1341
Mean	478	514	653	719	555	951	1148	1366
ss	101,752	105,818	83,068	103,614	209,938	162,216	583,986	206,628

[a] a_1 = target present; a_2 = target absent; b_1 = 5, b_2 = 10, b_3 = 15, and b_4 = 20.

24. There is some evidence that lab rats become addicted to drugs easily because there are raised in isolated cages and deprived of any other stimulation to compete with the drugs (Alexander, Coambs, & Hadaway, 1978). Bozarth, Murray, and Wise (1989) investigated the role of social isolation on rats' self-administration of heroin.

Bozarth et al. (1989) used 39 Long-Evans rats in their experiment. All rats were housed for the first 2 to 3 months of their lives in 24- × 24- × 9-inch cages with 25 rats per cage. Twenty-two of these rats were then housed for the next 6 to 8 weeks in small individual cages (about 7 × 10 × 7 inches), isolated from all interactions with other rats. The

iStock.com/fotografixx

remaining 17 rats were housed in larger cages (about 18 × 40 × 15 inches) with about 10 rats per cage. "[T]hese rats displayed normal play behavior, dominance struggles, and social grooming" (Bozarth et al., 1989, p. 904).

After about a week in their new cages, all rats were trained to bar press for a food reward. After 3 to 4 days of training, all rats underwent surgery to install catheters in their jugular veins. This allowed the researchers to later administer heroin directly into the rats' bloodstream.

For the next 5 weeks, each rat spent 2 hours a day in a small test cage (about 10 × 10 × 10 inches) with an operant lever (bar). Each time a rat pressed the bar, a small dose of heroin (0.1 mg/kg/injection) was delivered into the rat's bloodstream. Bozarth et al. (1989) measured the mean number of injections received per hour by each rat. Data that are qualitatively similar to those reported by Bozarth et al. (1989) are provided in the Bozarth.sav data file (available at study.sagepub.com/gurnsey). The means and standard deviations of these data are shown in the following table. Submit the data in the Excel file to a two-factor mixed ANOVA and report F, p, and $\hat{\eta}_p^2$ for A, B, and AB.

	a_1[a]					a_2				
	b_1	b_2	b_3	b_4	b_5	b_1	b_2	b_3	b_4	b_5
Mean	0.931	1.305	1.817	2.301	2.570	1.934	2.764	3.343	3.855	3.849
SD	1.124	1.092	1.121	1.174	2.086	1.403	2.031	1.998	1.794	2.104

[a] a_1 = animals housed in groups; a_2 = animals housed in isolation; b_1 = week 1, b_2 = week 2, b_3 = week 3, b_4 = week 4, and b_5 = week 5.

25. Glanzer and Cunitz (1966) investigated long- and short-term memory in a sample of 46 US Army personnel (all male). On each trial of the experiment, a subject was presented with a list of 15 one-syllable nouns. After presentation, the subject was given several minutes to recall as many words as possible from the list.

The subjects recalled the lists 0, 10, or 30 seconds after the last word was presented. For the 10- and 30-second delays, subjects counted from a randomly chosen number until the experimenter told

them to stop and write down as many words as he could remember from the list.

Each subject was presented with five lists at each of the three delays, and the researchers recorded the number of correct responses at each serial position (number correct at the first position, second position, up to the 15th position). The means and standard deviations of the 46 scores in each condition are shown in the following table. Raw data with characteristics similar to those reported by Glanzer and Cunitz (1966) are provided in the

Glanzer.sav data file (available at study.sagepub.com/gurnsey). Submit the data in the Excel file to a two-factor ANOVA and report F, p, and $\hat{\eta}_p^2$ for A, B, and AB, where A = delay (0, 10, and 30 seconds) and B represents serial position (1 to 15).

	a_1[a]														
	b_1	b_2	b_3	b_4	b_5	b_6	b_7	b_8	b_9	b_{10}	b_{11}	b_{12}	b_{13}	b_{14}	b_{15}
Mean	3.54	2.91	2.20	1.87	1.83	1.74	1.52	1.85	2.09	1.43	1.93	1.85	2.39	2.59	3.00
SD	0.98	1.15	1.17	1.19	1.02	1.02	1.01	1.09	1.17	1.07	1.14	1.05	1.22	1.24	1.14
	a_2														
Mean	3.57	2.85	2.20	2.00	1.83	1.54	1.50	1.76	1.59	1.72	1.85	1.74	1.91	1.87	2.04
SD	0.98	1.07	1.07	1.12	1.12	0.98	1.01	1.12	1.05	1.13	1.09	1.08	1.11	1.07	1.15
	a_3														
Mean	3.11	2.54	1.85	1.93	1.72	1.59	1.17	1.87	1.74	1.57	1.78	1.74	1.78	1.76	1.52
SD	1.12	1.11	1.13	1.08	1.11	1.17	0.88	1.09	1.08	1.05	1.07	1.08	1.05	1.08	0.98

[a] a_1 = immediate recall; a_2 = delayed recall (10), a_3 = delayed recall (30); $b_1 \ldots b_{15}$ = serial position in the list.

21 CONTRASTS IN TWO-FACTOR DESIGNS

INTRODUCTION

In Chapter 20, we saw how to conduct significance tests for omnibus effects in three versions of the two-factor ANOVA. We also pointed out that each of the orthogonal vectors coding for the omnibus effects could be tested for statistical significance. Throughout this book, we've seen that significance tests have many disadvantages that are largely avoided through the use of estimation. Therefore, in this chapter, we will see how to compute confidence intervals around contrasts that exist within two-factor ANOVAs.

The really good news about this chapter is that we are back on familiar ground. For the two-factor between- and within-subjects designs, we compute contrasts and confidence intervals exactly as in Chapters 18 and 19, respectively. For the mixed design, most contrasts of interest are computed exactly as in one-factor between- and within-subjects designs. There are a few special cases, but otherwise everything will feel very familiar.

AN OVERVIEW OF FIRST-ORDER AND SECOND-ORDER (INTERACTION) CONTRASTS

First-Order Contrasts

Before we begin, it's worth recalling that contrasts are computed as follows:

$$c = \sum_{i=1}^{k} w_i m_i \qquad (21.1)$$

when the weights (w) sum to zero and the absolute values of the weights ($|w|$) sum to 2. When the weights in the contrast are integers whose absolute values do not sum to 2, we divide the contrast $\sum_{i=1}^{k} w_i m_i$ by $\sum_{i=1}^{k} |w_i|/2$ to produce the same effect as equation 21.1. That is, we compute the contrast as

$$c = \frac{\sum_{i=1}^{k} w_i m_i}{\sum_{i=1}^{k} |w_i|/2}. \qquad (21.2)$$

The contrasts defined in equations 21.1 and 21.2 divide a collection of means into two groups associated with positive and negative weights. (Means associated with zeroes are not part of the contrast.) For simplicity, we will occasionally refer to these contrasts as *first-order contrasts*.

Second-Order (Interaction) Contrasts

In Chapter 20, we coded the omnibus interaction with $df_a df_b = (k_a-1)(k_b-1)$ product vectors, each of which produces an $\hat{\eta}^2_{source}$ that can be tested for statistical significance. Although we discussed interaction vectors briefly, we'll now have a closer look.

To explain interaction contrasts, we'll think about a 2×2 ANOVA, which involves four means ($m_{a_1b_1}$, $m_{a_1b_2}$, $m_{a_2b_1}$, and $m_{a_2b_2}$) as shown in Table 21.1 in the row labeled "Means." The row labeled "m" provides concrete values for these four means that we will use for illustration. The weights in a_w constitute a first-order contrast that compares the two means representing the main effect of A; i.e., $(20 + 30)/2 = 25$ and $(40 + 50)/2 = 45$. We compute this contrast as follows:

$$c = \frac{\sum_{i=1}^{k} a_{w_i} m_i}{\sum_{i=1}^{k} |a_{w_i}|/2} = \frac{-40}{2} = -20,$$

where k is the total number of means; i.e., $k_a k_b$. The weights in b_w constitute a first-order contrast that compares the two means representing the main effect of B; i.e., $(20 + 50)/2 = 35$ and $(20 + 40)/2 = 35$. We compute this contrast as follows:

$$c = \frac{\sum_{i=1}^{k} b_{w_i} m_i}{\sum_{i=1}^{k} |b_{w_i}|/2} = \frac{0}{2} = 0.$$

TABLE 21.1 ■ Interaction Contrasts

A	a_1		a_2	
B	b_1	b_2	b_1	b_2
Means	$m_{a_1b_1}$	$m_{a_1b_2}$	$m_{a_2b_1}$	$m_{a_2b_2}$
m	20	30	50	40
a_w	1	1	−1	−1
b_w	1	−1	1	−1
$a_w b_w$	1	−1	−1	1

Note: Interaction contrasts for a 2 x 2 design.

The interaction contrast is shown in the row labeled "$a_w b_w$." We will see that these weights necessarily compare four means, so the absolute values of the weights should sum to 4. Therefore, we compute the interaction contrast as follows:

$$c = \frac{\sum_{i=1}^{k} w_i m_i}{\sum_{i=1}^{k} |w_i|/4}. \tag{21.3}$$

As you can see, the only difference between equations 21.2 and 21.3 is that the sum of the absolute values of the weights is divided by 4 in the denominator of equation 21.3, rather than 2 in equation 21.2. Using equation 21.3, we can compute the contrast defined by the weights in $a_w b_w$ as follows:

$$c = \frac{\sum_{i=1}^{k} a_{w_i} b_{w_i} m_i}{\sum_{i=1}^{k} |a_{w_i} b_{w_i}|/4} = \frac{-20}{1} = -20.$$

We will now see that the interaction contrast tells us how a given effect in one variable changes across levels of the other variable. In our example, the interaction contrast assesses how different the effect of B is at the two levels of A. In other words, the contrast tells us how different ($m_{a_1b_1} - m_{a_1b_2}$) is from ($m_{a_2b_1} - m_{a_2b_2}$). To show that this is the case, when we substitute in the numbers from our example, we find that

$$c = (m_{a_1b_1} - m_{a_1b_2}) - (m_{a_2b_1} - m_{a_2b_2}) = (20-30)-(50-40) = -10-10 = -20,$$

A second-order (interaction) contrast is one that assesses how two first-order differences differ.

which is exactly what we got when we computed the contrast using equation 21.3. Because interaction contrasts assess the difference between two differences, we call them **second-order contrasts**.

To see the connection between $(m_{a_1b_1} - m_{a_1b_2}) - (m_{a_2b_1} - m_{a_2b_2})$ and contrast weights for a_wb_w, we can remove the parentheses to obtain the following:

$$c = (m_{a_1b_1} - m_{a_1b_2}) - (m_{a_2b_1} - m_{a_2b_2}) = m_{a_1b_1} - m_{a_1b_2} - m_{a_2b_1} + m_{a_2b_2}$$
$$= (1)m_{a_1b_1} + (-1)m_{a_1b_2} + (-1)m_{a_2b_1} + (1)m_{a_2b_2}.$$

That is, the weights (in blue) in the last line are exactly what is conveyed by contrast a_wb_w in Table 21.1.

Interestingly, interaction contrasts can also be seen as assessing how different the effect of A is at the two levels of B. In other words, the contrast tells us how different $(m_{a_1b_1} - m_{a_2b_1})$ is from $(m_{a_1b_2} - m_{a_2b_2})$. When we evaluate this contrast using the numbers in Table 21.1, we find that the following equation produces exactly the same contrast:

$$c = (m_{a_1b_1} - m_{a_2b_1}) - (m_{a_1b_2} - m_{a_2b_2}) = (20 - 50) - (30 - 40) = -30 + 10 = -20.$$

As before, we can remove the parentheses from this contrast to see its connection to a_wb_w:

$$c = (m_{a_1b_1} - m_{a_2b_1}) - (m_{a_1b_2} - m_{a_2b_2}) = m_{a_1b_1} - m_{a_2b_1} - m_{a_1b_2} + m_{a_2b_2}$$
$$= (1)m_{a_1b_1} + (-1)m_{a_2b_1} + (-1)m_{a_1b_2} + (1)m_{a_2b_2}.$$

Although the order of means is different from those in the previous example, each mean gets the weight defined in contrast a_wb_w. So again, second-order (interaction) contrasts compute a difference between two differences. In one case, the difference is the effect of B at two levels of A; in the second case, the difference is between the effect of A at two levels of B.

THE TWO-FACTOR, BETWEEN-SUBJECTS DESIGN

An Example

Let's consider the question of whether hunger has a detrimental effect on our ability to stay on task. This question may or may not have an answer in the literature, so we can use hypothetical data to illustrate contrasts in the two-factor between-groups design. We will imagine a 2 × 3 between-subjects design where one factor is *hours since breakfast* (2, 6, or 12 hours) and the other factor is *meal type* (a low-sugar high-protein breakfast, or a high-sugar low-protein breakfast).

Participants will be asked to spend 90 minutes studying the iTunes License Agreement and answering questions about what it implies. They will read the agreement and answer questions about it on an iPad connected to the Internet. Researchers would like to know how much time participants will spend on task (studying and answering questions about the iTunes License Agreement) as opposed to surfing the Internet. They expect that participants will have a harder time staying on task as the time since their last meal increases. They also think that those whose last meal was high in protein and low in sugar will find it easier to stay on task than whose last meal was low in protein and high in sugar. Therefore, the dependent variable will be *time on task*, measured in minutes. Participants will do the reading 2, 6, or 12 hours after their last meal (breakfast).

Sixty-six students are randomly selected from a departmental participant pool and randomly assigned to the six conditions of the experiment, with 11 participants per condition. On the day of the experiment, participants are brought to a research lab to have breakfast at 8:00 a.m. Half of the participants eat a low-sugar high-protein breakfast and

FIGURE 21.1 ■ Hypothetical Data

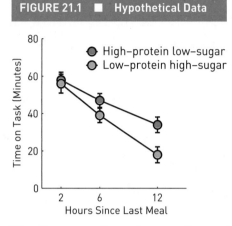

Plot of *time on task* as a function of type of last meal and time since last meal. Error bars represent ±*SEM*.

the other half eat a high-sugar low-protein breakfast. Two groups of 11 participants (one of each meal type) will do the reading task at 10:00 a.m. (2 hours after breakfast). Two groups will do the reading task at 2:00 p.m. (6 hours after breakfast). Two groups will do the reading task at 8:00 p.m. (12 hours after breakfast). Participants will have eaten only breakfast before doing the reading task.

The results of the hypothetical experiment are shown in Figure 21.1 and Table 21.2. The data show that *time on task* declined as time since the last meal increased. In addition, the decline is faster for the high-sugar meal than for the high-protein meal.

Confidence Intervals for Contrasts

Table 21.2 shows the mean and sum of squares (*ss*) for each of the six conditions of the experiment. Below the data are the five contrasts that will be assessed. Contrasts 1 to 3 assess the effect of *meal type* for each of the three times. The fourth contrast measures the change in *time on task* at 2 and 12 hours for the high-protein breakfast, and the fifth contrast measures the change in *time on task* at 2 and 12 hours for the high-sugar breakfast.

The sixth contrast (c_6) is a second-order contrast and has the same structure as the interaction contrast discussed Table 21.1. Contrast c_6 assesses how the difference between b_1 and b_3 changes from a_1 to a_2 (i.e., a difference between two differences). Contrast c_6 is the product of two contrasts: one that compares a_1 to a_2 averaged over B (i.e., [1, 1, 1, –1, –1, –1]) and one that compares b_1 to b_3 averaged over A (i.e., [1, 0, –1, 1, 0, –1]). The product of these two contrasts is [1, 0, –1, –1, 0, 1], as shown in the last row of Table 21.2.

Note that these six contrasts in Table 21.2 do not form an orthogonal set, but they answer questions that logically arise from the design of the experiment.

The contrasts in Table 21.2 make the important point that we don't need to think about a two-factor design as comprising two main effects and an interaction. Although in Chapter 20 we designed contrasts to encode the main effects and the interaction, there is no rule saying we have to do things this way. Rather, we have a collection of means and we can specify whatever contrasts interest us. Therefore, to compute contrasts in the two-factor between-subjects design, we simply treat the data set as a one-factor between-subjects design and proceed as we did in Chapter 18.

The first five contrasts in Table 21.2 are first-order contrasts, so we use equation 21.1 to compute them. When we are able to assume that all groups represent populations with the same variances (homoscedastic populations), the estimated standard errors for all contrasts are computed as

$$s_c = \sqrt{ms_{error} \sum_{i=1}^{k} \frac{w_i^2}{n_i}}. \tag{21.4}$$

Remember that the mean squared error (ms_{error}) is ss_{error}/df_{error}. In this design, $ss_{error} = \sum_{i=1}^{6} ss_i = 11,758$. (This is the sum of the numbers in row "ss" in Table 21.2.) Because there are 11 subjects in each condition, there are 10 degrees of freedom within each group, so $df_{error} = 6 * 10 = 60$. Therefore, $ms_{error} = ss_{error}/df_{error} = 11,758/60 = 195.97$. Because the first five contrasts involve the same two nonzero weights (1 and –1), $\sum_{i=1}^{k} w_i^2/n_i = 2/11 = 0.1818$ in all five cases. Therefore, the estimated standard error for each contrast is

TABLE 21.2   ■   Contrasts for a 2 × 3 Between-Subjects Design[a]

	a_1			a_2		
	b_1	b_2	b_3	b_1	b_2	b_3
	35	41	20	46	42	6
	69	65	15	68	43	7
	58	53	24	56	19	29
	70	64	28	69	41	19
	64	48	32	69	24	5
	67	36	30	27	58	11
	38	24	40	40	35	24
	45	48	33	38	60	5
	53	58	58	66	29	35
	62	42	36	82	44	10
	77	38	58	55	34	47
Mean	58	47	34	56	39	18
SS	1882	1564	1906	2780	1622	2004
s^2	188.2	156.4	190.6	278	162.2	200.4
c_1	1	0	0	−1	0	0
c_2	0	1	0	0	−1	0
c_3	0	0	1	0	0	−1
c_4	1	0	−1	0	0	0
c_5	0	0	0	1	0	−1
c_6	1	0	−1	−1	0	1

[a] a_1 = low-sugar high-protein; a_2 = low-protein high-sugar; b_1 to b_3 = 2, 6, and 12 hours, respectively.

$$s_c = \sqrt{ms_{\text{error}} \sum_{i=1}^{k} \frac{w_i^2}{n_i}} = \sqrt{195.97 \frac{2}{11}} = 5.969.$$

The sixth contrast involves four weights whose absolute values are 1. Therefore, $\sum_{i=1}^{k} w_i^2/n_i = 4/11 = 0.3636$, so the estimated standard error for contrast 6 is

$$s_{c_6} = \sqrt{ms_{\text{error}} \sum_{i=1}^{k} \frac{w_i^2}{n_i}} = \sqrt{195.97 \frac{4}{11}} = 8.442.$$

TABLE 21.3 ■ Statistics for Six Contrasts in a 2 × 3 Between-Subjects Design[a]

Contrast	c	s_c	df	Lower	Upper	t_{obs}	p
c_1	2	5.969	60	−9.94	13.94	0.335	.739
c_2	8	5.969	60	−3.94	19.94	1.340	.185
c_3	16	5.969	60	4.06	27.94	2.680	.009
c_4	24	5.969	60	12.06	35.94	4.021	.000
c_5	38	5.969	60	26.06	49.94	6.366	.000
c_6	−14	8.442	60	−30.89	2.89	−1.658	.102

[a] a_1 = low-sugar high-protein; a_2 = low-protein high-sugar; b_1 to b_3 = 2, 6, and 12 hours, respectively.

Table 21.3 shows the six contrasts and their associated statistics. The columns "Lower" and "Upper" show the 95% confidence limits for all contrasts. In all cases, $t_{\alpha/2}$ = 2.000; this was found using the **T.INV.2T** function in Excel.

The confidence intervals in Table 21.3 were not corrected for multiple comparisons. The Bonferroni-corrected intervals for six contrasts would have used $t_{\alpha/2}$ = 2.729 and thus produced confidence intervals that are 36% wider than those shown in Table 21.3. We find this in Excel using **T.INV.2T (.05/6, 60)**, where 6 is the number of contrasts.

The t_{obs} statistics in Table 21.3 were computed as $t_{obs} = c/s_c$ in all cases, and the p-values were calculated using **T.DIST.2T** in Excel. To obtain the Bonferroni-corrected p-values, simply multiply the p-values in the table by 6, the number of contrasts; if a corrected p-value exceeds 1, set it to 1.

APA Reporting

Sixty-six students volunteered to participate in an experiment assessing ability to stay on task (*time on task*, measured in minutes) as a function of *time since last meal* (2, 6, or 12 hours) and the type of meal last eaten (high-protein low-sugar vs. low-protein high-sugar). Eleven participants were randomly assigned to each of the six conditions of the experiment. The mean values of *time on task* for the six conditions are summarized in Figure 21.1.

At each of the three time points, those whose last meal was high in protein and low in sugar stayed on task longer than those whose last meal was low in protein and high in sugar. At 2 hours, the difference was 2, 95% CI [−9.94, 13.94]; at 6 hours, the difference was 8, 95% CI [−3.94, 19.94]; and at 12 hours, the difference was 16, 95% CI [4.06, 27.94]. Therefore, the effect of last meal type increased with time.

For both meal types, *time on task* decreased as *time since last meal* increased. The difference between the 2-hour and 12-hour conditions was 24 for the high-protein group, 95% CI [12.06, 35.94], and 38 for the high-sugar group, 95% CI [26.06, 49.94]. An interaction contrast that compared the difference at 2 and 12 hours for the two groups showed this second-order difference to be 24 − 38 = −14, 95% CI [−30.89, 2.89]. That is, the ability to stay on task fell more rapidly over time for those whose last meal was high in sugar than for those whose last meal was high in protein.

Computing Contrasts in SPSS

As noted earlier, contrast analysis can be carried out in SPSS by treating the experiment as a one-factor between-subjects design. I'll repeat that the concept of a factor can often be ignored in ANOVA. In reality, we simply have some number of means and a theoretical or practical interest in how they differ. Again, Table 21.2 specifies six contrasts, but there is no explicit analysis of factors.

Figure 21.2 shows how to set up the analysis. The first column in the data file shows *time on task* (Time_on_Task) for each subject in the experiment. The second column shows the condition of the experiment that each subject was in (Group); these are the numbers 1 to 6, although only the first condition is shown. The analysis is conducted through the One-Way ANOVA dialog, just as in Chapter 18. The contrasts were entered by hand to achieve the six contrasts shown in Table 21.2. (A seventh contrast, to be discussed later, illustrates the question of proper scaling of contrasts.)

Figure 21.2 shows that Time_on_Task is the dependent variable and Group contains codes for the six conditions of the experiment. Once the dependent and independent variables are entered and the contrasts are defined, we click OK to initiate the analysis. The results are shown in Figure 21.3.

Figure 21.3a shows is an ANOVA table showing that the independent variable explains a statistically significant portion of the variability in the dependent variable. Figure 21.3b shows the six contrasts described in Table 21.2 plus the seventh contrast (to be discussed later). This is an important box because it is easy to make mistakes when entering contrasts in SPSS. Therefore, you must always check to see that contrasts have been entered correctly.

The top part of Figure 21.3c shows the contrasts, their estimated standard errors, *t*-statistics, degrees of freedom, and *p*-values. These are identical to the corresponding quantities computed by hand in Table 21.2. As noted in Chapter 18, SPSS does not provide confidence intervals around the contrasts. However, once the contrasts and estimated standard errors have been computed, you can take the final step and turn these into confidence intervals.

The bottom part of Figure 21.3c shows the estimated standard errors, *t*-statistics, degrees of freedom, and *p*-values when equal variances are not assumed. The approach to confidence intervals in this situation is a more general version of the Welch-Satterthwaite

---

FIGURE 21.2 ■ Setting Up Planned Contrasts

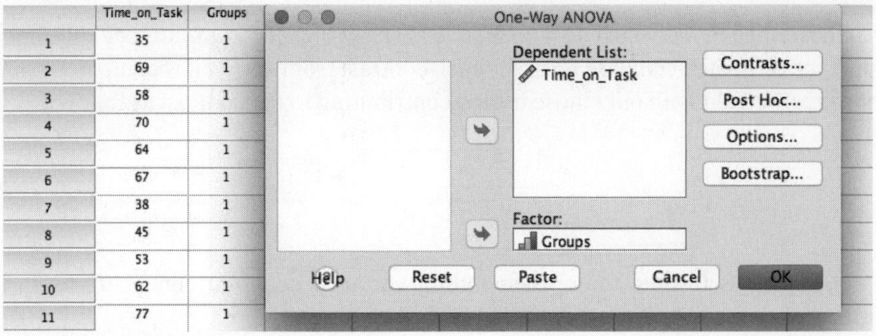

Setting up planned contrasts for a two-factor between-subjects design in SPSS. The design can be treated as a one-factor analysis with $2*3 = 6$ conditions. The dependent variable is *time on task*.

FIGURE 21.3 ■ Contrast Analysis

(a)

ANOVA

Time_on_Task

	Sum of Squares	df	Mean Square	F	Sig.
Between Groups	12386.000	5	2477.200	12.641	.000
Within Groups	11758.000	60	195.967		
Total	24144.000	65			

(b)

Contrast Coefficients

	Groups					
Contrast	1	2	3	4	5	6
1	1	0	0	−1	0	0
2	0	1	0	0	−1	0
3	0	0	1	0	0	−1
4	1	0	−1	0	0	0
5	0	0	0	1	0	−1
6	1	0	−1	−1	0	1
7	10	0	−10	−10	0	10

(c)

Contrast Tests

		Contrast	Value of Contrast	Std. Error	t	df	Sig. (2–tailed)
Time_on_Task	Assume equal variances	1	2.00	5.969	.335	60	.739
		2	8.00	5.969	1.340	60	.185
		3	16.00	5.969	2.680	60	.009
		4	24.00	5.969	4.021	60	.000
		5	38.00	5.969	6.366	60	.000
		6	−14.00	8.442	−1.658	60	.102
		7	−140.00	84.416	−1.658	60	.102
	Does not assume equal variances	1	2.00	6.510	.307	19.284	.762
		2	8.00	5.382	1.486	19.993	.153
		3	16.00	5.962	2.684	19.987	.014
		4	24.00	5.868	4.090	19.999	.001
		5	38.00	6.595	5.762	19.487	.000
		6	−14.00	8.828	−1.586	38.838	.121
		7	−140.00	88.276	−1.586	38.838	.121

Output of the contrast analysis for the completely between-subjects design. (a) The overall ANOVA. The treatment conditions explained 51.3% of the variability in the dependent variable, which corresponds to $F = 12.641$. (b) The contrasts that were entered. Each row corresponds to a contrast. (c) Contrast analyses when we assume equal variances (*top*) and when we do not (*bottom*).

method described in Appendix 11.4 (available at study.sagepub.com/gurnsey). Rather than compute the estimated standard error for each contrast from ms_{error}, it is computed from the variances associated with only those means contributing to the contrast as follows:

$$s_c = \sqrt{\sum_{i=1}^k w_i^2 \frac{s_i^2}{n_i}}. \tag{21.5}$$

If a weight (w_i) is 0, then the corresponding variance does not contribute to s_c. (The variances for each condition are shown in Table 21.2.)

For contrast 1, the estimated standard error computed in this way is $s_{c_1} = 6.51$. For contrast 6, it is $s_{c_6} = 8.828$. Remember that the coefficients for contrast 1 are $w_1 = [-1, 0, 0, 1, 0, 0]$ and the coefficients for contrast 6 are $w_6 = [1, 0, -1, -1, 0, 1]$. Therefore, with this information, we can determine that s_{c_1} is

$$s_{c_1} = \sqrt{\sum_{i=1}^{k} w_i^2 \frac{s_i^2}{n_i}} = \sqrt{(-1^2)\frac{188.2}{11} + (1^2)\frac{278}{11}} = 6.51$$

and s_{c_6} is

$$s_{c_6} = \sqrt{\sum_{i=1}^{k} w_i^2 \frac{s_i^2}{n_i}} = \sqrt{(1^2)\frac{188.2}{11} + (-1^2)\frac{190.6}{11} + (-1^2)\frac{278}{11} + (1^2)\frac{200.4}{11}} = 8.828.$$

The degrees of freedom associated with these estimated standard errors are based on the degrees of freedom of the conditions for which $w_i \neq 0$. For contrast 1, the adjusted degrees of freedom are based on conditions 1 and 4. For contrast 6, the adjusted degrees of freedom are based on conditions 1, 3, 4, and 6. For contrast 1, the adjusted degrees of freedom are $df_{c_1} = 19.284$. For contrast 6, the adjusted degrees of freedom are $df_{c_6} = 38.838$.

These estimated standard errors and corrected degrees of freedom can be used to compute 95% confidence intervals around the contrasts. However, the $t_{\alpha/2}$ values cannot be taken from the t-table because there are fractional parts to the degrees of freedom. Although the **T.INV.2T** function in Excel accepts fractional degrees of freedom, it simply ignores the fractional part and rounds down to the next integer; this is not much help. Fortunately, if you type

```
qt(p = .975, df = 19.284)
```

into **R**, it returns 2.09094, which is $t_{\alpha/2}$ for $df_{c_1} = 19.284$ when $\alpha = .05$. The value of p is $1-\alpha/2 = .975$, so qt returns the t score with 97.5% of the distribution below it when there are 19.284 degrees of freedom. Similarly, if you type

```
qt(p = .975, df = 38.838)
```

in **R**, it returns 2.022961, which, is $t_{\alpha/2}$ when $\alpha = .05$ and $df_{c_6} = 38.838$.

With this information, we can compute the 95% confidence interval for contrast 1 as

$$c_1 \pm t_{\alpha/2}(s_c) = 2 \pm 2.09094(6.51) = [-11.61, 15.61].$$

And the 95% confidence interval for contrast 6 is

$$c_6 \pm t_{\alpha/2}(s_c) = -14 \pm 2.022961(8.828) = [-31.86, 3.86].$$

These corrected confidence intervals are wider than those in Table 21.3, mainly because there are fewer degrees of freedom; hence, $t_{\alpha/2}$ is larger. However, it is possible for the estimated standard error of the contrast to be smaller when variances are not assumed to be equal, and this may make the confidence interval narrower.

Reminder: Scaling Contrasts

Contrasts 1 to 6 all involve weights of -1 and 1. This means the weights of the first-order contrasts (c_1 to c_5) sum to zero and their absolute values sum to 2. Therefore, these contrasts are properly scaled. For the second-order contrast (c_6), the weights sum to zero and their absolute values sum to 4. Therefore, these contrast weights are also properly scaled.

Scaling contrasts before entering them into SPSS can create problems because properly scaled contrast weights may end up with several digits after the decimal places. For example, properly scaling the first-order contrast $[-1, -1, -1, -1, -1, -1, 6]$ would produce $[-0.1667, -0.1667, -0.1667, -0.1667, -0.1667, -0.1667, 1]$. Because entering contrasts into SPSS can be a little tedious (and thus prone to error), it may be easier to use integer weights whose absolute values do not sum to 2, rather than real valued weights whose absolute values do sum to 2. The same applies to second-order (interaction) weights. That is, it may be easier to use integer weights whose absolute values do not sum to 4, rather than real valued weights whose absolute values do sum to 4.

Contrast 7 (c_7) in Figure 21.3 provides an example of a second-order contrast for which the absolute values of the weights do not sum to 4. In fact, the weights in c_7 are equal to the weights in c_6 multiplied by 10. As shown in Figure 21.3, contrasts c_6 and c_7 produce the same t-statistics (and p-values), so scaling can be ignored if your purpose is the test the null hypothesis that $\Psi = 0$. However, the computed contrasts differ by a factor of 10 ($c_6 = -14$ and $c_7 = -140$), and these clearly mean different things.

If you are going to use SPSS to compute a contrast and its estimated standard error, you can do the scaling after these have been computed. This means dividing the contrast and its estimated standard error by $\sum_{i=1}^{k} |w_i|/4$. In the case of c_7, this quantity is

$$\sum_{i=1}^{k} |w_i|/4 = [10+0+10+10+0+10]/4 = 10.$$

So, $c_7/10 = -140/10 = -14$, and $s_{c_7}/10 = 84.416/10 = 8.442$. By scaling the contrasts after they've been computed, we obtain exactly the same quantities that would have been obtained had the weights themselves been scaled.

Remember, for first-order contrasts, we scale the contrast and its estimated standard error by $\sum_{i=1}^{k} |w_i|/2$. For second-order contrasts, we scale the contrast and its estimated standard error by $\sum_{i=1}^{k} |w_i|/4$. Make sure to scale, and make sure to scale by the right quantity. Much confusion will ensue if this is not done.

THE TWO-FACTOR, WITHIN-SUBJECTS DESIGN

An Example

We've all had the experience of switching our attention from one task (e.g., studying for a final) to another (e.g., answering a text message). In such situations, we need to disengage from the first task and prepare for the next. A classic study by Rogers and Monsell (1995) showed that there are costs associated with switching attention from one task to another. In their experiment, subjects performed two tasks: one was to judge whether a letter presented on a computer screen was a consonant or a vowel, and the other was to judge whether a number was odd or even. We will call the letter task L and the number task N. The two tasks were presented in a repeated sequence such as *LLNNLLNNLLNN* . . . Subjects were to make the decision as quickly as possible, and reaction time (measured in milliseconds) was the dependent variable. (I'm leaving out some important details here, but the gist of the story is sufficient for this illustration.)

Rogers and Monsell found that subjects performed a task more quickly when they had done the same task on the immediately preceding trial (i.e., doing L after having just done L, or doing N after having just done N) versus when they had done a different task on the immediately preceding trial (i.e., doing L after having just done N, or doing N after having just done L). In other words, there was a cost to switching tasks on successive trials; in the literature, this has become known as a switch cost.

One might think that the switch cost would decrease the longer one has to prepare for a new task. To assess this possibility, Rogers and Monsell (1995) varied

LEARNING CHECK 1

1. The data in the following table are from a two-factor between-subjects design. There are three levels of A, three levels of B, and five subjects in each condition. Plot the means with levels of B on the x-axis and each level of A as a separate line. Use 95% confidence intervals as the error bars. Please note, each row represents a condition of the experiment. In other examples, each column represents a condition of the experiment.

A	B					
a_1	b_1	16	10	13	16	13
a_1	b_2	9	14	14	15	9
a_1	b_3	11	13	10	8	13
a_2	b_1	18	14	11	14	14
a_2	b_2	19	17	19	22	22
a_2	b_3	27	29	31	22	28
a_3	b_1	17	17	13	17	16
a_3	b_2	25	28	29	24	28
a_3	b_3	35	36	37	33	34

2. Use these data to compute the contrasts and their 95% confidence intervals for comparisons specified in a to f below. You may assume homoscedasticity of variances. If you do these calculations in SPSS, remember to scale the contrasts and standard errors before (or after) computing the confidence intervals. If you do these calculations in Excel, you must first compute ms_{error}. This is not an orthogonal set of contrasts.

   (a) Subtract the average score at a_1 from the average score at a_3.

   (b) Subtract the average score at b_1 from the average score at b_3.

   (c) Subtract the average score at a_2 from the average score at a_3.

   (d) Subtract the average score at b_2 from the average score at b_3.

   (e) Compute the second-order contrast that is the product of contrasts a and b.

   (f) Compute the second-order contrast that is the product of contrasts c and d.

   (g) Put into words what contrasts e and f accomplish.

Answers

1. The data plot is as shown to the right.

2. The following table shows the mean, sum of squares, and variance for each cell in the data set. Columns "a" to "f" show the contrasts required. The mean squared error is $ms_{error} = 5.76$. The contrasts and confidence intervals are also shown. (g) Contrast e measures the difference in the $b_3 - b_1$ contrast at a_1 and a_3. Contrast f measures the difference in the $b_3 - b_2$ contrast at a_2 and a_3.

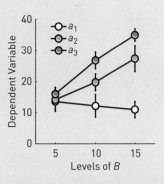

A	B	Mean	ss	Variance	a	b	c	d	e	f
a_1	b_1	13.6	25.2	6.3	−1	−1	0	0	1	0
a_1	b_2	12.2	34.8	8.7	−1	0	0	−1	0	0
a_1	b_3	11.0	18.0	4.5	−1	1	0	1	−1	0
a_2	b_1	14.2	24.8	6.2	0	−1	−1	0	0	0
a_2	b_2	19.8	18.8	4.7	0	0	−1	−1	0	1
a_2	b_3	27.4	45.2	11.3	0	1	−1	1	0	−1
a_3	b_1	16.0	12.0	3.0	1	−1	1	0	−1	0
a_3	b_2	26.8	18.8	4.7	1	0	1	−1	0	−1
a_3	b_3	35.0	10.0	2.5	1	1	1	1	1	1
				c	13.67	9.87	5.47	4.87	21.6	0.6
				Lower	11.89	8.09	3.69	3.09	17.24	−3.76
				Upper	15.45	11.65	7.25	6.65	25.96	4.96

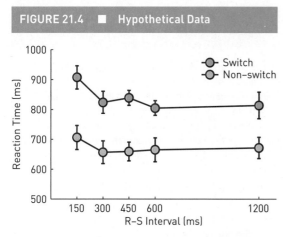

FIGURE 21.4 ■ Hypothetical Data

Response times (in milliseconds) for switch and non-switch trials as a function of Response-Stimulus (R-S) interval. Error bars represent ±*SEM*.

the interval between when a response was given on one trial and the appearance of the stimulus on the next trial. Because the sequence of trials was always the same (*LLNNLLNNLLNN* . . .), subjects knew which task they were to perform on each trial. (Also, the stimuli for the letter task appeared at different screen locations than the stimuli for the number task.) Therefore, a longer Response-Stimulus (R-S) interval between the end of one trial and the beginning of the next should allow subjects to better prepare for the new task, and thus reduce the switch cost.

Their experiment was run completely within subjects with 2 trial types (switch versus non-switch trials) × 5 R-S intervals (150, 300, 450, 600, and 1200 ms) for 10 conditions in total. The results shown in Figure 21.4 are simulated data that are similar to the results of Rogers and Monsell (1995; see also Nieuwenhuis & Monsell, 2002). The stimulated data are shown in Table 21.4; each row represents a simulated subject.

TABLE 21.4 ■ Contrasts for a 2 × 3 Within-Subjects Design[a]

	a_1					a_2				
	b_1	b_2	b_3	b_4	b_5	b_1	b_2	b_3	b_4	b_5
	884	732	817	767	795	679	538	630	599	645
	630	590	726	669	487	515	468	503	400	399
	1078	952	903	841	978	844	755	779	762	728
	927	919	890	898	932	770	777	729	719	780
	927	805	879	742	751	603	685	609	694	691
	868	801	820	795	774	647	619	575	618	701
	1041	949	975	896	950	947	857	839	883	748
	962	939	851	894	780	770	700	623	696	773
	830	764	719	732	885	647	557	651	600	601
	923	789	810	816	808	638	614	652	679	644
Mean	907	824	839	805	814	706	657	659	665	671
c_1	1	0	0	0	0	–1	0	0	0	0
c_2	0	1	0	0	0	0	–1	0	0	0
c_3	0	0	1	0	0	0	0	–1	0	0
c_4	0	0	0	1	0	0	0	0	–1	0
c_5	0	0	0	0	1	0	0	0	0	–1
c_6	1	–1	0	0	0	0	0	0	0	0
c_7	0	0	0	0	0	1	–1	0	0	0
c_8	4	–1	–1	–1	–1	0	0	0	0	0
c_9	0	0	0	0	0	4	–1	–1	–1	–1
c_{10}	4	–1	–1	–1	–1	–4	1	1	1	1

[a] a_1 = switch trials; a_2 = non-switch trials; b_1 to b_5 = 150, 300, 450, 600, and 1200 ms, respectively.

Confidence Intervals for Contrasts

The lower part of Table 21.4 shows 10 contrasts that we will examine. Of course, with 10 means, it is impossible for 10 contrasts to be orthogonal. The contrasts were chosen to examine several questions of interest in the data. The first five (c_1 to c_5) compare the switch versus non-switch trials for each of the five R-S intervals. You should be able to confirm that these five contrasts are orthogonal to each other.

Contrasts c_6 and c_7 compare R-S intervals 150 ms (b_1) and 300 ms (b_2) for the switch trials (c_6) and non-switch trials (c_7). These contrasts will tell us how much change in reaction time there is between R-S intervals 150 and 300 ms. Contrast c_8 compares the mean score at 150 ms (b_1) with the average of the remaining R-S intervals (b_2 to b_5) for the switch trials (a_1). Contrast c_9 compares the mean score at 150 ms (b_1) with the average of the remaining R-S intervals (b_2 to b_5) for the non-switch trials (a_2). These contrasts will tell us how much change in reaction time there is between R-S intervals 150 ms and the average of the remaining R-S intervals.

Contrasts c_1 to c_9 are first-order contrasts. Contrasts c_1 and c_7 are properly scaled (because they involve only 1 and −1), but c_8 and c_9 are not.

The final contrast (c_{10}) is a second-order (interaction) contrast. It asks how the difference between b_1 and the average of b_2 to b_5 differs between switch and non-switch trials. In other words, c_{10} simply compares c_8 and c_9. Contrast c_{10} also needs to be properly scaled.

Computing Contrasts in SPSS

Figure 21.5 shows the data from Table 21.3 in columns a1b1 to a2b5 of an SPSS Data Viewer file. Contrasts c_1 to c_{10} were computed using Transform→Compute Variable. . . . Contrasts were computed this way in Chapter 19. As noted, contrasts c_8 to c_{10} are not properly scaled. The contrasts could have been scaled when they were computed in Compare Variable . . . , and this is something you are of course free to do. However, for this exercise, we will scale the results rather than the contrasts weights.

The confidence intervals around the contrasts were computed using Analyze→Compare Means→One Sample T-Test. . . . There are other ways to do this (e.g., Analyze→Descriptive Statistics→Explore . . .), but the One Sample T-Test . . . dialog produces the most compact output. The results are shown in Figure 21.6 for each of the 10 contrasts. Our focus is on the contrasts in the column Mean Difference and the 95% confidence intervals shown in the last two columns. The second and fourth columns show the t-statistics and their corresponding p-values for each contrast, for those interested in testing the null hypothesis that $\Psi = 0$.

FIGURE 21.5 ■ Sample Data and Contrasts Scores

	a1b1	a1b2	a1b3	a1b4	a1b5	a2b1	a2b2	a2b3	a2b4	a2b5	c1	c2	c3	c4	c5	c6	c7	c8	c9	c10
1	884	732	817	767	795	679	538	630	599	645	205	194	187	168	150	152	141	425	304	121
2	630	590	726	669	487	515	468	503	400	399	115	122	223	269	88	40	47	48	290	−242
3	1078	952	903	841	978	844	755	779	762	728	234	197	124	79	250	126	89	638	352	286
4	927	919	890	898	932	770	777	729	719	780	157	142	161	179	152	8	−7	69	75	−6
5	927	805	879	742	751	603	685	609	694	691	324	120	270	48	60	122	−82	531	−267	798
6	868	801	820	795	774	647	619	575	618	701	221	182	245	177	73	67	28	282	75	207
7	1041	949	975	896	950	947	857	839	883	748	94	92	136	13	202	92	90	394	461	−67
8	962	939	851	894	780	770	700	623	696	773	192	239	228	198	7	23	70	384	288	96
9	830	764	719	732	885	647	557	651	600	601	183	207	68	132	284	66	90	220	179	41
10	923	789	810	816	808	638	614	652	679	644	285	175	158	137	164	134	24	469	−37	506

The data from Table 21.3 are shown in columns a1b1 to a2b5. The contrasts from Table 21.3 have been computed using Compute Variable, and the results are stored in columns c1 to c10.

FIGURE 21.6 ■ Contrast Analysis

One-Sample Test

					95% Confidence Interval of the Difference	
					Test Value = 0	
	t	df	Sig. (2–tailed)	Mean Difference	Lower	Upper
c1	9.001	9	.000	201.000	150.49	251.51
c2	11.438	9	.000	167.000	133.97	200.03
c3	9.129	9	.000	180.000	135.40	224.60
c4	5.821	9	.000	140.000	85.59	194.41
c5	5.176	9	.001	143.000	80.50	205.50
c6	5.257	9	.001	83.000	47.29	118.71
c7	2.476	9	.035	49.000	4.24	93.76
c8	5.713	9	.000	346.000	209.00	483.00
c9	2.530	9	.032	172.000	18.20	325.80
c10	1.842	9	.099	174.000	−39.72	387.72

The **One Sample T-Test** . . . dialog was used to compute the mean contrast and 95% confidence intervals for each of the 10 sets of contrast scores. The dialog also provides t-statistics and p-values for those who wish to test the null hypothesis that $\Psi = 0$.

To scale the first-order contrasts c_8 and c_9, we have to divide the computed contrasts by $\sum_{i=1}^{k} |w_i| / 2$, which in these cases produces $[4 + 1 + 1 + 1 + 1]/2 = 4$. We could also divide the estimated standard errors ($s_c = c/t_{obs}$) by 4 and then compute the 95% confidence intervals by hand or in Excel. However, simply dividing the confidence limits by 4 produces the same results. Therefore, the properly scaled version of c_8 is $346/4 = 86.5$, 95% CI [52.25, 128.75], and the properly scaled version of c_9 is $172/4 = 43$, 95% CI [4.55, 81.45].

To scale the second-order contrast c_{10}, we have to divide the computed contrasts by $\sum_{i=1}^{k} |w_i| / 4$, which in these cases produces $[2(4) + 8(1)]/4 = 4$. We can then scale the contrast and its confidence limits to produce c_{10}, which is $174/4 = 43.5$, 95% CI [−9.93, 96.93]. Remember that contrast c_{10} computed $c_8 - c_9$, meaning that c_{10} should be equal to $86.5 - 43.0 = 43.5$, which it is. Therefore, our scalings have computed the right things.

APA Reporting

Ten students volunteered to participate in an experiment assessing how reaction times change for switch and non-switch trials at Response-Stimulus (R-S) intervals of 150, 300, 450, 600, and 1200 ms. The 2 (switch versus non-switch) × 5 (R-S intervals) design was run completely within subjects. Figure 21.4 plots the mean response times on switch and non-switch trials for each of the five R-S intervals.

Planned contrasts (c_1 to c_5) show that reaction times were longer on switch trials than on non-switch trials for R-S intervals 150 to 1200 ms, respectively; see Table 21.5. It is worth noting that these differences decrease modestly as R-S intervals increase. The largest difference (201 ms) was found at R-S interval = 150 ms (c_1), meaning that switch cost was greatest at the shortest R-S interval.

TABLE 21.5 ■ Contrasts

	Contrast	s_c	Lower	Upper
c_1	201.0	22.33	150.49	251.51
c_2	167.0	14.60	133.97	200.03
c_3	180.0	19.72	135.40	224.60
c_4	140.0	24.05	85.59	194.41
c_5	143.0	27.63	80.50	205.50
c_6	83.0	15.79	47.29	118.71
c_7	49.0	19.79	4.24	93.76
c_8	86.5	15.14	52.25	120.75
c_9	43.0	17.00	4.55	81.45
c_{10}	43.5	23.62	−9.93	96.93

Contrasts c_6 and c_7 measured the mean difference in reaction times at R-S intervals 150 and 300 ms for switch (c_6) and non-switch (c_7) trials, respectively. This difference is greater for switch trials ($c_6 = 83$) than for non-switch trials ($c_7 = 49$), as expected. That is, increasing R-S intervals from 150 to 300 ms reduces reaction times more for switch trials than non-switch trials, but it should be noted that the corresponding confidence intervals overlap considerably.

Figure 21.4 shows that after a reduction of reaction time from R-S intervals of 150 to 300 ms, performance was essentially flat over the remaining intervals. Contrasts c_8 and c_9 compared mean reaction times at 150 ms with the average reaction time at 300 to 1200 ms, for both switch and non-switch trials. This difference is greater for switch trials ($c_8 = 86.5$, 95% CI [52.25, 120.75]) than for non-switch trials ($c_9 = 43$, 95% CI [4.55, 81.45]), consistent with contrasts c_6 and c_7.

Finally, an interaction contrast (c_{10}) compared contrasts c_8 and c_9 for switch and non-switch trials. The difference between the first (150 ms) and remaining (300 to 1200 ms) R-S intervals was greater for switch trials than for non-switch trials, $c_{10} = 43.5$, 95% CI [−9.93, 96.93].

Please note that the confidence intervals in Table 21.5 were not corrected for multiple contrasts. Had the Bonferroni correction been applied, all confidence intervals would be 1.63 times wider than those shown.

These results provide some support for the idea that increased preparation time reduces switch costs. However, even very long R-S intervals (1200 ms) do not eliminate switch costs, $c_5 = 143$, 95% CI [80.50, 205.50].

THE TWO-FACTOR MIXED DESIGN

An Example

When you have your hearing tested, an audiologist will present you with tones of different frequencies and will measure how intense (loud) each one must be for you to detect it. The frequency of a tone is measured in cycles per second, or hertz (Hz), and intensity is measured in decibels (dB). Low notes on a piano correspond to low frequencies, and high notes correspond to high frequencies. You can visit szynalski.com/tone-generator/ to find examples of tones of different frequencies.

The audibility curve is a plot of threshold (the lowest intensity that you can detect) at each frequency. Healthy young adults can hear frequencies ranging from 20 to 20,000 cycles per second (Hz), and their average audibility curve is used as a reference that defines normal hearing. Therefore, when your hearing is tested, the audiologist compares your thresholds to the reference thresholds.

An audiogram shows how many decibels higher (or lower) an individual's thresholds are relative to those of healthy young adults. To get a sense of your own audiogram, you can go to hearingtest.online/hearingTest_L.php and test your own hearing.

It is well known that our sensitivity to sounds decreases with age, but there is some question about how much of this hearing loss is an inevitable effect of aging, and how much is a consequence of the constant exposure to high levels of sound that we experience in industrialized societies. You may not notice how much noise there is in your daily life until you get into the country and experience real quiet. (I highly recommend doing this as much as possible.) Alternatively, think about how loud your usual daily activities would sound at 2:00 a.m. when everyone else in your home is asleep.

In this section, we will illustrate the two-factor mixed design using simulated data similar to those reported by Goycoolea, Goycoolea, Rodriguez, Farfan, Martinez, and Vidal (1986). Goycoolea et al. had the unique opportunity to study people from Easter Island, which lies about 3500 kilometers west of Chile in the Pacific Ocean. The society

LEARNING CHECK 2

1. The data in the following table are from a two-factor completely within-subjects design. There are three levels of A and three levels of B. Plot the mean with levels of A on the x-axis, and each level of B as a separate line. Use 95% confidence intervals as the error bars.

	a_1			a_2			a_3		
b_1	b_2	b_3	b_1	b_2	b_3	b_1	b_2	b_3
16	9	11	18	19	27	17	25	35
10	14	13	14	17	29	17	28	36
13	14	10	11	19	31	13	29	37
16	15	8	14	22	22	17	24	33
13	9	13	14	22	28	16	28	34

2. Use these data to compute the following contrasts and 95% confidence intervals for contrasts a to f. You should scale the trends so that the absolute values of the weights sum to 2. These calculations can be done in Excel or SPSS.

(a) Subtract the average score at a_1 from the average score at a_3.

(b) Subtract the average score at b_1 from the average score at b_3.

(c) Subtract the average score at a_2 from the average score at a_3.

(d) Subtract the average score at b_2 from the average score at b_3.

(e) Compute the second-order contrast that is the product of contrasts a and b.

(f) Compute the second-order contrast that is the product of contrasts c and d.

(g) Put into words what contrasts e and f accomplish.

Answers

1. The data plot is shown to the right.

2. The following table shows the contrast weights, contrast scores, contrasts, and confidence intervals for a to f. (g) Contrast e measures the difference in the $b_3 - b_1$ contrast at a_1 and a_3. Contrast f measures the difference in the $b_3 - b_2$ contrast at a_2 and a_3.

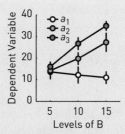

		a_1			a_2			a_3		
		b_1	b_2	b_3	b_1	b_2	b_3	b_1	b_2	b_3
		16	9	11	18	19	27	17	25	35
		10	14	13	14	17	29	17	28	36
		13	14	10	11	19	31	13	29	37
		16	15	8	14	22	22	17	24	33
		13	9	13	14	22	28	16	28	34
a		−1	−1	−1	0	0	0	1	1	1
b		−1	0	1	−1	0	1	−1	0	1
c		0	0	0	−1	−1	−1	1	1	1
d		0	−1	1	0	−1	1	0	−1	1
e		1	0	−1	0	0	0	−1	0	1
f		0	0	0	0	1	−1	0	−1	1
	a	b	c	d	e	f				
	13.67	7.33	4.33	6.67	23	2				
	14.67	12.33	7.00	6.33	16	−4				
	14.00	13.67	6.00	5.33	27	−4				
	11.67	5.33	5.33	0.67	24	9				
	14.33	10.67	4.67	5.33	18	0				
c	13.67	9.87	5.47	4.87	21.6	0.6				
Lower	12.20	5.56	4.14	1.86	16.01	−6.06				
Upper	15.13	14.17	6.79	7.87	27.19	7.26				

there is not industrialized, so its inhabitants are not regularly exposed to high levels of noise. Goycoolea et al. studied the hearing of three groups of older adults born on Easter Island. The three groups involved those who had (i) lived on the island all their lives, (ii) lived off the island in an industrialized society for 3 to 5 years, or (iii) lived off the island in an industrialized society for more than 5 years.

The average data shown in Figure 21.7 are similar to the average audiograms reported by Goycoolea et al. (1986) for the three groups of older adults. Notice that an audiogram shows threshold elevation (hearing level) on a reversed *y*-axis to indicate that higher threshold elevations mean poorer hearing. The *x*-axis shows frequency in kilohertz (kHz, thousands of cycles per second).

In the study by Goycoolea et al. (1986), the between-groups factor (*A*) was the three groups of Easter Islanders and the within-groups factor (*B*) is frequency. We will approach these data using trend analysis. In particular, we will assess the linear trend in *B* (frequency) for each participant within each of the three groups. Before we get into the details of the analysis, we will introduce two useful technical details having to do with (i) logarithms and (ii) the relationship between linear trends and the slope of the regression equation.

iStock.com/anharris

Logarithmic Scales

Remember that to conduct a trend analysis, the values of *x* must be equally spaced on a linear scale. In Figure 21.7, however, this is clearly not the case. Rather, the frequency increments are said to be octave increments, which means that frequency doubles from one step to the next (e.g., 1 = 2 * .5 and 8 = 2 * 4). Although they are not equally spaced on a linear scale, the values of *x* are equally spaced on a logarithmic scale.

You probably studied logarithms in high school, but a brief review will be useful. Logarithmic functions show the power to which some base would have to be raised to obtain that number. We can refer to log base 10 as $\log_{10}$. Therefore, $\log_{10}(100) = 2$ because $10^2 = 100$. And $\log_{10}(1000) = 3$ because $10^3 = 1000$. For our purposes, base 2 logarithms ($\log_2$) will be most useful. In base 2 logarithms, $\log_2(4) = 2$ because $2^2 = 4$. And $\log_2(8) = 3$ because $2^3 = 8$. [You can experiment with logarithmic transformations in Excel using the **LOG(x, base)** function. Typing in = **LOG(8, 2)** returns 3.]

When we transform our frequencies (0.5, 1, 2, 4, and 8 kHz) into logarithmic units, we find that

$$\log_{10}(0.5, 1, 2, 4, 8) = (-0.301, 0.000, 0.301, 0.602, 0.903)$$

and

$$\log_2(0.5, 1, 2, 4, 8) = (-1, 0, 1, 2, 3).$$

In both cases, the logarithmic units are equally spaced. Therefore, we can conduct a trend analysis when the values of *x* in kilohertz are transformed to logarithmic units. Although any logarithmic transformation of our frequencies will produce equally spaced units, the $\log_2$ transformation is useful because it provides an octave scale.

FIGURE 21.7 ■ Hypothetical Data

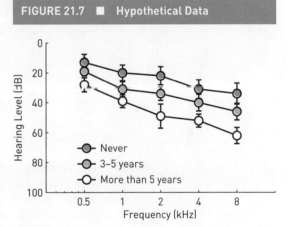

Data similar to those reported by Goycoolea et al. (1986). These average audiograms plot hearing level (deviation from normal hearing in decibels) for frequencies ranging from 500 to 8000 Hz (0.5 to 8 kHz). Error bars represent ±*SEM*.

Linear Contrasts and Linear Slopes

In earlier sections, we noted that scaling first- and second-order contrasts is necessary so that the contrasts relate to the original units of measurement. However, we didn't say much about scaling trend coefficients. If we were simply asking whether the linear trend in the data is statistically significant, then there would be no need to scale the trend coefficients. However, if we would like to know how threshold elevations change with frequency, then it would be important to determine the slope of the best-fitting line that relates frequency to threshold elevation. We will now see how to scale trend weights so that they yield the slope of the line relating the independent variable to the dependent variable.

In Figure 21.7, the average threshold elevations for those who had never left the island are $m = [13, 20, 22, 31, 34]$. As noted above, the frequencies expressed log base 2 are $x = [-1, 0, 1, 2, 3]$. If we were to compute the slope that relates m to x, we would find that it is $b = 5.3$. That is, our best estimate is that each octave increase in frequency produces a 5.3-dB increase in threshold elevation for this group.

To produce scaled trend coefficients, we first center the log-transformed frequencies by subtracting the mean from each. For our case, the centered, log-transformed frequencies are

$$x_c = x - m_x = (-1, 0, 1, 2, 3) - 1 = (-2, -1, 0, 1, 2).$$

We then scale x_c by the sum of squares of x_c, which is simply $ss_{x_c} = \sum x_c^2$. In this case, $ss_{x_c} = 10$. Finally, we can define our trend coefficients as

$$w = x_c / ss_{x_c} = (-2, -1, 0, 1, 2)/10 = (-0.2, -0.1, 0.0, 1.0, 2.0).$$

When we use these weights/coefficients to compute the linear trend in $m = [13, 20, 22, 31, 34]$, we find

$$c = \sum wm = \sum(-0.2, -0.1, 0.0, 0.1, 0.2)(13, 20, 22, 31, 34) = 5.3,$$

which is exactly the previously computed slope of $b = 5.3$.

With these preliminary points established, we can continue to explore contrasts for the mixed design. We will work with the data in Table 21.6, which shows three groups of five subjects. For each subject, we have an audiogram that shows threshold elevation (hearing level) at each of the five frequencies (0.5, 1, 2, 4, and 8 kHz). The means at each frequency within each group are shown in the *rows* labeled "Mean." These are the data plotted in Figure 21.7. Remember, the *y*-axis in Figure 21.7 is reversed to convey the degree of *hearing loss*. The sums of squared deviations (*ss*) are shown below each mean. The last row in Table 21.6, labeled "Means (frequency)," shows the mean threshold elevation, averaged over subjects and groups.

The column labeled "Mean" shows the mean hearing level for each subject, averaged over frequency. The rightmost column labeled "Means (groups)" shows the mean hearing level, averaged over subjects and frequencies within a given group.

Contrasts or Trends for *B* (Within) at Each Level of *A* (Between)

Figure 21.8 shows the data from Table 21.6 arranged in an SPSS data file. The first five columns show threshold elevations for each of the 15 subjects. The column labeled Group shows codes that distinguish members of the three groups of Islanders. The column labeled Linear shows the slope ($c = \sum wm$) relating $\log_2$ frequency to threshold elevation for each of the 15 subjects. The column labeled Mean shows the mean threshold elevation for each subject, averaged over frequency. Both Linear and Mean were computed using the Transform→Compute Variable . . . dialog.

TABLE 21.6 ■ Hearing Levels for Three Groups[a] at Five Frequencies

		Frequency (kHz)						
		0.5	1	2	4	8		
		$\log_2$ (kHz)						Means (Groups)
		−1	0	1	2	3	Mean	
a_1	s_1	−3	20	13	18	37	17	
a_1	s_2	16	11	12	24	22	17	
a_1	s_3	14	16	17	22	16	17	24
a_1	s_4	30	13	23	40	39	29	
a_1	s_5	8	40	45	51	56	40	
Mean		13	20	22	31	34		
SS		580	546	736	780	986	428	
a_2	s_6	13	32	26	28	36	27	
a_2	s_7	11	31	24	39	40	29	
a_2	s_8	25	22	33	29	36	29	34
a_2	s_9	21	21	38	46	54	36	
a_2	s_{10}	25	49	49	58	64	49	
Mean		19	31	34	40	46		
SS		176	506	406	626	624	328	
a_3	s_{11}	12	38	39	47	54	38	
a_3	s_{12}	26	43	26	38	47	36	
a_3	s_{13}	33	28	49	57	63	46	46
a_3	s_{14}	28	33	58	63	68	50	
a_3	s_{15}	41	53	73	55	78	60	
Mean		28	39	49	52	62		
SS		454	370	1286	376	582	376	
Means (frequency)		20	30	35	41	47		

[a] a_1 = Easter Islanders who never left the island; a_2 = Easter Islanders who lived on the continent for 3 to 5 years; a_3 = Easter Islanders who lived on the continent for more than 5 years.

FIGURE 21.8 ■ Simulated Data							
	F.5kHz	F1kHz	F2kHz	F4kHz	F8kHz	Group	Linear	Mean
1	−3	20.0	13	18	37	1	7.8	17
2	16	11.0	12	24	22	1	2.5	17
3	14	16.0	17	22	16	1	1.0	17
4	30	13.0	23	40	39	1	4.5	29
5	8	40.0	45	51	56	1	10.7	40
6	13	32.0	26	28	36	2	4.2	27
7	11	31.0	24	39	40	2	6.6	29
8	25	22.0	33	29	36	2	2.9	29
9	21	21.0	38	46	54	2	9.1	36
10	25	49.0	49	58	64	2	8.7	49
11	12	38.0	39	47	54	3	9.3	38
12	26	43.0	26	38	47	3	3.7	36
13	33	28.0	49	57	63	3	8.9	46
14	28	33.0	58	63	68	3	11.0	50
15	41	53.0	73	55	78	3	7.6	60

Simulated data showing threshold elevation for each of five frequencies, for three groups of subjects.

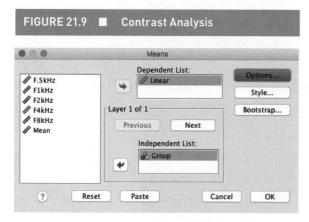

FIGURE 21.9 ■ Contrast Analysis

The Analyze→Compare Means→Means . . . dialog.

To compute the mean linear trend (slope) within each of the three groups, we will use the Analyze→Compare Means→One-Way ANOVA . . . dialog, as shown in Figure 21.9. We can obtain the mean contrast score for a within-groups contrast (the linear trend in this example) through the Contrasts . . . dialog. To do this, I entered the following weights: [1, 0, 0], [0, 1, 0], and [0, 0, 1]. Of course, these weights do not define actual contrasts because they involve only one mean. However, SPSS will use these weights to calculate the statistics necessary to put a confidence interval around the mean value of Linear within each of the three groups.

The results of this analysis are shown in Figure 21.10. Figure 21.10a shows the "contrasts" that were specified, and Figure 21.10b shows the mean value of each contrast and its estimated standard error. The estimated standard errors can be computed in two ways, based on whether we assume that the variances within groups are equal.

When variances are assumed to be equal, the estimated standard errors are computed using equation 21.4, as in the completely between-subjects design. In this case, ms_{error} is based on the sum of squares for the linear contrast scores within the three levels of Group. Therefore, the estimated standard error is associated with $k_a(n-1)$ degrees of freedom. The data in Figure 21.10b (equal variances assumed) provide the estimated standard errors and degrees of freedom required to compute 95% confidence intervals around the three mean values of the contrast scores:

$$5.3 \pm 2.179(1.427) = [2.19, 8.41],$$
$$6.3 \pm 2.179(1.427) = [3.19, 9.41], \text{ and}$$
$$8.1 \pm 2.179(1.427) = [4.99, 11.21].$$

FIGURE 21.10 ■ Contrasts and Standard Errors

(a)   **Contrast Coefficients**

Contrast	Group 1	Group 2	Group 3
1	1	0	0
2	0	1	0
3	0	0	1

(b)   **Contrast Tests**

		Contrast	Value of Contrast	Std. Error	t	df	Sig. (2–tailed)
Linear	Assume equal variances	1	5.300[a]	1.4268	3.715	12	.003
		2	6.300[a]	1.4268	4.416	12	.001
		3	8.100[a]	1.4268	5.677	12	.000
	Does not assume equal variances	1	5.300[a]	1.7661	3.001	4.000	.040
		2	6.300[a]	1.2178	5.173	4.000	.007
		3	8.100[a]	1.2268	6.603	4.000	.003

a. The sum of the contrast coefficients is not zero.

Statistics for the linear trends within the three groups. The means, estimated standard errors, and degrees of freedom can be used to compute confidence intervals around the contrasts, both when variances are assumed to be equal and when they're not.

For these confidence intervals, $t_{\alpha/2} = 2.179$, based on 12 degrees of freedom and $\alpha = .05$.

When variances are not assumed to be equal, the estimated standard errors are computed using equation 21.5. That is, they are based only on the variances of the groups contributing to the contrast. In this case, only one group is associated with the "contrast." The degrees of freedom are adjusted using an elaborated version of the Welch-Satterthwaite procedure described in Appendix 11.4. Because only one mean is involved in this contrast, the degrees of freedom are reduced to $(n-1)$. The data in Figure 21.10b (equal variances not assumed) provide the estimated standard errors and degrees of freedom required to compute 95% confidence intervals around the three mean values of the contrast scores:

$$5.3 \pm 2.776(1.766) = [0.40, 10.20],$$

$$6.3 \pm 2.776(1.218) = [2.92, 9.68], \text{ and}$$

$$8.1 \pm 2.776(1.227) = [4.69, 11.51].$$

For these confidence intervals, $t_{\alpha/2} = 2.776$, based on 4 degrees of freedom and $\alpha = .05$.

The same confidence intervals can be obtained through the Analyze→Descriptive Statistics→Explore . . . dialog. Of course, all of this could have been done just as easily in Excel.

Contrasts (Trends) for *B* (Within) Averaged Over Levels of *A* (Between)

In some cases, we may want to compute trends (or contrasts) based on the means of the within-subjects factor. In our example, this would mean analyzing the main effect of frequency; i.e., the means in the row labeled "Means (frequency)" in Table 21.6. To conduct this kind of contrast or trend analysis, we first compute a *contrast score* (*cs*) for each subject and each contrast, as we did in Figure 21.8 (see the column labeled Linear). We saw in Chapter 18 that the general form of this computation is

$$cs_j = \sum_{i=1}^{k_b} w_i y_{i,j},$$

(21.6)

where j is the jth subject, and k_b is the number of levels of B. As always, we assume that the contrast weights have been properly scaled. We then compute the mean contrast across levels of the between-subjects variable (A) as follows:

$$c = \frac{\sum_{i=1}^{k_a} m_{cs_i}}{k_a},$$

(21.7)

where k_a is the number of levels of A, and m_{cs_i} is the mean contrast score at the ith level of A. For our example, c is the average of the means for groups 1, 2, and 3 as shown in Figure 21.10. That is,

$$c = \frac{\sum_{i=1}^{k_a} m_{cs_i}}{k_a} = \frac{5.3 + 6.3 + 8.1}{3} = 6.567.$$

The estimated standard error for this kind of contrast is based on the sums of squares for the contrast scores within groups. The first step is to compute ms_{error}, which is done as follows:

$$ms_{\text{error}} = \frac{\sum_{i=1}^{k_a} ss_i}{\sum_{i=1}^{k_a} df_i},$$

where ss_i is the sum of squares of the contrast scores (cs) within the ith group. The sums of squares for our example are $ss_1 = 62.38$, $ss_2 = 29.66$, and $ss_2 = 30.10$, and they sum to $\Sigma ss_i = 122.14$. (These calculations are not shown, but you can verify them in Excel.) Because there are 4 degrees of freedom within each of the three groups, $ms_{\text{error}} = 122.14/12 = 10.178$.

Having computed ms_{error}, we compute the standard error of the contrast as

$$s_c = \sqrt{\frac{ms_{\text{error}}}{\sum_{i=1}^{k_a} n_i}},$$

(21.8)

which yields

$$s_c = \sqrt{\frac{ms_{\text{error}}}{\sum_{i=1}^{k_a} n_i}} = \sqrt{\frac{10.178}{15}} = 0.824.$$

To compute the 95% confidence interval around the mean contrast, $c \pm t_{\alpha/2}(s_c)$, we require $t_{\alpha/2}$, which is based on $\sum_{i=1}^{k_a} df_i = 12$ degrees of freedom in this example. Therefore, $t_{\alpha/2} = 2.179$, and the 95% confidence interval is

$$CI = c \pm t_{\alpha/2}(s_c) = 6.567 \pm 2.179(0.824) = [4.77, 8.36].$$

All of these computations can be done as easily in Excel as in SPSS.

We've seen before that SPSS will only test contrasts for statistical significance. If you run the data in Figure 21.8 through a mixed ANOVA and ask for polynomial trends for the within-subjects factor, you will find that the F-statistic for the linear trend in B, averaged over A, is $F_{obs} = 63.548$. The same result can be obtained from the quantities we computed above; that is,

$$F_{obs} = \left(\frac{c}{s_c}\right)^2 = \left(\frac{6.567}{0.824}\right)^2 = 63.548.$$

Contrasts (Trends) for A (Between)

There are several ways to use contrasts for the between-subjects factor. For example, one could compare the mean threshold elevation for groups 1 and 3 (i.e., two levels of the between-subjects factor, A) at a given frequency (i.e., a given level of the within-subjects factor, B). That is, we can compute a confidence interval for $m_{a_1b_1} - m_{a_3b_1}$. For our example, this means contrasting the mean hearing level at .5 kHz for those who never left the island and those who had been away for 5 or more years. Or we could compute the contrast $m_{a_1b_1} - (m_{a_2b_1} + m_{a_3b_1})/2$. That is, we could contrast the mean hearing level at .5 kHz for those who never left the island and those who had been away for 3 or more years. Or we could average over all levels of B (see Figure 21.8, column "Mean") and compute $m_{a_1} - (m_{a_2} + m_{a_3})/2$. That is, we could contrast the mean hearing level averaged over all frequencies for those who never left the island and those who had been away for 3 or more years.

In all cases, the contrasts are computed as in a one-factor between-subjects design. The only thing that changes from situation to situation is the dependent variable. As we've said, these could be the scores at a given level of B, the average scores across several levels of B, or the average scores over all levels of B. This means that the estimated standard error for each contrast is based on the mean squared error defined as

$$ms_{error} = \sum_{i=1}^{k_a} \frac{ss_i}{df_i}, \tag{21.9}$$

where ss_i is the sum of squares within groups for a specific level of the dependent variable. The estimated standard error is then computed exactly as in the one-factor between-subjects design; that is,

$$s_c = \sqrt{ms_{error} \sum_{i=1}^{k_a} \frac{w_i^2}{n_i}}. \tag{21.10}$$

We will consider two examples, starting with the contrast $m_{a_1b_1} - m_{a_3b_1}$. Looking back to Table 21.6, we see that $m_{a_1b_1} = 13$ and $m_{a_3b_1} = 28$. Therefore, $c = m_{a_1b_1} - m_{a_3b_1} = 13 - 28 = -15$. Table 21.6 also shows that the sum of squares for the three levels of A associated with b_1 (.5 kHz) are 580, 176, and 454. Therefore, from equation 21.9, we find that $ms_{error} = 100.883$; from equation 21.10, we find that $s_c = 6.351$. (Remember, because we are comparing $m_{a_1b_1}$ and $m_{a_3b_1}$, the contrast weights are $[1, 0, -1]$). From these numbers, we can compute the 95% confidence interval around the contrast as follows:

$$CI = c \pm t_{\alpha/2}(s_c) = -15 \pm 2.179(6.351) = [-28.837, -1.163].$$

As before, $t_{\alpha/2} = 2.179$ because we have 12 degrees of freedom.

The second contrast we'll consider is $m_{a_1} - m_{a_3}$, which is the difference between the mean of all scores in a_1 and all scores in a_3. Table 21.6 shows that $m_{a_1} = 24$ and $m_{a_3} = 46$, so $m_{a_1} - m_{a_3} = 24 - 46 = -22$. Table 21.6 also shows that the sum of squares for the three levels of A, averaged over B, are 428, 328, and 376. Therefore, from equation 21.9, we find that $ms_{\text{error}} = 94.333$; from equation 21.10, we find that $s_c = 6.143$. (Again, because we are comparing m_{a_1} and m_{a_3}, the contrast weights are [1, 0, −1]). From these numbers, we can compute the 95% confidence interval around the contrast as

$$\text{CI} = c \pm t_{\alpha/2}(s_c) = -22 \pm 2.179(6.143) = [-35.384, -8.616].$$

Again, $t_{\alpha/2} = 2.179$ because we have 12 degrees of freedom.

To conduct between-subjects contrasts in SPSS, we use the One-Way ANOVA dialog, just as in Chapter 18. Figure 21.11 shows this dialog, with seven variables moved into the Dependent List. The first five are columns that correspond to the five levels of B (frequencies), and the sixth is the mean score for each subject averaged over levels of B. (We will discuss the seventh column, Linear, in the following section.) Through the Contrasts dialog, I specified the contrast [1, 0, −1], which compares the first and third levels of A for each of the variables in the Dependent List.

The results of the contrast analysis are shown in Figure 21.12. Figure 21.12a shows a description of the contrast to be applied (i.e., [1, 0, −1]) to each variable in the Dependent List. Figure 21.12b shows the contrasts. As we noted before, SPSS only conducts significance tests for contrasts, but you can verify that for the contrasts on F.5kHz and Mean, the values of the contrasts (−15 and −22) and their estimated standard errors (6.351 and 6.143) are exactly as computed above. To complete the calculation of the confidence intervals, you only need to know $t_{\alpha/2}$, which (as before) is 2.179 because we have 12 degrees of freedom in all cases.

Also shown in Figure 21.12 are the standard errors and adjusted degrees of freedom used when the assumption of equal variances is not made. In these cases, the estimated standard errors are computed using equation 21.5 as follows:

$$s_c = \sqrt{\sum_{i=1}^{k} w_i^2 \frac{s_i^2}{n_i}}.$$

To find $t_{\alpha/2}$ for the adjusted degrees of freedom, we must use **R**, as described in the section on the two-factor between-subjects design.

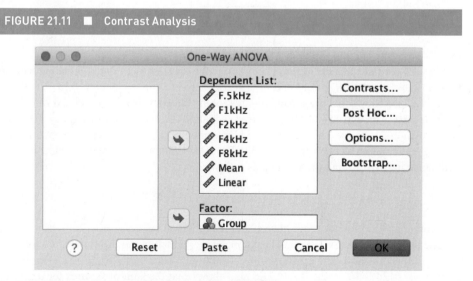

FIGURE 21.11 ■ Contrast Analysis

Conducting between-subjects contrasts in a mixed ANOVA.

FIGURE 21.12 ■ Contrast Analysis

(a)   **Contrast Coefficients**

Contrast	Group 1	Group 2	Group 3
1	1	0	−1

(b)   **Contrast Tests**

		Contrast	Value of Contrast	Std. Error	t	df	Sig. (2-tailed)
F.5kHz	Assume equal variances	1	−15.00	6.351	−2.362	12	.036
	Does not assume equal variances	1	−15.00	7.190	−2.086	7.883	.071
F1kHz	Assume equal variances	1	−19.000	6.8848	−2.760	12	.017
	Does not assume equal variances	1	−19.000	6.7676	−2.808	7.715	.024
F2kHz	Assume equal variances	1	−27.00	8.996	−3.001	12	.011
	Does not assume equal variances	1	−27.00	10.055	−2.685	7.449	.030
F4kHz	Assume equal variances	1	−21.00	7.707	−2.725	12	.018
	Does not assume equal variances	1	−21.00	7.603	−2.762	7.129	.028
F8kHz	Assume equal variances	1	−28.00	8.548	−3.276	12	.007
	Does not assume equal variances	1	−28.00	8.854	−3.162	7.502	.015
Mean	Assume equal variances	1	−22.00	6.143	−3.581	12	.004
	Does not assume equal variances	1	−22.00	6.340	−3.470	7.967	.008
Linear	Assume equal variances	1	−2.800	2.0178	−1.388	12	.190
	Does not assume equal variances	1	−2.800	2.1503	−1.302	7.131	.233

(a and b) Contrasts between a_1 and a_3 for each of the variables in the **Dependent List** from Figure 21.11.

Interaction Contrasts

Finally, we will consider interaction (second-order) contrasts for the mixed design. The simplest way to do this is to compute a new variable, representing a properly scaled contrast on the within-subjects variable, and then compute contrasts exactly as in the previous section across levels of the between-subjects variable. For example, we might ask how the linear trend in B changes between the first and last levels of A (i.e., between a_1 and a_3). That is, we might ask how the linear trend in frequency differs between those who never left the island and those who had been away for 5 or more years. We previously computed properly scaled linear trend scores for frequency (for each subject) and stored the results in the variable Linear. To assess the difference between the mean linear trend at a_1 and a_3, we simply conduct the analysis as in the previous section.

Remember that the mean linear trend at a_1 was 5.3 (see Figure 21.3) and the mean linear trend at a_3 was 8.1. Therefore, the difference between these mean trends is $5.3 - 8.1 = -2.8$, as shown in the last row of Figure 21.12. The standard error for this contrast can be computed in the usual way,

$$ms_{\text{error}} = \frac{\sum_{i=1}^{k_a} ss_i}{\sum_{i=1}^{k_a} df_i},$$

where ss_i is the sum of squares of the linear trend scores (cs) within the ith group. Earlier, when we computed the linear trend for the main effect of B, we found that $ms_{error} = 10.178$. Therefore, using the usual formula for a between-subjects contrast, we find the estimated standard error of the interaction contrast is

$$s_c = \sqrt{ms_{error} \Sigma_{i=1}^{k_a} \frac{w_i^2}{n_i}} = \sqrt{10.178 \frac{2}{5}} = 2.0178.$$

This is the standard error shown for the variable Linear in Figure 21.12 when equal variances are assumed. From the contrast $c = -2.8$ and its estimated standard error $s_c = 2.0178$, we can compute the 95% confidence interval as follows:

$$CI = c \pm t_{\alpha/2}(s_c) = -2.8 \pm 2.179(2.0178) = [-7.196, 1.596].$$

If we can't assume that variances are equal, we could compute the estimated standard error as follows:

$$s_c = \sqrt{\Sigma_{i=1}^{k} w_i^2 \frac{s_i^2}{n_i}} = \sqrt{1^2 \left(\frac{15.595}{5} \right) + 0^2 \left(\frac{7.415}{5} \right) + 1^2 \left(\frac{7.252}{5} \right)} = 2.1503,$$

where the variances are derived from ss and df_{error}. Once again, the fractional degrees of freedom (7.131 in this case) can be used to calculate $t_{\alpha/2}$ using **R**. For this contrast, $t_{\alpha/2} = 2.356$, and the corrected confidence interval would be

$$CI = c \pm t_{\alpha/2}(s_c) = -2.8 \pm 2.356(2.1503) = [-7.866, 2.266].$$

APA Reporting

Audiograms were obtained from three groups of older individuals born on Easter Island. One group ($n = 5$) had lived on the island all their lives. A second group ($n = 5$) had lived on the continent for 3 to 5 years. A third group ($n = 5$) had lived on the continent for more than 5 years. Hearing levels (threshold elevation in decibels) were obtained at 0.5, 1, 2, 4, and 8 kHz. The mean hearing levels measured at the five frequencies for the three groups are plotted in Figure 21.7.

For all three groups, hearing levels increased with frequency. The mean slopes of the linear functions relating hearing level to frequency were as follows for groups 1, 2, and 3, respectively: 5.3, 95% CI [0.40, 10.20]; 6.3, 95% CI [2.92, 9.68]; and 8.1, 95% CI [4.69, 1.51]. A second-order contrast comparing the mean linear slopes for groups 1 and 3 shows a mean difference of $5.3 - 8.1 = -2.8$, 95% CI [−7.20, 1.60], suggesting a faster hearing loss with frequency for those who had lived off-island for more than 5 years versus those who had never left the island.

Contrasts for mean hearing levels between groups 1 and 3 at each of the five frequencies showed differences at 0.5 to 8 kHz of −15, 95% CI [−28.84, −1.16]; −19, 95% CI [−34.00, −4.00]; −27, 95% CI [−46.60, −7.40]; −21, 95% CI [−37.79, −4.21]; and −28, 95% CI [−46.62, −9.38], respectively. Mean hearing levels, averaged over frequency, were 24 dB for those who had never left the island and 46 dB for those who had lived off-island for more than 5 years: $24 - 46 = -22$, 95% CI [−35.38, −8.62]. These data strongly suggest that living off-island leads to greater hearing loss because of exposure to greater levels of ambient noise. However, it must be kept in mind that subjects were not randomly assigned to conditions and other moderating factors we not controlled. Therefore, it is possible that some other factor explains this difference between groups.

LEARNING CHECK 3

1. The data in the following table are from a two-factor mixed design. There are three levels of A and three levels of B. A is a between-subjects variable and B is a within-subjects variable. B is a quantitative variable with values of 5, 10, and 15. Plot the means with levels of B on the x-axis and each level of A as a separate line. Use 95% confidence intervals as the error bars.

	a_1				a_2				a_3		
Subject	b_1	b_2	b_3	Subject	b_1	b_2	b_3	Subject	b_1	b_2	b_3
s_1	7	21	12	s_7	12	38	16	s_{13}	9	21	29
s_2	11	22	10	s_8	13	33	18	s_{14}	8	22	29
s_3	9	18	7	s_9	15	32	12	s_{15}	9	18	30
s_4	10	18	6	s_{10}	15	32	11	s_{16}	10	19	27
s_5	11	28	9	s_{11}	15	36	15	s_{17}	7	18	31
s_6	16	16	8	s_{12}	17	36	19	s_{18}	11	16	30

2. Use the data in the table to compute 95% confidence intervals for the following contrasts and linear trends. Scale the linear trends so that they compute the slope of the linear regression line relating y to x.

   (a) the linear trend in B averaged over levels of A,

   (b) the linear trend in B for each of the three levels of A, assuming equal variances,

   (c) the difference in the linear trend in B at levels a_1 and a_2, and

   (d) the difference in the linear trend in B at levels a_1 and a_3.

Answers

1. The data plot is shown to the right.

2. (a) $c = 0.63$, $s_c = 0.08$, 95% CI [0.46, 0.81].

   (b) $c_{L,a_1} = -0.20$, $s_c = 0.14$, 95% CI [−0.51, 0.11].
   $c_{L,a_2} = 0.07$, $s_c = 0.14$, 95% CI [−0.24, 0.37].
   $c_{L,a_3} = 2.03$, $s_c = 0.14$, 95% CI [1.73, 2.34].

   (c) $c = -0.27$, $s_c = 0.20$, 95% CI [−0.70, 0.17].

   (d) $c = -2.23$, $s_c = 0.20$, 95% CI [−2.67, −1.80].

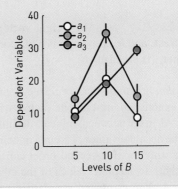

In conclusion, ANOVA is one of the most commonly used statistical analyses in psychology and related disciplines. In Chapters 18 to 20, we saw how to test the variability associated with omnibus effects for statistical significance. Although this is a routine part of reporting an ANOVA, it often serves no practical purpose. Rather, the researcher's interest is in how means differ, and the way to address this question is through contrasts of the sort discussed in this chapter. In my view, we should be thinking of the contrasts we wish to perform when designing experiments or quasi-experiments. Contrasts are conceptually simpler than omnibus effects and more directly related to our research questions. My hope is that analysis of omnibus effects will fall out of favor sooner rather than later.

SUMMARY

When there are two independent variables in an experiment or quasi-experiment, it is possible to partition variability into components associated with the two main effects and the interaction, and then test these for statistical significance through omnibus F-ratios. Unfortunately, these omnibus tests rarely provide answers to questions of interest to researchers. The questions of interest *are* answered by contrasts, and the relevant contrasts should be in the researcher's mind before data are collected. Assessing these planned contrasts is not dependent in any way on the results of omnibus tests.

For the *completely between-subjects design*, simple and complex contrasts are computed exactly as in the one-factor between-subjects design. In these cases, it is important that the contrast weights sum to 0 and that the absolute values of the contrast weights sum to 2. If contrasts represent the product of two contrast vectors, we call them *second-order* or *interaction contrasts*. In this case, it is important that the contrast weights sum to 0 and that their absolute values sum to 4. The estimated standard error for completely between-subjects contrasts is

$$s_c = \sqrt{ms_{error} \sum_{i=1}^{k} \frac{w_i^2}{n_i}},$$

where $ms_{error} = \Sigma ss_i / \Sigma df_i$, w is the weight vector that defines the contrast, and n_i is the number of scores in the ith condition. $t_{\alpha/2}$ is based on $df_{error} = df_{total} - df_a - df_b - df_{ab}$.

For the *completely within-subjects design*, contrasts are computed exactly as in the one-factor within-subjects design. As in the between-subjects design, it is important that the absolute values of simple or complex contrast weights sum to 2. For second-order contrasts, it is important that the absolute values of the contrast weights sum to 4. The estimated standard error for these contrasts is

$$s_c = \sqrt{\frac{s_{cs}^2}{n}},$$

where s_{cs}^2 is the variance of the contrast scores (cs), and n is the number of subjects. $t_{\alpha/2}$ is based on $df_{subjects} = n-1$.

For the *mixed design*, many contrast types are possible.

● Contrasts that compare means across levels of the between-subjects variable for a single level of the within-subjects variable are computed as in the one-factor between-subjects design, with $t_{\alpha/2}$ based on $df_{s/a} = k_a(n-1)$, where k_a is the number of levels of the between-subjects factor.

● Contrasts that compare means across levels of the within-subjects variable for a single level of the between-subjects variable are computed from contrast scores as in the one-factor within-subjects design. When the estimated standard error is computed from sums of squares for contrast scores within all levels of the between subjects factor, then $t_{\alpha/2}$ based on $\sum_{i=1}^{k_a} df_i$, where k_a is the number of levels of the between-subjects variable.

● Contrasts for the main effect of the between-subjects variable are computed from subject means over the within-subjects variable. Everything proceeds as in the case of the one-factor between-subjects design.

● Contrasts for the main effect of the within-subjects variable are computed from contrast scores for all subjects averaged over the between-subjects factor. The estimated standard error of the contrast is

$$s_c = \sqrt{\frac{ms_{error}}{n*k}},$$

where $ms_{error} = \Sigma ss_i / \sum_{i=1}^{k_a} df_i$ is computed from the sums of squares for the contrast scores within each level of the between-subjects factor. $t_{\alpha/2}$ is based on $df_{s/a} = \sum_{i=1}^{k_a} df_i$, where k_a is the number of levels of the between-subjects variable.

● Interaction (second-order) contrasts are computed from the means of contrast scores within levels of the between-subjects factor. The estimated standard error for these contrasts is

$$s_c = \sqrt{ms_{error} \sum_{i=1}^{k} \frac{w_i^2}{n_i}},$$

where $ms_{error} = \Sigma ss_i / \sum_{i=1}^{k_a} df_i$ is computed from the sums of squares for the contrast scores within each level of the between-subjects factor. $t_{\alpha/2}$ is based on $df_{s/a} = \sum_{i=1}^{k_a} n_i$.

KEY TERMS

second-order (interaction) contrast   594

EXERCISES

Definitions and Concepts

1. What is a second-order contrast?

True or False

State whether the following statements are true or false.

2. For a two-factor between-subjects design, the estimated standard error of a contrast is $s_c = \sqrt{ms_{error} \sum_{i=1}^{k} w_i/n_i}$.

3. To be a true contrast, the weights of a second-order contrast should sum to 4.

4. To be a true first-order contrast, the absolute values of the contrast weights should sum to 2.

5. Confidence intervals around between-subjects contrasts will generally be narrower when the equal variances assumption is violated.

6. If the three levels of a between-subjects variable are drug type, then linear and quadratic trends can be computed.

Calculations

7. The following data are from a two-factor ANOVA (available at study.sagepub.com/gurnsey.)

	b_1	b_2	b_3	b_4
a_1	9	9	20	18
a_1	10	13	11	14
a_1	8	16	12	24
a_1	5	10	17	20
a_2	9	14	8	13
a_2	4	8	7	9
a_2	6	6	9	7
a_2	5	8	12	11

(a) Assume that these data were derived from a two-factor, between-subjects design. Use Excel or SPSS to compute 95% confidence intervals for the following contrasts

(i) $b_1 - b_4$ at a_1,

(ii) $b_1 - b_4$ at a_2,

(iii) $a_1 - a_2$ at b_1,

(iv) $a_1 - a_2$ at b_4,

(v) $a_1 - a_2$ averaged over levels of B, and

(vi) the interaction of $(b_1 + b_2)/2 - (b_3 + b_4)/2$ with A.

(b) Compute the contrasts in part (a) assuming that the data were obtained in a two-factor within-subjects design.

(c) Compute the contrasts in (a) assuming that the data were obtained in a two-factor mixed design, where B is the within-subjects factor.

Scenarios

8. In Chapter 20, we described a visual search task in which participants are asked to report the presence or absence of a target as quickly as possible in a variable number of nontarget items. For example, in the display below, the target is Я and the nontarget items are Rs. To make the task more difficult, target and nontarget items are randomly rotated. In this task, the dependent variable is the *time* (measured in milliseconds) required to press a button, after display onset, to indicate that the target is present or absent. The two independent variables are (i) *display type* (i.e., target present or target absent) and (ii) *display size* (in this example, there were 5, 10, 15, or 20 items in each display).

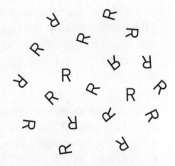

The data are presented in the following table (and are available at study.sagepub.com/gurnsey). The researcher expects that (i) subjects will be able to report the presence of the target faster than its absence, (ii) response times will increase with display size, and (iii) the difference in reaction times for target-present and target-absent conditions will increase with display size. Normally, the visual search task is run completely within subjects; for each of the exercises below, analyze the data in this way.

	a_1^a				a_2			
	b_1	b_2	b_3	b_4	b_1	b_2	b_3	b_4
	313	629	692	780	440	717	1041	1397
	675	598	871	903	778	1010	1419	1120
	560	702	722	548	539	1012	806	1731
	486	515	607	596	337	1049	1065	1290
	523	382	569	692	847	1202	1501	1388
	475	386	576	662	481	829	751	1295
	314	386	534	852	463	838	1453	1341
Mean	478	514	653	719	555	951	1148	1366
SS	101,752	105,818	83,068	103,614	209,938	162,216	583,986	206,628

[a] a_1 = target present; a_2 = target absent; b_1 = 5, b_2 = 10, b_3 = 15, and b_4 = 20.

(a) Plot the data in the table (using $\pm SEM$ as error bars) and comment on your initial observations.

(b) Use Excel or SPSS (where possible) to compute uncorrected 95% confidence intervals around the difference in reaction times for target-present versus target-absent conditions for each display size (i.e., $a_2 b_i - a_1 b_i$).

(c) Use Excel or SPSS (where possible) to compute uncorrected 95% confidence intervals around the linear trend in display size (B) for the two display types (A). Make sure to scale the linear trends so that they represent the slope of the linear regression line relating reaction time to display size.

(d) Create a new variable to represent the difference in slope between target-present and target-absent conditions for each subject. Use Excel or SPSS (where possible) to compute uncorrected 95% confidence intervals around the mean of these slope differences. Would you consider this a second-order contrast?

(e) What assumptions must be satisfied for these confidence intervals to be valid?

9. As we discussed in Chapter 20, there is some evidence that lab rats become addicted to drugs easily because they are raised in isolated cages and deprived of any other stimulation to compete with the drugs (Alexander, Coambs, & Hadaway, 1978). Bozarth, Murray, and Wise (1989) investigated the role of social isolation on rats' self-administration of heroin.

Bozarth et al. (1989) used 39 Long-Evans rats in their experiment. All rats were housed for the first 2 to 3 months of their lives in 24- × 24- × 9-inch cages with 25 rats per cage. Twenty-two of these rats were then housed for the next 6 to 8 weeks in small individual cages (about 7 × 10 × 7 inches), isolated from all interactions with other rats. The remaining 17 rats were housed in larger cages (about 18 × 40 × 15 inches) with about 10 rats per cage. "[T]hese rats displayed normal play behavior, dominance struggles, and social grooming" (Bozarth et al., 1989, p. 904).

iStock.com/fotografixx

After about a week in their new cages, all rats were trained to bar press for a food reward. After

3 to 4 days of training, all rats underwent surgery that allowed for heroin to be administered directly into their bloodstream.

For the next 5 weeks, each rat spent 2 hours a day in a small test cage (about $10 \times 10 \times 10$ inches) with an operant lever (bar). Each time a rat pressed the bar, a small dose of heroin (0.1 mg/kg/injection) was delivered into the bloodstream. Bozarth et al. (1989) recorded the mean number of injections/hour received by each rat, averaged over 5-day blocks. Data that are qualitatively similar to those reported by Bozarth et al. (1989) are available at study.sage pub.com/gurnsey The means and standard deviations of these data are shown in the following table.

	a_1[a]					a_2				
	b_1	b_2	b_3	b_4	b_5	b_1	b_2	b_3	b_4	b_5
Mean	0.931	1.305	1.817	2.301	2.570	1.934	2.764	3.343	3.855	3.849
SD	1.124	1.092	1.121	1.174	2.086	1.403	2.031	1.998	1.794	2.104

[a] a_1 = animals housed in groups; a_2 = animals housed in isolation; b_1 = week 1, b_2 = week 2, b_3 = week 3, b_4 = week 4, and b_5 = week 5.

(a) Plot the data from the table (using $\pm SEM$ as error bars) and comment on your initial observations.

(b) Compute the linear trend and the uncorrected 95% confidence interval for the within-subjects factor at both levels of the between-subjects factor. Make sure the trend weights are scaled so that they compute the slope of the linear regression equation relating weeks to the amount of heroin received. Assume equal variances.

(c) Compute the interaction (second-order) contrast on the linear trend for the within-subjects factor. Compute the uncorrected 95% confidence interval around the difference in the mean slopes for the grouped and isolated rats. Assume equal variances.

(d) Use Excel or SPSS (where possible) to compute uncorrected 95% confidence intervals around the difference in drug intake for grouped versus isolated rats for each week (i.e., $a_2b_i - a_1b_i$). Assume equal variances.

(e) The animals in the isolation group spent only about 6 weeks of their lives housed in isolation. What effect does this seem to have on their tendency to work for a drug reward?

10. As we described in Chapter 20, Glanzer and Cunitz (1966) investigated long- and short-term memory in a sample of 46 US Army personnel (all male). On each trial of the experiment, a subject was presented with a list of 15 one-syllable nouns. After presentation, the subject was given 1 to 5 minutes to recall as many words as possible from the list.

The subjects recalled the lists 0, 10, or 30 seconds after the last word was presented. For the 10- and 30-second delays, subjects counted from a randomly chosen number until the experimenter told them to stop and write down as many words as they could remember from the list. Counting was intended to prevent rehearsal.

Each subject was presented with five lists at each of the three delays, and the researchers recorded the number of correct responses at each serial position (number correct at the first position, second position, up to the 15th position). The means and standard deviations of the 46 scores in each condition are shown in the following table. Raw data with characteristics similar to those reported by Glanzer and Cunitz (1966) are available at study.sagepub.com/gurnsey.

	a_1[a]														
	b_1	b_2	b_3	b_4	b_5	b_6	b_7	b_8	b_9	b_{10}	b_{11}	b_{12}	b_{13}	b_{14}	b_{15}
Mean	3.54	2.91	2.20	1.87	1.83	1.74	1.52	1.85	2.09	1.43	1.93	1.85	2.39	2.59	3.00
SD	0.98	1.15	1.17	1.19	1.02	1.02	1.01	1.09	1.17	1.07	1.14	1.05	1.22	1.24	1.14
	a_2														
Mean	3.57	2.85	2.20	2.00	1.83	1.54	1.50	1.76	1.59	1.72	1.85	1.74	1.91	1.87	2.04
SD	0.98	1.07	1.07	1.12	1.12	0.98	1.01	1.12	1.05	1.13	1.09	1.08	1.11	1.07	1.15
	a_3														
Mean	3.11	2.54	1.85	1.93	1.72	1.59	1.17	1.87	1.74	1.57	1.78	1.74	1.78	1.76	1.52
SD	1.12	1.11	1.13	1.08	1.11	1.17	0.88	1.09	1.08	1.05	1.07	1.08	1.05	1.08	0.98

[a] a_1 = immediate recall; a_2 = delayed recall (10); a_3 = delayed recall (30); $b_1 \ldots b_{15}$ = serial m position in the list.

(a) Plot the data in the table with error bars representing ±SEM and put into words what the data seem to reflect about memory. (The data have many interesting features.)

(b) Compute the linear trends (and their uncorrected 95% confidence intervals) for the first five serial positions (i.e., 1 to 5) for each of the three delays. Scale the linear trend weights so that they show the slope of the regression line relating serial position to number recalled. Do these trends seem different to you?

(c) Compute the linear trends (and their uncorrected 95% confidence intervals) for the last five serial positions (i.e., 11 to 15) for each of the three delays. (Scale the trend weights as before.) Do these trends seem different to you?

(d) Compute the uncorrected 95% confidence interval for $a_1 - a_2$, $a_1 - a_3$, and $a_2 - a_3$ at b_{15}.

(e) Put into words what seem to be the most salient features of these data and explain what they mean without using the term *statistically significant*.

SELECTED ANSWERS TO CHAPTER EXERCISES

Please note that asterisks indicate questions relating to material covered in the appendixes or requiring insight beyond what was covered in the chapter.

CHAPTER 1

Definitions and Concepts

1. Variables are physical or abstract attributes, or quantities that we wish to measure. A variable can take on specific values. A score is the value that an individual has on a particular variable.

3. Qualitative variables have values that are qualities or categories. They are also referred to as nominal or categorical variables. There is no natural ordering of the values of such variables.

5. Equal interval scales have units of measurement, such as inches, meters, hours, degrees, pressure, or intensity.

7. The values of variables measured on an ordinal scale are qualitative and discrete, like qualitative variables, but they also have a natural ordering, like quantitative variables.

9. An operational measure is a tool used to measure a psychological construct. Very often, operational measures are derived from questionnaires. Operational measures may also derive from the speed and accuracy with which psychological tasks are completed.

11. A reliable measuring device gives very similar (if not identical) measurements each time it is applied to the same object.

13. A population comprises the scores of individuals (that share some characteristic of interest) on a variable of interest. A sample is any subset of a population.

15. Estimating the mean weight of all bluefin tuna in the Mediterranean Ocean from a random sample of all bluefin tuna in the Mediterranean Ocean is an instance of inferential statistics.

17. Sampling bias means that not all members of the population had an equal chance of being selected in a sample. Sampling bias can and should be avoided.

19. Sampling error is the difference between a statistic and the parameter it estimates. Sampling error cannot be avoided; it is an inevitable feature of random sampling.

True or False

21. False. There is no natural ordering to eye color.

23. True.

25. True.

27. False. One could be highly intelligent but not know the rules of Sudoku.

29. True.

31. False. This is an example of sampling bias.

33. False. A parameter is a numerical characteristic of a population.

35. True.

37. False. This is a biased sample because not all members of the population had an equal chance of being part of the sample.

Scenarios

39. There are 30 puzzle-solving times.

   (a) This is a discrete variable because all times have been measured to the nearest second.

   (b) This is a ratio scale because it has an absolute zero point and time is broken up into equal intervals of one second each.

   (c) This conclusion is not valid because the sample was drawn from the participant pool at the researcher's university. Therefore, the sample was not a random sample from the population of interest.

CHAPTER 2

Definitions and Concepts

1. A frequency table is a depiction of the number, or proportion of scores in a sample or population having each value (or range of values) of a variable.

3. A relative frequency is the proportion of scores (as opposed to the number of scores) having a particular value, or falling within a range of values, for a given a variable.

5. Relative frequency is the proportion of scores having some value or falling within a given interval. Cumulative relative frequency is the proportion of scores at or below a given value or interval of a quantitative variable.

7. The real limits of an interval are the minimum and maximum real values that define the interval. The real limits are defined as *midpoint ± width*/2.

9. A sampling experiment is the random selection of one of the possible values of a variable.

11. An outcome is a single value of a variable, whereas an event may involve many values of a variable. For example, an event might be "drawing an Ace or a Queen," whereas the Ace of Hearts is an outcome.

13. Mutually exclusive events are events that cannot co-occur.

15. A conditional probability is the probability of an event occurring, given that some condition is true.

True or False

17. True.

19. False. 19.4999 would fall in the interval 10–19; it is below the lower real limit of the interval 20–29.

21. True.

23. False. If our sample is large and there are many values of the variable, we can have a large number of intervals.

25. False. The lower score limit must be a multiple of 10.

27. False. For a histogram, there is a natural ordering to values on the x-axis and the bars for the histogram touch. For a bar graph, there is no natural ordering to values on the x-axis and the bars for the histogram do not touch.

29. False. The range in milliseconds is 1365, so an interval width of 20 would produce about 70 intervals.

Scenarios

31. The frequency table and bar graph are as follows:

Distribution of Political Affiliations in 165 Voters		
Value	f	p
Democrat	70	.424
Republican	63	.382
Independent	32	.194
	n = 165	

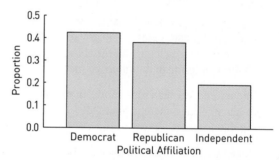

Bar graph showing the distribution of political affiliations in 165 American voters.

33. The range is 51.7. I would aim for approximately 10 intervals, which means that an interval width of 5 is about right. The lowest interval would be 10–14, and the highest interval would be 65–69. This means there would be 12 intervals. An interval width of 10 would also make sense. This will produce six intervals, which seems a little small given the number of scores we have.

35. A proportion is the number of times an event occurred divided by the number of sampling experiments. Probability is the proportion of times an event will occur in an infinite series of identical sampling experiments.

36. (a) $p = 4/52 = .0769$.
    (c) $p = 2/6 = .3333$.

(e) $p = 1/2 = .5$.

(g) $p = .382$.

(i) $p = 8/52 = .1538$.

*(k) $p = .5385$.

CHAPTER 3

Definitions and Concepts

1. Central tendency is a number that represents a typical score in a distribution.

3. The median is the number that divides a set of numbers into two groups of equal size.

5. Dispersion refers to how spread out scores are in a distribution of a quantitative variable.

7. A deviation score is the difference between a score and the mean of the sample or population from which the score came.

9. In a population, the variance is the average squared deviation from the mean $[\sigma^2 = \Sigma(y - \mu)^2/N]$.

11. The standard deviation is the square root of the variance.

13. A skewed distribution is an asymmetrical distribution, with one tail longer than the other.

15. A multimodal distribution has more than one peak.

True or False

17. True.

19. False. The variance is the sum of squares divided by $n-1$.

21. False. Kurtosis can be easily computed from scores.

23. False. Adding a constant will have no effect on the variance.

25. True.

27. False. The mean is based on all scores in a set.

29. False. Outliers will have more of an effect on the mean than the median.

Calculations

31. The calculations are as follows:

	Mean	Median	Variance	Standard Deviation
Sample 1	14.5	8.5	299.55	17.31
Sample 2	0	0.15	1.10	1.05
Sample 3	5	5.15	1.10	1.05
Sample 4	10.15	9.95	0.89	0.94
Sample 5	21.95	21.85	1.05	1.03

CHAPTER 4

Definitions and Concepts

1. A z-score, or standard score, expresses the distance of a score from the mean (μ) of its distribution in units of standard deviation (σ).

3. The standard normal distribution is a distribution of z-scores. The standard normal distribution has a mean of $\mu = 0$, a standard deviation of $\sigma = 1$, and variance of $\sigma^2 = 1$.

True or False

5. True.

7. False. Only the standard normal distribution has a mean of zero and standard deviation of 1.

9. True.

11. True.

13. False. $P(z_2) - P(z_1)$ is the proportion of a normal distribution within the interval x_1 to x_2.

Calculations

14. (a) .9452. (c) .9893.

15. (a) .0918. (c) .0314.

16. (a) .8574. (c) .9500. (e) .9902.

17. (a) .1636. (c) .8703.

18. (a) .02. (c) .50. (e) .98.

19. (a) .98. (c) .50. (e) .02.

20. (a) .14. (c) .82. (e) .96.

21. (a) .0778. (c) .5557. (e) .9554.

22. (a) .9222. (c) .4443. (e) .0446.

23. (a) .1833. (c) .6943. (e) .8776.

24. (a) 87.424. (c) 121.472. (e) 97.152.

25. (a) [74.112, 124.288]. (c) [76.8, 121.6]. (e) [66.048, 132.352].

CHAPTER 5

Definitions and Concepts

1. The distribution of means is a probability distribution of all possible values of a sample mean based on n scores. The distribution of means is also referred to as the sampling distribution of the mean.

3. μ_m is the mean of the distribution of means. μ_m always equals μ, the mean of the distribution of scores from which samples were drawn.

5. A statistic is unbiased when its expected value equals the parameter it estimates. The sample mean is an unbiased statistic.

7. σ_m^2 is that variance of the distribution of means. σ_m^2 always equals σ^2/n, the variance of the population of scores divided by sample size.

9. We use the term "standard error" when referring to the variability in the distribution of a statistic, and we use "standard deviation" when referring to the variability in a distribution of scores.

11. A sampling distribution is a probability distribution of all possible values of a sample statistic based on samples of the same size.

True or False

13. True.

15. True, because s^2 is an unbiased statistic.

17. False. We cannot determine the standard error without knowing sample size.

19. *False. The distribution of sample variances is generally right skewed.

21. *True.

Calculations

23. $\sigma_m = 3, z = 2, P(2) = .9772$.

25. $\sigma_m = 4, z = 1.55, P(1.55) = .9394$.

27. $\sigma_m = 2, z_1 = -1, z_2 = -.5, P(-.5) - P(-1) = .1498$.

29. $\sigma_m = .02, z_1 = .5, z_2 = 1, P(1) - P(.5) = .1498$.

Extra Practice

30. σ_m, z, and $P(z)$, respectively:
    (a) 1, −1, and ≈.16.
    (c) 1, 2, and ≈.98.
    (e) 2, 2, and ≈.98.

31. σ_m, z, and $1 - P(z)$, respectively:
    (a) 1, −1, and ≈.84.
    (c) 1, 2, and ≈.02.
    (e) 2, 2, and ≈.02.

32. σ_m, z_L, z_U, and $P(z_U) - P(z_L)$, respectively:
    (a) 1, −2, −1, and ≈.14.
    (c) 2, −2, 1, and ≈.82.
    (e) .2, 1, 2, and ≈.14.

33. σ_m, z, and $z(\sigma_m) + \mu$, respectively:
    (a) .2, −2, and 99.6.
    (c) .2, 1, and 100.2.
    (e) .2, 2, and 100.4.
    (f) .2, 1, and 100.2.

34. σ_m, z, and $P(z)$, respectively:
    (a) 1.5, −2, and .0228.
    (c) 1.5, 0, and .5.
    (e) 3, .67, and .7486.

35. σ_m, z, and $1 - P(z)$, respectively:
    (a) 1.5, −2, and .9772.
    (c) 1.5, 0, and .5.
    (e) 3, .67, and .2514.

36. σ_m, z, and $z(\sigma_m) + \mu$, respectively:
    (a) .2, −2.05, and 99.59.
    (c) .2, .99, and 100.198.
    (e) .2, 2.05, and 100.41.
    (f) .2, .99, and 100.198.

37. $z_{\alpha/2}$ and the two sample means:
    (a) 1.96, [7.02, 8.98].
    (c) 1.55, [7.6125, 8.3875].
    (e) 1.03, [7.485, 8.515].

CHAPTER 6

Definitions and Concepts

1. A point estimate is a single number that estimates a population parameter. In other words, a point estimate is a statistic.

3. An interval estimate is called a confidence interval when we can express the confidence we have that it contains the parameter.

5. $z_{\alpha/2}$ is a z-score for which $(\alpha)100\%$ of the standard normal distribution lies outside the interval $\pm z_{\alpha/2}$.

7. The four quantities are m, σ, n, and α. m is the center of the interval and has no effect on the width of the interval. Interval width increases as σ increases. Interval width decreases as n increases. Interval width decreases as α increases.

9. *Knowing only that $moe = 10$ does not tell you how much confidence you should have in an interval. moe depends on σ and n; without knowing these two things, we can't associate a confidence level with moe.

True or False

11. False. 95% of all intervals computed as $m \pm 1.96(\sigma_m)$ will contain μ.

13. True. See Figure 6.5.

15. True.

17. False. Exactly 5% of all $m \pm 1.96(\sigma_m)$ will fail to capture μ.

Calculations

19. $m \pm 1.64(\sigma_m)$.

20. α and $z_{\alpha/2}$, respectively: (a) .484 and 0.70. (c) .126 and 1.53.

21. $z_{\alpha/2}$ and α, respectively: (a) 2.1 and .036. (c) 1.2 and .230.

22. (a) 90% CI = [29.26, 34.74].

23. (b) 95% CI = [82.08, 89.92].

24. (a) 60% CI = [94.12, 105.88].

25. (b) Approximate 95% CI = [97.06, 102.94].

Scenarios

27. $m \pm z_{\alpha/2}(s_m) = 74{,}000 \pm 1.96(291.008) = [73{,}429.62, 74{,}570.38]$. We have to assume that the distribution of scores is normal. This is not a reasonable assumption because the distribution of incomes is skewed. However, the effects of violating assumptions decrease as sample size increases. In this case, our sample is very large ($n = 1600$), so the effects of violating the normality assumption are probably small.

29. $m \pm z_{\alpha/2}(s_m) = 11.2 \pm 1.96(.2) = [10.81, 11.59]$. We must assume random sampling and that the distribution of scores is normal. The question states that sampling was random. However, because the scores are a type of reaction time, the distribution may be skewed to the right. Therefore, it's possible that the normality assumption has been violated.

31. The calculations are as follows:

Scores	Deviations $(y - m)$	Squared Deviations
54	−2	4
67	11	121
27	−29	841
59	3	9
53	−3	9
36	−20	400
46	−10	100
53	−3	9
86	30	900
79	23	529
37	−19	361
80	24	576
57	1	1
49	−7	49
57	1	1
48	−8	64
50	−6	36
65	9	81
64	8	64
63	7	49
57	1	1
39	−17	289
56	0	0
65	9	81
53	−3	9
$m = \Sigma y/n$		$ss = \Sigma(y - m)^2$
56		4584

(i) $m = \Sigma y/n = 1400/25 = 56.$

(ii) $s^2 = \dfrac{ss}{n-1} = \dfrac{4584}{24} = 191.$

(iii) $s_m = \sqrt{\dfrac{s^2}{n}} = \sqrt{\dfrac{191}{25}} = 2.76$

(iv) $m \pm z_{\alpha/2}(s_m) = 56 \pm 1.96(2.76) = 56 \pm 5.41 = [50.58, 61.42].$

CHAPTER 7

Definitions and Concepts

1. The null hypothesis is a hypothesis about the distribution of scores from which a sample was drawn. The null hypothesis typically states that there is no effect of an experimental treatment on a measured variable.

3. A statistic is said to be statistically significant when its associated p-value under the null is small. When a statistic is statistically significant, we reject the null hypothesis.

5. A significance test is a statistical procedure for evaluating the plausibility of the null hypothesis. We compare a statistic computed from a sample with the distribution of the statistic under the null hypothesis. If our statistic is one that would occur very rarely when the null hypothesis is true, we reject the null hypothesis and declare our result to be statistically significant. Significance tests are often referred to as hypothesis tests, because we are in essence testing the plausibility of the null hypothesis.

7. A non-directional hypothesis predicts only that there is a difference between μ_0 and μ_1: $\mu_1 - \mu_0 \neq 0$.

9. Retaining H_0 means that it remains a plausible hypothesis about the mean of the distribution from which our sample was drawn. This does not mean that it is true or that we treat it as though it is true.

11. Practical significance relates to the interpretation and implications of a statistical analysis. The practical significance of a result rests in *how it matters*.

True or False

13. True.

15. False. When we reject the null hypothesis, we are saying that the observed statistic (or one more extreme) occurs infrequently when the null hypothesis is true. This is not the same thing as saying that the null hypothesis is false.

17. False. With this alternative hypothesis, this p-value is associated with $z_{obs} = 1$. The correct p-value is .1587.

19. False. Researchers typically believe that H_1 (the research hypothesis) is true.

Calculations

20. (a) $\sigma_m = 10/\sqrt{25} = 2, z_{obs} = (27 - 32)/2 = -2.5.$
    (b) $p = .0062.$

22. (a) $s_m = 16/\sqrt{64} = 2, z_{obs} = (58 - 54)/2 = 2.$
    (b) $p = .0456.$

Scenarios

23. (a) $H_0: \mu_{pink} - \mu_{white} = 0.$
    (b) $H_1: \mu_{pink} - \mu_{white} > 0.$
    (c) $z_{obs} = (68 - 75)/(3.5/7) = -7/.5 = -14.$
    (d) p is close to 1. The magnitude of this z-score is huge and its sign is negative. Therefore, almost 100% of the distribution is above it.
    (e) The result is not statistically significant. We can only reject H_0 if z_{obs} has a p-value less than .05.
    (f) Retain; z_{obs} is not statistically significant.
    (g) The researcher's hypothesis is not supported.
    (h) Any reasonable person would say that an exam printed on pink paper produces much lower grades than an exam printed on white paper. A sample mean of 68 would be *very* unusual if H_0 were true. However, because the instructor predicted the opposite effect, he is not allowed to say the result is statistically significant and is thus not allowed to reject H_0. This is one of the odd things that can happen with directional hypotheses.
    (i) My explanation would be that the exam was harder to read on pink paper than on white paper. This could have led to increased stress or possible failure to read the questions correctly. Maybe you have your own ideas.

25. (a) $\mu_0 = 900$ using a significance level of $\alpha = .05$.

   (i) H_0: $\mu_{training} - 900 = 0$.

   (ii) H_1: $\mu_{training} - 900 < 0$.

   (iii) $z_{obs} = (650 - 900)/(165/11) = -250/15 = -16.67$.

   (iv) $P(z_{obs}) \approx 0$.

   (v) Yes, z_{obs} is statistically significant.

   (vi) We should reject H_0.

   (vii) Well, there's no support for the claim that reading speeds double. However, I'm impressed. Three hours of training raised reading speeds from an average of 450 to an average of 650, with a margin of error of $\pm 1.96*15 = \pm 29.4$.

   (viii) This program seems promising. Of course, we'd have to be cautious and ask some important questions. For example, do readers sacrifice comprehension for speed? If you Google "Staples reading test" you will find that it asks a few basic questions that do not seem to be very challenging.

   (ix) Yes, I would. On the face of things, reading speeds improve quite dramatically following minimal training. However, I would caution my friend to pay close attention to whether she thinks she's sacrificing comprehension.

(b) $\mu_0 = 450$ using a significance level of $\alpha = .05$.

   (i) H_0: $\mu_{training} - 450 = 0$.

   (ii) H_1: $\mu_{training} - 450 \neq 0$. The student is not making a prediction about the direction of the effect and might imagine that training could improve or impair reading speed.

   (iii) Assuming sample size is 121, then $z_{obs} = (650 - 450)/(165/11) = -200/15 = 13.33$.

   (iv) The p-value is close to 0 because almost 100% of the z-distribution lies in the interval ± 13.333.

   (v) Yes, z_{obs} is statistically significant.

   (vi) We should reject H_0.

   (vii) See 25.a.viii.

26. Professor Hebb has committed the inverse probability error. He thinks $p(H_0|z_{obs}) = .048$.

CHAPTER 8

Definitions and Concepts

1. In significance tests, a Type II error is retaining the null hypothesis when it is false.

3. In significance tests, β is the probability of retaining the null hypothesis when it is false. That is, β is the probability of a Type II error. The probability of rejecting the null hypothesis when it is true is denoted α. That is, α is the probability of a Type I error.

5. Cohen's effect size classifications state that δ values of .2, .5, and .8, should be considered small, medium, and large, respectively.

True or False

7. True.

9. False. $\delta = (\mu_1 - \mu_0)/\sigma$.

11. False. Power is the probability of rejecting a false null hypothesis.

13. False. $\delta = 1$ could be meaningless in some situations.

15. False. $\beta = 1 - power$.

Calculations

17. β and power, respectively: (a) .3594 and .6406. (c) .1949 and .8051.

18. β and power, respectively: (a) .7995 and .2005. (c) .5199 and .4801.

19. σ and U_3, respectively: (a) 4 and .5987. (c) 1 and .8413.

20. δ, power, and n, respectively: (a) .8, .8, and 10. (c) .8, .2, and 155.

Scenarios

21. Pink paper and test grades.

   (a) He would need $n = 25$ scores to achieve power = .8.

   (b) $z_{obs} = (76.147 - 75)/(3.5/\sqrt{25}) = 1.64$, $p = .0505$. This result is not statistically significant.

   (c) The estimate is $(76.147 - 75)/3.5 = 0.33$.

   (d) U_3 computed from this estimate is .6293.

(e) $z_{obs} = (76.147 - 75)/(3.5/\sqrt{35}) = 1.94$, $p = .0262$. The conclusion would change because the result is statistically significant.

23. Seasonal affective disorder.

(a) The researcher would require $n = 7$ scores to achieve power $= .8$.

(b) $z_{obs} = (60.564 - 45)/(15/\sqrt{7}) = 2.75$. Yes, this is statistically significant.

(c) $p = .003$. (You will have to use Excel to compute this p-value.)

(d) The estimate δ is $(60.564 - 45)/15 = 1.04$. U_3 computed from this estimate is .8508.

(e) In this case, $n = 4$. $z_{obs} = (60.564 - 45)/(15/\sqrt{4}) = 2.08$, $p = .0188$. The conclusion would not change because the result remains statistically significant.

CHAPTER 9

Definitions and Concepts

1. A p-value is the probability of obtaining your statistic, or one more extreme, by chance when the null hypothesis is true.

3. The run-until-statistically-significant (RUSS) procedure chooses sample size based on the p-value rather than choosing a sample size and then computing a p-value. If you run until statistically significant, the p-value you report is much lower than it should be. Therefore, routine use of the RUSS procedure will increase the number of Type I errors in the literature.

5. If replications are not published, it would be impossible to determine when a Type I error has entered the literature.

7. The mean specified by the null hypothesis is only one of many means that are plausible hypotheses about the mean of the distribution from which the scores were drawn. Therefore, the mean specified by the null hypothesis is no more the "true mean" than any of these other plausible means.

True or False

9. False. Rejecting the null hypothesis could be a Type I error.

11. True.

13. False. Because $\mu_0 = 22$ is not in the interval, I should reject the null hypothesis.

15. False. $m - \mu_0 = 4$ (the center of the interval) and $\mu_0 = 5$. Therefore, $m = 9$.

Calculations

17. For $\mu_0 = 32$, $\sigma_0 = 10$, $m = 27$, and $n = 25$:

(a) 95% CI = [23.08, 30.92].

(b) 95% CI = [−8.92, −1.08].

(c) 95% CI = [−.892, −.108].

(d) The 95% confidence intervals around m, $m - \mu_0$, and d do not capture μ_0, 0, and 0, respectively; therefore, we can reject H_0 at the $p < .05$ criterion.

19. For $\mu_0 = 100$, $\sigma_0 = 15$, $m = 101$, and $n = 225$:

(a) 95% CI = [99.04, 102.96].

(b) 95% CI = [−.96, 2.96].

(c) 95% CI = [−.064, .197].

(d) The 95% confidence intervals around m, $m - \mu_0$, and d capture μ_0, 0, and 0, respectively; therefore, we cannot reject H_0 at the $p < .05$ criterion.

21. For $\mu_0 = 100$, $\sigma_0 = 15$, $m = 102$, and $n = 225$:

(a) 90% CI = [100.36, 103.65].

(b) 90% CI = [.36, 3.65].

(c) 90% CI = [.024, .243].

(d) The 90% confidence intervals around m, $m - \mu_0$, and d do not capture μ_0, 0, and 0, respectively. Furthermore, the difference is in the predicted direction. Therefore, we can reject H_0 at the $p < .05$ criterion.

(e) The difference is not in the predicted direction so it doesn't matter whether the 90% CIs capture μ_0, 0, and 0. We retain H_0.

Scenarios

23. Difference thresholds and hypnotism. The null hypothesis is that $\mu_1 - \mu_0 = 0$. $m - \mu_0 = 102.5 - 103 = -.5$, and the 95% confidence interval about this difference is [−0.89, −0.11]. This interval does not capture 0, so we can reject H_0. Hypnosis produces a statistically significant reduction in difference thresholds.

25. Social conformity. $m = 5.1$, and the approximate 95% confidence interval about this difference

is [4.51, 5.69]. The scenario suggests that had the participants not been influenced by the confederates, they would have given the incorrect answer 0 of 7 times. Therefore, μ_0 would be 0. We've computed an approximate confidence interval based on the sample standard deviation, $s = 1.2$. This study strongly suggests that individuals don't want to stand out as different from others and so modify their behaviors to conform. We saw in Chapter 6 that this interval is narrower than it should be because of the small sample size. Therefore, our concern is that we have less than 95% confidence in this interval. How much less remains to be seen in Chapter 10.

27. Memory for line drawings. $d = z_{obs}/\sqrt{n} = -2.4 / 6 = -.4$. Therefore, the 95% CI around d is [−.727, −.073]. The mean recognition score for older adults is $\mu_0 + d(\sigma)$ 75 + (−.4∗5) = 75 − 2 = 73. A decrease of 2 percentage points over a period of 47 years does not seem like a severe loss. (The actual percentage change is −2/75∗100 = −2.667.) Although the result is statistically significant, I doubt that such a change would have a meaningful impact on the daily lives of older adults.

29. Context and memory. $m - \mu_0 = 15.67 - 13.23 = 2.44$, 95% CI [0.38, 4.50]. These results are really interesting because they suggest that providing context for the passage yields more correctly recalled items, presumably because the context permitted subjects to make sense of the passage. Personally, I think this is a generally important result. Providing context before giving people a list of facts allows them to organize the material that they receive better than without context. A good teacher will provide you with a framework for what you're about to hear before filling in the details.

CHAPTER 10

Definitions and Concepts

1. A t-score is defined as $t = (m - \mu)/s_m$, which is the difference between a sample mean and the population mean, divided by the estimated standard error of the mean.

3. In this chapter, degrees of freedom refers to sample size minus 1 (i.e., $n-1$).

5. The proportion of a t-distribution falling above $t_{\alpha/2}$ is $\alpha/2$.

7. Yes this is possible. The sample mean is less than one standard error from the population mean, which is very close. Therefore, it seems odd that the 95% confidence interval does not capture μ. However, this can happen if by chance the sample standard deviation is very small. For example, if the sample standard deviation were 3.2, then $s_m = 3.2/\sqrt{16} = .8$, and the 95% confidence interval would be 98 ± 1.705 = [96.295, 99.705].

9. The scores must be a random sample from a normal distribution.

True or False

11. False. Distributions of t-scores are only called t-distributions if samples are drawn from a normal population.

13. False. It is possible for m to be the outside interval $\mu \pm 1.96(\sigma_m)$ and yet have its 95% confidence interval capture μ.

15. False. $t_{\alpha/2} = 2.131$.

17. True.

Calculations

19. (a) 2.262, 3.250, 1.833.
    (c) 2.030, 2.724, 1.690.
    *(e) 1.988, 2.634, 1.663.
    *(g) 1.962, 2.580, 1.646.

20. (a) 0.900.
    (c) 0.980.
    (e) 0.500.
    *(g) 0.920.

21. (b) $m \pm 2.064 (s_m)$.

23. $m \pm t_{\alpha/2}(s_m) = 22 \pm 2.69(.388) = [20.96, 23.04]$.

25. (a) $m \pm t_{\alpha/2}(s_m) = 10 \pm 2.365(1.165) = [7.24, 12.76]$.
    (c) $m \pm t_{\alpha/2}(s_m) = 5 \pm 2.16(1.512) = [1.73, 8.27]$.

26. (a) $m \pm t_{\alpha/2}(s_m) = [7.24, 12.76]$ does not capture 13; reject H_0.
    (c) $m \pm t_{\alpha/2}(s_m) = [1.73, 8.27]$ does not capture 13; reject H_0.

27. (a) $d \pm z_{\alpha/2}(s_d) = -.91 \pm 1.96(.429) = [-1.752, -.069]$.

    (c) $d \pm z_{\alpha/2}(s_d) = -1.414 \pm 1.96(.385) = [-2.169, -.659]$.

28. *(a) $n = 7$.

    (c) $n = 6$.

    (e) $n = 19$.

Scenarios

29. Facebook and emotional contagion.

    (a) $m \pm t_{\alpha/2}(s_m) = 5.29 \pm 1.96(.01) = [5.27, 5.31]$.

    (b) This 95% confidence interval does not contain 5.24, so we can reject the null hypothesis at the .05 level.

    (c) $t_{obs} = (5.29 - 5.24)/.01 = 4.78$. Yes, we can reject the null hypothesis at the .05 level.

    *(d) Exact p-value using Excel (requires Appendix 10.1): $p = 0.00000177$; remember, this is a two-tailed test because no prediction was made.

    (e) $d = (5.29 - 5.24)/4.12 = .012$.

    (f) $d \pm z_{\alpha/2}(s_d) = .012 \pm 1.96(.003) = [0.007, 0.017]$.

    (g) Receiving fewer negative posts leads to a statistically significant increase in positive posts. However, the effect size is tiny ($d = .012$), and it is difficult to see any practical significance in this result. This is an extreme example of a large sample size ($n = 155,000$) producing a statistically significant result.

30. Alone with our thoughts.

    (a) $m \pm t_{\alpha/2}(s_m) = 4.67 \pm 1.993(.209) = [4.25, 5.09]$. This confidence interval contains 5, so the difference is not statistically significant.

    (b) $t_{obs} = (4.67 - 5)/.209 = -1.58$. No, we cannot reject the null hypothesis at the .05 level.

    *(c) Requires Excel to compute exact p-value. Assuming a two-tailed test with $\alpha = .05$, $p = .12$.

    (d) $d = (4.57 - 5)/1.8 = -.183$.

    (e) $d \pm z_{\alpha/2}(s_d) = -.183 \pm 1.96(.117) = [-.413, .046]$.

    (f) Clearly subjects are not thrilled to be alone with their thoughts. Their mean rating is below the neutral level. However, the difference between their mean rating and a neutral rating is relatively small ($d = -.183$), and the measurement seems somewhat imprecise, approximate 95% CI [$-.413$, .046].

CHAPTER 11

Definitions and Concepts

1. An experiment involves the random assignment of participants to conditions. A quasi-experiment involves comparing samples drawn from existing populations (i.e., there was no random assignment of participants to conditions).

3. Independent samples comprise scores that are completely unrelated to each other. This means that knowing the score of an individual in one sample tells you nothing about the score of any individual in the other sample. Dependent samples comprise scores that are related to each other in some way; pairs of scores come either from the same individual or from individuals matched on some characteristic. This means that knowing the score of an individual in one sample tells you something about the score of the corresponding individual in the other sample.

True or False

5. True.

7. False. Think about bench-pressing 5-year-olds.

9. True.

11. True.

13. True.

Calculations

14. (a) $\mu_{m_1 - m_2} = -2, \sigma^2_{m_1 - m_2} = \dfrac{\sigma_1^2}{n_1} + \dfrac{\sigma_2^2}{n_2} = \dfrac{36}{15} + \dfrac{25}{15} = 4.07$.

    (c) $\mu_{m_1 - m_2} = 14, \sigma^2_{m_1 - m_2} = \dfrac{\sigma_1^2}{n_1} + \dfrac{\sigma_2^2}{n_2} = \dfrac{36}{15} + \dfrac{625}{4} = 158.65$.

    (e) $\mu_{m_1 - m_2} = -4, \sigma^2_{m_1 - m_2} = \dfrac{\sigma_1^2}{n_1} + \dfrac{\sigma_2^2}{n_2} = \dfrac{256}{32} + \dfrac{64}{64} = 9$.

15. (a) $m_1 = 10, m_2 = 12, s_1 = 6, s_2 = 5, n_1 = 15, n_2 = 15.$

(i) $(m_1 - m_2) \pm t_{\alpha/2}(s_{m_1 - m_2}) = -2 \pm 2.048$
$(2.017) = [-6.13, 2.13].$

(ii) $d \pm 1.96(s_d) = -0.362 \pm 1.96(0.368) =$
$[-1.084, .360].$

(iii) $t_{obs} = (m_1 - m_2)/s_{m_1 - m_2} = -2/2.017 =$
$-0.992.$

(iv) $r^2 = t_{obs}^2/(t_{obs}^2 + df_{within}) - 0.992^2/(-0.992^2$
$+ 28) = .034.$

(c) $m_1 = 8, m_2 = 7, s_1 = 4, s_2 = 3, n_1 = 9, n_2 = 15.$

(i) $(m_1 - m_2) \pm t_{\alpha/2}(s_{m_1 - m_2}) = 1 \pm 2.074$
$(1.433) = [-1.97, 3.97].$

(ii) $d \pm 1.96(s_d) = .294 \pm 1.96(0.424) =$
$[-.537, 1.125].$

(iii) $t_{obs} = (m_1 - m_2)/s_{m_1 - m_2} = 1/1.433 = 0.698.$

(iv) $r^2 = t_{obs}^2/(t_{obs}^2 + df_{within}) = 0.698^2/$
$(0.698^2 + 22) = .022.$

Scenarios

17. Comic Sans and reading.

(a) $r^2 = t_{obs}^2/(t_{obs}^2 + df_{within}) = -2.147^2/(-2.147^2 +$
$40) = .103.$

(b) $d \pm 1.96(s_d) = -0.663 \pm 1.96(0.317) =$
$[-1.285, -.041].$

(c) Yes, we can reject the null hypothesis. The researcher predicted a negative difference, which is what we found, and the confidence interval around d does not contain 0. Had we calculated the 90% confidence interval around d, the CI would have been even narrower and would clearly not capture 0. (We could have equally conducted a t-test to assess the null hypothesis.)

19. Like, no!

(a) $(m_{old} - m_{young}) \pm t_{\alpha/2}(s_{m_1 - m_2}) = 15 \pm$
$2.009(4.125) = [6.71, 23.29].$

(b) No. His hypothesis is $\mu_{old} - \mu_{young} < 0$, but he found that $m_{old} - m_{young} = 15$, which is opposite to what he predicted.

(c) $d \pm 1.96(s_d) = [0.409, 1.485].$

CHAPTER 12

Definitions and Concepts

1. Two populations of scores are dependent populations when each score in one population is paired with or is associated with a score in the other population. If there is no association between the scores in the two populations, then they are independent.

3. A repeated-measures design is one in which scores are obtained from participants on more than one occasion.

5. A carryover effect occurs in repeated-measures designs when obtaining measurements on previous trials affects the measurements made on subsequent trials.

7. σ_{m_D} is the standard error of the mean difference. s_{m_D} is the estimated standard error of the mean difference.

True or False

9. True.

11. True.

13. False. $\sigma_{m_D}^2$ is always smaller than σ_1^2.

15. True.

17. False. $d = m_D/s_D.$

19. False. A matched-samples design will generally yield wider confidence intervals than a repeated-measures design.

21. False. $\sigma_D \neq \sigma_{m_1 - m_2}.$

23. True.

Calculations

24. (a) $\mu_{m_D} = 10 - 12 = -2, \sigma_{m_D}^2 = 6^2/15 = 2.4.$

(c) $\mu_{m_D} = 25.6 - 21.9 = 3.7,$
$\sigma_{m_D}^2 = 8.8^2/121 = .64.$

(e) $\mu_{m_D} = 102 - 102 = 1, \sigma_{m_D}^2 = 15^2/225 = 1.$

25. (a) $m_1 = 18, m_2 = 15, s_D = 18, n = 9.$

(i) $(m_1 - m_2) \pm t_{\alpha/2}(s_{m_D}) = (18 - 15)$
$\pm 2.306(6) = [-10.84, 16.84].$

(ii) $d \pm z_{\alpha/2}(s_d) = (18 - 15)/18 \pm 1.96(.336)$
$= [-0.492, 0.825].$

(iii) $t_{obs} = m_D/s_{m_D} = (18-15)/6 = 0.5.$

(iv) $p > .05.$

(c) $m_1 = 88, m_2 = 101, s_D = 48, n = 64.$

    (i) $(m_1 - m_2) \pm t_{\alpha/2}(s_{m_D}) = (88 - 101)$:

       $\pm 2.00(6) = [-24.99, -1.01].$

    (ii) $d \pm z_{\alpha/2}(s_d) = (88 - 101)/48 \pm$
       $1.96(0.127) = [-0.520, -.021].$

    (iii) $t_{obs} = m_D/s_{m_D} = (88 - 101)/6 = -2.17.$

    (iv) $p < .05.$

(e) $m_1 = 99, m_2 = 98, s_D = 1, n = 25.$

    (i) $(m_1 - m_2) \pm t_{\alpha/2}(s_{m_D}) = (99-98)$
       $\pm 2.064(.2) = [0.59, 1.41].$

    (ii) $d \pm z_{\alpha/2}(s_d) = (99 - 98)/1 \pm 1.96(0.247)$
       $= [0.517, 1.483].$

    (iii) $t_{obs} = m_D/s_{m_D} = (99-98)/0.2 = 5.$

    (iv) $p < .05.$

26. (a) $\mu_{m_D} = 10 - 12 = -2,$
      $\sigma^2_{m_D} = 2[6^2 + 6^2 - 2*(4^2)]/25 = 3.2$

  (c) $\mu_{m_D} = 88 - 91 = -3,$
      $\sigma^2_{m_D} = 2[16^2 + 16^2 - 2(10^2)]/36 = 17.33.$

  (e) $\mu_{m_D} = 67 - 65 = 2,$
      $\sigma^2_{m_D} = 2[10^2 + 10^2 - 2(5^2)]/225 = 1.33$

27. (a) $m_1 = 10, m_2 = 12, s_1 = 6, s_2 = 6, s_S = 4, n = 25.$

    (i) $(m_1 - m_2) \pm t_{\alpha/2}(s_{m_D}) = (10-12)$
       $\pm 2.064(1.789) = [-5.69, 1.69].$

    (ii) $d \pm z_{\alpha/2}(s_d) = (10 - 12)/8.944 \pm$
       $1.96(0.203) = [-0.621, 0.173].$

    (iii) $t_{obs} = m_D/s_{m_D} = (10 - 12)/1.789$
       $= -1.12.$

    (iv) $p > .05.$

  (c) $m_1 = 41, m_2 = 43, s_1 = 24, s_2 = 24, s_S = 21,$
    $n = 144.$

    (i) $(m_1 - m_2) \pm t_{\alpha/2}(s_{m_D}) = (41 - 43)$:
       $\pm 1.984(1.936) = [-5.83, 1.83].$

    (ii) $d \pm z_{\alpha/2}(s_d) = (41 - 43)/23.238 \pm$
       $1.96(0.083) = [-0.250, 0.078].$

    (iii) $t_{obs} = m_D/s_{m_D} = (41-43)/1.936$
       $= -1.03.$

    (iv) $p > .05.$

  (e) $m_1 = 1234, m_2 = 999, s_1 = 300, s_2 = 300, s_S = 290, n = 225.$

    (i) $(m_1 - m_2) \pm t_{\alpha/2}(s_{m_D}) = (1234 - 999)$
       $\pm 1.972(10.242) = [214.81, \ 255.19]$

    (ii) $d \pm z_{\alpha/2}(s_d) = (1234 - 999)/153.623 \pm$
       $1.96(0.098) = [1.337, 1.722].$

    (iii) $t_{obs} = m_D/s_{m_D} = (1234 - 999)/10.242$
       $= 22.95.$

    (iv) $p < .05.$

Scenarios

29. Repeated measures.

  (a) 95% CI = [5.12, 22.88].

  (b) 95% CI = [0.340, 2.296].

  (c) $t_{obs} = 3.73, p < .05.$

  (d) The difference between m_D and 0 is statistically significant at the $p < .05$ level.

31. Motion aftereffect.

  (a) The independent variable is attention. The two levels are "attending to letters" and "not attending to letters."

  (b) Duration of the motion aftereffect (MAE).

  (c) This is a repeated-measures design.

  (d) The MAE is shorter (2.8 seconds on average) when attending to the letters than when not attending to the letters (6.4 seconds on average).

  (e) 95% CI = [0.02, 7.18].

  (f) Approximate 95% CI = [0.017, 2.482].

  (g) $t_{obs} = 2.794, p < .05.$

  (h) The difference between m_D and 0 is statistically significant at the .05 level because the 95% confidence interval does not include 0.

  (i) Yes, the results are meaningful. It would be worthwhile to read Chaudhuri's paper. The MAE is typically thought to arise from the fatigue of low-level motion selective mechanisms. After prolonged exposure to downward motion, the downward selective motion mechanisms become fatigued. When the motion stops, the spontaneous activity in the upward selective mechanisms is not balanced by the spontaneous activity in the downward selective mechanisms. Greater activity in the upward selective mechanisms than the downward selective mechanisms is the brain's code for upward motion. This standard account does not include a role for attention, and there is no reason to expect attention to affect the duration of the MAE. The results suggest that attention strongly affects the duration of the MAE, so this is important and new.

CHAPTER 13

Definitions and Concepts

1. We say that y is a function of x when for every value of x there is exactly one value of y.

3. An association between x and y is said to be linear when the changes in y with changes in x are best described by a straight line.

5. In this section of the book, we consider linear probabilistic associations.

7. *Grades* is the outcome variable and *hours* is the predictor variable. If I were to plot this association on a scatterplot, *hours* would be on the x-axis, and *grades* would be on the y-axis.

9. The correlation coefficient can take on a range of values from −1 to 1.

11. If all of the values in y were multiplied by 12.543, the correlation would remain unchanged because r is invariant under linear transformations of x and/or y.

True or False

13. True.

15. False. $r^2 = (ss_{total} - ss_{error})/ss_{total}$.

17. True.

19. False. This equation [$y = 5 + 3(x^2)$] is not a linear transformation because x is squared.

21. True.

Calculations

22. (a) $r = 1$. (c) $r = .5$.

23. (a) $\hat{y} = 16.08 - 0.34(x)$.
    (c) $\hat{y} = 4.67 + 0.33(x)$.

25. $r^2 = .25$.

Scenarios

27. Correlation and regression.
    (a) $sp_{xy} = 37$.
    (b) $r = 37/\sqrt{34 * 212} = .4358$.
    (c) $s_x = \sqrt{34/10} = 1.8439$.
    (d) $s_y = \sqrt{212/10} = 4.6043$.
    (e) $b = .4358 * 4.6043/1.8439 = 1.0882$.
    (f) $a = 21 - 1.0882(11) = 9.0298$.
    (g) $\hat{y} = 9.0298 + 1.0882(13) = 23.1765$.

x	y	$x - m_x$	$y - m_y$	$(x - m_x)(y - m_y)$
9	24	−2	3	−6
16	28	5	7	35
11	22	0	1	0
10	25	−1	4	−4
11	24	0	3	0
10	18	−1	−3	3
10	21	−1	0	0
11	16	0	−5	0
12	24	1	3	3
11	14	0	−7	0
10	15	−1	−6	6
m_x	m_y	$ss_x = \Sigma(x - m_x)^2$	$ss_y = \Sigma(y - m_y)^2$	$sp_{xy} = \Sigma(x - m_x)(y - m_y)$
11	21	34	212	37

CHAPTER 14

Definitions and Concepts

1. A bivariate distribution is a distribution of xy pairs.

3. Distributions are homoscedastic when they all have the same standard deviation.

5. The quantity $\sum(y - E(y|x))^2$ is called ss_{error}.

7. A marginal distribution of scores is a distribution that combines the scores from all conditional distributions.

9. A bivariate distribution is a distribution of xy pairs. The distribution of regression slopes (b) does not involve xy pairs.

11. The five properties of a bivariate normal distribution are as follows: (i) both the x and y distributions of scores (marginal distributions) are normal, (ii) for all values of x there is a distribution of y values (a conditional distribution), (iii) all conditional distributions are normal, (iv) all conditional distributions have the same standard deviation (homoscedasticity), and (v) the means of the conditional distributions of y values change linearly with x; i.e., $E(y|x) = \alpha + \beta(x)$.

13. To have 95% confidence in the interval $\hat{y} \pm t_{\alpha/2}(s_{\hat{y}})$ means knowing that 95% of all intervals computed this way will capture the population parameter, $E(y|x)$.

True or False

15. False. The regression model assumes scores were randomly sampled from conditional distributions associated with fixed values of x that are normal and homoscedastic.

17. False. The regression model makes no assumptions about the marginal distributions.

19. True.

21. True.

23. False. Regression analysis can plausibly imply a causal association between x and y if there has been random assignment to levels of x.

Calculations

25. Regression calculations.

   (a) $sp_{xy} = -68$.

   (b) $r = -.801$.

   (c) $s_x = 1.8439$.

   (d) $s_y = 4.6043$.

   (e) $s_{est} = 2.9059$.

   (f) $s_b = .4984$.

   (g) $\hat{y} = 43 - 2(x)$.

   (h) $s_b = 0.4984$, $t_{\alpha/2} = 2.262$, CI $= 2 \pm 2.262(0.4984) = [-3.1273, -0.8727]$.

   (i) $\hat{y} = 11$, $s_{\hat{y}} = 2.641$, $t_{\alpha/2} = 2.262$, CI $= 11 \pm 2.262(2.641) = [5.025, 16.975]$.

   (j) $\hat{y} = 11$, $s_{y_{next}} = 3.927$, $t_{\alpha/2} = 2.262$, PI $= 11 \pm 2.262(3.927) = [2.117, 19.883]$.

Computing the Regression Equation and Its Statistics				
x	y	$x - m_x$	$y - m_y$	$(x - m_x)(y - m_y)$
16	14	5	−7	−35
12	15	1	−6	−6
11	16	0	−5	0
11	18	0	−3	0
11	21	0	0	0
11	22	0	1	0
10	24	−1	3	−3
10	24	−1	3	−3
10	24	−1	3	−3
10	25	−1	4	−4
9	28	−2	7	−14
m_x	m_y	$ss_x = \sum(x - m_x)^2$	$ss_y = \sum(y - m_y)^2$	$sp_{xy} = \sum(x - m_x)(y - m_y)$
11	21	34	212	−68

Scenarios

27. Facebook and narcissism.

    (a) $\hat{y} = 4.5 + 0.002(x)$.

    (b) 95% CI $= .002 \pm 1.972(.0012) = [-0.0003, 0.0043]$.

    (c) 95% CI $= 5.2 \pm 1.972(0.155) = [4.973, 5.427]$.

    (d) 95% PI $= 5.2 \pm 1.972(1.997) = [1.263, 9.137]$.

    (e) The conclusion that having a large number of Facebook friends implies that you are narcissistic is wrong. First, we have to define what it means to have a large number of Facebook friends. The average number is 250 in this example, so one supposes a large number is some number greater than this. Let's say that a large number is one standard deviation above the mean (i.e., 350). Our best estimate of $E(y|350)$ is $\hat{y} = 5.2$, 95% CI [4.973, 5.427]. This is not very different from $m_{\text{NPI-16}} = 5$. Furthermore, the 95% prediction interval for a person with 350 Facebook friends is [1.263, 9.137], which is really wide. Two standard deviations around the sample mean is $5 \pm 4 = [1, 9]$. Therefore, knowing that someone has 350 Facebook friends tells you very little about their NPI-16 score. Of course, one would also have to know the clinical definition of narcissism and how it relates to scores on the NPI-16. We must be careful about throwing around terms like "narcissistic."

29. Video games and surgical skill.

    (a) $\hat{y} = 7.7625 - 0.0875(x)$.

    (b) 95% CI $= -0.0875 \pm 2.040(0.0421) = [-0.1733, -0.0017]$.

    (c) 95% CI $= 7.7625 \pm 2.040(0.2083) = [7.3378, 8.1872]$ and 95% CI $= 7.15 \pm 2.040(0.2361) = [6.6684, 7.6316]$.

    (d) 95% PI $= 7.7625 \pm 2.040(0.9743) = [5.7755, 9.7495]$ and 95% PI $= 7.15 \pm 2.040(0.9086) = [5.1501, 9.1499]$.

    (e) First and foremost, we must be careful about the inference that video game playing is the cause of the improved Top Gun performance. There is a very good chance that those who spend many hours playing video games do so because they are good at it. This may mean that many hours of video gaming and fewer errors on the Top Gun test both result from a preexisting condition of superior fine-motor control or selective attention.

    However, if we assume, for the moment, that there is a causal connection between video game playing and Top Gun scores, we can look at the results from two perspectives. Let's first consider the prediction intervals for those who didn't play video games (HPW = 0) and those who played 7 hours per week, which is one standard deviation above the mean. The two prediction intervals are [5.7755, 9.7495] and [5.1501, 9.1499], respectively. These two intervals overlap considerably. Therefore, if you were asked whether you wanted your laparoscopic surgery performed by a surgeon who played no video games in college or one who played 7 hours of video games per week in college, you would have no compelling statistical evidence to choose one over the other.

    We will now consider the results from the broader perspective of all possible laparoscopic surgeries to be performed this year. Let's think first about the confidence intervals around $\hat{y}$ for those who didn't play video games (HPW = 0) and those who played 7 hours per week, which is one standard deviation above the mean. The estimates of $E(y|0)$ and $E(y|7)$ are 7.7625 and 7.15, respectively, with confidence intervals of [7.3378, 8.1872] and [6.6684, 7.6316], respectively. Given that the standard deviation of the Top Gun scores was $s_{\text{TG}} = 1$, the difference between 7.7625 and 7.15 (i.e., .61) is about ⅗ standard deviation. This seems like a rather big difference and seems to suggest that there would be fewer errors made if all future laparoscopic surgeons played video games for 7 hours per week, which might lead to improved outcomes population-wise. However, this conclusion requires knowing the connection between errors on the Top Gun test and actual surgical errors. Furthermore, one would have to attach some measure of seriousness to these errors. Top Gun errors might correspond to small and inconsequential surgical errors. Above all, however, we have to keep in mind that proving a causal connection between video game playing and surgical skill is an important and challenging task.

CHAPTER 15

Definitions and Concepts

1. r is the correlation between x and y in a sample and ρ is the correlation between x and y in a population. r estimates ρ.

3. As sample size increases, the sampling distribution of r becomes increasingly normal.

5. The Fisher transformation transforms a correlation coefficient into a different score called z_r. The advantage of the transform is that the sampling distribution of z_r is approximately normal (with standard error of $1/\sqrt{n-3}$), whereas the sampling distribution of r is generally skewed. The transform allows us to use the normal distribution assumption to place a confidence interval around z_r, and the limits of this interval can be transformed back into the original units of r to provide a confidence interval around r.

True or False

7. False.

9. False.

11. True.

13. True.

15. True. Although the regression equation will change, it will still predict the group means.

Calculations

16. (a) 95% CI = [−.5028, .6373].
    (c) 95% CI = [−.4833, −.0915].
    (e) 95% CI = [.6085, .9034].
    (g) 95% CI = [.1452, .3864].
    (i) 95% CI = [−.2953, −.2035].

17. We can reject H_0 for intervals (c) to (h) because they do not include 0.

Scenarios

18. Physical activity and GPA.
    (a) $\hat{y} = 3.1374 + .0078(x)$.
    (b) Approximate 95% CI = [.04, .41].
    (c) There is a positive correlation between activity and GPA. Because our 95% confidence interval does not include 0, we can reject the null hypothesis that $\rho = 0$ and declare the result statistically significant. According to Cohen's classifications, $r = .23$ is between a small and a medium effect.

(i) The validity of our inferential statistics rests on the assumption that our sample was drawn from a bivariate normal distribution. Because we were not shown the original data, we are in no position to assess this claim.

(ii) If we can assume that our confidence interval is valid, then we can say that there is a modest correlation between physical activity and GPA, but we have no basis for judging the size of this correlation relative to an existing literature.

(iii) Of course, because this is a correlational study we can say nothing about causality. It is possible that some third variable explains the correlation between physical activity and GPA. From our regression equation we can say each 1-point increase in physical activity score leads to a .0078-point increase in GPA. Even if this were a causal association, we have no basis for assessing its practical significance.

CHAPTER 16

Definitions and Concepts

1. ε_i is the difference between y_i and $\hat{y}$ for the ith individual. ε_i is also called a residual or prediction error.

3. k is the number of predictor variables. In this case, $k = 3$.

5. $ss_{regression}$ is the sum of squared deviations of the predicted value of y ($\hat{y}$) from the mean of y. $ss_{regression}$ reflects the variability in y that is explained by y's association with x.

7. $ss_{regression} = .6 * 100 = 60$ and $ss_{error} = (1 − .6) * 100 = 40$.

9. It is difficult to compute an exact confidence interval around R^2 because the sampling distribution of R^2 is complex. There is no simple mathematical formula that describes it; therefore, approximate methods are used to construct confidence intervals around R^2.

11. An F-ratio is the product of the ratio of explained to unexplained variance $[R^2/(1 − R^2)]$ and the ratio of error to regression degrees of freedom ($df_{error}/df_{regression}$); $F = R^2/(1 − R^2) * df_{error}/df_{regression}$.

13. Homoscedasticity means that the distribution of prediction errors (ε) is the same for all combinations of the predictor variables.

True or False

15. False. R^2 associated with multiple predictors is always greater than or equal to an r^2 associated with a single predictor.

17. True.

19. True.

21. False. $F_{obs} = (.8/.2)*(10/20) = 2$.

23. False. If $df_{regression} = 10$, $df_{error} = 40$, and $\alpha = .01$, then $F_{critical} = 2.8$.

Calculations

25. R^2_{adj}, s_{est}, and approximate 95% CI, respectively: (a) .0719, 4.3084, and [.0029, .2118]. (c) .2528, 3.2343, and [.1086, .4192]. (e) .4667, 1.4606, and [.3322, .5904].

26. (a) $F_{critical} = 3.9$. (c) $F_{critical} = 2.2$. (e) $F_{critical} = 19.4$.

27. (a) $F_{obs} = 1.0370$, $p > .05$. (c) $F_{obs} = 27.0000$, $p < .05$. (e) $F_{obs} = 2.7500$, $p > .05$.

28. Hierarchical regression by hand.

(a)

Model	R	R^2	R^2_{adj}	s_{est}	Change Statistics				
					ΔR^2	ΔF	df_1	df_2	p
1	0.7071	0.5000	.4937	1.1251	0.5000	79.0000	1	79	.00
2	0.7906	0.6250	0.6154	0.9806	0.1250	26.0000	1	78	.00

(c)

Model	R	R^2	R^2_{adj}	s_{est}	Change Statistics				
					ΔR^2	ΔF	df_1	df_2	p
1	0.4472	0.2000	.1726	3.6980	0.2000	7.3125	4	117	.00
2	0.5000	0.2500	0.1969	3.6434	0.0500	1.8833	4	113	.12

Scenarios

29. Predictors of insomnia.

(a) Yes, the variables *caffeine consumption*, *anxiety*, and *depression* explain a statistically significant portion of the variability in *FIRST* scores. $R^2 = .618$, $F_{3,12} = 6.46$, $p = .008$.

(b) Approximate 95% CI = [0.09, 0.81].

(c) When *HF-HRV* is added as a predictor, R^2 increases from .618 to .733; $\Delta R^2 = .115$, $F_{1,11} = 4.74$, $p = .052$. Therefore, the result is not statistically significant.

(d) We assume $P^2 = .773$ and $\Delta P^2 = .115$; therefore, $f^2 = .4307$. In G*Power, we set $\alpha = .05$, power = .8, number of tested predictors = 1, and total number of predictors = 4. With these numbers, we find that we would need 21 scores in our sample to obtain power = .8.

(e) There is an obvious issue of cause and effect in this study. The term "produces" suggests that low *HF-HRV* causes insomnia; of course, we can't infer causation from correlation when we have random variables. In this case, *HF-HRV* could be a consequence of insomnia rather than a cause.

(f) The means and standard deviations for the variables *coffee*, *anxiety*, *depression*, *HF-HRV*, and *FIRST* are $m_{coffee} = 11.30$ (4.19), $m_{anxiety} = 20.01$ (6.48), $m_{depression} = 17.86$ (6.89), $m_{HF-HRV} = 15.81$ (6.12), and $m_{FIRST} = 23.78$ (7.55).

(i) $\hat{y}$ for an individual with scores equal to the means of the four predictors is 23.78. For this individual, the 95% confidence interval around $\hat{y} = 23.78$ is [21.27, 26.28], and the 95% prediction interval around $\hat{y} = 23.78$ is [13.44, 34.11].

(ii) An individual who scores one standard deviation below the mean of the *HF-HRV* scores has a score of $m_{HF-HRV} - s_{HF-HRV} = 15.81 - 6.12 = 9.69$. If this individual scores at the mean of the remaining predictors, then the 95% confidence interval around $\hat{y} = 26.68$ is [22.82, 30.53], and the 95% prediction interval around $\hat{y} = 26.68$ is [15.93, 37.42].

30. Mother's education and school performance.

First Analysis

(a) On step 1, $R^2 = .02$, approximate 95% CI [0, 0.079], $F_{1,177} = 3.69$, $p = .056$. Therefore, the result is not statistically significant.

(b) We assume $P^2 = .086$ and $\Delta P^2 = .065$; therefore, $f^2 = .071$. In G*Power, we set $\alpha = .05$, power = .8, number of tested predictors = 1, and total number of predictors = 2. With these numbers, we find that we would need 113 scores in our sample to obtain power = .8.

(c) The means and standard deviations for the variables *MEd*, *MInvolve*, and *SchoolPerf* are $m_{MEd} = 3.02$ (0.96), $m_{MInvolve} = 2.95$ (0.781), and $m_{SchoolPerf} = 4.55$ (0.688).

(i) When *MEd* = 3.02 and *MInvolve* = 2.95, the 95% confidence interval around $\hat{y} = 4.55$ is [4.45, 4.65], and the 95% prediction interval around $\hat{y} = 4.55$ is [3.24, 5.86]. This prediction interval is rather imprecise, as is to be expected.

(ii) A mother who scores one standard deviation above the mean of *MInvolve* has a score of $m_{MInvolve} + s_{MInvolve} = 2.95 + 0.78 = 3.73$. When *MEd* = 3.02 and *MInvolve* = 3.73, the 95% confidence interval around $\hat{y} = 4.74$ is [4.60, 4.89], and the 95% prediction interval around $\hat{y} = 4.74$ is [3.43, 6.06]. This prediction interval is also rather imprecise, as is to be expected, and overlaps considerably with the prediction interval for children whose mothers score at the mean of the predictors.

(d) On step 1, $b_{MEd} = 0.103$; on step 2, $b_{MEd} = 0.019$. This means that the contribution that *MEd* makes to explained variance in *SchoolPerf* is far less in the presence of *MInvolve* than on its own. We will return to this important point in the next chapter.

Second Analysis

(a) On step 1, $R^2 = .085$, approximate 95% CI [0.02, 0.176], $F_{1,177} = 16.482$, $p < .001$. Therefore, the result is statistically significant.

(b) We assume $P^2 = .086$ and $\Delta P^2 = .001$; therefore, $f^2 = .001$. In G*Power, we set $\alpha = .05$, power = .8, number of tested predictors = 1, and total number of predictors = 2. With these numbers, we find that we would need 7176 scores in our sample to obtain power = .8.

31. What predicts musical aptitude?

(a) On step 1, $R^2 = .382$, approximate 95% CI [0.22, 0.52], $F_{2,97} = 30.04$, $p < .001$. Therefore, the result is statistically significant.

(b) We assume $P^2 = .433$ and $\Delta P^2 = .051$; therefore, $f^2 = .0899$. In G*Power, we set $\alpha = .05$, power = .8, number of tested predictors = 1, and total number of predictors = 3. With these numbers, we find that we would need 90 scores in our sample to obtain power = .8.

(c) The means and standard deviations for the variables *music scores*, *digit span*, *spatial*, and *fluency* are $m_{MusicScores} = 3.02$ (1.035), $m_{DigitSpan} = 7.05$ (0.968), $m_{Spatial} = 10.21$ (3.086), and $m_{Fluency} = 2.97$ (1.029).

(i) For an individual who scores at the mean of the three predictors, the 95% confidence interval around $\hat{y} = 3.02$ is [2.86, 3.18] and the 95% prediction interval around $\hat{y} = 3.02$ is [1.44, 4.60].

(ii) An individual who scores one standard deviation above the mean of the *spatial* scores has a score of $m_{Spatial} + s_{Spatial} = 10.21 + 3.086 = 13.30$. $\hat{y}$ for this individual is 3.38. The 95% confidence interval around $\hat{y} = 3.38$ is [3.09, 3.68], and the 95% prediction interval around

$\hat{y} = 3.38$ is [1.79, 4.98]. This prediction interval is rather imprecise, as is to be expected, and overlaps considerably with the prediction interval for an individual who scored at the mean of all three predictors.

(d) No. High scores on the variable *spatial* predict poor musical ability. Remember low *MusicScores* mean strong musical ability. However, *MusicScores* and *spatial* are positively correlated, meaning that high scores on *spatial* are associated with high *MusicScore*, which means (again) poor musical ability.

CHAPTER 17

Definitions and Concepts

1. The sampling distribution of b_i is the distribution of all possible values of b_i computed from samples of the same size drawn from the multivariate distribution of interest.

3. When the association between x and y changes as a function of a third variable *mod*, we say that *mod* is a moderator variable. It moderates the association between x and y.

5. A partial correlation is the correlation between two sets of partial scores. That is, it is the correlation between two sets of scores when the effects of one or more other variables have been partialed out of each.

7. The variance inflation factor (*VIF*) is the factor by which $s_{b_i}^2$ is increased relative to $s_b^2 = s_{est}/\sqrt{ss_x}$, in which case, x_i is uncorrelated with the other predictors, $x_{j \neq i}$.

True or False

9. False. Random variables are typically correlated to some degree.

11. False. The correlation between two sets of partial scores is rarely 0.

13. True.

15. False. Bootstrapping can be used for all statistics.

17. True.

19. True.

21. False. There are nine defined by $x_{j \neq i}$.

23. False. $c = c' + ab$.

25. True.

27. False. In a moderation analysis, *mod* is a moderator variable.

Calculations

29. (a) $\hat{y} = 8.4 + 0.25(x_1)$.
    (c) $\hat{y} = 14.32 + 0.25(x_1)$.

30. (a) 95% CI = [0.5619, 1.8381].

31. (a) 95% CI = [−0.1124, 0.6124].
    (c) 95% CI = [−1.1678, 0.0478].

32. (a) Approximate 95% CI = [1.3360, 14.6640].
    (c) Approximate 95% CI = [−0.0556, 0.4576].

33. (a) $c' = 0.4$. (c) $b = -0.2$.

34. Direct effect and indirect effect, respectively: (a) $0.4(x)$ and $0.3(x)$. (c) $0.7(x)$ and $-0.1(x)$.

35. Computing a hierarchical regression by hand.

(a)

Model Summary									
					Change Statistics				
Model	R	R^2	R^2_{adj}	s_{est}	ΔR^2	ΔF	df_1	df_2	p
1	0.6502	0.4227	0.4154	4.1311	0.4227	57.8547	1	79	.00
2	0.6704	0.4494	0.4353	4.0603	0.0267	3.7808	1	78	.06

Coefficients

Model		b	S_{b_i}	Beta	t	p	95% Lower	95% Upper
1	b_1	0.1596	0.0210	0.6503	7.6072	.0000	0.1178	0.2014
2	b_1	0.1346	0.0243	0.5484	5.5406	.0000	0.0862	0.1830
	b_2	0.1025	0.0527	0.1925	1.9451	.0554	−0.0024	0.2074

(c)

Model Summary

Model	R	R^2	R^2_{adj}	s_{est}	ΔR^2	ΔF	df_1	df_2	p
						Change Statistics			
1	0.4790	0.2295	0.2164	3.1159	0.2295	17.5710	1	59	0.00
2	0.6871	0.4722	0.4540	2.6011	0.2427	26.6654	1	58	0.00

Coefficients

Model		b	s_{b_i}	Beta	t	p	95% Lower	95% Upper
1	b_1	0.3738	0.0892	0.4789	4.1909	.0001	0.1953	0.5523
2	b_1	0.2882	0.0763	0.3693	3.7782	.0004	0.1355	0.4409
	b_2	0.3781	0.0732	0.5049	5.1656	.0000	0.2316	0.5246

Scenarios

36. **Predictors of first-year GPA.**

(a) No, ΔR^2 is not statistically significant; $\Delta R^2 = 0.001$, $F_{1,22} = 0.017$, $p = .897$.

(b) On step 1, $b = 0.563$, Beta $= 0.522$. On step 2, $b = 0.641$, Beta $= 0.594$.

(c) Neither predictor is statically significant on step 2, despite the fact that the model as a whole is statistically significant. There is no contraction here. The two predictors are highly correlated, so the *VIF* is very large (10.3), meaning that the estimated standard errors of the predictors are inflated. The predictors don't need to be statistically significant for the model as a whole to be statically significant.

(d) The admissions committee could safely drop *HS_Ratings*. These scores are highly correlated ($r = .95$). The high school teachers are probably rating students based on their high school grades. Therefore, the two variables are measuring the same thing. Because *HS_GPA* is more highly correlated with *Univ_GPA* than is *HS_Ratings*, I would throw out *HS_Ratings*.

38. **Math scores and cognitive self-control.**

A mediation analysis was conducted with *math* as the outcome variable, *PPV* as the predictor variable, and *CSC* as the hypothetical mediator. The direct effect of *PPV* on *math* was $c' = 0.1704$, 95% CI [0.0862, 0.2546]. The indirect effect of *PPV* on *math* was $ab = 0.0944$, 95% bootstrapped CI [0.0434, 0.1459]. (Your bootstrapped confidence interval will be slightly different from mine.) This analysis shows partial mediation, because there is a substantial indirect effect; 0 seems to be an implausible estimate of the indirect effect because it is not included in the 95% bootstrapped confidence interval. However, the direct effect is stronger than the indirect effect, $c'/ab = 0.1744/0.0944 = 1.85$. Even though there is some evidence of statistical mediation, *CSC* is not a plausible mediator. Rather, *CSC* seems more likely to be an individual trait, rather than a result of kindergarten vocabulary. Statistical evidence of mediation does not support an implausible causal chain.

40. Parental pressure and grades.

    (a) A hierarchical regression analysis was conducted with *grade* as the outcome variable and *IQ* and *pressure* as predictors in block one. In block two, *IQ*pressure* was added as a predictor. Adding this interaction term produced a statistically significant increase in explained variance; $\Delta R^2 = .0269$, $F_{1,64} = 5.39$, $p = .02$. This is the conventional way to assess an interaction.

    (b) The interaction means that the association between *IQ* and *grade* changes as a function of *pressure*.

    (c) For $m_p - s = 6.94$, the slope is 0.67. For $m_p = 10.09$, the slope is 0.55. For $m_p + s = 13.24$, the slope is 0.43.

    (d) For $m_p - s$, 95% CI = [0.51, 0.83]. For m_p, 95% CI = [0.45, 0.65]. For $m_p + s$, 95% CI = [0.30, 0.55].

    (e) In this analysis, we ask PROCESS to print data for plotting. This gives us all combinations of $m_p - s$, m_p, $m_p + s$, $m_{IQ} - s$, m_{IQ}, and $m_{IQ} + s$. (We could have also calculated these quantities by hand by determining the mean and standard deviations from Descriptives in SPSS.) We then put these nine combinations of $m_p - s$, m_p, $m_p + s$, $m_{IQ} - s$, m_{IQ}, and $m_{IQ} + s$ in the appropriate columns of the SPSS data file. Finally, we rerun a regression analysis with *IQ*, *pressure*, and *IQ*pressure* all entered at once and obtain the confidence limits from the SPSS data file.

IQ	Pressure	Lower	Upper
86.73	6.94	11.34	16.55
99.85	6.94	20.96	24.50
112.98	6.94	28.65	34.39
86.73	10.09	11.34	14.90
99.85	10.09	19.08	21.56
112.98	10.09	25.74	29.32
86.73	13.24	9.95	14.64
99.85	13.24	16.15	19.68
112.98	13.24	21.08	25.99

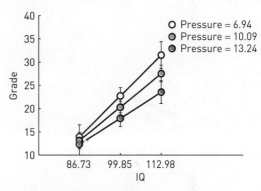

The interaction between *IQ* and *pressure* is shown for all combinations of $m_p - s$, m_p, $m_p + s$, $m_{IQ} - s$, m_{IQ}, and $m_{IQ} + s$. Error bars are 95% confidence intervals.

    (f) There seems to be relatively little effect of pressure on the average grades of students whose IQs are one standard deviation below the mean IQ. As the students' IQs increase, the effect of parental pressure seems to suppress grades. For example, for students whose IQs are one standard deviation above the mean, those with less parental pressure did better, on average, than those with more parental pressure.

42. Scientific literacy, movies, and global warming.

    (a) A hierarchical regression analysis was conducted with *AGW* as the outcome variable and *literacy* and *SciFi* as predictors in block one. In block two, *literacy*SciFi* was added as a predictor. Adding this interaction term did not produce a statistically significant increase in explained variance; $\Delta R^2 = .00$, $F_{1,55} = 0.00$, $p = .97$. This is the conventional way to assess an interaction.

(b) The first step of the regression analysis shows that *literacy* and *SciFi* were statistically significant predictors of *AGW*; $R^2 = .31$, approximate 95% CI [.11, .49], $F_{2,56} = 12.73, p < .001$. There is no evidence of an interaction between *literacy* and *SciFi*. This means that the association between *literacy* and *AGW* does not depend on the number of science fiction movies watched during the last year.

(c) For $m_{SF} - s = 6.85$, the slope is 1.44; 95% CI = [0.62, 2.26]. For $m_{SF} = 10.02$, the slope is 1.45; 95% CI = [0.86, 2.04]. For $m_{SF} + s = 13.19$, the slope is 1.47; 95% CI = [0.62, 2.31].

(d)

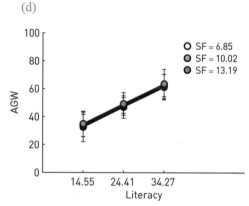

The interaction between *scientific literacy* and *SF* is shown for all combinations of $m_{Lit} - s$, m_{Lit}, $m_{Lit} + s$, $m_{SF} - s$, m_{SF}, and $m_{SF} + s$. Error bars are 95% confidence intervals.

(e) If this sample is a truly random sample from the participant pool, then (strictly speaking) those in the participant pool represent the population of interest. The student may have a larger population in mind, such as all university students or all people in their 20s. Further research would be needed to see whether the observed results generalize to these larger populations.

CHAPTER 18

Definitions and Concepts

1. ANOVA stands for ANalysis Of VAriance.

3. A condition of an ANOVA is a specific combination of the levels of the independent variables.

5. In a $5 \times 6 \times 3 \times 2$ between-subjects ANOVA, there are $5 \times 6 \times 3 \times 2 = 180$ conditions.

7. A mixed design involves at least one between-subjects factor and at least one within-subjects factor.

9. Having 95% confidence in an interval means knowing that 95% of all intervals computed in the same way will capture the parameter being estimated.

11. A simple contrast is the difference between two means. A complex contrast divides three or more means into two groups and computes the difference between the means of the two groups.

13. To control the family-wise Type I error rate, one must make the individual contrasts more conservative. This means making individual confidence intervals wider or *p*-values larger. Wider confidence intervals are more likely to capture 0, and larger *p*-values are more likely to exceed .05. Both make it harder to reject the null hypothesis, which means reducing power.

True or False

15. False. It would be far more likely to have neuroticism as the between-subjects factor and days as the within-subjects factor.

17. False. Homogeneity is another word for homoscedasticity.

19. True.

21. False. The w_i should be squared.

23. True.

Calculations

25. Assuming equal sample sizes, a one-factor between-subjects ANOVA with 10 levels of the independent variable and 120 participants would have 12 participants in each condition.

27. One-way between-subjects ANOVA.

(a) $m_y = 10$, $ss_{between} = 200$, $ss_{error} = 264$, $ss_{total} = 464$, $R^2 = .43$, $df_a = 4$, $df_{error} = 15$, $F = 2.84$.

(b) $\hat{y}_{i,j} = 10 - 0.5(x_1) + 0.167(x_2) + 1.833(x_3) + 3.5(x_4)$.

(c) 95% CI = [0.0, 0.58].

29. Contrasts in ANOVA.

(a) 95% CI = [−7.50, 2.50], t_{obs} = −1.07, $\hat{\eta}^2$ = .04, F = 1.14.

(b) 95% CI = [−0.47, 8.47], t_{obs} = 1.91, $\hat{\eta}^2$ = .14, F = 3.64.

(c) 95% CI = [−9.32, 3.32], t_{obs} = −1.01, $\hat{\eta}^2$ = .04, F = 1.02.

(d) 95% CI = [0.68, 13.32], t_{obs} = 2.36, $\hat{\eta}^2$ = .21, F = 5.57.

31. Yes. These numbers form an orthogonal set because all pairs of weights have products ($\Sigma w_i w_j = 0$).

Scenarios

33. Effectiveness of CBT and ST.

(a) Two planned contrasts.

(i) c_1 = 6, 95% CI [2.24, 9.76]. c_2 = 6, 95% CI [1.65, 10.35].

(ii) d_1 = 0.87, 95% CI [0.31, 1.43]. d_2 = 0.87, 95% CI [0.23, 1.51].

(iii) t_{obs} = 3.19, p = .0023. t_{obs} = 2.76, p = .0077.

(b) Omnibus statistics.

(i) R^2 = .24, 95% CI [0.06, 0.39].

(ii) F_{obs} = 8.91, p = .0004.

(c) Three possible contrasts between means.

(i) $m_1 − m_2$ = 3, 95% CI [−2.35, 8.35]. $m_1 − m_3$ = 9, 95% CI [3.65, 14.35]. $m_2 − m_3$ = 6, 95% CI [0.65, 11.35].

(ii) t_{obs} = 1.38, p = .517. t_{obs} = 4.15, p = .0003. t_{obs} = 2.76, p = .023.

(d) In this study, ST reduced scores on the NPI more than CBT. Taken on its own, this result might be seen as evidence favoring ST for the treatment of narcissistic personality disorder. Of course, there is always a question about how mean differences on instruments such as the NPI relate to the actual experiences of individuals. For example, you may score lower after than before treatment but still have the same life problems that led you to seek therapy. In terms of the dodo verdict, the current results produce a much larger effect size, d = 0.87, than in the results reported by Luborsky et al. (2002). However, this one study added to the previous 130 would make a small difference in the accumulated results; i.e., 1/131*(0.87)

+ 130/131*(0.02) = 0.205. Conversely, the approximate 95% CI around d = 0.87 was [0.23, 1.51], which does not include 0.2. This suggests that the current result differs quite a bit from the mean effect size from 130 previous studies. If this study were real, there might be methodological details that made it much more sensitive to differences in treatment outcomes. There is always the possibility that researchers will come up with a new method of analysis that renders all previous analyses obsolete. Taken on its own, however, the present study does not overthrow the dodo bird verdict.

35. Vision and auditory localization.

(a) The omnibus F = 7.45.

(b) The three contrasts are $c_1 = m_1 − m_2$, $c_2 = m_1 − m_3$, $c_3 = m_2 − m_3$, s_c = 12.89, t_{pc} = 2.55, and moe = 32.89. We thus find the following 95% CIs: c_1 = [−15.89, 49.89], c_2 = [−64.89, 0.89], and c_3 = [−81.89, −16.11].

(c) Violations of assumptions? There are very big differences in variances of the three groups, 160.89, 2256.89, and 73.33. The variance within the second group (congenitally blind participants) is a real outlier. If we had the original data, we would have to examine them closely to see whether there are either some outlying scores or perhaps distinct patterns of responses within the group; if you consult the original paper by Lessard et al. (1998), you will see that this is the case.

(d) The first contrast suggests that congenitally blind participants are better at monaural localization than sighted control participants because their average localization errors are (52 − 35)/53)*100% = 33% smaller than those of the control participants. This suggests that having to rely more heavily on audition throughout a lifetime improves auditory sensitivity. However, the 95% confidence interval is quite wide [−16.34, 49.45], so it would make sense to gather more data in order to obtain a more precise estimate of the difference between the means of the two populations. The true difference could be very small or very large. The second contrast shows a big difference in monaural localization errors between the sighted control participants and those with acquired vision loss (i.e., those that became blind later in life and still had some residual peripheral vision). The control participants are better at monaural localization than those with acquired vision loss because their average errors are $((84 - 52)/84){*}100\%$ = 38% smaller than those with acquired vision loss. The third contrast shows a very big difference between those who are congenitally blind and those with acquired vision loss, $c_3 = 35 - 84 = -48$, 95% CI $[-81.89, -16.11]$. The congenitally blind participants have average localization errors that are $((84 - 35)/84){*}100\% = 58\%$ smaller than those with acquired vision loss. This is an extremely interesting comparison because it shows that the simple loss of vision does not immediately lead to improved use of auditory information. It may be that with acquired vision loss, there is a period of neural reorganization during which both visual and auditory localization systems work suboptimally. However, one has to keep in mind that this was a quasi-experiment, so the characteristics of the samples may be quite different. For example, Lessard et al. (1998) did not provide information about the ages, sexes, or other demographic characteristics of their participants. Ensuring matched samples would be an important control before one could theorize in an informed way about why performance was so poor in the acquired vision loss group.

CHAPTER 19

Definitions and Concepts

1. ss_a is the sum of squares associated with condition means. It is equivalent to replacing each subject's scores with the condition means and then computing the total sum of squares. ss_a can be computed as $ss_a = n\sum_{i=1}^{k}(m_i - m_y)^2$, where n is the number of subjects. ss_s is the sum of squares associated with subjects. It is equivalent to replacing each subject's scores with his or her mean score and then computing the total sum of squares. ss_s can be computed as $ss_s = k\sum_{j=1}^{n}(s_j - m_y)^2$, where k is the number of

levels of the independent variable. ss_{error} is the residual sum of squares; $ss_{error} = ss_{total} - ss_a - ss_s$.

3. The omnibus F-statistic is defined as $F_{obs} = ss_a/ss_{error} {*} df_{error}/df_a$.

5. $\hat{\eta}^2 = ss_a/ss_{total}$ and $\hat{\eta}_p^2 = ss_a/(ss_a + ss_{error})$.

7. The sphericity assumption is that in the population, the difference scores for all possible simple contrasts have the same variance.

9. A contrast score is the sum of the products of a weight vector (contrast vector) and the scores of a subject at each level of the independent variable; i.e., $cs_j = \sum_{i=1}^{k} w_i y_{i,j}$.

11. s_c is the estimated standard error of a contrast and s_{cs} is the standard deviation of contrast scores.

13. $d = c/s_{cs}$ estimates $\delta = \psi/\sigma_{cs}$.

True or False

15. True. In a within-subjects design, $ss_{total} = ss_s + ss_a + ss_{error}$. So unless $ss_s = 0$, $ss_{total} > ss_a + ss_{error}$.

17. True

19. True. $\hat{\eta}_p^2 = 10/(10 + 40/4) = .5$.

21. False. Sphericity concerns the variance of difference scores, and homoscedasticity concerns raw scores.

23. False. d is always the center of the interval.

25. False. This is the inverse probability error.

27. True.

Calculations

29. If a one-factor within-subjects ANOVA has 10 levels of the independent variable and 100 participants, there are 100 subjects per condition.

31. One-way within-subjects trends.

(a) $m_y = 10$, $ss_a = 200$, $ss_s = 68.4$, $ss_{Total} = 464$, $df_a = 4$, $df_{error} = 12$, $F = 3.07$, $\hat{\eta}_p^2 = .51$.

(b) Trend analysis.
Linear: $t_{obs} = -1.23$, $F_{obs} = 1.52$, $p = .31$, $\hat{\eta}_p^2 = .34$.
Quadratic: $t_{obs} = -4.83$, $F_{obs} = 23.29$, $p = .02$, $\hat{\eta}_p^2 = .89$.
Cubic: $t_{obs} = -1.57$, $F_{obs} = 2.46$, $p = .21$, $\hat{\eta}_p^2 = .45$.
Quartic: $t_{obs} = 0.17$, $F_{obs} = 0.03$, $p = .87$, $\hat{\eta}_p^2 = .01$.

Scenarios

33. Exercise and school performance.

   (a) $m_y = 5, ss_a = 208, ss_s = 130, ss_{total} = 410,$
$df_a = 4, df_{error} = 28, F = 20.22, \hat{\eta}_p^2 = .74.$

   (b) Repeated contrasts in SPSS.

      1 versus 2: $F_{1,7} = 5.60, p = .050, \hat{\eta}_p^2 = .444$.

      2 versus 3: $F_{1,7} = 16.00, p = .005, \hat{\eta}_p^2 = .696.$

      3 versus 4: $F_{1,7} = 8.62, p = .022, \hat{\eta}_p^2 = .522.$

      4 versus 5: $F_{1,7} = 1.65, p = .240, \hat{\eta}_p^2 = .190.$

   (c) Repeated contrasts computed in Excel or SPSS.

      (i) c_1: [−2.00, 0.00]. c_2: [−3.18, −0.82]. c_3: [−3.61, −0.39]. c_4: [−2.84, 0.84].

      (ii) d_1: [−1.66, −0.02]. d_2: [−2.43, −0.40]. d_3: [−1.92, −0.16]. d_4: [−1.19, 0.28].

      (iii) d_1: [−1.63, −002]. d_2: [−2.39, −0.39]. d_3: [−1.89, −0.14]. d_4: [−1.17, 0.29].

   (d) First, there is a very clear improvement in mean ratings with exercise. The pattern is clear and impressive, given the small sample size. The improvements are greatest from 2 to 4 and from 4 to 8 minutes ToT. It seems as though the improvements level off between 8 and 16 minutes ToT. If we assume that we'd like to maximize the benefits of exercise and minimize the time spent exercising, then future research could be directed toward finding the optimal duration of exercise. Of course, this small study should be replicated on a much larger scale before the results could be used to make policy decisions.

   (e) I would be encouraged by the results. There is support from other sources for the idea that exercise makes a positive contribution to school performance. As an administrator, I would be willing to initiate a trial program in my school to determine if there are observable effects of exercise for my students. I would seek permission to select a random sample of students, put them on a regime of 10 minutes of treadmill walking each morning, and then compare their grades at the end of term to those of students who were not part of the study, to see how big a difference there is between their grades. I would try to run this as a matched-samples design to increase the precision of my estimates.

CHAPTER 20

Definitions and Concepts

1. Main effect refers to changes in the mean levels of a factor averaged across all levels of other variables.

3. These quantities are as follows: $\hat{\eta}_a^2$ is the proportion of total variability in the dependent variable explained by the main effect of independent variable A, $\hat{\eta}_b^2$ is the proportion of total variability in the dependent variable explained by the main effect of independent variable B, and $\hat{\eta}_{ab}^2$ is the proportion of total variability in the dependent variable explained by the interactive effect of independent variables A and B.

5. In a 2×2 ANOVA, there is $df = (2-1)(2-1) = 1*1 = 1$ degrees of freedom associated with the AB interaction.

7. An omnibus effect is a source of variance that is coded by two or more orthogonal vectors.

9. The seven sources of variance in a completely within-subjects ANOVA are $\hat{\eta}_a^2$, $\hat{\eta}_b^2$, $\hat{\eta}_{ab}^2$, $\hat{\eta}_s^2$, $\hat{\eta}_{sa}^2$, $\hat{\eta}_{sb}^2$, and $\hat{\eta}_{sab}^2$.

11. Omnibus main effects are of little value in an ANOVA because many different arrangements of factor means will produce the same omnibus $\hat{\eta}$ or F_{obs}. However, these different arrangements would have different interpretations. So it is the pattern of differences among means that has meaning, not the omnibus effect on its own.

True or False

13. True.

Calculations

15. A two-factor within-subjects ANOVA with 10 levels of both independent variables and 100 participants altogether will have 100 participants in each condition.

16. Two-factor, completely between-subjects design.

   (a) Degrees of freedom

      (i) $df_a = 4.$

      (ii) $df_b = 2.$

      (iii) $df_{ab} = 2*4 = 8.$ $df_{error} = df_{total} - df_a - df_b - df_{ab} = 149 - 4 - 2 - 8 = 135.$

(b) *F*-statistics

    (i) For *A*: $F_{obs} = \hat{\eta}_a^2/\hat{\eta}_{error}^2 * df_{error}/df_a = .2/.51*135/4 = 13.24, p < .01.$

    (ii) For *B*: $F_{obs} = \hat{\eta}_b^2/\hat{\eta}_{error}^2 * df_{error}/df_b = .05/.51*135/2 = 6.62, p < .01.$

    (iii) For *AB*: $F_{obs} = \hat{\eta}_{ab}^2/\hat{\eta}_{error}^2 * df_{error}/df_{ab} = .24/.51*135/8 = 7.94, p < .01.$

17. Two-factor, completely within-subjects design.

    (a) Degrees of freedom

        (i) $df_a = 2$.

        (ii) $df_b = 5$.

        (iii) $df_{ab} = 2*5 = 10$.

        (iv) $df_s = 10$.

        (v) $df_{sa} = 2*10 = 20$.

        (vi) $df_{sb} = 5*10 = 50$.

        (vii) $df_{sab} = 2*5*10 = 100$.

        (viii) $df_{total} = 2 + 5 + 10 + 10 + 20 + 50 + 100 = 197$.

        (ix) $n = 3*6*11 = 198$.

    (b) *F*-statistics

        (i) For *A*: $F_{obs} = \hat{\eta}_a^2/\hat{\eta}_{sa}^2 * df_{sa}/df_a = .2/.05*20/2 = 40, p < .01.$

        (ii) For *B*: $F_{obs} = \hat{\eta}_b^2/\hat{\eta}_{sb}^2 * df_{sb}/df_b = .05/.1*50/5 = 5, p < .01.$

        (iii) For *AB*: $F_{obs} = \hat{\eta}_{ab}^2/\hat{\eta}_{sab}^2 * df_{sab}/df_{ab} = .15/.2*100/10 = 7.5, p < .01.$

18. Two-factor mixed design.

    (a) Degrees of freedom

        (i) $df_a = 3$.

        (ii) $df_b = 5$.

        (iii) $df_{ab} = 3*5 = 15$.

        (iv) $df_{s/a} = 4*9 = 36$.

        (v) $df_{b*s/a} = 5*36 = 180$.

        (vi) $df_{total} = 3 + 5 + 15 + 36 + 180 = 239$.

        (vii) $n = 4*6*10 = 240$.

    (b) *F*-statistics

        (i) For *A*: $F_{obs} = \hat{\eta}_a^2/\hat{\eta}_{s/a}^2 * df_{s/a}/df_a = .15/.3*36/3 = 6, p < .01.$

        (ii) For *B*: $F_{obs} = \hat{\eta}_b^2/\hat{\eta}_{b*s/a}^2 * df_{b*s/a}/df_b = .25/.2 *180/5 = 45, p < .01.$

        (iii) For *AB*: $F_{obs} = \hat{\eta}_{ab}^2/\hat{\eta}_{b*s/a}^2 * df_{b*s/a}/df_{ab} = .1/.2 *180/15 = 6, p < .01.$

22. Partial eta squared.

    (a) Between: $\hat{\eta}_{p(A)}^2 = .024, \hat{\eta}_{p(B)}^2 = .348, \hat{\eta}_{p(AB)}^2 = .018.$

    (b) Within: $\hat{\eta}_{p(A)}^2 = .176, \hat{\eta}_{p(B)}^2 = .884, \hat{\eta}_{p(AB)}^2 = .742.$

    (c) Mixed: $\hat{\eta}_{p(A)}^2 = .123, \hat{\eta}_{p(B)}^2 = .394, \hat{\eta}_{p(AB)}^2 = .022.$

Scenarios

23. Visual search. $\hat{\eta}_{p(A)}^2 = .909, \hat{\eta}_{p(B)}^2 = .841, \hat{\eta}_{p(AB)}^2 = .522.$

24. Isolation and drug taking.

    (a) *A:* $F = 6.837, p = .013, \hat{\eta}_p^2 = .156.$

    (b) *B:* $F = 77.022, p = .000, \hat{\eta}_p^2 = .676.$

    (c) *AB:* $F = 1.836, p = .125, \hat{\eta}_p^2 = .047.$

25. List learning and delayed recall.

    (a) *A:* $F = 14.574, p = .000, \hat{\eta}_p^2 = .245.$

    (b) *B:* $F = 30.822, p = .000, \hat{\eta}_p^2 = .407.$

    (c) *AB:* $F = 1.999, p = .002, \hat{\eta}_p^2 = .043.$

CHAPTER 21

Definitions and Concepts

1. A second-order contrast is a contrast on a contrast. For example, one could have a contrast (c_1) between two levels of *B* at a_1 and the same two levels of *B* at a_2 (c_2). A contrast between c_1 and c_2 is a contrast-contrast, or a second-order contrast.

True or False

3. False. The absolute values of the weights ($|w_i|$) must sum to 4.

5. False. Confidence intervals around between-subjects contrasts will generally be wider when the equal variances assumption is violated.

Calculations

7. (a) A two-factor ANOVA.

        (i) $c = -11, s_c = 2.208,$ 95% CI [−15.557, −6.443].

        (ii) $c = -4, s_c = 2.208,$ 95% CI [−8.557, 0.557].

        (iii) $c = 2, s_c = 2.208,$ 95% CI [−2.557, 6.557].

        (iv) $c = 9, s_c = 2.208,$ 95% CI [4.443, 13.557].

(v) $c = 5$, $s_c = 1.104$,
95% CI [2.722, 7.278].

(vi) $c = -5$, $s_c = 2.208$,
95% CI [−9.557, −0.433].

(b) Two-factor, within-subjects design.

(i) $c = -11$, $s_c = 2.799$,
95% CI [−19.907, −2.093].

(ii) $c = -4$, $s_c = 1.080$,
95% CI [−7.437, −0.563].

(iii) $c = 2$, $s_c = 1.414$,
95% CI [−2.501, 6.501].

(iv) $c = 9$, $s_c = 2.828$,
95% CI [−0.001, 18.001].

(v) $c = 5$, $s_c = 1.080$,
95% CI [1.563, 8.437].

(vi) $c = -5$, $s_c = 2.483$,
95% CI [−12.903, −2.903].

(c) Two-factor, mixed-subjects design.

(i) $c = -11$, $s_c = 2.121$,
95% CI [−16.191, −5.809].

(ii) $c = -4$, $s_c = 2.121$,
95% CI [−9.191, 1.191].

(iii) $c = 2$, $s_c = 1.528$,
95% CI [−1.738, 5.738].

(iv) $c = 9$, $s_c = 2.449$,
95% CI [−3.006, 14.994].

(v) $c = 5$, $s_c = 1.155$,
95% CI [2.175, 7.825].

(vi) $c = -5$, $s_c = 2.582$,
95% CI [−11.318, 1.318].

Scenarios

8. Visual search.

(a) The results show strong linear trends in both the target-present and target-absent conditions. However, the increase in reaction time with display size is much greater in the target-absent condition than in the target-present condition. In other words, the difference between target-absent and target-present reaction times grows with display size. Furthermore, these data seem remarkably precise. The standard errors are so small that thvey are obscured in many cases by the plot symbols.

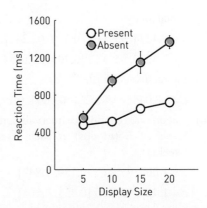

(b) The (uncorrected) 95% confidence intervals around the difference in reaction times for target-present versus target-absent conditions for each display size are as follows:

$c = 77$, 95% CI [−61.773, 215.773];

$c = 437$, 95% CI [231.959, 642.041];

$c = 495$, 95% CI [186.512, 803.488]; and

$c = 647$, 95% CI [379.256, 914.744].

(c) Trend analysis

$c = 17.240$, 95% CI [6.074, 28.406].

$c = 52.600$, 95% CI [38.861, 66.339].

(d) A contrast-contrast. $c = 35.360$, 95% CI [18.228, 52.492]. Yes, this is a second-order contrast. We can think of the slopes within the target-present and target-absent conditions as first-order contrasts. So this contrast is the difference between two first-order contrasts, making it a second-order contrast.

(e) We assume that the distributions of contrast scores are normal and randomly selected from the population of interest.

9. Isolation and drug taking.

(a) For both groups, the mean number of injections increases over time. However, there is strong evidence that rats housed in groups are less prone to self-administering heroin than those housed in isolation.

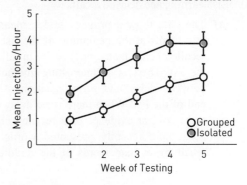

(b) Trend analysis.

$c = 0.4274$, 95% CI [0.329, 0.5247].

$c = 0.4920$, 95% CI [0.4055, 0.5784].

(c) Interaction contrast.

$c = -0.0646$, 95% CI [-0.0663, 0.1956].

(d) Uncorrected confidence intervals comparing levels of B (groups) at each level of A (weeks).

$c = 1.003$, 95% CI [0.159, 1.847].

$c = 1.459$, 95% CI [0.353, 2.565].

$c = 1.526$, 95% CI [0.430, 2.623].

$c = 1.554$, 95% CI [0.536, 2.573].

$c = 1.279$, 95% CI [-0.093, 2.650].

(e) The main message of these data seems to be that large increases in drug consumption result from relatively short periods of isolation. If we were to generalize from these data to humans, we might expect that individuals who are isolated may be at greater risk for drug addiction. This study does not say anything about whether or not the rats are addicted to heroin or have a physiological dependence on it. However, in humans, we know that continued use over time leads to addiction and physiological dependence. From the point of view of those interested in reducing drug addiction in human populations, the data strongly suggest that environmental factors play a critical role in continued drug use, and one would also expect environmental factors to play a large role in overcoming addition. Of course, even the non-isolated rats consume drugs, and their drug intake increases over time. Therefore, social isolation is not a necessary condition for increased drug consumption.

10. List learning and delayed recall.

(a) The data show a primacy effect, meaning that words at the beginning of the list are recalled better than words in the middle of the list. There is an interesting interaction between delay and serial position at the end of the list. When there is no delay (0), there is clear evidence of a recency effect (i.e., more recently presented words are recall better than words in the middle of

the list). However, with delayed recall that prevents rehearsal (10 and 30), the recency effect is greatly reduced or eliminated. This suggests that the delayed recall interferes with transfer to long-term memory.

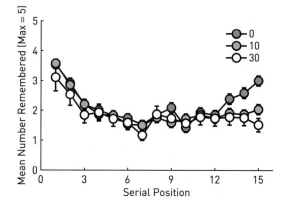

(b) Linear trends for first five serial positions

$c_0 = -0.448$, 95% CI [-0.542, -0.354].

$c_{10} = -0.433$, 95% CI [-0.539, -0.326].

$c_{30} = -0.339$, 95% CI [-0.437, -0.242].

These trends seem very similar, so delay had very little effect on items early in the list.

(c) Linear trends for last five serial positions

$c_0 = 0.287$, 95% CI [0.177, 0.397].

$c_{10} = 0.052$, 95% CI [-0.044, 0.148].

$c_{30} = -0.050$, 95% CI [-0.154, 0.054].

These trends seem very different, so delay had a very big effect on items late in the list. Performance increases over the last five conditions when there is no delay, but there is no change over positions for either the 10- or 30-second delays.

(d) Simple contrasts at the last serial position

$c_{1-2} = 0.957$, 95% CI [0.448, 1.465].

$c_{1-3} = 1.478$, 95% CI [1.015, 1.941].

$c_{2-3} = 0.522$, 95% CI [0.054, 0.989].

(e) The results show a primacy effect. There is a strong negative linear trend at each delay

($c_0 = -0.448$, 95% CI [-0.542, -0.354], $c_{10} = -0.433$, 95% CI [-0.539, -0.326], $c_{30} = -0.339$, 95% CI [-0.437, -0.242]).

These decreasing trends seem quite similar, suggesting that the delays did not affect the early parts of the three serial position curves. (Note, you could perform further analyses to measure differences in the early parts of the serial position curves to see how much recall delay affected them.)

The linear trends in the last five positions are quite different. There is a strong linear trend for the case of no delay ($c_0 = 0.287$, 95% CI [0.177, 0.397]), meaning that recall improves for words closer to the end of the list. The improvement is much less for the case of a 10-second delay ($c_{10} = 0.052$, 95% CI [−0.044, 0.148]) and the confidence interval around the linear contrast includes 0, so it is plausible that there is very little dependence on serial position. In the case of a 30-second delay, there is a weak negative trend in the data ($c_{30} = -0.050$, 95% CI [−0.154, 0.054]). Therefore, it is possible that items at the end of the list are recalled more poorly than those closer to the middle of the list.

In the final serial position (15), subjects recalled about one more word in the 0-delay condition than in the 10-second delay condition ($c_{1-2} = 0.957$, 95% CI [0.448, 1.465]) and about 1.5 more words than in the 30-second delay condition ($c_{1-3} = 1.478$, 95% CI [1.015, 1.941]). There was about a half a word difference in recall in the 10- and 30-second delay conditions ($c_{2-3} = 0.522$, 95% CI [0.054, 0.989]). These data show that there is a progressive effect of delay on the recall of the final list item.

These results strongly suggest that preventing rehearsal of the list items strongly influences the recency effect, with longer delays reducing recall. Of course, there are many questions that might follow from this study, not the least of which is whether the results replicate and are found in other populations. This study is 50 years old, and many subsequent studies have referenced it. If you're interested, you can read the literature and find out where this all led.

APPENDIX A

The *z*-Table for a Selection of *z*-Scores Ranging From −2.59 to 2.59									
	.00	.01	.02	.03	.04	.05	.06	.07	.08	.09
−2.5	.0062	.0060	.0059	.0057	.0055	.0054	.0052	.0051	.0049	.0048
−2.4	.0082	.0080	.0078	.0075	.0073	.0071	.0069	.0068	.0066	.0064
−2.3	.0107	.0104	.0102	.0099	.0096	.0094	.0091	.0089	.0087	.0084
−2.2	.0139	.0136	.0132	.0129	.0125	.0122	.0119	.0116	.0113	.0110
−2.1	.0179	.0174	.0170	.0166	.0162	.0158	.0154	.0150	.0146	.0143
−2.0	.0228	.0222	.0217	.0212	.0207	.0202	.0197	.0192	.0188	.0183
−1.9	.0287	.0281	.0274	.0268	.0262	.0256	.0250	.0244	.0239	.0233
−1.8	.0359	.0351	.0344	.0336	.0329	.0322	.0314	.0307	.0301	.0294
−1.7	.0446	.0436	.0427	.0418	.0409	.0401	.0392	.0384	.0375	.0367
−1.6	.0548	.0537	.0526	.0516	.0505	.0495	.0485	.0475	.0465	.0455
−1.5	.0668	.0655	.0643	.0630	.0618	.0606	.0594	.0582	.0571	.0559
−1.4	.0808	.0793	.0778	.0764	.0749	.0735	.0721	.0708	.0694	.0681
−1.3	.0968	.0951	.0934	.0918	.0901	.0885	.0869	.0853	.0838	.0823
−1.2	.1151	.1131	.1112	.1093	.1075	.1056	.1038	.1020	.1003	.0985
−1.1	.1357	.1335	.1314	.1292	.1271	.1251	.1230	.1210	.1190	.1170
−1.0	.1587	.1562	.1539	.1515	.1492	.1469	.1446	.1423	.1401	.1379
−0.9	.1841	.1814	.1788	.1762	.1736	.1711	.1685	.1660	.1635	.1611
−0.8	.2119	.2090	.2061	.2033	.2005	.1977	.1949	.1922	.1894	.1867
−0.7	.2420	.2389	.2358	.2327	.2296	.2266	.2236	.2206	.2177	.2148
−0.6	.2743	.2709	.2676	.2643	.2611	.2578	.2546	.2514	.2483	.2451
−0.5	.3085	.3050	.3015	.2981	.2946	.2912	.2877	.2843	.2810	.2776
−0.4	.3446	.3409	.3372	.3336	.3300	.3264	.3228	.3192	.3156	.3121
−0.3	.3821	.3783	.3745	.3707	.3669	.3632	.3594	.3557	.3520	.3483
−0.2	.4207	.4168	.4129	.4090	.4052	.4013	.3974	.3936	.3897	.3859
−0.1	.4602	.4562	.4522	.4483	.4443	.4404	.4364	.4325	.4286	.4247
−0.0	.5000	.4960	.4920	.4880	.4840	.4801	.4761	.4721	.4681	.4641
0.0	.5000	.5040	.5080	.5120	.5160	.5199	.5239	.5279	.5319	.5359
0.1	.5398	.5438	.5478	.5517	.5557	.5596	.5636	.5675	.5714	.5753
0.2	.5793	.5832	.5871	.5910	.5948	.5987	.6026	.6064	.6103	.6141
0.3	.6179	.6217	.6255	.6293	.6331	.6368	.6406	.6443	.6480	.6517
0.4	.6554	.6591	.6628	.6664	.6700	.6736	.6772	.6808	.6844	.6879
0.5	.6915	.6950	.6985	.7019	.7054	.7088	.7123	.7157	.7190	.7224
0.6	.7257	.7291	.7324	.7357	.7389	.7422	.7454	.7486	.7517	.7549
0.7	.7580	.7611	.7642	.7673	.7704	.7734	.7764	.7794	.7823	.7852
0.8	.7881	.7910	.7939	.7967	.7995	.8023	.8051	.8078	.8106	.8133
0.9	.8159	.8186	.8212	.8238	.8264	.8289	.8315	.8340	.8365	.8389
1.0	.8413	.8438	.8461	.8485	.8508	.8531	.8554	.8577	.8599	.8621
1.1	.8643	.8665	.8686	.8708	.8729	.8749	.8770	.8790	.8810	.8830
1.2	.8849	.8869	.8888	.8907	.8925	.8944	.8962	.8980	.8997	.9015
1.3	.9032	.9049	.9066	.9082	.9099	.9115	.9131	.9147	.9162	.9177
1.4	.9192	.9207	.9222	.9236	.9251	.9265	.9279	.9292	.9306	.9319
1.5	.9332	.9345	.9357	.9370	.9382	.9394	.9406	.9418	.9429	.9441
1.6	.9452	.9463	.9474	.9484	.9495	.9505	.9515	.9525	.9535	.9545
1.7	.9554	.9564	.9573	.9582	.9591	.9599	.9608	.9616	.9625	.9633
1.8	.9641	.9649	.9656	.9664	.9671	.9678	.9686	.9693	.9699	.9706
1.9	.9713	.9719	.9726	.9732	.9738	.9744	.9750	.9756	.9761	.9767
2.0	.9772	.9778	.9783	.9788	.9793	.9798	.9803	.9808	.9812	.9817
2.1	.9821	.9826	.9830	.9834	.9838	.9842	.9846	.9850	.9854	.9857
2.2	.9861	.9864	.9868	.9871	.9875	.9878	.9881	.9884	.9887	.9890
2.3	.9893	.9896	.9898	.9901	.9904	.9906	.9909	.9911	.9913	.9916
2.4	.9918	.9920	.9922	.9925	.9927	.9929	.9931	.9932	.9934	.9936
2.5	.9938	.9940	.9941	.9943	.9945	.9946	.9948	.9949	.9951	.9952

The left column shows *z*-scores rounded to one decimal, and the top row shows the value in the second decimal place. Each row/column combination specifies z to two decimal places. Each table entry shows $P(z)$, the proportion of the standard normal distribution falling below z. (Values were computed in Excel by the author.)

APPENDIX B

	Area in two tails (α)					
	.50	.20	.10	.05	.02	.01
	Area in one tail ($\alpha/2$)					
df	.250	.100	.050	.025	.010	.005
1	1.000	3.078	6.314	12.706	31.821	63.657
2	0.816	1.886	2.920	4.303	6.965	9.925
3	0.765	1.638	2.353	3.182	4.541	5.841
4	0.741	1.533	2.132	2.776	3.747	4.604
5	0.727	1.476	2.015	2.571	3.365	4.032
6	0.718	1.440	1.943	2.447	3.143	3.707
7	0.711	1.415	1.895	2.365	2.998	3.499
8	0.706	1.397	1.860	2.306	2.896	3.355
9	0.703	1.383	1.833	2.262	2.821	3.250
10	0.700	1.372	1.812	2.228	2.764	3.169
11	0.697	1.363	1.796	2.201	2.718	3.106
12	0.695	1.356	1.782	2.179	2.681	3.055
13	0.694	1.350	1.771	2.160	2.650	3.012
14	0.692	1.345	1.761	2.145	2.624	2.977
15	0.691	1.341	1.753	2.131	2.602	2.947
16	0.690	1.337	1.746	2.120	2.583	2.921
17	0.689	1.333	1.740	2.110	2.567	2.898
18	0.688	1.330	1.734	2.101	2.552	2.878
19	0.688	1.328	1.729	2.093	2.539	2.861
20	0.687	1.325	1.725	2.086	2.528	2.845
21	0.686	1.323	1.721	2.080	2.518	2.831
22	0.686	1.321	1.717	2.074	2.508	2.819
23	0.685	1.319	1.714	2.069	2.500	2.807
24	0.685	1.318	1.711	2.064	2.492	2.797
25	0.684	1.316	1.708	2.060	2.485	2.787
26	0.684	1.315	1.706	2.056	2.479	2.779
27	0.684	1.314	1.703	2.052	2.473	2.771
28	0.683	1.313	1.701	2.048	2.467	2.763
29	0.683	1.311	1.699	2.045	2.462	2.756
30	0.683	1.310	1.697	2.042	2.457	2.750
31	0.682	1.309	1.696	2.040	2.453	2.744
32	0.682	1.309	1.694	2.037	2.449	2.738
33	0.682	1.308	1.692	2.035	2.445	2.733
34	0.682	1.307	1.691	2.032	2.441	2.728
35	0.682	1.306	1.690	2.030	2.438	2.724
36	0.681	1.306	1.688	2.028	2.434	2.719
37	0.681	1.305	1.687	2.026	2.431	2.715
38	0.681	1.304	1.686	2.024	2.429	2.712
39	0.681	1.304	1.685	2.023	2.426	2.708
40	0.681	1.303	1.684	2.021	2.423	2.704
50	0.679	1.299	1.676	2.009	2.403	2.678
60	0.679	1.296	1.671	2.000	2.390	2.660
70	0.678	1.294	1.667	1.994	2.381	2.648
80	0.678	1.292	1.664	1.990	2.374	2.639
90	0.677	1.291	1.662	1.987	2.368	2.632
100	0.677	1.290	1.660	1.984	2.364	2.626
200	0.676	1.286	1.653	1.972	2.345	2.601
500	0.675	1.283	1.648	1.965	2.334	2.586
1000	0.675	1.282	1.646	1.962	2.330	2.581
10,000	0.675	1.282	1.645	1.960	2.327	2.576

The t-Table: t-Values Having a Proportion of .5 to .01 Outside the Interval ±t, for Selected df

Note: All values calculated by the author in Excel.

APPENDIX C

			The Fisher r to z_r Transform				
r	z_r	r	z_r	r	z_r	r	z_r
−.99	−2.65	−.49	−0.54	.01	0.01	.51	0.56
−.98	−2.30	−.48	−0.52	.02	0.02	.52	0.58
−.97	−2.09	−.47	−0.51	.03	0.03	.53	0.59
−.96	−1.95	−.46	−0.50	.04	0.04	.54	0.60
−.95	−1.83	−.45	−0.48	.05	0.05	.55	0.62
−.94	−1.74	−.44	−0.47	.06	0.06	.56	0.63
−.93	−1.66	−.43	−0.46	.07	0.07	.57	0.65
−.92	−1.59	−.42	−0.45	.08	0.08	.58	0.66
−.91	−1.53	−.41	−0.44	.09	0.09	.59	0.68
−.90	−1.47	−.40	−0.42	.10	0.10	.60	0.69
−.89	−1.42	−.39	−0.41	.11	0.11	.61	0.71
−.88	−1.38	−.38	−0.40	.12	0.12	.62	0.73
−.87	−1.33	−.37	−0.39	.13	0.13	.63	0.74
−.86	−1.29	−.36	−0.38	.14	0.14	.64	0.76
−.85	−1.26	−.35	−0.37	.15	0.15	.65	0.78
−.84	−1.22	−.34	−0.35	.16	0.16	.66	0.79
−.83	−1.19	−.33	−0.34	.17	0.17	.67	0.81
−.82	−1.16	−.32	−0.33	.18	0.18	.68	0.83
−.81	−1.13	−.31	−0.32	.19	0.19	.69	0.85
−.80	−1.10	−.30	−0.31	.20	0.20	.70	0.87
−.79	−1.07	−.29	−0.30	.21	0.21	.71	0.89
−.78	−1.05	−.28	−0.29	.22	0.22	.72	0.91
−.77	−1.02	−.27	−0.28	.23	0.23	.73	0.93
−.76	−1.00	−.26	−0.27	.24	0.24	.74	0.95
−.75	−0.97	−.25	−0.26	.25	0.26	.75	0.97
−.74	−0.95	−.24	−0.24	.26	0.27	.76	1.00
−.73	−0.93	−.23	−0.23	.27	0.28	.77	1.02
−.72	−0.91	−.22	−0.22	.28	0.29	.78	1.05
−.71	−0.89	−.21	−0.21	.29	0.30	.79	1.07
−.70	−0.87	−.20	−0.20	.30	0.31	.80	1.10
−.69	−0.85	−.19	−0.19	.31	0.32	.81	1.13
−.68	−0.83	−.18	−0.18	.32	0.33	.82	1.16
−.67	−0.81	−.17	−0.17	.33	0.34	.83	1.19
−.66	−0.79	−.16	−0.16	.34	0.35	.84	1.22
−.65	−0.78	−.15	−0.15	.35	0.37	.85	1.26
−.64	−0.76	−.14	−0.14	.36	0.38	.86	1.29
−.63	−0.74	−.13	−0.13	.37	0.39	.87	1.33
−.62	−0.73	−.12	−0.12	.38	0.40	.88	1.38
−.61	−0.71	−.11	−0.11	.39	0.41	.89	1.42
−.60	−0.69	−.10	−0.10	.40	0.42	.90	1.47
−.59	−0.68	−.09	−0.09	.41	0.44	.91	1.53
−.58	−0.66	−.08	−0.08	.42	0.45	.92	1.59
−.57	−0.65	−.07	−0.07	.43	0.46	.93	1.66
−.56	−0.63	−.06	−0.06	.44	0.47	.94	1.74
−.55	−0.62	−.05	−0.05	.45	0.48	.95	1.83
−.54	−0.60	−.04	−0.04	.46	0.50	.96	1.95
−.53	−0.59	−.03	−0.03	.47	0.51	.97	2.09
−.52	−0.58	−.02	−0.02	.48	0.52	.98	2.30
−.51	−0.56	−.01	−0.01	.49	0.54	.99	2.65
−.50	−0.55	.00	0.00	.50	0.55	.995	2.99

The F-Table: Critical Values of F Under the Null Hypothesis That $P^2 = 0$, for $\alpha = .05$ and $\alpha = .01$														
	$df_{regression}$													
df_{error}	1	2	3	4	5	6	7	8	9	10	15	20	25	30
1	161	200	216	225	230	234	237	239	241	242	246	248	249	250
	4052	5000	5403	5625	5764	5859	5928	5981	6022	6056	6157	6209	6240	6261
2	18.5	19.0	19.2	19.2	19.3	19.3	19.4	19.4	19.4	19.4	19.4	19.4	19.5	19.5
	98.5	99.0	99.2	99.2	99.3	99.3	99.4	99.4	99.4	99.4	99.4	99.4	99.5	99.5
3	10.1	9.6	9.3	9.1	9.0	8.9	8.9	8.8	8.8	8.8	8.7	8.7	8.6	8.6
	34.1	30.8	29.5	28.7	28.2	27.9	27.7	27.5	27.3	27.2	26.9	26.7	26.6	26.5
4	7.7	6.9	6.6	6.4	6.3	6.2	6.1	6.0	6.0	6.0	5.9	5.8	5.8	5.7
	21.2	18.0	16.7	16.0	15.5	15.2	15.0	14.8	14.7	14.5	14.2	14.0	13.9	13.8
5	6.6	5.8	5.4	5.2	5.1	5.0	4.9	4.8	4.8	4.7	4.6	4.6	4.5	4.5
	16.3	13.3	12.1	11.4	11.0	10.7	10.5	10.3	10.2	10.1	9.7	9.6	9.4	9.4
6	6.0	5.1	4.8	4.5	4.4	4.3	4.2	4.1	4.1	4.1	3.9	3.9	3.8	3.8
	13.7	10.9	9.8	9.1	8.7	8.5	8.3	8.1	8.0	7.9	7.6	7.4	7.3	7.2
7	5.6	4.7	4.3	4.1	4.0	3.9	3.8	3.7	3.7	3.6	3.5	3.4	3.4	3.4
	12.2	9.5	8.5	7.8	7.5	7.2	7.0	6.8	6.7	6.6	6.3	6.2	6.1	6.0
8	5.3	4.5	4.1	3.8	3.7	3.6	3.5	3.4	3.4	3.3	3.2	3.2	3.1	3.1
	11.3	8.6	7.6	7.0	6.6	6.4	6.2	6.0	5.9	5.8	5.5	5.4	5.3	5.2
9	5.1	4.3	3.9	3.6	3.5	3.4	3.3	3.2	3.2	3.1	3.0	2.9	2.9	2.9
	10.6	8.0	7.0	6.4	6.1	5.8	5.6	5.5	5.4	5.3	5.0	4.8	4.7	4.6
10	5.0	4.1	3.7	3.5	3.3	3.2	3.1	3.1	3.0	3.0	2.8	2.8	2.7	2.7
	10.0	7.6	6.6	6.0	5.6	5.4	5.2	5.1	4.9	4.8	4.6	4.4	4.3	4.2
12	4.7	3.9	3.5	3.3	3.1	3.0	2.9	2.8	2.8	2.8	2.6	2.5	2.5	2.5
	9.3	6.9	6.0	5.4	5.1	4.8	4.6	4.5	4.4	4.3	4.0	3.9	3.8	3.7
14	4.6	3.7	3.3	3.1	3.0	2.8	2.8	2.7	2.6	2.6	2.5	2.4	2.3	2.3
	8.9	6.5	5.6	5.0	4.7	4.5	4.3	4.1	4.0	3.9	3.7	3.5	3.4	3.3
16	4.5	3.6	3.2	3.0	2.9	2.7	2.7	2.6	2.5	2.5	2.4	2.3	2.2	2.2
	8.5	6.2	5.3	4.8	4.4	4.2	4.0	3.9	3.8	3.7	3.4	3.3	3.2	3.1
18	4.4	3.6	3.2	2.9	2.8	2.7	2.6	2.5	2.5	2.4	2.3	2.2	2.1	2.1
	8.3	6.0	5.1	4.6	4.2	4.0	3.8	3.7	3.6	3.5	3.2	3.1	3.0	2.9
20	4.4	3.5	3.1	2.9	2.7	2.6	2.5	2.4	2.4	2.3	2.2	2.1	2.1	2.0
	8.1	5.8	4.9	4.4	4.1	3.9	3.7	3.6	3.5	3.4	3.1	2.9	2.8	2.8
25	4.2	3.4	3.0	2.8	2.6	2.5	2.4	2.3	2.3	2.2	2.1	2.0	2.0	1.9
	7.8	5.6	4.7	4.2	3.9	3.6	3.5	3.3	3.2	3.1	2.9	2.7	2.6	2.5
30	4.2	3.3	2.9	2.7	2.5	2.4	2.3	2.3	2.2	2.2	2.0	1.9	1.9	1.8
	7.6	5.4	4.5	4.0	3.7	3.5	3.3	3.2	3.1	3.0	2.7	2.5	2.5	2.4
35	4.1	3.3	2.9	2.6	2.5	2.4	2.3	2.2	2.2	2.1	2.0	1.9	1.8	1.8
	7.4	5.3	4.4	3.9	3.6	3.4	3.2	3.1	3.0	2.9	2.6	2.4	2.3	2.3
40	4.1	3.2	2.8	2.6	2.4	2.3	2.2	2.2	2.1	2.1	1.9	1.8	1.8	1.7
	7.3	5.2	4.3	3.8	3.5	3.3	3.1	3.0	2.9	2.8	2.5	2.4	2.3	2.2
45	4.1	3.2	2.8	2.6	2.4	2.3	2.2	2.2	2.1	2.0	1.9	1.8	1.8	1.7
	7.2	5.1	4.2	3.8	3.5	3.2	3.1	2.9	2.8	2.7	2.5	2.3	2.2	2.1
50	4.0	3.2	2.8	2.6	2.4	2.3	2.2	2.1	2.1	2.0	1.9	1.8	1.7	1.7
	7.2	5.1	4.2	3.7	3.4	3.2	3.0	2.9	2.8	2.7	2.4	2.3	2.2	2.1
55	4.0	3.2	2.8	2.5	2.4	2.3	2.2	2.1	2.1	2.0	1.9	1.8	1.7	1.7
	7.1	5.0	4.2	3.7	3.4	3.1	3.0	2.9	2.7	2.7	2.4	2.3	2.1	2.1
60	4.0	3.2	2.8	2.5	2.4	2.3	2.2	2.1	2.0	2.0	1.8	1.7	1.7	1.6
	7.1	5.0	4.1	3.6	3.3	3.1	3.0	2.8	2.7	2.6	2.4	2.2	2.1	2.0
65	4.0	3.1	2.7	2.5	2.4	2.2	2.2	2.1	2.0	2.0	1.8	1.7	1.7	1.6
	7.0	4.9	4.1	3.6	3.3	3.1	2.9	2.8	2.7	2.6	2.3	2.2	2.1	2.0
70	4.0	3.1	2.7	2.5	2.3	2.2	2.1	2.1	2.0	2.0	1.8	1.7	1.7	1.6
	7.0	4.9	4.1	3.6	3.3	3.1	2.9	2.8	2.7	2.6	2.3	2.2	2.1	2.0
100	3.9	3.1	2.7	2.5	2.3	2.2	2.1	2.0	2.0	1.9	1.8	1.7	1.6	1.6
	6.9	4.8	4.0	3.5	3.2	3.0	2.8	2.7	2.6	2.5	2.2	2.1	2.0	1.9

Note: All values calculated in Excel by the author.

GLOSSARY

(1−α)100% confidence interval. A (1−α)100% confidence interval is an interval around a statistic. We have (1−α)100% confidence in this interval when we know that (1−α)100% of all such intervals will capture the parameter estimated by the statistic.

34-14-2 approximation. The 34-14-2 approximation summarizes the approximate proportion of normal distributions in intervals that are one standard deviation (σ) wide. Approximately 34% of a normal distribution falls between μ and μ + σ, approximately 14% of a normal distribution falls between μ + σ and μ + 2σ, and approximately 2% of a normal distribution lies above μ + 2σ.

95% confidence interval. A 95% confidence interval is an interval around a statistic. We have 95% confidence in this interval when we know that 95% of all such intervals will capture the parameter estimated by the statistic.

α. We use the symbol α to denote the probability of a Type I error.

Adjusted R^2 (R^2_{adj}). Adjusted R^2 (R^2_{adj}) is a statistic that reduces the positive bias in R^2.

Alternative hypothesis. The alternative hypothesis (also called the research hypothesis) is a hypothesis about the distribution of scores from which a sample was drawn. It is the alternative hypothesis that the researcher believes to be true.

Arguments. The inputs to mathematical functions are called arguments.

Association. Association means that the scores of one variable vary systematically but imperfectly with scores of the other variable. When there is an association between two variables, knowing an individual's score on one variable allows us to predict, with some degree of precision, his or her score on the other variable.

β. We use the symbol β to denote the probability of a Type II error.

Bar graph. A bar graph is a graphical depiction of the information in a frequency table. Each value of the variable is represented by a bar, and the height of each bar represents the number or proportion of scores having that value.

Between-group variability. Between-group variability reflects the variability within a collection of scores resulting from scores having been drawn from two or more populations.

Between-subjects factor. A between-subjects factor in an ANOVA involves scores on the dependent variable obtained from individuals that are unrelated to each other in any way. That is, scores are randomly selected from independent populations.

Biased statistic. A statistic is biased when its expected value (the mean of its sampling distribution) does not equal the parameter it estimates.

Bivariate distributions. Bivariate distributions are distributions of pairs of scores.

Bivariate normal distribution. In a bivariate normal distribution, x and y are normally distributed random variables and the conditional distributions are normal and homoscedastic, with expected values defined by the regression line.

Bonferroni correction. The Bonferroni correction for multiple contrasts achieves (1−$α_{fw}$)100% confidence in a family of n contrasts by setting the per-contrast α level ($α_{pc}$) to $α_{fw}/n$.

Carryover effect. A carryover effect occurs when obtaining measurements on previous trials affect the measurements made on subsequent trials.

Central limit theorem. The central limit theorem states (roughly) that the distribution of means will be a normal distribution if the population being sampled is normal. It also states that the distribution of sample means will converge on a normal distribution as sample size increases, irrespective of the shape of the distribution being sampled.

Central tendency. Central tendency is a number that is typical of the scores in a distribution.

Cohen's δ. Cohen's δ is an effect size. It is a standardized measure of difference between two population means in units of standard deviation. Cohen's δ is essentially a z-score.

Complex contrast. A complex contrast compares one or more means to another group of means. Properly constructed contrasts have negative weights that sum to −1 and positive weights that sum to +1.

Condition. Each condition of an ANOVA represents a specific combination of the levels of the independent variables. The number of conditions is the product of the number of levels in the factors.

Conditional distribution. In regression analysis, a conditional distribution is the distribution of y scores at a given value of x.

Confidence interval. An interval estimate is called a confidence interval when we can express the confidence we have that it contains the parameter. Confidence is expressed as a percentage.

Confounding variable. A confounding variable is an uncontrolled variable that affects the scores in our samples differently.

Continuous variables. Continuous variables have values that are real numbers.

Contrast score. A contrast score is the sum of the products of a weight vector (contrast vector) and the scores of a subject at each level of the independent variable; i.e., $cs_j = \sum_{i=1}^{k} w_i y_{i,j}$.

Convenience sample. A convenience sample is a sample that is conveniently available. It is the most common type of biased sample.

Correlation coefficient. The correlation coefficient is a quantitative measure of the precise degree of linear association between two variables.

Covariance. The covariance is the sum of products of deviations scores in x and y, divided by $n-1$, where n is the number of xy pairs. The covariance can be positive or negative, depending on whether there is a positive or negative linear association between x and y.

Cumulative frequency. A cumulative frequency is the number of scores at or below a given value of a variable.

Cumulative proportion (P). A cumulative proportion is the proportion of scores at or below a given value of a variable.

ΔR^2. $\Delta R^2 = R^2_{larger} - R^2_{smaller}$ is a statistic representing the difference in explained variance between two regression models. R^2_{larger} is the variance explained by a model with k predictors and $R^2_{smaller}$ is the variance explained by a model with a subset of these k predictors. ΔR^2 estimates $\Delta P^2 = P^2_{larger} - P^2_{smaller}$, the corresponding parameter.

Degrees of freedom (df). Degrees of freedom (df) refers to the number of independent variables required to account for a given source of variance (treatment, variable, or error) in a regression (or ANOVA) analysis.

Dependent samples. Dependent samples comprise scores that are related to each other in some way; either pairs of scores come from the same individual (repeated measures) or pairs of scores come from individuals matched on some characteristic (matched samples). This means that knowing the score of an individual in one sample tells you something about the score of the corresponding individual in the other sample.

Dependent variable. A dependent variable is the variable for which we obtain scores.

Descriptive statistics. Descriptive statistics are used to describe the characteristics of a collection of scores, without the goal of inferring something about a population parameter.

Deviation score. A deviation score is the difference between a score and the mean of the sample or population from which it came.

Direct effect (c'). The direct effect (c') is the effect of x on y while controlling for med; c' is the partial regression coefficient on x from a multiple regression analysis that predicts y from both x and med.

Directional tests. Directional tests make predictions about whether a statistic will be greater or less than a specified parameter. The p-value associated with a test statistic (such as z_{obs}) is the area in one tail of the distribution. Therefore, directional tests are also called one-tailed tests.

Discrete variables. Discrete variables have fixed values. Discrete quantitative variables have values that are integers or whole numbers.

Dispersion. Dispersion refers to how spread out scores are in a distribution of a quantitative variable.

Distribution of means. The distribution of means is a probability distribution of all possible values of a sample mean based on samples of the same size (n). The distribution of means is also referred to as the sampling distribution of the mean.

Empirical sampling distribution. An empirical sampling distribution is a histogram of a statistic computed from a large number of samples of the same size (n), drawn at random from the population of interest.

Event. An event is one or more of the possible outcomes of a sampling experiment.

Expected value (E). The expected value (E) of a statistic is the mean of its sampling distribution. We denote the expected value of a statistic as E (statistic).

Experiment. An experiment involves random assignment of individuals to treatment conditions. In an experiment, it is plausible to conclude that group differences on the dependent variable are caused by the different treatment conditions.

External validity. External validity refers to the validity of an inference from a sample to a population. External validity is high when a sample is a random sample from the population of interest, and lower when it is not.

Factor. A factor in an ANOVA is an independent variable. A one-factor ANOVA (or one-way ANOVA) has one independent variable, a two-factor ANOVA has two independent variables, a three-factor ANOVA has three independent variables, and so on.

File-drawer problem. The file-drawer problem refers to the large number of papers filed away in cabinets (or hard drives) because they were unpublished. As a consequence, many valid and worthwhile results are not available to guide and inform other researchers.

Fisher transformation. The Fisher transformation converts correlation coefficients (r), which are not generally normally distributed, into z_r statistics, which are approximately normally distributed. The transformation is simply $z_r = 0.5*\ln[(1+r)/(1-r)]$, where ln is the natural log function.

F-ratio. The F-ratio (or F-statistic) is the ratio of explained to unexplained variance multiplied by the ratio of error to regression degrees of freedom. The F-ratio increases as explained variance increases and as sample size increases.

Frequency table for a discrete quantitative variable. A frequency table for a discrete quantitative variable conveys the number or proportion of scores in a sample or population having specific values of the variable. The frequency table may also convey the number or proportion of scores at or below a given value of the variable.

Frequency table for a qualitative variable. A frequency table for a qualitative variable conveys the number or proportion of scores in a sample or population having each value of a variable.

F-statistic. See *F*-ratio.

Function. If there is exactly one value of *y* for each value of *x*, we say that *y* is a function of *x*. When *y* is a function of *x*, we can perfectly predict *y* from *x*.

Histogram. A histogram of a discrete quantitative variable is a graphical depiction of the number or proportion of scores in a set having a specific value of a variable.

Homogeneity of variance. Homogeneity of variance means that populations of interest have the same variance. This is exactly the same as saying the distributions are homoscedastic.

Homoscedasticity. We say that conditional distributions are homoscedastic when they all have the same standard deviation. Inferential statistics for regression requires the homoscedasticity assumption.

Hybrid model. The hybrid model combines elements of Fisher's significance tests and the Neyman-Pearson decision-making model. The hybrid model is referred to as null hypothesis significance testing (NHST).

Hyperplane. A hyperplane is a plane in three or more dimensions. It is the regression surface when there are three or more predictor variables.

Hypothetical population. A hypothetical population is one that does not exist, but which could exist. Individuals in an experimental group that undergo a novel treatment can be considered a sample from a hypothetical population of all individuals that undergo the same treatment.

Independent events. Events are independent if the occurrence of one does not affect the probability of the other occurring.

Independent samples. Independent samples comprise scores that are completely unrelated to each other. This means that knowing the score of an individual in one sample tells you nothing about the score of any individual in the other sample.

Independent variable. An independent variable is a qualitative or quantitative variable whose values define the two (or more) groups of interest.

Indirect effect (ab). The indirect effect (*ab*) is the product of the effect of *x* on *med* (*a*) and the effect of *med* on *y*, controlling for *x* (*b*). That is, *a* is the simple regression coefficient on *x* that predicts *med* from *x*, and *b* is the partial regression coefficient on *med* in a multiple regression analysis that predicts *y* from both *x* and *med*.

Inferential statistics. Inferential statistics is the act of inferring population parameters from sample statistics.

Interaction. An interaction between two independent variables means that the effect of one independent variable on the dependent variable changes as a function of the other independent variable.

Interaction vectors. Interaction vectors in a two-factor ANOVA are the products of orthogonal vectors coding for the two main effects. Because df_a vectors are required to capture the main effect of *A* and df_b vectors are required to capture the main effect of *B*, we require $df_a * df_b$ vectors to capture the *AB* interaction.

Interval. An interval is a range of values of a quantitative variable.

Interval estimate. An interval estimate is a range of values believed to contain the population parameter.

Interval midpoint. An interval midpoint is the middle value of an interval.

Interval scales. Interval scales have equal units of measurement but no absolute zero point.

Inverse Fisher transformation. The Inverse Fisher transformation converts z_r statistics to correlation coefficients (*r*). The transformation is simply $r = [exp(2z_r) - 1]/[exp(2z_r) + 1]$, where exp is a function that raises the base of the natural logarithm (≈ 2.7182) to some power *x*; i.e., $exp(x) \approx 2.7182^x$.

Inverse probability error. The inverse probability error is the mistaken belief that the *p*-value associated with our obtained statistic reflects the probability that H_0 is true.

Kurtosis. The kurtosis of a distribution refers to how sharp or flat its peak is.

Least-squares fit. The least-squares fit means the fit (line) through a scatter of points that produces the smallest sum of squared deviations of data points from the line.

Left skewed. A skewed distribution is an asymmetrical distribution, with one tail longer than the other. When the longer tail is on the left of the peak, it is said to be left skewed, or negatively skewed. Left skewed distributions are a consequence of ceiling effects.

Levels. Each independent variable (factor, way) in an ANOVA has two or more discrete values. These values are said to be levels of the independent variable (factor, way).

Linear regression line. The linear regression line is the best-fitting straight line through a scatter of *xy* points.

Linear transformation. A linear transformation of a set of scores (*x*) is any combination of additions or multiplications that can be reduced to $a + b(x)$.

$m_{critical}$. $m_{critical}$ is the value that our sample mean must *exceed* in order for us to reject the null hypothesis.

Main effect. The main effect of an independent variable refers to changes in the means of its levels averaged across all levels of other variables.

Marginal distribution. A marginal distribution is a distribution of the *x* scores from all *xy* pairs or the *y* scores from all *xy* pairs.

Matched-samples design. The matched-samples design creates dependent samples without measuring individuals twice. Individuals in two samples are matched on features that might be associated with the dependent variable, such as age, sex, socioeconomic status, IQ, health status, income, height, weight, or empathy, to name just a few that might be important in psychological research.

Mean of the distribution of means (μ_m). μ_m is the mean of the distribution of means. μ_m always equals μ, the mean of the distribution from which samples were drawn.

Mean-centering. Subtracting the mean from a set of scores produces a new set of scores that have a mean of 0. This new set of scores is said to be a mean-centered version of the original set of scores.

Mean square. Mean square is the sum of squares divided by its associated degrees of freedom.

Measurement error. Measurement error refers to the fact that each time something is measured a slightly different score will be obtained.

Median. The median is the number that divides an ordered set of numbers into two groups of equal size.

Mediation. When variable x affects variable y indirectly through its effect on variable *med*, we say that *med* mediates the association between x and y.

Meta-analysis. Meta-analysis is a quantitative method of data analysis that combines the results of many individual studies to obtain a more precise estimate of a population parameter.

Mixed design. A mixed ANOVA involves one or more between-subjects factors and one or more within-subjects factors.

Mode. The mode is the most frequently occurring score in a set.

Moderation. When the association between x and y changes as a function of a third variable *mod*, we say that *mod* moderates the association between x and y, or that there is an interaction between x and *mod*.

Multimodal distribution. A multimodal distribution has more than one peak.

Multiple correlation coefficient. The multiple correlation coefficient in a population is the correlation between y and $E(y|x_1, x_2, \ldots, x_k)$; it is denoted P. The multiple correlation coefficient in a sample is the correlation between y and $\hat{y}$; it is denoted R.

Mutually exclusive events. Mutually exclusive events are events that cannot co-occur.

Negatively skewed. *See* Left skewed.

Nominal scale. The values of variables measured on a nominal scale are qualitative and discrete and have no natural ordering.

Non-directional tests. Non-directional tests make no prediction about whether a statistic will be greater or less than a specified parameter. The p-value associated with the test statistic (such as z_{obs}) is the area in two tails of the distribution. Therefore, non-directional tests are also called two-tailed tests.

Normal distribution. A normal distribution is a unimodal, symmetrical distribution whose cross-section resembles the cross-section of a bell. Hence, normal distributions are often called bell curves.

Null hypothesis. The null hypothesis is a hypothesis about the distribution of scores from which a sample was drawn.

Null hypothesis significance testing (NHST). *See* Hybrid model.

Omnibus F-statistic. The omnibus F-statistic in a multiple regression is an F-statistic computed with $df_{regression} > 1$. The omnibus F-statistic in ANOVA is an F-statistic computed with $df_{source} > 1$.

One-tailed tests. *See* Directional tests.

Operational measure. An operational measure is a tool used to measure a psychological construct. Very often, operational measures are derived from questionnaires. Operational measures may also derive from the speed and accuracy with which psychological tasks are completed.

Ordinal scale. The values of variables measured on an ordinal scale are qualitative and discrete and have a natural ordering, but they do not have a unit of measurement.

Orthogonality. Two vectors are orthogonal if their correlation coefficient is $r = 0$.

Orthogonal set. A set of contrasts forms an orthogonal set if all pairs of contrast weight vectors are orthogonal. For k sample means, there are only $k - 1$ orthogonal contrasts.

Outcome. The outcome of a sampling experiment is the value of the variable that was selected.

Outliers. Outliers are scores that deviate by an unusual amount from the central tendency of a distribution.

P^2. P^2 is the proportion of variability in y explained by $x_1, x_2, \ldots, x_k$ in a multivariate population.

$P(z)$. $P(z)$ denotes the proportion of the standard normal distribution falling below a given z-score.

Parameter. A parameter is a numerical characteristic of a population.

Partial correlation coefficient. A partial correlation coefficient is the correlation between x_i and y when the effects of all other predictors ($x_{j \neq i}$) have been removed from the association between x_i and y.

Partial eta squared ($\hat{\eta}_p^2$). Partial eta squared is an effect size. It is the variance associated with a source, divided by itself plus its associated error variance.

Partial regression coefficient. A partial regression coefficient is the regression coefficient relating x_i to y when the effects of all other predictors ($x_{j \neq i}$) have been removed from the association between x_i and y.

Percentile rank. The percentile rank of a score is its cumulative proportion multiplied by 100; i.e., $P*100\%$.

Perfect correlation. There is a perfect correlation between x and y when $y = a + b(x)$. In this case, r will be -1 or 1 depending on the sign of b, the slope of the linear transformation.

p-hacking. The term p-hacking refers to a multitude of questionable practices in which researchers make undisclosed adjustments to their data analysis procedure in efforts to produce statistically significant results.

Point-biserial correlation coefficient (r_{pb}). The point-biserial correlation coefficient (r_{pb}) is the correlation coefficient that results from a dichotomous predictor variable (coded by any two numbers) and a quantitative outcome variable.

Point estimate. A point estimate is a single number that estimates a population parameter. In other words, a point estimate is a statistic.

Polynomial trend. A polynomial trend is a pattern of change across levels of the independent variable. The order of a polynomial trend is related to which coefficients in a polynomial equation can have nonzero weight.

Pooled variance (s^2_{pooled}). Pooled variance is computed as a weighted sum of two separate estimates of σ^2.

Population. A population comprises the scores on a variable of interest, obtained from individuals that share some characteristic of interest.

Positively skewed. *See* Right skewed.

Post hoc contrasts. Unplanned contrasts are also called post hoc contrasts. They are contrasts conducted after a statistically significant omnibus effect has been found.

Power. Power is the probability of rejecting the null hypothesis when it is false. That is, power is the probability of correctly rejecting the null hypothesis.

Practical significance. Practical significance relates to the interpretation and implications of a statistical analysis. The practical significance of a result rests on *how it matters*.

Prediction interval. A prediction interval is an interval that we can say with some confidence will contain the next value of *y* for a given value of *x*.

Probability. From the frequentist perspective, the probability of an event is the proportion of times the event would occur if the same sampling experiment were repeated infinitely many times.

Probability density function. A probability density function plots the density of scores at each value of a continuous variable. In statistics, density is defined at a point.

Probability distribution. A probability distribution conveys the probability that a randomly selected score will have a given value or fall in a given interval.

Proportion of successes. The proportion of successes in a sequence of sampling experiments is the number of successes divided by the number of sampling experiments, or trials: $p = N_{success}/N_{SE}$.

Psychological construct. A psychological construct is a hypothetical attribute (such as intelligence, introversion, or happiness) that is thought to explain some aspect of behavior, but which cannot be measured directly with a physical measuring device.

Publication bias. Publication bias "occurs whenever the research that appears in the published literature is systematically unrepresentative of the population of completed studies" (Rothstein, Sutton, & Borenstein, 2006). One form of publication bias occurs when journals, editors, reviewers, and even authors favor publication of results that achieve statistical significance.

***p*-value.** A *p*-value expresses the probability of a statistic, or one more extreme, occurring by chance when the null hypothesis is true. ("*More extreme*" means further from the mean of the distribution.)

Qualitative variables. Qualitative variables have values that are qualities or categories. They are also referred to as nominal or categorical variables.

Quantitative (scale) variables. Quantitative (scale) variables have values that are numbers. They often reflect how much of some quantity an individual possesses.

Quasi-experiment. A quasi-experiment compares two naturally occurring groups (i.e., groups not formed as a result of random assignment). In a quasi-experiment, it is not always plausible to conclude that group differences on the dependent variable are caused by group membership.

***r*.** *r* is the slope of the regression line predicting z_y from z_x.

Random variable. A variable is said to be a random variable when scores on the variable arise from random selection.

Range. The range is the difference between the largest and smallest scores in a set.

Ratio scales. Ratio scales have units of measurement and an absolute zero. The ratio of two values on a ratio scale expresses their relative distances from 0.

Raw frequency counts. Raw frequency counts (or tallies) represent the number of scores in a sample or population having a particular value or falling in a given interval.

Real limits. The real limits of an interval are the minimum and maximum real values of scores that define the interval.

Relative frequency. Relative frequencies represent the proportion of scores in a sample or population having a particular value or falling in a given interval.

Reliable. A reliable measuring device gives very similar (if not identical) measurements each time it is applied to the same object.

Repeated-measures design. A repeated-measures design involves scores obtained from participants on more than one occasion. That is, we've taken repeated measures from the same participants.

Researcher degrees of freedom. Researcher degrees of freedom refers to the number of undisclosed decisions that researchers make about what statistics are reported.

Research hypothesis. *See* Alternative hypothesis.

Retaining the null hypothesis. Retaining the null hypothesis means that it remains a plausible hypothesis about the distribution from which our sample was drawn. Retaining the null hypothesis does not mean that it is true or that we treat it as though it is true.

Right skewed. A skewed distribution is an asymmetrical distribution, with one tail longer than the other. When the longer tail is on the right of the peak, it is said to be right skewed, or positively skewed. Right skewed distributions are a consequence of floor effects.

Sample. A sample is a subset of a population.

Sample variance (s^2). The sample variance (s^2) is the sum of squares (ss) divided by $n-1$.

Sampling bias. Sampling bias means that not all members of the population had an equal chance of being selected in the sample. Sampling bias can and should be avoided.

Sampling distribution. A sampling distribution is a probability distribution of all possible values of a sample statistic based on samples of the same size.

Sampling distribution of b. The sampling distribution of b is a probability distribution for all possible values of b, for samples of size n, drawn from a homoscedastic bivariate distribution with fixed values of x.

Sampling distribution of the difference between two means. The sampling distribution of the difference between two means is a probability distribution of all possible differences between two sample means, m_1 and m_2, of size n_1 and n_2, respectively, drawn at random from two independent populations. In other words, it is the sampling distribution of all possible values of $m_1 - m_2$.

Sampling distribution of the mean. The distribution of means is a probability distribution of all possible values of a sample mean based on samples of the same size (n).

Sampling error. Sampling error is the difference between a statistic and the parameter it estimates. Sampling error cannot be avoided; it is an inevitable feature of random sampling.

Sampling experiment. A sampling experiment is the random selection of one of the possible values of a variable. We could also call a sampling experiment a trial.

Sampling with replacement. Sampling with replacement means that each time a score is drawn from a population, it is returned to the population before the next score is drawn.

Sampling without replacement. Sampling without replacement means that when a score is drawn from a population, it is not returned to the population before the next score is drawn.

Scatterplot. A scatterplot plots pairs of xy scores representing either two scores drawn from the same individual or two scores taken from individuals who are related in some way.

Score. A score is the value that an individual has on a particular variable.

Score limits. The score limits of an interval are the minimum and maximum whole values of the units of measurement that define the interval.

Second-order (interaction) contrast. A second-order (interaction) contrast is one that assesses how two first-order differences differ.

Shape. The shape of a distribution refers to how density (or frequency) changes as a function of the values of the variable.

Significance level. The significance level of a test defines the largest p-value that will be considered statistically significant; that is, p-values less than the significance level are statistically significant, and those that are greater are not. The significance level is typically set at .05.

Significance test. A significance test is a statistical procedure for evaluating the plausibility of the null hypothesis.

Simple contrast. A simple contrast compares two means. Properly constructed contrasts have negative weights that sum to -1 and positive weights that sum to $+1$.

Simple random sampling. Simple random sampling means that all members of the population had an equal chance of being selected in the sample.

Skewed distribution. A skewed distribution is an asymmetrical distribution, with one tail longer than the other.

Source table. A source table shows the sources of variance, degrees of freedom, and some indication of the relative importance of each source.

Sphericity. The sphericity assumption is that the variances of difference scores for all possible simple contrasts are the same in the population.

Squared deviation score. A squared deviation score is the square of a deviation score.

Standard deviation. The standard deviation is the square root of the variance. The standard deviation is roughly the average distance (deviation) of scores from the mean.

Standard error of estimate (σ_{est}). The standard error of estimate (σ_{est}) is the square root of the average squared deviation from the regression line in a population. σ_{est} is the standard deviation common to all conditional distributions.

Standard error of the mean. The standard error of the distribution of means (also called the standard error of the mean, or SEM) is denoted σ_m. The value of σ_m always equals $\sigma/\sqrt{n}$, which is the standard deviation of the population of scores divided by the square root of the sample size.

Standard normal distribution. The standard normal distribution is a distribution of z-scores. The standard normal distribution has a mean of $\mu = 0$, a standard deviation of $\sigma = 1$, and a variance of $\sigma^2 = 1$.

Standard score. A z-score, or standard score, expresses the distance of a score from the mean (μ) of its distribution in units of standard deviation (σ).

Statistic. A statistic is a numerical characteristic of a sample.

Statistical control. Statistical control refers to partialing out the effect of one variable (e.g., z) from the association between two others (e.g., x and y). When this is done, we can talk about the association between x and y while controlling for z.

Statistically significant. A statistic is said to be statistically significant when its associated p-value under the null is small.

Subjective probability. Subjective probability expresses an individual's subjective judgment about the likelihood that some event will occur.

Subject variability. Subject variability is the variability of individuals' scores around the group mean.

Sum of squares (ss). The sum of squares (ss) is the sum of all squared deviation scores in a sample or population.

t-distribution. A *t*-distribution is a distribution of *t*-scores $[t = (m - \mu)/s_m]$ computed from all possible samples of size *n, drawn from a normal distribution*. If the distribution from which the samples were drawn is not normal, we do not call the resulting distribution of *t*-scores a *t*-distribution.

Test statistic. A test statistic (such as z_{obs}) is computed from a sample statistic, such as the sample mean.

Total effect. The total effect of *x* on *y* is related to the strength of association between *x* and *y* without controlling for other variables. The total effect is denoted with *c*, which is defined as the slope of the simple regression equation that predicts *y* from *x*.

Trial. In the context of probability, a trial is the random selection of one of the possible values of a variable.

Two-tailed tests. See Non-directional tests.

Type I error. A Type I error occurs if the null hypothesis is rejected when it is true. This means that a sample was drawn from the distribution of scores specified by the null hypothesis, yet the statistic was statistically significant.

Type II error. A Type II error occurs if the null hypothesis is retained when it is false. This means that a sample was not drawn from the distribution of scores specified by the null hypothesis, yet the null hypothesis was retained.

U_3. U_3 is the proportion of one distribution of scores that falls below (or above) the mean of a second distribution.

Unbiased statistic. A statistic is unbiased when its expected value equals the parameter it estimates.

Unplanned contrasts. Unplanned contrasts are also called post hoc contrasts. They are contrasts conducted after a statistically significant omnibus effect has been found.

Valid. A measuring device is valid if it measures what it is supposed to measure.

Variables. Variables are physical or abstract attributes or quantities that we wish to measure. A variable can take on specific values.

Variance. In a population, the variance is the mean squared deviation from the mean.

Variance of the distribution of means (σ_m^2). σ_m^2 is the variance of the distribution of means. σ_m^2 always equals σ^2/n, the variance of the population of scores divided by sample size.

Weighted sum. A weighted sum is a way of computing the mean of two or more statistics by multiplying each statistic by a weight related to sample size and then summing the products.

Within-group variability. Within-group variability is the variability about the mean of a sample or population.

Within-subjects factor. A within-subjects factor in an ANOVA involves scores obtained from the same individuals at different levels of the independent variable. Therefore, scores are randomly selected from dependent populations.

$z_{\alpha/2}$. $z_{\alpha/2}$ is a *z*-score for which $(\alpha)100\%$ of the standard normal distribution lies outside the interval $\pm z_{\alpha/2}$.

$z_{critical}$. $z_{critical}$ is the *z*-score that z_{obs} must *exceed* for us to reject the null hypothesis. By *exceed* we mean further from 0 than the criterion. For example, 2.3 exceeds 1.645, and −2.3 exceeds −1.645.

z-score. *See* Standard score.

z-table. The *z*-table shows the proportion of the standard normal distribution falling below each of a large number of *z*-scores.

REFERENCES

Aldhous, P. (2011). Journal rejects studies contradicting precognition. *New Scientist*. Retrieved from http://www.newscientist.com

Alexander, B. K., Coambs, R. B., & Hadaway, P. F. (1978). The effect of housing and gender on morphine self-administration in rats. *Psychopharmacology, 58*(2), 175–179. doi:10.1007/BF00426903

Algina, J., Keselman, H. J., & Penfield, R. J. (2008). Note on a confidence interval for the squared semipartial correlation coefficient. *Educational and Psychological Measurement, 68*(5), 734–741. doi:10.1177/0013164407313371

American Psychological Association. (2010). *Publication manual of the American Psychological Association* (6th ed.). Washington, DC: American Psychological Association.

Ames, D. R., Rose, P., & Anderson, C. P. (2006). The NPI-16 as a short measure of narcissism. *Journal of Research in Personality, 40*(4), 440–450. doi:10.1016/j.jrp.2005.03.002

Anscombe, F. J. (1973). Graphs in statistical analysis. *The American Statistician, 27*(1), 17–21. doi:10.2307/2682899

Anvari, S. H., Trainor, L. J., Woodside, J., & Levy, B. A. (2002). Relations among musical skills, phonological processing, and early reading ability in preschool children. *Journal of Experimental Child Psychology, 83*(2), 111–130. doi:10.1016/S0022-0965(02)00124-8

Arent, S. M., & Landers, D. M. (2003). Arousal, anxiety, and performance: A reexamination of the inverted-U hypothesis. *Research Quarterly for Exercise and Sport, 74*(4), 436–444. doi:10.1080/02701367.2003.10609113

Asch, S. E. (1951). Effects of group pressure upon the modification and distortion of judgments. In H. Guetzkow (Ed.), *Groups, leadership, and men* (pp. 177–190). Oxford, England: Carnegie Press.

Bakan, D. (1966). The test of significance in psychological research. *Psychological Bulletin, 66*(6), 423–437. doi:10.1037/h0020412

Bargh, J. A., Chen, M., & Burrows, L. (1996). Automaticity of social behavior: Direct effects of trait construct and stereotype-activation on action. *Journal of Personality and Social Psychology, 71*(2), 230–244. doi:10.1037/0022-3514.71.2.230

Baron, R. M., & Kenny, D. A. (1986). The moderator–mediator variable distinction in social psychological research: Conceptual, strategic, and statistical considerations. *Journal of Personality and Social Psychology, 51*(6), 1173–1182. doi:10.1037/0022-3514.51.6.1173

Bayley, N. (2006). *Bayley Scales of Infant and Toddler Development*. San Antonio, TX: The Psychological Corporation.

Beck, A. T., Epstein, N., Brown, G., & Steer, R. A. (1988). An inventory for measuring clinical anxiety: Psychometric properties. *Journal of Consulting and Clinical Psychology, 56*(6), 893–897. doi:10.1037/0022-006X.56.6.893

Bem, D. J. (2011). Feeling the future: Experimental evidence for anomalous retroactive influences on cognition and affect. *Journal of Personality and Social Psychology, 100*(3), 407–425. doi:10.1037/a0021524

Bird, K. D. (2004). *Analysis of variance via confidence intervals*. London, UK: SAGE.

Box, J. F. (1978). *R. A. Fisher: The life of a scientist* (p. 146). New York, NY: Wiley.

Bozarth, M. A., Murray, A., & Wise, R. A. (1989). Influence of housing conditions on the acquisition of intravenous heroin and cocaine self-administration in rats. *Pharmacology, Biochemistry and Behavior, 33*(4), 903–907. doi:10.1016/0091-3057(89)90490-5

Buhr, K., & Dugas, M. J. (2002). The Intolerance of Uncertainty Scale: Psychometric properties of the English version. *Behaviour Research and Therapy, 40*(8), 931–945. doi:10.1016/S0005-7967(01)00092-4

Button, K. S., Ioannidis, J. P., Mokrysz, C., Nosek, B. A., Flint, J., Robinson, E. S., & Munafo, M. R. (2013). Power failure: Why small sample size undermines the reliability of neuroscience. *Nature Reviews: Neuroscience, 14*(5), 365–376. doi:10.1038/nrn3475

Cain, S. (2012) *Quiet: The power of introverts in a world that can't stop talking*. New York, NY: Crown Publishing.

Carney, D. R., Cuddy, A. J. C., & Yap, A. J. (2010). Power posing brief nonverbal displays affect neuroendocrine levels and risk tolerance. *Psychological Science, 21*(10), 1363–1368. doi:10.1177/0956797610383437

Carroll, L. (1865). *Alice's adventures in wonderland*. London, UK: Macmillan.

Carver, R. P. (1978). The case against statistical significance testing. *Harvard Educational Review, 48*(3), 378–399.

Chabris, C., & Simons, D. (2011). *The invisible gorilla: And other ways our intuitions deceive us*. New York, NY: Harmony Books.

Chamorro-Premuzic, T., Furnham, A., & Ackerman, P. L. (2006). Incremental validity of the typical intellectual engagement scale as predictor of different academic performance measures. *Journal of Personality Assessment, 87*(3), 261–268. doi:10.1207/s15327752jpa8703_07

Chaudhuri, A. (1990). Modulation of the motion aftereffect by selective attention. *Nature, 344*(6261), 60–62. doi:10.1038/344060a0

Chen, C. Y., & Hong, R. Y. (2010). Intolerance of uncertainty moderates the relation between negative life events and anxiety. *Personality and Individual Differences, 49*(1), 49–53. doi:10.1016/j.paid.2010.03.006

Cohen, J. (1962). The statistical power of abnormal-social psychological research: A review. *The Journal of Abnormal and Social Psychology, 65*(3), 145–153. doi:10.1037/h0045186

Cohen, J. (1988). *Statistical power analysis for the behavioral sciences.* Hillsdale, NJ: Erlbaum.

Cohen, J. (1994). The earth is round (p < .05). *American Psychologist, 49*(12), 997–1003. doi:10.1037/0003-066X.49.12.997

Cohen, J., & Cohen, P. (1983). *Applied multiple regression/correlation analysis for the behavioral sciences.* Hillsdale, NJ: Erlbaum.

Cumming, G. (2012). *Understanding the new statistics effect sizes, confidence intervals, and meta-analysis.* New York, NY: Routledge, Taylor & Francis Group.

Dawes, R. M. (1988). *Rational choice in an uncertain world.* San Diego, CA: Harcourt Brace Jovanovich.

DeBruine, L., & Jones, B. (2017). Face Research Lab London Set. *figshare.* doi:10.6084/m9.figshare.5047666.v3

Dooling, D. J., & Lachman, R. (1971). Effects of comprehension on retention of prose. *Journal of Experimental Psychology, 88*(2), 216–222. doi:10.1037/h0030904

Dugas, M. J., Gosselin, P., & Ladouceur, R. (2001). Intolerance of uncertainty and worry: Investigating specificity in a nonclinical sample. *Cognitive Therapy and Research, 25*(5), 551–558. doi:10.1023/A:1005553414688

Echambadi, R., & Hess, J. D. (2007). Mean-centering does not alleviate collinearity problems in moderated multiple regression models. *Marketing Science, 26*(3), 438–445. doi:10.1287/mksc.1060.0263

Ellemberg, D., & St-Louis-Deschênes, M. (2010). The effect of acute physical exercise on cognitive function during development. *Psychology of Sport and Exercise, 11*(2), 122–126. doi:10.1016/j.psychsport.2009.09.006

Ellis, P. D. (2010). *The essential guide to effect sizes: Statistical power, meta-analysis, and the interpretation of research results.* New York, NY: Cambridge University Press.

Falk, R., & Greenbaum, C. W. (1995). Significance tests die hard: The amazing persistence of a probabilistic misconception. *Theory & Psychology, 5*(1), 75–98. doi:10.1177/0959354395051004

Faul, F., Erdfelder, E., Buchner, A., & Lang, A. G. (2009). Statistical power analyses using G*Power 3.1: Tests for correlation and regression analyses. *Behavior Research Methods, 41*(4), 1149–1160. doi:10.3758/BRM.41.4.1149

Faul, F., Erdfelder, E., Lang, A. G., & Buchner, A. (2007). G*Power 3: A flexible statistical power analysis program for the social, behavioral, and biomedical sciences. *Behavior Research Methods, 39*(2), 175–191. doi:10.3758/BF03193146

Field, A. P. (2013). *Discovering statistics using IBM SPSS Statistics* (4th ed.). Thousand Oaks, CA: SAGE.

Fisher, R. A. (1925). *Statistical methods for research workers* (1st ed.). London, England: Oliver & Boyd.

Fisher, R. A. (1935). *The design of experiments.* Edinburgh, Scotland: Oliver and Boyd.

Fisher, R. A. (1939). Student. *Annals of Eugenics, 9*(1), 1–9. doi:10.1111/j.1469-1809.1939.tb02192.x

Fisher, R. A. (1955). Statistical methods and scientific induction. *Journal of the Royal Statistical Society, Series B (Methodological), 17*(1), 69–78.

Fisher, R. A. (1956). *Statistical methods and scientific inference.* Oxford, England: Hafner Publishing Co.

Galak, J., LeBoeuf, R. A., Nelson, L. D., & Simmons, J. P. (2012). Correcting the past: Failures to replicate psi. *Journal of Personality and Social Psychology, 103*(6), 933–948. doi:10.1037/a0029709

Gigerenzer, G. (1993). The superego, the ego, and the id in statistical reasoning. In G. Keren & C. E. Lewis (Eds.), *A handbook for data analysis in the behavioral sciences: Methodological issues* (pp. 311–339). Hillsdale, NJ: Erlbaum.

Gigerenzer, G. (2004). Mindless statistics. *The Journal of Socio-Economics, 33,* 587–606.

Glanzer, M., & Cunitz, A. R. (1966). Two storage mechanisms in free recall.

Journal of Verbal Learning and Verbal Behavior, 5(4), 351–360. doi:10.1016/S0022-5371(66)80044-0

Goff, M., & Ackerman, P. L. (1992). Personality-intelligence relations: Assessment of typical intellectual engagement. *Journal of Educational Psychology, 84*(4), 537–552. doi:10.1037/0022-0663.84.4.537

Gouin, J. P., Wenzel, K., Boucetta, S., O'Byrne, J., Salimi, A., & Dang-Vu, T. T. (2015). High-frequency heart rate variability during worry predicts stress-related increases in sleep disturbances. *Sleep Medicine, 16*(5), 659–664. doi:10.1016/j.sleep.2015.02.001

Goycoolea, M. V., Goycoolea, H. G., Rodriguez, L. G., Farfan, C. R., Martinez, G. C., & Vidal, R. (1986). Effect of life in industrialized societies on hearing in natives of Easter Island. *The Laryngoscope, 96*(12), 1391–1396. doi:10.1288/00005537-198612000-00015

Gurnsey, R., & Browse, R. A. (1987). Micropattern properties and presentation conditions influencing visual texture discrimination. *Perception & Psychophysics, 41*(3), 239–252. doi:10.3758/BF03208222

Haller, H., & Krauss, S. (2002). Misinterpretations of significance: A problem students share with their teachers. *Methods of Psychological Research, 7*(1), 1–20.

Harlow, L. L., Mulaik, S. A., & Steiger, J. H. (2016). *What if there were no significance tests?* (classic ed.). New York, NY: Routledge.

Harwood, E., & Naylor, G. (1969). Recall and recognition in elderly and young subjects. *Australian Journal of Psychology, 21*(3), 251–257.

Hayes, A. F. (2013). *Introduction to mediation, moderation, and conditional process analysis: A regression-based approach.* New York, NY: Guilford Press.

Hedges, L. V., & Olkin, I. (1995). *Statistical methods for meta-analysis.* London, England: Academic Press.

Hetherington, A. W., & Ranson, S. W. (1940). Hypothalamic lesions and adiposity in the rat. *The Anatomical Record, 78*(2), 149–172. doi:10.1002/ar.1090780203

Hillman, C. H., Pontifex, M. B., Raine, L. B., Castelli, D. M., Hall, E. E., & Kramer, A. F. (2009). The effect of acute treadmill walking on cognitive control and academic achievement in preadolescent children. *Neuroscience, 159*(3), 1044–1054. doi:10.1016/j.neuroscience.2009.01.057

Hunter, J. E., & Schmidt, F. (1990). *Methods of meta-analysis.* Newbury Park, CA: SAGE.

Ioannidis, J. P. (2005). Why most published research findings are false. *PLoS Medicine, 2*(8), e124. doi:10.1371/journal.pmed.0020124

Julesz, B. (1981). Textons, the elements of texture perception, and their interactions. *Nature, 290*(5802), 91–97. doi:10.1038/290091a0

Kelley, K. (2007). Methods for the behavioral, educational, and social sciences: An R package. *Behavior Research Methods, 39*(4), 979–984. doi:10.3758/BF03192993

Kline, R. B. (2004). *Beyond significance testing: Reforming data analysis methods in behavioral research* (1st ed.). Washington, DC: American Psychological Association.

Kline, R. B. (2013). *Beyond significance testing: Statistics reform in the behavioral sciences* (2nd ed.). Washington, DC: American Psychological Association.

Kline, R. B. (2015). The mediation myth. *Basic and Applied Social Psychology, 37*(4), 202–213. doi:10.1080/01973533 2.015.1049349

Kohn, P. M., Lafreniere, K., & Gurevich, M. (1990). The Inventory of College Students' Recent Life Experiences: A decontaminated hassles scale for a special population. *Journal of Behavioral Medicine, 13*(6), 619–630. doi:10.1007/BF00844738

Kramer, A. D., Guillory, J. E., & Hancock, J. T. (2014). Experimental evidence of massive-scale emotional contagion through social networks. *Proceedings of the National Academy of Sciences, 111*(24), 8788–8790. doi:10.1073/pnas.1320040111

Kruskal, W. (1980). The significance of Fisher: A review of *R. A. Fisher: The life of a scientist,* by Joan Fisher

Box. *Journal of the American Statistical Association, 75*(372), 1019–1030. doi:10.2307/2287199

Lambdin, C. (2012). Significance tests as sorcery: Science is empirical—significance tests are not. *Theory and Psychology, 22*(1), 67–90. doi:10.1177/0959354311429854

Langlois, J. H., & Roggman, L. A. (1990). Attractive faces are only average. *Psychological Science, 1*(2), 115–121. doi:10.1111/j.1467-9280.1990.tb00079.x

Lansford, J. E., Miller-Johnson, S., Berlin, L. J., Dodge, K. A., Bates, J. E., & Pettit, G. S. (2007). Early physical abuse and later violent delinquency: A prospective longitudinal study. *Child Maltreatment, 12*(3), 233–245. doi:10.1177/1077559507301841

Lehmann, E. L. (1994). Jerzy Neyman 1894–1981: A biographical memoir. National Academy of Sciences. Retrieved from http://www.nasonline.org/member-directory/deceased-members/20000800.html

Lessard, N., Paré, M., Lepore, F., & Lassonde, M. (1998). Early-blind human subjects localize sound sources better than sighted subjects. *Nature, 395*(6699), 278–280. doi:10.1038/26228

Loftus, E. F., & Palmer, J. C. (1974). Reconstruction of auto-mobile destruction: An example of the interaction between language and memory. *Journal of Verbal Learning and Verbal Behavior, 13*(5), 585–589. doi:10.1016/S0022-5371(74)80011-3

Luborsky, L., Rosenthal, R., Diguer, L., Andrusyna, T. P., Berman, J. S., Levitt, J. T., Seligman, D. A., & Krause, E. D. (2002). The dodo bird verdict is alive and well—mostly. *Clinical Psychology: Science and Practice, 9*(1), 2–12. doi:10.1093/clipsy.9.1.2

Luce, R. D. (1988). The tools-to-theory hypothesis. Review of G. Gigerenzer and D. J. Murray, "Cognition as intuitive statistics." *Contemporary Psychology, 33,* 582–583.

Lykken, D. T. (1968). Statistical significance in psychological research. *Psychological Bulletin, 70*(3), 151–159. doi:10.1037/h0026141

MacKinnon, D. P., Warsi, G., & Dwyer, J. H. (1995). A simulation study of mediated effect measures. *Multivariate Behavioral Research, 30*(1), 41–62. doi:10.1207/s15327906mbr3001_3

Meehl, P. E. (1978). Theoretical risks and tabular asterisks: Sir Karl, Sir Ronald, and the slow progress of soft psychology. *Journal of Consulting and Clinical Psychology, 46*(4), 806–834. doi:10.1037/0022-006X.46.4.806

Meehl, P. E. (1990). Why summaries of research on psychological theories are often uninterpretable. *Psychological Reports, 66*(1), 195–244. doi:10.2466/pr0.1990.66.1.195

Minium, E. W., Clarke, R. C., & Coladarci, T. (1999). *Elements of statistical reasoning* (2nd ed.). Hoboken, NJ: John Wiley & Sons.

Nash, R. A., Nash, A., Morris, A., & Smith, S. L. (2015). Does rapport-building boost the eyewitness eye-closure effect in closed questioning? *Legal and Criminological Psychology, 21*(1), 305–318. doi:10.1111/lcrp.12073

Nasreddine, Z. S., Phillips, N. A., Bédirian, V., Charbonneau, S., Whitehead, V., Collin, I., Cummings, J. L., & Chertkow, H. (2005). The Montreal Cognitive Assessment, MoCA: A brief screening tool for mild cognitive impairment. *Journal of the American Geriatrics Society, 53*(4), 695–699. doi:10.1111/j.1532-5415.2005.53221.x

Nelson, R. A., & Strachan, I. (2009). Action and puzzle video games prime different speed/accuracy trade-offs. *Perception, 38*(11), 1678–1687. doi:10.1068/p6324

Neyman, J. (1937). Outline of a theory of statistical estimation based on the classical theory of probability. *Philosophical Transactions of the Royal Society of London, Series A, Mathematical and Physical Sciences, 236*(767), 333–380.

Neyman, J. (1956). Note on an article by Sir Ronald Fisher. *Journal of the Royal Statistical Society, Series B, 18*(2), 288–294.

Neyman, J. (1961). Silver jubilee of my dispute with Fisher. *Journal of the Operations Research Society of Japan, 3*(4), 145–154.

Neyman, J. (1974). *The heritage of Copernicus: Theories "pleasing to the mind."* Cambridge, Massachusetts: MIT Press.

Neyman, J., & Pearson, E. S. (1933). On the problem of the most efficient tests of statistical hypotheses. *Philosophical Transactions of the Royal Society of London, Series A, Containing Papers of a Mathematical or Physical Character, 231,* 289–337.

Nieuwenhuis, S., & Monsell, S. (2002). Residual costs in task switching: Testing the failure-to-engage hypothesis. *Psychonomic Bulletin & Review, 9*(1), 86–92. doi:10.3758/BF03196259

Nuzzo, R. (2014). Scientific method: Statistical errors. *Nature, 506*(7487), 150–152. doi:10.1038/506150a

Oakes, M. (1986). *Statistical inference: A commentary for the social and behavioral sciences.* New York, NY: Wiley.

Oliva, A. (2005). Gist of the scene. *Neurobiology of Attention, 696*(64), 251–258.

Pearson, E. S. (1939). "Student" as Statistician. *Biometrika, 30*(3/4), 210–250.

Pearson, E. S. (1955). Statistical concepts in their relation to reality. *Journal of the Royal Statistical Society, Series B (Methodological), 17*(2), 204–207.

Peterson, D. (2016). The baby factory: Difficult research objects, disciplinary standards, and the production of statistical significance. *Socius: Sociological Research for a Dynamic World, 2.* doi:10.1177/2378023115625071

Posner, M. I. (1980). Orienting of attention. *Quarterly Journal of Experimental Psychology, 32*(1), 3–25.

Ranehill, E., Dreber, A., Johannesson, M., Leiberg, S., Sul, S., & Weber, R. A. (2015). Assessing the robustness of power posing no effect on hormones and risk tolerance in a large sample of men and women. *Psychological Science, 26*(5), 653–656. doi:10.1177/0956797614553946

Raskin, R., and Terry, H. (1988). A principal-components analysis of the Narcissistic Personality Inventory and further evidence of its construct validity. *Journal of Personality and Social Psychology, 54*(5), 890–902. doi:10.1037/0022-3514.54.5.890

Rayner, K., Foorman, B. R., Perfetti, C. A., Pesetsky, D., & Seidenberg, M. S. (2001). How psychological science informs the teaching of reading. *Psychological Science in the Public Interest, 2*(2), 31–74. doi:10.1111/1529-1006.00004

Reid, C. (1998). *Neyman.* New York, NY: Springer-Verlag.

Rensink, R. A., O'Regan, J. K., & Clark, J. J. (1997). To see or not to see: The need for attention to perceive changes in scenes. *Psychological Science, 8*(5), 368–373.

Roberts, D. C., & Bennett, S. A. (1993). Heroin self-administration in rats under a progressive ratio schedule of reinforcement. *Psychopharmacology, 111*(2), 215–218. doi:10.1007/BF02245526

Rogers, R. D., & Monsell, S. (1995). Costs of a predictable switch between simple cognitive tasks. *Journal of Experimental Psychology: General, 124*(2), 207–231. doi:10.1037/0096-3445.124.2.207

Rosenzweig, S. (1936). Some implicit common factors in diverse methods of psychotherapy. *American Journal of Orthopsychiatry, 6*(3), 412–415. doi:10.1111/j.1939-0025.1936.tb05248.x

Rosser, J. C., Jr., Lynch, P. J., Cuddihy, L., Gentile, D. A., Klonsky, J., & Merrell, R. (2007). The impact of video games on training surgeons in the 21st century. *Archives of Surgery, 142*(2), 181–186; discussion 186. doi:10.1001/archsurg.142.2.181

Rothstein, H. R., Sutton, A. J., & Borenstein, M. (2006). *Publication bias in meta-analysis: Prevention, assessment and adjustments.* New York, NY: John Wiley & Sons.

Rozeboom, W. W. (1960). The fallacy of the null-hypothesis significance test. *Psychological Bulletin, 57*(5), 416–428. doi:10.1037/h0042040

Sachs, J. S. (1967). Recognition memory for syntactic and semantic aspects of connected discourse. *Perception & Psychophysics, 2*(9), 437–442. doi:10.3758/BF03208784

Salsburg, D. (2001). *The lady tasting tea: How statistics revolutionized science in the twentieth century.* New York, NY: Henry Holt and Company.

Savage, L. J. (1961). The foundations of statistics reconsidered. In *Proceedings of the Fourth Berkeley Symposium on Mathematical Statistics and Probability: Vol. 1. Contributions to the theory of statistics* (pp. 575–586). Berkeley, CA: University of California Press.

Sedki, F., D'Cunha, T., & Shalev, U. (2013). A procedure to study the effect of prolonged food restriction on heroin seeking in abstinent rats. *Journal of Visualized Experiments,* (81), e50751. doi:10.3791/50751

Sedlmeier, P., & Gigerenzer, G. (1989). Do studies of statistical power have an effect on the power of studies? *Psychological Bulletin, 105*(2), 309–316. doi:10.1037/0033-2909.105.2.309

Simmons, J. P., Nelson, L. D., & Simonsohn, U. (2011). False-positive psychology: Undisclosed flexibility in data collection and analysis allows presenting anything as significant. *Psychological Science, 22*(11), 1359–1366. doi:10.1177/0956797611417632

Simon, H. A. (1992). What is an "explanation" of behavior? *Psychological Science, 3,* 150–161. doi:10.1111/j.1467-9280.1992.tb00017.x

Skinner, B. F. (1972). *Cumulative record.* New York, NY: Appleton-Century-Crofts.

Smith, M. L., & Glass, G. V. (1977). Meta-analysis of psychotherapy outcome studies. *American Psychologist, 32*(9), 752–760. doi:10.1037/0003-066X.32.9.752

Sobel, M. E. (1982). Asymptotic confidence intervals for indirect effects in structural equation models. *Sociological Methodology, 13,* 290–312. doi:10.2307/270723

Stevens, S. S. (1946). On the theory of scales of measurement. *Science, 103*(2684), 677–680. doi:10.1126/science.103.2684.677

Stevens, S. S. (1960). The predicament in design and significance. *Contemporary Psychology, 5*(9), 273–276. doi:10.1037/006366

Stevenson, D. L., & Baker, D. P. (1987). The family-school relation and the child's school performance.

Child Development, 58(5), 1348–1357. doi:10.2307/1130626

Stolley, P. D. (1991). When genius errs: R. A. Fisher and the lung cancer controversy. *American Journal of Epidemiology, 133*(5), 416–425. doi:10.1093/oxfordjournals.aje.a115904

Thompson, B. (Ed.). (2003). *Score reliability: Contemporary thinking on reliability issues.* Thousand Oaks, CA: SAGE.

Tomopoulos, S., Dreyer, B. P., Berkule, S., Fierman, A. H., Brockmeyer, C., & Mendelsohn, A. L. (2010). Infant media exposure and toddler development. *Archives of Pediatrics and Adolescent Medicine, 164*(12), 1105–1111. doi:10.1001/archpediatrics.2010.235

Turner, E. H., Matthews, A. M., Linardatos, E., Tell, R. A., & Rosenthal, R. (2008). Selective publication of antidepressant trials and its influence on apparent efficacy. *New England Journal of Medicine, 358*(3), 252–260. doi:10.1056/NEJMsa065779

Wasserstein, R. L., & Lazar, N. A. (2016). The ASA's statement on *p*-values: Context, process, and purpose. *The American Statistician, 70*(2), 129–133. doi:10.1080/00031305.2016.1154108

Wilson, T. D., Reinhard, D. A., Westgate, E. C., Gilbert, D. T., Ellerbeck, N., Hahn, C., Brown, C. L., & Shaked, A. (2014). Just think: The challenges of the disengaged mind. *Science, 345*(6192), 75–77. doi:10.1126/science.1250830

Yerkes, R. M., & Dodson, J. D. (1908). The relation of strength of stimulus to rapidity of habit-formation. *Journal of Comparative Neurology and Osychology, 18*(5), 459–482. doi:10.1002/cne.920180503

Zabell, S. L. (2008). On Student's 1908 article: "The Probable Error of a Mean." *Journal of the American Statistical Association, 103*(481), 1–7. doi:10.1198/016214508000000030

INDEX